The Marriage and Family Experience

10TH EDITION

The Marriage and Family Experience

Intimate Relationships in a Changing Society

Bryan Strong
Formerly of University of California, Santa Cruz

Christine DeVault
Cabrillo College

Theodore F. Cohen
Ohio Wesleyan University

THOMSON
———✦———
WADSWORTH

Australia · Brazil · Canada · Mexico · Singapore · Spain
United Kingdom · United States

THOMSON
™
WADSWORTH

The Marriage and Family Experience: Intimate Relationships in a Changing Society, **Tenth Edition**
Bryan Strong, Christine DeVault, and Theodore F. Cohen

Acquisitions Editor: Chris Caldeira
Development Editor: Sherry Symington
Assistant Editor: Christina Ho
Editorial Assistant: Tali Beesley
Technology Project Manager: Dave Lionetti
Marketing Manager: Michelle Williams
Marketing Assistant: Emily Elrod
Marketing Communications Manager:
    Darlene Amidon-Brent
Project Manager, Editorial Production: Cheri Palmer
Creative Director: Rob Hugel

Art Director: John Walker
Print Buyer: Becky Cross
Permissions Editor: Bob Kauser
Production Service: Graphic World Inc.
Text Designer: Diane Beasley
Photo Researcher: Sarah Evertson
Cover Designer: Yvo Riezebos
Cover Images: Randy Faris/Corbis
    TV: John Springer Collection/Corbis
Compositor: Graphic World Inc.
Printer: R.R. Donnelley/Willard

Library of Congress Control Number: 2006935083

Student Edition:
ISBN-13: 978-0-534-62424-8
ISBN-10: 0-534-62424-3

Loose-leaf Edition:
ISBN-13: 978-0-495-50083-4
ISBN-10: 0-495-50083-6

Thomson Higher Education
10 Davis Drive
Belmont, CA 94002-3098
USA

For more information about our products, contact us at:
Thomson Learning Academic Resource Center
1-800-423-0563

For permission to use material from this text or product, submit a request online at
http://www.thomsonrights.com.
Any additional questions about permissions can be submitted by email to **thomsonrights@Thomson.com.**

I have been blessed to share my own marriage and family experience with two beautiful and remarkable women, the late Susan Jablin Cohen and Julie Pfister Cohen. In both looking back and looking ahead, I know how much my life has been enriched because it has been spent with them. With loving gratitude, I dedicate this edition to them.

—Ted Cohen

Brief Contents

Contents

CHAPTER 3   Differences: Historical and Contemporary Variations in American Family Life   68

UNIT 2   INTIMATE RELATIONSHIPS

CHAPTER 4   Gender and Family   114

CHAPTER 5 Friendship, Love, and Intimacy 148

CHAPTER 6 Understanding Sex and Sexualities 188

CHAPTER 7   Communication, Power, and Conflict   234

UNIT 3 FAMILY LIFE

CHAPTER 8 Singlehood, Pairing, and Cohabitation 278

CHAPTER 9 Marriages in Societal and Individual Perspective 320

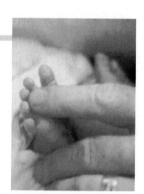

Chapter 10 Should We or Shouldn't We: Choosing Whether and How to Have Children 360

CHAPTER 11 Experiencing Parenthood: Roles and Relationships of Parents and Children 386

UNIT 4 FAMILY CHALLENGES AND STRENGTHS

CHAPTER 12 Marriage, Work, and Economics 420

CHAPTER 13   Intimate Violence and Sexual Abuse   454

CHAPTER 14   Coming Apart: Separation and Divorce   488

CHAPTER 15   New Beginnings: Single-Parent Families, Remarriages, and Blended Families   524

Boxes

Exploring Diversity

Popular Culture

Issues and Insights

Real Families

Understanding Yourself

Preface

This is the third time I have had the pleasure to revise and update *The Marriage and Family Experience: Intimate Relationships in a Changing Society*. Since it was first written by Bryan Strong, for many years and through many editions, it has appealed to teachers and students in a number of different types of institutions and across a range of academic and applied disciplines. The book recognizes that in whatever form(s) we experience them, our families shape who we are and who we become, provide us with our most intimate and loving relationships, and need to be valued and supported. The current edition retains that focus, as it broadens and updates coverage of changing family patterns, variations of family experience, and how family experiences are shaped both from within and by external influences over which we have less control. Once again, the diverse and dynamic nature of families are major emphases.

When I first began working on the eighth edition of this book, I was fortunate to be in a strong and stable marriage of more than twenty years. I anticipated it lasting at least thirty more. I was the father of two young teenagers who were the center of my hectic life. Between the eighth and ninth editions, my family was tragically transformed by my wife Susan's fourteen-month-long struggle with brain cancer and her subsequent death. I was a widower and a single parent. My children were making their way into and out of high school. Between that edition and this tenth edition, my family world has once again been transformed. A long-distance relationship culminated in a remarriage two years later. My two children are now both in college and I have three stepchildren, the youngest in kindergarten, the oldest about to enter high school. All of these experiences—marriage, fatherhood, caregiving, widowerhood, single parenting, remarriage, stepfatherhood—not only strengthened my belief in the importance of the kinds of relationships addressed in this book, but also heightened my awareness of how dynamic and diverse family experience is, even within a single lifetime, and how none of us can necessarily anticipate or fully control the directions our families may take. I don't pretend that my experience of change strengthens my expertise on these various matters, but it has certainly raised my sensitivity to the human experiences found in so many of the issues you will find in this book.

Once again, I have drawn from various disciplines to update, broaden, and illustrate the many issues this book covers. In trying to retain the book's wonderful balance between an academic and more functional approach, I have again drawn most heavily from recent research in sociology, family studies, and psychology.

New to This Edition

Returning users will notice a number of noteworthy changes from the last edition. Along with new features, there is a new chapter (Chapter 9), *Marriages in Societal and Individual Perspective*. This chapter addresses the competing viewpoints about the status of marriage in the United States today, trends in marriage patterns, how those trends vary among different segments of society, and what those trends might mean. The chapter also retains, but greatly expands, material on marriage across the life cycle. Additionally, the sequence of chapters has been altered so that the chapter on sex and sexuality (retitled, "Understanding Sex and Sexualities") appears earlier in the book. With its significantly expanded coverage of material on sexual orientation and on sexuality in relationships, it more appropriately follows the chapters about gender (Chapter 4) and love and intimacy (Chapter 5). Also, Chapter 16 has been dropped, and as many reviewers suggested, some of that material has been moved into other chapters.

New or Expanded Topics

Chapter 1, *The Meaning of Marriage and the Family.* The chapter now begins with a series of recent news stories reflecting underlying debates about what families are or should be and what rights we ought to have as spouses and parents. This fits the chapter's enlarged emphasis on different and competing viewpoints about the meaning of family and the interpretation of changing family patterns. The chapter includes a new discussion of religious differences on family issues, and now concludes by identifying the major themes of the text formerly found in Chapter 16.

Chapter 2, *Studying Marriages and Families.* Modest changes were made to this chapter, mostly to use new examples of theories (for example, social exchange theory) or to clarify or expand certain points (such as expanding the discussion of independent, dependent, and intervening variables by including a fourth diagram depicting a more complex causal hypothesis). Two features that were retained have been updated and broadened.

Chapter 3, *Differences: Historical and Contemporary Variations in American Family Life.* Along with much new data on racial and ethnic differences in social and family characteristics, there is a new section on families of Middle Eastern immigrants (replacing the section on Amish families). The discussion of poverty has been moved from a later chapter to the section on social class.

Chapter 4, *Gender and Family.* This chapter now includes data on the wage gap between women and men and an updated discussion of gender and education that reflects current concerns about boys.

Chapter 5, *Friendship, Love, and Intimacy.* There are new sections on the need for intimacy, qualities we seek in friends and lovers, and displays of love among different racial groups and among intra- versus interracial couples. Discussions of gender differences in love and friendship, stalking, infidelity, and jealousy have all been enlarged and updated.

Chapter 6, *Understanding Sex and Sexualities.* There is expanded coverage of sexual orientation and much new data on sexual behavior (nonmarital, marital, and extramarital), especially how it changes as we age. New material includes comparisons of gay, lesbian, and heterosexual relationships, new data on anti-gay prejudice, and an expanded section on bisexuality. There are also new sections on adolescent and young adult sexual behavior, the meanings and experiences of virginity and virginity loss among females and males, changes in sexual desire and the experience of sexual problems as we age, and new data on marital and extramarital sex.

Chapter 7, *Communication, Power, and Conflict.* There is new material on gender and cultural differences in nonverbal communication, topical areas that create communication problems between partners, premarital factors that shape conflict management styles, consequences of conflict, and forgiveness.

Chapter 8, *Singlehood, Pairing, and Cohabitation.* Notable changes include new material on interracial relationships, marital and family history homogamy, the causes and consequences of heterogamy, gender-based dating scripts, and the process and consequences of breakups. The coverage of cohabitation now includes causes and outcomes of different types of cohabitation, the impact of children on cohabiting relationships, and the impact of cohabitation on health.

Chapter 9, *Marriages in Societal and Individual Perspective.* Along with the material on the debate about the status of marriage in the United States today, the chapter also considers legal issues surrounding marriage, reasons people marry, and the benefits of marriage. Much of the "life cycle" material from prior editions has been retained to illustrate how marriage changes across time, but it is supplemented by new material on premarital factors that affect marital outcomes, the meaning and symbolism of weddings, the early stages of marriage relationships, and the effects of social stress and social context on marital relationships.

Chapter 10, *Should We or Shouldn't We: Choosing Whether and How to Have Children.* The chapter contains new material on choices people make about whether and when to have children, as well as broadened coverage of women's and men's reactions to childbirth, women's assessments of the birth experience and hospital care, and issues surrounding adoption.

Chapter 11, *Experiencing Parenthood: Roles and Relationships of Parents and Children.* There are new sections on what fatherhood means to men, how parenthood affects marriage and mental health, how marital status affects parenting, non-parental childcare, and gay men and lesbians as parents. The discussion of who cares for children has been updated and expanded.

Chapter 12, *Marriage, Work, and Economics.* The chapter includes new material on work induced time strains, workplace related stress, family to work spillover, emotion work, shift work, and factors that lead to male participation in household labor.

Chapter 13, *Intimate Violence and Sexual Abuse.* The chapter includes more recent data on the prevalence of intimate violence experienced by males and females as well as across categories of social class, race, ethnicity, and sexual orientation. Additionally, there is new material on legal and police response to intimate partner violence.

Chapter 14, *Coming Apart: Separation and Divorce.* This chapter now includes new data and discussions of the prevalence of divorce, factors that increase or decrease the risk of divorce, the effects of prior divorce on marital stability, the intergenerational transmission of divorce, the reality and policies around child support, and the effects of divorce on children.

Chapter 15, *New Beginnings: Single-Parent Families, Remarriages, and Blended Families.* This material has been updated with more recent statistical information on gender differences in single parenting, remarriage rates, and discussions of the role of private safety nets and social fathers, the effects of single parenthood on children, the effects of cohabitation, race, gender, and parenthood on one's likelihood of remarriage. More attention is paid to gender differences in step-parenting.

Pedagogy

What Do You Think? Self-quiz chapter openers let students assess their existing knowledge of what will be discussed in the chapter. We have found these quizzes engage the student, drawing them into the material and stimulating greater interaction with the course.

Chapter Outlines. Each chapter contains an outline at the beginning of the chapter to allow students to organize their learning.

Exploring Diversity. These boxes let students see family circumstances from the vantage point of other cultures, other eras, or within different lifestyles in the contemporary United States. There are new diversity features on Iranian families, Asian American sexuality, the contributions of daughters in low-income single parent households, the expe-

riences of young African American single fathers, and how different cultures view love.

Understanding Yourself. These boxed inserts use research topics, findings, and instruments to stimulate students to examine their own family experiences or expectations. They help students see the personal meaning of otherwise abstract material. Among the new features are "Does the 'F-word' Fit You?" examining feminist attitudes and self-identities; "What Kind of Touching Makes You Feel Loved?" exploring the importance we place on different displays of physical affection; and "How Nuclear is Your Family?" asking students to consider the roles played by grandparents in their lives.

Issues and Insights. These boxes focus on high-interest topics. New to this edition, examples include features on the multiple meanings of virginity loss, gender ideals in middleclass Black families, the topics most couples fight about, and the familial aftermath of 9/11 and Hurricane Katrina.

Popular Culture. Also new to this edition, these features discuss the ways family issues are portrayed through various forms of popular culture. Topics include the depiction of fathers and mothers in comic strips, the portrayal of love relationships in Disney films, the effect of Valentine's Day on breakups, the sexual content of web logs, and distortions in the portrayal of adoption in the news.

Real Families. The final new feature, these provide first person accounts of issues raised in the text as they are experienced by people in their everyday lives. Examples include an autobiographical account of growing up gay, marital conflict experienced by Korean immigrants, heterosexuals who choose to become domestic partners, and stepfathers who "claim" their stepchildren as their own.

Matter of Fact items provide statistics and quick data relating to the concepts in the chapter.

Reflections bring students closer to the material by encouraging them to consider their own ideas and beliefs.

Each chapter also has a *Chapter Summary* and a list of *Key Terms,* all of which are designed to maximize students' learning outcomes. The Chapter Summary reviews the main ideas of the chapter, making review easier and more effective.

The Key Terms are boldfaced within the chapter and listed at the end, along with the page number where the term was introduced. Both Chapter Summaries and Key Terms assist students in test preparation.

Glossary. There is a comprehensive glossary of key terms included at the back of the textbook.

Appendixes. There are appendixes on sexual structure and the sexual response cycle, fetal development, and managing money.

Website Resource Center. Material that had been included in the Resource Center in prior editions is still available online at the Wadsworth website. This includes a self-help directory and practical information on financial, health, sexual, and consumer matters that affect families. There are also study guides and topically organized lists of websites.

Supplements and Resources

The Marriage and Family Experience, Tenth Edition, is accompanied by a wide array of supplements prepared for both the instructor and student. Some new resources have been created specifically to accompany the Tenth Edition, and all of the continuing supplements have been thoroughly revised and updated.

Supplements for the Instructor

Classroom Activities and Lecture Ideas for Marriage and Family

This collection of ideas for classroom activities and lectures is made up of contributions from Marriage and Family instructors around the country. Free to adopters of *The Marriage and Family Experience,* this supplement will provide you with some creative ideas to enhance your lectures.

Instructor's Edition of *The Marriage and Family Experience,* Tenth Edition

The Instructor's Edition contains a visual preface, a walk-through of the text that provides an overview of its key features, themes, and supplements.

Instructor's Resource Manual with Test Bank

This manual will help the instructor to organize the course and to captivate students' attention. The manual includes a chapter focus statement, key learning objectives, lecture outlines, in-class discussion questions, class activities, student handouts, extensive lists of reading and online resources, and suggested Internet sites and activities. The Test Bank portion includes approximately 40 to 50 multiple choice, 20 true/false, 10 short answer, and 5 to 10 essay questions, all with answers and page references, for each chapter of the text.

ExamView® Computerized Testing

Create, deliver, and customize tests and study guides (both print and online) in minutes with this easy-to-use assessment and tutorial system. *ExamView* offers both a Quick TestWizard and an Online Test Wizard that guide you step-by-step through the process of creating tests. The test appears on the screen exactly as it will print or display online. Using *ExamView*'s complete word processing capabilities, you can enter an unlimited number of new questions or edit existing questions.

Transparency Masters

A selection of masters consisting of tables and figures from this text is available to help prepare lecture presentations. Free to qualified adopters.

Wadsworth Sociology Video Library

This large selection of thought-provoking films, including many from the Films for the Humanities collection, is available to adopters who meet adoption criteria. Please contact your local sales representative for more information.

ABC®Video: Marriage and Family, Volume 2

Launch your lectures with riveting footage from ABC, a leading news television network. ABC Videos allow you to integrate the newsgathering and programming power of ABC into the classroom to show students the relevance of course topics to their everyday lives. Organized by topics covered in a typical course, these videos are divided into short segments—perfect for introducing key concepts. The wide selection includes thought-provoking clips such as "David Reimer: Raised as a Girl," "All-Female Fire Department," and "Effect of Holding Hands." High-interest clips are followed by questions designed to spark class discussion.

Marriage and Family DVD (also available on VHS)

Enhance your classroom lecture with unique, Wadsworth-owned video clips on the following topics: gender, relationships, and AIDS.

JoinIn™ on TurningPoint®

This interactive classroom tool turns your ordinary Microsoft®PowerPoint® application into powerful audience-response software, allowing you to take attendance, poll students on key issues to spark discussion, check student comprehension of difficult concepts, collect student demographics to better assess student needs, and even administer quizzes without collecting papers or grading. Polling Questions—mini-quizzes students can complete to measure a particular aspect of themselves or their relationship with their partners—are now available through JoinIn on TurningPoint, allowing instructors to use this feature as a lecture launcher or to stimulate class discussion. This will also have Marriage and Family Video Clips with accompanying multiple choice questions—designed to test comprehension and illustrate certain key points of the clip.

Families and Society: Classic and Contemporary Readings

This reader is designed to promote a sociological understanding of families, at the same time demonstrating the diversity and complexity of contemporary family life. The different sections of the reader are designed to "map" onto most textbooks and course syllabi relating to sociology of the family. Edited by Scott L. Coltrane, University of California, Riverside.

The Marriage and Families Activities Workbook

This workbook of interactive self-assessments was written by Ron J. Hammond and Barbara Bearnson, both of Utah Valley State College. They present questions such as: What are your risks of divorce? Do you have healthy dating practices? What is your cultural and ancestral heritage, and how does it affect your family relationships? The answers to these and many more questions are found in this workbook of nearly a hundred interactive self-assessment quizzes designed for students studying marriage and family. These self-awareness instruments, all based on known social science research studies, can be used as in-class activities or homework assignments.

Supplements for the Student

Study Guide

For each chapter of the text, this student study tool contains a chapter focus statement, key learning objectives, key terms, chapter outlines, assignments, Internet activities and websites, and practice tests containing 20 multiple choice and 15 true/false with answers and page references, and 5 short answer questions with page references.

Audio Downloads

Thomson iAudio features MP3-ready chapter review content for quick study. Whether walking to class, doing laundry, driving to work, or studying at their desk, students now have the freedom to choose when, where, and how they interact with their audio-based educational media. Students may purchase access to Thomson Audio Study Products for this text online at www.thomsonedu.com.

Online Resources

Companion Website

www.thomsonedu.com/sociology/strong

Students can gain an even better understanding of the material by using the additional study resources at the book companion website. For example, a Pre- and Post-Test quizzing tool helps gauge understanding of chapter topics, and flash cards help master key terms. Visit the Marriage and Family Resource Center on the site. You'll also find special features such access to InfoTrac® College Edition (a database that allows you access to more than 18 million full-length articles from 5,000 periodicals and journals), as well as GSS Data and Census information to help with research projects and papers.

Acknowledgments

Many people deserve to be recognized for the roles they played in the revision and production of this edition. I thank the following reviewers, whose comments and reactions were encouraging and whose suggestions were helpful: Penny Bove, Walsh University; Naima Brown, Santa Fe Community College; Wanda Clark, South Plains College; Preston M. Dyer, Baylor University; Patricia Gibbs, Foothill College; Deborah Helsel, California State University, Fresno; Joy Jacobs, Michigan State University; Lori Maida, Westchester Community College; and Lorraine M. Perry, Cameron University.

This book remains the product of many hands. Bryan Strong and Christine DeVault created a wonderful book from which to branch. Their strong convictions about the meaning and importance of families, their well-conceived organization, and their reader-friendly prose make it obvious why this book has appealed to so many for so long. I am proud to, once again, follow on their heels.

A number of people at Wadsworth Publishing deserve thanks. Editor-in-chief Eve Howard, publisher Michele Sordi, and Sociology editor Chris Caldeira, showed considerable faith in and continued, enthusiastic support for this book. Chris brought many fresh ideas and wise suggestions. I hope that I have done at least some of them justice. My developmental editor, Sherry Symington was extraordinary. She offered encouragement, expressed enthusiasm, reminded me of deadlines (and helped me meet them) and helped immensely with the editing and rewriting. This book would not be what it is without her efforts. I am enormously grateful and truly fortunate to have worked with her.

Tali Beesley, editorial assistant, performed tremendously in helping assemble the manuscript for production; Christina Ho, the assistant editor, has put together the strong ancillary package that accompanies the Tenth Edition; Dave Lionetti, the technology project manager, is responsible for the state of the art technology and website supporting the text; while Michelle Williams, marketing manager, and Darlene Amidon-Brent, marketing communications manager, have skillfully directed the introduction of the Tenth Edition to adopters and prospective adopters. I want to extend my thanks to Cheri Palmer, the senior production project manager at Wadsworth, who oversaw the complex production process with great skill. Thanks, too, to Bob Kauser in the permissions department at Wadsworth. Anne Williams and Graphic World Inc. were tremendously helpful and highly competent in the copyediting and production phases. The text looks and reads better just because of their involvement. My appreciation again goes to Sarah Evertson, photo editor and researcher, for finding such good examples of what were occasionally vaguely requested subjects.

Once again, I wish to express appreciation to my colleagues and friends at both Ohio Wesleyan University and Rowan University for the support they provided me. Ohio Wesleyan allowed me to take an unpaid leave of absence, without which I would have been unable to invest the time and energy I have in this book. Rowan University has been a comfortable place to be during the 2004-2005 and 2006-2007 academic years. When I return to my good friends and colleagues at Ohio Wesleyan, I will miss my new friends and colleagues at Rowan. My Ohio Wesleyan colleagues Jan Smith, Mary Howard, Akbar Mahdi, Jim Peoples, John Durst, and Pam Laucher, along with the dozen colleagues I have enjoyed at Rowan, make me realize that I have been very fortunate. I appreciate their friendship and support.

I want again to express my appreciation to my family. My parents, Kalman and Eleanor Cohen, and sisters, Laura Cohen and Lisa Merrill, continue to give me a solid foundation on which to stand. My children, Danny and Allison, continue to bring me unimaginable joy as they blossom into adulthood. I am more proud of them than I can ever adequately show. They are wonderful legacies to their beautiful mother and invaluable friends to me. My stepchildren, Daniel, Molly, and Brett have reminded me what it is that children can bring to one's life. I feel honored to be a part of their lives and to share mine with them.

I want to express special thanks to my wife, Julie, who first entered my life as a supportive friend and has become my partner in life. She continuously offers me empathy, wise and practical advice, steady encouragement, and a truly unimaginable love. More than even that, despite each having lived lives with too much sadness and loss, she has given me a reason to look forward to the rest of my life, to imagine a future that is brighter and more beautiful because it will be spent with her.

About the Author

THEODORE COHEN is a professor in the sociology/anthropology department at Ohio Wesleyan University. He earned his M.A. and Ph.D. in sociology from Boston University. His research and teaching specializations center around gender and family life, with special attention to men's family lives and to emergent family lifestyles. He is the editor of *Men and Masculinity: A Text Reader* (Wadsworth, 2000).

The Meaning of Marriage and the Family

Outline

What Do YOU Think?

Are the following statements TRUE or FALSE?
You may be surprised by the answers (see answer key on the following page).

T F **1** Most American families are traditional nuclear families in which the husband works and the wife stays at home caring for the children.

T F **2** Families are easy to define and count.

T F **3** No U.S. state prohibits interracial marriage.

T F **4** All cultures traditionally divide at least some work into male and female tasks.

T F **5** In the United States, all states recognize same-sex "civil unions."

T F **6** There is widespread agreement about the nature and causes of change in American family patterns.

T F **7** Most cultures throughout the world prefer monogamy—the practice of having only one husband or wife.

T F **8** Married men tend to live longer than single men.

T F **9** Most people who divorce eventually marry again.

T F **10** Nuclear families, single-parent families, and stepfamilies are equally valid family forms.

A course in marriage and the family is unlike almost any other course you are likely to take. At the start of the term—before you purchase any books, before you attend any lectures, and before you take any notes—you may believe you already know a lot about families. Indeed, each of us acquires much firsthand experience of family living before being formally instructed about what families are or what they do. Furthermore, each of us comes to this subject with some pretty strong ideas and opinions about families: what they're like, how they should live, and what they need. Our personal beliefs and values shape what we think we know as much as our experience in our families influences our thinking about what family life is like. But if pressed, how would we describe American family life? Are our families "healthy" and stable? Is marriage important for the well-being of adults and children? Are today's fathers and mothers sharing responsibility for raising their children? How many cheat on each other? What happens when people divorce? Do stepfamilies differ from biological families? How common are abuse and violence in families? Questions such as these will be considered throughout this book; they encourage us to think about what we know about families and where our knowledge comes from.

In this chapter, we examine how marriage and family are defined by individuals and society, paying particular attention to the discrepancies between the realities of family life as uncovered by social scientists and the impressions we have formed elsewhere. We then look at the functions that marriages and families fulfill and examine extended families and kinship. We close by introducing the themes that will be pursued through the remaining chapters.

Personal Experience, Social Controversy, and Wishful Thinking

Experience versus Expertise

As we begin to study family patterns and issues, we need to understand that our attitudes and beliefs about families may affect and distort our efforts. In contemplating the wider issues about families that are the substance of this book, it is likely that we will consider our own households and family experiences. How we respond to the issues and information presented over the 600-plus pages and 14 chapters that follow will be influenced by what we have experienced in and come to believe about families. For some of us, those experiences have been largely loving and the relationships have remained stable. For others, family life has been characterized by conflict and bitterness, separations and reconfigurations. Most people have experienced both sides of family life, the love and the conflict, whether their families remained intact or not.

The temptation to draw conclusions about families from personal experiences of particular families is understandable. Thinking that experience translates to expertise, we may find ourselves tempted to generalize from what we experience to what others must also encounter in family life. The dangers of doing that are clear; although the knowledge we have about our own families is vividly real, it is also highly subjective. We "see" things, in part, as we want to see them. Likewise, we overlook some things because we don't want to accept them. Perhaps, we want to pretend they don't exist. The meanings we attach to our experiences are affected by the emotions we feel within the relationships that comprise our families. Our family members are likely to have different perceptions and attach different meanings to even those same relationships. Thus, the understanding we have of our families is very likely a distorted one.

Furthermore, no other family is exactly like your family. We don't all live where you live or how you live, and we don't all possess the same financial resources, draw from the same cultural backgrounds, and build on the same sets of experiences that make your family unique. As well as we might think we know our families, they are poor sources of more general knowledge about the wider marital or family issues that are the focus of this book.

Ongoing Social Controversy

Learning about marriage and family relationships is challenging for another reason. Few areas of social life are more controversial than family matters. Just consider the following news stories. Can you identify any underlying issues involved? What is your position on such issues?

- On September 9, 2005, Texas juvenile court judge Carl Lewis ordered that 13-year-old Katie Wernecke receive chemotherapy to treat her Hodgkin's disease, despite her parents' opposition. Custody of Katie was taken from her parents, Edward and Michelle Wernecke, after Michelle left the state with Katie. The Werneckes did not oppose medical treatment on religious grounds but rather opposed the high-dosage treatment because they felt it posed other medical problems (Associated Press, June 16, 2005). On October 31, 2005, state district court judge Jack Hunter returned Katie, whose health was declining, to the custody of her parents, who still wanted to seek alternative, mostly vitamin, treatments. Doctors estimated that Katie's chances for survival had dropped from 80% to 20% because of the incomplete treatment she received (Brezosky 2005).

- In May 2001, 52-year-old Tom Green became the first Utah man in more than 50 years to be prosecuted for bigamy. Green, a fundamentalist Mormon, proudly proclaimed that his family of five wives and their 33 children was an expression of his devout Mormon faith. For nonsupport and multiple counts of bigamy, he was convicted and sentenced to 5 years in prison. Subsequently, Green was then tried for child rape for having had sex with a 13-year-old girl who later became one of his wives and who gave birth to seven of his children. He was further sentenced to 5 years to life. At present, an estimated 30,000–50,000 people live in polygamous families in Utah.

- The 11 adopted special needs children of Michael and Sharen Gravelle were taken from their custody after it was discovered that 8 of the 11 were kept in "enclosures" or wooden cages, without pillows or mattresses, either overnight or as discipline (Associated Press, January 9, 2006). The children have disorders

AP Images/Lynn Ischay/THE PLAIN DEALER

■ *Michael Gravelle stands next to a bunk bed Sunday, October 23, 2005, built over a clothing storage area in the room where four of his adopted children slept in cage-like enclosures. Gravelle and his wife Sharen lost custody of their eleven adopted special needs children when it was discovered that they made some sleep in cages.*

such as fetal alcohol syndrome, autism, human immunodeficiency virus, and pica, a disorder that involves eating nonfood items. Although the Gravelles claimed to stand behind their childrearing practices, they said they would give up the enclosures and be more lenient in their discipline to get their children back. Meanwhile, the Ohio agency responsible for overseeing children's needs came under severe criticism for allowing the situation to go unnoticed.

- On June 17, 2005, Tina Burch, lesbian partner of the late Christina Smarr, was awarded custody of Smarr's 5-year-old son by West Virginia's highest court. The women had been life partners for 4 years, had planned and arranged Smarr's pregnancy, and were raising Smarr's son. Smarr was killed in an auto accident in June 2002. A family court judge awarded Burch custody, only to have it overturned by a Clay County circuit court judge on the grounds that West Virginia law doesn't give gay partners the right to legal guardianship of a former partner's child. A divided West Virginia Supreme Court overturned the circuit court decision and declared that a "psychological parent" could be a biological, adoptive, foster, or stepparent. The court decision was a milestone: for the first time a same-sex partner was accorded psychological parent status (Ramsey 2005).

Each of the preceding cases contain underlying family issues that spark considerable disagreement and social controversy. In determining both the medical care and the disciplinary methods to which children will be subjected, how much freedom and latitude should parents enjoy? How much should the state restrict people's choices of whom they wish to marry? How far do the rights of gays and lesbians extend in areas of marriage and parenting? This is but a partial list of the issues and implications of the aforementioned cases. And these cases are but a sampling of ongoing controversies to which we could add, for example, grandparents' rights, implications of advances in reproductive technology, divorce-related policy initiatives, custody and child support, and social policies and personal strategies of juggling paid work and family life. As a society, we are often divided, sometimes deeply and bitterly, on such family issues. That we are so deeply invested in certain values regarding family life makes a course about families a different kind of learning experience than if you were studying material to which you were less connected. Ideally, as a result, you will find yourself more engaged, even provoked, to think about and question things you take for granted. At minimum, you will be exposed to information that can help you more objectively understand the realities behind the more vocal debates.

families, cohabiting adults, child-free families, families headed by gay men or by lesbians, and so on. With such variety, how can we define family? What are the criteria for identifying these groups as families?

For official counts of the numbers and characteristics of American families, we can turn to the U.S. Census Bureau. The Census Bureau defines a **family** as "a group of two or more persons related by birth, marriage, or adoption and residing together in a household" (U.S. Census Bureau 2001). A distinction is made between a family and a **household.** A household consists of "one or more people—everyone living in a housing unit makes up a household" (Fields 2003). Single people who live alone, roommates, lodgers, and live-in domestic service employees are all counted among members of households, as are family groups. *Family households* are those in which at least two members are related by birth, marriage, or adoption (Fields 2003). Thus, the U.S. Census reports on characteristics of the nation's households *and* families (Figure 1.1). Of the 111,278,000 households in the United States in 2003, 75,596,000, or 68%, were family households (Fields 2003). Among family households, 76% (57,320,000) consisted of married couples, either with or without children (Fields 2003).

In individuals' perceptions of their own life experiences, *family* has a less precise definition. For example, when we asked our students whom they included as family members, their lists included such

What Is Family?
What Is Marriage?

To accurately understand marriage and family, it is important to define these terms. Before reading any further, think about what the words *marriage* and *family* mean to you. As simple and straightforward as this may seem, as you attempt to systematically define these words, you may be surprised at the complexity involved.

Defining Family

As contemporary Americans, we live in a society composed of many kinds of families—married couples, stepfamilies, single-parent families, multigenerational

Figure 1.1 ■ **Household Composition, 2003**

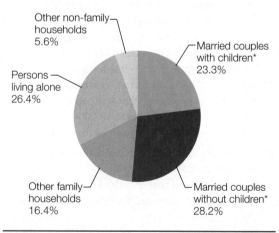

Other non-family households 5.6%

Married couples with children* 23.3%

Persons living alone 26.4%

Other family households 16.4%

Married couples without children* 28.2%

SOURCE: U.S. Census Bureau, *Current Population Survey,* March, 2nd Annual Social and Economic Supplements: 1970–2003

expected relatives as mother, father, sibling, spouse, as well as the following:

best friend	lover	priest
boyfriend	minister	rabbi
girlfriend	neighbor	and teacher.
godchild	pet	

Most of those designated as family members are individuals related by descent, marriage, remarriage, or adoption, but some are **affiliated kin**—unrelated individuals who feel and are treated as if they were relatives.

Reflections

Think about all the people you consider your family. What criteria—biological, legal, affectional—did you use? Did you exclude any biological or legal family? If so, whom and why?

Furthermore, being related biologically or through marriage is not always sufficient to be counted as a family member or kin. One researcher (Furstenberg 1987) found that 19% of the children with biological siblings living with them did not identify their brothers or sisters as family members. Sometimes an absent or divorced parent was not counted as a relative. Stepparents, stepsiblings, or stepchildren were the most likely not to be viewed as family members (Furstenberg 1987; Ihinger-Tallman and Pasley 1987). Emotional closeness may be more important than biology or law in defining family.

There are also ethnic differences as to what constitutes family. Among Latinos, for example, *compadres* (or godparents) are considered family members. Similarly, among some Japanese Americans the *ie* (pronounced "ee-eh") is the traditional family. The *ie* consists of living members of the extended family (such as grandparents, aunts, uncles, and cousins), as well as deceased and yet-to-be-born family members (Kikumura and Kitano 1988). Among many traditional Native American tribes, the **clan,** a group of related families, is regarded as the fundamental family unit (Yellowbird and Snipp 1994).

A major reason we have such difficulty defining *family* is that we tend to think that the "real" family is the **nuclear family,** consisting of mother, father, and children. The term "nuclear family" is less than 60 years old, coined by anthropologist Robert Murdock in 1949 (Levin 1993). What most Americans consider to be the **traditional family** is a mostly middle-class version of the nuclear family in which women's primary roles are wife and mother and men's primary roles are

Photofest

■ *The strength and vitality of kin ties was a major theme in the popular 2002 movie,* My Big Fat Greek Wedding. *The film has grossed more that $240 million.*

husband and breadwinner. As shown in Chapter 3, the traditional family exists more in our imaginations than it ever did in reality.

Because we believe that the nuclear or traditional family is the real family, we compare all other family forms against these models. To include these diverse forms, the definition of family needs to be expanded beyond the boundaries of the "official" census definition. A more contemporary and inclusive definition describes family as "two or more persons related by birth, marriage, adoption, or choice. Families are further defined by socio-emotional ties and enduring responsibilities, particularly in terms of one or more members' dependence on others for support and nurturance" (Allen, Demo, and Fine 2000). Such a definition more accurately and completely reflects the diversity of contemporary American family experience.

Defining Marriage

More than half of the population of the United States, age 15 and older, is married (U.S. Census Bureau 2004). Among males, 55% are currently married and 68% have at least experienced marriage (that is, are married, separated, divorced, or widowed). Although a smaller percentage of females is currently married (51.6%), 75% of females 15 and older, are or have been married (Fields 2003) (see Figure 1.2).

With marriage being such a central part of adult life for so many, it seems marriage would be an easy phenomenon to define and understand.

Figure 1.2 ■ Marital Status of U.S. Population

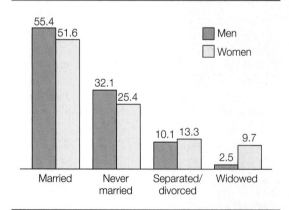

SOURCE: Fields, Jason. 2003. *America's Families and Living Arrangements: 2003*. Current Population Reports, P20-553. U.S. Census Bureau, Washington, DC. (Figure 6) http://www.census.gov/prod/2004pubs/p20-553.pdf.

A **marriage** is a legally recognized union between two people, generally a man and a woman, in which they are united sexually, cooperate economically, and may give birth to, adopt, or rear children. The union is assumed to be permanent (although it may be dissolved by separation or divorce). As simple as such a definition may make marriage seem, it differs among cultures and has changed considerably in our society.

With one exception, the Na of China, marriage has been a universal institution throughout recorded history (Coontz 2005). Despite the universality of marriage, widely varying rules across time and cultures dictate whom one can, should, or must marry; how many spouses one may have at any given time; and where married couples can and should live—including whether husbands and wives are to live together or apart, whether resources are shared between spouses or remain the individual property of each, and whether children are seen as the responsibility of both partners or not (Coontz 2005).

Among non-Western cultures, who may marry whom and at what age varies greatly from our society. In some areas of India, Africa, and Asia, for example, children as young as 6 years may marry other children (and sometimes adults), although they may not live together until they are older. In many cultures, marriages are arranged by families who choose their children's partners. In many such societies, the "choice" partner is a first cousin. And in one region of China, marriages are sometimes arranged between unmarried young men and women who are dead.

Considerable cultural variation exists in what societies identify as the essential characteristics that define couples as married. In many societies, marriage entails an elaborate ceremony, witnessed and legitimated by others, which then bestows a set of expectations, obligations, rights, and privileges on the newly married. Far from this relatively familiar construction of marriage, Stephanie Coontz notes that in some "small-scale societies" the act of eating alone together defines a couple as married. In such instances, as found among the Vanatinai of the South Pacific, for example, dining together alone has more social significance than sleeping together (Coontz 2005). Anthropological study of Sri Lanka revealed that when a woman cooked a meal for a man, this indicated that the two were married. Likewise, if a woman stopped cooking for a man, their marriage might be considered a thing of the past.

Although cultural and historical variation abounds, the following seem to be shared among all arrangements defined as marriages (Coontz 2005):

- Marriage typically establishes rights and obligations connected to gender, sexuality, relationships with kin and in-laws, and legitimacy of children.

- Marriage establishes specific roles within the wider community and society. It specifies the rights and duties of husbands and wives, as well as of their respective families, to each other and makes such duties and responsibilities enforceable by the wider society.

- Marriage allows the orderly transfer of wealth and property from one generation to the next.

Many Americans believe that marriage is divinely instituted; others assert that it is a civil institution involving the state. The belief in the divine institution of marriage is common to religions such as Christianity, Judaism, and Islam and to many tribal religions throughout the world. But the Christian church only slowly became involved in weddings; early Christianity was at best ambivalent about marriage, despite being opposed to divorce (Coontz 2005). Over time, as the church increased its power, it extended control over marriage. Traditionally, marriages had been arranged between families (the father "gave away" his daughter in exchange for goods or services); by the tenth century, marriages were valid only if they were performed by priests. By the thirteenth century, the ceremony was required to take place in a church (Gies and Gies 1987). As states competed with organized religion for power, governments began to regulate marriage. In the United States today, for marriages to be legal—whether they are performed by ministers, priests, rabbis, or imams—they must be validated through government-issued marriage licenses. This is a right for which many gay men and lesbians continue to fight.

Matter of Fact

In 2003, 58.8% of the adult population in the United States (age 18 and older) were married. This includes 60.7% of men and 57.1% of women (U.S. Census Bureau 2004-2005, Tables 51 and 53).

Who May Marry?

Who may marry has changed over the last 150 years in the United States. Laws once prohibited enslaved African Americans from marrying because they were regarded as property. Marriages between members of different races were illegal in more than half the states until 1966, when the U.S. Supreme Court declared such prohibitions unconstitutional. Each state enacts its own laws regulating marriage, leading to some discrepancies from state to state. For example, in some states, first cousins may marry; other states prohibit such marriages as incestuous. We will more fully explore such legal aspects of marriage (such as the age at which one can marry, whom one may marry, and so on) in Chapter 9.

The greatest current controversy regarding legal marriage is over the continuing question of same-sex marriage. As you read this book, we remain amid potentially revolutionary change. Before we look at current developments, let's glance back at the recent past.

Beginning in the 1990s, countries such as Germany, France, the Netherlands, Sweden, and Norway enacted legislation extending marital rights or marriage-like protections to gay couples. Some stopped short of allowing gay or lesbian couples to legally marry, but in the Netherlands, Belgium, Canada, and Spain, as well as in Massachusetts in the United States, the right to marry extends to same-sex couples. Sometime in 2006 (as this book is in production), South Africa will extend the right to marry to gay couples. In addition, a number of countries—including Denmark, Norway, Sweden, Switzerland, Iceland, France, Germany, Finland, Luxembourg, Britain, Portugal, Slovenia, Australia and New Zealand, as well as the states of Connecticut and Vermont in the United States—grant recognition to same-sex couples who register as "domestic partners" or enter "civil unions." With the issue in such a dynamic state of change, by the time you read this that list may well have grown.

In the United States, the issue of gay marriage has been in flux for more than a decade. In the 1990s, U.S. courts rendered decisions that seemed to pave the way toward American legalization of same-sex marriage. The two most notable cases were in Hawaii and Vermont. In 1993, the Hawaii Supreme Court ruled that denying gay men and lesbians the right to marry was unconstitutional in that it violated the equal protection clause of the state constitution. This decision led many to anticipate the eventual legalization of same-sex marriage. It also caused opponents of gay marriage to take action. A number of state legislatures, along with the federal government, passed laws that declared marriage to be the union of one man and one woman, which prevented the forced acceptance of gay or lesbian marriages should the Hawaiian decision stand up to an appeal.

In 1996, Congress passed the **Defense of Marriage Act,** and President Bill Clinton signed it into law. This act denied federal recognition to same-sex couples and gave states the right to legally ignore gay or lesbian marriages should they gain legal recognition in Hawaii or any other state. But the earlier Hawaiian decision did not stand. In a November 1998 ballot, 69% of Hawaiian voters chose to amend the state constitution, giving lawmakers the power to block same-sex marriage and limit legal marriage to heterosexual couples.

Similar laws were passed in more than half of the 50 states by November 1998. As 1999 drew to a close, the state of Vermont took a major step toward what some believed would be the eventual legal recognition of gay marriage.

There, three same-sex couples filed lawsuits, challenging a 1975 state ruling prohibiting same-sex couples from marrying. On December 20, 1999, the Vermont Supreme Court ruled that the state legislature had to either grant marriage rights to same-sex couples or assure them a legal equivalent to marriage, providing them the same range of state benefits enjoyed by married heterosexuals.

On April 26, 2000, Vermont Governor Howard Dean signed into law legislation recognizing same-sex "civil unions." Although they are not marriages, "civil unions" are officially entered, offer the same rights and protections as marriages, and must be officially dissolved when they fail. As of January 2005, more than 7,500 such civil unions had been recorded in Vermont, more than 1,100 between state residents and another 6,400 involving residents of almost every state, the nation's capital, and several other countries, including Canada (Vermont Guide to Civil Unions, http://www.sec.state.vt.us/otherprg/civilunions/civilunions.html).

In October 2001, California passed Chapter 893, a law granting gay or lesbian domestic partners many benefits (including tax benefits, stepparent adoption, sick leave, and permission to make medical decisions) otherwise restricted to married couples. Although far less sweeping in scope than Vermont's civil union legislation, Chapter 893 provided same-sex couples more benefits than found anywhere in the United States other than Vermont (Vermont Guide to Civil Unions 2005). In June 2002, Connecticut passed more limited legislation, giving gay or lesbian couples certain partnership rights and responsibilities.

On June 26, 2003, in the case of *Lawrence and Garner v. Texas*, the U.S. Supreme Court ruled 6–3 that

■ *Same sex marriage is now legal in the U.S., but as of 2006, only in the Commonwealth of Massachusetts.*

Issues and Insights The Rights and Benefits of Marriage

Lieutenant Laurel Hester, 49 years old and a long-time officer for the Ocean County, New Jersey, county prosecutor's office, was dying of lung cancer. For months, as her disease spread, Hester fought against the Ocean County freeholders, seeking the right to have her pension benefits pass to her same-sex partner, Stacie Andree. For a variety of expressed reasons, the freeholders rejected her plea. In fact, had she worked in a different New Jersey County or for the New Jersey state government, she would have had the right to leave her benefits to a domestic partner. But pension benefits for those in the Police and Fire Retirement System could only be passed to spouses (Bonafide 2006), and New Jersey state law does not allow same-sex couples to marry. Just weeks prior to her death, Ocean County freeholders relented. Facing considerable public outcry, Ocean County freeholders voted to allow the $13,000 benefit to be transferred to Andree.

Heterosexuals rarely stop to think about the privileges that sexual orientation offers. One such privilege is the right to marry. Those couples who do marry receive many more rights and protections than couples who don't marry.

For heterosexual cohabitants, this is a matter of choice; they do so because they prefer the more informal arrangement. For many same-sex couples, the historical *inability* to

marry has cost them many protections, some of which are listed below. It is the lack of these rights and protections that state courts in the United States (Hawaii, Vermont, and Massachusetts, for example) have found unconstitutional.

- Accidental death benefit for the surviving spouse of a government employee
- Appointment as guardian of a minor
- Award of child custody in divorce proceedings
- Burial of service member's dependents
- Control, division, acquisition, and disposition of community property
- Death benefit for the surviving spouse for a government employee
- Disclosure of vital statistics records
- Division of property after dissolution of marriage
- Funeral leave for government employees
- Income tax deductions, credits, rates exemption, and estimates
- Legal status with partner's children
- Partner medical decisions
- Nonresident tuition deferential waiver
- Payment of worker's compensation benefits after death
- Permission to make arrangements for burial or cremation
- Proof of business partnership
- Public assistance from the Department of Human Services
- Qualification at a facility for the elderly

- Right of survivorship to custodial trust
- Right to change names
- Right to enter into a premarital agreement
- Right to file action for nonsupport
- Right to inherit property
- Right to support after divorce
- Right to support from spouse
- Spousal privilege and confidential marriage communications
- Spousal immigration benefits
- Status of children
- In vitro fertilization coverage

There are also potential personal and emotional benefits related to the right to marry. Knowing that the wider society recognizes, accepts, or respects a relationship may cause feelings of greater self-validation and comfort within the relationship. On the other hand, knowing that people do not respect, accept, or recognize a commitment may cause additional emotional suffering and personal anguish for the partners involved. Opposition to same-sex marriage is rarely based on issues such as legal rights. Opponents most often question the moral acceptability of gay or lesbian relationships. They often refer to religious grounds for their rejection of gay marriage. Morality is harder to address objectively than the question of legal rights. Clearly, those who can and do marry receive substantial privileges and protections that those who don't or can't must live without.

SOURCE: Parents, Families, and Friends of Lesbians and Gays, "What Is Marriage, Anyway?" http://www.pflag.org/education/marriage.html.

existing laws against sodomy, in Texas and 12 other states, were illegal invasions of privacy. The ruling, which struck down the 13 remaining state sodomy statutes, stemmed from a 1998 arrest of two Houston men, John Lawrence and Tyron Garner, who were having sex when police entered their home on a false emergency call. The men were arrested, jailed overnight, and fined $200 under the Texas sodomy statute. Texas was one of four states whose sodomy statute pertained only to same-sex relations. The remaining nine statutes pertained to heterosexuals and homosexuals. All 13 were nullified with the Court's decision. Although the ruling was about private, consensual sex, not about same-sex marriage, many perceived it as a potential step further down the path toward gay marriage.

Of greatest significance, the Massachusetts Supreme Court ruled November 18, 2003, that the state's ban of same-sex marriage was unconstitutional. The ruling gave the state legislature 6 months to remedy the situation. Although Vermont's response was to create civil unions that provided the same rights and benefits as legal marriage, the Massachusetts court's decision specified the *right to marry* (that is, not the right to enter something similar to marriage). Although the Massachusetts legislature and governor remained opposed to same-sex marriage, on February 4, 2004, the state supreme court ruled 4–3 that a "civil union" solution was unacceptable in that it would constitute "an unconstitutional, inferior, and discriminatory status for same-sex couples." Writing in the *Boston Globe,* journalist Raphael Lewis quoted the court: "For no rational reason the marriage laws of the Commonwealth discriminate against a defined class; no amount of tinkering with language will eradicate that stain. . . . The [civil unions] bill would have the effect of maintaining and fostering a stigma of exclusion that the Constitution prohibits" (Lewis 2004).

As you read these words, Massachusetts has had more than 2 years of fully legal gay marriage recognized in the United States for the first time. In the first 16 months of the law, 6,500 gay or lesbian couples married in Massachusetts.

It is difficult to predict what level of opposition to gay marriage will continue in Massachusetts or what effect it will have. It is also difficult to predict what may happen elsewhere in the United States. Some states may eventually recognize civil unions performed in Vermont or same-sex marriages performed in Massachusetts. Other state legislatures might create their own domestic partner legislation. In January 2006, five states—New Jersey, New York, Washington, Iowa, and California—had cases pending much like the Hawaii case that led to civil union legislation there (http://www.lambdalegal.org, 2006). Also, some form of civil union or domestic partnership is available to same-sex couples in six states: Hawaii, Vermont, Connecticut, New Jersey, Maine, and California. It is hard to predict how many additional states may enact similar legislation.

We have witnessed continued reluctance to legalize gay marriage as more than forty other states have since enacted legislation modeled on the Defense of Marriage Act, passing constitutional amendments limiting marriage to heterosexual couples. Ohio, for example, enacted some of the most restrictive defense-of-marriage legislation in the country. The bill, signed by Governor Bob Taft on February 5, 2004, explicitly defined and prohibited gay marriage as "against the strong public policy of the state." It further denies benefits to state employees' unmarried partners, whether they be heterosexuals, gay men, or lesbians. Without reciprocal recognition (i.e., other states acknowledging and supporting same-sex marriages performed in Massachusetts), more civil suits are certain to follow.

Forms of Marriage

In Western cultures such as the United States, the only legal form of marriage is **monogamy,** the practice of having only one spouse at one time. Monogamy is also the only form of marriage recognized in *all* cultures. Interestingly, and possibly surprisingly, it is not the *preferred* form of marriage in most other cultures. Among world cultures, only 24% of the known cultures perceive monogamy as the ideal form of marriage (Murdock 1967). The preferred marital arrangement worldwide is **polygamy,** the practice of having more than one wife or husband. One study of 850 non-Western societies found that 84% of the cultures studied (representing, nevertheless, a minority of the world's population) practiced or accepted **polygyny,** the practice of having two or more wives (Gould and Gould 1989). **Polyandry,** the practice of having two or more husbands, is actually quite rare: where it does occur, it often coexists with poverty, a scarcity of land or property, and an imbalanced ratio of men to women.

Even within polygynous societies, monogamy is the *most widely practiced* form of marriage. In such societies, plural marriages are in the minority, primarily for simple economic reasons: they are a sign of status that relatively few people can afford and require wealth that few men possess. As we think about polygyny, we may imagine high levels of jealousy and conflict among wives. Indeed, problems of jealousy may and do arise in plural marriages—the Fula in Africa, for example, call the second wife "the jealous one." Based on data from 69 polygynous societies (56% of which were in Africa), Jankowiak, Sudakov, and Wilreker suggest that co-wife conflict and competition for access to the husband is common, but there are also circumstances that reduce conflict (for example, when the wives are sisters, when one is fertile and one barren or postmenopausal). For both the men and the women involved, polygyny brings higher status.

Even though conflict and competition among co-wives is often found in polygynous societies, the level is probably less than would result if our monogamous society was to suddenly allow people multiple spouses. In part because of our culture's traditional roots in Christianity, polygamy has been illegal in the United States since a U.S. Supreme Court decision in 1879. Polygamy was prohibited because it was considered a potential threat to public order (Tracy 2002). As a result, polygamy was looked on as strange or exotic. However, it may not seem so strange if we look at actual American marital practices. Considering the high divorce and remarriage rates in this country, monogamy may no longer be the best way of describing our marriage forms. For many, our marriage system might more accurately be called **serial monogamy** or **modified polygamy,** a practice in which one person may have several spouses over his or her lifetime although no more than one at any given time. In our nation's past, enslaved Africans tried unsuccessfully to continue their traditional polygamous practices when they first arrived in America; these attempts, however, were rigorously suppressed by their masters (Guttman 1976). Members of the Church of Jesus Christ of Latter-day Saints, more commonly known as Mormons, practiced polygamy from the 1830s until the late nineteenth century, when they officially abandoned the practice as a condition of Utah's becoming a state. The U.S. Supreme Court decision *Reynolds v. the United States* asserted that polygamy was not protected by the Constitution. Just four years later, in 1882, Congress passed the Edmunds Act, making "bigamous cohabitation" a crime. In 1890, the leadership of the Church of Jesus Christ of Latter-day Saints formally advised members to refrain from polygamy because it was in violation of the law. After numerous warnings, the church leadership began excommunicating members who continued to practice polygamy. These excommunicated Mormons became the fundamentalists that continue, even through to today, to practice plural marriage and live polygamously. The offices of Utah's and Arizona's attorneys general jointly report that there are at least a dozen fundamentalist Mormon groups, ranging in size from 100 to 10,000, living polygamously in parts of the southwest. The two largest, the Fundamentalist Church of Jesus Christ of Latter-day Saints and the Apostolic United Brethren, each claim perhaps as many as 10,000 members. The former lives a fairly isolated and secluded lifestyle. The latter tends to be integrated into the wider society (Shurtleff and Goddard 2005). As explained in a manual jointly published by the Utah and Arizona attorney general's offices, rather than crack down on the polygamy itself (which is criminal, as seen earlier in the case of Tom Green), most law enforcement efforts are focused on crimes committed within polygamous communities, such as tax evasion, child or spouse abuse, sexual assault, or fraud (Shurtleff and Goddard 2005). Although all wives may live as married in polygamous unions, only the first wife has legal status as a wife.

Functions of Marriages and Families

Whether it is the mother–father–child nuclear family, a married couple with no children, a single-parent family, a stepfamily, a dual-worker family, or a cohabiting family, the family generally performs four important functions: (1) it provides a source of intimate relationships; (2) it acts as a unit of economic cooperation and consumption; (3) it may produce and socialize children; and (4) it assigns social roles and status to individuals. Although these are the basic functions that families are "supposed" to fulfill, families do not have to fulfill them all (as in families without children), nor do they always fulfill them well (as in abusive families).

Intimate Relationships

Intimacy is a primary human need. Human companionship strongly influences rates of illnesses such as cancer or tuberculosis, as well as suicide, accidents, and mental illness.

Studies consistently show that married couples and adults living with others are generally healthier and have a lower mortality rate than divorced, separated, and never-married individuals (Ross, Mirowsky, and Goldsteen 1991). Although some of this difference results from what is known as the *selection* factor—wherein healthier people are more likely to marry or live with someone—both marriage and cohabitation yield benefits to health and well-being. This holds true for Caucasians and African Americans (Broman 1988). Chapter 9 will consider in more detail whether it is the selection into marriage of healthier individuals or the protective benefits of marriage that accounts for the health benefits of marriage.

Family Ties

Marriage and the family usually furnish emotional security and support. This has probably been true from the earliest times. Thousands of years ago, in the Judeo-Christian Bible, the book of Ecclesiastes (4:9–12) emphasized the importance of companionship:

> Two are better than one, because they have a good reward for their toil. For if they fall, one will lift up his fellow; but woe to him who is alone when he falls and has not another to lift him up. Again if two lie together, they are warm; but how can one be warm alone? And though a man might prevail against one, two will withstand him. A three-fold cord is not quickly broken.

It is in our families we generally seek and find our strongest bonds. These bonds can be forged from love, attachment, loyalty, obligation, or guilt. The need for intimate relationships, whether they are satisfactory or not, may hold unhappy marriages together indefinitely. Loneliness may be a terrible specter. Among the newly divorced, it may be one of the worst aspects of the marital breakup.

Since the nineteenth century, marriage and the family have become even more important as sources of companionship and intimacy. They have become "havens in a heartless world" (Lasch 1977). As society has become more industrialized, bureaucratic, and impersonal, it is within the family that we increasingly seek and expect to find intimacy and companionship. In the larger world around us, we are generally seen in terms of status. A professor may see us primarily as students; a used-car salesperson relates to us as potential buyers; a politician views us as voters. Only among our intimates are we seen on a personal level, as Jen or Matt. Before marriage, our friends are our intimates. After marriage, our spouses are expected to be the ones with whom we are *most* intimate. With our spouses we disclose ourselves most completely, share our hopes, rear our children, and hope to grow old together.

Pets and Intimacy

The need for intimacy is so powerful that many rely on pets as additional or even substitute sources for satisfaction of those needs. Animals have been important human companions since prehistoric times (Siegel 1993). They have been important emotional figures in our lives, especially if our other relationships are not fulfilling. Unmarried adults, for example, are more attached to their pets than are married men and women (Stallones et al. 1990).

■ *A major function of marriages and families is to provide us with intimacy and social support, thus protecting us from loneliness and isolation.*

This does not mean, however, that we reject Fido or Fluffy when we become romantically involved or married. What often happens is that the pet may become less important—he or she becomes more an "animal" and less "someone" to whom we are emotionally attached.

However, we should neither overstate this change nor assume that it is inevitable. After all, studies of the role of pets in human relationships suggest that the most prized aspects of pets, especially dogs and cats, are their attentiveness to their owners, their welcoming and greeting behaviors, and their role as confidants—qualities valued in our intimate relationships with humans. Pets give children an opportunity to nurture, and they provide a best friend, someone to love. As recent developments in family law reveal, for many people the relationships with their pets outlast their marriages, even becoming the source of custody disputes between divorcing spouses. Consider the case of a San Diego couple who spent close to $150,000 in their efforts to resolve their "custody" dispute over their dog, Gigi. As reported in the *Seattle Times,* to resolve the dispute the judge called on the expertise of an animal behaviorist and viewed a videotape, depicting a "Day in the Life of Gigi" (Gigi seen under the couch, around her bowl, romping through water). The video was designed to help the judge determine whether the dog's lifestyle was better suited for life with the husband or the wife.

Such custody cases, now part of the growing legal subspecialty of animal law, are described by Adam Karp, a lawyer who specializes in them, as more bitterly fought, with more "dirt" thrown back and forth, than even child custody cases (Aviv 2004). More than three dozen law schools, including those at Harvard University and Yale University, now offer animal law courses, most of which cover the issues surrounding pet custody (Aviv 2004). Although there is interesting research on the ways in which we attach human qualities and familial connections to pets, the remainder of this text will consider human experiences in their intimate relationships and their interactions in families.

Economic Cooperation

The family is a unit of economic cooperation that traditionally divides its labor along gender lines—that is, between males and females (Fox and Murry 2000; Ferree 1991). Although a division of labor by gender is characteristic of virtually all cultures, the work that males and females perform varies from culture to culture. Among the Nambikwara in Africa, for example, the fathers take care of the babies and clean them when they soil themselves; the chief's concubines, secondary wives in polygamous societies, prefer hunting over domestic activities. In American society, from the last century until recently, men were expected to work away

■ *Pets are often considered to be members of the family. They often provide their owners with comfort and a sense of intimacy.*

from home whereas women were to remain at home caring for the children and house.

Such tasks are assigned by culture, not biology. Only a man's ability to impregnate and a woman's ability to give birth and produce milk are biologically determined. And some cultures practice *couvade*, ritualized childbirth in which a male gives birth to the child's spirit and his partner gives physical birth.

We commonly think of the family as a consuming unit, but it also continues to be an important producing unit. The husband is not paid for building a shelf or bathing the children; the wife is not paid for fixing the leaky faucet or cooking.

Although children contribute to the household economy by helping around the house, they generally are not paid (beyond an "allowance") for such things as cooking, cleaning their rooms, or watching their younger brothers or sisters (Coggle and Tasker 1982; Gecas and Seff 1991). Yet they are all engaged in productive labor.

Over the past decade, economists have begun to reexamine the family as a productive unit (Ferree 1991). If men and women were compensated monetarily for the work done in their households, the total would be equal to the entire amount paid in wages by every corporation in the United States.

As a service unit, the family is dominated by women. Because women's work at home is unpaid, the productive contributions of homemakers have been overlooked (Ciancanelli and Berch 1987; Walker 1991). Yet women's household work is equal to about 44% of the gross domestic product, and the value of such work is double the reported earnings of women. If women were paid wages for their labor as mothers and homemakers according to the wage scale for chauffeurs, physicians, babysitters, cooks, therapists, and so on, many women would make more for their work in the home than most men do for their jobs outside the home. One economic estimate of a typical homemaker's work placed the yearly value at more than $60,000 (Crittenden 2001). Because family power is partly a function of who earns the money, paying the stay-at-home partner for household work might significantly affect marital relations.

Reproduction and Socialization

The family makes society possible by producing (or adopting) and rearing children to replace the older members of society as they die off. Traditionally, reproduction has been a unique function of the married family. But single-parent and cohabiting families also perform reproductive and socialization functions. As we will look at in some detail in Chapter 10, technological change has also affected reproduction. Developments in contraception, artificial insemination, and in vitro fertilization have separated reproduction from sexual intercourse.

Depending on their contraceptive choices, couples can engage in sexual intercourse with relatively high confidence that they will not become parents.

Innovations in reproductive technology permit many infertile couples to give birth. Such techniques have also made it possible for lesbian couples to become parents.

The family traditionally has been responsible for **socialization**—the shaping of individual behavior to

{"image_ref_note":"© David Young Wolff/PhotoEdit"}

■ *Much childhood socialization occurs in nonfamily settings such as preschools or day-care centers.*

conform to social or cultural norms. Children are help-less and dependent for years following birth. They must learn how to walk and talk, how to take care of themselves, how to act, how to love, and how to touch and be touched. Teaching children how to fit into their particular culture is one of the family's most important tasks.

This socialization function, however, often includes agents and caregivers outside of the family. The involvement of nonfamily in the socialization of children need not indicate a lack of parental commitment to their children or a lack of concern for the quality of care received by their children. Still, nonparental sources of childrearing, which will be addressed in Chapter 11, may be one of the most significant societal changes in our lifetimes. Since the rise of compulsory education in the nineteenth century, the state has become responsible for a large part of the socialization of children older than age 5. Increasing numbers of dual-earner households and employed single mothers have resulted in placing many infants, toddlers, and small children under the care of nonfamily members, thus broadening the role of others (such as neighbors, friends, or paid caregivers) and reducing the family's role in childrearing.

Assignment of Social Roles and Status

We fulfill various social roles as family members, and these roles provide us with much of our identities.

During our lifetimes, most of us will belong to two families: the family of orientation and the family of procreation. The **family of orientation** (sometimes called the *family of origin*) is the family in which we grow up, the family that orients us to the world. The family of orientation may change over time if the marital status of our parents changes. Originally, it may be an intact nuclear family or a single-parent family; later it may become a stepfamily. We may even speak of *binuclear families* to reflect the experience of children whose parents separate and divorce. With parents maintaining two separate households and one or both possibly remarrying, children of divorce are members in two different, parentally based nuclear families.

The common term for the family formed through marriage and childbearing is **family of procreation.** Because many families have stepchildren, adopted children, or no children, we can use a more recent term—**family of cohabitation**—to refer to the family formed

through living or cohabiting with another person, whether we are married or unmarried. Most Americans will form families of cohabitation sometime in their lives.

Much of our identity is formed in the crucibles of families of orientation, procreation, and cohabitation. In a family of orientation, we are given the roles of son or daughter, brother or sister, stepson or stepdaughter. We internalize these roles until they become a part of our being. In each of these roles, we are expected to act in certain ways. For example, children obey their parents, and siblings help one another.

Sometimes our feelings fit the expectations of our roles; other times they do not.

Our family roles as offspring and siblings are most important when we are living in a family of orientation. After we leave home, these roles gradually diminish in everyday significance, although they continue throughout our lives. In relation to our parents, we never cease being children; in relation to our siblings, we never cease being brothers and sisters. The roles simply change as we grow older.

As we leave a family of orientation, we usually are also leaving adolescence and entering adulthood. Being an adult in our society is defined in part by entering new family roles—those of husband or wife, partner, father or mother. These roles formed in a family of procreation take priority over the roles we had in a family of orientation. In our nuclear family system, when we marry we transfer our primary loyalties from our parents and siblings to our partners.

Later, if we have children, we form additional bonds with them. When we assume the role of spouse or bonded partner, we assume an entirely new social identity linked with responsibility, work, and parenting.

In earlier times, such roles were considered lifelong. Because of divorce or separation, however, these roles today may last for considerably less time.

The status or place we are given in society is acquired largely through our families. Our families place us in a certain socioeconomic class, such as blue collar (working class), middle class, or upper class. We learn the ways of our class through identifying with our families. As shown in Chapter 3, different classes experience the world differently. These differences include the ability to satisfy our needs and wants but may extend to how we see men's and women's roles, how we value education, and how we rear our children (Lareau 2003; Rubin 1976, 1994).

Our families also give us our ethnic identities as African American, Latino, Jewish, Irish American, Asian American, Italian American, and so forth. Families also provide us with a religious tradition as Protestant, Catholic, Jewish, Greek Orthodox, Islamic, Hindu, or Buddhist—as well as agnostic, atheist, or New Age. These identities help form our cultural values and expectations. These values and expectations may then influence the kinds of choices we make as partners, spouses, or parents.

Why Live in Families?

As we look at the different functions of the family we can see that, at least theoretically, most of them can be fulfilled outside the family. For example, artificial insemination permits a woman to be impregnated by a sperm donor and embryonic transplants allow one woman to carry another's embryo. Children can be raised communally, cared for by foster families or childcare workers, or sent to boarding schools. Most of our domestic needs can be satisfied by microwaving prepared foods or going to restaurants, sending our clothes to the laundry, and hiring help to clean our bathrooms, cook our meals, and wash the mountains of dishes accumulating (or growing new life-forms) in the kitchen. Friends can provide us with emotional intimacy, therapists can listen to our problems, and sexual partners can be found outside of marriage. With the limitations and stresses of family life, why bother living in families?

Sociologist William Goode (1982) suggests that there are several advantages to living in families:

■ *Families offer continuity as a result of emotional attachments, rights, and obligations.* Once we choose a partner or have children, we do not have to search continually for new partners or family members who can better perform a family task or function such as cooking, painting the kitchen, providing companionship, or bringing home a paycheck. We expect our family members—whether partner, child, parent, or sibling—to participate in family tasks over their lifetimes. If at one time we need to give more emotional support or attention to a partner or child than we receive, we expect the other person to reciprocate at another time. We further expect that we can enjoy the fruits of our labors together. We count on our family members to be there for us in multiple ways. We rarely have the same extensive expectations of friends.

■ *Families offer close proximity.* We do not need to travel across town or the country for conversation or help. With families, we do not even need to leave the house; a husband or wife, parent or child, or brother or sister is often at hand (or underfoot). This close proximity facilitates cooperation and communication.

■ *Families offer an abiding familiarity with others.* Few people know us as well as our family members, because they have seen us in the most intimate circumstances throughout much of our lives. They have seen us at our best and our worst, when we are kind or selfish, and when we show understanding or intolerance. This familiarity and close contact teach us to make adjustments in living with others. As we do so, we expand our knowledge of ourselves and others.

■ *Families provide many economic benefits.* They offer us economies of scale. Various activities, such as laundry, cooking, shopping, and cleaning, can be done almost as easily and with less expense for several people as for one. As an economic unit, a family can cooperate to achieve what an individual could not. It is easier for a working couple to purchase a house than an individual, for example, because the couple can pool their resources. Because most domestic tasks do not take great skill (a corporate lawyer can mop the floor as easily as anyone else), most family members can learn to do them. As a result, members do not need to go outside the family to hire experts. For many family tasks—from embracing a partner to bandaging a child's small cut or playing peekaboo with a baby—there are no experts to compete with family members.

These are only some of the theoretical advantages families offer to their members. Not all families perform all of these tasks or perform them equally well. But families, based on mutual ties of feeling and obligation, offer us greater potential for fulfilling our needs than do organizations based on profit (such as corporations) or compulsion (such as governments).

Extended Families and Kinship

Society "created" the family to undertake the task of making us human. According to some anthropologists, the nuclear family of man, woman, and child is universal, either in its basic form or as the building block for other family forms (Murdock 1967). Other anthropologists disagree that the father is necessary, arguing that the basic family unit is the mother and child *dyad*, or pair (Collier, Rosaldo, and Yanagisako 1982). The use of artificial insemination and new reproductive technologies, as well as the rise of female-headed single-parent families, are cited in support of the mother–child model.

Extended Families

The **extended family,** as already described, consists not only of the cohabiting couple and their children but also of other relatives, especially in-laws, grandparents, aunts and uncles, and cousins. In most non-European countries, the extended family is often regarded as the basic family unit.

For many Americans, especially those with strong ethnic identification and those in certain groups (discussed in Chapter 3), the extended family takes on great importance. Sometimes, however, we fail to recognize the existence of extended families because we assume the nuclear family model as our definition of family. We may even be blind to the reality of our own family structure.

When someone asks us to name our family members, if we are unmarried, most of us will probably name our parents, brothers, and sisters. If we are married, we will probably name our husband or wife and children. Only if questioned further will some bother to include grandparents, aunts or uncles, cousins, or even friends or neighbors who are "like family." We may not name all our blood relatives, but we will probably name the ones with whom we feel emotionally close, as shown earlier in the chapter.

Although most family households in the United States are nuclear in structure, there are more than 4 million multigenerational households in the United States. Looking only at the three most common multigenerational households (householder–householder's parent–householder's child, householder–householder's child–householder's grandchild, householder–householder's parent–householder's child–householder's grandchild), Census 2000 reported that such households represent 3.7% of all U.S. households (Simmons and O'Neil 2001). Such extended family households are somewhat more common among immigrants and where economic necessity dictates. They can be found in greater proportion in states where there are large concentrations of certain ethnic populations. For example, in Hawaii, which has a large Asian population, more than 8% of households are multigenerational. Among families in California, where there is a large Hispanic population, close to 6% of households fall under this arrangement (Max 2004).

The most common type of multigenerational household is one in which the householder lives with both his or her child or children and grandchild or grandchildren. In 2000, these 2.6 million households made up nearly two-thirds of all multigenerational households. Another third, or 1.3 million households, consist of the householder, sandwiched between his or her child or children and parent (or parent-in-law). Only 2% of multigenerational households, numbering 78,000, contain four generations living under one roof (Simmons and O'Neil 2001). But even in the absence of multigenerational households, many Americans maintain what have been called *modified extended families,* in that care and support are shared among extended family members even though they don't share a residence.

Kinship Systems

The **kinship system** is the social organization of the family. It is based on the reciprocal rights and obligations of the different family members, such as those between parents and children, grandparents and grandchildren, and mothers-in-law and sons-in-law.

Conjugal and Consanguineous Relationships

Family relationships are generally created in two ways: through marriage and through birth. Family relationships created through marriage are known as **conjugal relationships.** (The word *conjugal* is derived from the Latin *conjungere,* meaning "to join together.") In-laws, such as mothers-in-law, fathers-in-law, sons-in-law, and daughters-in-law, are created by law—that is, through marriage. **Consanguineous relationships** are created through biological (blood) ties—that is, through birth. (The word *consanguineous* is derived from the Latin *com-,* "joint," and *sanguineous,* "of blood.")

Families of orientation, procreation, and cohabitation provide us with some of the most important roles we will assume in life. The nuclear family roles (such as parent, child, husband, wife, and sibling) combine with extended family roles (such as grandparent, aunt, uncle, cousin, and in-law) to form the kinship system.

Kin Rights and Obligations

In some societies, mostly non-Western or nonindustrialized cultures, kinship obligations may be more extensive than they are for most Americans in the twenty-first century. In cultures that emphasize wider kin groups, close emotional ties between a husband and a wife are viewed as a threat to the extended family. A remarkable form of marriage that illustrates the precedence of the kin group over the married couple is the institution of spirit marriage, which continues today in Canton, China. According to anthropologist Janice Stockard (1989), a **spirit marriage** is arranged by two families whose son and daughter died unmarried. After the dead couple is "married," the two families adopt an orphaned boy and raise him as the deceased couple's son to provide family continuity.

In another Cantonese marriage form, women do not live with their husbands until at least 3 years after marriage, as their primary obligation remains with their own extended families. Among the Nayar of India, men have a number of clearly defined obligations toward the children of their sisters and female cousins, although they have few obligations toward their own children (Gough 1968).

In American society, the basic kinship system consists of parents and children, but it may include other relatives as well, especially grandparents. Each person in this system has certain rights and obligations as a result of his or her position in the family structure. Furthermore, a person may occupy several positions at the same time. For example, an 18-year-old woman may simultaneously be a daughter, a sister, a cousin, an aunt, and a granddaughter.

Each role entails different rights and obligations. As a daughter, the young woman may have to defer to certain decisions of her parents; as a sister, to share her bedroom; as a cousin, to attend a wedding; and as a granddaughter, to visit her grandparents during the holidays.

In our culture, the nuclear family has many norms regulating behavior, such as parental support of children and sexual fidelity between spouses, but the rights and obligations of relatives outside the basic kinship system are less strong and less clearly articulated. Because neither culturally binding nor legally enforceable norms exist regarding the extended family, some researchers suggest that such kinship ties have become voluntary. We are free to define our kinship relations much as we wish. Like friendship, these relations may be allowed to wane (Goetting 1990).

Despite the increasingly voluntary nature of kin relations, our kin create a rich social network for us. Studies suggest that most people have a large number of kin living in their areas (Mancini and Blieszner 1989). Adult children and their parents often live close to one another; make regular visits; and help one another with childcare, housework, maintenance, repairs, loans, and gifts. The relations among siblings also are often strong throughout the life cycle (Lee, Mancini, and Maxwell 1990).

We generally assume kinship to be lifelong. In the past, if a marriage was disrupted by death, in-laws generally continued to be thought of as kin. But today, divorce is as much a part of the American family system as marriage. Although shunning the former spouse may no longer be appropriate (or polite), no new guidelines on how to behave have been developed. The ex-kin role is a role with no clearly defined rules.

The Major Themes of This Text

Throughout the many chapters and pages that follow, as we examine in detail intimate relationships, marriage, and family in the United States, we will introduce a range of theories, provide much data, and look at a number of family issues and relationships in ways you may never have considered before. As we do so, we will visit and revisit the following points:

Families Are Dynamic

As we will see shortly (in Chapter 3) and throughout the text, the family is a dynamic social institution that has undergone considerable change in its structure and functions. Similarly, values and beliefs about families have changed over time. We are more accepting of divorce, employed mothers, and cohabitation. We expect men to be more involved in hands-on childcare. We place more importance on individual happiness than on self-sacrifice for family.

In Chapter 3, we explore some of the major changes that have occurred in how Americans experience families. Then, throughout the text, as we address topics such as marriage, divorce, cohabitation, raising children, and managing employment and family, we ask, In what ways have things changed, and why? What consequences and implications result from these changes? Because familial change is often differently perceived and interpreted (see the final theme in this section), we also present different possible interpretations of the meaning of change. Are families merely changing, or are they declining?

Throughout much of the text we also look at how individual family experience changes over time. From the formation of love relationships, the entry into marriage, the bearing, raising, and aging of children, the aging and death of spouses, families are ever changing.

Families Are Diverse

Not all families experience things the same way. Beginning with Chapter 3, we look closely at a variety of factors that create differences in family experience. We consider, especially, the following major sources of patterned variation in family experience: race, ethnicity, gender, social class, sexuality and lifestyle choice.

"Race" and Ethnicity

There were more than 240 different native cultures that lived in what is now the United States when the colonists first arrived (Mintz and Kellogg 1988). Since then, American society has housed immigrant groups from the world over who bring with them some of the customs, beliefs, and traditions of their native lands, including those about families. Thus, we can speak of African American families, Latino families, Asian families, Native American families, European families, Middle Eastern families, and so on. In Chapter 3, we provide a brief sketch of the major characteristics of the family experiences of each of these racial or ethnic groups. As we proceed from there, we compare and contrast, where relevant and possible, major differences in family experiences across racial and ethnic lines.

Social Class

Different social classes (categories of individuals and families that share similar economic positions in the wider society) have different experiences of family life. Because of both the material and the symbolic (including cultural and psychological) dimensions of social class, our chances of marrying, our experiences of marriage and parenthood, our ties with kin, our experience of juggling work and family, and our likelihood of experiencing violence or divorce all vary. And this is but a partial list of major areas of family experience that differ among social classes.

Gender

Although gender roles have changed considerably over time, gender differences still surface and loom large in each area of marriage and family on which we touch. Love and friendship, sexual freedom and expression, marriage responsibilities and gratifications, involvement with children, experience of abuse, consequences of divorce and becoming a single parent, and chances for remarriage all differ between women and men.

Sexuality and Lifestyle Variation

A striking difference between twenty-first-century families and early American families is the diversity of family lifestyles that people choose or experience. There is no family form that encompasses most people's aspirations or experiences. Statistically, the dual-earner household is the most common form of family household with children, but there is considerable variation among dual-earner households and between such households traditional or single-parent families.

Increasingly, people choose to cohabit our experiences of marriage and parenthood, increasing numbers of couples choose not to have children, and increasing numbers of others choose expensive procedures to assist their efforts and enable them to bear children. This diversity of family types and lifestyles will not soon abate. In the chapters that follow specific attention is directed at singles (with and without children), cohabitation, childless or child-free couples, and role-reversed households. In addition, we examine sexual orientation and, where data are available, compare and contrast how experiences of such things as intimacy, sexual expression, parenting, abuse, and separation differ among heterosexuals, gay men, and lesbians.

Family Experience Is Influenced by Social Institutions and Forces Outside of the Family

This book takes a mostly sociological approach to relationships, marriage, and families, in that we repeatedly stress the outside forces that shape family experiences. The family is one of the core social institutions of society, along with the economy, religion, the state, education, and health care. As such, the shape and substance of family life is heavily affected by the needs of the wider society in which it is located. In addition, other social institutions influence how we experience our families.

Similarly, cultural influences in the wider society, such as the values and beliefs about what families are or should be like and the norms (or social rules) that distinguish acceptable from unacceptable behavior, guide how we choose to live in relationships and families. Thus, although each of us as an individual makes a series of decisions about the kinds of family lives we want, the choices we make are products of the societies in which we live.

In addition, options available to each of us may not reflect what we would freely choose if we faced no constraints on our choices. So, for example, parents who might prefer to stay at home with their children might find such a choice impractical to impossible because economic necessity forces them to work outside the home. Working parents may find the time they spend with their children more a reflection of the demands of their jobs and the inflexibility of their workplaces than of their own personal preferences, just as some at-home parents might prefer to be employed, but find that their children's needs, the cost and availability of quality childcare, the jobs available to them, and the demands and benefits contained in those jobs push them to stay home.

Even the decision to marry requires a pool of potential and suitable spouses from which to choose and the preferred marital choice to be accepted within the society in which we live. We cannot marry if there are no "marriageable options available" (as may be the case in many inner-city, low-income areas) or if our choice of spouse is not legally allowed (as in gay or lesbian relationships).

Our familial lives reflect decisions we face, choices we make, and the opportunities and/or constraints we confront. In the wider discourse about families, we tend to encounter mostly individualistic explanations for what people experience, focusing sometimes exclusively on personal choices. Throughout this text, we examine the wider environments within which our family choices are made and the ways in which some of us are given more opportunities whereas others face limited choices.

Healthy Families Are Essential to Societal Well-Being, and Societal Support Is Essential to Familial Well-Being

Family is the irreplaceable means by which most of the social skills, personality characteristics, and values of individual members of society are formed. Hope, purpose, and general attitudes of commitment, perseverance, and well-being are nurtured in the family. Indeed, even the rudimentary maintenance and survival care provided by families is no small contribution to the well-being of a community.

Some of the services provided by families are such a basic part of our existence that we tend to overlook them. These include such essentials as the provision of food and shelter—a place to sleep, rest, and play—as well as caretaking, including supervision of health and hygiene, transportation, and the accountability of family members involving their activities and whereabouts. Without families, communities would have to provide extensive dormitories and many personal-care workers with different levels of training and responsibility to perform the many activities in which families are engaged. On a more emotional level, without families individuals must look elsewhere to satisfy basic needs for intimacy and support. We marry or form marriage-like cohabiting relationships, have children, and maintain contact with other kin (adult siblings, aging parents, and extended kin) because such relationships retain importance as bases for our identities and sources of social and emotional sustenance. We bring to these relationships high affective expectations.

When our intimacy needs are not met (in marriage or long-term cohabitation), we terminate those relationships and seek others that will provide them. We believe, however, that those needs are best met in families.

To function effectively, if not optimally, families need outside assistance and support. Better childcare, more flexible work environments, economic assistance for the neediest families, protection from violent or abusive partners or parents, and a more effective system for collecting child support are just some examples we consider of where families clearly have needs for greater societal or institutional support. The health and stability of our society depend largely on strong and stable families. When families fail, individuals must turn to society for assistance; social institutions must be designed to fill the voids left by failing families, and the pathologies created by weak family structures make society a less livable place. There are enormous costs that result from neglecting the needs of America's families and children.

Family Patterns Can Be Interpreted Differently Depending on Individual Values

As we noted at the outset of this chapter, marriage and family issues inspire much debate. With so much "noise" in the wider society around what family life is and should be like, how families are changing and whether those changes are good or bad, it may be difficult to know what conclusions to draw about family issues.

Many of the so-called culture wars over such "hot button" issues as the status of women, abortion, the effects of divorce, nonmarital births, and gay rights may really be conflicts over differing conceptions of family (Benokraitis 2000; Glenn 2000; Hunter 1991).

For instance, those who believe that families of male providers, female homemakers, and their dependent children living together, 'til death do they part, are what families *should be* cannot be encouraged by the continued high rates of divorce and cohabitation or by the declining rates of marriage or full-time motherhood. Those on the "other" side who claim that there are basic inequities within the traditional family especially regarding the status of women, will not mourn the diminishing numbers of breadwinner–housewife families. Similarly, the question of gay marriage will divide those who believe marriage *must* be a relationship between a man and a woman from those who believe that we *must* recognize and support all kinds of families.

Given the lack of societal consensus, it is easy to become confused or be misled about what American families are really like. There is undeniable evidence that family life has changed, repeatedly and dramatically, throughout history, as familial "change . . . not stability . . . has been the norm" since colonial days (Mintz and Kellogg, 1988). But not everyone sees change through the same lens. To some, contemporary family life is weaker because of cultural and social changes and is now, to some extent, endangered (Wilson 2002; Popenoe 1993).

More optimistic interpretations of changing family patterns celebrate the increased domestic diversity of numerous family types and the richer range of choices now available to Americans (Coontz 1997; Stacey 1993). Like the proverbial glass, some see the family as "half empty" others see it as "half full." What makes the "half full, half empty" metaphor so apt is that even when looking at the same phenomenon or the same trend, some interpret it as evidence of the troubled state U.S. families are in and others see today's families as different or changing. So, for instance, although the rates of divorce and marriage, the numbers of children in nonparental childcare, or the extent of increase in cohabitation can, like the volume of liquid in a partially filled glass, be objectively measured, the meaning of those measures can vary widely depending on perspective.

The following example nicely illustrates this. In October 2005, PBS (the public broadcasting network) conducted a poll of American attitudes and opinions on a host of family issues. Sampling 1,130 American adults for the program "Religion and Ethics Newsweekly," the pollsters asked about a number of family issues. The survey garnered interesting results. Consider a few:

- 80% of respondents agreed that it is better for children if their parents are married.

- 71% believe that "God's plan for marriage is one man, one woman, for life."

- 49% agree that it is okay for a couple to live together without intending to marry.

- 52% agree that divorce is the best solution for a couple who cannot work out their marriage problems.

- 55% agree that "Love makes a family . . . and it doesn't matter if parents are gay or straight, married or single."

- When asked if the government should play a role in encouraging people to marry and stay married or the government should stay out, more than three-fourths say stay out.

- 73% agree that a "working mother" can have just as warm and secure a relationship with her children as a stay-at-home mother.

Interestingly, within each of these items there were big differences in attitudes based on respondents' religious backgrounds. Look again at some of these same items, comparing respondents of different religious backgrounds (see Table 1.1).

What the data clearly show is that big differences exist between those of more traditional or conservative Christian backgrounds and "mainline Protestants" or liberal Catholics (no other religious groups were included in the sample). Overall, the most liberal attitudes are held by those who identified themselves as having no religious preference (or as atheists or agnostics). Note, however, the differences between traditional and liberal Catholics and between evangelical Christians and mainline Protestants. Such differences are often obscured when we look at overall attitudes of Americans or even at attitude differences between Protestants and Catholics. Clearly, religious affiliation and degree of identification are among the sources of difference in attitudes about families.

This divisiveness is neither new nor unique to the United States. In the early twentieth century we witnessed considerable pessimism about whether families would survive the changing and liberalizing culture of sexuality, the increasing numbers of women delaying marriage for educational or occupational reasons, the declining birthrate and increases in divorce. In considering the same sorts of changes, others advocated that these trends were positive signs of families adapting to changes in the wider society (Mintz and Kellogg 1988).

In recent years, many other countries have faced similar cultural clashes over trends and changes in family life. In Spain, for example, there is a dispute pitting the Spanish socialist government against the Catholic Church, as governmental initiatives to legalize same-sex marriage, and make abortion and divorce easier or quicker have met with strong and vocal opposition from the church. Whereas some in the Spanish Socialist Party or among its allies such as the United Left Party believe Spain hasn't gone far enough in recognizing and embracing change, organizations aligned with the church, such as the Institute of Family Policy, consider the climate in Spain "family phobic" (Fuchs 2004).

Ultimately, the ways we view families depend on what we conceive of *as* families. Such disagreements reflect both different definitions of family and different value orientations about particular kinds of families. Often the product of personal experience as much

Table 1.1 ■ **Religious Differences in Attitudes toward Family Issues: Results from PBS "Faith and Family" Survey, October 2005**

Item	Total	Evangelical Christian (%)	Mainline Protestant (%)	Traditional Catholic (%)	Liberal Catholic (%)	No Preference/ Atheist/Agnostic (%)
Better for children if parents are married	80	86	82	88	75	58
God's plan for marriage. . .	71	92	62	91	60	31
Divorce is usually best	52	48	61	46	63	50
All right to live together	49	21	57	38	72	78
Love is what makes a family	55	33	62	41	77	80

as of religious background, these value positions reflect what we want families to be like and, thus, what we come to believe about the kinds of issues that are raised throughout this book.

In the wider, societal discourse about families, we can see opposing ideological positions on the well-being of families (Glenn 2000). The two extremes, which sociologist Norval Glenn calls conservative and liberal, are like the half empty–half full disagreement, a difference between pessimistic and optimistic viewpoints. Conservatives are fairly pessimistic about the state of today's families. To **conservatives,** cultural values have shifted from individual self-sacrifice toward personal self-fulfillment. This shift in values is seen as an important factor in some major changes in family life that occurred in the last 3 or 4 decades of the twentieth century (especially higher divorce rates, more cohabitation, and more births outside of marriage).

Furthermore, conservatives believe that as a result of such changes, today's families are weaker and less able to meet the needs of children, adults, or the wider society (Glenn 2000). Conservatives therefore recommend social policies to reverse or reduce the extent of such changes (recommendations to repeal no-fault divorce and the introduction of covenant marriage are two examples we examine later).

Compared with conservatives, liberals are more optimistic about the status and future of family life in the United States. **Liberals** tend to believe that the changes in family patterns are just that—changes, not signs of familial decline (Benokraitis 2000). The liberal position also portrays these changing family patterns as products of and adaptations to wider social and economic changes rather than a shift in cultural values (Benokraitis 2000; Glenn 2000). Such changes in family experience create a wider range of contemporary household and family types and require greater tolerance of such diversity. Placing great emphasis on economic issues, liberal family policies are often tied to the economic well-being of families. Additional examples would include supportive policies for the increasing numbers of employed mothers and two-earner households.

According to Glenn, there is a third position in the discourse about families. **Centrists** share aspects of both conservative and liberal positions. Like conservatives, they believe that some familial changes have had negative consequences. Like liberals, they identify wider social changes (for example, economic or demographic) as major determinants of the changes in family life, but they assert greater emphasis than liberals do on the importance of cultural values. They note that too many people are too absorbed in their careers or too quick to surrender in the face of marital difficulties (Benokraitis 2000; Glenn 2000).

The assumptions within and the differences between these positions are more important than they might first appear to be. The perceptions we have of what accounts for the current status of family life or the directions in which it is heading influence what we believe families *need*. These, in turn, influence social policies regarding family life. As Nijole Benokraitis states, "Conservatives, centrists, liberals, and feminists who lobby for a variety of family-related 'remedies' affect our family lives on a daily basis" (2000, 19).

It should be noted that a similar difference of interpretation can be seen among social scientists who study families. In other words, changing family patterns, and trends in marriage, divorce, parenting, and childcare, are explained and interpreted differently even by the experts who study them. Consider, for example, the following two statements about the effects of divorce on children. The first, is by Constance

Popular Culture   The Family Values Debate Captures SpongeBob SquarePants

Who or what is SpongeBob? Whether or not you have ever been a regular viewer, many of you are likely familiar with the cartoon *SpongeBob SquarePants*, one of the most popular cartoons in recent memory.

Nearly 60 million people, 35% adults age 18–49 and another 23% teenagers, tune in each month to watch SpongeBob's antics. However, you might be wondering why are you reading about him here? The answer is a little complicated but nicely illustrates the point made earlier about clashing views on families and family values.

SpongeBob is among a number of characters—besides Barney, the purple dinosaur; Kermit the Frog; and Winnie the Pooh—singing the disco-era hit "We are Family" in a video. The video was produced by the We Are Family Foundation and was de-

signed to be used in elementary schools around the country to teach tolerance, cooperation, unity, and appreciation of diversity (http://www.wearefamilyfoundation.org).

In January 2005, the video and organization that produced it became the target of Dr. James Dobson, the 70-year-old founder and board chairperson of Focus on the Family, a nonprofit evangelical Christian organization Dobson started in 1977.

Dobson has been sought out by politicians such as Jimmy Carter, George W. Bush and Trent Lott, is heard on the radio daily in nearly 100 countries, and is seen on television on 100 stations throughout the United States. He is also the author of some three dozen books on parenting and marriage, including *Dare to Discipline* (which has sold more than 4.5 million copies).

To Dobson, the "We Are Family" video was an attempt by a gay-supporting organization to convince children to accept homosexuality, although no mention of homosexuality can be found in the video.

Addressing a George W. Bush presidential inauguration dinner, Dobson mentioned SpongeBob as he offered his warning regarding the "We Are Family" video. Although he later acknowledged that both the cartoon sponge and the "We Are Family" video are harmless, he was disturbed by the use for which he believed the video was intended. He claimed that the We Are Family Foundation's efforts to use the video to teach "tolerance" and recognize "diversity" extended to teaching children that homosexuality is an acceptable lifestyle. Dobson contends that "tolerance and diversity . . . are almost always buzzwords for homosexual advocacy" (Dobson 2005).

After he made his remarks the issue exploded into controversy. Numerous media outlets reported that Dobson accused SpongeBob of being gay or that Dobson saw the *SpongeBob SquarePants* cartoon as promoting homosexuality or as a threat to the family. On the Focus on the Family website, Dobson is explicit: "One more time let me say that the

Ahrons, a noted scholar on family relationships and author of numerous articles and books on divorce. She is professor emerita of sociology at the University of Southern California. A member of the Council on Contemporary Families, Ahrons offers these encouraging words:

The good news about divorce is that the vast majority of children develop into reasonably competent individuals, functioning within a normal range. Studying the long-term effects of parental divorce on children is very complex and many of the research findings are equivocal. A review of the research literature reveals a strong bias towards using a deficit approach that focuses on the problems cre-

ated by divorce and relies on the "intact family" as the reference point. However, a small group of studies is emerging that explores the effects of divorce from a "strength and resilience" perspective. This perspective represents an important shift in our thinking. It will direct our attention to the life course of postdivorce families and those factors that mediate between the divorce and its long-term implications.

Overall, the findings thus far clearly indicate that it is not the divorce per se, but the quality of the relationship between divorced parents that has an important long term impact on adult children's lives. Good or "good enough" divorces (characterized by parents who are able to minimize their

problem is not with SpongeBob or the other cartoon characters (in the video). It is with the way they will be used in the classroom as part of an effort that threatens the well-being of American families." So here is an issue that ensnared one of the most popular cartoon characters in a net of controversy. The mere fact that a controversy arose and that it pitted those with more conservative views against those with more liberal views demonstrates that family issues are differently defined and interpreted, often in highly divisive and heated ways.

■ *SpongeBob*

© Paramount Pictures/Bureau L.A.Collections/CORBIS

conflict and continue to share parenting, even if minimally) maintain the bonds of family and extended kinship ties.

Contrast Ahrons' comments with the following from David Popenoe, also a well-known sociologist, author, and/or editor of numerous books about contemporary American families. Popenoe, a Rutgers University sociologist and codirector of the National Marriage Project, provides a different perspective:

Divorce increases the risk of interpersonal problems in children. There is evidence, both from small, qualitative studies and from large-scale, long-term empirical studies, that many of these problems are long lasting. In fact, they may even become worse in adulthood. . . . While it found that parents' marital unhappiness and discord have a broad negative impact on virtually every dimension of their children's well-being, so does the fact of going through a divorce. In examining the negative impacts on children more closely, the study discovered that it was only the children in high-conflict homes who benefited from the conflict removal that divorce may bring. In lower-conflict marriages that end in divorce—and the study found that perhaps as many as two-thirds of the divorces were of this type—the situation of the children was made much worse following a divorce.

Based on the findings of this study, except in the minority of high-conflict marriages, it is better for the children if their parents stay together and work out their problems than if they divorce.

Although there are ways to reconcile the two seemingly contrary points of view, clearly they come from different overall perspectives about marriage, divorce, and the well-being of children. Thus, it is important to realize that, just as the wider society and culture is fraught with conflicting opinions and values about marriage and family relationships, the academic disciplines that study family life suffer lack of consensus.

As we set off on our exploration of marriage and family issues, it is important to realize that many of the topics we cover are part of the ongoing debates about families. As you try to make sense of the material we introduce, we do not require you to take a particular viewpoint but rather to keep in mind that multiple interpretations are possible. Where different interpretations are particularly glaring (as in the many issues surrounding divorce, for example), we present them and allow you to decide which better fits the evidence presented.

Hopefully, as you now begin studying marriage and the family, you will see that such study is both abstract and personal. It is abstract insofar as you will learn about the general structure, processes, and meanings associated with marriage and the family, especially within the United States. In the chapters that follow, the things that you learn should also help you better understand your own family, how it compares to other families, and why families are the way they are. In other words, as we address family more generally, in some ways it is *your* present, *your* past, and *your* future that you are studying. By providing a wider sociological context to marriage, family, and intimate relationships, we show you how and where your experiences fit and why.

Summary

- Marriage is a legally recognized union between a man and a woman in which they are united sexually; cooperate economically; and may give birth to, adopt, or rear children. Marriage differs among cultures and has changed historically in our own society. In Western cultures, the preferred form of marriage is *monogamy,* in which there are only two spouses, the husband and wife. *Polygyny,* the practice of having two or more wives, is commonplace throughout many cultures in the world.

- Legal marriage provides a number of rights and protections to spouses that couples who live together lack.

- The current legal definitions of marriage are changing in the United States and in many other countries. The greatest change relates to same-sex marriage.

- Defining the term *family* is complex. Most definitions of family include individuals related by descent, marriage, remarriage, or adoption; some also include affiliated kin. *Family* may be defined as one or more adults related by blood, marriage, or affiliation who cooperate economically, who may share a common dwelling, and who may rear children. There are also ethnic differences as to what constitutes family.

- Four important family functions are (1) the provision of intimacy, (2) the formation of a cooperative economic unit, (3) reproduction and socialization, and (4) the assignment of social roles and status, which are acquired both in a *family of orientation* (in which we grow up) and in a *family of cohabitation* (which we form by marrying or living together).

- Advantages to living in families include (1) continuity of emotional attachments, (2) close proximity, (3) familiarity with family members, and (4) economic benefits.

- The *extended family* consists of grandparents, aunts, uncles, cousins, and in-laws. It may be formed *conjugally* (through marriage), creating in-laws or step-kin, or *consanguineously* (by birth) through blood relationships.

- The *kinship system* is the social organization of the family. In the *nuclear family,* it generally consists of parents and children, but it may also include members of the extended family, especially grandparents, aunts, uncles, and cousins. Kin can be *affiliated,* as when a nonrelated person is considered "kin," or a relative may fulfill a different kin role, such as a grandmother taking the role of a child's mother.

- Unmarried *cohabitation* is a relationship that occurs when a couple lives together and is sexually involved.

Key Terms

affiliated kin 6

centrists 25

clan 7

conjugal relationship 20

consanguineous relationship 20

conservatives 25

extended family 19

family 6

family of cohabitation 17

family of orientation 17

family of procreation 17

household 6

kinship system 20

liberals 25

marriage 8

modified polygamy 13

monogamy 12

nuclear family 7

polyandry 12

polygamy 12

polygyny 12

serial monogamy 13

socialization 16

spirit marriage 20

traditional family 7

Resources on the Internet

Companion Website for This Book

http://www.thomsonedu.com/sociology/strong

Gain an even better understanding of this chapter by going to the companion website for additional study resources. Take advantage of the Pre- and Post-Test quizzing tool, which is designed to help you grasp difficult concepts by referring you back to review specific pages in the chapter for questions you answer incorrectly. Use the flash cards to master key terms and check out the many other study aids you'll find there. Visit the Marriage and Family Resource Center on the site. You'll also find special features such as access to Info-Trac© College Edition (a database that allows you access to more than 18 million full-length articles from 5,000 periodicals and journals), as well as GSS Data and Census information to help you with your research projects and papers.

CHAPTER 2

Studying Marriages and Families

Outline

What Do YOU Think?

Are the following statements **TRUE** or **FALSE**?
You may be surprised by the answers (see answer key on the following page.)

T F   **1** To answer questions about families, we need to rely most on our "common sense."

T F   **2** "Everyone should get married" is an example of an objective statement.

T F   **3** Many researchers believe that both love and conflict are normal features of families.

T F   **4** Stereotypes about families, ethnic groups, and gays and lesbians are easy to change.

T F   **5** We tend to exaggerate how much other people's families are like our own.

T F   **6** Family researchers formulate generalizations derived from carefully collected data.

T F   **7** Every method of collecting data on families is limited in some way.

T F   **8** A belief that our own ethnic group, nation, or culture is innately superior to another is an example of an ethnocentric fallacy.

T F   **9** According to some scholars, in marital relationships we tend to weigh the costs against the benefits of the relationship.

T F   **10** It is impossible to observe family behavior.

A word of warning: The subjects covered in this book come up often and unexpectedly in everyday experience. You may be reading the paper or watching television and come upon some news about research on the effects of divorce or day care on children. You might be having lunch with friends or dinner with your parents, and before you know it someone may make claims about what marriages need or lack or how some kinds of families are better or stronger than others. The following hypothetical situation is not an uncommon or unrealistic one.

Imagine having coffee with a close friend. She confides that she is worried about her relationship with her boyfriend of 2 years now that they are separated by nearly 600 miles while at different colleges. You feel for your friend, sensing the seriousness of her anxiety and the depth of her fears. You think hard about her predicament and, wanting to be a supportive friend, smile reassuringly. She shares the following: "I don't know, I guess I sometimes think I'm worrying too much. After all, how many times have I heard, 'absence makes the heart grow fonder'? Everyone knows that. Maybe my relationship will actually get stronger and deeper through this separation." Before you can reply, she continues: "But then I guess I do think too much. Don't they say, 'Out of sight, out of mind.'? Maybe it's just a matter of time before this relationship is history. In fact, I wonder if I should just prepare myself, even start looking for someone new. Now." In obvious distress and confusion, she looks to you for advice: "Hey, you're in a family class. So which is it? Will 'absence make his heart grow fonder' or now that I am 'out of sight' am I soon to be 'out of mind'?"

How would you answer your friend? Both reactions can't be true. Moreover, how can "everyone know" one thing even though "they" say the opposite? Surely, there must be a way to resolve such a contradiction.

In this chapter we examine how family researchers attempt to explore issues such as the one posed here. In that sense, this chapter differs from all of the others. Instead of presenting material about different aspects of the marriage and family experience, it explains and illustrates how we learn the information about relationships and families found in the rest of the book. However, it will enable you to understand better and appreciate more how much our knowledge and understanding of families is enriched by the theories and research procedures we introduce. In learning *how information is obtained and interpreted*, we set the stage for the in-depth exploration of family issues in the chapters that follow.

How Do We Know?

As sociologist Earl Babbie suggests, social research is one way we can learn about things (Babbie 2002). However, most of what we "know" about the social world we have "learned" elsewhere through other less systematic means (Babbie 2002; Neuman 2000). In the previous chapter we noted the dangers inherent in generalizing from personal experiences. We all do this. If you or someone you know had an unfavorable experience with a long-distance relationship, you probably favor the "out of sight, out of mind" response more than the optimistic one your friend is hoping to hear.

The opening scenario illustrates the difficulty involved in relying on what are often called common sense–based explanations or predictions (Neuman 2000). Commonsense understanding of family life may be derived from "tradition," what everyone knows because it has always been that way or been thought to be that way; from "authority figures," whose expertise we trust and whose knowledge we accept; or from various media sources.

The mass media are so pervasive that they become invisible, almost like the air we breathe. Yet they affect us. Popular culture, in all its forms, is a key source of both information and misinformation about families. Cumulatively, television, popular music, the Internet, magazines, newspapers, and movies help shape our attitudes and beliefs about the world in which we live. On average, each of us spends more than 3,400 hours a year using one of these media (U.S. Census Bureau 2001, Table 1125). Television has a particularly powerful effect on our values and beliefs (see the "Popular Culture" box on families in the media). Popular culture conveys images, ideas, beliefs, values, myths,

Answer Key for What Do You Think

1 False, see p. 32;   2 False, see p. 36;   3 True, see p. 49;   4 False, see p. 36;   5 True, see p. 36;   6 True, see p. 37;   7 True, see p. 55;   8 True, see p. 37;   9 True, see p. 44;   10 False, see p. 57.

and stereotypes about every aspect of life and society, including the family.

Because so much of the day-to-day stuff of family life (for example, caring for children, arguing, dividing chores, and engaging in sexual behavior) takes place in private, behind closed doors, we do not have access to what really goes on. But we are privy to those behaviors on television and in movies and magazines. Thus, those depictions can influence what we *assume* happens in real families. If you have seen a movie or television show or read magazine articles in which couples in long-distance relationships thrived despite distance, those sources will likely influence you toward reassuring your nervous friend.

Cumulatively, the multiple forms of commonsense knowledge (experience, tradition, authority, and media) are typically poor sources of accurate and reliable knowledge about social and family life. Often, what we consider and accept as common sense is fraught with the kinds of contradictions depicted previously (or, for example, "birds of a feather flock together" but "opposites attract"). Even in the absence of contradiction, many commonsense beliefs are simply untrue. Thus, if we "really want to know" about how families work or what people in different kinds of family situations or relationships experience, we would be better informed by seeking and acquiring more trustworthy information.

Thinking Critically about Marriage and the Family

Before we examine the specific theories and research techniques used by family researchers, it is important to emphasize that the attitudes of the researcher (or you, as you read research) are important. To obtain valid research information, we need to keep in mind the rules of critical thinking. The term *critical thinking* is another way of saying "clear and unbiased thinking."

We all have perspectives, values, and beliefs regarding marriage, family, and relationships. These can create blinders that keep us from accurately understanding the research information. We need instead to develop a sense of **objectivity** in our approach to

■ *Television sitcoms, such as the popular* Everybody Loves Raymond *or* King of Queens, *influence our beliefs and attitudes about marriage and family. What messages and expectations do these programs convey?*

Popular culture in all its forms is a key source of information and misinformation about families. Often, critics point to the pervasive influence of television and its distortions of reality, familial and otherwise. For example, prime-time television, in both dramas and situation comedies, unrealistically depicts married life, understates the unique issues faced by various ethnic families, inaccurately depicts single-parent family life, inaccurately portrays the relative sexual activity levels of marrieds and singles, and portrays conflict as something easily resolved within 20 minutes, often with humor.

The combined portrayal of family life on daytime television that results from soap operas and talk shows is unrealistic and highly negative. Those who have scrutinized daytime soap operas note the extremely high rates of conflict, betrayal, infidelity, and divorce that afflict soap opera families (Pingree and Thompson 1990; Benokraitis and Feagin 1995). Characters go through multiple marriages, often carrying "deep, dark secrets" that they keep from their spouses. Soaps often stereotype women as starry-eyed romantics or scheming manipulators of men. Particularly unrealistic is the way soap operas portray sex, leading viewers to envision exaggerated estimates of how much sex does and should occur within relationships, as well as how often people sexually stray outside of their marriages and relationships (Lindsey 1997). Daytime talk shows, from *Jerry Springer* and *Maury* to *The Montel Williams Show* contribute to the idea that American family life is deeply dysfunctional, that parents are anything from

"irresponsible fools" to "in-your-face monsters" (Hewlett and West 1998), that spouses and partners routinely cheat on each other and often strike each other, and that teenagers are recklessly out of control. Half-naked fisticuffs on *Jerry Springer* and contested allegations and paternity tests on *Maury* are especially distasteful distortions of families and relationships.

As you will see throughout this book, although families are not without their share of serious problems, daily family life is as poorly represented by daytime television as by prime-time programming.

(Un)reality Television

The newest genre of television programming is what has been termed *reality television*. Operating without scripts or professional actors, reality television typically puts "ordinary people" into situations or locations that require them to meet various challenges.

Much of what is considered reality television has nothing to do with relationships, marriages, or families. However, there have been a number of reality programs that focus directly on relationships and family life, including the following current or cancelled shows: *Who Wants to Marry a Millionaire?, Trading Spouses, Who's Your Daddy?, Supernanny, Brat Camp, Meet Mr. Mom, Boy Meets Boy,* and *Renovate My Family.* Note that these represent just a fraction of the reality genre. Whether these shows match and/or marry people or showcase aspects of families or relationships, they hardly represent what their genre claims as its name. By highlighting extreme cases or introducing artificial circumstances and/or competitive goals, these shows are no more representative of familial reality than the daytime talk shows. It would be dangerous to draw gen-

eralizations from shows such as *Supernanny* or *Brat Camp* and conclude that "kids today" are disrespectful or out of control. Although you may consider yourself too sophisticated to make such a generalization, millions of others watch programs such as these. Are all of them equally sophisticated?

Advice and Information

This media genre transmits information and conveys values and norms—cultural rules and standards—about marriage and family, often disguised as information and intended as entertainment. A veritable industry exists to support the advice and information genre. It produces self-help and childrearing books, advice columns, radio and television shows, and numerous articles in magazines and newspapers.

In newspapers in the past, this genre was represented by such popular "advice columnists" as Abigail Van Buren (real name Pauline Esther Friedman, whose column "Dear Abby" is now written by her daughter), Dan Savage (whose sex-advice column "Savage Love" is syndicated in 70 newspapers), and the late Ann Landers (Abby's twin sister, Esther Pauline Friedman).

Newer, Web-based columnists such as Alison Blackman Dunham and her late twin sister Jessica Blackman Freedman, the self-proclaimed "Advice Sisters," (or "Ann and Abby for the new millennium") helped carry this genre to the Internet. Radio therapists, such as Dr. Joy Browne and Dr. Laura Schlessinger, have daily callers seeking advice or information about relationships, family crises, and so on.

On television, Dr. Philip McGraw's *Dr. Phil* has become a ratings success. McGraw, a psychologist of some 25 years, was featured often on *The Oprah Winfrey Show* before landing his own talk show in 2002. His shows cover a range of personal and family issues. In a recent 2-week period, for example, episodes included "The Stepford Family," "Is This Normal?" "Wifestyles," "Extreme Parenting," "Nasty Custody Battles," and "Pressured Into Marriage." Dr. Phil is also the author of a number of best-selling books, including *Self Matters: Creating Your Life from the Inside Out,* and *Relationship Rescue: Seven Steps for Reconnecting With Your Partner,* and has a website from which visitors can obtain a variety of suggestions for how to deal with the kinds of relationship and personal issues featured on his show or in his books.

Evaluating the Advice and Information Genre

The various radio or television talk shows, columns, articles, and advice

■ *Dr. Philip McGraw is a licensed clinical psychologist who in addition to his television program has authored six* New York Times *No. 1 best-selling books.*

■ *Montel Williams is among a number of television and radio talk show hosts who often focus on family issues, although he approaches them with more seriousness than most others.*

books have several things in common. First, their primary purpose is to sell books or periodicals or to raise program ratings. They must capture the attention of viewers, listeners, or readers. In contrast, the primary purpose of scholarly research is the pursuit of knowledge.

Second, the media must entertain while disseminating information about relationships and families. Thus, the information and advice must be simplified. Complex explanations and analyses must be avoided because they would interfere with the entertainment purpose. Furthermore, the genre relies on high-interest or shocking material to attract readers or viewers. Consequently, we are more likely to read or view stories about finding the perfect mate or protecting our children from strangers than stories about new research methods or the process of gender stereotyping.

Third, the advice and information genre focuses on how-to information or morality. The how-to material advises us on how to improve our relationships, sex lives, childrearing abilities, and so on. Advice and

normative judgments (evaluations based on norms) are often mixed together. Advice columnists act as moral arbiters, much as do ministers, priests, rabbis, and other religious leaders.

Fourth, the genre uses the trappings of social science without its substance. Writers and columnists interview social scientists and therapists to give an aura of scientific authority to their material. They rely especially heavily on therapists with clinical rather than academic backgrounds. Because clinicians tend to deal with people with problems, they often see relationships as problematical.

To reinforce their authority, the media also incorporate statistics, which are key features of social science research. But Susan Faludi (1991) offers this word of caution:

> The statistics that the popular culture chooses to promote most heavily are the very statistics we should view with the most caution. They may well be in wide circulation not because they are true but because they support widely held media preconceptions.

With the media awash in advice and information about relationships, marriage, and family, how can we evaluate what is presented to us? Here are some guidelines:

■ *Be skeptical.* Remember: Much of what you read or see is meant to entertain you. Are the sources scholarly or popular? Do they rely on self described "experts" or "victims"? How representative are the people interviewed?

■ *Search for biases, stereotypes, and lack of objectivity.* Information is often distorted by points of view. What conflicting information may have been omitted? Does the media's idea of family include diverse family forms and experiences?

Continues

- *Look for moralizing.* Many times what passes as fact is disguised moral judgment. What are the underlying values of the article or program?
- *Go to the original source or sources.* The media simplify. Find out for yourself what the studies said. How valid were their methodologies? What were their strengths and limitations?

- *Seek additional information.* The whole story is probably not told. In looking for additional information, consider information in scholarly books and journals, reference books, or college textbooks.

Throughout this book you will be exposed to a variety of information or data about families. This information may or may not reflect your experiences, but its value is this: It will enable you to learn about how other people experience family life. This knowledge and the results of different kinds of responses to family situations enable a more informed understanding of families in general and of yourself as an individual. Finally, such information is important and necessary for a variety of professionals and practitioners, especially those who provide social services, medical care, or legal assistance, as they deal with family-related issues.

---

information—to suspend the beliefs, biases, or prejudices we have about a subject until we understand what is being said (Kitson et al. 1996). We can then take that information and relate it to the information and attitudes we already have. Out of this process a new and enlarged perspective may emerge.

One area in which we may need to be alert to maintaining an objective approach is that of family lifestyle. The values we have about what makes a successful family can cause us to decide ahead of time that certain family lifestyles are "abnormal" because they differ from our experience or preference. We may refer to single-parent families as "broken" or say that adoptive parents are "not the real parents."

A clue that can sometimes help us "hear" ourselves and detect whether we are making value judgments or objective statements is as follows: A **value judgment** usually includes words that mean "should" and imply that our way is the correct way. An example is, "Everyone should get married." This text presents information based on scientifically measured findings—for example, concluding that "about 90% of Americans marry."

Opinions, biases, and stereotypes are ways of thinking that lack objectivity. An **opinion** is based on our experiences or ways of thinking. A **bias** is a strong opinion that may create barriers to hearing anything contrary to our opinion. A **stereotype** is a set of simplistic, rigidly held, and overgeneralized beliefs about the personal characteristics of a group of people. They form the "glasses" with which we "see" people and groups. Stereotypes are fairly resistant to change. Furthermore, stereotypes are often negative. Common stereotypes related to marriages and families include the following:

- Nuclear families are best.
- Stepfamilies are unhappy.
- Lesbians and gay men cannot be good parents.
- Latino families are poor.
- Women are instinctively nurturing.
- People who divorce are selfish.

We all have opinions and biases; most of us, to varying degrees, think stereotypically. But the commitment to objectivity requires us to become aware of these opinions, biases, and stereotypes and to put them aside in the pursuit of knowledge.

Fallacies are errors in reasoning. These mistakes come as the result of errors in our basic presuppositions. The *gambler's fallacy*, for example, is based on the belief that following a stretch of bad luck at cards or dice the next hand or roll has to be better. Or, having been "hot," the gambler should quit because luck has or will soon "run out." However, every roll of two dice or hand of cards dealt is independent of whatever came before. Statistically, there is no truth to the gambler's fallacy.

Two common types of fallacies that especially affect our understanding of families are egocentric fallacies and ethnocentric fallacies. The **egocentric fallacy** is the mistaken belief that everyone has the same experiences and values that we have and therefore should think as we do. The **ethnocentric fallacy** is the belief that our ethnic group, nation, or culture is innately superior to others. In the next chapter, when we consider the differences and strengths of families from different ethnic and economic backgrounds, you need to keep both of these fallacies from distorting your understanding.

From the day of your birth you have been forming impressions about human relationships and developing ways of behaving based on these impressions. Hence, you might feel a sense of "been there, done that" as you read about an aspect of personal development or family life. However, your study of the information in this book will provide you the opportunity to reconsider your present attitudes and past experiences and relate them to the experiences of others. As you do, this you will be able to use the logic and problem-solving skills of critical thinking so that you can effectively apply that which is relevant to your life.

Theories and Research Methods

Family researchers come from a variety of academic disciplines—from sociology, psychology, and social work to communication and family studies (sometimes known as "family and consumer sciences").

Although these disciplines may differ in terms of the specific questions they ask or the objectives of their research, they are unified in their pursuit of accurate and reliable information about families through the use of social scientific theories and research techniques. Scholarly research about the family brings together information and formulates generalizations about certain areas of experience. These generalizations help us predict what happens when certain conditions or actions occur.

Family science researchers use the **scientific method**—well-established procedures used to collect information about family experiences. With scientifically accepted techniques, they analyze this information in a way that allows other people to know the source of the information and to be confident of the accuracy of the findings. Much of the research family scientists do is shared in specialized journals (for example, *Journal of Marriage and the Family, Journal of Family Issues*) or in book form. By communicating their results through such channels, other researchers can build on, refine, or further test research findings. Much of the information contained in this book originally appeared in scholarly journals.

Theories of Marriage and Families

One of the most important differences between the knowledge about marriage and family derived from family research and that acquired elsewhere is that family research is influenced or guided by **theories**—sets of general principles or concepts used to explain a phenomenon and to make predictions that may be tested and verified experimentally. Although researchers collect and use a variety of kinds of data on marriages and families, these data alone do not automatically convey the meaning or importance of the information gathered. Concepts and theories supply the "story line" for the information we collect.

Concepts are abstract ideas that we use to represent the reality in which we are interested. We use concepts to focus our research and organize our data. Many examples of concepts—for example, nuclear families, monogamy, and socialization—were introduced in the previous chapter. Family research involves the processes of **conceptualization,** the specification and definition of concepts used by the researcher, and of **operationalization,** the identification and/or development of research strategies to observe or measure concepts. For example, to study the relationship between social class and childrearing strategies, we need to define and specify how we are going to identify and measure a person's social class position and childrearing strategies.

In **deductive research,** concepts are turned into **variables,** concepts that can vary in some meaningful way. Marital status is an example of a variable used by family researchers. We may be married, divorced, widowed, or never married. As researchers explore the causes and/or consequences of marital status, they may formulate **hypotheses,** or predictions, about the relationships between marital status and other variables. We might hypothesize that race or social class influences whether someone is married or not. In such an example, race is an **independent variable** and marital status the **dependent variable** in that race is thought to influence the likelihood of becoming or staying married. Marital status, on the other

Figure 2.1 ■ Marital Status as a Dependent, Independent, and Intervening Variable

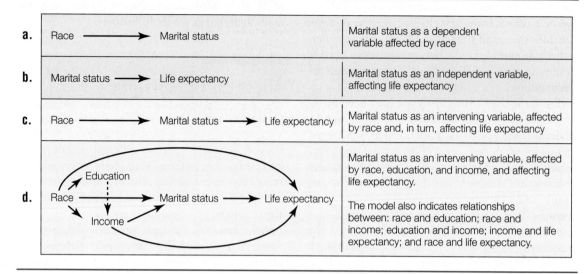

hand, may be a causal or independent variable in a hypothesized relationship between being married and life expectancy. Finally, marital status might be hypothesized as an **intervening variable,** affected by the independent variable, race, and in turn affecting the dependent variable, life expectancy. In that instance, the hypothesis suggests that race differences in marital status account for race differences in life expectancy (Figure 2.1).

Rarely do researchers construct theories with only two or three variables. They may hypothesize multiple independent and intervening variables and seek to identify those having the greatest effect on the dependent variable (Neuman 2000). In Figure 2.1, panel d is an illustration of this. Race is hypothesized to have direct and indirect effects on marital status. Race is alleged to have effects on both income and education, which—in turn—are hypothesized to affect marital status. And, finally, race, income, and marital status are all hypothesized to have effects on life expectancy.

Inductive research is not hypothesis testing research. Instead, it begins with a topical interest and perhaps some vague concepts. As researchers gather their data, typically in the form of field observations or interviews, they refine their concepts, seek to identify recurring patterns out of which they can make generalizations, and, perhaps, end by building a theory (or asserting some hypotheses) based on the data collected. Theory that emerges in this inductive fashion is often referred to as **grounded theory,** in that it is

grounded or "rooted in observations of specific, concrete details" (Neuman 2000).

Theoretical Perspectives on Families

On a more abstract level of theory, we can identify major theoretical frameworks or perspectives that guide much of the research about families. These perspectives (sometimes called *paradigms*) are sets of concepts and assumptions about how families work and how they fit into society. Theoretical frameworks guide the kinds of questions we raise, the types of predictions we make, and where we look to find answers or construct explanations (Babbie 1992).

In this section we discuss several of the most influential theories sociologists and psychologists use to study families, including: ecological, symbolic interaction, social exchange, developmental, structural functional, conflict, and family systems theory. We also look at the influence of feminist perspectives on family studies. As you examine them, notice how the choice of a theoretical perspective influences the way data are interpreted. Furthermore, as you read this book, ask yourself how different theoretical perspectives would lead to different conclusions about the same material.

Family Ecology Theory

The emphasis of **family ecology theory** is on how families are influenced by and in turn influence the wider environment. The theory was introduced in the late nineteenth century by plant and human ecologists. German biologist Ernst Haeckel first used the term *ecology* (from the German word *oekologie,* or "place of residence") and placed conceptual emphasis on **environmental influences.** This focus was soon picked up by Ellen Swallows Richards, the founder and first president of the American Home Economics Association (now known as the American Association of Family and Consumer Sciences). An MIT-trained chemist, Richards believed that scientists needed to focus on home and family, "for upon the welfare of the home depends the welfare of the commonwealth" (quoted in White and Klein 2002).

The core concepts in ecological theory include **environment** and **adaptation.** Initially used to refer to the adaptation of plant and animal species to their physical environments, these concepts were later extended to humans and their physical, social, cultural, and economic environments (White and Klein 2002). As applied to family issues, the family ecology perspective asks: How is family life affected by the environments in which families live?

We use the plural *environments* to reflect the multiple environments that families encounter. In Urie Brofenbrenner's ecologically based theory of human development, the environment to which individuals adapt as they develop consists of four levels: (1) microsystem, (2) mesosystem, (3) exosystem, and (4) **macrosystem.** Cumulatively, these levels make up the environments in which we live. The *microsystem* contains the most immediate influences with which individuals have frequent contact. For example, in adolescence our microsystem could include our families, peers, schools, and neighborhoods. In each of these, roles and relationships exert influence over how we develop. The *mesosystem* consists of the interconnections between microsystems—for example, the ways school experiences and home experiences influence each other. The *exosystem* consists of settings in which the individual does not actively participate but which nonetheless affect his or her development. Parental work experiences—everything from salaries to schedules to continued employment—will influence adolescent development. Finally, the *macrosystem* operates at the broadest level, encompassing the laws, customs, attitudes, and belief systems of the wider society, all of which influence individual development and experience (Rice and Dolgin 2002).

Similarly, in constructing an ecological framework to better understand marriage relationships, Ted Huston illustrated how marital and intimate unions are "embedded in a social context" (Huston 2000). This social context includes the *macroenvironment*—the wider society, culture, and physical environment in which a couple lives—and their particular *ecological niche*—the behavior settings in which they function on a daily basis (for example, a poor, urban neighborhood as opposed to a small town or suburb). Also included in the social context is the marriage relationship itself, especially as it is affected by a larger network of relationships. The final key element in Huston's ecological approach contains the physical, psychological, and social attributes of each spouse, including attitudes and beliefs about their relationship and each other. As illustrated in Figure 2.2, each of these environments influences and is influenced by the others (Huston 2000). We cannot fully understand marriage without exploring the interconnections among these three elements.

In a study of work–family stresses and problem drinking, Joseph Grzywacz and Nadine Marks (2000) applied an ecological approach, wherein problem drinking is seen as a consequence of "negative person–environment interactions," including, especially, high levels of work or family stress or issues arising from the mesosystem of work and family. Ecological factors, then, operate "above and beyond" individual factors in accounting for problem drinking. Negative "spillover" from work to home includes such things as job-induced irritability and fatigue inhibiting home involvement, and job worries that lead to distraction at home. All of these were factors that elevated the likelihood of problem drinking. Furthermore, positive person–environment interactions, such as positive work–family spillover, were associated with reduced likelihood of problem drinking.

As evident, ecological approaches examine how family experience is affected by the broader social environment. In many ways, much of what we examine in subsequent chapters has at least this level of ecological focus. We cannot understand what happens within families without considering the wider cultural, social, and economic environments within which family life takes place.

For many of us, even years later, images are still vividly with us and recalled anytime we hear the date 9/11. We can remember where we were on that date in 2001 when we first heard the news, or saw the footage, of the planes flying into the World Trade Center, into the Pentagon, and into the ground in Pennsylvania. The twin towers of the World Trade Center, which took 6 years and 8 months to build, collapsed less than 2 hours after being struck by the hijacked planes (St. Petersburg Times, September 8, 2002). More than 1.6 million tons of debris and nearly 20,000 body parts were removed from the site. But the memories of planes striking buildings, of the two massive towers collapsing to the ground, and of the smoke and debris and chaos on the streets of New York City are not easily removed.

Neither are the images and memories of the more recent tragedy in the Gulf Coast states of Mississippi, Louisiana, Alabama, and Florida from Hurricane Katrina, an eventual Category 5 hurricane that hit land nearly 2 weeks shy of the fourth anniversary of the destruction of the World Trade Center. With winds that occasionally reached 170 miles an hour Katrina devastated the region. Hardest hit was New Orleans, where 80% of the city was submerged under water, but Biloxi and Gulfport, Mississippi, and parts of Mobile, Alabama suffered similarly. More than 1.7 million people lost power, damage estimates exceed $100 billion, and the future of the region, particularly New Orleans, faces challenges (National Climatic Data Center, http://www.ncdc.noaa.gov/oa/climate/research/2005/katrina.html, December 29, 2005).

The human cost of both tragedies was enormous. More than 1,300 people, in five states, died from Hurricane Katrina. More than twice as many died in the September 11, 2001, attack on the World Trade Center.

■ *Disasters such as Hurricane Katrina and the terrorist attack on the World Trade Center on September 11, 2001 often throw families into extreme situations of ambiguous loss.*

Clearly, both events also brought great suffering to families. Parents, spouses, children and siblings—as well as extended family and friends—suffered sudden and unanticipated loss. Thousands of families were forced to cope and grieve. Husbands and wives became widowers and widows, children faced life without mothers or fathers. Parents faced the terrible reality of losing children. Brothers and sisters were left without their sisters and brothers. In addition, many thousands of families, for at least a time, were left in limbo, without news about the whereabouts of missing loved ones. Did they survive? Where were they? How could they be reunited? It is hard to imagine how such uncertainty weighs on the families of the missing. More than 4 years after 9/11, all but two dozen "missing" were accounted for. Five months after Hurricane Katrina, more than 3,000 of the nearly 11,500 people reported missing were still missing (Associated Press, January 19, 2006). What must families feel in such situations? Extraordinary as these events are, can we make any sense of the familial aftermath?

Pauline Boss, has spent more than 30 years studying families dealing with either physically missing or psychologically missing members. Beginning in the early 1970s by looking at *psychological father absence,* wherein fathers were present but distant, Boss broadened her interest to include situations in which any family member might be said to be "there, but not there." She labeled such circumstances *ambiguous loss* (Boss 2004). Ultimately, she defined **ambiguous loss** as "a situation of unclear loss resulting from not knowing whether a loved one is dead or alive, absent or present" (Boss 2004, 554). Such loss, she suggests, is the

most stressful because it remains unresolved and creates lasting confusion "about who is in or out of a particular family" (p. 553). There is no death certificate, no funeral, no opportunity to honor the deceased or bury remains. It prevents family members from reaching psychological closure, and it leaves families in a situation of boundary ambiguity—unable to carry out expected roles, manage daily tasks, or make necessary decisions. As a result, families are immobilized, roles are confused, and tasks remain undone.

Boss considers two situations of ambiguous loss. First is the ambiguous loss of "there, but not there," of "physical presence and psychological absence" mentioned previously and applicable in unexpected situations such as when a family member suffers from dementia (including Alzheimer's disease), depression, or addictions, and in more common situations, such as preoccupation with work; obsessive involvement with the Internet, or divorce followed by remarriage. In the second form of ambiguous loss, members remain psychologically present despite physical absence. This "not there (physically), but there (psychologically)" version of ambiguous loss can be found in tragic situations of war (for families of soldiers missing in action), among families of incarcerated inmates, in families where a member deserts, and in such events as occurred on 9/11 or in the Gulf states, especially if no body is recovered. Even more common versions of "not there but there" can occur after divorce or adoption, work relocations, and children leaving home and the "nest" emptying. We can face both types of ambiguous loss simultaneously, as Boss describes in the case of a woman who, after 9/11, had a

physically missing husband while caring for her psychologically missing mother, who was suffering from Alzheimer's disease (Boss 2004).

Not all situations of ambiguous loss result in the same outcomes or suffering. Some families manage to redraw otherwise ambiguous boundaries (such as when an aunt or uncle steps in and is viewed as a parent). As Boss notes, "longtime partners of missing workers perceived themselves as wives and then widows, challenging the officials in charge of remunerations" (555). It appears as if some people have higher tolerance for ambiguity and therefore may be more resilient in instances such as Hurricane Katrina or the World Trade Center aftermath.

Individuals may suffer many emotional or psychological wounds after a tragedy such as 9/11 or Katrina. Indeed, surviving family members may also suffer from post-traumatic stress disorder (PTSD). But ambiguous loss is not the same as PTSD. PTSD treatment focuses on individuals, not families as whole systems. Also, PTSD is a pathology, a psychological illness. Ambiguous loss is a situation of stress that can lead to individual suffering but needs to be understood *on the familial level* (Boss 2004).

More than 8,000 of those reported missing in the Gulf Coast after Katrina have been found or their bodies have been identified. Still, 3,200 or more families struggle to find closure and come to terms with what the storm took from them. Using concepts such as *ambiguous loss* enables us to better understand what they suffer from and why. Such understanding won't alter their suffering or reduce the pain of their losses, but it may make it possible to be more effective in any efforts to help them move on.

Figure 2.2 ■ A Three-Level Model for Viewing Marriage

The various contexts and environments in which families live influence each other. Macroenvironment (A): spouses' ecological niche (a1) and macrosocietal context (a2). Individuals (B): spouses' beliefs and attitudes, (b1) spouses' psychological and physical makeup, and (b2) marital behavior in context. (C) Marital dyad: (c1) social network context (c2).

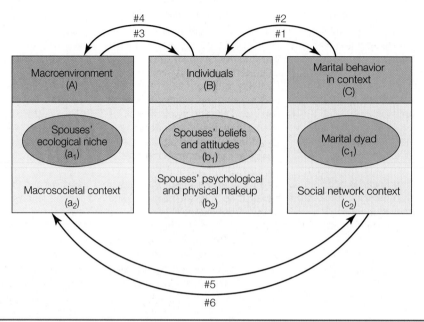

SOURCE: Huston 2000.

There have been a variety of criticisms of ecological theory (White and Klein 2002). Here we note two. It is often unclear which level of analysis is most appropriate—individual, group, or population—to account for the behavior we attempt to explain. In addition, there is a lack of specificity as to the process through which families are affected and what specifically is responsible for the outcomes we seek to explain. Also, some criticize the perspective for seeming to apply more easily to development and growth rather than decline or degeneration. Yet families are prone to decline and degeneration as much as they are to development and growth.

Symbolic Interaction Theory

Symbolic interaction theory looks at how people interact with one another. An **interaction** is a reciprocal act, the everyday words and actions that take place between people. For an interaction to occur, there must be at least two people who both act and respond to each other. When you ask your sister to pass the potatoes and she does it, an interaction takes place. Even if she intentionally ignores you or tells you to "get the potatoes yourself," an interaction occurs (even if it is not a positive one). Such interactions are conducted through symbols, words, or gestures that stand for something else.

Interaction consists of more than merely reacting to others. To interact, we interpret or define the meaning of their words, gestures and actions. If your sister did not respond to your request for the potatoes, what did her nonresponse mean or symbolize? Hostility? Rudeness? A hearing problem? We interpret the meaning and act accordingly. If we interpret the nonresponse as not hearing, we may repeat the request. If we believe it symbolizes hostility, or rudeness, we may become angry.

Symbolic interactionists, like the rest of us, are concerned with relationships. When we feel that our partner does (or does not) understand us, that we communicate well (or poorly), that our relationship can (or cannot) withstand the difficulties created by long distance, we are expressing feelings at the heart of symbolic interaction research. Symbolic interactionists study the interactions that make up a relationship.

Family as the Unity of Interacting Personalities

In the 1920s, Ernest Burgess defined the family as a "unity of interacting personalities" (1926). This definition has been central to symbolic interaction theory and in the development of marriage and family studies. Marriages and families consist of individuals who interact with one another over time. Such interactions and relationships define the nature of a family: a loving family, a dysfunctional family, a conflict-ridden family, an emotionally distant family, a high-achieving family, and so on.

In marital and family relationships, our interactions are partly structured by **social roles**—established patterns of behavior that exist independently of a person, such as the role of wife or husband existing independently of any particular husband or wife. Each member in a marriage or family has one or more roles, such as husband, wife, mother, father, child, or sibling. These roles help give us cues as to how we are supposed to act. When we marry, for example, these roles help us "become" wives and husbands; when we have children, they help us "become" mothers and fathers.

Symbolic interactionists study how the sense of self is maintained in the process of acquiring these roles. We are, after all, more than simply the roles we fulfill. There is a core self independent of our being a husband or wife, father or mother, son or daughter. Symbolic interactionists ask how we fulfill our roles and continue to be ourselves and, at the same time, how our roles contribute to our sense of self. Our identities as humans emerge from the interplay between our unique selves and our social roles.

Only in the most rudimentary sense are families created by society. According to symbolic interactionists, families are "created" by their members. Each family has its own unique personality and dynamics created by its members' interactions. To classify families by structure, such as nuclear family, stepfamily, and single-parent family, misses the point of families. Structures are significant only insofar as they affect family dynamics. It is what goes on inside families, the construction, communication, and interpretation of shared meanings that is important.

This is nicely illustrated in a widely acclaimed book, *The Second Shift,* by sociologist Arlie Hochschild. Hochschild interviewed 50 dual-earner couples to see how they divided housework and childcare. She noted that only 20% of her sample couples shared housework responsibilities equally. In 70% of her sample couples, men did between one-third and one-half of the housework, and in the remaining 10% of sample households, men did less than one-third of the household tasks.

But Hochschild went further and deeper. She examined what happened in households where what couples did (their actual behavior) conflicted with what each partner believed they should do (their "gender ideologies"). She described the strategic use of **family myths,** views of reality that together couples construct and apply to account for why their domestic arrangement is other than they expected (Hochschild 1989).

The clearest example of the workings of such myths can be found among a couple Hochschild calls Evan and Nancy Holt. After repeated but unsuccessful efforts on Nancy's part to convince husband Evan to share more of the housework, Nancy considered the possibility of a divorce. Unwilling to end her marriage "over a dirty frying pan," she and Evan arrived at a "solution," which Hochschild calls the "upstairs–downstairs" myth. Under this version of domestic reality, Nancy notes that she does the "upstairs" and Evan has taken responsibility for, and freed her from, the "downstairs." Hochschild points out that although portrayed by the Holts as "sharing," this solution leaves much unequal. The "upstairs" included the living room, dining room, kitchen, two bedrooms, and two bathrooms; whereas the "downstairs" amounted to the garage, which included responsibility for the car and the dog. Nevertheless, by constructing and believing in the idea that they "share," Nancy was able to live with their arrangement. Thus, the meanings Nancy attached to their arrangement ("I do the upstairs, he does the downstairs; we share"), what we might consider her "definition of her marital situation" became more important than their actual division of responsibilities.

Family myths were used in the opposite direction in Hochschild's sample as well. In other words, couples who believed that traditional divisions of labor (male breadwinner, female homemaker) were best but who could not financially afford such an arrangement often constructed myths that explained away their failure to achieve them. In one such case, Carmen Delacorte was considered an at-home wife even though she brought in one-third of the household income by providing childcare in her home.

Critique

Although symbolic interaction theory focuses on the daily workings of the family, it suffers from several drawbacks. First, the theory tends to minimize the role of power in relationships. If a conflict exists, it may reveal more than differences in meaning and it may take more than simply communicating to resolve it. If one partner strongly wants to pursue a career in Los Angeles and the other just as strongly wants to pursue a career in Boston, no amount of communication and role adjustment may be sufficient to resolve the conflict. The partner with the greater power in the relationship may prevail.

Second, symbolic interaction does not fully account for the psychological aspects of human life, especially personality and temperament. It sees us more in terms of our roles, thus neglecting the self that exists independently of our roles and limiting our uniqueness as humans.

Perhaps most important, the theory does not place marriage or family within a larger social context. It thereby disregards or minimizes the forces working on families from the outside, such as economic or legal discrimination against minorities and women.

Social Exchange Theory

According to **social exchange theory,** we measure our actions and relationships on a cost–benefit basis, seeking to maximize rewards and minimize costs by employing our resources to gain the most favorable outcome. An outcome is basically figured by the equation *Reward − Cost = Outcome.*

How Exchange Works

At first glance, exchange theory may be the least attractive theory we use to study marriage and the family. It seems more appropriate for accountants than for lovers. But all of us use a cost–benefit analysis to some degree to measure our actions and relationships.

One reason many of us do not recognize our use of this interpersonal accounting is that we do much of it unconsciously. If a friend is unhappy with a partner, you may ask, "What are you getting out of this relationship? Is it worth it?" Your friend will start listing pluses and minuses: "On the plus side, I get company and a certain amount of security; on the minus side, I don't get someone who really understands me." When the emotional costs outweigh the benefits of the relationship, your friend will probably end it. This weighing of costs and benefits is social exchange theory at work.

One problem many of us have in recognizing our exchange activities is that we think of rewards and costs as tangible objects, like money. In personal relationships, however, resources, rewards, and costs are more likely to be things such as love, companionship, status, power, fear, and loneliness. As people enter into relationships, they have certain resources—either tangible or intangible—that others consider valuable, such as intelligence, warmth, good looks, or high social status. People consciously or unconsciously use their various resources to obtain what they want, as when they

■ How family members interact with one another is partly determined by how they define their roles and by the meanings they attach to such behaviors as housework and childcare.

"turn on" the charm. Most of us have had friends, for example, whose relationships are a mystery to us. We may not understand what our friend sees in his or her partner; our friend is so much better looking and more intelligent than the partner. (Attractiveness and intelligence are typical resources in our society.) But it turns out that the partner has a good sense of humor, is considerate, and is an accomplished musician, all of which our friend values highly.

Equity

A corollary to exchange is **equity**: exchanges that occur between people have to be fair, to be balanced. We are always exchanging favors: you do the dishes tonight and I'll take care of the kids. Often we do not even articulate these exchanges; we have a general sense that they will be reciprocated. If, in the end, we feel that the exchange was not fair, we are likely to be resentful and angry. Some researchers suggest that people are most happy when they get what they feel they deserve in a relationship (Hatfield and Walster 1981). Oddly, both partners feel uneasy in an inequitable relationship:

> While it is not surprising that deprived partners (who are, after all, getting less than they deserve) should feel resentful and angry about their inequitable treatment, it is perhaps not so obvious why their overbenefited mates (who are getting more than they deserve) feel uneasy, too. But they do. They feel guilty and fearful of losing their favored position.

When partners recognize that they are in an inequitable relationship, they generally feel uncomfortable, angry, or distressed. They try to restore equity in one of three ways:

- They attempt to restore actual equity in the relationship.
- They attempt to restore psychological equity by trying to convince themselves and others that an obviously inequitable relationship is actually equitable.
- They decide to end the relationship.

Society regards marriage as a permanent commitment. Because marriages are expected to endure, exchanges take on a long-term character. Instead of being calculated on a day-to-day basis, outcomes are judged over time.

An important ingredient in these exchanges is whether the relationship is fundamentally cooperative or competitive. In cooperative exchanges, both husbands and wives try to maximize their "joint profit" (Scanzoni 1979). These exchanges are characterized by mutual trust and commitment. Thus, a husband might choose to work part-time and care for the couple's infant so that his wife may pursue her education. In a competitive relationship, however, each is trying to maximize individual profit. If both spouses want the freedom to go out whenever or with whomever they wish, despite opposition from the other, the relationship is likely to be unstable.

Applying Exchange Theory to Marital Outcomes

Exchange theory has been applied to a number of areas of marriage and family including mate selection or partner choice, transition to parenthood, and decisions to divorce. Looking more closely at the latter, the theory suggests the following:

- *Attractiveness of relationship.* A relationship's attractiveness depends on its relative rewards and costs. A relationship's rewards include love, support, security, and sexual intimacy, as well as material goods and services that marriage allows us to obtain or enjoy. Costs associated with marriage may include being and staying in a relationship that causes us emotional or physical suffering, increased and unequal responsibility, lack of freedom, or absence of rewards (Knoester and Booth 2000). A marriage in which we obtain more rewards than costs likely will be attractive and satisfying.

- *Attractiveness of alternatives.* Exchange theory suggests that we are always comparing our relationship outcomes to what we perceive as the sum of rewards and costs in available alternatives. In these terms, alternatives can be a new partner—whether in marriage or something more casual, greater freedom as a single person, or even the chance to focus on a career instead of remaining married. The logic of the theory suggests that if we perceive greater rewards in some alternative or alternatives, we will think about and/or seek a divorce.

- *Barriers to divorce.* Chris Knoester and Alan Booth (2000) note that the final piece in this exchange theory approach to divorce is the presence or absence of barriers to divorce. In some ways, barriers to divorce may be understood as costs associated with leaving the marriage. Even if the rewards of marriage are low and less than could be found outside of the marriage, we may have barriers to overcome if we are to leave our relationships. If the costs of leaving are greater than the rewards of leaving and/or the costs of staying, exchange theory would predict that we would stay, even unhappily, married (Knoester and Booth 2000; Levinger 1976).

Knoester and Booth (2000) tested the importance of eight perceived barriers to divorce: financial security, not wanting to leave the residence, the spouse's dependency on the respondent, the respondent's dependency on the spouse, importance of religious beliefs, concern about the children suffering, concern about losing (custody or contact with) a child, and disapproval of family or friends.

For each barrier, respondents were asked, "How important is (blank) in keeping your marriage together?" Possible responses ranged from "very important" to "somewhat important" to "not very important."

In order from highest to lowest, the most important barriers and the percentage of respondents who identified them as "very important" are as follows:

Barrier	"Very Important" (%)
Child suffering	50.1
No loss of child	46.0
Religious beliefs	41.4
Dependence on spouse	32.9
Spouse's dependence	30.5
Financial security	24.1
Reluctance to leave residence	18.7
Family and friends	11.6

Knoester and Booth note differences in men's and women's answers. Men were more likely than women to attach greater importance to the threat of losing a child and the influence of family and friends. Women placed greater importance on dependence on spouse and religious beliefs. Also, although financial security was considered "very important" by only one-fourth of the sample, another 50% considered it "somewhat important" (not shown), meaning that it is at least a consideration to 75% of the sample.

Knoester and Booth (2000) determine that perceived barriers are mostly ineffective as deterrents to divorce, despite the logic behind believing them to be so. They suggest that such barriers may have once been more important factors in the divorce process but that in an era in which so many marriages end in divorce, the idea of barriers to divorce keeping people married is no longer useful.

Critique

Social exchange theory assumes that we are all rational, calculating individuals, weighing the costs and rewards of our relationships and making cost–benefit comparisons of all alternatives. In reality, sometimes we are rational and sometimes we are not. Sometimes we act altruistically without expecting any reward. This is often true of love relationships and parent–child interactions.

Social exchange theory also has difficulty ascertaining the value of costs, rewards, and resources. If you want to buy eggs, you know they are a certain price per dozen and you can compare buying a dozen eggs with spending the same amount on a notebook. But how does the value of an outgoing personality compare with the value of a compassionate personality? Is 1 pound of compassion equal to 10 pounds of enthusiasm? Compassion may be the trait most valued by one person but may not be important to another. The values that we assign to costs, rewards, and resources may be highly individualistic.

Family Development Theory

Of all the theories discussed here, **family development theory** is the only one exclusively directed at families (White and Klein 2002). It emphasizes the patterned changes that occur in families through stages and across time. In its earliest formulations, family development theory borrowed from theories of individual development and identified a set number of stages that all families pass through as they are formed: growth with the birth of children, change during the raising of children, and contract as children leave and spouses die. Such stages created the *family life cycle.* Eventually, other concepts were introduced to replace the idea of a family life cycle. Roy Rodgers (1973) and Joan Aldous (1978, 1996) proposed the notion of the *family career,* which was said to consist of subcareers like the marital or the parental career, which themselves were affected by an educational or occupational career. Most recently, the idea of the *family life course* has been used to examine the dynamic nature of family experience.

The family life course consists of "all the events and periods of time (stages) between events traversed by a family" (White and Klein 2002). Because all of these concepts emphasize the change and development of families over time, they are complementary and overlapping.

Family development theory looks at the changes in the family that typically commence in the formation of the premarital relationship, proceed through marriage, and continue through subsequent sequential stages. The specification of stages may be based on family economics, family size, or developmental tasks that families encounter as they move from one stage to the next. The stages are identified by the primary or orienting event characterizing a period of the family history. An eight-stage family life cycle might consist of the following: (1) beginning family, (2) childbearing family, (3) family with preschool children, (4) family with schoolchildren, (5) family with adolescents, (6) family as launching center, (7) family in middle years, and (8) aging family.

As we grow, each of us responds to certain universal developmental challenges (Person 1993). For example, all people encounter normative age-graded influences, such as the biological processes of puberty and menopause or sociocultural markers such as the beginning of school and the advent of retirement. Normative history-graded influences come from historical facts that are common to a particular generation, such as the political and economic influences of wars and economic depressions that are similar for individuals in a particular age group (Santrock 1995).

The life-cycle model gives us insights into the complexities of family life and the different tasks that families perform. This model describes the interacting influences of changing roles and circumstances through time and how such changes produce corresponding changes in family responsibilities and needs. Planning that uses the developmental model alerts the family to seek resources appropriate to the upcoming needs and to be aware of vulnerabilities associated with each family stage (Higgins, Duxbury, and Lee 1994).

There are a variety of developmental theories that examine the stages involved in specific family phenomena such as "falling in love," choosing a spouse, or experiencing divorce. Instead of attempting to depict all stages families might encounter, these theories look at the unfolding of specific aspects of family life across stages. You will find such approaches in a number of later chapters.

Critique

An important criticism sometimes made of family development theory is that it assumes the sequential processes of intact, nuclear families. It further assumes that all families go through the same process of change across the same stages. Thus, the theory downplays both the diversity of family experience and the experiences of those who divorce, remain childless, or bear children but never marry (Winton 1995). For example, lesbian-headed families are likely to experience a life-cycle pattern quite different from the traditional one (Slater 1995). Similarly, stepfamilies experience different stages and tasks (Ahrons and Rogers 1987). Nevertheless, the universality of the family life cycle may transcend the individuality of the family form. Single-parent and two-parent families go through many of the same development tasks and transitions. They may differ, however, in the timing and length of those transitions.

A related criticism points out that gender, race, ethnicity, and social class all create variations in how we experience family dynamics. The very sequence of stages may reflect a middle- to upper-class family reality. Many lower- and working-class families do not have lengthy periods of early childless marriage. The transitions to marriage and parenthood may be encountered simultaneously or in reverse of what the stages specify. In neglecting these sorts of variations, the developmental model can appear overly simplistic.

Structural Functionalism Theory

Structural functionalism theory explains how society works, how families work, and how families relate to the larger society and to their own members. The theory is used largely in sociology and anthropology, disciplines that focus on the study of society rather than of individuals. When structural functionalists study the family, they look at three aspects: (1) what functions the family serves for society (discussed in Chapter 1), (2) what functional requirements family members perform for the family, and (3) what needs the family meets for its individual members.

Society as a System

Structural functionalism is deeply influenced by biology. It treats society as if it were a living organism, like a person, animal, or tree. The theory sometimes uses the analogy of a tree in describing society. In a tree, there are many substructures or parts, such as the trunk, branches, roots, and leaves. Each structure has a function. The roots gather nutrients and water from the soil, the leaves absorb sunlight, and so on. Society is like a tree insofar as it has different structures that perform functions for its survival. These structures are called **subsystems.**

The subsystems are the major institutions, such as the family, religion, government, and the economy. Each of these structures has a function in maintaining society, just as the different parts of a tree serve a function in maintaining the tree. Religion gives spiritual support, government ensures order, and the economy produces goods. The family provides new members for society through procreation and socializes its members so that they fit into society. In theory, all institutions work in harmony for the good of society and one another.

The Family as a System

Families may also be regarded as systems. In looking at families, structural functionalists examine how the family organizes itself for survival and what functions the family performs for its members. For the family to survive, its members must perform certain functions, which are traditionally divided along gender lines. Men and women have different tasks: Men work outside the home to provide an income, whereas women perform household tasks and childrearing.

According to structural functionalists, the family molds the kind of personalities it needs to carry out its functions. It encourages different personality traits for men and women to ensure its survival. Men develop instrumental traits, and women develop expressive traits. *Instrumental traits* encourage competitiveness, coolness, self-confidence, and rationality—qualities that will help a person succeed in the outside world. *Expressive traits* encourage warmth, emotionality, nurturing, and sensitivity—qualities appropriate for someone caring for a family and a home.

Such a division of labor and differentiation of temperaments is seen as efficient because it allows each spouse to specialize, thus minimizing competition and reducing ambiguity or uncertainty over such things as who should work outside the home or whose outside employment is more important. For these reasons, such role allocation may be deemed functional.

Critique

Although structural functionalism has been an important theoretical approach to the family, it has declined in significance in recent decades for several reasons. First, because the theory cannot be empirically tested, we'll never know if it is "right" or "wrong." We can only discuss it theoretically, arguing whether it accounts for what we know about the family.

Second, it is not always clear what function a particular structure serves. "The function of the nose is to hold the *pince-nez* [eyeglasses] on the face," remarked the eighteenth-century philosopher François Voltaire. What is the function of the traditional division of labor along gender lines? Efficiency, survival, or the subordination of women?

If interdependence, specialization, and clarity of role responsibilities are what make breadwinner–homemaker households most "functional," those same objectives could be met by household arrangements wherein men stay home, rear kids, and tend house and women earn incomes. In some relationships these role reversals might be more functional. There are women who earn higher incomes than their husbands, are in jobs with greater opportunities for advancement, and are more dedicated to their careers than are their husbands. If their husbands are frustrated by or stagnated at work but have developed or discovered a deeper-than-anticipated fulfillment from children, a reversal of the male provider–female homemaker household would be most functional for them.

Third, how do we know which family functions are vital? The family, for example, is supposed to socialize children, but much socialization has been taken over by the schools, peer groups, and the media. Is this "functional"?

Fourth, structural functionalism has a conservative bias against change. Aspects that reflect stability are called functional, and those that encourage instability (or change) are called dysfunctional. Traditional roles are functional, but nontraditional ones are dysfunctional. Employed mothers are viewed as undermining family stability because they should be home caring for the children, cleaning house, and providing emotional support for their husbands. But in reality, employed mothers may be contributing to family stability by earning money; their income often pushes their families above the poverty line.

Finally, structural functionalism looks at the family abstractly. It looks at it formally, from a distance far removed from the daily lives and struggles of men, women, and children. It views the family in terms of functions and roles. Family interactions, the lifeblood of family life, are absent. Because of its formalism, structural functionalism often has little relevance to real families in the real world.

Conflict Theory

Where structural functionalists assert that existing structures benefit society, conflict theorists ask, "Who benefits?" **Conflict theory** holds that life involves discord. Conflict theorists see society not as basically cooperative but as divided, with individuals and groups in conflict with one another. They try to identify the competing forces.

Sources of Conflict

How can we analyze marriages and families in terms of conflict and power? Such relationships are based on love and affection, aren't they? Conflict theorists agree that love and affection are important elements in marriages and families, but they believe that conflict and power are also fundamental. Marriages and families are composed of individuals with different personalities, ideas, values, tastes, and goals. Each person is not always in harmony with every other person in the family.

Imagine that you are living at home and want to do something your parents don't want you to do, such as spend the weekend with a friend they don't like. They forbid you to carry out your plan. "As long as you live in this house, you'll have to do what we say." You argue with them, but in the end you stay home. Why did your parents win the disagreement? They did so because they had greater power, according to conflict theorists.

Conflict theorists do not believe that conflict is bad; instead, they think it is a natural part of family life. Families always have disagreements, from small ones, such as what movie to see, to major ones, such as how to rear children. Families differ in the number of underlying conflicts of interest, the degree of underlying hostility, and the nature and extent of the expression of conflict. Conflict can take the form of competing goals, such as a husband wanting to buy a new CD player and a wife wanting to pay off credit cards. Conflict can also occur because of different role expectations: An employed mother may want to divide housework 50–50, whereas her husband insists that household chores are "women's work."

Sources of Power

When conflict occurs, who wins? Family members have different resources and amounts of power. There are four important sources of power: legitimacy, money, physical coercion, and love. When arguments arise in a family, a man may want his way "because I'm the head of the house" or a parent may argue "because I'm your mother." These appeals are based on legitimacy—that is, the belief that the person is entitled to prevail by right. Money is a powerful resource in marriages and families. "As long as you live in this house. . ." is a directive based on the power of the purse. Because men tend to earn more than women, they have greater economic power; this economic power translates into marital power. Physical coercion is another important source of power. "If you don't do as I tell you, you'll get a spanking" is one of the most common forms of coercion of children. But physical abuse of a spouse is also common, as we will see in a later chapter. Finally, there is the power of love. Love can be used to coerce someone emotionally, as in "If you really loved me, you'd do what I ask." Or love can be a freely given gift, as in the case of a person giving up something important, such as a plan, desire, or career, to enhance a relationship.

Everyone in the family has power, although the power may be different and unequal. Adolescent children, for example, have few economic resources, so they must depend on their parents. This dependency gives the parents power. But adolescents also have power through the exercise of personal charm, ingratiating habits, temper tantrums, wheedling, and so on.

Families cannot live comfortably with much open conflict. The problem for families, as for any group, is how to encourage cooperation yet allow for differences. Because conflict theory sees conflict as normal, the theory seeks to channel it and to seek solutions through communication, bargaining, and negotiations. We return to these items in Chapter 5 in the discussion of conflict resolution.

Critique

A number of difficulties arise in conflict theory. First, conflict theory derives from politics, in which self-interest, egotism, and competition are dominant elements. Yet is such a harsh judgment of human nature justified? People's behavior is also characterized by self-sacrifice and cooperation.

Love is an important quality in relationships. Conflict theorists do not often talk about the power of love or bonding; yet the presence of love and bonding may distinguish the family from all other groups in society. We often will make sacrifices for the sake of those we love. We will defer our wishes to another's desires;

■ *According to conflict theory, conflict is a normal and natural feature of family life. Families often have disputes over such things as the division of responsibilities, allocation of resources, and levels of commitment.*

we may even sacrifice our lives for a loved one. Second, conflict theorists assume that differences lead to conflict. Differences can also be accepted, tolerated, or appreciated. Differences do not necessarily imply conflict. Third, conflict in families is not easily measured or evaluated. Families live much of their lives privately, and outsiders are not always aware of whatever conflict exists or how pervasive it is. Also, much overt conflict is avoided because it is regulated through family and societal rules. Most children obey their parents, and most spouses, although they may argue heatedly, do not employ violence.

Family Systems Theory

Family systems theory combines two of the previous sociological theories, structural functionalism and symbolic interaction, to form a psychotherapeutic theory. Mark Kassop (1987) notes that family systems

theory creates a bridge between sociology and family therapy.

Structure and Patterns of Interaction

Like functionalist theory, family systems theory views the family as a structure of related parts or subsystems. Each part carries out certain functions. These parts include the spousal subsystem, the parent–child subsystem, the parental subsystem (husband and wife relating to each other as parents), and the personal subsystem (the individual and his or her relationships). One of the important tasks of these subsystems is maintaining their boundaries. For the family to function well, the subsystems must be kept separate (Minuchin 1981). Husbands and wives, for example, should prevent their conflicts from spilling over into the parent–child subsystem. Sometimes a parent will turn to the child for the affection that he or she ordinarily receives from a spouse. When the boundaries of the separate subsystems blur, as in incest, the family becomes dysfunctional.

As in symbolic interaction, interaction is important in systems theory. A family system consists of more than simply its members. It also consists of the pattern of interactions of family members: their communication, roles, beliefs, and rules. Marriage is more than a husband and wife; it is also their pattern of interactions. The structure of marriage is determined by how the spouses act in relation to each other over time (Lederer and Jackson 1968). Each partner influences, and in turn is influenced by, the other partner. And each interaction is determined in part by the previous interactions. This emphasis on the pattern of interactions within the family is a distinctive feature of the systems approach.

Virginia Satir (1988) compared the family system to a hanging mobile. In a mobile, all the pieces, regardless of size and shape, can be grouped together and balanced by changing the relative distance between the parts. The family members, like the parts of a mobile, require certain distances between one another to maintain their balance. Any change in the family mobile—such as a child leaving the family, family members forming new alliances, and hostility distancing the mother from the father—affects the stability of the mobile. This disequilibrium often manifests itself in emotional turmoil and stress. The family may try to restore the old equilibrium by forcing its "errant" member to return to his or her former position, or it may adapt and create a new equilibrium with its members in changed relations to one another.

Analyzing Family Dynamics

In looking at the family as a system, researchers and therapists believe the following:

- *Interactions must be studied in the context of the family system.* Each action affects every other person in the family. The family exerts a powerful influence on our behaviors and feelings, just as we influence the behaviors and feelings of other family members. On the simplest level, an angry outburst by a family member can put everyone in a bad mood. If the anger is constant, it will have long-term effects on each member of the family, who will cope with it by avoidance, hostility, depression, and so on.

- *The family has a structure that can only be seen in its interactions.* Each family has certain preferred patterns of acting that ordinarily work in response to day-to-day demands. These patterns become strongly ingrained "habits" of interactions that make change difficult. A warring couple, for example, may decide to change their ways and resolve their conflicts peacefully. They may succeed for a while, but soon they fall back into their old ways. Lasting change requires more than changing a single behavior; it requires changing a pattern of relating.

- *The family is a purposeful system; it has a goal.* In most instances, the family's goal is to remain intact as a family. It seeks **homeostasis**, or stability. This goal of homeostasis makes change difficult, for change threatens the old patterns and habits to which the family has become accustomed.

- *Despite resistance to change, each family system is transformed over time.* A well-functioning family constantly changes and adapts to maintain itself in response to its members and the environment. The family changes through the family life cycle—for example, as partners age and as children are born, grow older, and leave home. The parent must allow the parent–child relationship to change. A parent must adapt to an adolescent's increasing independence by relinquishing some parental control. The family system adapts to stresses to maintain family continuity while making restructuring possible. If the primary wage earner loses his or her job, the family tries to adapt to the loss in income; the children may seek work, recreation may be cut, or the family may be forced to move.

Although it has been applied to a variety of family dynamics, systems theory has been particularly influential in studying *family communication* (White and Klein 2002). As applied by systems theorists, interaction and communication between spouses are the kinds of systems wherein a husband's (next) action or communication toward his wife depends on her prior message to him. But through research in family communications, we recognize that marital communication is more complex than a simple *quid pro quo* or reciprocity expectation, such as "if she is nasty, he is nasty." John Gottman has explored marital communication patterns that differentiate distressed from nondistressed couples. He identifies the importance of nonverbal communication over that of verbal messages spouses send (White and Klein 2002). As shown in later chapters, certain nonverbal messages are especially useful predictors of the eventual success or failure of a relationship (Gottman et al. 1998; Gottman and Levenson 1992).

Critique

It is difficult for researchers to agree on exactly what family systems theory is. Many of the basic concepts are still in dispute, even among the theory's adherents, and the theory is sometimes accused of being so abstract that it loses any real meaning (Melito 1985; White and Klein 2002).

Family systems theory originated in clinical settings in which psychiatrists, clinical psychologists, and therapists tried to explain the dynamics of dysfunctional families. Although its use has spread beyond clinicians, its greatest success is still in the analysis and treatment of dysfunctional families. As with clinical research, however, the basic question is whether its insights apply to healthy families, as well as to dysfunctional ones. Do healthy families, for example, seek homeostasis as their goal, or do they seek individual and family well-being?

Feminist Perspectives

As a result of the feminist movement of the past two decades, new questions and ways of thinking about the meaning and characteristics of families have arisen. Although there is not a unified "feminist family theory," **feminist perspectives** share a central concern regarding family life.

Blending some central ideas of conflict theory with those of interactionist theory, feminists critically examine the ways in which family experience is shaped by **gender**—the social aspects of being female or male. This is the orienting focus that unifies most feminist writing, research, and advocacy. Feminists maintain that family and gender roles have been constructed by society and do not derive from biological or absolute conditions. They believe that family and gender roles have been created by men to maintain power over women. Basically, the goals of the feminist perspective are to work to accomplish changes and conditions in society that remove barriers to opportunity and oppressive conditions and are "good for women" (Thompson and Walker 1995).

Gender and Family: Concepts Created by Society

Who or what constitutes a family cannot be taken for granted. The "traditional family" is no longer the predominant family lifestyle. Today's families have great diversity. What we think family should be is influenced by our own values and family experiences. Research demonstrates that couples actually may construct gender roles in the ongoing interactions that make up their marriages (Zvonkovic et al. 1996).

Are there any basic biological or social conditions that require the existence of a particular form of family? Some feminists would emphatically say no. Some object to efforts to study the family because to do so accepts as "natural" the inequalities built into the traditional concept of family life. Feminists urge an extended view of family to include all kinds of sexually interdependent adult relationships regardless of the legal, residential, or parental status of the partnership. For example, families may be formed of committed relationships between lesbian or gay individuals, with children obtained through adoption, from previous marriages, or through artificial insemination.

Feminist Agenda

Feminists strive to raise society's level of awareness regarding the oppression of women. Furthermore, some feminists make the point that all groups defined on the basis of age, class, race, ethnicity, disability, or sexual orientation are oppressed; they extend their concern for greater sensitivity to all disadvantaged groups (Allen and Baber 1992). Feminists assume that the experiences of individuals are influenced by the social system in which they live. Therefore, the experiences

of each individual must be analyzed to form the basis for political action and social change. The feminist agenda is to attend to the social context as it affects personal experience and to work to translate personal experience into community action and social critique.

Feminists believe that it is imperative to challenge and change the system that exploits and devalues women. They are aware of the dangers of speaking out but feel their integrity will be threatened if they fail to do so. Some feminists have described themselves as having "double vision"—the ability to be successful in the existing social system and simultaneously work to change oppressive practices and institutions.

Men as Gendered Beings

Inspired and influenced by the writing and research of feminist scholars, many social scientists now focus on how men's experiences are shaped by cultural ideas about masculinity and by their efforts to either live up to or challenge those ideas (Kimmel and Messner 1998; Cohen 2001). Instead of assuming that gender only matters to or includes women, this perspective looks at men as men, or as "gendered beings," whose experiences are shaped by the same kinds of forces that shape women's lives (Kimmel and Messner 1998).

With increased attention on gender courtesy of feminist scholars, and a more recent refocusing of attention to men as "gendered beings," we now have a greatly enlarged and still growing body of literature about men as husbands, fathers, sexual partners, ex-spouses, abusers, and so on (for example, see Cohen 1987; Coltrane 1996; Daly 1993; Gerson 1993; LaRossa 1988; Marsiglio 1998; and Johnson 1996). Throughout this book, we explore how gender shapes women's and men's experiences of the family issues we examine.

Critique

The feminist perspective is not a unified theory; rather, it represents thinking across the feminist movement. It includes a variety of viewpoints that have, however, an integrating focus relating to the inequity of power between men and women in society and especially in family life (MacDermid et al. 1992).

Some family scholars who conceptualize family life and work as a "calling" have taken issue with feminists' focus on power and economics as a description of family. This has created a moral dialogue concerning the place of family life and work in "the good society" (Ahlander and Bahr 1995; Sanchez 1996). Feminists

■ *Surveys are often used to look at how daily housework, such as cooking, is divided between marriage partners.*

© David Young-Wolff/PhotoEdit

today recognize considerable diversity within their ranks, and the ideas of feminist theorists and other family theorists often overlap.

Applying Theories to Long-Distance Relationships

Although the preceding theories were illustrated with numerous examples, it is worthwhile to return to the opening scenario and look at some of the questions each of the theories might pose about long-distance relationships, given their major assumptions. These examples, illustrated in Table 2.1, are not intended to exhaust all possible questions suggested by each theory, nor do they necessarily favor *either* "absence makes the heart grow fonder" or "out of sight, out of mind." They are meant merely as examples of how each theory's core ideas might apply to long-distance relationships.

Conducting Research on Families

In gathering their data, researchers use a variety of techniques. Some researchers ask the same set of questions of great numbers of people. They collect information from people of different ages, sexes, living

situations, and ethnic backgrounds. This is known as "representative sampling." In this way researchers can discover whether age or other background characteristics influence people's responses. This approach to research is called **quantitative research** because it deals with large quantities of information that is analyzed and presented statistically. Quantitative family research often uses sophisticated statistical techniques to assess the relationships between variables. Survey research and, to a lesser extent, experimental research (discussed in the following sections) are examples of quantitative research.

Other researchers study smaller groups or sometimes individuals in a more in-depth fashion. They may place observers in family situations, conduct intensive interviews, do case studies involving information provided by several people, or analyze letters, diaries, or other records of people whose experiences represent special aspects of family life. This form of research is known as **qualitative research** because it is concerned with a detailed understanding of the object of study. The sections on observational research illustrates qualitative research (Ambert, Adler, and Detzner 1995).

In addition to using information provided specifically by people participating in a research project, researchers use information from public sources. This research is called **secondary data analysis.** It involves reanalyzing data originally collected for another purpose. Examples might include analyzing U.S. Census data and official statistics, such as state marriage, birth, and divorce records. Secondary data analysis also

Table 2.1 ■ Applying Theories to Long-Distance Relationships

Theory	Assumptions about Families	Applied to Long-Distance Relationships
Ecological	Families are influenced by and must adapt to environments.	How do the characteristics of each partner's different living environments affect their abilities to maintain their commitments to the relationship? How does the physical separation place the partners in somewhat different ecological niches, which in turn may be more or less conducive to maintaining the relationship? How does the cultural exosystem impose certain beliefs or expectations that might influence the stability of these relationships?
Symbolic Interaction	Family life acquires meaning for family members and depends on the meanings they attach.	What meaning do couples attach to being separated? How does this alter their perceptions of the relationship? Does separation prevent or inhibit the construction of a shared definition of the relationship?
Social Exchange	Individuals seek to maximize rewards, minimize costs, and achieve equitable relationships.	How do both partners define the costs and rewards associated with their relationship? If the rewards of continuing the relationship are felt to be greater than the costs associated with their physical separation, they will maintain their relationship. If either perceives the costs of being apart as too great, or finds another more rewarding relationship, the long-distance relationship will end.
Family Development	Families undergo predictable changes over time and across stages.	How do couples handle the transition to a long-distance relationship? What are the stages or phases that couples encounter as they adjust to being separated? What are the key tasks that must be accomplished at each stage for the relationship to survive?
Structural Functionalism	The institution of the family contributes to the maintenance of society. On a familial level, roles and relationships within the family contribute to its continued well-being.	How does physical separation function to maintain or threaten the stability of the relationship? What benefits does separation have for the individual partners and for the couple's relationship?
Conflict	Family life is shaped by social inequality. Within families, as within all groups, members compete for scarce resources (for example, attention, time, power, and space).	To what extent does one partner benefit more from being apart? Assuming that one partner has a greater commitment to the relationship, how does physical separation create inequality between partners? How does separation prevent couples from effectively managing and resolving conflict?
Family Systems	Families are systems that function and must be understood on that level.	How does being physically separated make it difficult for the couple to communicate effectively? What difficulties does separation create for maintaining the equilibrium of the relationship? How are boundaries between the family system and the wider society altered by being separated?
Feminist	Gender affects our experiences of and within families. Gender inequality shapes how women and men experience families. Families perpetuate gender difference.	How are women and men differently affected by separation? Do long-distance heterosexual relationships create a gender-unequal relationship? Does separation lead men to exploit women by expecting women—but not men—to remain faithful or monogamous? Do women bear more of the burden of managing and maintaining the relationship?

includes content analysis of various communication media such as newspapers, magazines, letters, and television programs.

Family science researchers conduct their investigations using **ethical guidelines** agreed on by professional researchers. These guidelines protect the privacy and safety of people who provide information in the research. For example, any research conducted with college students requires the investigator to present the plan and method of the research to a "human subjects review committee." This ensures that subjects' participation is voluntary and that their privacy is protected. To protect the privacy of participants, researchers promise them either anonymity or confidentiality. **Anonymity** insists that no one, including the researcher, can connect particular responses to the individuals who provided them. Much questionnaire research is of this kind, providing that no identifying information is found on the questionnaires. According to the rules of **confidentiality,** the researcher knows the identities of participants and can connect what was said to who said it but promises not to reveal such information publicly.

To protect the safety of research participants, researchers design their studies with the intent to minimize any possible and controllable harm that might come from participation. Such harm is not typically physical harm but rather embarrassment or discomfort. Much of what family researchers study is ordinarily kept private. Talking about personal matters with an interviewer or answering a series of survey questions may create unintended anxiety on the part of the participants. At best, researchers carefully design their studies to reduce the extent and likelihood of such reactions. Unfortunately, they cannot always be completely prevented (Babbie 2002).

Research ethics also require researchers to conduct their studies and report their findings in ways that assure readers of the accuracy, originality, and trustworthiness of their reports. Falsifying data, misrepresenting patterns of findings, and plagiarizing the research of others are all unethical.

What researchers know about marriage and the family comes from four basic research methods: survey research, clinical research, observational research, and experimental research. There is a continual debate as to which method is best for studying marriage and the family. But such arguments may miss an important point: each method may provide important and unique information that another method may not (Cowan and Cowan 1990).

Survey Research

The **survey research** method, using questionnaires or interviews, is the most popular data-gathering technique in marriage and family studies. Surveys may be conducted in person, over the telephone, or by written questionnaires. Typically, the purpose of survey research is to gather information from a smaller, representative group of people and to infer conclusions valid for a larger population. Questionnaires offer anonymity, may be completed fairly quickly, and are relatively inexpensive to administer.

Quantitative questionnaire research is an invaluable resource for gathering data that can be generalized to the wider population. Because researchers who use such techniques typically draw or use *probability-based random samples,* they can estimate the likelihood that their sample data can be safely inferred to the pop-

ulation in which they are interested. Furthermore, preestablished response categories or existing scales or indexes used by all respondents allow more comparability across a particular sample and between the sample data and related research.

For example, Chloë Bird's 1997 study examined the psychological distress associated with the burdens of parenting, as they vary by gender. Using data from 1,601 men and women under age 60 who participated in the U.S. Survey of Work, Family, and Well-Being, she contrasted the levels of distress experienced by parents with those of nonparents, and—among parents—compared mothers with fathers.

Although the details of her analysis are too complex to be dealt with here, she determined that, on average, parents report higher levels of distress than do people without children, and mothers report higher levels of distress than do fathers (Bird 1997). Women with children under age 18 living at home reported experiencing the highest levels of distress. From her carefully controlled analysis, Bird determined that it is not children but rather increased social and economic burdens that accompany children that seem to create the psychological outcomes she identified.

Questionnaires usually do not allow in-depth responses, however; a person must respond with a short answer, a *yes* or *no,* or a choice on a scale of, for example, 1 to 10, from *strongly agree* to *strongly disagree,* from *very important* to *unimportant,* and so on. Unfortunately, marriage and family issues are often too complicated for questionnaires to explore in depth.

Interview techniques avoid some of this shortcoming of questionnaires because interviewers are able to probe in greater depth and follow paths suggested by the interviewee. They are also typically better able to capture the particular meanings or the depth of feeling people attach to their family experiences.

Consider these two examples, each of which conveys reactions to the life changes associated with becoming or being parents. The first comes from Sharon Hays's interview study of 38 mothers of 2- to 4-year-old children (1996).

In describing how priorities are restructured when a woman becomes a mother, one of Hays's informants offered this comment:

> I think the reason people are given children is to realize how selfish you have been your whole life—you are just totally centered on yourself and what you want. And suddenly here's this helpless thing that needs you constantly. And I kind of think that's

why you're given children, so you kinda think, okay, so my youth was spent for myself. Now, you're an adult, they come first. . . . Whatever they need, they come first.

The second example comes from research conducted by one of the authors on men's experiences becoming and being fathers (Cohen 1993). Here, a 33-year-old municipal administrator describes how becoming a father changed his life:

> I think everything in a personal relationship a baby changes. . . . It's just fantastic . . . it knocked me for a loop. Something creeps into your life and then all of a sudden it dominates your life. It changes your relationship to everybody and everything, and you question every value you ever had. . . . And you say to yourself, "This is a miracle."

These examples of narrative data convey much about the experience of parenthood, including a depth of feeling and degree of nuance that quantitative questionnaire data cannot. By having respondents circle or check the appropriate preestablished response categories to a researcher's questions, we may never identify what that response means to the respondent or how it fits within the wider context of her or his life. However, interviewers are less able to determine how commonly such experiences or attitudes are found. Interviewers may also occasionally allow their own preconceptions to influence the ways in which they frame their questions and to bias their interpretation of responses.

There are problems associated with survey research, whether done by questionnaires or interviews. First, how representative is the sample (the chosen group) that volunteered to take the survey? In the case of a probability-based sample this is not a concern. Self-selection (volunteering to participate) also tends to bias a sample. Second, how well do people understand their own behavior? Third, are people underreporting undesirable or unacceptable behavior? They may be reluctant to admit that they have extramarital affairs or that they are alcoholics, for example. If for any reason people are unable or unwilling to answer questions honestly, the survey technique will produce misleading or inaccurate data.

Nevertheless, surveys are well suited for determining the incidence of certain behaviors or for discovering traits and trends. Much of the research that family scientists conduct and use—on topics as far reaching as the division of housework and childcare, the frequency of and satisfaction with sex, or the effect of divorce on children or adults—is derived from interview or questionnaire data. Surveys are more commonly used by sociologists than by psychologists, because they tend to deal on a general or societal level rather than on a personal or small-group level. But surveys are not able to measure well how people interact with one another or what they actually do. For researchers and therapists interested in studying the dynamic flow of relationships, surveys are not as useful as clinical, experimental, and observational studies.

Secondary Analysis

As mentioned earlier, many researchers use a technique known as secondary data analysis. Because of the various costs associated with conducting surveys on large, nationally representative samples, researchers often turn to one of the available survey data sets such as the General Social Survey (GSS) conducted by the National Opinion Research Center at the University of Chicago. The GSS includes many social science variables of interest to family researchers. Family researchers also often use data issued by the U.S. Census Bureau, which include many descriptive details about the U.S. population, including characteristics of families and households.

Additional examples of available survey data of particular value to family researchers include the National Survey of Families and Households (NSFH) and the National Health and Social Life Survey (NHSLS). The NSFH has provided much information about a range of family behaviors including the division of housework, the frequency of sexual activity, and the relationships between parents and their adult children. The NHSLS is based on a representative sample of 3,432 Americans, aged 18 to 59, and contains much useful data about sexual behavior (Christopher and Sprecher 2000).

The major difficulty associated with secondary data analysis is that the material collected in the original survey may "come close to" but not be exactly what you wanted to examine. Perhaps you would have worded it differently to capture the essence of what you are interested in. Likewise, perhaps you would have asked additional questions to further or more deeply explore your topical interest (Babbie 2002). This disadvantage, although real, does not negate the benefits associated with secondary analysis.

Clinical Research

Clinical research involves in-depth examination of a person or a small group of people who come to a psychiatrist, psychologist, or social worker with psychological or relationship problems. The **case-study method,** consisting of a series of individual interviews, is the most traditional approach of all clinical research; with few exceptions, it was the sole method of clinical investigation through the first half of the twentieth century (Runyan 1982).

Clinical researchers gather a variety of additional kinds of data, including direct, first-hand observation or analysis of records. Rather than a specific technique of data collection, clinical research is distinguished by its examination of individuals and families that have sought some kind of professional help. The advantage of clinical approaches is that they offer long-term, in-depth study of various aspects of marriage and family life. The primary disadvantage is that we cannot necessarily make inferences about the general population from them. People who enter psychotherapy are not a representative sample. They may be more motivated to solve their problems or have more intense problems than the general population (Kitson et al. 1996).

One of the more widely cited and celebrated clinical studies is Judith Wallerstein's longitudinal study of 60 families who sought help from her divorce clinic. Wallerstein has published three books, *Surviving the Breakup: How Children and Parents Cope With Divorce; Second Chances: Men, Women, and Children a Decade After Divorce;* and *The Unexpected Legacy of Divorce: The 25 Year Landmark Study,* following the experiences of most of the children in these families (she has retained 93 of the original 131 children that she first interviewed in 1971) at 5, 10, and 25 years after divorce (Wallerstein 1980, 1989, 2000). All three books are sensitively written and richly convey the multitude of short- and long-term effects of divorce in the lives of her sample. Her critics have questioned whether findings based on such a clinically drawn sample (60 families from Marin County, California, who sought help as they underwent divorce) apply to divorced families more generally (Coontz 1998).

Clinical studies, however, have been fruitful in developing insight into family processes. Such studies have been instrumental in the development of family systems theory, discussed earlier in this chapter. By analyzing individuals and families in therapy, psychiatrists, psychologists, and therapists such as R. D. Laing, Salvador Minuchin, and Virginia Satir have been able to understand how families create roles, patterns, and rules that family members follow without being aware of them.

Observational Research

Observational research and experimental studies (discussed in the next section) account for less than 5% of recent research articles (Nye 1988). In **observational research,** scholars attempt to study behavior systematically through direct observation while remaining as unobtrusive as possible. To measure power in a relationship, for example, an observer researcher may sit in a home and videotape exchanges between a husband and a wife. The obvious disadvantage of this method is that the couple may hide unacceptable ways of dealing with decisions, such as threats of violence, when the observer is present. Individuals within families, as well as families as groups, are concerned with appearances and the impressions they make.

Another problem with observational studies is that a low correlation often exists between what observers see and what the people observed report about themselves (Bray 1995). Researchers have suggested that self-reports and observations measure two different views of the same thing: A self-report is an insider's view, whereas an observer's report is an outsider's view (Jacob et al. 1994). Some observational research involves family members being given structured activities to carry out. These activities involve interaction that can be observed between family members (Milner and Murphy 1995). They may include problem-solving tasks, putting together puzzles or games, or responding to a contrived family dilemma. Different tasks are intended to elicit different types of family interaction, which provide the researchers with opportunities to observe behaviors of interest.

A third problem that observational researchers encounter involves the essentially private nature of most family relationships and experiences. Because we experience most of our family life "behind closed doors," researchers typically cannot see what goes on "inside," without being granted access. For more public family behavior, observational data can be effectively used.

For example, an observational study by sociologist Paul Amato examined the question, "Who takes care

Understanding Yourself

What Do Surveys Tell You about Yourself?

Survey questionnaires are the leading source of information about marriage and the family. The questionnaire that follows was developed by Don Martin to gain information about attitudes toward marriage and the family. On a scale of 1 to 5 as shown, indicate your response for each statement.

The Marriage and Family Life Attitude Survey	Strongly agree (1)	slightly agree (2)	neither agree nor disagree (3)	slightly disagree (4)	strongly disagree (5)
I. Cohabitation and Premarital Sexual Relations					
1. I have or would engage in sexual intercourse before marriage.	1	2	3	4	5
2. I believe it is acceptable to experience sexual intercourse without loving one's partner.	1	2	3	4	5
3. I want to live with someone before I marry him or her.	1	2	3	4	5
4. If I lived intimately with a member of the opposite sex, I would tell my parents.	1	2	3	4	5
II. Marriage and Divorce					
5. I believe marriage is a lifelong commitment.	1	2	3	4	5
6. I believe divorce is acceptable except when children are involved.	1	2	3	4	5
7. I view my parents' marriage as happy.	1	2	3	4	5
8. I believe I have the necessary skills to make a good marriage.	1	2	3	4	5
III. Childhood and Childrearing					
9. I view my childhood as a happy experience.	1	2	3	4	5
10. If both my spouse and I work, I would leave my child in a day care center while at work.	1	2	3	4	5
11. If I have a child, I feel only one parent should work so that the other can take care of the child.	1	2	3	4	5
12. The responsibility for raising a child is divided between both spouses.	1	2	3	4	5
13. I believe I have the knowledge necessary to raise a child properly.	1	2	3	4	5
14. I believe children are not necessary in a marriage.	1	2	3	4	5
15. I believe two or more children are desirable for a married couple.	1	2	3	4	5
IV. Division of Household Labor and Professional Employment					
16. I believe household chores and tasks should be equally shared between marital partners.	1	2	3	4	5
17. I believe there are household chores that are specifically suited for men and others for women.	1	2	3	4	5
18. I believe women are entitled to careers equal to those of men.	1	2	3	4	5
19. If my spouse is offered a job in a different locality, I will move with my spouse.	1	2	3	4	5
V. Marital and Extramarital Sexual Relations					
20. I believe sexual relations are an important component of a marriage.	1	2	3	4	5
21. I believe the male should be the one to initiate sexual advances in a marriage.	1	2	3	4	5
22. I do not believe extramarital sex is wrong for me.	1	2	3	4	5

After you have completed this questionnaire, ask yourself the following questions:

- Were the questions correctly posed so that your responses adequately portrayed your attitudes?
- Were questions omitted that are important for you regarding marriage and the family? If so, what were they?
- Do your attitudes reflect your actual behavior?

The Marriage and Family Life Attitude Survey	Strongly agree (1)	slightly agree (2)	neither agree nor disagree (3)	slightly disagree (4)	strongly disagree (5)
VI. Privacy Rights and Social Needs					
23. I believe friendships outside of marriage with the opposite sex are important in a marriage.	1	2	3	4	5
24. I believe the major social functioning in a marriage should be with other couples.	1	2	3	4	5
25. I believe married couples should not argue in front of other people.	1	2	3	4	5
26. I want to marry someone who has the same social needs as I have.	1	2	3	4	5
VII. Religious Needs					
27. I believe religious practices are important in a marriage.	1	2	3	4	5
28. I believe children should be made to attend church.	1	2	3	4	5
29. I would not marry a person of a different religious background.	1	2	3	4	5
VIII. Communication Expectations					
30. When I have a disagreement in an intimate relationship, I talk to the other person about it.	1	2	3	4	5
31. I have trouble expressing what I feel toward the other person in an intimate relationship.	1	2	3	4	5
32. When I argue with a person in an intimate relationship, I withdraw from that person.	1	2	3	4	5
33. I would like to learn better ways to express myself in a relationship.	1	2	3	4	5
IX. Parental Relationships					
34. I would not marry if I did not get along with the other person's parents.	1	2	3	4	5
35. If I do not like my spouse's parents, I should not be obligated to visit them.	1	2	3	4	5
36. I believe each spouse's parents should be seen an equal amount of time.	1	2	3	4	5
37. I feel parents should not intervene in any matters pertaining to my marriage.	1	2	3	4	5
38. If my parents did not like my choice of a marriage partner, I would not marry this person.	1	2	3	4	5
X. Professional Counseling Services					
39. I would seek premarital counseling before I got married.	1	2	3	4	5
40. I would like to attend marriage enrichment workshops.	1	2	3	4	5
41. I will seek education and/or counseling to learn about parenting.	1	2	3	4	5
42. I feel I need more education of what to expect from marriage.	1	2	3	4	5
43. I believe counseling is only for those couples in trouble.	1	2	3	4	5

The comparative data included here about the division of household labor are representative of the kind of data family researchers gather through survey instruments. Presented in table format, as data typically are throughout this book, they can be used for many purposes. Look closely at Table 2.2. What do the numbers in the table actually represent?

Using data from the International Social Justice Project, a multination study of perceptions of social and economic justice, Shannon Davis and Theodore Greenstein examine perceptions of the division of household labor. Although they look at a number of issues in their analysis, we look here only at gender differences in perceptions of who bears responsibility for housework.

Their measure of the division of household labor is based on answers to the following question (translated into each country's native language): "Please tell me how the following responsibilities are divided. Are they always done by yourself, usually by yourself, equally between yourself and your partner, usually by your partner, or always by your partner? First of all, housework such as cooking, cleaning and laundry?"

Data from 10,153 respondents, 5,104 men and 5,049 women, are presented. All respondents were married and living with their spouses at the time of the interviews. The sample does not contain married *couples*, as only one spouse in a household was interviewed. Answers were recoded into the following categories: always the wife, usually the wife, shared equally, usually the husband, always the husband. Individuals who

gave answers other than these (as in paid help or someone else in the family) were excluded from the following findings.

In the following table, data are presented by country, first for men and then for women (with numbers of men and women in the country samples provided). We can compare countries or, by comparing the two rows within each country, see the extent to which gender differences separate men's and women's answers in each country. Study the table. What interesting things do you notice?

We can use these data in the table to note a number of different things. First, in each country, by both men's and women's accounts, women are responsible for housework. Similarly, looking at all countries together, 65.8% of males and 72.7% of females say housework is usually and always done by wives (columns 3 and 4).

Second, in each country, men and women differ in their responses about who does the housework in their households, with men nearly always indicating somewhat greater sharing than women credit men with. Perhaps this doesn't surprise you.

Third, the gender gap in asserting that housework is "always" done by wives is often wide. Only in the Czech Republic and Russia is it less than 10%. In countries such as Poland, East and West Germany, and the United Kingdom, it is nearly or in excess of 20%. Conversely, greater percentages of husbands than wives report that housework is "usually done by wives" everywhere but Russia and the United States. Combining the "always wife" and "usually wife" categories (into "usually or always wife") reduces the gender difference considerably and reveals what percentages of women and men attribute housework to

women. With a single category of "usually or always wife," the gender difference is reduced to an average of 8% (ranging from 1.4% between male and female respondents in Bulgaria to a 13.3% difference between male and female West German respondents).

Fourth, combining the categories "always wife" and "usually wife" shows large variation across the 13 countries. Women's reports range from Russia, where women report housework is "usually" or "always" done by the wife in 37.4% of households, to Japan, where women report that they always or usually do the housework in 97.8% of households. Men's reports vary similarly across countries. Men report housework is "usually" or "always" done by wives in 30.3% of Russian households to 92.6% of Japanese households.

Fifth, by combining the last three columns, we can see the percentages of women and men who report that men do about half or more of the housework. There is a cross-national range here, too, from Japan, where 7.3% of men and 2.2% of women report men doing at least "almost equal" amounts of housework, to Russia, where 69.8% of the men and 61.7% of the women report men's involvement as "almost equal" or greater. In the United States, 43.2% of male respondents and 33.1% of female respondents said husbands shared "about equally" or usually or always did the housework.

Together, these data reveal that, in the United States and abroad, responsibility for domestic work rests heavily on women's shoulders. Along with other comparative survey research such as Jean Baxter's 5-country comparison and Makiko Fuwa's 22-country analysis, we can use the Davis and Greenstein data in the following table to demonstrate that in all countries studied the

Table 2.2 ■ Men's and Women's Responses across 13 Countries to Who Does the Housework?

Country	n = x	Always Wife (%)	Usually Wife (%)	Equal (%)	Usually Husband (%)	Always Husband (%)
Bulgaria						
Husband	497	20.3	49.3	25.8	2.2	2.4
Wife	482	37.6	33.4	27.6	1.2	0.2
Czech Republic						
Husband	372	19.6	48.1	30.6	0.8	0.8
Wife	359	22.0	38.7	37.3	1.9	0.0
Estonia						
Husband	279	7.9	43.4	44.1	4.3	0.4
Wife	275	21.5	40.7	35.3	1.5	1.1
West Germany						
Husband	428	22.2	51.6	21.5	1.2	3.5
Wife	356	50.6	36.5	11.0	1.4	0.6
East Germany						
Husband	292	12.3	52.4	31.2	2.1	2.1
Wife	297	37.7	37.7	23.9	0.7	0.0
Hungary						
Husband	314	26.1	37.9	30.9	1.9	3.2
Wife	309	39.8	30.7	26.9	1.9	0.6
Japan						
Husband	258	62.8	29.8	5.0	0.4	1.9
Wife	276	79.3	18.5	2.2	0.0	0.0
Netherlands						
Husband	608	22.4	51.2	25.3	1.0	0.2
Wife	510	39.8	41.8	17.6	0.8	0.0
Poland						
Husband	502	32.9	43.0	18.7	2.0	3.4
Wife	471	52.2	30.1	13.6	2.8	1.3
Russia						
Husband	499	3.8	26.5	66.7	2.6	0.4
Wife	494	9.3	28.1	60.1	1.4	0.2
Slovenia						
Husband	385	29.1	46.5	17.9	3.4	3.1
Wife	480	45.6	36.3	17.7	0.4	0.0
United Kingdom						
Husband	311	21.5	42.8	29.9	4.2	1.6
Wife	356	41.0	35.4	21.9	0.8	0.8
United States						
Husband	359	14.8	42.1	40.4	1.7	1.1
Wife	384	24.7	42.2	30.5	1.6	1.0
All Nations						
Husband	5,104	22.0	43.8	30.3	2.1	1.8
Wife	5,049	37.8	34.9	25.6	1.3	0.4

From Davis and Greenstein 2004.

Continues

responsibility for housework falls most heavily on women's shoulders (Baxter 1997; Fuwa 2004; Davis and Greenstein 2004).

Keep in mind that these data, like all questionnaire data, report only what people say; we do not have behavioral indicators of what they actually do. Furthermore, from these data alone we do not know why household chores are divided as they are. Nor do we know whether women and/or men object to this allocation of responsibilities; for that we would need more and different data.

Different theories, such as those raised earlier in this chapter, offer a range of explanations as to why tasks become divided by gender and what implications such divisions have. These sorts of issues are raised, and survey data such as these are used, throughout this book.

of children in public places?" Amato suggested that using "naturalistic observations," wherein people are unaware that they are being watched, eliminated the concern about potential face-saving or impression-making distortions to people's "real behavior." Using researchers strategically stationed in a variety of public places (for example, parks, shopping malls, and restaurants) in San Diego, California, and Lincoln, Nebraska, Amato compiled 2,500 observations of children with their male and/or female caretakers. He used such observations to test five hypotheses about adult male–child interaction (Amato 1989).

Overall, Amato found that 43% of the young children observed were cared for by men. His specific findings indicated that boys were more likely than girls to be looked after by a man; preschool children were most likely and infants were least likely to have male caretakers; male caretaking was highest in recreational settings and lowest in restaurants; male caretaking rates were higher among men who were accompanied by women than among men by themselves; and there were only modest differences between the California and the Nebraska locations. In addition to its substantive contributions, Amato's research showed that

■ *There are aspects of family life that can be easily observed, such as care for children in public.*

■ *What goes on at home, behind closed doors, may not be easily accessible to observational researchers.*

although not widely popular among family researchers, observational research can be used to study certain family phenomena.

Because a major limitation of strictly observational data is identifying meanings people attach to their behavior or attributing motives for why people are doing what they are observed doing, researchers often combine observational data with other sorts of data in a process known as **triangulation.**

As an example of triangulation, Jan R. Gerris, Maja Dekovic, and Jan M. Janssens (1997) examined whether social class affected the childrearing values and behaviors of a sample of 237 Dutch mothers and fathers. Researchers interviewed participants; administered a 25-minute "family interaction task" (puzzle solving by both parents and a target child), which they observed; and asked participants to complete a questionnaire detailing their childrearing techniques. The observational data were recorded on tape and later analyzed, along with the interview and questionnaire data, to identify a variety of ways in which social class effects surfaced in childrearing.

Experimental Research

In **experimental research,** researchers isolate a single factor under controlled circumstances to determine its influence. Researchers are able to control their experiments by using *variables*, aspects or factors that can be manipulated in experiments. Recall the earlier discussion of types of variables, especially independent and dependent variables. In experiments, independent variables are factors manipulated or changed by the experimenter; dependent variables are factors affected by changes in the independent variable.

Because it controls variables, experimental research differs from the previous methods we have examined. Clinical studies, surveys, and observational research are correlational in nature. Correlational studies measure two or more naturally occurring variables to

determine their relationship to one another. Because correlational studies do not manipulate the variables, they cannot tell us which variable causes the others to change. But because experimental studies manipulate the independent variables, researchers can reasonably determine which variables affect the other variables.

Experimental findings can be powerful because such research gives investigators control over many factors and enables them to isolate variables. Researchers believing that stepmothers and stepfathers are stigmatized, for example, tested their hypothesis experimentally (Ganong, Coleman, and Kennedy 1990). They devised a simple experiment in which subjects were asked to evaluate 20 traits of a person in a family who was described in a short paragraph.

The person was variously identified as a father or mother in a nuclear family, a biological father or mother in a stepfamily, or a stepfather or stepmother in a stepfamily. When identified as a biological parent in either a nuclear family or a stepfamily, the individual was rated more favorably than when identified as a stepfather or a stepmother. This paper-and-pencil experiment confirmed the researchers' hypothesis that stepparents are stigmatized.

The obvious problem with such studies is that we respond differently to people in real life than we do in controlled situations, especially in paper-and-pencil situations. We may not stigmatize a stepparent in real life. Experimental situations are usually faint shadows of the complex and varied situations we experience in the real world.

Differences in sampling and methodological techniques help explain why studies of the same phenomenon may arrive at different conclusions. They also help explain a common misperception many of us hold regarding scientific studies. Many of us believe that because studies arrive at different conclusions, none are valid. What conflicting studies may show us, however, is that researchers are constantly exploring issues from different perspectives as they attempt to arrive at a consensus.

Researchers may discover errors or problems in sampling or methodology that lead to new and different conclusions. They seek to improve sampling and methodologies to elaborate on or disprove earlier studies. In fact, the very word *research* is derived from the prefix *re-*, meaning "over again," and *search*, meaning "to examine closely." And that is the scientific endeavor: searching and re-searching for knowledge.

Researching Long-Distance Relationships

We have come almost full circle. At the beginning of this chapter we posed a scenario in which "common sense" failed to resolve contradictory advice to a friend about the likely outcome of her long-distance relationship. We have revisited this issue throughout the chapter. It is now time to look at what researchers who have approached this phenomenon have learned.

Long-distance relationships are increasingly common, both within the college context and in the adult world (especially of two earner couples). In the 1990s, studies estimated that anywhere from 33% to more than 40% of romantic relationships among undergraduates were long-distance relationships (Sahlstein 2004).

Unfortunately, much of the research supports a pessimistic view of how long-distance relationships fare (Knox et al. 2002; Van Horn et al. 1997; Guldner 1996; Stafford and Reske 1990; Schwebel et al. 1992). Although they may not fail because once "out of sight, (we are) out of mind," they don't appear to hold up especially well over time. Using survey research techniques with samples of college students, K. Roger Van Horn (1997) and colleagues found that partners in long-distance romantic relationships reported less companionship, less disclosure, less satisfaction, and less certainty about the future together compared to partners in geographically close relationships. Comparing 164 students in long-distance relationships with 170 in geographically proximal relationships, Gregory Guldner (1996) reported that the separated partners showed more depressive symptoms. Andrew Schwebel and colleagues (1992) studied 34 men and 55 women in relationships in which they were separated by at least 50 miles from their partners. Within 9 weeks, nearly a quarter of the relationships had ended. Finally, David Knox and colleagues (2002) surveyed 438 undergraduates to test their belief in the "out of sight, out of mind" idea and to gauge their experiences of such relationships. Nearly 20% of the sample reported being in a long-distance relationship, here meaning separated by 200 miles or more. Of the sample, 37% reported having been in a long-distance relationship that ended. Although more than half of the sample with experience of a long-distance relationship had phoned and/or e-mailed several times a week, more than 40%

felt that the distance had worsened (20%) or ended (21.5%) their relationship. Conversely, 18% said that it "improved" their relationship. Having experienced a long-distance relationship made respondents more likely to believe "out of sight, out of mind."

The most optimistic findings suggest that long-distance relationships are not especially different from proximal relationships (Guldner and Swensen 1995). Comparing 194 students in long-distance relationships with 190 who were in geographically close relationships, Gregory Guldner and Clifford Swensen found that the two types of relationships were rated with about the same levels of self-reported relationship satisfaction and similar levels of intimacy, trust, and degree of relationship progress.

Although long-distance relationships clearly face obstacles that proximal relationships don't (for example, lack of time together and pressure to maximize quality of time partners do spend together), they also benefit from a determination to make their time special, to value each other's company in ways that couples who see each other easily and often may not (Sahlstein 2004). This hardly constitutes "making the heart grow fonder," but it is less negative than the other research.

Did *any* research support "absence makes the heart grow fonder?" The answer is yes, yet that itself may be problematic for healthy relationship development. Comparing 34 "geographically close" couples and 37 long-distance couples (separated by an average of 421.6 miles), Laura Stafford and James Reske found that long-distance couples were more satisfied with their relationships and with the level of communication they had. They also were by their assessments "more in love." Acknowledging the possibility that the long-distance relationships were "better" than the geographically closer relationships, Stafford and Reske go on to suggest that a process of *idealization* occurs in long-distance relationships, largely because of their more restricted communication (more phone calls and letters as opposed to face-to-face interaction and less overall interaction). As a consequence of this idealization, long-distance couples set themselves up for later problems that couples with less restricted communication (that is, geographically closer couples) avoid. They "may have little idea of how idealized and inaccurate their images (of their relationships) are" (Stafford and Reske 1990).

It is worth noting that even if all existing research painted a negative picture of what happens in long-distance relationships, that does not mean that any particular relationship (that is, yours or your friend's) is destined to meet that unfortunate outcome. Family scientists seek to identify and account for patterns in social relationships. There are always going to be exceptions to any identified pattern. This is important for two reasons. First, don't assume that patterns reported in this book will happen in your life. Your experience may constitute an exception to the more general pattern. Second, and equally important, don't dismiss findings reported here because they don't fit your experiences or those of people you may know. Instead, try to account for why your experience

departs from the more generally observed social regularities.

By using critical thinking skills and by understanding something about the methods and theories used by family researchers, we are in a position to more effectively evaluate the information we receive about families. We are also better able to step outside our personal experience, go beyond what we've always been told, and begin to view marriage and family from a sounder and broader perspective. In Chapters 3 and 4, we take such steps and explicitly examine the factors and forces that create differences in family experience.

Summary

- We need to be alert to maintain *objectivity* in our consideration of different forms of family lifestyle. *Opinions, biases,* and *stereotypes* are ways of thinking that lack objectivity.

- *Fallacies* are errors in reasoning. Two common types of fallacies are *egocentric fallacies* and *ethnocentric fallacies:* the belief that all people are, or should be, the same as we are or that our way of living is superior to all others.

- *Theories* attempt to provide frames of reference for the interpretation of data. Theories of marriage and families include family ecology, symbolic interaction, social exchange, family development, structural functionalism, conflict, and family systems.

- Theories are built from concepts, abstract ideas about reality. Conceptualization is the process of identifying and defining the concepts we are studying, and operationalization is the development of research strategies to observe our concepts.

- Deductive research tests hypotheses, statements in which we turn our concepts into variables and specify how variables are related to each other. An independent variable is a variable that influences or shapes our dependent variable. Intervening variables are those that follow our independent variables and have direct effects on dependent variables.

- Inductive research does not test hypotheses. It begins with a more general interest. As data is collected, concepts are specified in more detail

leading to the development of hypotheses and to grounded theory.

- *Family ecology theory* examines how families are influenced by and, in return, influence the wider environments in which they function.

- *Symbolic interaction theory* examines how people interact and how we interpret or define others' actions through the symbols they communicate (their words, gestures, and actions). Symbolic interactionists study how social roles and personality interact.

- *Social exchange theory* suggests that we measure our actions and relationships on a cost–benefit basis. People seek to maximize their rewards and minimize their costs to gain the most favorable outcome. A corollary to exchange is *equity:* exchanges must balance out or hard feelings are likely to ensue. Exchanges in marriage can be either cooperative or competitive.

- *Family development theory* looks at the changes in the family, beginning with marriage and proceeding through seven sequential stages reflecting the interacting influences of changing roles and circumstances through time.

- *Structural functionalism theory* looks at society and families as though they were organisms containing different structures, each of which has a function. Structural functionalists study: (1) the functions the family serves for society, (2) the functional requirements performed by the family for its survival, and (3) the needs of individual members that are met by the family. Family functions are usually divided along gender lines.

- *Conflict theory* assumes that individuals in marriages and families are in conflict with one another. Power is often used to resolve the conflict. Four important sources of power are legitimacy, money, physical coercion, and love.

- *Family systems theory* approaches the family in terms of its structure and pattern of interactions. Systems analysts believe that (1) interactions must be studied in the context of the family, (2) family structure can be seen only in the family's interactions, (3) the family is a system purposely seeking homeostasis (stability), and (4) family systems are transformed over time.

- *Feminist perspectives* provide an orienting focus for considering gender differences relating to family and social issues. In the writing, research, and advocacy of the feminist movement, the goals are to help clarify and remove oppressive conditions and barriers to opportunities for women. Recently the feminist perspective has been expanded to include constraints affecting black–white and gay–straight dichotomies. Such attention to gender gave rise to men's studies, a field in which scholars examine how masculinity and male socialization shape men's experiences, including their family lives.

- Family researchers apply the *scientific method*—well-established procedures used to collect information.

- Professional family researchers follow ethical principles to protect participants from having their identities revealed and to minimize the discomfort the subjects experience from their participation in the research.

- Research data come from surveys, clinical studies, and direct observation, in which naturally occurring variables are measured against one another. Data are also obtained from experimental research.

- *Survey research* uses questionnaires and interviews. They are more useful for dealing with societal or general issues than for personal or small-group issues. Limits of the method include (1) volunteer bias or an unrepresentative sample, (2) individuals' lack of self-knowledge, and (3) underreporting of undesirable or unconventional behavior.

- Frequently researchers conduct secondary analyses on already existing data. This allows researchers to examine large representative samples at little cost of time or resources.

- *Clinical research* involves in-depth examinations of individuals or small groups that have entered a clinical setting for the treatment of psychological or relationship problems. The primary advantage of clinical studies is that they allow in-depth case studies; their primary disadvantage is that the people coming into a clinic are not representative of the general population.

- In *observational research*, interpersonal behavior is examined in a natural setting, such as the home, by an unobtrusive observer. Major difficulties include the possibility that participants behave unnaturally, hide less acceptable behavior from researchers, and the fact that most family behavior is highly private. Further, one may not be able to know what the observed behavior means to those engaged in it.

- In *experimental research,* the researcher manipulates variables. Such studies are of limited use in marriage and family research because of the difficulty of controlling behavior and duplicating real-life conditions.

- To overcome limitations with any particular method, researchers often engage in triangulation, the use of multiple methods and/or multiple sources of data.

- Family researchers strive to identify and account for patterns of behavior. There will be exceptions to all patterns. Exceptions do not negate the importance or validity of research conclusions.

Key Terms

adaptation 39	equity 45
ambiguous loss 41	ethical guidelines 54
anonymity 54	ethnocentric fallacy 37
bias 36	experimental research 63
case-study method 57	fallacies 36
clinical research 57	family development theory 46
conceptualization 37	
confidentiality 54	family ecology theory 39
conflict theory 49	family myths 43
deductive research 37	family systems theory 50
dependent variable 37	feminist perspectives 51
egocentric fallacy 36	gender 52
environment 39	grounded theory 38
environmental influences 39	homeostasis 51
	hypotheses 37

Resources on the Internet

Companion Website for This Book

http://www.thomsonedu.com/sociology/strong

Gain an even better understanding of this chapter by going to the companion website for additional study resources. Take advantage of the Pre- and Post-Test quizzing tool, which is designed to help you grasp difficult concepts by referring you back to review specific pages in the chapter for questions you answer incorrectly. Use the flash cards to master key terms and check out the many other study aids you'll find there. Visit the Marriage and Family Resource Center on the site. You'll also find special features such as access to Info-Trac© College Edition (a database that allows you access to more than 18 million full-length articles from 5,000 periodicals and journals), as well as GSS Data and Census information to help you with your research projects and papers.

Differences: Historical and Contemporary Variations in American Family Life

Outline

What Do YOU Think?

Are the following statements **TRUE** or **FALSE**?
You may be surprised by the answers (see answer key on the following page.)

T F   1 Compared with contemporary families, colonial family life was considered more private.

T F   2 Industrialization transformed the role families played in society, as well as the roles women and men played in families.

T F   3 Slavery destroyed the African American family system.

T F   4 Compared with what came both before and after, families of the 1950s were unusually stable.

T F   5 Within upper-class families, husbands and wives are relatively equal in their household roles and authority.

T F   6 Lower-class families are the most likely to be single-parent families.

T F   7 Family relationships can suffer as a result of either downward or upward mobility.

T F   8 Compared with Caucasian families, relationships between African American husbands and wives are more traditional.

T F   9 Asian American or Latino families show much variation within each group, depending on the country from which they came, why they left, and when they arrived in the United States.

T F   10 European ethnic groups are as different from one another as they are from African Americans, Latinos, Asian Americans, or Native Americans.

One thing you can almost always count on is that sometime during the term or semester, whether in class or in conversation, someone will make the oft-heard statement, "Well, all families are different." There is a lot of truth to that sentiment. For example, your family is not like your best friend's family in every way. Furthermore, assuming your best friend is someone a lot like you (which, as you've probably noticed, is common among people who become best friends), the differences between your families likely understate how richly variable family experience actually is.

Although it is true that in some ways *every family is different*, over the next few chapters we look closely at some patterned variations that separate and diversify family experiences. Although there are a number of factors that we could include as sources of such variation, the current chapter is concerned with the following four: time, social class, race, and ethnicity. In subsequent chapters we also look at how gender and sexual orientation shape people's experiences of relationships and families. Then, throughout the remainder of the book, we draw comparisons and make contrasts among different types of households and families—singles, cohabiting and married couples, parents and nonparents, single-parent households and two-parent households, dual earners, male-breadwinner–female homemaker households and "role reversers," first marriages and remarriages, and step relationships in blended families and blood relationships in birth families. Therefore, the task we start here won't end until you finish this book.

We begin by detailing the historical development of the kinds of families that predominate in the United States today, noting key transformations and the forces that created them. This accomplishes two things: It gives you a better sense of where today's American families have come from, and it enables you to see how different family life has been across generations, even within the same families. We then shift our attention to some major racial, ethnic, and economic variations that diversify contemporary American families.

Answer Key for What Do You Think

1 False, see p. 72;   2 True, see p. 73;   3 False, see p. 75;   4 True, see p. 79;   5 False, see p. 90;   6 True, see p. 89;   7 True, see p. 94;   8 False, see p. 98;   9 True, see p. 104;   10 False, see p. 110.

American Families across Time

American marriages and families are dynamic and must be understood as the products of wider cultural, demographic, and technological developments (Mintz and Kellogg 1988). Although we tend to emphasize the more familiar changes that have occurred over the past half century or so (post–World War II), those changes represent only more recent instances of more than 300 years of change that comprise the history of American family life from the colonial period through the twentieth century. Armed with this brief history, we can recognize and make connections between changes in society and changes in families. In addition, we will be better positioned to assess the meaning of some of the more dramatic changes that have occurred recently in American family life.

Finally, on a more personal level, you can better understand your own genealogies and family histories by recognizing the shifting stage on which they were played out.

The Colonial Era

The colonial era is marked by differences among cultures, family roles, customs, and traditions. These families were the original crucible from which our contemporary families were formed.

Native American Families

The greatest diversity in American family life probably existed during our country's earliest years, when 2 million Native Americans inhabited what is now the United States and Canada. There were more than 240 groups with distinct family and kinship patterns. Many groups were **patrilineal:** rights and property flowed from the father. Others, such as the Zuni and Hopi in the Southwest and the Iroquois in the Northeast, were **matrilineal:** rights and property descended from the mother.

Native American families tended to share certain characteristics, although it is easy to overgeneralize. Most families were small. There was a high child mortality rate, and mothers breastfed their infants; during breastfeeding, mothers abstained from sexual intercourse.

Children were often born in special birth huts. As they grew older, the young were rarely physically disciplined. Instead, they were taught by example. Their families praised them when they were good and publicly shamed them when they were bad. Children began working at an early stage. Their play, such as hunting or playing with dolls, was modeled on adult activities. Ceremonies and rituals marked transitions into adulthood. Girls underwent puberty ceremonies at first menstruation. For boys, events such as growing the first tooth and killing the first large animal when hunting signified stages of growing up. A vision quest often marked the transition to manhood.

Reflections

How far back can you trace your family's history? What would you like to know about it? What values, traits, or memories do you wish to pass on to your descendants?

Marriage took place early for girls, usually between 12 and 15 years; for boys, it took place between 15 and 20 years. Some tribes arranged marriages; others permitted young men and women to choose their partners.

Most groups were monogamous, although some allowed two wives. Some tribes permitted men to have sexual relations outside of marriage when their wives were pregnant or breastfeeding.

Colonial Families

From earliest colonial times, America has been an ethnically diverse country. In the houses of Boston, the mansions and slave quarters of Charleston, the mansions of New Orleans, the haciendas of Santa Fe, and the Hopi dwellings of Oraibi (the oldest continuously inhabited place in the United States, dating back to A.D. 1150), American families have provided emotional and economic support for their members.

THE FAMILY. Colonial America was initially settled by waves of explorers, soldiers, traders, pilgrims, servants, prisoners, farmers, and slaves. In 1565, in St. Augustine, Florida, the Spanish established the first permanent European settlement in what is now the United States. But the members of these first groups came as single men—as explorers, soldiers, and exploiters.

In 1620, the leaders of the Jamestown colony in Virginia, hoping to promote greater stability, began importing English women to be sold in marriage. The European colonists who came to America attempted to replicate their familiar family system. This system, strongly influenced by Christianity, emphasized **patriarchy** (rule by father or eldest male), the subordination of women, sexual restraint, and family-centered production.

The family was basically an economic and social institution, the primary unit for producing most goods and caring for the needs of its members. The family planted and harvested food, made clothes, provided shelter, and cared for the necessities of life.

As a social unit, the family reared children and cared for the sick, infirm, and aged. Its responsibilities included teaching reading, writing, and arithmetic because there were few schools. The family was also responsible for religious instruction: it was to join in prayer, read scripture, and teach the principles of religion.

Unlike New Englanders, the planter aristocracy that came to dominate the Southern colonies did not give high priority to family life; hunting, entertaining, and politics provided the greatest pleasure. The planter aristocracy continued to idealize gentry ways until the Civil War destroyed the slave system upon which the planters based their wealth.

MARITAL CHOICE. Romantic love was not a factor in choosing a partner; one practical seventeenth-century marriage manual advised women that "this boiling affection is seldom worth anything" (Fraser 1984). Because marriage had profound economic and social consequences, parents often selected their children's mates. In the seventeenth century, 8 of the 13 colonies had laws *requiring parental approval* and imposed sanctions as harsh as imprisonment or whipping on men who "insinuated" themselves into a woman's affections without her parents' approval (Coontz 2005). Even in instances without such restrictions, in which individuals were "free to choose," children rarely went against their parents' wishes. If parents disapproved, their children typically gave up out of fear of the social and financial consequences of defying their parents (Coontz 2005). Love was not irrelevant but came after marriage. It was a person's duty to love his or her spouse. The inability to desire and love a marriage partner was considered a defect of character.

Although the Puritans prohibited premarital intercourse, they were not entirely successful. **Bundling**, the New England custom in which a young man and woman spent the night in bed together, separated by a wooden bundling board, provided a courting couple with privacy; it did not, however, encourage restraint. An estimated one-third of all marriages in the eighteenth century took place with the bride pregnant (Smith and Hindus 1975).

FAMILY LIFE. The colonial family was strictly patriarchal, and such paternal authority was reinforced by both the church and the community (Mintz 2004). Steven Mintz describes the range of fathers' influence (2004, 13). Fathers were

> responsible for leading their households in daily prayers and scripture reading, catechizing their children and servants, and teaching household members to read so that they might study the Bible. . . . Childrearing manuals were thus addressed to men, not their wives. They had an obligation to help their sons find a vocation or calling, and a legal right to consent to their children's marriage. Massachusetts Bay Colony and Connecticut underscored the importance of paternal authority by making it a capital offense (that is, punishable by death) for youths sixteen or older to curse or strike their father.

The authority of the husband/father rested in his control of land and property. In an agrarian society such as colonial America, land was the most precious resource. The manner in which the father decided to dispose of his land affected his relationships with his children. In many cases, children were given land adjacent to the father's farm, but the title did not pass into their hands until the father died. This power gave fathers control over their children's marital choices and kept them geographically close.

This strongly rooted patriarchy called for wives to submit to their husbands. The wife was not an equal but was a helpmate. This subordination was reinforced by traditional religious doctrine. Like her children, the colonial wife was economically dependent on her husband. Upon marriage, she transferred to her husband many rights she had held as a single woman, such as the right to inherit or sell property, to conduct business, and to attend court.

For women, marriage marked the beginning of a constant cycle of childbearing and childrearing. On average, colonial women had six children and were consistently bearing children until around age 40. In addition to their maternal responsibilities, colonial women were expected to do a wide range of chores from cooking and cleaning to spinning, sewing, gardening, keeping chickens, and even brewing beer. (Mintz and Kellogg 1988).

CHILDHOOD AND ADOLESCENCE. The colonial conception of childhood was radically different from ours. First, children were believed to be evil by nature The community accepted the traditional Christian doctrine that children were conceived and born in sin.

Second, childhood did not represent a period of life radically different from adulthood. Such a conception is distinctly modern (Aries 1962; Meckel 1984; Vann 1982). In colonial times, a child was regarded as a small adult. From the time children were 6 or 7 years old, they began to be part of the adult world, participating in adult work and play.

Third, children between the ages 7 and 12 were often "bound out" or "fostered" as apprentices or domestic servants (Mintz 2004). They lived in the home of a relative or stranger where they learned a trade or skill, were educated, and were properly disciplined. **Adolescence**—the separate life stage between childhood and adulthood—did not exist. They went from a shorter childhood (than what we are accustomed to) to adulthood (Mintz and Kellogg 1988; Mintz 2004). Thus, our contemporary notions of a rebellious life stage filled with inner conflicts, youthful indiscretions, and developmental crises do not fit well with the historical record of Plymouth Colony (Demos 1970; Mintz 2004).

African American Families

In 1619, a Dutch man-of-war docked at Jamestown in need of supplies. Within its cargo were 20 Africans who had been captured from a Portuguese slaver. The captain quickly sold his captives as indentured servants. Among those first Africans was a woman known by the English as Isabella and a man known as Antony; their African names are lost. In Jamestown, Antony and Isabella married. After several years, Isabella gave birth to William Tucker, the first African American child born in what is today the United States. William's birth marked the beginning of the African American family, a unique family system that largely grew out of the African adjustment to slavery in America. By 1664, when the British gained what had been Dutch governed New Amsterdam, 40% of the colony's population consisted of African slaves.

■ *Strong family ties endured in enslaved African American families. The extended family, important in West African cultures, continued to be a source of support and stability.*

During the seventeenth and much of the eighteenth centuries, enslaved Africans and their descendants faced difficulty forming and maintaining families. It was hard for men, who often outnumbered women 60% to 40% or worse, to find wives. Enslaved African Americans were more successful in continuing the traditional African emphasis on the extended family, in which aunts, uncles, cousins, and grandparents played important roles. Although slaves were legally prohibited from marrying, they created their own marriages.

Childhood experience was often bitter and harsh. It was common for children to be separated from their parents because of a sale, a repayment of a debt, or a plantation owner's decision to transfer slaves from one property to another (Mintz 2004). Despite the hardships placed on them, enslaved Africans and African Americans developed strong emotional bonds and family ties. Slave culture discouraged casual sexual relationships and placed a high value on marital stability. On the large plantations, most enslaved people lived in two-parent families with their children. To maintain family identity, parents named their children after themselves or other relatives or gave them African names. In the harsh slave system, the family provided strong support against the daily indignities of servitude. As time went on, the developing African American family blended West African and English family traditions (McAdoo 1996).

Nineteenth-Century Marriages and Families

In the nineteenth century, the traditional colonial family form gradually vanished and was replaced by the modern family.

Industrialization and the Shattering of the Old Family

In the nineteenth century, the industrialization of the United States transformed the face of America. It also transformed American families from self-sufficient farm families to wage-earning, increasingly urban families. As factories began producing gigantic harvesters, combines, and tractors, significantly fewer farm workers were needed. Looking for employment, workers migrated to the cities, where they found employment in the ever-expanding factories and businesses. Because goods were now bought rather than made in the home, the family began its shift from being primarily a production unit to being more of a consumer- and service-oriented unit. With this shift, a radically new **division of labor** arose in the family. Men began working outside the home in factories or offices for wages they then used to purchase the family's necessities and other goods. Men became identified as the family's sole provider or breadwinner. Their work was given higher status than women's work because it was paid in wages. Men's work began to be increasingly identified as "real" work, distinct from the unpaid domestic work done by women.

Marriage and Families Transformed

Without its central importance as a work unit, and less and less the source of other important societal functions (for example, education, religious worship, protection, and recreation), the family became the focus and abode of feelings. The emotional support and well-being of adults and the care and nurturing of the young became the two most important family responsibilities.

THE POWER OF LOVE. This new affectionate foundation of marriage brought love to the foreground.

Love as the basis of marriage represented the triumph of individual preference over family, social, or group considerations. Stephanie Coontz reports that "By the middle of the nineteenth century there was near unanimity in the middle and upper classes throughout western Europe and North America that the love-based marriage, in which the wife stayed home and was protected and supported by her husband, was a recipe for heaven on earth" (Coontz 2005, 162).

Women now had a new degree of power: they were able to choose whom they would marry. Women could rule out undesirable partners during courtship; they could choose mates with whom they believed they would be compatible. Mutual esteem, friendship, and confidence became guiding ideals. Without love, marriages were considered empty shells.

CHANGING ROLES FOR WOMEN. The two most important family roles for middle-class women in the nineteenth century were that of housewife and mother. As there was a growing emphasis on domesticity in family life, the role of housewife increased in significance and status. Home was the center of life, and the housewife was responsible for making family life a source of fulfillment for everyone. For many women, especially middle class, this "doctrine of separate spheres" was wholeheartedly accepted and enthusiastically embraced (Coontz 2005).

Women also increasingly focused their identities on motherhood. The nineteenth century witnessed the most dramatic decline in fertility in American history. Between 1800 and 1900, fertility dropped by 50%. Where at the beginning of the nineteenth century American mothers typically gave birth to between 7 and 10 children, beginning "in her early twenties and (giving birth) every two years or so until menopause," by 1900 the average number of births had fallen to just 3 (Mintz 2004).

Women reduced their childbearing by insisting that they, not men, control the frequency of intercourse. Childrearing rather than childbearing became one of the most important aspects of a woman's life. Having fewer children, and having them in the early years of marriage, allowed more time to concentrate on mothering and opened the door to greater participation in the world outside the family. This outside participation manifested itself in women's heavy involvement in abolition, prohibition, and women's emancipation movements.

CHILDHOOD AND ADOLESCENCE. A strong emphasis was placed on children as part of the new conception of the family. A belief in childhood innocence replaced the idea of childhood corruption. A new sentimentality surrounded the child, who was now viewed as born in total innocence. Protecting children from experiencing or even knowing about the evils of the world became a major part of childrearing.

The nineteenth century also witnessed the beginning of adolescence. In contrast to colonial youths, who participated in the adult world of work and other activities, nineteenth-century adolescents were kept economically dependent and separate from adult activities and often felt apprehensive when they entered the adult world. This apprehension sometimes led to the emotional conflicts associated with adolescent identity crises.

Education also changed as schools, rather than families, became responsible for teaching reading, writing, and arithmetic, as well as educating students about ideas and values. Conflicts between the traditional beliefs of the family and those of the impersonal school were inevitable. At school, the child's peer group increased in importance.

The African American Family: Slavery and Freedom

Although there were large numbers of free African Americans—100,000 in the North and Midwest and 150,000 in the South—most of what we know about the African American family before the Civil War is limited to the slave family.

THE SLAVE FAMILY. By the nineteenth century, the slave family had already lost much of its African heritage. Under slavery, the African American family lacked two key factors that helped give free African American and Caucasian families stability: autonomy and economic importance. Slave marriages were not recognized as legal. Final authority rested with the owner in all decisions about the lives of slaves. The separation of families was a common occurrence, spreading grief and despair among thousands of slaves. Furthermore, slave families worked for their masters, not themselves. It was impossible for the slave husband/father to become the provider for his family. The slave women worked in the fields beside the men. When an enslaved woman was pregnant, her owner determined her care during pregnancy and her relation to her infant after birth.

Slave children endured deep and lasting deprivation. Often shoeless, sometimes without underwear or adequate clothing, hungry, underfed and undernourished, and forced into hard physical labor as young as age 5 or 6, slave children suffered considerably. Rates of illness and death in infancy and childhood were high. Furthermore, family life was fragile and often disrupted. Steven Mintz reports that separation of children from parents, especially fathers, was so common that at least half of all enslaved children experienced life separate from their father, because he died, lived on another plantation, or was a white man who declined to acknowledge that they were his children. By their late teens, either temporary or permanent separation from their parents was something virtually all slave children had suffered (Mintz 2004).

Still, it is important to reiterate that slavery did not destroy all aspects of slave families. Despite the intense oppression and hardship to which they were subjected, many slaves displayed resilience and survived by relying on their families and by adapting their family system to the conditions of their lives (Mintz and Kellogg 1988). This included, for example, relying on extended kinship networks and, where necessary, on unrelated adults to serve as surrogates for parents absent because of the forced breakup of families.

Furthermore, enslavement did not forever destroy the African American family system. In no way does saying this diminish the horrors of slavery. Instead, it acknowledges the resilience of those who survived enslavement, and it illustrates how family systems may be pivotal sources of support and key mechanisms of surviving even the most extraordinary distress.

AFTER FREEDOM. When freedom came, the formerly enslaved African American families had strong emotional ties and traditions forged from slavery and from their West African heritage (Guttman 1976; Lantz 1980). Because they were now legally able to marry, thousands of former slaves formally renewed their vows. The first year or so after freedom was marked by what was called "the traveling time," in which African Americans traveled up and down the South looking for lost family members who had been sold. Relatively few families were reunited, although many continued the search well into the 1880s.

African American families remained poor, tied to the land, and segregated. Despite poverty and continued exploitation, the Southern African American family usually consisted of both parents and their children. Extended kin continued to be important.

Immigration: The Great Transformation

THE OLD AND NEW IMMIGRANTS. In the nineteenth and early twentieth centuries, great waves of immigration swept over America. Between 1820 and 1920, 38 million immigrants came to the United States. Historians commonly divided them into "old" immigrants and "new" immigrants. The old immigrants, who came between 1830 and 1890, were mostly from western and northern Europe. During this period, Chinese also immigrated in large numbers to the West Coast. The new

National Park Service: Statue of Liberty National Monument

■ *Except for Native Americans, most of us have ancestors who came to America—voluntarily or involuntarily. Between 1820 and 1920, more than 38 million immigrants came to the United States.*

immigrants, who came from eastern and southern Europe, began to arrive in great numbers between 1890 and 1914 (when World War I virtually stopped all immigration).

Japanese also immigrated to the West Coast and Hawaii during this time. Today, Americans can trace their roots to numerous ethnic groups.

As the United States expanded its frontiers, surviving Native Americans were incorporated. The United States acquired its first Latino population when it annexed Texas, California, New Mexico, and part of Arizona after its victory over Mexico in 1848.

THE IMMIGRANT EXPERIENCE. Most immigrants were uprooted; they left only when life in the old country became intolerable. The decision to leave their homeland was never easy. It was a choice between life and death and meant leaving behind ancient ties.

Most immigrants arrived in America without skills. Although most came from small villages, they soon found themselves in the concrete cities of America. Again, families were key ingredients in overcoming and surviving extreme hardship. Because families and friends kept in close contact even when separated by vast oceans, immigrants seldom left their native countries without knowing where they were going—to the ethnic neighborhoods of New York, Chicago, Boston, San Francisco, Vancouver, and other cities. There they spoke their own tongues, practiced their own religions, and ate their customary foods. In these cities, immigrants created great economic wealth for America by providing cheap labor to fuel growing industries.

In America, kinship groups were central to the immigrants' experience and survival. Passage money was sent to their relatives at home, information was exchanged about where to live and find work, families sought solace by clustering together in ethnic neighborhoods, and informal networks exchanged information about employment locally and in other areas.

The family economy, critical to immigrant survival, was based on cooperation among family members. For most immigrant families, as for African American families, the middle-class idealization of motherhood and childhood was a far cry from reality. Because of low industrial wages, many immigrant families could survive only by pooling their resources and sending mothers to work and even sending their children to work in the mines, mills, and factories.

Most groups experienced hostility. Crime, vice, and immorality were attributed to the newly arrived ethnic groups; ethnic slurs became part of everyday par-

lance. Strong activist groups arose to prohibit immigration and promote "Americanism." Literacy tests required immigrants to be able to read at least 30 words in English. In the early 1920s, severe quotas were enacted that slowed immigration to a trickle.

It is interesting to note what crucial roles families played in enabling people to survive the oppression of enslavement, the difficulties of immigration, and the impoverishment induced by industrialization.

Reflections

As you read through these historical perspectives, what are your feelings about such struggles and triumphs? How does knowledge of your family history affect you, your values, and your behavior?

Twentieth-Century Marriages and Families

The Rise of Companionate Marriages: 1900–1960

By the beginning of the twentieth century, the functions of American middle-class families had been dramatically altered from earlier times. Families had lost many of their traditional economic, educational, and welfare functions. Food and goods were produced outside the family, children were educated in public schools, and the poor, aged, and infirm were increasingly cared for by public agencies and hospitals. The primary focus of the family was becoming even more centered on meeting the emotional needs of its members. In time, cultural emphasis would shift from self-sacrificing familism to more self-centered individualism, and individuals' sense of their connections and obligations to their families would be greatly transformed.

THE NEW COMPANIONATE FAMILY. Beginning in the 1920s, a new ideal family form was beginning to emerge that rejected the "old" family based on male authority and sexual repression. This new family form was based on the **companionate marriage.**

There were four major features of this companionate family (Mintz and Kellogg 1988): (1) Men and women were to share household decision making and tasks. (2) Marriages were expected to provide romance, sexual fulfillment, and emotional growth. (3) Wives were no longer expected to be guardians of virtue and sexual restraint. (4) Children were no longer to be protected from the world but were to be given greater freedom to explore and experience the world; they were to be treated more democratically and encouraged to express their feelings.

Through the Depression and World Wars

The history of twentieth-century family life cannot be told without considering how profoundly family roles and relationships were affected by the Great Depression and two world wars. Although many different connections could be drawn, two seem particularly significant: changes in the relationship between the family and the wider society and changes in women's and men's roles in and outside of the family.

LINKING PUBLIC AND PRIVATE LIFE. The economic crisis during the Depression was staggering in its scope. Unemployment jumped from less than 3 million in 1929 to more than 12 million in 1932, and the rate of unemployment rose from 3.2% to 23.6%.

Over that same span of time, average family income dropped 40% (Mintz and Kellogg 1988). To cope with this economic disaster, families turned inward, modifying their spending, increasing the numbers of wage earners to include women and children, and pooling their incomes. Often it was a broadened "inward" to

which they turned, because people often took in relatives or relied on kinship ties for economic assistance (Mintz and Kellogg 1988).

Ultimately, these more personal, intrafamilial efforts proved insufficient. President Franklin Roosevelt's New Deal social programs attempted to respond to the social and economic despair that more localized efforts were unable to alleviate. Farm relief, rural electrification, Social Security, and a variety of social welfare provisions were all implemented in the hope of doing what local communities and individual families could not. Such federal initiatives reflected a dramatic ideological shift wherein government now bore responsibility for the lives and well-being of families (Mintz and Kellogg 1988).

Precipitated by the mass entrance into the workforce of millions of previously unemployed women, including many with young children, there was a clear need and opportunity for public resources to be committed to childcare. Unfortunately, the federal government's response was slow and inadequate given the sudden and dramatic increase in need and demand (Filene 1986; Mintz and Kellogg 1988; Mintz 2004). Most mothers who entered the labor force had to rely on neighbors and grandparents to provide childcare. When such supports were unavailable, many had no choice but to turn their children into "latch-key" kids

■ *During Word War II women were urged to enter the labor force and especially to enter nontraditional occupations left vacant by the deployment of men overseas. The images here illustrate the kinds of messages women received and the kinds of jobs they helped fill.*

fending for themselves (Mintz 2004). Unlike some of our European allies who invested more heavily in policies and services to accommodate employed mothers (Mintz and Kellogg 1988), it took the federal government 2 years to "appropriate funds to build and staff day-care centers, and the funds were sufficient for only one-tenth of the children who needed them" (Filene 1986). Despite having engineered a propaganda campaign to entice women into jobs vacated by the 16 million men who entered the service, the government remained ambivalent about welcoming mothers of young children into those positions. However inadequate or slow their efforts were, they were still more ambitious than what followed for most of the rest of the century.

GENDER CRISES: THE GREAT DEPRESSION AND WORLD WARS. Both the Depression and the two world wars (especially World War II) reveal much about the gender foundation on which twentieth-century families rested. During the Depression, it was men whose gender identities and family statuses were threatened by their lost status as providers. During each world war, women were the ones who faced challenges that required them to abandon their gender socialization and step into roles and situations that fell outside their traditional familial roles. In each instance, the familial gender roles and identities had to be altered to match extraordinary circumstances.

What is especially striking about men's reactions to their job loss is their internalization of fault for what was a society-wide economic crisis. Given how widespread unemployment was, we might think that men would take some comfort in knowing that the predicaments they faced were not of their own making. Yet they had so deeply internalized their sense of themselves as providers that their identities, family statuses, and sense of manhood were all invested in wage earning and providing. When unable to provide, many men were deeply shaken. Some were even driven to the point of emotional breakdown or suicide by their sense of economic failure (Filene 1986).

For many families, survival depended on the efforts of wives or the combination of women's earnings, children's earnings, assistance from kin, or some kind of public assistance. For those who depended at least somewhat on women's earnings, there were other gender consequences of running the household. Sometimes, men were pressed by their wives to contribute domestically in the women's "absence." Although some did, many others resisted (Filene 1986). Sometimes

women displayed ambivalence about the meaning of male unemployment and male housework. Whereas 80% of the women who were surveyed in 1939 by the *Ladies' Home Journal* thought an unemployed husband should do the domestic work in the absence of his employed wife, 60% reported they would lose respect for men whose wives out-earned them (Filene 1986).

If the Depression illustrates male anxiety about their familial roles as providers, we see in women's experiences during World Wars I and II that gender crises were not limited to men. Both wars share that, in the absence of millions of men, women were pressed to step into their vacant shoes and participate in wartime production. During World War I, 1.5 million women entered the wartime labor force, many in jobs previously held largely by men (Filene 1986). During World War II, the number of employed women rose dramatically. Between 1941 and 1945, the numbers of employed women increased by more than 6 million to a wartime high of 19 million (Degler 1980; Lindsey 1997). Furthermore, "nearly half of all American women held a job at some time during the war" (Mintz and Kellogg 1988). Whereas single women had long worked, and poor or minority women had worked even after marriage, the biggest change in women's labor force participation during World War II was among married, middle-class women. Thus, despite the strong and widely held cultural emphasis on the special nurturing role of women and the belief that the home was a woman's "proper place," American society needed women to take over for the absent men.

Once enticed into nontraditional female employment, women received both material and nonmaterial benefits that were hard for many to surrender once the war ended and men returned.

Materially, women in traditionally male occupations received higher wages than they had in their past, more sex-segregated work experiences. As important, they also found a sense of gratification and enhanced self-esteem that were often missing from the jobs they were more accustomed to. However reluctant they may have been to take on such work, many were clearly more than a little ambivalent to leave it.

To assist women in their departures from these jobs, pro-family rhetoric and a new ideology extolling the value and importance of women's roles as mothers and caregivers were broadly conveyed by a variety of sources (for example, popular media, social workers, and educators).

Families of the 1950s

In the long history of American family life, no other decade has come to symbolize so much about that history, despite actually representing relatively little of it. (Mintz and Kellogg 1988; Coontz 1997). In many ways, the 1950s appear to be a period of unmatched family stability. Marriage and birthrates were unusually high, divorce rates were uncharacteristically low, and the economy enabled many to buy houses with only one wage-earning spouse.

During the 1950s, marriage and family seemed to be central to American lives. It was a time of youthful marriages, increased birthrates, and a stable divorce rate. Most families were comprised of male breadwinners and female homemakers. Traditional gender and marital roles mostly prevailed. Man's place was in the world and woman's place was in the home. Women were expected to place motherhood first and to sacrifice their opportunities for outside advancement to ensure the success of their husbands and the well-being of their children.

Given the meaning often invested in this era, it is important to understand that the 1950s were unique. Compared with both what came before and what followed, families of the 1950s were exceptional. This is important: It means that anyone who uses this decade as a baseline against which to compare more recent trends in such family characteristics as birth, marriage, or divorce rates starts with a faulty assumption about how representative it is of American family history. Looking at those same trends with a longer view reveals that the changes that followed the 1950s were more consistent with some patterns evident in the nineteenth and earlier part of the twentieth century (Mintz and Kellogg 1988).

For example, the trend since the Civil War had been an increase in the divorce rate of about 3% per decade until the 1950s. During the 1950s, the divorce rate increased less than in any other decade of the twentieth century. Similarly, after more than 100 years of declining birthrates and shrinking family sizes, during the 1950s "women of childbearing age bore more children, spaced . . . closer together, and had them earlier and faster" than had previous generations (Mintz and Kellogg 1988). After all, this was the height of the baby boom; married couples had more children than either those that preceded them or those that followed.

Much familial experience of the 1950s was created and sustained by the unprecedented economic growth and prosperity of the postwar economy (Coontz 1997).

The combination of suburbanization and economic prosperity, supplemented by governmental assistance to veterans, allowed many married couples to achieve the middle-class family dream of home ownership while raising their children under the loving attention of full-time caregiving mothers. We must be careful, though, not to oversimplify family experience of the 1950s. Americans did not all benefit equally from the economic prosperity and opportunity of the decade. Thus, overgeneralizations would leave out the experiences of poor and working-class families and racial minorities for whom neither full-time mothering nor home ownership were commonplace (Coontz 1997). In addition, many women found that the ideal lifestyle of the period left them longing for something more (Friedan 1963).

When we look at family changes that occurred in subsequent decades, we need to recognize that economic factors, again, were among the most important determinants of some more dramatic departures from the 1950s model. This especially pertains to the emergence of the dual-earner household. As Stephanie Coontz points out, "By the mid-1970s, maintaining the prescribed family lifestyle meant for many couples giving up the prescribed family form. They married later, postponed children, and curbed their fertility; the wives went out to work" (Coontz 1997). They did this not in rejection of the family lifestyle of the 1950s but in the pursuit of central features of that lifestyle, such as home ownership.

Aspects of Contemporary Marriages and Families

The remaining 12 chapters of this book look closely at families of the latter decades of the twentieth century and the beginning of the twenty-first century. The characteristics displayed by these families did not emerge suddenly but were established over years. Beginning with the latter years of the 1950s and escalating through and then beyond the 1960s and 1970s, some striking family trends surfaced. These trends persisted through and beyond the end of the twentieth century, leaving marriages and families reshaped and the meaning and experience of family life significantly altered.

Birthrates dropped, people delayed and departed marriage as almost never before, and individuals increasingly were drawn to cohabitation. The median age for marriage began to climb in the 1960s, by 1996 reaching the highest it had been in more than 100 years. Even after a slight drop in the last few years of the 1990s, the age at entering first marriage climbed more in the first years of this century and remains 4 years older for men and nearly 5 years for women than the 1960 ages (Table 3.1).

Marriage and divorce rates rose and fell, the prevalence of cohabitation substantially increased, and birthrates dropped. But even across this shorter historical span family trends are not linear; they go up and then drop (Table 3.2). Thus, it appears that, even in the short term, the only constant in family life is change (Mintz and Kellogg 1988).

Although the trends shown in Tables 3.2 and 3.3 are not the only dimensions of family life that have seen major change, they are important indicators that the family is a dynamic institution. Such trends are also often the sources of much controversy over their larger meaning. The debates about what is happening to family life in the United States that we depicted in Chapter 1 often focus on these very swings. It is obvious that such trends depict change, but what is less clear is what those changes say about the vitality of the family.

As we noted, some argue that changes such as these are worrisome signs of family decline (Popenoe 1993). With fewer people marrying, more people divorcing, and more people living together or by themselves outside of marriage, the importance of the family—as reflected in the stability or desirability of marriage—appears to be declining, and the future of family life

Table 3.1 ▪ Median Age at First Marriage, 1960-2003

Year	Males (age)	Females (age)
1960	22.8	20.3
1970	23.2	20.8
1980	24.7	22.0
1990	26.1	23.9
2000	26.8	25.1
2003	27.1	25.3

SOURCE: Fields 2003.

Table 3.2 ▪ Couples and Children: 1970-2000

	1970	1980	1990	2000
Married couples	44,728,000	49,112,000	52,317,000	55,311,000
Married couples with children	25,541,000	24,961,000	24,537,000	25,248,000
Percentage of all married couples with children	57%	51%	47%	46%
Unmarried couple households	523,000	1,589,000	2,856,000	4,486,000
Unmarried couples with children	196,000	431,000	891,000	1,563,000
Children living with two parents	59,681,000	47,543,000	46,820,000	49,688,000
Children living with one parent	8,426,000	12,349,000	15,842,000	19,227,000
Births to unmarried women	399,000	666,000	1,165,000	1,308,000
As percentage of all births	11%	18%	28%	33%

SOURCES: U.S. Census Bureau 2000, Table 77; Jason Fields 2002; National Vital Statistics Reports 2001; Fields and Casper 2000.

Table 3.3 ▪ Trends in Marriages, Divorces, and Births: 1970-2005

	1970	1980	1990	2000	2005
Marriages	2,159,000	2,390,000	2,443,000	2,329,000	2,230,000
Marriage rate	10.6%	10.6%	9.8%	8.5%	7.4%
Divorces	708,000	1,189,000	1,182,000	1,135,000	NA
Divorce rate	2.2%	3.5%	5.2%	4.7%	3.7%
Births	3,731,000	3,612,000	4,158,000	4,063,000	4,143,000
Birth rate[*]	18.4%	15.9%	16.7%	14.5%	14.0%

SOURCES: U.S. Census Bureau 2002; Munson and Sutton, National Center for Health Statistics, 2005.
NA means data not available.
[*]Rate per 1,000 people.

is in some doubt. Others take the more liberal position that change is not a bad thing and that with these changes come more choices for people about the kinds of families they wish to create and experience (Mintz and Kellogg 1988; Coontz 1997). Certainly, today's families do reflect considerable diversity of structure. In painting a picture of today's families, we would include many categories: breadwinner–homemaker families with children, two-earner couples with children, single-parent households with children, marriages without children, cohabiting couples with or without children, blended families, role-reversed marriages, and gay and lesbian couples with or without children.

Whereas American families have from their beginnings been diverse entities, with varying cultural and economic backgrounds (Mintz and Kellogg 1988), what distinguishes contemporary families is the diversity represented by the range and spread of people across these varying chosen lifestyles.

Factors Promoting Change

Marriages and families are shaped by a number of different forces in society. In looking over the major changes to American families, we can identify four important factors that initiated these changes: (1) economic changes, (2) technological innovations, (3) demographics, and (4) gender roles and opportunities for women.

Economic Changes

As noted earlier, over time, the family has moved from being an economically productive unit to a consuming, service-oriented unit. Where families once met most needs of their members—including providing food, clothing, household goods, and occasionally surplus crops that it bartered or marketed—most of today's families must purchase what they need.

Economic factors have been responsible for major changes in the familial roles played by women and men. Inflation, economic hardship, and an expanding economy have led to married women entering the labor force in unprecedented numbers. Even women with preschool-aged children are typically employed outside the home (Figure 3.1). As a result, the dual-earner marriage and the employed mother have become commonplace features of contemporary families. As women have increased their participation in the

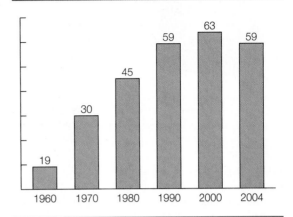

Figure 3.1 ■ **Percentage of Married Women Employed outside the Home Who Have Children 6 Years Old or Younger**

SOURCE: U.S. Census Bureau 2006: Table 586.

paid labor force, other familial changes have occurred. For instance, women are less economically dependent on either men or marriage. This provides them greater legitimacy in attempts to exercise marital power. It has also increased the tension around the division of household chores and raised anxiety and uncertainty over who will care for children.

Technological Innovations

Before you read further, stop and think about how your family routines and relationships are affected by various technological innovations and devices. Can you imagine how your family experiences would be different *without* these devices? You may be wondering what things we're talking about. Consider this: The family has been affected by most major innovations in technology—from automobiles, telephones, cell phones, televisions, DVD players, and microwaves to personal computers and the Internet. These devices were not designed or invented to transform families but to improve transportation, enhance communication, expand choices and quality of entertainment, and maximize efficiency. Nevertheless, they have had major repercussions in how family life is experienced.

For example, older devices such as automobiles and telephones, as well as more recent innovations such as personal computers, have aided families in maintaining contact across greater distances, thus allowing extended families to sustain closer relations and nuclear family members to stay available to one another through school- and job-related travel or relocation.

Before we begin examining marriages and families in detail, consider some changes that have occurred over the past 4 decades, sparking so much debate.

- *Cohabitation.* In its technical sense, **cohabitation** refers to individuals sharing living arrangements in an intimate relationship, whether these individuals are married or unmarried. In common usage, however, cohabitation refers to relationships in which unmarried individuals share living quarters and are sexually involved. (*Cohabitation* and *living together* are often used interchangeably.) A cohabiting relationship may be similar to marriage in many of its functions and roles, but it does not have equivalent legal sanctions or rights. Cohabitation has increased dramatically over the past 40 years. In addition to the almost 5 million heterosexual couples, there are an additional 600,000 to 1 million–plus same-sex couples living together outside of marriage.

- *Marriage.* A combination of factors including the women's movement, shifting demographics, family policy, and changing values, particularly as they relate to sexuality, have altered the meaning of marriage and the role it plays in people's lives. Still, between 80% and 90% of young unmarried women and men *will marry* at least once in their lifetimes.

- *Separation and divorce.* Separation occurs when two married people

no longer live together. It may or may not lead to divorce. Many more people separate than divorce. Divorce is the legal dissolution of a marriage. Over the last 50 years, divorce has changed the face of marriage and the family in America. At present, among adults 18 and over, there are nearly 20 million who are divorced. The divorce rate is two to three times what it was for our parents and grandparents.

Slightly less than half of all those who marry will divorce within 7 years. These trends in prevalence of divorce have led to what might be considered "the normalization of divorce." Divorce has become so widespread that many scholars view it as one variation of the normal life course of American marriages (Coontz 1997). The high divorce rate does not indicate that Americans devalue marriage, however. Paradoxically, Americans may divorce because they value marriage so highly. If a marriage does not meet their standards, they divorce to marry again. They hope that their second marriages will fulfill the expectations that their first marriages failed to meet (Furstenberg and Spanier 1987).

- *Remarriages, stepfamilies, and single-parent families.* Contemporary divorce patterns are largely responsible for three related versions of American marriages and families: single-parent families, remarriages, and stepfamilies. Because of their widespread incidence, these variations are becoming part of our normal marriage and family patterns.

- *Single-parent families.* As many as one out of six U.S. households consist of single mothers or single

fathers and their dependent children. Single-parent households represented 32% of all households with children under 18. Most such households are created by divorce, but some are the product of nonmarital births, widowerhood, or widowhood. There were 12.4 million single-parent households in the United States in 2003; 18% (2.3 million) were male-headed households and 82% were female-headed families. Of note, important differences can be observed between single-father-headed and single-mother-headed households in terms of standard of living, numbers of children, and marital histories (Fields 2003).

- *Remarriage.* Half of all recent marriages are remarriages for at least one partner (Coleman, Ganong, and Fine 2000). Most individuals who divorce tend to remarry. Rates differ between men and women and across different ethnic groups. Those who remarry are usually older, have more experience in both life and work, and have different expectations than those who marry for the first time. Remarriages also may create stepfamilies. When remarriages include children, a person may become not only a husband or a wife but also a stepfather or a stepmother.

- *Stepfamilies.* An estimated one-third of children will reside in a stepfamily household before reaching adulthood (Coleman et al. 2000). Ironically, despite the hopes and experience of those who remarry, their divorce rate is at least as much as that of those who marry for the first time.

The proliferation of automobiles also altered the residential and relationship experiences of many Americans, making it possible for people to live greater distances from where they work—thus contributing to the suburbanization of America—and to experience premarital relationships away from more watchful adult supervision.

Televisions, and more recently the Internet, have altered the recreation and socialization activities in which families engage, with both beneficial and negative consequences. Sitting and watching television programs together gives family members the opportunity for shared experiences. As important as the entertainment function of both television and the Internet are, they also operate as additional socialization agents, beyond parents and other relatives.

What we watch on television, or view and read on the Internet, helps shape our values and beliefs about the world around us. As shown in a subsequent chapter, the Internet has also greatly expanded our options for meeting potential partners and spouses. Finally, cell phones, e-mail, and instant messaging have altered the ways in which parents monitor children and family members remain in contact with one another.

The range of domestic appliances—from washing machines and dishwashers to microwaves—has altered how the tasks of housework are done. Although we might be tempted to conclude that such devices free people from some time- and labor-intensive burdens associated with maintaining homes, historical research has shown that this is not automatically so. For instance, as technology made it possible to more easily wash clothes, the standards for cleanliness increased. In the case of microwaves, the time needed for tasks associated with meal preparation has been reduced, freeing people to spend more time in other activities (not necessarily as families, and often away from their families—at work, for example).

Finally, revolutions in contraception and biomedical technology have reshaped the meaning and experience of sexuality and parenthood. Much of what we call the "sexual revolution" in the 1970s and beyond was fueled partly by safer and more certain methods of preventing pregnancy, such as the birth control pill. Regarding parenthood, people who in the past would have been unable to become parents have the opportunity to enjoy childbearing and rearing as a result of **assisted reproductive technologies**—including medical advances such as in vitro fertilization, as well as surrogate motherhood and sperm donation. Such developments have thus altered the meaning of parenthood, as multiple individuals may be involved in any single conception, pregnancy, and eventual birth. Sperm and/or egg donors, surrogate mothers, and the parent or parents who nurture and raise the child all can claim in some way to have reared the child in question. Such changes have complicated the social and legal meanings of parenthood as that they have opened the possibility of parenthood to previously infertile couples or same-sex couples.

Reflections

As you stop and think about your family routines and relationships, how much do they seem to be affected by the kinds of technological innovations discussed here? How would your experiences be *without* these devices?

Demographics

The family has undergone dramatic demographic changes in areas that include family size, life expectancy, divorce, and death. Three important changes have emerged:

- *Increased longevity.* As people live longer, they are experiencing aspects of family life that few experienced before. In colonial times, because of a relatively short life expectancy, husbands and wives could anticipate a marriage lasting 25 years. Today, couples can remain married 50 or 60 years. Today's couples can anticipate living many years together after their children are grown; they can also look forward to grandparenthood or great-grandparenthood. Since men tend to marry women younger than themselves and on average die younger than women do, American women can anticipate a prolonged period of widowhood.

- *Increased divorce rate.* The increased divorce rate, beginning in the late nineteenth century (even before 1900, the United States had the highest divorce rate in the world), has led to the rise of single-parent families and stepfamilies. In this way, it has dramatically altered the experience of both childhood and parenthood and has altered our expectations of married life.

- *Decreased fertility rate.* As women bear fewer children, they have fewer years of childrearing responsibility.

With fewer children, partners are able to devote more time to each other and expend greater energy on each child. Children from smaller families benefit in a variety of ways from the greater levels of parental attention, although they may lack the advantages of having multiple siblings. From the adults' perspective, smaller families afford women greater opportunity for entering the workforce.

Gender Roles and Opportunities for Women

Changes in gender roles are the fourth force contributing to alterations in American marriages and families. The history summarized earlier indicated some major changes that took place in women's and men's responsibilities and opportunities. These gender shifts then directly or indirectly led to changes in both the ideology surrounding and the reality confronting families.

The emphasis on childrearing and housework as women's proper duties lasted until World War II, when, as we saw, there was a massive influx of women into factories and stores to replace the men fighting overseas. This initiated a trend in which women increasingly entered the labor force, became less economically dependent on men, and gained greater power in marriage.

The feminist movement of the 1960s and 1970s led many women to reexamine their assumptions about women's roles. Betty Friedan's *The Feminine Mystique* challenged head-on the traditional assumption that women found their greatest fulfillment in being mothers and housewives. The women's movement emerged to challenge the female roles of housewife, helpmate, and mother, appealing to some women as it alienated others.

More recently, the dual-earner marriage made the traditional division of roles an important and open question for women. Today, contemporary women have dramatically different expectations of male–female roles in marriage, childrearing, housework, and the workplace than did their mothers and grandmothers. Changes in marriage, birth, and divorce rates, and in the ages at which people enter marriage, have all been affected by women's enlarged economic roles.

We have also witnessed changes in what men expect and are expected to do in marriage and parenthood. Although it may still be assumed that men will be "good providers," that is no longer enough. Married men face greater pressure to share housework and participate in childcare. Although they have been slow to increase the amount of housework they do, there

has been more acceptance of the idea that greater father involvement benefits both children and fathers. New standards and expectations of paternal behavior and more participation by fathers in raising children help explain the ongoing changes—from how dual-earner households function to why we are more accepting of fathers staying home to care for their young children.

The U.S. Census Bureau estimates that there are 1 million fathers (and 7 million mothers) of children age 15 or younger who were out of the labor force and home full-time for all of 2003. Of these, 160,000 (16%) of the fathers said that the "primary reason" they were home was to care for home and family. Another 45% of the fathers were home because of illness or disability. Among at-home mothers, 6 million of the 7 million (88%) said they were home to care for home and family (Fields 2004). Furthermore, 2 million preschoolers are cared for by their fathers while their mothers are at work. This is more than the numbers cared for by any other noninstitutionalized caregiver (U.S. Census Bureau 2003).

Gender issues are so central to family life that they are the subject of the entire next chapter. It is not an exaggeration to say that we cannot truly understand the family without recognizing the gender roles and differences on which it rests.

Cultural Changes

We can, in conclusion, point to a shift in American values from an emphasis on obligation and self-sacrifice to individualism and self-gratification (Bellah et al. 1985; Mintz and Kellogg 1998; Coontz 1997). The once strong sense of **familism,** in which individual self-interest was expected to be subordinated to family well-being, has given way to more open and widespread individualism, in which even families can be sacrificed for individual happiness and personal fulfillment.

This shift in values has had consequences for how people weigh and choose among alternative lifestyle paths. For example, complex decisions—about whether and how much to work, whether to stay married or to divorce, how much time and attention to devote to children or to spouses—are increasingly made against a backdrop of pursuing self-gratification and individual happiness. Values alone have not changed families, but such shifts in values have contributed to the choices people make, out of which new family forms predominate (Coontz 1997).

How Contemporary Families Differ from One Another

The preceding discussion traced some ways families have changed throughout history and why. In that sense, it has led us to family life of today. But today's families differ from one another, a topic we now explore. We look first at economic factors that differentiate families and then at cultural characteristics, social class, and race and ethnicity.

Economic Variations in Family Life

A **social class** is a category of people who share a common economic position in the stratified (that is, unequal) society in which they live. We typically identify classes using economic indicators such as ownership of property or wealth, amount of income earned, the level of prestige accorded to work, and so forth. Social class has both a structural and a cultural dimension. Structurally, social class reflects the occupations we hold (or depend on), the income and power they give us, and the opportunities they present or deny us. The cultural dimension of social class refers to any class-specific values, attitudes, beliefs, and motivations that distinguish classes from one another. Cultural aspects of social class are somewhat controversial, especially when applied to supposed "cultures of poverty"—an argument holding that poor people become trapped in poverty because of the values they hold and the behaviors in which they engage (Harrington 1962; Lewis 1966). What is unclear regarding "cultures of class" is how much difference there is in the values and beliefs of different classes and whether such differences cause or follow the more structural dimensions that separate one class from another.

To an extent, there is also a psychological aspect to social class. By this we mean the internalization of economic status in the self-images we form and the self-esteem we possess. These may also be seen as consequences of other aspects of class position, such as the self-identity that results from the prestige accorded to work or the respect paid to accomplishments. Like the structural and cultural components of social class, these are brought home and affect our experiences in our families.

The effect of social class is far reaching and deep. In an article about how social class affects marriage,

New York Times reporter Tamara Lewin quotes one of her sources, Della Mae Justice, of Piketown, Kentucky. Justice grew up in the coal-mining world of Appalachia, in a house without indoor plumbing. Having put herself through college and later law school, she is now solidly and unambiguously middle class. Justice says, "I think class is everything, I really do. When you're poor and from a low socioeconomic group, you don't have a lot of choices in life. To me, being from an upper class is all about confidence. It's knowing you have choices, knowing you have connections" (Lewin 2005).

Clearly, many facets of our lives (often referred to by sociologists as **life chances**) are affected by our **socioeconomic status,** including our health and well-being, safety, longevity, religiosity, and politics. A host of family experiences also vary up and down the socioeconomic ladder. For instance, class variations can be found in such family characteristics as age at marriage, age at parenthood, timing of marriage and parenthood, division of household labor, ideologies of gender, socialization of children, meanings attached to sexuality and intimacy, and likelihood of violence or divorce.

Conceptualizations of social class vary in how class is defined and how many classes are identified and counted in American society. In some formulations of social class, it is a person's relationship to the means of production that defines class position. In other models, people are grouped into classes because of similar incomes, amounts of wealth, degrees of occupational status, and years of education. Whether we claim that the United States has two (owners and workers), three (upper, middle, and lower), four (upper, middle, working, and lower), six (upper-upper, lower-upper, upper-middle, lower-middle, upper-lower, and lower-lower), or more classes, the important point about the concept of social class is that life is differently experienced by individuals across the range of identified classes and similarly experienced by people within any one of the class categories.

Using a fairly common model, we can describe these classes as follows.

Upper Classes

Roughly 7% to 10% of the population occupies an "upper class" position. The uppermost level of this class represents approximately 3% of the population (Renzetti and Curran 1998; Curry, Jiobu, and Schwirian 2002). They own 25% to 30% of all

Popular Culture

Can We See Ourselves in "Zits"? Comic Strips and Changes in Family Life

Every day, millions of people in hundreds of countries open thousands of newspapers and pause to read comic strips. The list of more popular strips will have familiar names on it for just about everyone: *Calvin & Hobbes*, *Cathy*, *Doonesbury*, *Dilbert*, *Sally Forth*, and *Peanuts*, which is so popular that it has run repeated comics for years since the retirement and later death of creator Charles Schulz.

Many comic strips focus on family life, typically featuring a couple and their young to adolescent children. Such is the case with two popular award-winning strips, *Baby Blues* and *Zits*, both of which are written by

Jerry Scott although they have different illustrators (Rick Kirkman for *Baby Blues* and Jim Borgman for *Zits*).

Baby Blues portrays what happens to Wanda and Darryl MacPherson when, in their 30s, they embark on life's great adventure of parenthood with the arrival of Zoe, Hammish, and Wren. On the *Baby Blues* website (http://www.babyblues.com/family _tree/familytree.htm), Wanda is

private wealth and 60% to 70% of all corporate wealth. They also receive as much as 25% of all yearly income. They are sometimes referred to as the "upper-upper class" or the "ruling class" or "elite." Their "extraordinary wealth" often takes them into the hundreds of millions if not billions of dollars (Curry, Jiobu, and Schwirian 2002).

The rest of the upper class live on yearly incomes ranging from hundreds of thousands to billions of dollars, own substantial amounts of wealth, and enjoy

described as having "traded working full-time as a public relations executive for working **fuller** time as a mother." She is "a complex person . . . part nurturing mother, part dynamic organizer, and part exhausted woman." Husband Darryl is depicted as "the consummate involved parent." Although he finds fatherhood "the hardest work he's ever done," Darryl is described as loving it. A March 6, 2005, strip depicted some ways in which the demands of parenthood come to dominate life, especially for mothers:

Zits looks at the life of 15-year-old Jeremy Duncan, "high school freshman with, thank God, four good friends, but other than that a seriously boring life in a seriously boring town made livable only by the knowledge that someday in the far-off future at least this will all be over" (http://www.kingfeatures.com/ features/comics/zits/about.htm). His parents, mother Connie, a frustrated novelist, and father Walt, an orthodontist, struggle to find ways to communicate with Jeremy, a brooding "handful" of adolescent hormones and moods.

Comics like *Zits* and *Baby Blues* allow us to laugh at some familiar, occasionally exaggerated situations and conversations, but they do much more. They offer us a window

through which to explore the wider cultural attitudes and values about family life. Sociologist Ralph LaRossa, one of the leading experts on changes in parenthood and especially fatherhood through the twentieth century, has studied the portrayal of gender and parental roles across six decades (1940–1999) of popular comic strips (LaRossa et al. 2000, 2001). Some comics that they examined included: *Blondie, Cathy, Dennis the Menace, For Better or Worse, Hi and Lois, Garfield,* and *Ziggy* (LaRossa et al. 2000). They looked to see how prominently the Mother's Day or Father's Day theme was represented, what activities fathers and mothers were portrayed doing, whether the fathers and mothers were portrayed as incompetent, whether they were mocked or made to look foolish, and whether father and mother characters were engaged in nurturant behaviors such as expressing affection toward, caring for, comforting, listening to, teaching, or praising a child or children. The data revealed fluctuating portrayals of fathers as nurturant or competent. Looking across the 6 decades (in 5-year increments), revealed a U-shaped curve; in the late 1940s and early 1950s there were high percentages of nurturant fathers unmatched until the 1990s. However, an increase in father nurturance can

be seen beginning in the 1980s. This may be surprising to students who think that only late in the twentieth century did nurturing qualities become valued or expected of fathers. At least comic strip fathers of the late 1940s and early 1950s were often nurturing and supportive toward their children. Nurturant portrayals of mothers "spiked" in the late 1950s and late 1970s and were consistently high from the mid-1980s to the end of the century. LaRossa and colleagues note that, although cartoonists seem to have tried to acknowledge the new ideology of nurturant fathers, they did not do it "at the expense" of mothers. "Indeed, if anything, they seemed to pay homage to fatherhood and motherhood at the end of the millennium" (LaRossa et al. 2000, 385).

In Chapter 11 we more explicitly examine the "culture" and "conduct" of fatherhood and motherhood. For now, we can use the research on comic strips to reiterate and illustrate that cultural changes have occurred in our ideas about families, in this case our expectations of fathers. We can also better appreciate how much Walt Duncan and Darryl MacPherson fit the wider context of involved, if exasperated, comic strip fathers.

much prestige. Some members of the lower-upper class may be wealthier than their elite counterparts, living well in large private homes in exclusive communities and enjoying considerable privilege. The major distinction typically drawn between the elite and the lower-upper class is between "old" and "new" money (Steinmetz, Clavan, and Stein 1992; Langman 1988). In other words, the clearest distinction we can draw between them is in how they achieved and how long they have enjoyed their affluence.

Middle Classes

In some analyses, the middle class is considered the largest class, representing between 45% and 50% of the population (Curry, Jiobu, and Schwirian 2002; Renzetti and Curran 1998). Often, the middle class is subdivided into two groupings: the upper-middle class and the lower-middle class.

The **upper-middle class** consists of highly paid professionals (for example, lawyers, doctors, and engineers) who have annual incomes that may reach into the hundreds of thousands of dollars (Renzetti and Curran 1998). They are typically college educated, although they may not have attended the same elite colleges as the upper-upper class (Curry, Jiobu, and Schwirian 2002). Women and men of the upper-middle class have incomes that allow them luxuries such as home ownership, vacations, and college educations for their children. The **lower-middle class** comprises a larger portion of the population. Although it is impossible to specify an exact income threshold that separates the lower from the upper middle class, the lower middle class is comprised of white-collar service workers who live on less income and have less education (or less prestigious degrees) and social standing than their professional and managerial counterparts (for example, physicians, attorneys, managers) in the upper-middle class. They own or rent more modest homes and purchase more affordable automobiles than their upper-middle class counterparts, and they hope, but with less certainty, to send their children to college.

Working Class

About a third of the U.S. population is considered **working class.** Members of this class tend to work in blue-collar occupations (as skilled laborers, for example), earn between $15,000 and $25,000, and have high school or vocational educations. The working class lives somewhat precariously, with little savings and few liquid assets should illness or job loss occur (Rubin 1994). They also have difficulty buying their own homes or sending their children to college (Curry, Jiobu, and Schwirian 2002).

Lower Class

The lower class consists of those who live in poverty. Despite an official estimate of 12.5% of the population being below the "poverty line" (U.S. Bureau of Labor Statistics 2005), a more accurate assessment might indicate that closer to 20% of Americans are poor (Seccombe 2000). As originally established, the *poverty line* was determined by calculating the annual costs of a "minimal food budget" multiplied by three, since 1960s survey data estimated that families spent one-third of their budgets on food (Seccombe 2000). In 2005, the "poverty line" for a family of four with two children was drawn at $19,350 (see Table 3.4). Families whose incomes are just *$1 above* the threshold for their size are not officially classified as poor.

Poverty is consistently associated with marital and family stress, increased divorce rates, low birth weight and infant deaths, poor health, depression, lowered life expectancy, and feelings of hopelessness and despair. Poverty is a major contributing factor to family dissolution.

Poor families are characterized by irregular employment or chronic underemployment. Individuals work at unskilled jobs that pay minimum wage and offer little security or opportunity for advancement (Renzetti and Curran 1998). Although many lower-class individuals rent substandard housing, we also find a homelessness problem among poor families. Karen Seccombe (2000) effectively describes the problems: "Poverty affects one's total existence. It can impede adults' and children's social, emotional, biological, and intellectual growth and development." She further notes that over a year, most poor families experience one or more of the following: "eviction, utilities disconnected, telephone disconnected, housing with upkeep problems, crowded housing, no refrigerator, no stove, or no telephone" (Seccombe 2000).

Despite stereotypes of the poor being African Americans and Latinos, most poor families—and of those who receive assistance—are Caucasian. However, African Americans and Hispanics or Latinos are more likely to experience poverty than are Caucasians (U.S. Bureau of Labor Statistics 2005). Those living in poverty, like their upper- and middle-class counterparts, can be subdivided.

THE WORKING POOR. Since 1979, there have been large increases in the proportion of the population who, despite paid employment, live in poverty. The label *working poor* refers to people who spent at least 27 weeks in the labor force but whose incomes fell below the poverty threshold (U.S. Bureau of Labor Statistics

Table 3.4 ■ 2005 Federal Poverty Guidelines

People in Family Unit	48 Contiguous States and D.C.	Alaska	Hawaii
1	$ 9,570	$11,950	$11,010
2	12,830	16,030	14,760
3	16,090	20,110	18,510
4	19,350	24,190	22,260
5	22,610	28,270	26,010
6	25,870	32,350	29,760
7	29,130	36,430	33,510
8	32,390	40,510	37,260
For each additional person, add	3,260	4,080	3,750

SOURCE: *Federal Register* 2005, 8,373–8,375.

2005). Factors such as low wages, occupational seg-regation, and the dramatic rise in single-parent fam-ilies account for why having a job and an income may not be enough to keep people out of poverty (Ellwood 1988).

Based on a U.S. Bureau of Labor Statistics 2005 report, "A Profile of the Working Poor: 2003," we can make the following statements about the working poor:

- Of the nation's poor, 20% can be classified as "work-ing poor." This amounts to 7.4 million people and 4.2 million families.

- Single-parent families are more likely to be among the working poor than are families of married cou-ples. More than one of every five (22.5%) single-female-headed families are "working poor," compared to 13.5% of single-male-headed families and 8.4% of families of married couples.

- Certain categories of people are more vulnerable to being among the working poor—younger workers, people who fail to finish high school, and people who work part-time. Women are more likely to be among the working poor than are older workers, college grad-uates, full-time workers, and men.

Although their family members may be working or looking for work, these families cannot earn enough to raise themselves out of poverty. An individual work-ing full-time at minimum wage simply does not earn enough to support a family of three. Thus, this kind of poverty results from problems in the economic structure—low wages, job insecurity or instability, or lack of available jobs.

THE GHETTO POOR. The homeless and *ghetto poor*—inner-city residents, disproportionately African Americans and Latinos, who live in poverty—are deeply disturb-ing counterpoints to wider cultural values and be-liefs that are definitive features of American life. Their lifestyles and circumstances challenge cherished im-ages of wealth, opportunity, and economic mobility. It is not clear exactly who the ghetto poor are. They are primarily a phenomenon of the ghettos and bar-rios of decaying cities, where poor African Americans and Latinos are overrepresented.

The behaviors, actions, and problems found among the ghetto poor are often responses to lack of oppor-tunity, urban neglect, and inadequate housing and schooling. With the flight of manufacturing, few job opportunities exist in the inner cities; the jobs that do exist are usually service jobs that fail to pay their work-ers sufficient wages to allow them to rise above poverty. Schools are substandard. The infant death rate approaches that of third world countries, and HIV infection and AIDS are epidemic. The housing projects are infested with crime and drug abuse, turn-ing them into kingdoms of despair. Gunfire often punctuates the night. A woman addicted to crack ex-plained, "I feel like I'm a different person when I'm not here. I feel good. I feel I don't need drugs. But being in here, you just feel like you're drowning. It's like being in jail. I hate the projects. I hate this rat hole" (DeParle 1991).

SPELLS OF POVERTY. Most of those who fall below the poverty threshold tend to be there for spells of time rather than permanently (Rank and Cheng 1995).

About a quarter of the American population may require some form of assistance at one time during their lives because of changes in families caused by divorce, unemployment, illness, disability, or death. About half of our children are vulnerable to poverty spells at least once during their childhood. Many families who receive assistance are in the early stages of recovery from an economic crisis caused by the death, separation, divorce, or disability of the family's major wage earner. Many who accept government assistance return to self-sufficiency within a year or two. Most children in these families do not experience poverty after they leave home.

Two major factors are related to the beginning and ending of spells of poverty: changes in income and changes in family composition. Many poverty spells begin with a decline in earnings of the head of the household, such as a job loss or a cut in work hours. Other causes include a decline in earnings of other family members, the transition to single parenting, the birth of a child to a single mother, and the move of a youth to his or her own household.

POOR WOMEN AND CHILDREN. The **feminization of poverty** is a painful fact that has resulted primarily from high rates of divorce, increasing numbers of unmarried women with children, and women's lack of economic resources (Starrels, Bould, and Nicholas 1994). When women with children divorce, their income and standard of living fall, often dramatically. By family type, 26.5% of single-mother families are below the poverty line (U.S. Census Bureau 2003).

In 2004, 13 million children, 17.8% of children under 18, were poor. The rate is higher among younger children; 18.6% of children under age 6, living in families, were poor (Proctor and Dalaker 2003). Like their parents, they move in and out of spells of poverty, depending on major changes in family structure, employment status of family members, or the disability status of the family head (Duncan and Rodgers 1988). These variables affect ethnic groups differently and account for differences in child poverty rates. African Americans, for example, have significantly higher unemployment rates and numbers of never-married single mothers than do other groups. As a result, their childhood poverty rates are markedly higher. More than a third of African American children are poor, as are nearly 30% of Hispanic children. In contrast, 10.5% of Caucasian children and 9.8% of Asian American children lived in poverty in 2004 (U.S. Census Bureau 2004, Report P60, n. 229, Table B-2, pp. 52–57). Being poor puts the most ordinary needs—from health care to housing—out of reach.

Class and Family Life

Working within this framework, we can note some ways in which family life is differently experienced by each of the four classes. Although there are a number of family characteristics we could consider (including divorce, domestic violence, and the division of labor), we look briefly at class-based differences in marriage relationships, parent–child relationships, and ties between nuclear and extended families.

Marriage Relationships

Within upper-class families we tend to find sharply sex-segregated marriages in which women are subordinated to their husbands. Upper-class women often function as supports for their husbands' successful economic and political activities, thus illustrating the **two-person career** (Papanek 1973).

Although their supportive activities may be essential to the husbands' success, such wives are neither paid nor widely recognized for their efforts. Rather than having their own careers, they often volunteer within charitable organizations or their communities. They are free to pursue such activities because they have many servants—from cooks to chauffeurs to nannies—who do the domestic work and some childcare or supervision.

Middle-class marriages tend to be *ideologically* more egalitarian and are often two-career marriages. In fact, middle-class lifestyles increasingly require two incomes. This creates both benefits and costs for middle-class women. The benefits include having more say in family decision making and greater legitimacy in asking for help with domestic and childrearing tasks. The costs include the failure to receive the help they request. Because working wives likely earn less than their husbands, the strength of their role in family decision making may still be less than that of their husbands. We say they are "ideologically" more egalitarian because middle-class couples more highly value and more readily accept the ideal of marriage as a sharing, communicating relationship in which spouses function as "best friends."

 Family experiences are affected by such variables as social class and ethnicity.

Once more explicitly traditional, working-class marriages are becoming more like their middle-class counterparts. Whereas such marriages in the past were clearly more traditional in both rhetoric and division of responsibilities, in recent years they have moved toward a model of sharing both roles and responsibilities (Komarovsky 1962; Rubin 1976, 1994). The sharply segregated, traditional marriage roles evident even just 2 decades ago have given way to two-earner households, increasingly driven by the need for two incomes.

Especially among those working-class couples who work "opposite" shifts, we find higher levels of sharing domestic and childcare responsibilities, as well as greater male involvement in home life (Rubin 1994). The reality of being the only parent home forces men to take on tasks that otherwise might be done by wives. Necessity, not ideology, creates this outcome. The meaning of male participation in home life may vary more than actual behavior or vary differently than levels of actual involvement. Male involvement may have greater "value" in the circles in which middle-class men live and work but be more of a practicality or necessity for working-class men. Thus, working-class men may understate, and middle-class men may exaggerate, what and how much they do.

Marriages among the lower class are the least stable marriages. Men are often absent from day-to-day family life. Resulting from the combination of high divorce rates and widespread nonmarital childbearing, a third of single mothers and their children are poor, roughly six times the rate of poverty among married couple families with children. Furthermore, although

they represent only about a fourth of all families in the United States, they are nearly half of the 6 million poor families (Lichter and Crowley 2002). The cultural association of men's wage earning with fulfillment of their family responsibilities subjects lower-class men to harsher experiences within families. They are less likely to marry. If married, they are less likely to remain married, and when married they derive fewer of the benefits that supposedly accrue in marriage.

Catherine Ross and her colleagues (1991) account for the connection between poverty and divorce as follows:

> It is in the household that the larger social and economic order impinges on individuals, exposing them to varying degrees of hardship, frustration, and struggle. The struggle to pay the bills and to feed and clothe the family on an inadequate budget takes its toll in feeling run-down and tired, having no energy, and feeling that everything is an effort, that the future is hopeless, that you can't shake the blues, that nagging worries make for restless sleep, and that there isn't much to enjoy in life.

When marriages cross class lines, other problems can arise. People may find themselves feeling out of step, as if they are in a world where there are different, perhaps dramatically different, assumptions about how to discipline and raise children, where to go and what to do on vacation, and how to save or spend money (Lewin 2005). It is more difficult to measure than interracial marriage or religious intermarriage, but using education as an indicator of class, there appear to be less cross-class marriages than in the past. Most of those marriages that do cross class lines are now between women with more education marrying men with less. This combination does not bode especially well for the future stability of the marriages (Lewin 2005).

Parents and Children

The relationships between parents and children vary across social lines, but most research has focused on the middle and working classes (Kohn 1990). Among the upper class, some hands-on childrearing may be done by nannies or au pairs. Certainly, mothers are involved, and relationships between parents and children are loving, but parental involvement in economic and civic activities may sharply curtail time with children (Langman 1987). For upper-class parents, an important objective is to see that children acquire the appropriate understanding of their social standing and that they cultivate the right connections with others like themselves. They may attend private and exclusive boarding schools and later join appropriate clubs and organizations. Their eventual choice of a spouse receives especially close parental scrutiny.

A considerable amount of research indicates that working- and middle-class parents socialize their children differently and have different objectives for childrearing (Kohn 1990; Rubin 1994; Hays 1996; Lareau 2003). Although all parents want to raise happy and caring children, middle-class parents tend to emphasize autonomy and self-discipline and working-class parents tend to stress compliance (Kohn 1990, Hays 1996). In her 1996 study of mothers, Sharon Hays identified differences between what middle- and working-class mothers believed made for a "good mother" and what they thought children most need. Whereas working-class mothers saw and therefore stressed education as essential for their children's later life chances, middle-class mothers took for granted that their children would receive good-quality educations and emphasized, instead, the importance of building children's self-esteem. And although both classes of mothers acknowledged using spanking to discipline their children, middle-class mothers spanked more selectively and favored other methods of discipline (for example, "timeout") (Hays 1996).

One of the more recent and fascinating class comparisons is Annette Lareau's *Unequal Childhoods* (2003). Lareau contends that "social class does indeed create distinctive parenting styles . . . that parents differ by class in the ways they define their own roles in their children's lives as well as in how they perceive the nature of childhood" (Lareau 2002, 748). Lareau introduces the concepts of *concerted cultivation* and *accomplishment of natural growth* to represent class-based differences in philosophy of childrearing.

Middle class families engage in concerted cultivation. Parents enroll their children in numerous extracurricular activities, from athletics to art and music, that come to dominate their children's lives, as well as the life of the whole family. Through these activities, however, children partake in and enjoy a wider range of outside activities and interact with a range of adults in authoritative positions, giving them experiences and expertise that can serve them well later. Because of the way household life tends to center around children's schedules and activities, the other members of these middle-class families (parents and siblings) are forced to endure a frenzied pace and a shortage of family time (See the "Real Families" box).

Working-class parents, lacking the material resources to enroll their children in such activities, tend to focus less on developing their children and more on letting them grow and develop naturally, play freely in unsupervised settings, and spend time with relatives and in the neighborhood. For a sense of how these two approaches may have been experienced by children, see the "Real Families" box.

Lower-class families are the most likely of all families to be single-parent families. Single parents, in general, may suffer stresses and experience difficulties that parents in two-parent households do not, but this situation is exacerbated for low-income single parents (McLanahan and Booth 1989). Parent–child relationships suffer from a variety of characteristics of lower-class life: unsteady, low-pay employment; substandard housing; and uncertainty about obtaining even the most basic necessities (food, clothing, and so forth). All of these can affect the quality of parent–child relationships and the ability of parents to supervise and control what happens to and with their children.

Extended Family Ties

Links between nuclear family households and extended kin vary in kind and meaning across social class. By some measures, the least closely connected group may be the middle class, which, because of the geographic mobility that accompanies their economic status, may find themselves the most physically removed from their kin. As Matthijs Kalmijn observed among the upper-middle-class families he studied in the Netherlands, they live almost three times as far from their siblings,

Middle-Class Parenting, Middle-Class Childhood

Louise and Don Tallinger are proud parents of three boys, 10-year-old Garrett, 7-year-old Spencer, and 4-year-old Sam. They are also busy professionals; Louise is a personnel consultant, and Don is a fund-raising consultant. Between them, they earn $175,000, making them comfortably upper-middle class. They travel a lot for work; Don is out of town an average of three days a week, and Louise, four or five times a month, flies out of state early in the morning and returns sometime after dinner. Don often doesn't return home from work until 9:30 p.m. This middle-class family of five is one of the families studied by Annette Lareau in her fascinating class comparison, *Unequal Childhoods*.

With 10-year-old Garrett's involvements in baseball, soccer, swim team, piano, and saxophone lessons, his schedule dictates the pace and routines of the household. Lareau describes what life is like for Don and Louise (2003, 42):

Rush home, rifle through the mail, prepare snacks, change out of . . . work clothes, make sure the children are appropriately dressed and have the proper equipment for the upcoming activity, find the car keys, put the dog outside, load the children and equipment into the car, lock the door and drive off.

The Tallingers epitomize the middle-class childrearing strategy Lareau called concerted cultivation. This lifestyle, dedicated as it is to each child's individual development and enrichment, is exhausting just to read about. It may be familiar to you as an extreme example of your experiences; Lareau argues that among the middle class it is not uncommon.

The brothers hardly go long stretches without some scheduled activity, most of which require adult-supplied transportation, adult supervision, and adult planning and scheduling. Rarely can they count on playing outside all day like children in working and lower-class families might.

Of course, working and lower-class children could not participate in all of the activities that the Tallinger boys do. By the Tallingers' estimate, Garrett's activities alone cost more than $4,000 a year. Despite this obvious advantage, Garrett feels disadvantaged because he cannot attend the private school he once attended.

Although Lareau is careful to illustrate what middle-class children like the Tallinger boys miss out on—free play, closer connections to relatives, more time for themselves away from adult supervision and control, comfort with and ability to amuse themselves, and less fatigue—she also illustrates the many benefits they receive beyond involvement in activities that they enjoy. The Tallingers believe that all the activities that their boys participate in teach them to work as part of a team, to perform on a public stage and in front of adults, to compete, to grow familiar with the many performance-based assessments that will come at them through school and work

experiences, and to prioritize. The children travel to tournaments, eat in restaurants, and stay in hotels; they may fly to summer camps or special programs out of state or overseas. Indeed, Lareau suggests that children like the Tallinger boys may travel more than working-class and poor adults (2003, 63). These experiences, combined with what the Tallingers teach the boys at home, promote skills that enhance their chances of staying or even moving higher up in the middle class.

In lifestyles such as this one, family life is organized and ruled by large calendars that detail the children's sports, play activities, music, and scouting events. It then falls on the parents to see that their children arrive at these activities, often directly from one to another. As Lareau somberly puts, "At times, middle-class homes seem to be little more than holding places for the occupants during the brief periods when they are between activities" (2003, 64).

For Further Consideration

1. What is your reaction to the Tallingers' lifestyle? Is it at all familiar to you from either firsthand experience or experiences of those you know?

2. As you see it, what are the biggest benefits and costs associated with this way of life?

3. What do you see as the effect on Garrett? His parents? His brothers?

SOURCE: Lareau 2003.

and more than three times as far from their parents and their (adult) children, as does the lowest-educated class (Kalmijn 2004). Similar class differences can be observed in the United States.

Middle-class families do visit kin or phone regularly and are available to exchange aid when needed. Still, the emphasis is on the conjugal family of spouses and children.

Closer connections may be found among both the working and the upper classes, although the reasons differ. In the case of working-class families, there are often both the opportunity and the need for extensive familial involvement. Opportunity results from lesser levels of geographic mobility, which results in closer proximity and allows more continuous contact to result. The need for involvement is created by the pooling of resources and exchange of services (for example, childcare) that often result between adults and their parents or among adult siblings. Intergenerational upward mobility may lessen the reliance on extended families (see discussion later in this chapter).

Upper-class families, especially among the "old" upper class, highly value the importance of family name and ancestry. They tend to maintain strong and active kinship groups that exert influence in the mate selection processes of members and monitor the behavior of members. Inheritance of wealth gives the kin group more than symbolic importance in their ability to influence behavior of individual members.

Among the lower class, kin ties—both real and fictive—may be essential resources in determining economic and social survival. Grandparents, aunts, and uncles may fill in for or replace absent parents, and multigenerational households (for example, children living with their mothers and grandmothers) are fairly common. **Fictive kin ties** refer to the extension of kinship-like status to neighbors and friends, thus symbolizing both an intensity of commitment and a willingness to help one another meet needs of daily life (Stack 1974; Liebow 1967).

The Dynamic Nature of Social Class

Like other aspects of family life, social class position is not set in stone. Individuals may experience **social mobility,** movement up *or* down the social class ladder. Either kind of social mobility can affect family relationships, especially, although not exclusively, intergenerational relationships (Kalmijn 2004; Newman 1988; Sennett and Cobb 1972). For example, children who see their parents "fall from grace," through job loss and dwindling assets look differently at those parents. Fathers who once seemed heroic may become the source of concern, and even resentment, as their job loss threatens the lifestyle of the family on which children depend (Newman 1988). Children who in adulthood climb upward occasionally find their relationships with their parents suffering as a result. As

they are exposed to new values and ideas that differ from those held by their parents, generational tension and social distance may follow. Furthermore, as they move into a new social circle, parents (as well as less mobile siblings) may appear to fit less well with their new life circumstances. The more they strive to fit into new circles and circumstances accompanying their increased social standing, the less well they may fit comfortably within their ongoing family relationships.

Aside from the difficulty fitting parents and siblings into a new social standing, practical considerations, imposed by a job, may create obstacles preventing individuals from maintaining closer relationships. As is true elsewhere, ascending the ladder to a higher rung may require geographic relocation. Such jobs may also impose greater demands on the individual time. To these constraints of time and distance we can add that as someone establishes new friendships and participates in leisure activities, further reductions in opportunity and availability may result (Kalmijn 2004).

Marital relationships, too, may be altered by either downward or upward mobility. Research indicates that some men who lose their jobs and "slide downward" react to their economic misfortune by abusing their spouses, turning to alcohol or other substances, withdrawing emotionally, or leaving the home (Rubin 1994; Newman 1988). Changes in the marriage are not entirely of men's doing; after an initial period of sympathy and support, wives may grow impatient with their husbands' unemployment or alter their positive views of the husbands' dedication as a worker or job seeker. In addition, as couples are forced to scale back their accustomed lifestyle, tensions may rise and resentment and distance may grow.

Upward mobility may also transform marriage relationships. We are familiar with the situation faced by women who, after sacrificing to help launch their husbands' careers by supporting them through school, are left by those same husbands once they have achieved their career goals. With their own increasing economic opportunity, some women find that marriage becomes less desirable because of the constraints it continues to impose on their career development.

Racial and Ethnic Diversity

The United States is a richly diverse society. This is not news to you; we pride ourselves on our multicultural mix of groups, whether we see them "melting" together into one large pot or, like a salad bowl, retaining their uniqueness even when tossed together. As we begin to look at the racial and ethnic variations in family experience, we need to first note the multiplicity of different groups that make up the U.S. population. To get at this, the U.S. Census Bureau asked the following question:

> What is this person's ancestry or ethnic origin? (For example, Italian, Jamaican, African American, Cambodian, Cape Verdeian, Norwegian, Dominican, French Canadian, Haitian, Korean, Lebanese, Polish, Nigerian, Mexican, Taiwanese, Ukranian, and so on.)

The U.S. Census Bureau goes on to define *ancestry* as any of the following: "where their ancestors are from, where they or their parents originated, or simply how they see themselves ethnically" (Brittingham and de la Cruz 2004). The census also contains items about a person's race and whether she or he is of Hispanic origin. Thus, there are multiple attempts to get at the diversity of the population. This creates some inconsistency or incompatibility in the data, however. Although both African American and Mexican are options for people to select as their "ancestry," many fewer people in both groups identified themselves in terms of these ancestry categories than answered that their race was African American or that they were Hispanic of Mexican origin. In the census, 12 million fewer people answered that their ancestry was African American than answered that their race was African American. In addition, 2 million fewer people listed Mexican ancestry than answered that way on the question about Hispanic origin. Thus, the ancestry data need to be approached with some caution when dealing with groups that surface on more than one question (for example, African Americans, Chinese, Mexican, and American Indian).

Of the population, 80% identified one or more ancestries, with 58% specifying one ancestry group and another 22% specifying two. Of the remainder, 19% did not report any ancestry and 1% reported some otherwise unclassifiable category such as "a mixture" (Brittingham and de la Cruz 2004).

Seven different ancestries were reported by at least 15 million people each. Most common was German. Almost 43 million people identified themselves as German or part German, nearly one out of six people or 15% of the population. The other six ancestries that were selected by at least 15 million people were as shown in Table 3.5.

Table 3.5 ■ Most Common Ancestries

Group	Number of People	Percentage of Population
German	43.0 million	15
Irish	30.5 million	10.8
African American*	24.9 million	8.8
English	24.5 million	8.7
American	20.2 million	7.2
Mexican*	18.4 million	6.5
Italian	15.6 million	5.6

*Remember, these are undercounts compared with what other census questions yield.

In addition, there are eight other ancestries that represent at least 4 million people each: Polish, French, American Indian, Scottish, Dutch, Norwegian, Scotch-Irish, and Swedish.

Race, Ethnicity, and Minority Groups

Before we begin to look more closely at diversity in family experience, we need to define several important terms. A **race** or **racial group** is a group of people, such as whites, blacks, and Asians, classified according to their **phenotype**—their anatomical and physical characteristics. Racial groups share common phenotypical characteristics, such as skin color and facial structure. The concept of race is often misused and misunderstood. We should neither assume a purity or homogeneity within racial groupings (in skin color, facial features, and so on) nor treat racial groups as superior or inferior in comparison to one another. In either of those biological applications, the concept of race is clearly a myth (Henslin 2000). Socially, however, we perceive or identify ourselves within racial classifications and are treated and act toward others on the basis of race, which makes it a highly significant factor in shaping our life experiences. Although its biological importance may be doubtful, its social significance remains great.

An **ethnic group** is a set of people distinct from other groups because of cultural characteristics. Such things as language, religion, and customs are shared within and allow us to differentiate among ethnic groups. These cultural characteristics are transmitted from one generation to another and may then shape how each person thinks and acts—both inside and outside of families.

Either a racial or an ethnic group can be considered a **minority group** depending on social experience. Minority groups are so designated not because of their numerical size in the wider population but because of their status (position in the social hierarchy), which places them at an economic, social, and political disadvantage (Taylor 1994b). Thus, African Americans are simultaneously an ethnic, a racial, and a minority group in the United States (as well as an *ancestry category* as shown previously). The term *African American,* used increasingly instead of *black,* reflects the growing awareness of the importance of ethnicity (culture) in contrast to race (skin color) (Smith 1992; but see Taylor 1994b).

As we will soon see, ethnic and/or racial differences are often difficult to untangle from social class differences. It may be that some differences in family patterns reflect cultural background factors or distinctive values. However, it is equally plausible that ethnic or racial differences in family patterns reflect the different socioeconomic circumstances under which different groups live (Aponte, Beal, and Jiles 1999).

According to recent census data (U.S. Census Bureau 2000), more than 30% of the U.S. population are people of color: 13% are African American, 13% are Hispanic, 4% are Asian/Pacific Islander, and 1% are Native American. By 2050, the population is expected to be just over 50% Caucasian, 24% Hispanic, 13% African American, 9% Asian, and 1% Native American.

As we embark on our discussion of race and ethnicity in family life, it is important to be aware of the danger of thinking in terms of ethnocentric fallacies (a term introduced in Chapter 2), beliefs that your ethnic group, nation, or culture is innately superior to others. In the following sections we consider briefly some distinctive characteristics and strengths of families from various ethnic and cultural groups.

We also need to keep in mind that until the last 35 years most research about American marriages and

families tended to be limited to the white, middle-class family. The nuclear family was the norm against which all other families, including single parent and stepfamilies were evaluated and often viewed as pathological because they differed from the traditional norm. A similar distortion also has influenced our understanding of African American, Latino, Asian American, and Native American families. Instead of recognizing the strengths of diverse ethnic family systems, misguided researchers viewed these families as "tangles of pathology" for failing to meet the model of the traditional nuclear family (Moynihan 1965). Part of this distortion resulted from the long-term scarcity of studies on families from African American, Latino, Asian American, Native American, and other ethnic groups. Furthermore, many earlier studies focused on weaknesses rather than strengths, giving the impression that all families from a particular ethnic group were riddled by problems (Dilworth-Anderson and McAdoo 1988; Taylor 1994a, 1994b; Taylor et al. 1991).

The "culture of poverty" approach, for example, sees African American families as being deeply enmeshed in illegitimacy, poverty, and welfare as a result of their slave heritage. As one scholar notes, the culture of poverty approach "views black families from a white middle-class vantage point and results in a pejorative analysis of black family life" (Demos 1990). This approach ignores most families that are intact or middle class. It also fails to see African American family strengths, such as strong kinship bonds, role flexibility, love of children, commitment to education, and care for the elderly.

America is a pluralistic society. Thus, it is important that students and researchers alike reexamine diversity among our different ethnic groups as possible sources of strength rather than pathology (DeGenova 1997). For instance, cultures may vary widely in how the best interests of the child are defined (Murphy-Berman, Levesque, and Berman 1996). Differences may not necessarily be problems but solutions to problems; they may be signs of adaptation rather than weakness (Adams 1985). As two family scholars pointed out, "Whether a phenomenon is viewed as a problem or a solution may not be objective reality at all but may be determined by the observer's values" (Dilworth-Anderson and McAdoo 1988).

AFRICAN AMERICAN FAMILIES. According to the 2000 census, the more than 34 million African Americans in the United States represented 12.2% of the population. If we include those who consider themselves

Matter of Fact

According to the United States Census Bureau, nearly fifty-two million Americans, more than 19%, speak languages other than English at home. Of those, more than 32 million speak Spanish. Nearly eight million people, 15% of the population, speak Asian and Pacific Island languages, the most common being Chinese, which is spoken by more than 2.2 million people in the United States (SOURCE: U.S. Census Bureau, 2005 American Community Survey).

biracial, in this case black combined with one or more other races, the total reaches 36.2 million people, or 12.9% of the population of the United States (McKinnon and Bennet 2005).

Compared with the total U.S. population, African Americans are younger and less likely to be married (see Figure 3.2) Although they are no more likely to be divorced or widowed, a much greater percentage of blacks than whites have never married (43% versus 25%). Blacks are more likely to bear children outside of marriage and more likely to live in single-parent, mostly mother-headed, families. These patterns continued to increase throughout the past decade but even more so among the general population than among African Americans (McLoyd et al. 2000b).

Although African Americans are as likely as the general population to live in family households, their households differ from the family households in the general population. A third of black households are headed by married couples. In the wider population, 53% of households are headed by married couples (McKinnon and Bennett 2005, Figure 4). Because of high rates of divorce and of births to unmarried women, in 2002 53% of African American children

lived in households headed by single mothers (48%) or single fathers (5%) (U.S. Census Bureau 2003). More than 30% of black households are headed by women with no husbands present compared to 12% in the population overall. Fewer than 6% of black households and households overall are headed by men with no wives in the home (McKinnon and Bennet 2005).

Considering families rather than households, data from 2002 reveal that 48% of African American families were married-couple families, 43% were headed by single women, and 9% were headed by single men. The equivalent percentages for whites show that more than 80% of white families are headed by married couples, 13% by women without husbands, and 5% by men without wives (McKinnon 2003).

In addition, we can note the following:

- Compared to the general population, African Americans are less likely to have completed college (17% versus 29%) (McKinnon 2003).

- Black women are slightly more likely than black men to have completed college (18% versus 16%) (McKinnon 2003).

Figure 3.2 ■ Marital Status of People 15 Years and Older by Race and Hispanic Origin: 2002

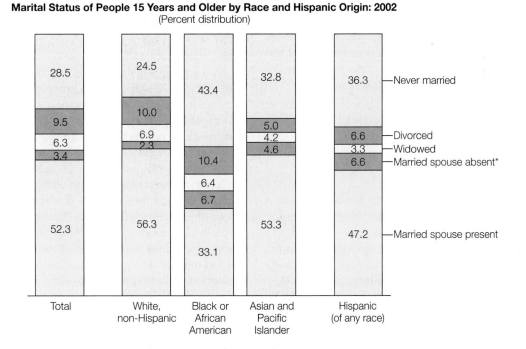

Marital Status of People 15 Years and Older by Race and Hispanic Origin: 2002
(Percent distribution)

SOURCE: U.S. Census Bureau, Population of the United States: Dynamic Version; Families and Living Arrangements in 2002

- Overall, African Americans are somewhat less likely than the general population to be employed (60% versus 64%), but this difference is really a reflection of male employment patterns. Black women are actually *slightly more likely* to be employed than are women overall (59.6% versus 57.5%). Among men, a significant 10% difference separates blacks from the general population of men (60.9% versus 70.7%) (McKinnon and Bennett 2005).

- Median earnings for African Americans who are employed full-time year-round were $27,264 in 2000. This amounts to approximately 85% of the median for all workers ($32,098). Black men's median earnings were 81% of the median for men overall. Black women's earnings were much closer (94%) to earnings among all women. The gender wage gap among blacks is narrower than it is among the general population; black women's median income is 85% of the median among black men. In the general population, the gender wage gap was 73% in 2000 (McKinnon and Bennett 2005).

- The median income of black families was $33,300 in 2000. This is two-thirds the median among all families ($50,000). If we look only at married couple families, the gap between blacks and the general population is smaller; black married couples earned 89% of the median income of all married couples (McKinnon and Bennett 2005).

- Of African Americans, 25% live below the poverty line. The percentage of impoverished blacks is nearly three times the percentage of poor whites (8%) and almost twice the percentage of the general population who are poor (12%) (McKinnon 2003).

- Related to the previous point, 16% of children under 18 live in poverty. The poverty rate for black children is nearly twice as high at 30%, and black children have three times the rate of poverty as white children (10%) (McKinnon 2003).

There are several noteworthy features of African American families. First, African American families, in contrast to Caucasian families, have a long history of being dual-earner families as a result of economic need. As a consequence, employed women have played important roles in the African American family. They also have more egalitarian family roles. Black men have more positive attitudes toward working wives, take on a slightly larger share of household labor, and spend more time on domestic tasks and childcare activities (McLoyd et al. 2000b). Second, marital relations more often show signs of greater distress than is true of the general population. Some evidence indicates a greater likelihood of spousal violence and lower levels of reported marital happiness among African American marriages (McLoyd et al. 2000b). Third, kinship bonds are especially important, because they provide economic assistance and emotional support in times of need (Taylor 1994c; Taylor et al. 1991). Fourth, African Americans have a strong tradition of familism (emphasis on family and family loyalty), with an important role played by intergenerational ties. Fifth, the African American community values children highly. Finally, African Americans are much more likely than Caucasians to live in **extended households,** households that contain several different generations (Taylor 1994c). Black children are more likely than other children to live in their grandparent's household or to have a grandparent living with them in their parent's household. Typically, this grandparent is a grandmother (U.S. Census Bureau 2003).

Many of these characteristics are often associated with poverty and thus may not be features inherent in African American families. When divorce rates are adjusted according to socioeconomic status, racial differences are minimal. Poor African Americans have divorce rates similar to poor Caucasians, and middle-class African Americans have divorce rates similar to middle-class Caucasians (Raschke 1987). Thus, understanding socioeconomic status, especially poverty, is critical in examining African American life (Bryant and Coleman 1988; Julian, McKenry, and McKelvey 1994; Wilkinson 1997).

As the preceding data reveal, African Americans and their families are at a clear economic disadvantage relative to the wider population. Compared to Caucasians, they have more than twice the unemployment rate, nearly three times the poverty rate, and two-thirds the median income (see Table 3.6).

These economic indicators point out the potential difficulty of comparing black and white family characteristics. Combined with the tendency of upper-status African American families (that is, middle and upper-middle class) to be as stable as Caucasian families of comparable status, these economic indicators suggest that much of what we may assume to be race differences are confounded by economic differences or may be social class differences *masquerading as race ones.*

This more economic argument pertains especially well to an understanding of race differences in marriage rates, divorce rates, and the numbers of single-mother-headed families. The most widely applied

Table 3.6 ■ Race, Ethnicity and Socioeconomic Status: 2003-2004

	Total	Whites	African Americans	Latinos	Asians
Median Family Income	$52,680	$55,768	$34,369	$34,272	$63,251
Percentage Unemployed	5.5	4.8	10.4	7.0	4.4
Percentage of families in poverty	10.3	8.1	22.3	20.8	12.2
Percentage of children in poverty	17.2	13.9	33.6	29.5	12.1

SOURCE: U.S. Census Bureau 2006. Tables 578, 679, 694, 698.

argument is that blacks "marital prospects" have shifted dramatically, especially among the poor (Aponte, Beal, and Jiles 1999). Wilson's notion of the "male marriageable pool index" emphasizes the importance of male employment to their "marriageability" (Wilson 1987). Downward shifts in male employment patterns would then account for some decline in marriage rates and the increase in single-mother-headed families. Not only are African Americans unlikely to devalue marriage, they may actually more highly value marriage than do other groups.

Despite the benefits of linking class and race in our efforts to understand family diversity, we cannot simply interpret all race differences as economic in nature. Don't forget that a major feature of race in American society is that it determines much treatment we receive from others. Thus, the opportunities we are offered or refused, and whether others insult, avoid, or think less of us, are all affected by race. The interpretation of race differences as only (or even largely) class differences unfortunately minimizes or ignores such expressions of racism and discrimination and fails to acknowledge patterns that may have cultural origins to them—such as greater emphasis on extended family ties or gender equality.

LATINO FAMILIES. Latinos (or Hispanics) are now the largest ethnic group in the United States, as well as the fastest growing. The 2000 census reported 35 million Hispanics, representing 12.5% of the U.S. population. Furthermore, it is projected that by 2050, at least 25% of the population will be of Hispanic origin. These increases result from both immigration and higher birthrate among Latinos (U.S. Census Bureau 1996; Vega 1991).

Currently, 65.8% of Latinos are of Mexican descent, 9.4% are Puerto Rican, and another 4% are Cuban. The remaining 21% includes 7.8% from Central

■ *Latino culture emphasizes the family as a basic source of emotional support for children.*

©Tony Freeman/PhotoEdit

American countries and 5.2% from South American countries (see Figure 3.3). Overall, more than three-fourths of Hispanics live in western and southern states, with California and Texas, together, accounting for more than half the Latino population in the United States. Latinos account for 24% of the population in the western United States, a proportion nearly twice their national level. Latinos, mostly of Mexican and Central American descent, are concentrated in California and the Southwest. Latinos of Puerto Rican descent are concentrated in the Northeast, especially New York. The greatest numbers of Cuban Americans are found in Florida. There are also significant Latino populations in Illinois, New Jersey, and Massachusetts (U.S. Census Bureau 2001).

Continued immigration has transformed the nature of Latino culture in the United States. First, immigration makes both Latino culture and the larger society a "permanently unfinished" society. The newer immigrants are urban and overwhelmingly workers and laborers rather than professionals. Second, in some areas, immigration is changing the proportion of U.S.-born and foreign-born Latinos. In 1960 in California, for example, four out of five Mexicans were born in the United States; today, because of the massive influx of immigrants, only about half are born here (Zinn 1994).

It is important to remember that there is considerable diversity among Latinos in terms of ethnic heritage (such as Mexican, Cuban, or Puerto Rican), socioeconomic status (Sanchez 1997; Walker 1993), and family characteristics. Tables 3.7 and 3.8 show how marital status and types of households vary between Hispanics and the wider population, as well as among different Hispanic groups.

As the data reveal, there are differences between Hispanics and non-Hispanics, as well as among Hispanics. Generally, Hispanics are less likely than both the overall population and non-Hispanic whites to divorce or to be married. With the exception of Cubans, they are more likely than the population overall and non-Hispanic whites to have female-headed households. Regardless of Hispanic ethnicity, they are less likely than whites and the general population to maintain families headed by married couples.

Across the various Hispanic categories there is considerable social and economic variation. For example, Cubans and South Americans have the highest socioeconomic status, as indicated by incomes, poverty rates, home ownership, and educational attainment. Puerto Ricans and Mexicans and Central Americans

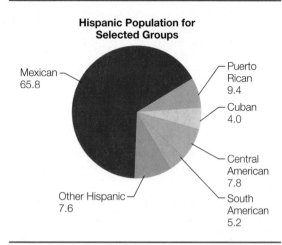

Figure 3.3 ■ **U.S. Hispanic Population for Selected Groups: 2004**

Hispanic Population for Selected Groups

Mexican 65.8
Puerto Rican 9.4
Cuban 4.0
Central American 7.8
South American 5.2
Other Hispanic 7.6

SOURCE: U.S. Census Bureau, Current Population Survey, Annual Social and Economic Supplement, 2004, Population Division, Ethnic and Hispanic Statistics Branch.

tend to have similar characteristics, with Mexican American families being slightly more likely to be poor except for female-headed families (42% of Puerto Rican female-headed families are below the poverty line compared to 39.6% of Mexican and 35% of Central American female-headed families). These differences, although real, are not as distinctive as the ways in which Cubans, and South Americans, differ from the other Hispanic groups (see Table 3.9).

We can combine the familial characteristics with the economic ones and note interesting connections. The more affluent Hispanic groups, especially Cubans, are among the most likely to have their families be married couple families (along with Mexicans), and they are less likely to have their families be female-headed, single-parent families.

Research also indicates that the percentage of children born to unmarried mothers ranges from a low of 27% among Cubans, to 41% among Mexicans, to a high of 60% among Puerto Ricans. Similarly, the percentage of births to teenage mothers ranges from 7.5% among Cubans to 20% among Puerto Ricans. Puerto Rican women are more likely to have their first child before marriage, and Mexican American women tend to have their first child after marriage. Cuban women tend to marry later and have the lowest fertility rates among Hispanic women (McLoyd et al. 2000b). This diversity is not merely because of economics. It is further accentuated by the varying proportions of U.S.-born and foreign-born Latinos in each group. Finally,

Table 3.7 ■ **Marital Status of Hispanics in 2004: Comparisons to Non-Hispanic Population and across Hispanic Groups**

Marital Status	U.S. White	Non-Hispanic	Hispanic	Mexican	Puerto Rican	Cuban	Central American	South American
Married	53.3%	57.0%	49.9%	51.5%	41.9%	55.6%	48.2%	51.0%
Widowed	6.1	6.7	3.3	3.0	3.8	8.4	2.3	2.6
Divorced	9.6	10.1	7.1	5.9	10.1	10.7	5.4	9.3
Separated	2.0	1.4	3.5	3.4	4.8	2.8	3.7	3.4
Never married	29.0	24.9	36.2	36.2	39.4	22.5	40.4	33.7

SOURCE: U.S. Census Bureau 2004, http://www.census.gov/population/www/socdemo/hispanic/ho04.html, Tables 2.1 and 2.2.

Table 3.8 ■ **Household Type: Comparisons of Hispanics and Non-Hispanics, 2004**

Household Type	U.S. White	Non-Hispanic	Hispanic	Mexican	Puerto Rican	Cuban	Central American	South American
Family households as % of households	68.1	66.0	79.3	81.0	77.0	72.8	81.3	76.7
Married couple as % of family households	75.6	82.5	67.0	70.0	57.0	77.0	63.0	68.0
Male head, no spouse, as % of all families	6.2	5.0	10.0	10.0	8.0	8.9	12.0	8.0
Female head, no spouse, as % of all families	18.1	13.3	23.0	20.0	34.6	13.6	25.0	23.5
Nonfamily households (%)	31.9	34.0	20.7	19.0	23.0	27.2	18.7	23.3

SOURCE: U.S. Census Bureau, "Household Type by Hispanic Origin and Race of Householder: 2004," Current Population Survey, Annual Social and Economic Supplement, 2004, Ethnicity and Ancestry Branch, Population Division.

Table 3.9 ■ **Selected Socioeconomic Characteristics of the Hispanic Population: 2004**

	All	Mexican Hispanic	Puerto Rican	Cuban	Central American	South American
% families in poverty	20.8	22.9	21.1	11.3	18.8	11.4
Married couples	15.7	18.4	9.0	8.8	14.7	7.6
Female headed	37.1	39.6	42.0	19.4	35.0	25.0
% unemployed	7.6	8.3	8.2	4.0	5.7	4.7
% earning < $25,000	34.8	36.7	36.0	26.0	33.4	24.3
% earning > $75,000	15.8	13.0	19.2	24.5	16.2	24.9
% living in own home	51.2	52.0	45.7	71.2	38.8	53.6
% 25 and older						
% with < high school	41.6	48.1	28.2	27.9	49.7	17.3
% with college or >	12.1	7.9	14.1	24.0	10.5	33.0

SOURCE: U.S. Census Bureau 2004, http://www.census.gov/population/www/socdemo/hispanic/ho04.html, Tables 6.2, 9.2, 13.2, 15.2, and 17.2.

keep in mind that characterizations of Mexican, Puerto Rican, Cuban, or any other Latino family types must avoid overgeneralization. None of these groups have a singular family system. More specifically, in Mexico, Cuba, or Puerto Rico there is much diversity, some of which results from socioeconomics, some from rural versus urban living, some from religion, and so on (Aponte, Beal, and Jiles 1999).

Traditional Mexican and Puerto Rican families can be characterized by two distinctive cultural traits: devotion to family (that is, *familism*) and male dominance (that is, *machismo*). *La familia* is based on the nuclear family, but it also includes the extended family of grandparents, aunts, uncles, and cousins. All tend to live close by, often in the same block or neighborhood. There is close kin cooperation and mutual assistance, especially in times of need, when the family bands together. Family unity and interdependence, sometimes extended to include fictive kin (for example, Cuban *compadres* and *comadres*—godparents), reflect the importance of extended kin ties. Male dominance, as suggested, although often exaggerated in the misuse or misunderstanding of *machismo,* is part of traditional Latino family systems but has declined, as has familism, especially among dual-earner couples. Migration and mobility disrupt traditional Latino family forms and lead to change. This change can be seen as part of a wider process of "convergence," in which distinctive ethnic traits diminish over time (Aponte, Beal, and Jiles 1999).

Children are especially important. Fertility rates are still relatively higher among Hispanics than among the general U.S. population, although they are dropping. Because Spanish is important in maintaining ethnic identity, many Latinos, as well as educators, support bilingualism in schools and government. Catholicism is also an important factor in Latino family life. Although there has been a tradition of male dominance, current day-to-day living patterns suggest noteworthy change has occurred.

Women have gained power and influence in the family as they have increased their participation in paid employment. When wives are co-providers, Hispanic men spend more time on household tasks (Aponte, Beal, and Jiles 1999; McLoyd et al. 2000b).

ASIAN AMERICAN FAMILIES. As of 2002, Asian Americans and Pacific Islanders made up more than 4% of the U.S. population (Reeves and Bennett 2003). The most complete data we have on where the Asian American population comes from is from the 2000 census. As revealed in Figure 3.4, Asian Americans are especially diverse, comprising Chinese, Filipino, Asian Indians, Japanese, Vietnamese, Cambodian, Hmong, Thai, and other groups.

In the 2000 census, questions about race were modified to allow individuals to identify whether they were Asian alone or Asian with some other category. In 2000, the census reported 10.2 million people identifying themselves as "Asian alone" and an additional 1.7 million who reported themselves as Asian with some other racial group. Figure 3.4 represents the population of selected Asian groups that results from combining the "Asian alone" and "Asian, in combination" categories into a population numbering nearly 11.9 million people." As can be seen, the largest Asian American groups are Chinese Americans, Filipino Americans, Asian Indians, Koreans, Vietnamese, and Japanese Americans. Five groups—Asian Indians, Chinese, Filipinos, Koreans, and Vietnamese—account for at least 1 million people each.

Groups such as Cambodians, Laotians, and Hmong are more recent arrivals, first coming to this country in the 1970s as refugees from the upheavals resulting from the Vietnam War. In the 1980s, Koreans, Filipinos, and Asian Indians began immigrating in larger numbers. Half of all Asian Americans live in the western United States. More than half of the Asian and Pacific Islander population in the United States lives in just 3 states: California, New York, and Hawaii. Furthermore, just 10 states—California, New York, Hawaii, Texas, New Jersey, Illinois, Washington, Florida, Virginia, and Massachusetts—accounted for 75% of the Asian population. These same 10 states represent 47% of the overall population, indicating a greater tendency among Asians to cluster in these states. Of the Asian and Pacific Islander population, 95% lives in metropolitan areas, compared to 78% of non-Hispanic whites (Barnes and Bennett 2002).

General comparisons show that in many key ways Asian Americans are less like other racial or ethnic minorities than they are like Caucasians. They are as likely

Figure 3.4 ■ Selected Asian Groups

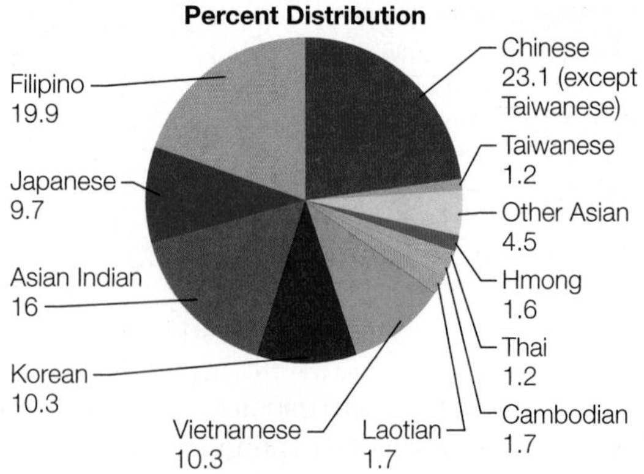

Percent Distribution

Chinese 23.1 (except Taiwanese)
Taiwanese 1.2
Other Asian 4.5
Hmong 1.6
Thai 1.2
Cambodian 1.7
Laotian 1.7
Vietnamese 10.3
Korean 10.3
Asian Indian 16
Japanese 9.7
Filipino 19.9

SOURCE: Barnes and Bennett 2002.

Exploring the Factors That Have Shaped Your Family

What we experience in our family relationships is partly a product of when we are born and live. This, like your race, ethnicity, and social class, is something over which you neither have control nor exercise choice, yet it limits or offers you choices and constrains or opens opportunities.

One way to illustrate this is to gather information on your own family, its history, its socioeconomic status, and its ethnic and/or racial background. Examining how your family has changed over time; how it has prospered, struggled, or held its ground economically; and how it has maintained or minimized the importance of its ethnic origins will go a long way toward helping you understand why your family has experienced the things it has. Such an analysis will not include everything that influences family life but it will move you in a sociologically enlightening direction, supplying a wider context to the particulars of your family.

If you carefully map your family's history and compare it with some historical patterns discussed in this chapter, you will likely see connections between these broader patterns and your family's story. You will then be better able to both see the larger picture and understand your family's unique experiences across generations.

What do you know of your family's history? Where does your family originally come from? How, why, and when did your family members come to the United States? Where did they settle? How did they survive economically? What sorts of work did earlier

generations of your family do? How much education did they receive? Over time and generations, how did their educational and economic experiences change?

Depending on whom you might be able to get information from, consider the following questions: How did your grandparents first meet? When did they marry? How many children did they have? Where did they live and what did they do? Of your parents, how did they get together? When did they first meet? What attracted them to each other and motivated them to start a family? How many siblings do you have? How many did your parents and your grandparents have?

Did your mother work outside the home when you were younger? Were your grandmothers employed when your parents were children?

Comparing across generations, how many, if any, divorces have occurred in your family? When was the first one?

There are many ways to explore your family's history. You can examine family photographs, read letters and diaries, or interview living members to learn what happened, when, and why. Interviews need not be formal. We highly recommend learning as much as you can about your families from surviving members of your families. These may be opportunities to hear family stories and correct any misunderstanding you have had about your families that otherwise might be forever lost as people age and pass away.

Family photographs can reveal much about the relationships among members. If you can, gather photographs of your immediate family your grandparents, great-grandparents, and so on. Identify who you can. Look at such details as facial expressions, and positioning of family members relative to one another. Are family members clustered closely together or far apart? Is someone standing off from the others?

After you gather information about your relatives, see what aspects of the family discussed in this chapter apply to your family. Was a great-great-great-grandmother a slave? How did your family weather the Depression? How did relatives go about their daily household tasks? If you can, interview members of your family about what they know. Try to find out stories about the oldest family members.

Where did they come from? What did they pass down—love of learning, ambition, money, pride? Did they speak a language other than English? Did they have to learn English? What important historical events occurred during their lifetimes? In which ones did they actually participate? What were their experiences of joy and sorrow?

Such family histories will better enable you to understand where you come from and what factors have shaped the family experiences you have had. Connecting your family history to the wider history of American families is a first step toward a better understanding of both.

---

as Caucasians to be married (57% for each group) but only half as likely to be divorced (5% versus 10% of non-Hispanic whites). They are also less likely to be widowed but more likely to have never married (33%

versus 25% among non-Hispanics whites). Almost three-fourths (73%) of Asian American households are family households, a level greater than found among Caucasians (66%) (Reeves and Bennett 2003).

Typically, Asian Americans have fewer children, have them within marriage, and have them later than do other ethnic groups. Where 10% of European American, 18% of Hispanic, and 23% of African American births occur to women under age 20, only 6% of Asian American births occur to teenage mothers (McLoyd et al. 2000a).

Values that continue to be important to Asian Americans in general include a strong sense of importance of family over the individual, self-control to achieve societal goals, and appreciation of cultural heritage. Chinese Americans tend to exercise strong parental control while encouraging their children to develop a sense of independence and strong motivation for achievement (Ishii-Kuntz 1997; Lin and Fu 1990).

Almost 90% of Asian Americans graduated from high school, a rate comparable to that of non-Hispanics whites (89%). However, Asians are more likely than Caucasians to graduate from college. More than half of Asian men (52%) and 44% of Asian women earned at least a B.A. degree. These are significantly higher percentages than found among Caucasian men (32%) and women (27%). At the other end of educational attainment, Asians are more likely to have less than 9 years of schooling than are non-Hispanics whites (7% versus 4%).

Economically, Asians are an unusual minority in that they often exceed the economic status and earnings of the dominant majority. In 2001, for example, Asian families were more likely to earn at least $75,000 than were Caucasian families (40% versus 35%). Unemployment rates were nearly the same (6% among Asians, 5% among non-Hispanics whites), and within employment categories, Asian men and women were more likely than Caucasian men and women to be employed in managerial and professional occupations. It is also true, however, that the poverty rate among Asians was slightly higher than among non-Hispanics whites (10% versus 8%) (Reeves and Bennett 2003).

As with Hispanics, there is noteworthy variation among different Asian American groups. In marital status, for example, although Asians were less likely than the general population to be separated, widowed, or divorced, there was much variation among them. Two thirds of Asian Indians and Pakistanis were married, but less than half of all Cambodians were married. Only 6% of Asian Indians or Pakistanis were separated, divorced, or widowed compared with Cambodians, Filipinos, Koreans, Laotians, and Thai, who

ranged between 10% and 15% in these marital statuses. The highest percentages of widowed, separated, or divorced were found among the Japanese, at 14.8%. However, this is potentially misleading because Japanese widows and widowers make up nearly half of that percentage (7.1%). The divorce rate among Japanese was 6.7%, still 3% less than among the total population (9.7%) and lower than the 7.4% found among Thai (Reeves and Bennett 2004, Figure 4).

Educationally and economically, there was also much variation among Asians. Asian Indians had the highest percentage earning a B.A. degree (64%), followed by Pakistanis (54%) and Chinese (48.1%). Meanwhile, about 50% of Laotians and Cambodians and 60% of Hmong had not completed high school. The Japanese, at 91%, had the highest percentage to have completed high school (Reeves and Bennett 2004, Figure 9). Asian Indian, Japanese, and Chinese women and men had the highest median earnings. Hmong, Cambodian, and Laotian men's and women's incomes were at the opposite end. At $51,904, Asian Indian men had the highest median income found among Asian men. Japanese women, with median earnings of $35,998, had the highest median income among Asian women (Reeves and Bennett 2004, Figure 12).

The median family income among Asians ranged from a low of $32,384 among Hmong to a high of $70,849 earned by Japanese families. Asian Indians were a close second at $70,708. Along with Hmong families, Cambodian, Korean, Laotian, Pakistani, Thai, and Vietnamese families all had median incomes "substantially lower" than the median for all Asian families ($59,324). Finally, poverty rates varied quite a bit. The lowest poverty rates were found among Filipinos (6.3%), Japanese (9.7%), and Asian Indians (9.8%). At the other end, the poverty rate for Hmong (37.8%) and Cambodians (29.3%) were 2.5 to 3 times the rate among Asians overall (Reeves and Bennett 2004, Figures 13 and 14).

Clearly, much diversity can be observed within Asian American families based on where they're from, time of arrival in the United States, and reasons for coming to this country (for example, political versus economic). More recent immigrants retain more culturally distinct characteristics, such as family structure and values, than do older groups, such as Chinese Americans and Japanese Americans. Asian American families tend to be slightly larger than the average U.S. family (U.S. Census Bureau 1996), although there is wide variation between older and more recent immigrants. Among the more assimilated Japanese, the average family has 2.5 members. Among more recent Asian immigrants (for example, Cambodians, Laotians, Vietnamese, and Hmong), families average between 4 and 5.1 members (McLoyd et al. 2000b). The greater family size reflects the presence of extended kin.

Migration and assimilation alter many traditional Asian family patterns. For example, among Japanese families there are considerable differences among the *Issei* (immigrant generation), the *Nisei* (first-generation American-born), and the *Sansei* and subsequent generations on such family characteristics as the relative importance of marriage over extended kin ties, the role of love in the choice of a spouse, and the relationship between the genders (Kitano and Kitano 1998). Similarly, we can draw distinctions between traditional Vietnamese families and American-born Vietnamese. Attitudes toward marriage and family, changes in familial gender roles, increased prevalence of divorce, and single-parent households all separate the generations.

We can also see marked change between parents' and children's attitudes about individualism and self-fulfillment versus family obligation and self-sacrifice (Tran 1998).

The most dramatic change affecting Chinese Americans has been their sheer increase in numbers over the last 30 years. The Chinese American population increased from 431,000 to 2.7 million between 1970 and 2000. More recent immigrants tend to be from Taiwan or Hong Kong rather than mainland China (Glenn and Yap 1994). Because of the large numbers of new immigrants, it is important to distinguish between American-born and foreign-born Chinese Americans; little research is available concerning the latter. Contemporary American-born Chinese families continue to emphasize familism, although filial piety and strict obedience to parental authority have become less strong. Chinese Americans tend to be better educated, have higher incomes, and have lower rates of unemployment than the general population. Their sexual values and attitudes toward gender roles tend to be more conservative. Chinese American women are expected to be employed and to contribute to the household income. More than 1.2 million speak Chinese at home.

NATIVE AMERICAN FAMILIES. More than 4 million Americans identify themselves as being of native descent, as American Indian or Alaska Native. This includes 2.5 million Americans who identify themselves as American Indian or Alaska Native alone and an additional 1.6 million who identify themselves as American Indian/Alaska Native, as well as one or more other races. Cumulatively, this population represents 1.5% of the 2000 population of the United States.

The increase in native population between 1990 and 2000 was greater than the increase in the entire U.S. population. Considering those who identify themselves as American Indian or Alaska Native alone, the increase was 26%, twice the size of the 13% increase in the entire U.S. population. Looking at those who identified themselves as native Americans in combination with one or more other races, the increase was a staggering 110% increase, an increase of more than 2.2 million people between 1990 and 2000.

Those who continue to be deeply involved with their own traditional culture give themselves a tribal identity, such as Dine (Navajo), Lakota, or Cherokee (Kawamoto and Cheshire 1997). The largest tribal groups include the Cherokee, Navajo, Latin American Indian, Choctaw, Sioux, and Chippewa. Together, these six tribal groups account for more than 40% of the American Indian population (see Table 3.10). Among Alaska Native tribal groups, there were 54,761 Eskimos, making them the largest group (Ogunwole 2002).

Those who are more acculturated, such as urban dwellers, tend to give themselves an ethnic identity as Native Americans or Indians. Most Americans of native descent consider themselves members of a tribal group rather than an ethnic group. According to John Price (1981), "Specific tribal identities are almost universally stronger and more important than identity as a Native American."

The American Indian population is unevenly distributed throughout the United States: 43% live in the West, 31% in the South, 17% in the Midwest, and only 9% in the Northeast. California and Oklahoma, together, account for nearly one-fourth of the American Indian population. Along with these eight other states, they lay claim to more than half of the American Indian population: Arizona, Texas, New Mexico, New York, Washington, North Carolina, Michigan, and Alaska (Ogunwole 2002).

There has been a considerable migration of Native Americans to urban areas since World War II because of poverty on reservations and pressures toward acculturation. Today, 1.2 million Americans of native descent live outside tribal lands; most live in cities, where they are separated from their traditional tribal

Table 3.10 ■ Largest American Indian Tribal Groupings, Census 2000

Tribal Group	# Identifying American Indian Alone	# Identifying American Indian in Combination	Total
Total	2,475,956	1,643,345	4,119,301
Tribe specified	1,963,996	1,098,848	3,062,844
Cherokee	281,069	448,464	729,533
Navajo	269,202	28,995	298,197
Latin American Indian	104,354	76,586	180,940
Choctaw	87,349	71,425	158,774
Sioux	108,272	45,088	153,360
Chippewa	105,907	43,762	149,669

SOURCE: Ogunwole 2000; U.S. Census Bureau 2002.

cultures and may experience great cultural conflict as they attempt to maintain traditional values. Not surprisingly, those in the cities are more acculturated than those remaining on the reservations. Urban Native Americans may attend powwows, intertribal social gatherings centering on drumming, singing, and traditional dances. Powwows are important mechanisms in the development of the Native American ethnic identity in contrast to the tribal identity. Urban Native Americans, however, may visit their home reservations regularly.

Based on data from the 2000 report "Census of Population and Housing Characteristics of American Indians and Alaska Natives by Tribe and Language," we can make the following points regarding the American Indian and Alaska Native populations (U.S. Census Bureau 2003):

■ Of the 770,334 American Indian and Alaska Native households, 563,651, or 73%, were family households, of which 59% had children under age 18 living with them.

■ Of the family households, 61% were married couple families, of which 57% had children under 18. Another 28% were female-headed families with no husband, of which 64% had a child or children under 18.

■ The American Indian and Alaska Native populations were less likely to complete high school and college and more likely to drop out of school before completing high school than were the general population. Where 80% of the general population, age 25 and older, had completed high school, 71% of the American Indian and Alaska Native populations had. Of the American Indian and Alaska Native populations, 11.5% had completed college, less than half the percentage of the general population (24.4%). At the other end, whereas 9.8% of 16–19 year olds in the United States had not graduated high school but were also no longer enrolled, among American Indian and Alaska Native populations the percentage was 16.1%.

■ Although similar percentages of American Indian and Alaska Native populations were in the labor force as the general population, with similar levels of female employment and employment of mothers with children under 6, the American Indian and Alaska Native populations were twice as likely to be unemployed, twice as likely to be below poverty, and had substantially lower median household and family incomes (see Table 3.11).

Although there is considerable variation among different tribal groups, and hence no single type of American Indian or Alaska Native family, three aspects of Native American families are important. First, extended families are significant. These extended families may be different from what the larger society regards as an extended family (Wall 1993). They often revolve around complex kinship networks based on clan membership rather than birth, marriage, or adoption. Concepts of kin relationships may also differ. A child's "grandmother" may be an aunt or great-aunt in a European-based conceptualization of kin (Yellowbird and Snipp 1994).

Second, increasingly large numbers of Native Americans are marrying non-Indians. Among married Native Americans, more than half have non-Indian spouses. With such high rates of intermarriage, a key question is whether Native Americans can sustain their ethnic identity. Michael Yellowbird and Matthew Snipp (1994) wonder if "Indians, through their spousal choices, may accomplish what disease, Western civilization, and decades of federal Indian policy failed to achieve."

Table 3.11 ■ Comparative Measures of Economic Well-Being: American Indian/Alaska Native and U.S. Overall: 2000

	U.S. Overall	American Indian/Alaska Native	American Indian as % of U.S. overall
% unemployed	5.8	12.4	
Median household income	$41,994	$30,599	73
Median family income	$50,046	$33,144	66
% of households with incomes > $100,000	12.3%	5.4	
Median earnings: male	$37,057	$28,919	78
Median earnings: female	$28,919	$22,834	79
% in poverty: total	12.4	25.7	
% in poverty: children	16	31	
People > 65 years old	9.9	23.5	
Families in poverty	9.2	21.8	
Female-headed families in poverty	34.3	45.7	

Source: U.S. Census Bureau, 2000 Census of Population and Housing Characteristics of American Indians and Alaskan Natives by Tribe and Language: PHC-5, Washington, D.C., 2003; Tables 9, 10, 12, 13.

Third, family characteristics are affected by the economic status of American Indians and Alaska Natives. Given the higher levels of unemployment and poverty, and lower overall earnings and educational attainment, once again social class may be confounding our attempt to look at patterns of family living.

FAMILIES OF MIDDLE EASTERN BACKGROUND. People of Middle Eastern ethnic backgrounds living in the United States are among the fastest growing ethnic minority in the country. Estimates of the population vary, depending on such issues as what countries are included and whether we count only naturalized citizens or includes all immigrants, legal and illegal, temporary (for example, students and guest workers) and permanent (Camarota 2002; Brittingham and de la Cruz 2005). Furthermore, the census provides more detailed analysis of people of Arab ancestries than people whose ancestry is Middle Eastern. Thus, estimates from the census tend to undercount the overall population of Middle Eastern background.

As defined by Steven Camarota of the Center for Immigration Studies, "Middle Eastern" includes people whose backgrounds can be traced to one of the following: Pakistan, Bangladesh, Afghanistan, Turkey, the Levant, the Arabian peninsula, and Arab North Africa. In terms of specific countries, the designation "Middle Eastern" encompasses Afghanistan, Bangladesh, Pakistan, Iran, Iraq, Israel, Jordan, Kuwait, Lebanon, Syria, Turkey, Oman, Qatar, Bahrain, Saudi Arabia, United Arab Emirates, Yemen, Algeria, Egypt, Libya, Morocco, Sudan, Tunisia, West Sahara, and Mauritania (Camarota 2002).

The Middle Eastern immigrant population is relatively recent and very diverse. The population includes non-Arab countries such as Israel, Iran, Turkey, and Pakistan, representing half of the top eight Middle Eastern countries of origin in 2000. Further complicating counts, among those Middle Eastern immigrants from Arab countries we find many non-Arabs. Similarly, many immigrants from non-Arab countries, such as Israel, for example, are Arabs.

The U.S. census, counting the more narrowly defined Arab population, estimates that 1.2 million people claim some Arab ancestry, either alone or in combination. Meanwhile, the Center for Immigration Studies states that Middle Eastern immigrants numbered closer to 1.5 million in 2000, with 40% of Arab background. The center further estimates that within a decade (that is, by 2010) the number is likely to be 2.5 million or more. Putting aside the question of counts, the U.S. census, in two separate reports on the Arab population in the United States, provides the following profile:

■ Three ancestry groups, Lebanese, Syrian, and Egyptian, account for 60% of the Arab population. The largest is Lebanese, representing 37% of the U.S. Arab population (Syrian and Egyptian account for 12% each).

- The Arab population, spread fairly evenly across the four regions of the United States, is disproportionately found in just five states, California, Florida, Michigan, New Jersey, and New York. Cumulatively they account for nearly half (48%) of the Arab population. The city with the largest Arab population is New York City, with 69,985 people of Arab ancestry. Second is Dearborn, Michigan, with 29,181. Interestingly, the Arab population in New York accounts for less than 1% of the city's population, whereas the Arab population in Dearborn is 30% of the city's population.

- In comparison to the general population, the Arab population is disproportionately male. Males comprise 57% of the population, compared to 49% of the total U.S. population. Furthermore, 31% of the Arab population consisted of men age 20–49. This same demographic group represented 22% of the total U.S. population (Brittingham and de la Cruz 2005).

- The Arab American population is more likely than the total population to be married and less likely to be widowed, separated, or divorced. Where 54% of the total U.S. adult population is married, 61% of the Arab population is married. As was true of Asians and Latinos, much variation exists among Arab ethnicities. Moroccans are the least likely to be married (53.4%) and Jordanians the most likely (67%). Nearly one out of five adults in the U.S. population is separated, widowed, or divorced; among Arab Americans, 13% fall into those categories (Brittingham and de la Cruz 2005).

- Compared to the total population, a greater proportion of Arab households consisted of married couples, with or without children, in 2000. Married couples made up 60% of Arab households, compared to 53% of all U.S. households. Among Palestinians and Jordanians, the percentage of married couples reached 70%. Meanwhile, where more than 1 in 10 (12%) U.S. households was headed by a woman with no husband present, only about 1 in 5 (6%) of Arab American households were headed by a woman (Brittingham and de la Cruz 2005).

- Arab Americans tend to be highly educated, employed, and have higher incomes than the total population. However, Arab women are much less likely to be in the labor force than are women overall (see Table 3.12).

- Obscured by the data in Table 3.12 are the differences among Arabs. For example, 94% of Egyptians graduated from high school compared to 73% of Iraqis in the United States. Similarly, 64% of Egyptians 25 and older had B.A. degrees compared to 36% of Iraqis (which still surpassed the 24% in the total population). Median family income ranged from a low of $41,277 among Moroccans to a high of $60,677 among Lebanese. Poverty rates ranged from 11% among Lebanese and Syrians to a high of 25% among Iraqis.

- Although the Middle East is approximately 98% Muslim, immigrants to the United States from the region historically were not. In the past, most were Christian. This changed in the 1990s, and estimates are that nearly three-fourths of the Middle Eastern immigrant population is Muslim (Camarota 2002). The fact of their Muslim faith may influence certain family patterns, although as happens to other ethnic groups

Table 3.12 ■ Socioeconomic Differences: Arab Americans and U.S. Overall

		Arab	U.S.
With ≥ high school		84%	80%
With ≥ undergraduate degree		40%	24%
In labor force	Male	73%	71%
	Female	46%	58%
Median earnings	Male	$41,700	$37,100
	Female	$31,800	$27,200
Median family income		$52,318	$50,046
In poverty	Total	17%	12%
	Children	22%	16.6%

Among Middle Eastern ethnic groups in the United States, Iranians are one of the faster growing and more successful (Mostashari and Khodamhosseini 2004). Although the population is estimated to be between 319,000 and 371,000, the Iranian Studies Group at the Massachusetts Institute of Technology speculates that the true population may be closer to 690,000 but is undercounted because of reluctance to identify oneself as Iranian out of fear of "adverse effects" (Mostashari 2004).

Iranians are highly educated, with 57% completing college compared to only 24% among the overall U.S. population. A greater proportion of Iranians have graduate degrees than the wider population has B.A. degrees. Iranian Americans have family incomes 38% higher than the median family income for the United States, own homes valued at 2.5 times the value of an average American home, and are twice as likely to have family incomes in excess of $100,000 as the general population (Mostashari and Khodamhosseini 2004). More than 80% of employed Iranians work in professional, managerial, sales, or office positions compared to 60% of the total population (Mostashari and Khodamhosseini 2004).

Iranian sociologist Ali Akbar Mahdi undertook a survey comparing the division of household labor for Iranian married couples in the United States and those in Iran. Mahdi focused on the women in a sample of 149 couples in the United States and 514 couples living in Iran. His U.S. sample was more highly educated and affluent than the general profile of Iranian American families. More than half of the women had graduate or professional degrees and another third had "just" B.A. degrees. His sample of couples living in Iran was also highly educated (60% had attended college) and comfortable (45% middle class, 26% upper-middle class). Mahdi compared how the two samples differed in their allocation of 10 household tasks. Compared to the women living in Iran, the immigrant women in the U.S. were less likely to bear responsibility for childcare and for domestic tasks that included cleaning the house or apartment; sewing, ironing and laundry. They reported their husbands as more likely to take responsibility for cooking, cleaning, and childcare than did the women living in Iran (Mahdi 2001). Although there was no typically female domestic task for which most immigrant women claimed that their husbands were now responsible, there had been some movement from tradition in the immigrant sample. As Mahdi notes about the immigrant couples (Mahdi 2001, 184):

> Men are taking a more active role in the household chores . . . women also are participating actively in the roles traditionally performed by men, such as managing family finances, attending to family business, and even caring for the family car. Iranian women are seeking open equality in doing household chores, in child-rearing, decision making, ownership of family property . . . even in their sexual relationship.

Transitions such as Mahdi depicts do not come without some difficulty. He notes that in his immigrant sample some men felt, for at least a time, a loss of the traditional privilege and higher status that men in Iran expect to enjoy in marriage. There is also stress and confusion felt by husbands and wives as they attempt to renegotiate and redefine their respective places in marriage and the family. Compounding this is the absence of the wider kin network that, in Iran, may have buffered couples from some conflict (or reinforced a particular way of living and thus prevented changes of this kind). He observes that in the new setting, each spouse has to play the role of intimate partner and also, in many cases, the role of an absent father, mother, or brother" (Mahdi 2001, 187).

Mahdi suggests that although wives and husbands share in the economic gains, the social gains have been unequal. Men surrendered privileges that they previously enjoyed (or that they were raised to expect). They lost authority and the automatic respect within their marriages and kin networks that men traditionally commanded. Women, on the other hand, escaped some oppressive features of the society they left behind and gained independence; autonomy; individuality; a new, more equal identity; and a "clearer sense of their sexuality" (Mahdi 2001, 190).

assimilation operates against strict adherence to even religiously reinforced customs. Such is the case, for example, with both dating and mate selection. Sharply sex-segregated customs surrounding dating and a preference toward arranged marriage are characteristic of Muslim family life, yet both undergo considerable challenge from sons and especially daughters who are exposed to and may come to value Western notions of love, marriage, and family life (Zaidi and Shuraydi 2002). As Arshia Zaidi and Muhammad Shuraydi report from their examination of Pakistanis, "Families, depending on their educational, religious, economic, and social backgrounds, are coping with these changes by modifying the traditional authoritarian structure of the family system and their attitudes."

EUROPEAN ETHNIC FAMILIES. The sense of ethnicity among Americans of European descent grew in recent decades. This is especially true among working-class Germans, Italians, Greeks, Poles, Irish, Croats, and Hungarians. This increasing awareness seems to be part of a general rise in ethnic identification over the last 30 years (Rubin 1994). Earlier, members of European ethnic groups sought to assimilate—to adopt the attitudes, beliefs, and values of the dominant culture. Most white ethnic groups have assimilated to a considerable degree—they have learned English, moved from their ethnic neighborhoods, and married outside their group, but many continue to be bound emotionally to their ethnic roots. These roots are psychologically important, giving them a sense of community and a shared history. This common culture is manifested in shared rituals, feast days, and saint's days, such as St. Patrick's Day.

Except for some West Coast enclaves, such as Little Italy in San Francisco, white ethnicity is strongest in the East and Midwest. The Irish neighborhoods of Boston, the Polish areas of Chicago, and the Jewish sections of Brooklyn, for example, have strong ethnic identities. Common languages and dialects are spoken in the homes, stores, and parks. Traditional holidays are celebrated; the foods are prepared from recipes passed down through generations. Elders speak of the old country and their villages—even if it was their parents or grandparents who immigrated.

As is true of some non-European ethnic groups, as children grow up and move from their neighborhoods, their ethnic identity often becomes weaker in terms of language and marriage to others within their group—but they may retain some elements of ethnic pride. Their ethnicity is what Herbert Gans (1979) calls *symbolic ethnicity*—an ethnic identity that's used only when the individual chooses. Symbolic ethnicity has little effect on day-to-day life. It is not linked to neighborhoods, accents, the use of a foreign language, or working life. Others cannot easily identify the person's ethnicity; he or she "looks" American. Nevertheless, for many Americans, ethnicity has emotional significance. A person is Irish, Jewish, Italian, or German, for example—not only an American.

European ethnic groups differ from one another in many ways. However, a major study of contemporary American ethnic groups (Lieberson and Waters 1988) found that European ethnic groups are more similar to one another than they are to African Americans, Latinos, Asian Americans, and Native Americans. The researchers concluded that a European–non-European distinction remains a central division in our society. There are several reasons for this. First, most European ethnic groups no longer have **minority status**—that is, unequal access to economic and political power. Some scholars suggest that what separates ethnic groups into distinctive lifestyles is their social placement. As groups become more similar in their access to opportunities, their family lifestyles may "converge" toward a common pattern, one that includes smaller families, increased divorce, less interdependent ties with extended families, and less male dominance (Aponte, Beal, and Jiles 1999). Second, because most European ethnic groups are not physically distinguishable from other white Americans, they are not discriminated against racially.

This chapter has covered much ground. As we have now seen, in a host of ways, American families are diverse. They vary across time and, within any given period, between racial, ethnic, and socioeconomic groups. Family diversity is reflected throughout subsequent, more specialized chapters as relevant variations by race, class, or ethnicity are discussed. Thus, our goal of understanding American families will be made more complete and representative.

Acknowledging the diversity that exists across families has personal consequences as well. It ought to make us a bit more cautious in generalizing from our particular set of family experiences to what others "must also experience." In addition, in noting how historical, economic, and cultural factors shape our families, we link our personal experiences to broader societal forces. In that way, we are better able to apply "sociological imaginations" to family experiences, identifying how our private and personal family worlds are largely products of when, where, and how we live (Mills 1959). Simply put, if we come of age during a period of great economic upheaval, we may put off marrying, bearing children, or divorcing because of the opportunities and constraints we face. Similarly, the kinds of family experiences we are able to have are limited or enhanced by the economic resources at our disposal, regardless of what we might otherwise choose to do.

Despite the extent to which the factors discussed in this and the next chapter may limit your opportunities or narrow your range of choices, remember that you do and will make choices about what kind of family you wish to create. You decide whether or not to marry, whether or not to bear children, how to rear your children, whether to stay married, and so on. A major goal of this book is to equip you with a foundation of accurate information about family issues from which you can make sound choices more effectively.

Summary

- In the early years of colonization, there were 2 million Native Americans in what is now called the United States. Many families were *patrilineal*; rights and property flowed from the father. Other tribal groups were *matrilineal*. Most families were small.

- Diverse groups settled America, including English, Germans, and Africans. In colonial America, marriages were arranged. Marriage was an economic institution, and the marriage relationship was *patriarchal*.

- African American families began in the United States in the early seventeenth century. They continued the African tradition that emphasized kin relations. Most slaves lived in two-parent families that valued marital stability.

- In the nineteenth century, industrialization revolutionized the family's structure; men became wage earners, and women, once they married, became housewives. Childhood was sentimentalized, and adolescence was invented. Marriage was increasingly based on emotional bonds.

- The stability of the African American enslaved family suffered because it lacked autonomy and had little economic importance. Enslaved families were broken up by slaveholders, and marriage between slaves was not legally recognized. African American families formed solid bonds nevertheless.

- Beginning in the twentieth century, *companionate marriage* became an ideal. Men and women shared household decision making and tasks, marriages were expected to be romantic, wives were expected to be sexually active, and children were to be treated more democratically.

- The 1950s, the golden age of the companionate marriage, was an aberration. It was an exception to the general trend of rising divorce and nontraditional gender roles. Prosperity was unusually high; suburbanization led to increased isolation.

- The terms *ethnic group, racial group,* and *minority group* are conceptually distinct. An ethnic group is a group of people distinct from other groups because of cultural characteristics. A *racial group* is a group of people, such as whites, blacks, or Asians, classified according to phenotype, as well as anatomical and physical characteristics. A *minority group* is a group whose status (position in the social hierarchy) places its members at an economic, social, and political disadvantage.

- African Americans are the second largest ethnic group in the United States. *Socioeconomic status* is an important element in understanding African American families.

- Because of economic necessity, African American women traditionally have been employed, which has given them important economic roles in the family and more egalitarian relationships. Kinship bonds and intergenerational ties are important sources of emotional and economic assistance in times of need. African Americans are much more likely than Caucasians to live in *extended households*.

- Latinos are now the largest ethnic group as a result of immigration and a higher birthrate than the general population. There is considerable ethnic and economic diversity among Latinos. Latinos emphasize extended kin relationships, cooperation, and mutual assistance. *La familia* includes not only the nuclear family but also the extended family.

- Asian Americans are the third largest ethnic group in the United States. Immigration has contributed heavily to the dramatic recent increase in the Asian American population. The largest Asian American groups are Chinese Americans, Filipino Americans, and Asian

Indians. More recent immigrants retain more culturally distinct characteristics, such as family structure and values, than do older groups. There are differences between Asian ethnic groups, much of which results from their socioeconomic position in U.S. society.

- More than 4 million Americans identify themselves as American Indians or Alaska Natives. Tribal identity remains a key part of their identity. More than half of Native Americans live in cities, although many remain in contact with their home reservation. Extended families are important and are often based on clan membership. About 53% of Native Americans are married to non-Indians.

- In recent years, increasing numbers of people from Middle Eastern countries have come to the United States. Overall, people of Middle Eastern background are economically better off than the general population, more highly educated, and more likely to live in married-couple headed households, though there is much social, economic and familial diversity within the Middle Eastern population.

- Ethnic identity among Americans of European descent has been growing, especially among working-class families. For many, their ethnicity is symbolic and has little effect on day-to-day life. Most members of European ethnic groups are physically indistinguishable from other white Americans and no longer have minority status.

Resources on the Internet

Companion Website for This Book

http://www.thomsonedu.com/sociology/strong

Gain an even better understanding of this chapter by going to the companion website for additional study resources. Take advantage of the Pre- and Post-Test quizzing tool, which is designed to help you grasp difficult concepts by referring you back to review specific pages in the chapter for questions you answer incorrectly. Use the flash cards to master key terms and check out the many other study aids you'll find there. Visit the Marriage and Family Resource Center on the site. You'll also find special features such as access to InfoTrac© College Edition (a database that allows you access to more than 18 million full-length articles from 5,000 periodicals and journals), as well as GSS Data and Census information to help you with your research projects and papers.

Key Terms

adolescence 72

assisted reproductive technologies 83

bundling 71

companionate marriage 76

ethnic group 95

extended households 98

familism 84

feminization of poverty 90

fictive kin ties 94

life chances 85

lower-middle class 88

matrilineal 70

minority group 96

minority status 110

patriarchy 71

patrilineal 70

phenotype 95

racial group (race) 95

social class 85

social mobility 94

socioeconomic status 85

two-person career 90

upper-middle class 88

working class 88

Gender and Family

Outline

What Do YOU Think?

Are the following statements TRUE or FALSE?
You may be surprised by the answers (see answer key on the following page.)

T   F   **1** Gender roles reflect the instinctive nature of males and females.

T   F   **2** Gender roles are influenced by ethnicity.

T   F   **3** The only universal feature of gender is that all societies sort people into only two categories.

T   F   **4** Parents are not always aware that they treat their sons and daughters differently.

T   F   **5** Peers are the most important influence on gender-role development from adolescence through old age.

T   F   **6** Both boys and girls suffer from gender-related problems in school.

T   F   **7** For African Americans, the traditional female gender role includes both employment and motherhood.

T   F   **8** Research shows it is possible for women and men to establish work or family roles that are counter to their socialization.

T   F   **9** Compared with traditional roles, contemporary male gender roles place more emphasis on the expectation that men will be actively involved with their children.

T   F   **10** Men's and women's movements have consistently stressed the importance of family.

Did you ever stop to consider how similar or different your life might be if you had been born the opposite sex? Would you be the kind of person you are? Participate in the same activities? Have the same friends? Have the same roles and relationships within your families? Would your goals be the same as they are now? Would you be enrolled in the same college? Take the same courses? Be reading this book? What about your expectations for relationships? Would you envision the same familial future? Asking ourselves such questions reminds us that much of what we do, who we are, what we expect, and what happens to us is influenced by gender. In this chapter we examine how deeply interconnected family experience is with gender. It is no exaggeration to say that we cannot fully understand one without taking the other into account.

The traditional view of gender depicts male and female, masculinity and femininity, men and women as polar opposites. Our gender stereotypes fit this pattern of polar differences: we believe that if men are aggressive, women are passive; if men are instrumental (task oriented), women are expressive (emotion oriented); if men are rational, women must be irrational; if men want sex, women want love (Duncombe and Marsden 1993; Lips 1997).

As shown in this chapter, this *perception* of male–female differences is greater than the actual differences (Hare-Mustin and Marecek 1990b). We may be accustomed to thinking that we are as different as Martians would be from Venusians, but both women and men inhabit Earth (Kimmel 2000). At the same time, our family experience *is* highly "gendered" (that is, differently experienced for women and men). Marriages might be said to consist of "two marriages, his and hers," that are not entirely the same (Bernard 1982). Similarly, we could argue that there are "two courtships," "two parenthoods," "two divorces," and so on. In each area of marriage and family life, we often observe differences in what women and men experience. Some data suggest that men and women may define and experience love differently, enter marriage with different emphases and expectations, react to the onset of parenthood and relate to their children differently, divorce for different reasons and with different consequences, and so on. The chapters that follow identify and illustrate some of these gender differences.

In this chapter we examine some gender and socialization theories and illustrate how much our families influence how we learn to act masculine and feminine. Next, we explore some areas of family experience that have been and remain differently experienced by women and men. Finally, we discuss changing gender roles and consider some gender-based social movements of the past 5 decades.

Understanding Gender and Gender Roles

Studying Gender

Before we commence, we need to define several key terms useful in building an understanding of the importance of gender in family life. These terms include *sex, role, gender role, gender-role stereotype, gender-role attitude,* and *gender-role behavior.* **Sex** refers here to the biological aspect of being male or female. As such, it includes chromosomal, hormonal, and anatomical characteristics that differentiate females from males. In general, a **role** consists of culturally defined expectations that an individual is expected to fulfill in a given situation in a particular culture. A **gender role** is a role that a person is expected to perform as a result of being male or female in a particular culture. (The term *gender role* is a more recent concept that has largely replaced the traditional term *sex role.*) A **gender-role stereotype** is a rigidly held and oversimplified belief that all males and females, as a result of their sex, possess distinct psychological and behavioral traits. Stereotypes tend to be false not only for the group but also for any individual member of the group. Even if the generalization is statistically valid in describing a group average, such as males are taller than females, we cannot necessarily predict whether Jason will be taller than Tanya. **Gender-role attitude** refers to the beliefs we have regarding appropriate male and female personality traits and activities. **Gender-role behavior** refers to the actual activities or behaviors we engage in as males and females. When we discuss

Answer Key for What Do You Think

1 False, see p. 119;   2 True, see p. 127;   3 False, see p. 117;   4 True, see p. 124;   5 False, see p. 129;   6 True, see p. 128;   7 True, see p. 128;   8 True, see p. 131;   9 True, see p. 131;   10 False, see p. 144.

gender roles, it is important not to confuse stereotypes with reality or to confuse attitudes with behavior.

Historically, most gender-role studies focused on the Caucasian middle class. This made it difficult to know whether and how gender roles may have differed among African Americans, Latinos, Asian Americans, and other ethnic groups. Students and researchers must be just as careful not to project onto other groups the gender-role concepts or aspirations characteristic of their own groups. Too often such projections can lead to distortions or moral judgments. Although we may come to accept one particular standard of behavior as more "appropriate" masculinity or femininity, there are actually **multiple masculinities and femininities,** out of which emerges a version that is expected or accepted (Connell 1995; Kimmel 2000; Messerschmidt 1993).

These dominant or **hegemonic models of gender** are held up as the standards for all women and men to emulate (Kimmel 2000). They are also dynamic and culturally variable. They change over time (Kimmel 1996), differ across space (Gilmore 1990), and—within a given time and place—are challenged for cultural dominance by those who advocate other versions of masculinity or femininity (Kimmel 1994; Connell 1995).

Gender and Gender Roles

Gender is simultaneously experienced on both personal and political levels. At birth, we are identified as either male or female. This identification, based essentially on inspection of genitalia, typically leads to the self-identity or **gender identity** we form of ourselves as females or males. We say "typically" because there are individuals who for a combination of reasons are categorized as *transsexuals*—males and females who develop self-identities that differ from the gender category into which they have been placed. They opt for reconstructive surgery to bring their biology into line with the identity they have developed, seeking "to become—physically, socially and legally—the sex they have always been psychologically. If they succeed in doing so, they typically consider themselves simply as members of their new sex, rejecting any significance to how they arrived there" (Coombs 1997).

Increasingly, we see the term *transgenderism* being used to refer to a range of situations in which a person's gender identity or gender presentation (whether individuals present themselves as a male or as a female) do not match what would be expected by wider soci-ety for someone with the anatomical characteristics she or he possesses. This would include cross-dressers (transvestites), transsexuals, non-operative transsexuals (individuals who identify as opposite their biological sex but do not seek to undergo sex reassignment surgery), and individuals Mary Coombs refers to as "bigendered," the gender equivalent of bisexuals, who choose at times to present themselves as male and at other times as female (Coombs 1997).

We acquire our gender identities at a young age. Furthermore, gender identity may well be the deepest concept we hold of ourselves. The psychology of insults reveals this depth; few things offend a person, especially a male, as much as to be tauntingly characterized as a member of the "opposite" sex. Gender identity determines many of the directions our lives will take—for example, whether we will fulfill the role of husband or wife, father or mother. When the scripts are handed out in life, the one you receive depends largely on your gender.

At the same time that it denotes how we perceive ourselves, gender is a basis for the assignment of social roles, the distribution of rewards, and the exercise of power. Most societies are **patriarchal societies,** in which males dominate political and economic institutions and exercise power in interpersonal relationships. Although many societies have been identified as more **egalitarian** (in which women and men enjoy similar amounts of power and neither dominates the economic or political institutions), truly **matriarchal societies** have not been evident. Within patriarchal societies, families tend to be male dominated. That is to say, in daily decision making and the division of responsibilities, men have privileges that women do not (for example, freedom from domestic work). The familial power that men have stems from various sources, including the marriage contract and their wage-earning roles. Later chapters explore in more detail how gender and power are connected within households and families.

Each culture determines the content of gender roles in its own way. In some cultures, there are more than two gender categories. Among some Asian and Native American societies, for example, men or women become *berdaches*. They then live as members of the opposite sex. The Hua of Papua, New Guinea, perceive gender as fluid, capable of changing over the individual's life span. In other societies, alternative categories (for example, the Hjira of India) are socially recognized for individuals who are *neither* male nor female (Renzetti and Curran 1999; Nanda 1990).

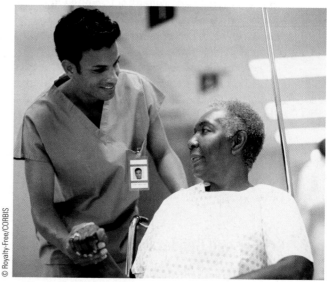

■ *Generally, the only limit on the jobs that women or men hold is social custom, not biology or individual ability. Even sex-segregated jobs such as nursing and firefighting can be performed by either gender.*

We can identify less extreme cultural variations in conceptualizations of gender. Among the Arapesh of New Guinea, both males and females possess what we consider feminine traits. Men and women alike tend to be passive, cooperative, peaceful, and nurturing. The father is said to "bear a child," as well as the mother; only the father's continual care can make a child grow healthily, both in the womb and in childhood. Eighty miles away, the Mundugumor live in remarkable contrast to the peaceful Arapesh. Margaret Mead (1975) offered this observation:

> Both men and women are expected to be violent, competitive, aggressively sexed, jealous, and ready to see and avenge insult, delighting in display, in action, in fighting. . . . Many, if not all, of the personality traits which we have called masculine or feminine are as lightly linked to sex as are the cloth-

ing, the manners, and the form of headdress that a society at a given period assigned to either sex.

Biology creates males and females, but culture creates masculinity and femininity.

Masculinity and Femininity: Opposites or Similar?

Until the last generation, a **bipolar gender role** was the dominant model used to explain male–female differences. In this model, males and females are seen as polar opposites, with males possessing exclusively instrumental traits and females possessing exclusively expressive ones. Sandra Bem (1993) describes the culture of the United States as one that looks at gender through a series of "lenses," including the belief that males and females are fundamentally different. She calls this assumption *gender polarization*. Our entire society is organized around such supposed differences (Renzetti and Curran 2003). In light of the widespread acceptance of this viewpoint and its immense popularization through John Gray's *Men Are from Mars, Women Are from Venus,* Michael Kimmel (2000) cleverly calls this viewpoint the "interplanetary theory of gender."

Traditional views of masculinity and femininity as opposites have several implications. First, if a person differs from the male or female stereotype, he or she is seen as being more like the other gender. If a woman is sexually assertive, for example, she is not only less feminine but also is believed to be more masculine. Similarly, if a man is nurturing, he is not only less masculine but also is seen as more feminine. Second, because males and females are perceived as opposites, they cannot share the same traits or qualities. A "real man" possesses exclusively masculine traits and behaviors, and a "real woman" possesses exclusively feminine traits and behaviors. A man is assertive, and a woman is receptive; in reality, both men and women are often both assertive and receptive. Third, because males and females are viewed as opposites, they are believed to have little in common with each other, and a "war of the sexes" is alleged as the norm. Men and women can't understand each other, nor can they expect to do so. Difficulties in their relationships are attributed to their "oppositeness."

The fundamental problem with the view of men and women as opposites is that it is erroneous. As men and women we are significantly more alike than we are different.

Our culture, however, has encouraged us to look for differences and, when we find them, to exaggerate their degree and significance. It has taught us to ignore the most important fact about males and females: that we are both human. As humans, we are significantly more alike biologically and psychologically than we are different. As men and women, we share similar respiratory, circulatory, neurological, skeletal, and muscular systems. (Even the penis and the clitoris evolved from the same undifferentiated embryonic structure.) Hormonally, both men and women produce androgens and estrogen (but in different amounts). Where men and women biologically differ most significantly is in terms of their reproductive functions: men impregnate, whereas women menstruate, gestate, and nurse. Beyond these reproductive differences, biological differences are not great. In terms

of social behavior, studies suggest that men are more aggressive both physically and verbally than women; the gender difference, however, is not large. Most differences can be traced to gender-role expectations, male–female status, and gender stereotyping.

Although we are more similar than different in our attributes and abilities, large and meaningful differences do exist in the *statuses* (or positions in various groups and organizations) we occupy and the privileges and responsibilities these carry. Although either gender may have the *ability* to nurture children, support families, clean, or cook, these tasks are assumed to be more appropriate for one gender than the other. Although women and men may possess the *ability* to do many kinds of jobs, the labor force is sex-segregated into jobs that are disproportionately male or female. Men's jobs typically carry more prestige, earn higher salaries, and offer more opportunity for advancement than do women's jobs.

We often refer to these differences as "gaps." The "wage gap" refers to the difference between what men tend to earn and what women tend to earn. Recent data indicate that when we compare the median weekly earnings of women and men employed full-time, women earn 75.7% of what men earn. As Table 4.1 shows, in 2003 white women earned 75.6% of what white men earned.

We can also speak of "prestige gaps" or "mobility gaps." Jobs that tend to be among the most highly respected jobs (typically, jobs such as physician, attorney, and engineer) tend to be held disproportionately by men. Jobs held largely by women (such as types of clerical work, elementary and preschool teaching, household service, and nursing) are often undervalued (and, not surprisingly, underpaid). We should not assume that "men's jobs" are highly paid and highly respected and "women's jobs" are devalued and underpaid. Instead, the point is that in those jobs rewarded with higher levels of prestige and higher salaries, we tend to find more men than women. Finally, compared to jobs in which we find mostly men, jobs typically held by women may offer only limited

Table 4.1 ■ The Wage Gap by Gender and Race

Median annual earnings of black men and women, Hispanic men and women, and white women as a percentage of white men's median annual earnings.

Year	White men	Black men	Hispanic men	White women	Black women	Hispanic women
1970	100%	69.0%	NA	58.7%	48.2%	NA
1975	100	74.3	72.1	57.5	55.4	49.3
1980	100	70.7	70.8	58.9	55.7	50.5
1985	100	69.7	68.0	63.0	57.1	52.1
1990	100	73.1	66.3	69.4	62.5	54.3
1992	100	72.6	63.3	70.0	64.0	55.4
1994	100	75.1	64.3	71.6	63.0	55.6
1995	100	75.9	63.3	71.2	64.2	53.4
1996	100	80.0	63.9	73.3	65.1	56.6
1997	100	75.1	61.4	71.9	62.6	53.9
1998	100	74.9	61.6	72.6	62.6	53.1
1999	100	80.6	61.6	71.6	65.0	52.1
2000	100	78.2	63.4	72.2	64.6	52.8
2003	100	78.2	63.3	75.6	65.4	54.3

SOURCE: National Committee on Pay Equity, http://www.infoplease.com.
NA means data not available.

levels of upward mobility (movement "up" in income and position). Cumulatively, we may refer to these economic gaps and inequalities as indicators of *gender stratification,* a term that denotes that in economic, as well as many social and political, ways, men are "on top" in society. This is similar to the way that the upper and middle classes are "above" the working and lower classes. Although women are not literally beneath men, on average they earn less and wield less political power. In the "gender hierarchy," men are superordinates and women subordinates (Goode 1980).

Despite possessing traits of both genders, most of us feel either masculine or feminine; we usually do not doubt our gender (Heilbrun 1982). Unfortunately, when people believe that individuals should *not* have the attributes culturally identified or associated with the other gender, females suppress their **instrumental traits** (perceived as their "masculine side") and, to an even greater extent, males suppress their **expressive traits** (perceived as their "feminine side"). As a result, the range of possible human behaviors is further reduced and limited by expectations attached to gender roles. As psychologist Sandra Bem (1975) pointed out more than 30 years ago, "Our current system of sex role differentiation has long since outlived its usefulness, and . . . now serves only to prevent both men and women from developing as full and complete human beings."

When we initially meet a person, we unconsciously note whether the individual is male or female (a process called *gender attribution*) and respond accordingly (Skitka and Maslach 1990). But what happens if we cannot immediately classify a person as male or female? Many of us feel uncomfortable because we don't know how to act if we don't know the gender. This is true even if gender is irrelevant, as in a bank transaction, walking past someone on the street, or answering a query about the time. ("Was that a man or woman?" a person may ask in exasperation, although it really makes no difference.) An inability to tell a person's gender may provoke a hostile response. As Hilary Lips (1997) writes:

> It is unnerving to be unsure of the sex of the person on the other end of the conversation. The labels *female* and *male* carry powerful associations about what to expect from the person to whom they are applied. We use the information the labels provide to guide our behavior toward other people and to interpret their behavior toward us.

Our need to classify people as male or female and its significance is demonstrated in the well-known Baby X experiment (Condry and Condry 1976). In this experiment, three groups played with an infant known as Baby X. The first group was told that the baby was a girl, the second group was told that the baby was a boy, and the third group was not told what gender the baby was. The group that did not know what gender Baby X was felt extremely uncomfortable, but the group participants then made a decision based on whether the baby was "strong" or "soft." When the baby was labeled a boy, its fussing behavior was called "angry"; when the baby was labeled a girl, the same behavior was called "frustrated." Once the baby's gender was determined (whether correctly or not), a train of responses followed that could have profound consequences in his or her socialization. The study was replicated numerous times with the same general results. Even birth congratulations cards reflect gender stereotyping of newborns (Bridges 1993).

A review of studies on infant labeling found that gender stereotyping is strongest among children, adolescents, and college students (Stern and Karraker 1989). Stereotyping diminishes among adults, especially among infants' mothers (Vogel et al. 1991).

Gender and Sexual Orientation

Often we assume that the way an individual acts out his or her gender (gender display or presentation) is a sign of their **sexual orientation,** or the nature of someone's sexual preference, be it for partners of the same or opposite sex or both. In other words, we link characteristics of gender with assumptions of sexual preference. Although we often dichotomize sexual preference into a duality of homosexuality and heterosexuality, the universe of sexual orientation is more diverse and wide ranging (encompassing bisexuality and situational sexuality). We need to sever this almost automatic assumption. We assume that women who depart from the variety of behavioral norms associated with femininity and female roles must be lesbians; we assume that men who depart from masculinity and reject male roles ("feminine" men) must be gay men. Neither is true. Sexual preference cannot be "read" by demeanor or role behavior. Men who fit within norms of "masculine behavior" may be heterosexual, bisexual, or homosexual. Men whose behavior seems "feminine" by wider cultural standards may be gay, bisexual, or heterosexual. The same holds true for women.

On a second level, we often make connections between gender and sexual orientation by raising doubts and suspicions about the sexual orientation of those who depart from gender expectations. In this way,

gender norms are bolstered and reinforced. Men, especially, may monitor and restrict their behavior so as to avoid the disparaging and unwanted sort of question, "What are you anyway, a fag?" These potential doubts accomplish the feat of keeping people conforming to gender roles and expectations.

In various ways, gender transcends sexual orientation. There are similarities that exist between heterosexual and gay men (for example, in areas like acceptance of nonmonogamous relationships) because they are men (for example, men typically are more tolerant of and interested in infidelity than are women).

Gender and Gender Socialization

There are several prominent theories used to explain the significance of gender in our culture and how we learn what is expected of us. These include gender theory, social learning theory, and cognitive development theory.

Gender Theory

In studying gender, feminist scholars begin with two assumptions: (1) that male–female relationships are characterized by power issues and (2) that society is constructed in such a way that males dominate females. They argue that on every level, male–female relationships—whether personal, familial, or societal—reflect and encourage male dominance, putting females at a disadvantage. Male dominance is neither natural nor inevitable, however. Instead, it is created by social institutions, such as religious groups, government, and the family (Acker 1993; Ferree 1991). The question is: How is male–female inequality created?

Social Construction of Gender

In the 1980s, gender theory emerged as an important model explaining inequality. According to this theory, gender is a **social construct,** an idea or concept created by society through the use of social power. **Gender theory** asserts that society may be best understood by how it is organized according to gender and that social relationships are based on the *socially perceived* differences between females and males that are used to justify unequal power relationships (Scott 1986; White 1993). Imagine, for example, an infant crying in the night. In the mother–father parenting relationship, which parent gets up to take care of the baby? In most cases, the mother does because women are socially perceived to be nurturing and it's the woman's "responsibility" as mother (even if she hasn't slept in two nights and is employed full-time).

Gender theory focuses on (1) how specific behaviors (such as nurturing or aggression) or roles (such as childrearer, truck driver, or secretary) are defined as male or female; (2) how labor is divided into man's work and woman's work, both at home and in the workplace; and (3) how different institutions bestow advantages on men (such as male-only clergy in many religious denominations or women receiving less pay than men for the same work).

Central to the creation of gender inequality are the belief that men and women are fundamentally different and the fact that the differences between the genders—in personalities, abilities, skills, and traits—are unequally valued: reason and aggression (defined as male traits) are considered more valuable than sensitivity and compliance (defined as female traits). Making men and women appear to be opposite of and of unequal value requires the suppression of natural similarities by the use of social power. The exercise of social power might take the form of greater societal value being placed on looks than on achievement for women, of sexual harassment of women in the workplace or university, of patronizing attitudes toward women, and so on.

"Doing Gender"

Some gender scholars emphasize the situational nature of gender: how it is reproduced or constructed in everyday social encounters. They argue that more than what we *are*, gender is something that we *do* (West and Zimmerman 1987; Risman 1998). As Greer Fox and Velma Murry (2000) explain it, "men and women not only vary in their degree of masculinity or femininity but have to be constantly persuaded or reminded to be masculine and feminine. That is, men and women have to 'do' gender rather than 'be' a gender."

We "do gender" whenever we take into account the gendered expectations in social situations and act accordingly. We don't so much perform an internalized role as tailor our behaviors to convey our suitability as a woman or a man in the particular situation in

which we find ourselves (West and Zimmerman 1987). To fail to conform to the expectations for someone of our gender in a given situation exposes us to potential criticism, ridicule, or rejection as an incompetent or immoral man or woman (Risman 1998). But in living up to or within those social expectations, we help create and sustain the idea of gender difference. According to Michael Kimmel (2000, 104), "successfully being a man or a woman simply means convincing others that you are what you appear to be."

Although we see the social construction or "doing" of gender in all kinds of social settings, the family is a particularly gendered domain (Risman 1998). There are cultural expectations about how wage earning, housework, childcare, and sexual intimacy should be allocated and performed between women and men. Thus, much of the experience that people have in their families is understandable as both an exercise in and a consequence of how they and others "do gender."

Gender as Social Structure

Another key idea shared by many gender theorists is the notion that gender is a social structure that constrains behavior by the opportunities it offers or denies us (Risman 1987, 1998; Lorber 1994; Connell 1987). The consequences of the different opportunities afforded women and men can be seen at the *individual* level in the development of gendered selves, at the *interactional* level in the cultural expectations and situational meanings that shape how we "do gender," and at the *institutional* level in such things as sex-segregated jobs, a wage gap, and other economic and institutional realities that differentiate women's and men's experiences (Risman 1998). Although we may more often focus on individuals making choices that reflect their internalization of gender expectations, situations and institutions also shape behavior.

Gender Socialization through Social Learning Theory

Many theorists see gender like any other socially acquired role. They stress that we have to be socialized to act according to the expectations attached to our status as female or male. The emphasis on socialization has been considerable, although consensus on the process of socialization has not. In other words, there is considerable agreement that we undergo gender socialization, but there are different theories of how such socialization proceeds. **Social learning theory** is derived from behaviorist psychology and its emphasis on observable events and their consequences rather than internal feelings and drives. According to behaviorists, we learn attitudes and behaviors as a result of social interactions with others (hence, the term *social learning*).

The cornerstone of social learning theory is the belief that consequences control behavior. Acts regularly followed by a reward are likely to occur again; acts regularly punished are less likely to recur. Girls are rewarded for playing with dolls ("What a nice mommy!"), but boys are not ("What a sissy!").

This behaviorist approach has been modified recently to include **cognition**—that is, mental processes (such as evaluation and reflection) that intervene between stimulus and response. The cognitive processes involved in social learning include our ability to use language, anticipate consequences, and make observations. These cognitive processes are important in learning gender roles. By using language, we can tell our daughter that we like it when she does well in school and that we don't like it when she hits someone. A person's ability to anticipate consequences affects behavior. A boy does not need to wear lace stockings in public to know that such dressing will lead to negative consequences. Finally, children observe what others do. A girl may learn that she "shouldn't" play video games by seeing that the players in video arcades are mostly boys.

We also learn gender roles by imitation, according to social learning theory. Learning through imitation is called **modeling.** Most of us are not even aware of the many subtle behaviors that make up gender roles—the ways in which men and women use different mannerisms and gestures, speak differently, and so on. We don't "teach" these behaviors by reinforcement. Children tend to model friendly, warm, and nurturing adults; they also tend to imitate adults who are powerful in their eyes—that is, adults who control access to food, toys, or privileges. Initially, the most powerful models that children have are their parents. Reflecting on your own family, you might examine the division of labor in your household. How is housework divided? How is unpaid household work valued in comparison with employment in the workplace?

As children grow older and their social world expands, so do the number of people who may act as their role models: siblings, friends, teachers, media figures, and so on. Children sift through the various demands and expectations associated with the different models to create their unique selves.

■ *Playing "dress up" is one way children model the characteristics and behaviors of adults. It is part of the process of learning what is appropriate for someone of their gender.*

Cognitive Development Theory

In contrast to social learning theory, **cognitive development theory** focuses on the child's active interpretation of the messages he or she receives from the environment. Whereas social learning theory assumes that children and adults learn in fundamentally the same way, cognitive development theory stresses that we learn differently, depending on our age. Swiss psychologist Jean Piaget (1896–1980) showed that children's abilities to reason and understand change as they grow older.

Lawrence Kohlberg (1969) took Piaget's findings and applied them to how children assimilate gender-role information at different ages. At age 2, children can correctly identify themselves and others as boys or girls, but they tend to base this identification on superficial features, such as hair and clothing. Girls have long hair and wear dresses; boys have short hair and never wear dresses. Some children even believe they can change their sex by changing their clothes or hair length. They don't identify sex in terms of genitalia, as older children and adults do. No amount of reinforcement will alter their views because their ideas are limited by their developmental stage.

When children are 6 or 7 years old and capable of grasping the idea that basic characteristics do not change, they begin to understand that gender is permanent. A woman can be a woman even if she has short hair and wears pants. Oddly enough, although children can understand the permanence of sex, they tend to insist on rigid adherence to gender-role stereotypes. Even though boys can play with dolls, children of both sexes believe they shouldn't because "dolls are for girls." Researchers speculate that children exaggerate gender roles to make the roles "cognitively clear."

According to social learning theory, children learn appropriate gender-role behavior through reinforcement and modeling. But according to cognitive development theory, once children learn that gender is permanent, they independently strive to act like "proper" girls or boys. They do this on their own because of an internal need for congruence, the agreement between what they know and how they act. Also, children find that performing the appropriate gender-role activities is rewarding. Models and reinforcement help show them how well they are doing, but the primary motivation is internal.

How Family Matters: Learning Gender Roles

Although biological factors, such as hormones, clearly are involved in the development of male and female differences, the extent of biological influences is not well understood. Moreover, it is difficult to analyze the relationship between biology and behavior because learning begins at birth. In this section, we explore gender-role learning from infancy through adulthood, emphasizing the influence of our families in the construction of our ideas about gender.

Childhood and Adolescence

In our culture, infant girls are usually held more gently and treated more tenderly than boys, who are ordinarily subjected to rougher forms of play. As early as the first day after birth, parents tend to describe their daughters as soft, fine featured, and small and their sons as hard, large featured, big, and attentive. Fathers tend to stereotype their sons more extremely than mothers do (Fagot and Leinbach 1987). Although it is impossible for strangers to know the gender of a diapered baby, once they learn the baby's gender, they respond accordingly. Such gender-role socialization occurs throughout our lives. By middle childhood, although conforming to gender-role behavior and attitudes becomes increasingly important, there is still considerable flexibility (Absi-Semaan, Crombie, and Freeman 1993). It is not until late childhood and adolescence that conformity becomes most characteristic. The primary agents forming our gender roles are parents. Eventually, teachers, peers, and the media also play important roles.

Parents as Socialization Agents

During infancy and early childhood, a child's most important source of learning is the primary caretaker—often both parents, but also often just the mother, father, grandmother, or someone else. Most parents may not be aware of how much their words and actions contribute to their children's gender-role socialization (Culp et al. 1983). Nor are they aware that they treat their sons and daughters differently because of their gender. Although parents may recognize that they respond differently to sons than to daughters, they usually have a ready explanation—the "natural" differences in the temperament and behavior of girls and boys. Parents may also believe that they adjust their responses to each particular child's personality. In an everyday living situation that involves changing diapers, feeding babies, stopping fights, and providing entertainment, it may be difficult for harassed parents to recognize that their own actions may be largely responsible for the differences they attribute to nature.

The role of nature cannot be ignored completely, however. Temperamental characteristics may be present at birth. Also, many parents who have conscientiously tried to raise their children in a nonsexist way have been frustrated to find their toddler sons shooting each other with carrots or their daughters primping in front of the mirror. Indeed, it is increasingly likely that some gender differences are influenced by hormones and/or chromosomes. At the same time, it is undeniable that children are socialized differently based on their gender.

Childhood gender socialization occurs in many ways. Children's literature, for example, typically depicts girls as passive and dependent, whereas boys are instrumental and assertive (Kortenhaus and Demarest 1993). In the more than 4,000 children's books published annually, females are rarely portrayed as brave or independent and are typically presented in supporting roles (Renzetti and Curran 2003). Children's toys and clothing also reinforce gender differences. In general, children are socialized by their parents through four subtle processes: manipulation, channeling, verbal appellation, and activity exposure (Oakley 1985):

- *Manipulation.* From infancy onward parents treat daughters more gently (telling them how beautiful they are, advising them that nice girls do not fight,

■ *Generally, daughters are given more responsibilities than are sons.*

Exploring Diversity The Work Daughters Do to Help Families Survive

Listen as two of the teenage girls Lisa Dodson and Jillian Dickert studied describe their contributions to their families.

> I have to take care of the house and take care of the kids and I don't go outside. I have to stay home. They have to work and so I take over.
>
> 15-year-old Ella
>
> I have to clean up the kitchen in the morning before school and then do whatever shopping or whatever on the way home. I cook for the kids (younger sister and cousin) before I start my (home)work.
>
> 16-year-old Anita

Ella and Anita carry heavy family responsibilities. Think back to your own childhood and adolescence. Because of a tendency to focus either on middle-class families or on younger children, the importance of children's contributions to household labor has been minimized and misunderstood (Gager, Cooney, and Call

1999). In many families, however, especially low-income or single-parent families, the contributions made by the children, particularly daughters, become part of a "survival strategy" without which their families would suffer greatly (Dodson and Dickert 2004).

Although both sons and daughters often contribute labor to the household, what they do, how much they do, and the consequences of their labor—both for themselves and for their families—greatly differ (Gager, Cooney, and Call 1999). Using data on 825 high school students who were part of the larger Youth Development Study, researchers Constance Gager, Teresa Cooney, and Kathleen Thiede Call compared the household labor of sons and daughters when they were in ninth and later twelfth grade. Among their findings were the following:

- As ninth graders, boys spent only 87% as much time as girls in housework. By twelfth grade, boys spent only 68% as much time as girls. They also differed in what tasks they were involved in.

- In ninth grade, girls averaged more than 2 hours per week in household tasks beyond the time boys spent on average (17 hours for girls to nearly 15 hours for boys). As twelfth graders, the gap had practically doubled (13 hours to 9 hours).

- Boys spent more time than girls on "male tasks." However, fewer household tasks are predominantly male. Such tasks—doing yard work, shoveling snow, and taking out the trash—tend to be less repetitive than stereotypical female tasks. These were the only tasks boys reported doing more often than girls. Female tasks included cooking, setting the table, washing dishes, doing laundry, cleaning, shopping for groceries, and caring for other family members.

- Twelfth graders living with single parents devoted the most time to housework—3 hours more per week than children living with both biological parents.

- On top of doing greater amounts of housework, girls devoted more time to homework, paid work, and volunteering than boys did, result-

and so on) and sons more roughly (telling them how strong they are, advising them not to cry, and so on). Eventually, children incorporate such views as integral parts of their personalities. Differences in girls' and boys' behaviors may result from parents expecting their children to behave differently (Connors 1996, cited in Renzetti and Curran 2003).

- *Channeling.* Children are directed toward specific objects and activities and away from others. Toys, for example, are differentiated by gender and are marketed with gender themes, as can be seen in toy ads and displays in retail stores. Parents purchase different toys for their daughters and sons, who—influenced by advertising, the reinforcement by their parents, and the enthusiasm of their peers—

are attracted to gendered toys (Renzetti and Curran 2003).

- *Verbal appellation.* Parents use different words with boys and with girls to describe the same behavior. A boy who pushes others may be described as "active," whereas a girl who does the same may be called "aggressive."

- *Activity exposure.* Both genders are usually exposed to feminine activities early in life, but boys are discouraged from imitating their mothers, whereas girls are encouraged to be "mother's little helpers." Chores are categorized by gender (Gager, Cooney, and Call 1999; Dodson and Dickert 2004). Boys' domestic chores take them outside the house, whereas girls' chores keep them in it—another rehearsal for traditional adult life.

ing in an adolescent version of a leisure gap between the genders.

- Summing up their findings, Gager, Cooney, and Call report, "when we consider all household tasks, teenage girls are more likely to pick up the slack when the need arises."

Daughters' contributions to their households become even more evident in the research reported by Dodson and Dickert (2004). Girls like Ella and Anita are not merely helping out; they are indispensable ingredients in their families' survival. Dodson and Dickert specifically note the ways teenage daughters in low-income families take responsibility for household tasks, including caring for younger siblings, freeing their frequently single and employed mothers from either additional and burdensome childcare costs or reduced income (from having to miss work or cut back hours). Driven by economic necessity, low-income parents, especially single mothers, are pushed to depend on their daughters to do what they, themselves, are unavailable to do. This includes caregiving and domestic work. In caring for younger siblings, girls may feed and wash them, help them with schoolwork, monitor their activities, and put them to bed. Household chores might include cooking, cleaning, laundry, shopping, and even household maintenance. In short, daughters do what mothers are unable to do, either because of employment-induced absence (entering the labor force, working increased hours, or commuting greater distances) or familial circumstance (birth, adoption, maternal illness, or illness or death of a former childcare provider) (Dodson and Dickert 2004). Though essential, such contributions may carry great costs for the daughters. As Dodson and Dickert (2004, 326) put it bluntly, daughters "lose the opportunity to focus on their own young lives."

The opportunity costs that daughters suffer include sacrifices they make in their own educations so as to care for younger siblings or meet the familial needs they are asked to satisfy. Middle-schooler Davida is chronically late for school because she has to drop off her baby sister at day care before going to school herself. As described by a teacher, "She never says why, she just takes the punishment . . . she doesn't want to tell." Instead, she lives with the reputation of a careless, uninterested student (Dodson and Dickert 2004).

In keeping their families going, caring for siblings, or doing significant amounts of housework, there was often little time left to devote to schoolwork or to guarantee punctual and consistent attendance. After-school extracurricular activities such as homework clubs, sports programs, and theater and arts programs were luxuries that their lifestyles did not allow them.

Sons, too, may help, but daughters are perceived as more responsible and more "naturally inclined" to provide effective care for home and siblings. To Dodson and Dickert, the combination of educational inattentiveness and extracurricular uninvolvement that results leaves such young women less able to develop talents and abilities, discover interests, and build the confidence and competence they might need to find a way to improve the economic position from which they start. Instead, daughters of low-income families may become low-income mothers themselves (Dodson and Dickert 2004).

Although it is generally accepted that parents socialize their children differently according to gender there are differences between fathers and mothers. Fathers pressure their children more to behave in gender-appropriate ways. Fathers set higher standards of achievement for their sons than for their daughters, play more interactive games with their sons, and encourage them to explore their environments (Renzetti and Curran 2003). Fathers emphasize the interpersonal aspects of their relationships with their daughters and encourage closer parent–child proximity. Mothers also reinforce the interpersonal aspect of their parent–daughter relationships (Block 1983). They typically engage in more "emotion talk" with their daughters than with their sons, and—unsurprisingly—as early as first-grade girls are more adept at monitoring emotion and social behavior (Renzetti and Curran 2003).

Both parents of teenagers and the teenagers themselves believe that parents treat boys and girls differently. It is not clear, however, whether parents are reacting to existing differences or creating them (Fagot and Leinbach 1987). It is probably both, although by that age, gender differences are fairly well established in the minds of adolescents.

Various studies have indicated that ethnicity and social class are important in socialization (Renzetti and Curran 2003; Zinn 1990; see Wilkinson, Chow, and Zinn 1992 for scholarship on the intersection of ethnicity, class, and gender). Among Caucasians, working-class families tend to differentiate more sharply than middle-class families between boys and girls in terms of appropriate behavior; they tend to place more restrictions on girls. African American families tend to socialize their children toward more egalitarian gender roles (Taylor 1994c). There is evidence that African American families socialize their daughters to be more independent than Caucasian families do. Indeed, among African Americans, the "traditional" female role model may never have existed. The African American female role model in which the woman is both wage earner and homemaker is more typical and more accurately reflects the African American experience (Lips 1997).

Other Sources of Socialization

Although primary, both in importance and in exposure, families are not the only influences on the ideas we acquire about gender. Our early lives are lived in the company of many others who shape our ideas about men and women, femininity and masculinity. As children grow even just a little older, their social world expands and so do their sources of learning.

SCHOOL. Around the time children enter day care centers or kindergarten, teachers (and peers, discussed next) become important influences. Day care centers, nursery schools, and kindergartens are often a child's first experience in the wider world outside the family. Teachers become important role models for their students. Because most day care, nursery school, kindergarten, and elementary schoolteachers are women, children tend to think of child–adult interactions as primarily taking place with women. Teachers also monitor children's behavior, reinforcing gender differences along the way.

A decade or so ago we could paint the following picture of gendered school experience. Classroom observations documented that boys were louder, more demanding, and received a disproportionate amount of the teacher's attention. Teachers called on boys more often, were more patient with boys in their explanations, and more generous toward them with their praise. Girls, praised for their appearance and the neatness of their work more than its substance or quality, grew more tentative and hesitant as they approached and entered middle school. By high school, they suffered drops in their self-esteem and self-confidence, prefacing their answers with disclaimers: "I'm probably wrong, but . . ." or "I'm not sure, but . . ." Intelligent girls often found that they were devalued by boys. Only in all-girl schools, argued Myra and

David Sadker (1994), did female students assert themselves vigorously in class. The Sadkers believed that girls benefited from gender-segregated schools and classes by not having to compete with boys for the teacher's attention, not becoming overly concerned with their appearance, and not having to fear that their intelligence would make them undesirable as dates. The picture in coeducational settings was bleaker; coed schools had "failed at fairness," and girls suffered the harsher consequences (Sadker and Sadker 1994).

Fast forward to 2006. From kindergarten through high school, we are increasingly finding that it is boys, not girls, whose performance lags. Girls generally excel over boys in all areas during grade school. They have less difficulty learning to read, learn to read earlier, are more likely to recognize words by sight by the second half of first grade, score higher on fourth-grade standardized reading and writing tests, and are less likely to be diagnosed with learning or speech problems or to repeat a grade. Boys are twice as likely to be diagnosed with learning disabilities or be placed in special-education classes (Tyre 2006). In middle school, girls score higher than boys on eighth-grade standardized reading and writing tests. In high school, girls take more advanced placement or honors biology classes, are more likely to plan on attending college, and are less likely to drop out. Twelfth-grade girls score, on average, 16 points higher on standardized reading tests and 24 points higher on standardized writing tests than twelfth-grade boys. Unsurprisingly, between 1980 and 2001 the number of boys who say they dislike school increased by nearly 75% (Tyre 2006).

Girls have long performed better than boys on standardized tests of verbal or writing ability but tended to lag, sometimes far behind, in math and science. More recent examination of math and science scores shows that the differences have greatly diminished. The Third International Mathematics and Science Study, one of the largest international comparisons of academic performance, examined math and science performance across 21 countries: Australia, Austria, Canada, Cyprus, the Czech Republic, Denmark, France, Germany, Hungary, Iceland, Italy, Lithuania, the Netherlands, New Zealand, Norway, the Russian Federation, Slovenia, South Africa, Sweden, Switzerland, and the United States. The United States was one of only three countries in which there was no significant gender difference in math scores. Although there was a gap between male and female science scores (in which males performed better than females), the U.S. gap was smaller than that for 19 of the other 20 countries in the comparison. However, in physics and advanced math, U.S. male twelfth graders outperformed females, as was also true in most other countries.

As a result of the variety of trends noted here, increased attention and concern are being directed at what boys experience in school, why, and with what consequences (Pollack 1998; Sadker and Sadker 1994). For example, although boys have long commanded more teacher time and classroom attention than girls, the attention boys receive is not always positive—they are subject to more discipline and receive more of the teacher's anger than do girls, even when the disruptiveness of their behavior is similar. Furthermore, their academic performance often suffers, as indicated by their rates of failing, acting up, and/or dropping out (Sadker and Sadker 1994; Renzetti and Curran 1995; Pollack 1998). As school curricula become more rigid, more focused on assessment and demonstrating proficiency, teachers have less leeway to teach to the student's strengths or needs and less tolerance for the typically boy style of learning—disorganized, distracted, high energy, and potentially disruptive. Boys are also often unwilling to seek help and admit weakness. Much as the earlier call for all-girl schools was seen by some as a remedy for girl's school problems, it is now being embraced by some as a solution for what ails boys (Tyre 2006).

Gender doesn't operate alone in shaping school experiences. Race and class matter, too. In schools, black males face especially difficult circumstances and receive the most unfavorable teacher treatment when compared with white males, white females, or black females (Sadker and Sadker 1994; Basow 1992). They receive the most recommendations for special education and are subjected to low expectations by teachers. Teachers describe black males as having the worst work habits, and they predict lower levels of academic success for them, *regardless of their actual behavior* (Basow 1992).

PEERS. A child's age-mates, or **peers,** become especially important when the child enters school. By granting or withholding approval, friends and playmates have great influence on us. They may affect what games we play, what we wear, what music we listen to, what television programs we watch, and even what cereal we eat or beverage we drink. Peer influence is so pervasive that it is hardly an exaggeration to say that in some cases children's peers tell them what to think, feel, and do.

Peers also provide standards for gender-role behavior in several ways (Carter 1987), such as through the play activities they engage in, the toys with which they play, and the approval or disapproval they display, verbally or nonverbally, toward others' behavior. Children's perceptions of their friends' gender-role attitudes, behaviors, and beliefs encourage them to adopt similar ones so that they are accepted. If a girl's female friends play soccer, she is more likely to play soccer. If a boy's male friends display feelings, he is more likely to display feelings.

During adolescence, peers continue to have a strong influence, one that often leaves parents feeling helpless and as though their importance has been reduced in guiding or shaping their sons and daughters. But research indicates that parents can be more influential than peers (Gecas and Seff 1991). Parents influence their adolescent's behavior primarily by establishing norms, whereas peers influence others through modeling behavior. Even though parents tend to fear the worst from their children's peers, peers provide important positive influences. It is within their peer groups, for example, that adolescents learn to develop intimate relationships (Gecas and Seff 1991). Also, adolescents tend to be more egalitarian in gender roles than parents do, especially fathers (Thornton 1989).

POPULAR CULTURE AND MASS MEDIA. In all its forms, the mass media depict females and males quite differently. We can safely assert that the media typically have "ignored, trivialized, or condemned women," a process known as *symbolic annihilation* (Renzetti and Curran 2003).

Much of television programming promotes or condones negative stereotypes about gender, ethnicity, age, and gay men and lesbians. Women are significantly underrepresented on television (Media Report to Women 1999; Signorielli 1997). Through the 1970s, men outnumbered women on prime-time television three to one. Even on *Sesame Street*, 84% of the characters were male in 1992, compared with 76% 5 years earlier ("Muppet Gender Gap" 1993). Recent data reveals that nearly two-thirds of all prime-time television characters are male (65% versus 35% female), including 59% of the characters featured in programs' opening credits, an indication of characters of importance. Consistently, since 1999, female characters have been outnumbered by almost 2:1 (Children Now 2004).

The women depicted on television represent women less than the men depicted represent men. A 2003 study of gender and age of characters revealed that female characters continue to be younger than male characters. The largest percentage of female characters was in the 20- to 29-year-old range, and the largest age range among male characters was 30 to 39. Furthermore, males are twice as likely as females (16% to 8%) to be in their 50s and 60s (Children Now 2004). Almost half of female characters are "thin and attractive"; only 16% of men are "thin or very thin" (Renzetti and Curran 2003; Signiorelli 1997). Television women are portrayed as emotional and needing emotional support; they are also sympathetic and nurturing. Not surprisingly, women are often portrayed as wives, mothers, or sex objects (Vande Berg and Streckfuss 1992). Occupationally, although both male and female characters are displayed in a range of both high- and low-status jobs, looking at characters by their jobs reveals that attorneys, physicians, executives, and elected or appointed officials are usually male characters, whereas two-thirds or more of characters who are domestic workers, clerical workers, and nurses are female (Children Now 2004).

On television, male characters are shown as more aggressive and constructive than female characters. They solve problems and rescue others from danger. Only more recent prime-time series have portrayed males in emotional, nurturing roles. Still, 100% of characters who are full-time homemakers are female (Children Now 2004). Although things have improved, ethnic and sexual stereotypes continue to be commonly found in television.

Gender Development in Adulthood

Although more attention has been directed at early experiences and socialization in childhood and adolescence, gender development doesn't stop there. Many life experiences that we have in adulthood alter our ideas about and actions as males and females. Again, families loom large in reshaping our gendered ideas and behaviors. From a 1970s perspective known as *role transcendence,* an individual goes through three stages in developing his or her gender-role identity: (1) undifferentiated stage, (2) polarized stage, and (3) transcendent stage (Hefner, Rebecca, and Oleshansky 1975).

Young children have not clearly differentiated their activities into those considered appropriate for males or females. As children enter school, however, they begin to identify behaviors as masculine or feminine. They tend to polarize masculinity and femininity as they test the appropriate roles for themselves. As they enter young adulthood, they slowly begin to shed the

rigid male–female polarization as they are confronted with the realities of relationships. As they mature and grow older, men and women transcend traditional masculinity and femininity. They combine masculinity and femininity into a more complex role.

More recent research details some of the ways adult life experiences can transform how we act as a male or female (Gerson 1985, 1993; Risman 1986, 1987, 1988, 1998). These more structural analyses have shown how adult life experiences both inside and outside of families have the potential to restructure our identity, redefine our role responsibilities, and take us in directions quite different from those suggested by our early gender socialization. In adulthood, new or different sources of gender-role learning may include marriage and parenthood, as well as college and experiences in the workplace.

COLLEGE. Within the past 30 years, the undergraduate student population has shifted from being 58% male to 56% female (Tyre 2006). Unlike high school, in the college setting, many young adults learn to think critically, to exchange ideas, and to discover the bases for their actions. There, many young adults first encounter alternatives to traditional gender roles, either in their personal relationships or in their courses. A longitudinal study of gender roles found that traditional and egalitarian gender-role attitudes affected dating relationships in college but had little effect on later life (Peplau, Hill, and Rubin 1993).

MARRIAGE. Marriage is an important source of gender-role learning because it creates the roles of husband and wife. For many individuals, no one is more important than a partner in shaping gender-role behaviors through interaction. Our partners have expectations of how we should act as a husband or wife, and these expectations are important in shaping behavior.

Husbands tend to believe in innate gender roles more than wives do. This should not be especially surprising, because men tend to be more traditional and less egalitarian about gender roles. Husbands stand to gain more in marriage by believing that women are "naturally" better at cooking, cleaning, shopping, and caring for children.

PARENTHOOD. For most men and women, motherhood alters life more significantly and visibly than fatherhood does. For some men, fatherhood may mean little more than providing for their children. It is unlikely to find many who would associate motherhood only with providing. As parents, mothers do more, are expected to do more, and are expected to juggle this "more" with their paid employment. As a consequence, fatherhood does not typically create the same degree of work–family conflict that motherhood does. Fathers who strive to be a fully or nearly equal co-parent will, however, discover the ways in which the demands of parenthood clash with demands of the workplace. Whereas traditionally a man's work role allowed him to fulfill much of his perceived parental obligation, we now expect more out of fathers.

Not only have our expectations shifted toward more nurturing versions of fatherhood, but where traditional fatherhood was tied to marriage, today a third of all current births occur outside of marriage and nearly half of all current marriages end in divorce. What, then, is the father's role for a man who is not married to his child's mother or who is divorced and does not have custody? What are his role obligations as a single father as distinguished from those of married fathers? For many men, the answers are painfully unclear, as evidenced by the low rates of contact between unmarried or divorced fathers and their children.

Women today have somewhat greater latitude as wives. It is now both accepted and expected that women will work outside the home at least until they become mothers and more than likely that they will continue or return to paid employment sometime after they have children. Even with increases in the numbers of women who remain childless, women may be expected to become mothers and be subjected to social pressure toward motherhood. Once children are born, roles tend to become more traditional, even in previously nontraditional marriages. Often, the wife remains at home, at least for a time, and the husband continues full-time work outside the home. The woman must then balance her roles as wife and mother against her needs and those of her family.

THE WORKPLACE. It is well established that men and women are psychologically affected by their occupations (Menaghan and Parcel 1991; Schooler 1987). Work that encourages self-direction, for example, makes people more active, flexible, open, and democratic; restrictive jobs tend to lower self-esteem and make people more rigid and less tolerant. If we accept that sex-segregated female occupations are often of lower status with little room for self-direction, we can understand why some women are not as achievement oriented as men. With different opportunities for promotion, men and women may express different attitudes toward achievement. Women may downplay their desire for promotion, suggesting that promotions would interfere with their family responsibilities. But this really may be related to a need to protect themselves from frustration because many women are in jobs where promotion to management positions is unlikely.

Household work affects women psychologically in many of the same ways that paid work affects them in female-dominated occupations, such as clerical and service jobs (Schooler 1987). Women in both situations feel greater levels of frustration because of the repetitive nature of the work, time pressures, and being held responsible for things outside their control. Such circumstances do not encourage self-esteem, creativity, or a desire to achieve.

Remaking Women and Men

Focusing on adulthood is important because it reveals the gaps that often exist between earlier gender socialization and adult experiences. The lives we lead are often different from those we were raised to lead or expected to lead (Gerson 1993; Risman 1987, 1998). To some scholars, this diminishes the importance of socialization and discredits theories that deterministically link early socialization to later life outcomes (Gerson 1993). In some ways, those theories are no better than *biological determinism,* in which we are limited to those behaviors that our genetic or hormonal characteristics allow. They simply substitute socialization for biology (Risman 1989).

Socialization is important, especially in affecting our expectations and offering us role models for lives we might live. But life is more circuitous than linear. Unanticipated twists and turns often take us in directions we neither expected nor intended. Research on women's and men's career and family experiences bear this out. For example, Kathleen Gerson's research on women's and men's career and family choices reveals that many people develop commitments to either careers or parenting that stem from their experiences in jobs and relationships (Gerson 1985, 1993). Some women and men who anticipate "traditional" adult outcomes move in nontraditional directions based on

Popular Culture

Video Gender: Gender, Music Videos, and Video Games

Over the past 25 years, recreation and entertainment, especially for young people, increasingly encompass video images and technologies. Popular music was revolutionized by the "invention" of the music video and by the inception of MTV, which premiered in 1981. An estimated 350 million households worldwide tune in to MTV, and three-fourths of all 12–19 year-old females and males watch MTV regularly, averaging more than 6 hours a week (6.2 for females, 6.6 for males) (National Institute on Media and the Family 2001).

Meanwhile, the video game industry has revolutionized "play" for millions of young Americans, especially males. Billions of dollars and countless hours have been spent on arcade or home video games (Dietz 1998). Since his first appearance in 1981 Super Mario has become a fixture in millions of households. Mario is the central figure in the various *Super Mario Brothers* games, which sold a combined 184 million copies between 1983 and 2005. Together these media have also altered the experience of gender socialization.

Video Games

Although the average age of video game players is now 29 (Gentile and Gentile 2005), concern is perhaps greatest regarding the quantity and quality of exposure of younger populations. Research indicates that the average 2- to 17-year-old in the United States plays video games for 7 hours a week (Gentile and Walsh 2002). Such a figure is a bit misleading, however, because it masks the sizable differences between the genders. For example, Douglas Gentile and colleagues (2004) found that, in their study of video games and aggressive behavior, the average time males spent playing video games was more than 2.5 times the average for females (13 hours a week compared to 5 hours for females). Elementary and middle school–age girls play an average of 5.5 hours compared to 13 hours for their male peers. Children 2 to 7 years old play an average of 43 minutes a day. Other research, looking exclusively at the youngest children (2 to 5 years old), reports that they play an average of 28 minutes daily (see Gentile and Gentile 2005; Gentile and Anderson, 2006). Douglas Gentile and J. Ronald Gentile (2005) found that 15% of their eighth- and ninth-grade subjects and 5% of their college-age subjects could be classified as "addicted" to video games; 86% of the addicted adolescents were males.

Aggression and violence are major components of many games. Content analyses document that 89% of video games feature some violent content, and half the time serious violence is directed at other game characters (Children Now 2001), in which such characters suffer serious injury or die (Gentile and Gentile 2005). Male characters mostly are the perpetrators of video game violence, and their targets are generally other male characters or some nonhuman characters (such as monsters, aliens, creatures, or animals). Occasionally (20% in one study of a sample of 33 Nintendo or Sega games), violence is directed at female characters, although that is not typical (Dietz 1998). Overall, violent themes and aggressive action are commonplace.

In most video games, when females are present, they are most often either victims ("damsels in distress") or sex objects ("visions of beauty with large breasts and thin hips"); they are rarely heroes or action characters. In the typical video game, females are absent (Dietz 1998). In the earliest story lines of some of the most popular series, the main male characters are trying to save a female. Later "chapters" alter this plot but keep intact the male hero trying to save his village, city, or world. The *Zelda* and *Mario* series are tame, however, in comparison to a more disturbing trend in games geared for teens and older players.

In their 2002 annual "video game report card," the National Institute on Media and the Family, describe a "growing tendency" to portray females as targets or recipients of "graphic violence" in some of the best-selling and most popular games. For a particularly disturbing example,

---

the levels of fulfillment and opportunity at work, the experiences and aspirations of their partners, and their experiences with children. Similarly, men and women who aspire to nontraditional outcomes (career attachment for women, involved fatherhood for men)

may "reluctantly" abandon those directions as a result of firsthand experiences at home and work.

Barbara Risman's research on single custodial fathers pointed to similar adult development. Men who reluctantly found themselves as lone, custodial parents

in *Grand Theft Auto: Vice City*, one of the most popular games, players are rewarded for kicking a prostitute to death (http://www.mediafamily.org).

Music Videos

With innovations such as the iPod revolutionizing the experience of listening to music, listening continues to be important in our lives—adolescents, for example, listen to about 10,500 hours of it between seventh and twelfth grade. Still, for 25 years now, music has come to be "seen," as well as heard. Visual images are as important as the music and the lyrics; indeed, the images may even be more important than the music.

Studies of gender stereotypes and ratios of males and females portrayed in music videos have consistently found males featured more extensively and portrayed more widely than females (Seidman 1999). Of the "characters" featured (including performers, dancers, and any characters in more storytelling videos), 63% were male. Steven Seidman (1999) further reported that in examining characters in MTV videos, there was much gender stereotyping: more than 90% of the occupational roles that we would typically classify as male (for example, manual laborer, physician, and mechanic) were portrayed by male actors and 100% of the stereotypically female occupational roles (for example, secretary, librarian, and phone opera-

tor) featured females. Some specific, and striking, distributions: more than 90% of manual laborers, police personnel, photographers, and soldiers and all stage hands, criminals, and politicians were males. Females made up 85% of the dancers and, as characters, all domestic cleaners, fashion models, and prostitutes (Seidman 1999).

Although there is considerable verbal or physical aggressiveness in music and video games (Kalis and Neuendorf 1989), there is a lesser level of violence in music videos than in video games. A study of four major music television networks found a range of violent videos from 11% to 22%. In one content analysis study of 391 acts of music video violence, males were the aggressors in 78% of the incidents. Females were victims or targets 46% of the time (Rich et al. 1998).

In music videos, female aggression is often provoked by jealousy. Male aggression is often unprovoked. Aggression is often a part of male swagger—the assertion of power and status—especially in heavy metal and rap videos. Critic James Twitchell (1992) contends that music videos "are rife with adolescent misogyny, homophobia, and threats of violence. They are rude, bawdy, boastful, with a kind of 'in your face' aggression . . . characteristic of insecure masculinity."

Most music videos are dominated by male singers or male groups, and women may be present mostly to provide erotic backdrop or vocal backup (Seidman 1992; Sommers-Flanagan, Sommers-Flanagan, and Davis 1993). Often, women are depicted as sex objects, pictured condescendingly, are provocatively dressed, or all of these. One study found that adolescent or male viewers generally rated music videos, especially sexually provocative ones, more positively than did older or female viewers (Greeson 1991). Another study found that both male and female undergraduates responded with positive emotions to music videos with sexual content; they responded negatively to those with violence. The music videos declined in appeal when sex and violence were combined (Hansen and Hansen 1990).

Cumulatively, video games and music videos become part of the gender socialization process. Their themes—male as aggressive and violent, females as sex objects and victims—fit, both with each other and with other popular media content (such as television, film, cartoons, and advertising). Clearly, no single game or video will determine a person's attitudes toward women or propensity toward violence. Collectively, however, such images help shape and reinforce traditional gender attitudes and make aggressive outcomes more likely.

developed nurturing abilities that their socialization had not included. More important than how they were raised was how they interacted with their children, as well as the lack of a female in their lives to whom caregiving tasks could be assigned. Thus, these single fa-

thers "mothered" their children in ways that were more like women's relationships with children than what we would have predicted (Risman 1986). Importantly, socialization contributes to but neither guarantees us nor restricts us to any particular family outcome.

Gender Matters in Family Experiences

Within the past generation, there has been a significant shift from traditional toward more egalitarian gender roles (Brewster and Padavic 2000). Women have changed more than men, but men are changing. These changes seem to affect all classes, although not to the same extent. Also, there is still resistance to change as those from more conservative religious groups, such as Mormons, Catholics, and fundamentalist and evangelical Protestants, continue to adhere more strongly to traditional roles (Jensen and Jensen 1993).

Contemporary gender-role attitudes have changed partly as a response to the steady increase in women's participation in the labor force. Although this increase was especially evident in the 1970s and 1980s, as was the move toward more egalitarian attitudes, it continued through the 1990s. College-educated women and men, especially, are considerably less likely to hold traditional ideas about gender, work, and family roles (Brewster and Padavic 2000).

Within the family, although attitudes toward gender roles have become more liberal, in practice, gender roles continue to place women at a disadvantage, especially by making them responsible for housekeeping and childcare activities (Atkinson 1987; Coltrane 2000; Hochschild 1989). Some of the most important changes affecting men's and women's roles in the family are briefly described in the following sections.

Men's Roles in Families and Work

In traditional gender-role stereotypes, many of the traits ascribed to one gender are not ascribed to the other. Theoretically, men's instrumental traits complemented women's expressive ones, much as (Hort, Fagot, and Leinbach 1990) women's and men's traditional family roles complemented each other.

Central features of the traditional male role, whether among Caucasians, African Americans, Latinos, or Asian Americans, include dominance, work, and family. Males are generally regarded as being more power oriented than females. Statistically, men demonstrate higher degrees of aggression, especially violent aggression (such as assault, homicide, and rape); seek to dominate and lead; and show greater competitiveness. Although aggressive traits are thought to be useful in the corporate world, politics, and the military, such characteristics are rarely helpful to a man in fulfilling marital and family roles requiring understanding, cooperation, communication, and nurturing.

Traditionally, across ethnic and racial lines, male roles have centered on providing, and the centrality of men's work identity affected their family roles as husbands and fathers. Men's identity as providers take precedence over all other family functions, such as nurturing and caring for children, doing housework, preparing meals, and being intimate. Because of this focus, traditional men may become confused by their spouses' expectations of intimacy; they believe that they are good husbands simply because they are good providers (Rubin 1983). When circumstances render them unable to provide, the blow to their self-identities can be quite powerful (Rubin 1994).

The somewhat traditional gender rhetoric of the 1990s Million Man March on Washington, D.C., by African American men was not that far from the more explicitly traditional rhetoric espoused by the Christian Promise Keepers. Both groups implored men to live up to their responsibilities to their families and communities, and central to the familial responsibilities was to lead and provide.

However, because race, ethnicity, and economic status often overlap, certain categories of men face more difficulty meeting the expectations of the traditional provider role. Because African Americans and Latinos often fare less well economically, men often are left unable to lay claim to the household status and power that traditional masculine roles promise.

Occasionally, characterizations of Latino families have exaggerated the extent of male dominance, as suggested by the notion of *machismo*. Although such a notion may have been somewhat more accurate in depicting gender ideologies of rural Mexico and the Caribbean in the first half of the twentieth century, it is inaccurately applied to contemporary Latino families (McLoyd et al. 2000b). Both African American and Chicano men have more positive attitudes toward employed wives. Ethnic differences in traditional notions of masculinity and men's roles are more evident among older and less educated African Americans and among Mexican Americans not born in the United States (McLoyd et al. 2000b).

Because the key assumption about male gender roles has been the centrality of work and economic success, many earlier researchers failed to look closely at how men interacted within their families. Over the past 2 decades, as part of a closer examination of men's lives, we witnessed a dramatic increase in the popular and scholarly attention paid to men's family lives (see, for example, Cohen 1987, 1993; Coltrane 1996; Gerson 1993; Daly 1993, 1996; LaRossa 1988; Popenoe 1996; and Marsiglio 1998).

Researchers finally began to ask about men's lives some of the same questions previously asked about women, looking at whether and how men juggle paid work and family and maintain sufficient involvement in each (Gerson 1993; Daly 1996; Coltrane 1996). Although we may not yet treat working fathers with the same concern we bring to working mothers, we have made strides in examining how men experience conflicts between work and parenting.

In addition, research indicates that men consider their family role to be much broader than that of family breadwinner (Cohen 1993; Gerson 1993; Coltrane 1996). Other dimensions of men's experiences include emotional, psychological, community, and legal dimensions; they also include housework and childcare activities (Goetting 1982). Later chapters will look at men's experiences of marriage, parenting, and the division of household labor.

Still, even with enlarged emphasis on men's more nurturing qualities, men continue to be expected to work and to support or help support their families. Although their financial contributions may be no less essential to maintain their family standard of living or even remain out of poverty, women are not judged as successful wives and mothers based on whether they succeed at paid employment. As a result, men have less role freedom than women to choose whether to work (Russell 1987; Cohen and Durst 2001). When a man's roles of worker and father come into conflict, usually it is the father role that suffers. A father may want to spend time with his children, but his job does not allow flexibility. Because he must provide income for his family, he will not be able to be more involved in parenting. In a familiar scene, a child comes into the father's home office to play, and the father says, "Not now. I'm busy working. I'll play with you later." When the child returns, the "not now, I'm busy" phrase is repeated. The scene recurs as the child grows up, and one day, as his child leaves home, the father realizes that he never got to know him or her.

Many men strive to avoid this potential nightmare and prevent the father–child estrangement that they may remember experiencing as children. They go out of their way to be more involved and more nurturing

with their children. However, what they learn is that the complexity of juggling work and family is not restricted to women. Men who attempt the same juggling act often experience similar role strain and role overload (Gerson 1993).

In addition, men continue to have greater difficulty expressing their feelings than do women (Real 1997). Men tend to cry less and show love, happiness, and sadness less. When men do express their feelings, they are more forceful, domineering, and boastful; women, in contrast, tend to express their feelings more gently and quietly. When a woman asks a man what he feels, a common response is "I don't know" or "Nothing." Such men have lost touch with their inner lives because they have repressed feelings that they have learned are inappropriate. This male inexpressiveness often makes men strangers to both themselves and their partners.

Men continue to expect and, in many cases, are expected to be the dominant member in a relationship. Unfortunately, the male sense of power and command often does not facilitate personal

■ As contemporary male gender roles allow increasing expressiveness, men are encouraged to nurture their children.

relationships. Without mutual respect and equality, genuine intimacy is difficult to achieve. We cannot control another person and at the same time be intimate with that person.

Women's Roles in Families and Work

Although the main features of traditional male gender roles vary more by class than ethnicity, there are more striking ethnic differences in traditional female roles.

Traditional white female gender roles center on women's roles as wives and mothers. When a woman leaves adolescence, she is expected to either go to college or to marry and have children. Although a traditional woman may work before marriage, she is not expected to defer marriage for work goals, and soon after marriage she is expected to be "expecting." Within the household, she is expected to subordinate herself to her husband. Often this subordination is sanctioned by religious teachings.

We still know relatively less about the lives of married African American women, as most research focuses more upon unmarried mothers and the poor (Wyche 1993). Yet we know that the traditional Caucasian female gender role does not extend to African American women. This may be attributed to a combination of the African heritage, slavery (which subjugated women to the same labor and hardships as men), and economic discrimination that pushed women into the labor force. Karen Drugger (1988) notes:

> A primary cleavage in the life experiences of Black and White women is their past and present relationship to the labor process. In consequence, Black women's conceptions of womanhood emphasize self-reliance, strength, resourcefulness, autonomy, and the responsibility of providing for the material as well as emotional needs of family members. Black women do not see labor-force participation and being a wife and mother as mutually exclusive; rather, in Black culture, employment is an integral, normative, and traditional component of the roles of wife and mother.

One study (Leon 1993) found that African American women appear more instrumental than either Caucasian or Latina women; they also have more flexible gender and family roles. African American men are generally more supportive than Caucasian or Latino men of egalitarian gender roles.

In traditional Latina gender roles, the notion of *Marianismo* has been the cultural counterpart to *machismo*. Drawn from the Catholic ideal of the Virgin Mary, *Marianismo* stresses women's roles as self-sacrificing mothers suffering for their children (McLoyd et al. 2000b). Thus, traditional Latino women are expected to subordinate themselves to males (Vasquez-Nuttall, Romero-Garcia, and De Leon 1987). But this subordination is based more on respect for the male's role as provider than on subservience (Becerra 1988). It also appears to be waning. Latina women are increasingly adopting values incompatible with a belief in male dominance and female subordination. They also display higher levels of marital satisfaction and less depression when their husbands share more of the domestic work (McLoyd et al. 2000b). Wives have greater equality if they are employed; they also have more rights in the family if they are educated (Baca Zinn 1994).

Latino gender roles, unlike those of Anglos, are strongly affected by age roles in which the young subordinate themselves to the old. In this dual arrangement, notes Rosina Becerra (1988), "females are viewed as submissive, naive, and somewhat childlike. Elders are viewed as wise, knowledgeable, and deserving of respect." As a result of this intersection of gender and age roles, older women are treated with greater deference than younger women.

Even though the traditional roles for white women have typically been those of wife and mother, increasingly over the past few decades an additional role has been added: employed worker or professional. It is now generally expected that most women will be employed at various times in their lives. Women generally attempt to reduce the conflict between work and family roles by giving family roles precedence. As a result, they tend to work outside the home in greatest numbers before motherhood and after divorce, when single mothers generally become responsible for supporting their families. After marriage, most women are employed even after the arrival of the first child. Regardless of whether a woman is working full-time, she almost always continues to remain responsible for housework and childcare.

Cultural expectations impose high standards of devotion and labor-intensive self-sacrifice on women who become mothers, what is described as the **intensive mothering ideology**—the belief that children need full-time, unconditional attention from moth-

Issues and Insights Gender Ideals in Middle-Class Black Families

[When I decide to get married, my mate] has to have fatherhood qualities. He has to like kids. He needs to be ambitious and motivated. He has to have a set of career aspirations. . . . He needs to be employed. If he lost his job, then, you know, we'll cope with that for a couple of months. But he's got to go out there and get a job.

Such are the words of Ms. Morgan, a 31-year-old single African American woman, interviewed by sociologist Faustina Haynes. Haynes undertook an exploratory study of the gender ideals evident among a sample of middle-class African Americans. Asking her sample of middle-class black men and women what they expect or expected from marital life, Haynes challenges what she sees as overgeneralized portraits of black families.

Based on a tendency in the research literature on African American family life to focus on working-class and lower-class black families, certain impressions have been formed and stereotypes perpetuated. Among these is the idea that black families are *egalitarian,* embodying a more equal division of domestic and paid labor than their counterparts among other racial groups, or *matriarchal,* with black men being relatively absent and unimportant because of their economic difficulties and failures. Noting that, traditionally, the dominant culture in the United States has defined men as household heads and providers, Haynes suggests that characterizations of black families as *either* egalitarian *or* matriarchal makes them seem

deviant, as they are said to depart from this long-standing white, middle-class norm.

Based on her interviews with a small sample of 19 black female and 15 black male high schoolteachers, Haynes found the following attitudes to be prevalent:

- Respondents possess what Haynes refers to as *neopatriarchal* gender ideals. That is, the females see men with expressive qualities and egalitarian ideas as attractive but not if they lack instrumental characteristics, especially those associated with successful providing. Similarly, men see women with instrumental characteristics as appealing, but women who lack expressive characteristics as well are not desirable as potential spouses. They further expect to pass these beliefs to their children. Girls will be raised to be feminine, little ladies and womanly. Boys will be raised to be masculine and manly.

- Both the male and the female schoolteachers contend that they have always anticipated egalitarian relationships. Single women and men say they expect to share tasks and finances equally. Married respondents expected before they married to share, and they continue to expect to share.

- Despite stated desires to share, household activities and family roles are still perceived to be gender specific. Although they are not traditional, they also are not egalitarian. Haynes calls them *transitionalists* (Hochschild 1989), in that they neither identify fully with traditional roles nor completely embrace the idea

that women and men are fully equal.

- Men and women reject the idea that wives are subservient to their husbands. However, both female and male respondents believe that "men, especially Black men, have to and should be in the provider roles in their families to feel 'like men'" (Haynes 2000, 834).

- To account for less than equal sharing, female respondents suggest that *competence* determines actual task allocation; a household task should be done by whoever is better at that task. Haynes notes that the desire for egalitarian household roles is thus thwarted by experience. Raised in more traditional households, females have "become better" at domestic tasks than men have. As a consequence, women carry more responsibility for household tasks because they "are good at them." Meanwhile, men continue to suffer from the demands of the provider role because they are expected to be providers for their families.

Haynes reminds us of the dangers in characterizing group differences without attention to the multiple factors (for example, race, gender, and social class) that shape both gender beliefs and familial behavior. Although other research supports the generalization that African American husbands perform a statistically significant greater amount of housework (Greenstein 1996), Haynes' study reminds us this need not mean that they depart entirely or even widely from some long-standing gender expectations.

SOURCE: Haynes 2000, 811–837.

ers to develop into healthy, well-adjusted people. This puts all mothers in a demanding position, but it creates a particularly difficult dilemma for mothers who also choose or need to work outside the home (Hays 1996). It leads increasing numbers of women to question whether they should have children and, if they do, how much of their time and attention their children need.

Women from ethnic and minority groups, however, are less likely than Caucasians to view motherhood as an impediment. African American women and Latinas tend to place greater value on motherhood than the Caucasian or Anglo majority. For African Americans, tradition has generally combined work and motherhood; the two are not viewed as necessarily antithetical (Basow 1993). For Latinas, the cultural and religious emphasis on family, the higher status conferred on motherhood, and their own familial attitudes have contributed to high birthrates (Jorgensen and Adams 1988).

Although husbands were once the final authority, wives have greatly increased their power in decision making. Today they are expected not to be submissive but to have significant, if not equal, input in marital decision making. This trend toward equality is limited in practice by an unspoken rule of marital equality: "Husbands and wives are equal, but husbands are more equal." In practice, husbands may continue to have greater power than wives, becoming what sociologist John Scanzoni (1982) once described as the "senior partner" of the marriage.

This is not absolute or inevitable, however. Interesting exceptions to this pattern exist, especially among some dual-earner couples. Some couples develop and act upon an ideology of sharing and fairness, valuing and pursuing such relationship characteristics as equality and equity (Schwartz 1994; Risman and Johnson-Sumerford 1997). Although these "peer" and **postgender relationships** (that is, relationships lived outside the constraints of gender expectations) are not yet the norm, they reflect the most concerted efforts to establish greater equality in marriage.

Breakdown of the Instrumental and Expressive Dichotomy

The identification of masculinity with instrumentality and femininity with expressiveness appears to be breaking down. Men perceive themselves to be more instrumental than do women and women perceive themselves as being more expressive than do men. A substantial minority of both genders is relatively high in both instrumentality and expressiveness or is low in both. It is interesting that the instrumental and expressiveness ratings men and women give each other

have little to do with how they rate themselves as masculine or feminine (Spence and Sawin 1985).

Constraints of Contemporary Gender Roles

Even though substantially more flexibility is offered to men and women today, contemporary gender roles and expectations continue to limit our potential. Indeed, there is considerable evidence that some stereotypes about gender traits are still very much alive. Men are perceived as having more undesirable self-oriented traits (such as being arrogant, self-centered, and domineering) than women. Women are viewed as having more traits reflecting a lack of a healthy sense of self (such as being servile and spineless).

Research suggests that the traditional female gender role does not facilitate self-confidence or mental health. Both men and women tend to see women as being less competent than men. A study by Lyn Brown and Carol Gilligan (1992) revealed that the self-esteem of adolescent girls plummeted between the age of 9 and the time they started high school.

The combination of gender-role stereotypes and racial or ethnic discrimination tends to encourage feelings of both inadequacy and lack of physical attractiveness among African American women, Latinas, and Asian American women (Basow 1993).

The situation of contemporary women in dual-earner households imposes its own constraints on women's lives. Because they continue to shoulder the bulk of responsibility for housework and childcare *on top of full-time jobs,* they often experience fatigue, stress, resentment, and lack of leisure (Hochschild 1989). Especially for women who try to be "supermoms," the volume and complexity of work and family can force them to cut back on their aspirations or compromise their expectations for marriage and motherhood (Hochschild 1989, 1997). Significantly, despite the ongoing stresses, women who "juggle" are less distressed and more fulfilled than full-time homemakers (Crosby 1991).

Finally, there is still a "double standard of aging" that treats men and women differently. As women grow older, they tend to be regarded as more masculine and as unattractive. As men age, they become distinguished; women simply become older. Masculinity is associated with independence, assertiveness, self-control, and physical ability; with the exception of physical ability, none of these traits necessarily decreases with age. Because older women are considered to have lost their attractiveness and because they have fewer potential partners, they are less likely to marry.

Resistance to Change

We may think that we want change, but both men and women reinforce traditional gender-role stereotypes among themselves and each other (Hort, Fagot, and Leinbach 1990). Both genders react more negatively to men displaying so-called female traits (such as crying easily or needing security) than to women displaying male traits (such as assertiveness or worldliness), and both define male gender-role stereotypes more rigidly than they do female stereotypes. Men, however, do not define women as rigidly as women do men. And both men and women describe the ideal female in androgynous terms (Hort, Fagot, and Leinbach 1990).

Despite the limitations that traditional gender roles may place on us, changing them is not easy. Gender roles are closely linked to self-evaluation. Our sense of adequacy often depends on gender-role performance as defined by parents and peers in childhood ("You're a good boy" or "You're a good girl"). Because gender roles often seem to be an intrinsic part of our personality and temperament, we may defend these roles as being natural, even if they are destructive to a relationship or to ourselves. To threaten an individual's gender role is to threaten his or her gender identity as male or female because people do not generally make the distinction between gender role and gender identity. Such threats are an important psychological mechanism that keeps people in traditional roles.

Furthermore, the social structure reinforces traditional gender norms and behaviors and makes change more difficult. Some religious groups, for example, strongly support traditional gender roles. The Catholic Church, conservative Protestantism, Orthodox Judaism, and fundamentalist Islam, for example, view traditional roles as being divinely ordained. Accordingly, to violate these norms is to violate God's will. The marketplace also helps enforce traditional gender roles. The wage disparity between men and women (remember, women earn about 75% of what men earn) is a case in point. Such a significant difference in income makes it "rational" that the man's work role takes precedence over the woman's work role. If someone needs to remain at home to care for the children or an elderly relative, it makes "economic sense" for a heterosexual woman to quit her job because her male partner probably earns more money.

Gender Movements and the Family

Gender issues have been the source of much collective action and the focus of a number of social movements that press for change. These movements include the

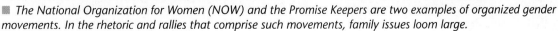

■ *The National Organization for Women (NOW) and the Promise Keepers are two examples of organized gender movements. In the rhetoric and rallies that comprise such movements, family issues loom large.*

Real Families   Degendering Marriage and Family

Making Gender Matter Less

When asked by sociologist John Durst to reflect on her situation, then 32-year-old Karen Wilson described having what she considered an almost ideal life. "Oh, how much time do you have. . . . Do you have like three hours? I love it! I love it! God, there's just so many things about it." Karen is a success in her career in sales and promotions for a communications company. She has a husband she loves and two children she adores, a 3-year-old daughter and a 4-month-old son. After having had an on-again, off-again work history, followed by a stint as a stay-at-home mom, an opportunity presented itself for her to become the full-time breadwinner and for her husband, Kevin, to stay home with their two young children. Here's her description of how she and Kevin reached the arrangement they have and what she most enjoys about it:

You know what, we talked about this before we got married. I spent all this money on my education [earning both a bachelor's and a master's degree] so that I could go to work. I planned on working

after I had kids. [I told Kevin] . . . "if you want to stay home, great, but I can't stand it!" And he said, "Yeah, we can do that." Then, BAM! This opportunity came along for me to make more money than he was and I said, "You wanna do it? You wanna live the dream?"

Pushed to identify what she sees as the biggest positives of her lifestyle, she enthusiastically replied:

I like being able to get away between 8 and 5 and to have a lot more control over my life without having to worry about two other responsibilities (son and daughter) and Kevin, too. I should say all three of them. I like that. I like being able to turn it off and just go, but I like coming back and having my daughter's little face pressed against the window (waiting), Kevin standing there with a beer in his hand, the dog running around me, it's really nice to come home to. . . . I love bringing home the paycheck and telling Kevin, "Here, honey, split it up. . . ." I love that. I love contributing; I just think it's the ultimate.

I love not having all the responsibilities he has. I hated cooking. I hated the dishes, the laundry—I felt like it was the least rewarding

job anybody could have because you never get any pats on the back. I like having a title and being able to say, "This is what I do. I'm contributing to my family." What else? I like being able to go out to lunch and playing with the "big dogs." I like doing that.

I feel like I paid for my education and I deserve that, to try that, to work on it. I like the intellectual stimulation I get from doing that. . . . And I like that Kevin's just so calm and relaxed and really laid back. The kids keep him moving constantly yet at the end of the day he's still relaxed enough to talk to me. I think it's been really wonderful.

Research on intimate relationships, marriage, and family consistently reveals the importance of gender in dividing up domestic responsibilities and shaping personal and familial experiences. Women perform two to three times as much housework as men, and employed wives experience greater stress and enjoy less leisure than their husbands (Coltrane 2000). The consistency with which such inequalities are reported may give the impression of inevitability, that they are somehow unavoidable parts of marriage and parenthood, but couples such as the Wilsons offer a more hopeful scenario to those who might

---

range of perspectives within the contemporary women's movement but also various "men's movements," that, although less visible, have organized to change aspects of men's lives. We look briefly here at some of the ways these movements have framed and acted on family matters.

A complete history of American feminism is beyond the scope of this book. In the eighteenth, nineteenth, and twentieth centuries, women organized around issues such as economic justice, abolition of slavery, temperance, and women's suffrage. In their antislavery activity during the nineteenth century, many

women were sensitized to the extent of their own oppression and disadvantage, which helped energize their pursuit of voting rights (Renzetti and Curran 1999; Lindsey 1997). After gaining the right to vote with the passage of the Nineteenth Amendment in 1919, many women withdrew from active feminist involvement because they thought they had reached equality with men (Renzetti and Curran 1999).

During the 1960s, feminism resurfaced dramatically. Catalyzed by the publication of Betty Friedan's *The Feminine Mystique*, many women began to look critically at the sources of their "problems with no

wish to someday depart from the norm, whether to create more equal partnerships or, more dramatically, reverse roles.

Sociologists Barbara Risman and Danette Johnson-Sumerford interviewed their own sample of 15 couples who explicitly reject conventional conceptions of gender, opting instead for more gender-neutral relationships. That is, they carefully and intentionally share responsibility for paid work and share responsibility as caregivers for their children. At minimum, they "changed how gender works in their families." Furthermore, "in the negotiation of marital roles and responsibilities, they have moved beyond using gender as their guidepost" (Risman and Johnson-Sumerford 1998, 24).

Let's look briefly at the different paths couples took to construct their "postgender" marriages.

- *Dual-career couples.* The most common path to "postgender marriage" begins with a marriage of two career-oriented professionals in which at least the wife, but preferably the husband as well, values equality and is committed to sharing. Both partners retain strong career commitments, although both scale back to achieve the lifestyle they desire.

- *Dual nurturers.* Dual-nurturer couples place their priorities on home and family, not careers. Their work is for money to enable them to spend their time together and with their children. In one dual-nurturer couple, neither spouse had consistently held a full-time job. Instead, they pieced together part-time work, seeing to it that they weren't both working simultaneously each day.

- *Posttraditionalists.* This path begins with a traditional arrangement, meaning a gender-based division of household roles and labor, although not necessarily male breadwinner–female home-maker. Couples found themselves dissatisfied in gender-based arrangements, whether in their current or a former marriage and were strongly motivated to avoid the sort of unfairness that often plagues dual-earner couples.

- *External forces.* This path consists of couples "pushed" by circumstance (for example, economic factors such as a wife's higher salary and less flexible work schedule than her husband or an illness) toward more equal domestic arrangements. Whatever the circumstances, they came to recognize and appreciate the gender equality that resulted.

Regardless of the route couples took to arrive at their postgender family arrangements, they used criteria other than gender to organize their daily activities. They have rejected the ideas that "wifehood involves a script of domestic service or that breadwinning is an aspect of successful masculinity" (Risman and Johnson-Sumerford 1998). Such couples are still rare and their lifestyles may require high levels of female income and professional autonomy if women are to be able to move beyond male dominance or privilege. Furthermore, all but two couples employ paid help with domestic tasks such as cleaning, dusting, bathrooms, and yard work, which made life easier and made fairness more achievable. However, couples who used paid help reported that domestic responsibilities had been shared even before they started paying for housekeeping services.

The significance of couples like the Wilsons or Risman and Johnson-Sumerford's postgender couples is that they reveal a wider range of possible marital outcomes than most literature reports. There is no inevitable inequality that engulfs married couples. Equality and fairness take work and persistence but are possible for those who seek them.

names," and the family was seen as a major culprit. In addition, wage inequality was made a public issue through President John F. Kennedy's Commission on the Status of Women in 1961 and the passage of the Equal Pay Act of 1963. Then in 1966, the National Organization for Women (NOW) was established. Over the last 40-plus years, this liberal, reform-oriented feminist organization has grown to include more than half a million members in its more than 500 chapters throughout the United States. It is the largest, although not the only, organized plank of the women's movement, and its philosophy represents one of a number of "feminisms" (Renzetti and Curran 2003; Lindsey 1997). Contemporary feminist positions range across a spectrum of perspectives, including liberal, socialist, radical, lesbian, multiracial, and postmodern feminism (Lorber 1998; Renzetti and Curran 1999). Each has a specific emphasis on issues and advocates different strategies to improve women's lives.

Judith Lorber sorts the various feminist perspectives into three broader categories: **Gender-reform feminism** is geared toward giving women the same rights and opportunities that men enjoy; **gender-resistant feminism** advocates more radical, separatist

Think about the material presented throughout this chapter. Clearly, women and men do not have identical experiences in families, as we continue to examine throughout this text. But we can probably all agree that gender matters in shaping what we do and don't experience in our relationships and our families, as well as in school, the workplace, and wider society. As you think about gender issues, ask yourself the following three questions, answering each one "Yes" or "No" (or agree/disagree):

1. Girls and women have not been treated as well as boys and men in our society.

2. Women and men should be paid equally for the same work.

3. Women's unpaid work (for example, housework and childcare) should be more socially valued.

How did you answer? Did you agree with (answer "Yes" to) one, two, or all three of the questions? Or did you answer "No" to all of them?

What can answers to the preceding items tell us? Together the three are used as measures or attitudinal indicators of feminist identification, in that they are among the "cardinal beliefs of feminism" (Zucker 2004). In other words, these items assess whether you agree with the most fundamental tenet of feminism: equality between the sexes. As conceptualized by researcher Alyssa Zucker, "feminists" would answer "Yes" to all three questions, demonstrating consistent agreement with feminist ideals.

Studying 333 alumnae from the University of Michigan (drawn from graduating classes of 1951 or 1952, 1972, and 1992) Zucker found the following:

4 women (1%) rejected all three beliefs

19 women (6%) rejected two of the three beliefs

81 women (24%) rejected one of the three beliefs

219 women (66%) accepted all three beliefs

Does this mean that two-thirds of this sample are feminists? Not so fast. The picture is about to become complicated. After all, we need to take into account how people perceive themselves. Think about yourself for the moment. How would you answer one last question:

4. Do you consider yourself a feminist?

This question measures your self-labeling or "acceptance of the feminist label." When Zucker asked this of her subjects (by having those who considered themselves feminist to complete certain other questions and those who didn't consider themselves feminist to proceed to others) 152 women, 46% of her sample, indicated that they saw themselves as feminists. Another 138 women, 41% of the sample, indicated that they did not identify themselves as feminists. Finally, 3% (11 women) could not decide whether they considered themselves feminists or not and 10% (32 women) didn't complete the questionnaire.

With both attitude and identification items in hand, Zucker determined that 123 of her subjects were feminists in that they agreed with all three items and considered themselves to be feminists. Interest-

---

strategies for women out of the belief that their subordination is too embedded in the existing social system; and **gender-rebellion feminism** tends to emphasize overlapping and interrelated inequalities of gender, sexual orientation, race, and class (Lorber 1998; Renzetti and Curran 1999).

Given this diversity of opinion, it is difficult to characterize *one* "feminist" position on families. Furthermore, such attempts occasionally exaggerate or simplify complex positions. In her critique of American feminism, for example, economist Sylvia Hewlett notes that neither liberal feminism ("equal rights" feminism) nor radical feminist positions have recognized the commitments women feel toward their families and the consequences of those commitments. By stressing equal rights and full equality with men, liberal feminism may have downplayed the responsibilities women carry within families and not recognized that women may need different supports than those needed by men (Hewlett 1986). Some of the more radical feminist positions articulated in the 1960s and 1970s may have been fairly anti-marriage or anti-motherhood, as either or both have at times been seen as relationships that oppress women and keep them from achieving their full capabilities.

Hewlett compares both approaches to a movement more characteristic of European feminist activity: **social feminism**—the belief that workplace and fam-

Table 4.2 ■ Age Effects among Study Subjects

Class of	Feminist (%)	Egalitarian (%)	Nonfeminist (%)
1951/1952	27	39	34
1972	51	30	19
1992	58	22	20

ingly, Zucker found that 84 women who endorsed all three feminist beliefs didn't consider themselves feminists (or couldn't choose). These are the people who might be heard in conversation to say, "I'm not a feminist but . . ." and then proceed to assert some point of view clearly in keeping with feminism. Zucker labeled them *egalitarians*. She found that 65 women were "nonfeminists" in that they rejected at least one of the cardinal beliefs of feminism and rejected the label. (The remaining women were not part of this analysis because of incomplete data or because despite disagreement with at least one feminist belief they considered themselves feminists.) This left the following percentage distribution in her sample:

45% feminists

31% egalitarians

24% nonfeminists

Zucker determined that there were some interesting age effects, as could be seen in the differences among the three subsamples:

Clearly, younger women were more likely than older women to consider themselves feminists. There were some other noteworthy differences within her sample. Feminists were significantly more likely than either egalitarians or nonfeminists to have feminist family members or to have had relationships with more feminists. They were also more likely than the other two groups to mention suffering, either personally or of

someone close to them, the consequences of sexism.

What are we to make, however, of women who agree with feminism but distance themselves from the label? Zucker contends that feminism is a social identity that "is both concealable and often stigmatized or socially devalued, and thus public identity as a feminist is both optional and potentially costly." Given the understandable desire to avoid stigma and other social costs, perhaps it is even expected that women would understate or deny being what their attitudes suggest.

Questions to Consider

1. How well does the preceding discussion fit your answers to the questions?

2. What reasons can you think of for why someone who agrees with feminist ideals would reject or avoid the feminist label?

3. Why do you think "feminist" has negative connotations for some people?

SOURCE: Zucker 2004.

---

ily supports are essential if women are to experience a high quality of life (Hewlett 1986). Feminist critics of Hewlett rightly point out that the greatest activism on behalf of public support for families has and continues to come most strongly from women; thus, her characterization is said to be unfair. Although American feminists have been active at the forefront of pushing for parental leave, childcare, and so forth, organizations such as NOW still stress abortion rights, reproductive freedom, opposing bigotry against lesbians and gays, and ending violence against women more heavily than specifically family-focused issues.

Divisions of opinion and multiple perspectives on gender inequality constitute a basic similarity between

women and men. Just as there is no one perspective on how women should be or what they should do, neither is there unanimity about men's lives. Just as there are multiple feminisms, each with its own agenda, there are different viewpoints on whether, in what direction, and how men ought to change (Clatterbaugh 1997; Messner 1997; Renzetti and Curran 1995).

In recent years, at one time or another, we have witnessed a variety of "men's movements": the mythopoeic men's movement, the men's rights and fathers' rights advocates, the Christian men's movement (for example, Promise Keepers), and the profeminist, gay-affirmative men's organization, the National Organization of Men Against Sexism. Each represents just a part among many movements. Many of these movements differ in what they see as men's roles in and responsibilities to their families.

Central to a **profeminist men's movement** is the issue of *fairness*. Profeminist men believe that men ought to share responsibilities within their households and that women and men ought to be equal partners. Also, profeminists argue that men and children would both benefit from closer connections between fathers and their children.

Both the Promise Keepers and the organizers of the 1995 Million Man March and rally in Washington, D.C., by African American men also stressed the idea of men's responsibilities to their families, although their versions of responsibilities included more traditional notions of men's roles as the heads of their households. They also argued that men needed to be more accountable to spouses and children. Finally, the men's rights movement has stressed supposed discrimination that men face, in and out of family matters. They note, for example, that only men can be subject to compulsory military service. They also look at what they believe are inequalities in areas of divorce settlements and custody or visitation arrangements (Farrell 2001).

It is interesting to note the different positions taken on the family by the various feminist and men's movements. Although it is inaccurate and overstated to suggest that feminists are antifamily, the resurgent women's movement of the 1960s did grow partly out of the articulation of discontent. Similarly, early "second wave" feminists (1960s–1970s) attempted to sever the automatic connections typically made among women, children, and families as a way of liberating women to pursue other aspirations.

Conversely, across most men's movements there is a sense that men need to enlarge their family role, live up to or "honor" their commitments to their families, and/or share in caring for children and households.

Such involvement is often seen as potentially "liberating" for men, because it reconnects them to their emotional sides and broadens their lives beyond wage earning.

Looked at more closely, these movements are really not as different as they seem. What feminists railed against was not the *family* but the *gendered family*. They were less antagonistic to what women felt toward and did in the family than what men did not. Because of the differential burden carried by women in households, family life imposed constraints on women's opportunities for outside involvements in ways it did not on men's. More recently, the various men's movements have acknowledged men's lack of involvement or weaker commitments and opposed defining men solely in terms of what they do away from the family.

Contemporary gender roles are still in flux. Few men or women are entirely egalitarian or traditional. Even those who are androgynous or who have egalitarian attitudes, especially males, may be more traditional in their behaviors than they realize. Few with egalitarian or androgynous attitudes, for example, divide all labor along lines of ability, interest, or necessity rather than gender. Also, marriages that claim to be traditional rarely have wives who submit to their husbands in all things. Among contemporary men and women, women find that their increasing access to employment puts them at odds with their traditional (and personally valued) role as mother. Women continue to feel conflict between their emerging equality in the workplace and their continued responsibilities at home. Within marriages and families, the greatest areas of gender inequality continue to be the division of housework and childcare. But change continues to occur in the direction of greater gender equality, and this equality promises greater intimacy and satisfaction for both men and women in their relationships.

Summary

- A *gender role* is the role a person is expected to perform as a result of being male or female in a particular culture. *Gender-role stereotypes* are rigidly held and oversimplified beliefs that males and females possess distinct psychological and behavioral traits. *Gender identity* refers to the sense of being male or female.

- Men and women are not "opposites," they are actually more similar than different. Innate gender differences are generally minimal; differences are encouraged by socialization.

- Within any given society, there are multiple versions of masculinity and femininity, one of which comes to dominate our thinking about gender. Across societies, much variation exists in how gender is perceived, including the perception of how many gender categories there are.

- Gender relations are also power relations. *Patriarchal societies* are social structures in which men dominate. Logically, *matriarchal societies* would be societies in which women dominate political and economic life. Researchers have not found any society that truly embodies a matriarchal social structure.

- According to *gender theory*, social relationships are based on the socially perceived differences between males and females that justify unequal power relationships.

- Symbolic interactionists view gender as something we actively create or "do" in everyday situations and relationships, not an internalized set of behavioral and personal attributes.

- Two important socialization theories are social learning theory and cognitive development theory. *Social learning theory* emphasizes learning behaviors from others through rewards, punishments, and *modeling*. *Cognitive development theory* asserts that once children learn that gender is permanent, they independently strive to act like "proper" boys or girls because of an internal need for congruence.

- Parents, teachers, and *peers* (age-mates) are important agents of socialization during childhood and adolescence. Ethnicity and social class also influence gender roles. Among African Americans, strong women are important female role models.

- After many years of evidence showing how schools disadvantage female students, recent evidence indicates that males are lagging behind educationally.

- The media tend to portray traditional stereotypes of men and women, as well as of ethnic groups. For students, colleges and universities are important sources of gender-role learning, especially for nontraditional roles. Marriage, parenthood, and the workplace also influence the development of adult gender roles.

- The gender roles we play in adulthood are affected by situations, opportunities, and constraints, which can alter the path established by socialization.

- Traditional male roles emphasize dominance and work. For women, there is greater role diversity according to ethnicity.

- Contemporary gender roles are more *egalitarian* than the traditional ones of the past. They reflect: (1) the acceptance of women as workers and professionals; (2) increased questioning of motherhood as a core female identity; (3) greater equality in marital power; (4) the breakdown of the instrumental and expressive dichotomy; and (5) the expansion of male family roles.

- Changing gender-role behavior is often difficult because (1) each sex reinforces the traditional roles of its own and the other sex; (2) we evaluate ourselves in terms of fulfilling gender-role concepts; (3) gender roles have become an intrinsic part of ourselves and our roles; and (4) the social structure reinforces traditional roles.

- There have been various social movements dedicated to challenging or changing women's or men's roles, including various feminisms and various "movements" and perspectives on men and masculinity. Ironically, whereas early 1960s and 1970s feminists often rallied against women being associated with family responsibilities, most of the current men's movements attempt to reconnect men with families.

Key Terms

Resources on the Internet

Companion Website for This Book

http://www.thomsonedu.com/sociology/strong

Gain an even better understanding of this chapter by
going to the companion website for additional study
resources. Take advantage of the Pre- and Post-Test
quizzing tool, which is designed to help you grasp difficult concepts by referring you back to review specific
pages in the chapter for questions you answer incorrectly. Use the flash cards to master key terms and check
out the many other study aids you'll find there. Visit
the Marriage and Family Resource Center on the site.
You'll also find special features such as access to Info-
Trac[©] College Edition (a database that allows you access to more than 18 million full-length articles from
5,000 periodicals and journals), as well as GSS Data
and Census information to help you with your research
projects and papers.

CHAPTER 5

Friendship, Love, and Intimacy

Outline

What Do YOU Think?

Are the following statements TRUE or FALSE?
You may be surprised by the answers (see answer key on the following page).

T F **1** A high value on romantic love is unique to the United States.

T F **2** The development of mutual dependence is an important factor in love.

T F **3** Love and commitment are inseparable.

T F **4** Friendship and love share many characteristics.

T F **5** Men fall in love more quickly than do women.

T F **6** Heterosexuals, gay men, and lesbians are equally likely to fall in love.

T F **7** In many ways, love is like the attachment an infant experiences for a parent or primary caregiver.

T F **8** A high degree of jealousy is a sign of true love.

T F **9** Partners with different styles of loving are likely to have more satisfying relationships because their styles are complementary.

T F **10** Love is something experienced and expressed similarly by people regardless of their ethnic or racial backgrounds.

"Y"ou had me at hello."

So says Renee Zellweger's character, Dorothy Boyd, to Tom Cruise's character, Jerry Maguire, near the end of the 1996 movie of the same name. It is the defining moment in the movie, when Jerry has arrived to re-state his love for Dorothy. As he stammers and stumbles, searching for the right words with which to articulate his feelings and to explain his return, she cuts him off with her simple but moving pronouncement: "You had me at hello."

It appears as though Americans share the sentiment first expressed by American businessman Franklin P. Jones: "Love doesn't make the world go round, love is what makes the ride worthwhile." We are, it seems, in love with love. We can see this in the ways we live our daily lives, especially in the kinds of relationships we want, seek, and make and in the steps we take to find and keep them. It is also evident in the popular culture that we produce and consume. There, we can see our love affair with love in everything from the things we read and watch to the music we listen to. Love is *the* dominant theme of popular music, where song titles and lyrics are typically testimonies to the power, pleasure, and pain associated with falling in and out of love. Among book genres, romance novels sell widely, accounting for more than half of all popular mass-market fiction and 40% of all fiction sold in the United States. According to Romance Writers of America, nearly 65 million Americans read romance novels, and the genre had annual sales revenue of $1.2 billion in 2004 (Romance Writers of America 2006, https://www.rwanational.org). However, more than in music or books, our devotion to love stands out especially well in movies.

Romantic movies, *love stories* as they are often appropriately called, provide us with vivid scenes and memorable lines filled with heartfelt, often poignant declarations of the depth of a character's love. Often scenes stay with us, even coming to symbolize our very idea of true love. Sometimes it is love lost, as in the 1993 movie, *Sleepless in Seattle,* when Sam Baldwin, played by Tom Hanks, describes for a radio talk show host what it was about his late wife that made him love her:

> It was a million tiny little things that, when you added them all up, they meant we were supposed to be together . . . and I knew it. I knew it the very first time I touched her. It was like coming home . . . only to no home I'd ever known. . . . I was just taking her hand to help her out of a car and I knew. It was like . . . magic.

Sometimes it is love found, even if it is found many years into an on-again, off-again friendship, as it was between Billy Crystal's Harry and Meg Ryan's Sally in the 1989 film *When Harry Met Sally:*

> I love that you get cold when it's 71 degrees out. I love that it takes you an hour and a half to order a sandwich. I love that you get a little crinkle in your nose when you're looking at me like I'm nuts. I love that after I spend the day with you, I can still smell your perfume on my clothes. And I love that you are the last person I want to talk to before I go to sleep at night. And it's not because I'm lonely, and it's not because it's New Year's Eve. I came here tonight because when you realize you want to spend the rest of your life with somebody, you want the rest of your life to start as soon as possible.

And sometimes, as in *Jerry Maguire*, it is love reclaimed:

> Hello? Hello. I'm lookin' for my wife. Wait. Okay . . . okay . . . okay. If this is where it has to happen, then this is where it has to happen.
>
> I'm not letting you get rid of me. How about that?
>
> This used to be my specialty. You know, I was good in a living room. They'd send me in there, and I'd do it alone. And now I just . . .
>
> But tonight, our little project, our company had a very big night—a very, very big night.
>
> But it wasn't complete, wasn't nearly close to being in the same vicinity as complete, because I couldn't share it with you. I couldn't hear your voice or laugh about it with you.
>
> I miss my—I miss my wife.
>
> We live in a cynical world, a cynical world, and we work in a business of tough competitors.
>
> I love you. You—complete me. And I just had—

Courtesy of Everett Collection

■ *Popular films, such as* Jerry Maguire, *reflect how much American popular culture emphasizes romantic love.*

To which Renee Zellweger's character, Dorothy, interrupts:

> Shut up. Just shut up. . . . You had me at hello. You had me at hello.

We want to feel "magic." We want someone with whom we can spend the rest of our lives. We want to be and feel "completed." We want to make "the ride" worthwhile. In short, we want to be in love.

Much like the culture that surrounds them, American families place high value on love. Decisions about entering or exiting a marriage, assessments of the quality and success of any particular marriage, and devotion between spouses or parents and children all come down to love. On both an individual level and a familial level, then, it is important to consider the role love plays in our lives. This chapter is devoted to such consideration. However, before we turn to love, we need to consider the broader phenomenon of intimacy, including the intimacy of friendship.

The Need for Intimacy

Humans require other humans with whom we feel close and to whom we can commit. We need to form relationships in which we can share ourselves with others, exchange affection, and feel connected. In the developmental model formulated by psychologist Erik

Erikson, this was the great task facing us in young adulthood—intimacy versus isolation; either we satisfy our need for intimacy or we remain socially and emotionally isolated (Hook et al. 2003). But what exactly is intimacy and why is it so important?

In its most general sense, **intimacy** refers to closeness between two people. Sometimes we associate "intimacy" or "being intimate" with sexual relations. Certainly, sexual relations are part of physical intimacy, as are kisses, caresses, and hugs. However, it is more the emotional intimacy, having someone to talk to, to share ourselves with that is such an important part of our social and psychological well-being.

Reviewing research and theory on intimacy, Misty K. Hook and colleagues (2003) suggest that intimacy consists of four key features: the presence of *love and/or affection, personal validation, trust,* and *self-disclosure.* The more we feel as though another person likes us or loves us, the more comfortable we will be sharing our innermost feelings and revealing our most personal thoughts. When we feel as though someone understands and appreciates us, we feel accepted and more freely open ourselves to this person. We feel safe in the thought that we will be neither judged nor betrayed. Finally, to be intimate entails **self-disclosure,** the sharing of both the facts of our lives and our deeper feelings (Hook et al. 2003).

Intimate relationships provide us with a variety of benefits. They buffer us against loneliness, provide us with positive feelings about ourselves and others, give us confidence that our needs will be fulfilled in the future, and enhance our self-esteem. Intimate relationships are connected to happiness, contentment, and a sense of well-being. They also offer protection from some stress-related symptoms and reduce our likelihood of illness, depression, and accidents. People who lack satisfying, positive intimate relationships are at greater risk of illness; once ill, they recover more slowly and have higher susceptibility to relapse or recurrence of their illness. If we "cannot connect in a positive, intimate way with another human being, then physical, interpersonal, and emotional difficulties will ensue" (Hook et al. 2003, 463).

In a relationship, intimacy can be expressed in a variety of ways—talking together, listening to each other, making time for each other, being open and honest with each other, and trusting each other. As a determinant of relationship satisfaction, the degree of intimacy is more important than independence (autonomy, individuality, freedom), agreement (harmony, few quarrels), or sexuality (sexual harmony and satisfaction, physical contact). The importance of intimacy in defining relationship quality cannot be stressed strongly enough. This holds true in but also beyond the United States. In comparative research using German and Canadian samples, intimacy was the factor most highly correlated with relationship satisfaction in both countries and for both males and females (although it was somewhat more strongly correlated with women's than with men's relationship quality and may have different meanings for females and males) (Hasselbrauck and Fehr 2002; Hook et al. 2003).

The Intimacy of Friendship and Love

Although both have proved difficult to define with precision or consistency, friendship and love are among the most important sources of intimacy we have. They bind us together, provide emotional sustenance, buffer us against stress, and help preserve our physical and mental well-being. The loss of a friend and especially a loved one can lead to illness and even suicide.

Reflections

What are the ideas you associate with love? With friendship? How do they overlap? Have you ever mistaken one for the other?

Friendship often supplies the foundation for a strong love relationship. Shared interests and values, acceptance, trust, understanding, and enjoyment are at the root of friendship and form a basis for love. As much as they may benefit us similarly, love and friendship are not the same thing. One way to see the differences between love and friendship is to look at the qualities we value and seek in a friend as opposed to a romantic partner. Do we want the same things in our friends as we do our romantic partners?

The evidence is mixed. There are more similarities in what we want from friends and lovers than there are differences (Sprecher and Regan 2002). For example, trust, enjoyment, acceptance, kindness, and warmth are valued in both friends and romantic partners. A study by Mary Laner and J. Neil Russell (1998) found that when college students were asked what qualities they'd most want in a best friend and a spouse,

answers overlapped quite a bit. Both men and women included qualities such as "sensitive/warm," "open/honest," "trustworthy," and "communicative." Given that romantic partners are would-be spouses and we expect our spouses to be our closest intimates—our best friends—this overlap is not entirely unexpected. Yet other research suggests that we differentiate either in kind or degree between those qualities we seek in a close or best friend and what we desire in a lover or romantic partner (Sprecher and Regan 2002; Cann 2004).

Unlike both more formal role relationships (such as between boss and employee, teacher and student, or coworkers), or the more intense relationship between romantic partners in a relationship, the role of friend and the qualities sought in friends are more ambiguous (Cann 2004). Unlike more formal relationships, there is no specific task or purpose we seek to satisfy with friends aside from finding pleasure in our interactions. Unlike romantic love relationships, we have less at stake and are therefore less certain about criteria we desire in selecting friends, aside from shared interests, kindness, and loyalty. Potential friends may be deemed desirable based on their specific combination of unique attributes and how those attributes match our needs and wants at a given point in time. Romantic partners, on the other hand, are more carefully selected, evaluated as desirable, based on possession of certain qualities or attributes that might indicate their commitment to the relationship, their potential reproductive success, and their eventual attachment to offspring. In terms introduced by John Scanzoni and colleagues, romantic partners are selected based on their seeming ability to satisfy multiple needs that are products of the multiple "interdependencies" two people share. Interdependencies consist of shared activities, statuses, and patterned exchanges between two people. Romantic partners are expected to be able to satisfy four types of interdependencies: intrinsic (for example, emotional support), extrinsic (for example, money or services), sexual (sexual activity), and formal (shared legal status). Friends, however, typically provide only intrinsic resources (Scanzoni et al. 1989; Sprecher and Regan 2002). In addition, because romantic partners are potential spouses and our spouses are expected to be our closest emotional support, qualities such as affection, kindness, and sensitivity take on greater importance in choosing romantic partners than in choosing friends.

In a study comparing the importance of 34 different qualities in friends, romantic partners, bosses and employees, Arnie Cann (2000) found that respondents rated those qualities associated with intimacy, achievement, dependability, and kindness higher in romantic partners than in friends. For qualities associated with intimacy (such as physical attractiveness, sensitivity, being affectionate, and gentleness), respondents rated romantic partners higher than they did the other three relationships, suggesting such qualities are especially desired in romantic partners. Respondents rated friends and bosses similarly, indicating that although they desire to have emotional connections with friends, the connection is not significantly more important to them than what they desire from their bosses. Achievement qualities (such as intelligent, analytic, competitive, and good earning potential) were considered less important in close friends than in the other three relationships. Although qualities associated with dependability (reliable, truthful, helpful, efficient, confident, and ambitious) were rated highly for all four types of relationships, again they were lowest for friends. Finally, qualities associated with kindness (compassionate, sincere, tactful, and conscientious) were rated important for all four relationships. Romantic partners were rated highly for all four composite qualities, reflecting the importance of such a choice. We are evaluated and judged not just by our own qualities but also by our partner's qualities. His or her strengths and weaknesses become our strengths and weaknesses.

Research by Susan Sprecher and Pamela Regan also looked at similarities and differences in qualities desired in their romantic partners and friends by a sample of 700 students at a large, midwestern university. Specifically, they explored the importance of 14 qualities that might be desired in a casual sex partner, dating partner, marriage partner, same-sex friend, and opposite-sex friend. Qualities included attractiveness, intelligence, warmth, earning potential, sense of humor, exciting personality, and similarity on interests, leisure activities, social skills, and background characteristics (such as race or social class). For romantic or sexual partners (for example, casual sex partner, dating partner, and marriage partner), prior sexual experience and sexual passion were also included. Looking across all five relationship types, certain qualities stood out as most desirable regardless of type of relationship. Warmth and kindness, openness, expressiveness, and a sense of humor were judged most desirable and most important. However, romantic partners were subjected to higher standards for these attributes, suggesting that such qualities are more important in romantic partners than in friends. Although intrinsic attributes may be desired in all relationships, they take on particular importance in romantic relationships. Meanwhile, romantic partners were expected to display extrinsic attributes (for example, qualities associated with appearance or social status) as well, whereas friends were not (Sprecher and Regan 2002).

The Importance of Love

Love is essential to our lives. Love binds us together as partners, spouses, parents and children, and friends and relatives. The importance of romantic love cannot be overstated. We make major life decisions, such as marrying, on the basis of love. Love creates bonds that we hope will enable us to endure the greatest hardships, suffer the severest cruelty, and overcome any distance. Because of its significance, we may even torment ourselves with doubts about the sincerity ("Is it really love?") or mutuality ("Do you love me as much as I love you?") of love.

Love is both a feeling and an activity. We feel love for someone and act in a loving manner. But the paradox of love is that it encompasses opposites, including both affection and anger, excitement and boredom, stability and change, bonds and freedom. Its paradoxical quality makes some ask whether they are really in love when they are not feeling "perfectly" in love or when their relationship is not going smoothly. Love does not give us perfection; however, it does give us meaning.

We can look at love in many ways besides through the eyes of lovers, although other ways may not be as entertaining. Whereas love was once the province of lovers, madmen, poets, and philosophers, social scientists have also taken a look at love. Although there is something to be said for the mystery of love, understanding how love works in the day-to-day world may help us keep our love vital and growing.

Love and American Families

Romantic love is the basis for family formation in the United States, as it has been for most of the last two centuries (Coontz 2004). Although American marriages were never quite as formally arranged as they have been in other places in the world, throughout the eighteenth century they were guided by more practical considerations and subject to more parental,

■ *While it is difficult to come up with a formal definition of love, we usually know what we mean when we tell someone we love them. Such feelings are important at the individual, relationship, and institutional level.*

©Esbin Anderson/The Image Works

especially paternal, control. By the end of the nineteenth century, however, most active parental involvement in their children's marriage choices had dissipated (Coontz 2004; Mintz 2004). Economic developments had decreased the dependency of adult children on their parents; increasingly, economic opportunity could be found without parental assistance, which freed people from worrying about the consequences of parental disapproval of their choice of mate. With increasing economic activity among women, a spread of legal and social recognition of women's rights, and enhanced opportunity for young people to meet and mingle, American courtship was further transformed (Mintz and Kellogg 1988; Murstein 1986). Love, as experienced, perceived, and pursued by individuals, became the vehicle that drove mate selection.

In the early decades of the twentieth century, new ideals about marriage and family emerged. Although American family life had already shifted from an economic to an emotional emphasis with the appearance of the democratic family, this was extended even further with the emergence and celebration of **companionate marriage,** wherein spouses were to be each other's best friends, confidants, and romantic partners (Mintz and Kellogg 1988). Love was the foundation upon which marriage was built and the criterion by which spouses were chosen.

Selecting a spouse on the basis of romantic love has consequences. It may lead to a greater tendency to idealize the partner, display affection toward the partner, and to attach more importance to sexual intimacy (Medora et al. 2002). Ironically, perhaps, the high emphasis we place on love as the basis for spousal choice contributes to the American patterns of divorce and remarriage. The qualities we "fall in love" with may not be easy to sustain across the lifelong duration of a marriage. Thus, we are more likely to perceive our marriages as "failures" when we sense that those qualities are gone or diminished. We then seek those same idealized qualities from subsequent marriage partners.

Within our marriage practices we find a number of distinct but related cultural beliefs about the character and place of love, including (1) that love is the criterion for choosing a spouse ("love and marriage, love and marriage, go together like a horse and carriage . . ." or "First comes love, then comes marriage . . ."), and (2) that love is uncontrollable and irrational ("Love is blind"). However, as much data show, Americans tend to follow a marriage pattern known as **homogamy**—the tendency to marry people much like themselves. The prevalence of homogamy casts some doubt on some of our ideas about love and marriage.

Perhaps love is more controllable and rational than we pretend (and therefore not blind), because we seem to fall in love with people like ourselves. On the other hand, if love *is* blind (that is, uncontrollable and irrational), it must not be the only determinant of mate selection. In other words, if—as the song lyrics suggest—love and marriage go together like a horse and carriage, we are selective in which horses we harness to our carriages. We don't marry simply because we've fallen in love and probably recognize some "loves" as unwise marriages. Finally, the social circles within which we live and move limit love. Thus, our "one and only" is drawn from a smaller pool than what the romantic mystique surrounding love suggests. With these qualifications in mind, we should still remember that most Americans who marry say they are marrying because they are in love.

In addition to the above, there are other beliefs comprising the ideology of romanticism in U.S. culture (see the "Exploring Diversity" box). Many Americans believe that love strikes powerfully upon first sight, that each of us has one and only one "one and only," and that as long as we love each other everything else will work out. As we will see, these beliefs are not as widely shared in other cultures as they are in the U.S. and western European societies.

Love across Cultures

Neither "falling in love" nor the experience of romantic love are unique to Americans; 90% of the 166 societies examined by William Jankowiak and Edward Fischer recognize and value love as an important element in building intimate relationships (Jankowiak and Fischer 1992). But love appears to have a more central role in American mate selection than in other Western societies (Goode 1982; Peoples and Bailey 2006). It fits well with and helps reinforce other features of American families and society. Love-based marriage validates the importance of individual autonomy and freedom from parental intervention and control, establishes the relative independence of the **conjugal family** from the extended family, and fits with the wider social freedoms granted to adolescents and young adults (Goode 1982). All of these make romantic love functional in industrial societies (Goode 1977). Conversely, in societies in which nuclear families are deeply embedded in extended families, or in which it is important for economic or political reasons to create alliances and exchanges through marriages, romantic love is not a central factor in mate selection. In such societies it may be entirely irrelevant (Medora, et al. 2002).

Love reflects the positive factors, such as caring and attraction, that draw two people together and sustain them in a relationship. Related to love, **commitment** reflects the stable factors, including not only love but also obligations and social pressure, that help maintain a relationship, for better or for worse. Although love and commitment are related, they are not inevitably connected.

It is possible to love someone without being committed, without making the sacrifices and adjustments needed to sustain the relationship. It is also possible to be committed to someone without loving that person. We might remain in a relationship, such as marriage, because of perceived obligation, for the sake of the children, or because of the fear of how other aspects of our life might be negatively affected. Yet, when all is said and done, most of us long for a love that includes commitment and a commitment that encompasses love.

Gender and Intimacy: Men and Women as Friends and Lovers

As shown in the last chapter, many areas of our lives are gendered, meaning they are experienced differently by males and females. Love, friendship, and intimacy are just such areas. In most scientific literature, there is a recurring theme highlighting men's supposed shortcomings as friends and partners. Unlike women, who are said to relate more easily and deeply with others and who develop a greater capacity for disclosing and sharing their inner selves, men maintain greater emotional distance, even as they experience their closest relationships.

Francesca Cancian (1985) argued that there is a gender bias in our cultural constructions of love that distorts our understanding of how both men and women love. Through the **feminization of love,** in other words, by defining or "seeing" love in largely expressive terms (telling each other how you feel), we ignore important qualities or aspects of both women's and men's intimacy. For example, much of what women do as expressions of love (for spouses and children, especially) is **instrumental,** consisting of tasks associated with nurturing and caregiving, more than **expressive** displays, such as telling others how much we care about or love them. Although done *out of love,* such activities may not be seen as *displays of love.* Likewise,

Although most cultures recognize and value love, the meanings and expectations attached to love vary, sometimes greatly. In individualistic cultures, such as the United States, people value **passionate love**, the kind experienced as an "intense longing for another," a "lovesickness" that often takes us on "a roller coaster of elation and despair, thrills and terror" (Kim and Hatfield 2004). As a student once put it to one of this book's authors, such love sometimes feels "like you've been run over by a truck—but in a good way." If reciprocal, passionate love brings us ecstasy and fulfillment; if unrequited, it can bring us emptiness and sadness. In individualistic cultures, it is expected that people will marry out of such an intense love, which is to be the most important factor in finding a spouse. This is part of the greater **romanticism** found in such societies (Medora et al. 2002). Prime importance is given to the affective element of relationships, and there is a stronger belief in each of the following components associated with romanticism:

1. Love conquers all.
2. For each person there is "one and only one" romantic match.
3. Our beloved should and will meet our highest ideals.
4. Love can and often most powerfully does strike "at first sight."
5. We should follow the heart not the mind when choosing a partner.

In collectivist cultures, including many Asian societies such as Japan, China, India, and Korea, individual happiness is subordinated to group well-being. Loyalty, especially to the wider kin group and extended family,

dictates decisions people make about entering marriage and who they shall marry. Higher value is placed on what Elaine Hatfield and Richard Rapson (1993) call **companionate love,** a less intense emotion in which warm affection and tenderness is felt and expressed toward those to whom our lives are deeply connected. Importance is placed on shared values, commitment, intimacy, and trust. Passionate love and marriage based on romantic love are seen negatively as potential threats to family approved and/or arranged marriages, associated with sadness and jealousy, and thought to interfere with family closeness and kin obligations (Kim and Hatfield 2004). More traditional and less developed collectivist eastern cultures, such as China and India, are reported to attach the least importance to romantic love. The idea of baring the soul, sharing or confiding innermost and heartfelt feelings to a partner receives more cultural validation in the United States and other individualistic cultures than in collectivist cultures (Kito 2005).

Additional cross-cultural research compared the attitudes toward romantic love of college undergraduates from the United States, Turkey, and India. The United States is an individualistic society in which romanticism is idealized and topics such as love, dating, and finding a partner are openly and frequently discussed, often becoming subjects of considerable media attention. To the contrary, India is a sexually conservative and more collectivist society in which family stability is valued above individual gratification and autonomy and marriages are frequently arranged. Turkey is a society "in transition." The ideal of romantic love was introduced as part of the processes of westernization and secularization. Although families may still

"assist" in the process of finding a spouse, formal arranging is uncommon. Comparing the attitudes of college undergraduates from the three countries, researchers found the students from the United States to be most and the Indian students to be least romantic (Medora et al. 2002). Using a 29-item, 5-point scale (from 1 = Strongly disagree to 5 = Strongly agree), individuals could score between a low of 29 and a high of 145, with higher scores indicating more romanticism. Items included statements such as the following:

"Somewhere there is an ideal mate for most people. The problem is just finding that one."

"Love at first sight is often the deepest and most enduring type of love."

"Common interests are really unimportant; as long as each of you is truly in love, you will adjust."

The average scores were as follows:

	U.S.	Turkey	India
	N = 200 (86 male, 114 female)	N = 223 (114 male, 114 female)	N = 218 (98 male, 120 female)
Mean score	86.09	74.92	70.33
Standard deviation	15.6	13.6	14.4

In all three national subsamples, females scored higher than males. Overall, the gender difference was as shown below:

	Male	Female
Mean score	74.63	79.81
Standard deviation	15.5	16.0

From Medora et al. 2002.

because men believe they "show" or express love by *what they do* more than by *what they say,* conceptualizing or recognizing love largely in terms of things said renders men's sincere attempts to show intimacy invisible and leaves them looking especially inadequate as intimate partners (Hook et al. 2003).

Misty Hook and colleagues note the following gender differences in intimacy (2003). To women intimacy means sharing love and affection and expressing warm feelings toward someone. To men, being intimate may mean engaging in sexual behavior and being physically close. Women display intimacy in their verbal exchanges, which can become "negotiations for closeness, during which people try to reach agreement and both give and receive support" (Hook et al. 2003, 464). Women express more empathy, being more likely than men are to come to an understanding of what others are feeling. Men are more likely to react to disclosures of negative or problematic emotions by trying to solve a supposed problem. Men also associate intimacy with "doing" things together or for another person and often find women's need or desire to "talk things through" puzzling. Although men may feel as though they show intimacy by sharing activities and interests, telling stories, and even sitting together in silence, women associate intimacy with being together and sharing themselves with another (Hook et al. 2003). To reinforce Cancian's critique of feminized conceptions of love, and to extend it more broadly to conceptualizations of intimacy, which of the preceding descriptions seems like "true" or "real" intimacy, women's or men's?

Gender and Friendship

The critique of the cultural feminization of love applies as well to friendship. We tend to conceptualize "real" or "true" friendship by such qualities as emotional support and self-disclosure—telling each other innermost feelings and attitudes and sharing personal experiences (Sprecher and Hendrick 2004). Friends share their inner lives with each other; sharing how they feel, including how they feel about each other. The closer the friend, the more personal and more frequent the disclosures. This conceptualization measures friendship against a standard more consistent with female friendships and may underestimate the "real" intimacy that men's friendships contain, especially if such closeness is expressed in other, more covert ways (Swain 1989; Twohey and Ewing 1995, Hook et al. 2003).

Indeed, there are gender differences in disclosure in same-sex friendships. If intimacy means self-disclosure, as early as age 6, female friendships are more intimate. This gender difference is accentuated in adolescence and persists into and through adulthood (Benenson and Christakos 2003). Women experience and express "closeness" with each other through conversation, disclosing more of both a positive and a negative nature (Hook et al. 2003; Sprecher and Hendrick 2004). Given the expectation and opportunity for greater sharing and disclosure between female friends, we might predict that their friendships would protect females from depression and emotional difficulty more than male friendships would protect boys and men. However, research shows that females more than males experience depression in adolescence. Psychologist Amanda Rose explains these seemingly contradictory findings (females disclose and share more but are more depressed) through the concept of **co-rumination.** Co-rumination may be thought of as *excessive disclosure* or sharing of personal problems—as in either discussing the same problem repeatedly, speculating about problems, mutually encouraging each other to talk about problems and, generally, "focusing on negative feelings" (Rose 2002, 1,830). Rose uses as an example, "talking at length about whether the ambiguous behavior of a boyfriend or girlfriend is signaling the demise of the relationship" (1,830). Co-rumination points to the possibility that disclosure that is excessive and/or focuses too much on negative topics may not benefit the friends sharing such disclosures.

There are other noteworthy differences in the number and nature of male and female friendships. Males reportedly have more friends (Dolgin 2001). In childhood and adolescence, boys spend more of their time in groups and in group activities, especially physical activities, games, and sports; girls spend more time in dyads and engage in more mutual disclosure. Looking at "closest friendships," girls' closest friendships tend to exist "in isolation," boys' closest friendships tend to be embedded in a larger group context. As a consequence, when conflict arises between close friends, males may have an easier time reaching resolution. Within a group context we can draw others in, drawing upon third parties to act as mediators, serve as allies, or even become alternate partners. With more loyalty to the larger group, one-on-one conflict may be kept to a lower level (Benenson and Christakos 2003).

Boys spend less time sharing, less time co-ruminating, but as a consequence may be spared some

Popular Culture   Love in the World of Disney

Do you remember Belle? How about Ariel? Both are characters in animated Disney films where love relationships take center stage. The films in which we meet them, Belle in *Beauty and the Beast* (1991) and Ariel in *The Little Mermaid* (1989), are among 26 full-length animated Disney movies analyzed by Lisa Tanner, Shelley Haddock, Toni Schindler Zimmerman, and Lori Lund in research published in *The American Journal of Family Therapy* (2003).

Beginning with an assertion that children gain information and develop their understanding of couples and families from numerous sources *other than their own families,* the authors set out to identify dominant messages and themes found in the medium of animated Disney films. Noting that children use media, popular stories, myths, and fairy tales to make sense of themselves and their social environment, Tanner and colleagues turn their attention to Disney, "a major contributor to most avenues of children's media . . . (including) a major television network, cable television networks, and radio stations . . . children's books, cartoons, movies, videos, computer software and games . . . backpacks, lunch boxes, and clothing" (2003, 356). Using the 26-film sample, they set out to "identify the prominent themes about family relationships" (357).

The films studied included early Disney classics such as *Snow White and the Seven Dwarves* (1937), *Pinocchio* (1940), and *Lady & the Tramp* (1955), as well as more recent films such as *Tarzan* (1999), *Mulan* (1998), and the two previously mentioned films, *Beauty and the Beast* (1991) and *The Little Mermaid* (1989). Although their article also looks at portrayals of families (for example, "Who comprises a family?"

■ *The animated love story,* Beauty and the Beast, *is one of 26 Disney films analyzed for its messages about couples and relationships. It was also the first animated film nominated for a Best Picture Oscar.*

Courtesy of Everett Collection

of the emotional fallout from dwelling on problems. In addition, men display less affection, using either words or touch, than women do toward their friends (Dolgin 2001). Yet female friendships appear to be more fragile. With increasingly intense sharing comes more opportunity for misunderstanding or even for conflict. Furthermore, when females' closest friendships end they are more likely to "find themselves alone" (Benenson and Christakos 2003).

Men are more open and intimate in cross-sex relationships than in their friendships with other men (Dolgin 2001). Wives or romantic partners are often the closest confidants in men's lives. In those relationships, men find themselves reaping the benefits that come from greater disclosure, even if the levels at which they disclose don't match what their partners desire. Certainly, the tendency to funnel their intimacy into one relationship, especially marriage, is consistent with the cultural expectations of marriage as best-friendship. But even outside marriage, the depth of men's disclosure to women stands in contrast to the male–male style, suggesting not so much inability as unwillingness at or discomfort with male–male intimacy.

and "How are families created and maintained?") and parents (for example, "Which parents are present?" and "What is the nature of mothers and fathers?"), here we concern ourselves with portrayals of couple relationships in keeping with the theme of the present chapter.

Who Comprises a Couple?

Unsurprisingly, all couples shown in the Disney films were heterosexual. The researchers note that in *Mulan,* as long as Mulan was thought to be a man, she and Lee Shang were only friends. As soon as he discovered that Mulan was a woman, they fell in love.

How Are Couple Relationships Created?

Because 3 of the 26 films provided no information, this analysis was based on the remaining 23 movies. In 78% of the films (18 films) we find the notion of "love at first sight." It typically took just minutes for couples to fall in love. In *Pocahontas,* John Smith and Pocahontas fell in love despite not being able to speak the same language. Appearance alone was enough to bring them together. In *The Little Mermaid,* Ariel falls in love with Prince Eric at first sight; he falls in love with her voice.

How Are Couple Relationships Maintained?

In most of the movies, there was a "happily ever after" theme; couples fell in love, married, and lived together easily, as well as happily, ever after. Tanner and colleagues cite the example of Snow White, who managed to fall in love while asleep and who, when asked if it was hard to fall in love, replied, "It was easy." Commonly, in the selected Disney films, falling in love seemed to follow too quickly and easily upon a man and a woman meeting. In only three (13%) of the sampled films (*Rescuers Down Under, Mulan,* and *Tarzan*) did falling in love take time, at least enough time for the couple to get to know each other.

What Are Couple Relationships Like?

Although many films provided too little information to generalize from, in 8 of the 23 relevant movies (34.8%), couples were unequal in the amount of power that each partner had. Of these eight, only one (*Alice in Wonderland*) depicted the female (in this case, the Queen of Hearts) as more powerful than the male. Three movies (*101 Dalmatians, The Rescuers Down Under,* and *Tarzan*) depicted couple relationships

between equals who shared power in their relationships.

Traditional Gender Representations Predominate

The authors concluded that most couples in the world of Disney are portrayed in relatively traditional ways. Men and women fall in love at first sight and live happily ever after, and the films stress appearance as the most important factor in selecting a partner and entering a relationship. Marriage and children are presented as the life goal, even though in portraying marriage and motherhood women are often powerless and marginalized. This gives girls a mixed message—strive for something that, once obtained, will not treat you fairly.

Movies are but one element of popular culture to which children are exposed. It is hardly likely that just from watching one or a few (or even 26) Disney films they will expect reality to fit the animated images and themes. But these images and themes do have an effect, especially when they are consistent with other elements of popular culture to which children are exposed.

SOURCE: Tanner et al. 2003, 355–373.

Gender and Love

With regard to love, the genders differ in a number of ways. Men fall in love more quickly than women, describe more instrumental styles of love (that is, love as "doing"), and are more likely to see sex as an expression of love. Because men have fewer deeply intimate, self-disclosing friendships, when they find this quality in a relationship they are more likely to perceive that relationship as special. Having more intimates with whom they can share their feelings, women are less likely to be as quick to characterize a particular relationship as love. In addition, traditionally, women could do so less safely unless other, economic, criteria were also met. Thus, men could afford to be more romantic, and women needed to be more realistic (Knox and Schacht 2000). In a study of 147 never-married undergraduate students designed to look for and at gender differences in timing and reason for saying "I love you," David Knox, Marty Zusman, and Vivian Daniels (2002) found the following:

■ In heterosexual relationships, males say "I love you" before their partners do.

- Males say "I love you" in part to increase the likelihood that their partner will agree to have sex with them.

Other gender differences surface in the connection between love and sex. Although men are often depicted as easily separating sex and love, there is evidence that within relationships men see sex as a means of expressing or showing love (Rubin 1983; Cancian 1987). Women's experiences of love and sexuality are different. Although sexual scripts have been changed in the direction of more open and acceptable expressions of female sexuality, to feel loved requires more than sexual expression.

Gender differences may be more exaggerated in *what people say* than in *what they do*. This is certainly the case with friendship, where bigger differences show up in how the genders talk *about* friendship than in what they experience as friends (Walker 1994). Karen Walker discovered that although her male and female informants validated the more common characterizations of how women's and men's friendships differ, on talking about *their friends,* they revealed more complex, often gender "inappropriate" patterns of relating. Thus, men had male friends to whom they disclosed personal information, and women had some relationships that resembled men's friendship patterns (Walker 1994). Similarly, despite the gendered expectations and definitions of love, within the context of heterosexual, romantic love relationships, significant gender differences in self-disclosure are absent; men and women disclose similarly (Sprecher and Hendrick 2004).

In identifying the factors that shape men's and women's intimate relationships, most researchers point to aspects of gender socialization (McGill 1985; Basow 1992). Some emphasize dominant cultural constructions of masculinity and femininity, wherein men are inexpressive, competitive, rational, and uncomfortable with revealing their innermost feelings, especially of vulnerability or of affection toward other males (Bell 1981; Rubin 1985; McGill 1985; Stein 1986). Women are allowed and encouraged to express a wider range of feelings without concern for the consequences.

Other researchers suggest that gender-specific relationship styles emerge because of differences in how males and females resolve the developmental task of early childhood identity formation (Chodorow 1978; Rubin 1985). As a result of being "mothered," and having the closest early relationship be with a female, the genders develop different ways of relating. Females develop "permeable ego boundaries" open to relationships with others, and they retain a strong connection with their mothers. Males are forced to separate from their mothers, identify with absent or less present fathers, and build boundaries around themselves in relation to their most nurturant caregivers. This haunts them throughout their later relationships, because it makes them less able to "connect" intimately with others (Rubin 1985). Women experience themselves in the context of relationships, whereas men—depicted as "selves in separation"—remain oriented more toward independence and task completion (Kilmartin 1994).

We might also emphasize the role-model consequences of being "mothered" but not "fathered." Without a loving, attentive, nurturing presence from fathers or other male role models, boys come to inhibit their own emotional expressiveness, identifying such behavior as typical of mothers (and women in general) and to be avoided. Because of the relative involvement of mothers versus fathers in caring for young children, and the greater prevalence of single-mother over single-father households, boys have fewer available role models for intimacy. Furthermore, what role models they have are products of gender socialization and carry a style of relating that results from that socialization. Girls have the opportunity to observe up close a caring, loving female role model from which they learn how to relate and express love.

Finally, still others stress evolutionary explanations for gender differences. Beginning with the idea that each gender has different "reproductive strategies," differences in intimacy are linked to such sex-specific goals. For males, the objective is to reproduce as widely as possible, seeing that their genetic material is spread widely in multiple offspring. For women, the objective is to see that each child successfully survives to a healthy adulthood. Such a difference is said to explain numerous other differences, especially in areas of intimacy, love, and sexuality. For example, from an evolutionary perspective on qualities desired in romantic and sexual partners, females will desire males of high status who are ambitious and dependable. Males will desire physically attractive females (Sprecher and Regan 2002).

Exceptions: Love between Equals

The gender differences depicted earlier, although common in the literature on love and intimacy, are not inevitable. As noted in the previous chapter, there are marriages and intimate heterosexual relationships that depart from traditional gender patterns (Schwartz 1994; Risman and Johnson-Sumerford 1997). Pepper Schwartz's research on peer couples illustrates loving relationships that avoid the aforementioned gender patterns. Schwartz conceptualized **peer marriage** as a relationship built on principles of equity, equality, and **deep friendship.** The emphases on equity and equality resulted in shared chores, equal say in decision making, and equal involvement in childrearing. More important for now is the element of deep friendship. These couples most valued an intense companionship, "a collaboration of love and labor in order to produce profound intimacy and mutual respect" (Schwartz 1994).

In peer marriages spouses become more alike over time, and thus both husbands and wives are more likely to display and value a blend of female and male styles of intimacy. Women value and appreciate the instrumental displays of love from their partners (for example, finding her husband has had her car serviced) because they know what it is like to make or take time from their demanding daily routines to attend to such things. Because husbands are more involved in daily domestic and childrearing routines, they share interests and concerns with their wives that traditional spouses do not. With enlarged identities outside the marriage and home, peer wives also need less of the conventional, conversational demonstrations of love and affection. They and their husbands have "learned love on each other's terms" (Schwartz 1994).

Schwartz's peer couples, like Risman and Johnson-Sumerford's post-gender couples, are uncommon. They represent what is possible in marriage, but creating such a lifestyle requires both an ideological commitment to sharing and equality and an ability to withstand scrutiny and curiosity from more typical couples. Most such lifestyles also require each spouse to have a job or career that the other recognizes as equal in importance to his or her own.

Showing Love: Affection and Sexuality

Within relationships based on romantic or passionate love, the emotional connection between partners is expressed in many ways, including typically through displays of affection and through sexual desire and activity. The state of "being in love" is assumed by most people to include sexual desire. Two people in a relationship absent sexual desire are assumed to not be in love (Regan 2000). Psychologist Lisa Diamond challenges this assumption, noting that sexual desire often occurs in the absence of romantic or passionate love and, more controversially, that romantic love, even in its earliest and most passionate stage, does not require sexual desire (Diamond 2003).

Although love and sex are separate phenomena, recent research shows that for both men and women sex often includes intimacy and caring, key aspects of love, and love is most often expected to include sexual desire. Men and women who feel the greatest sexual desire for dating partners are also likely to report the strongest feelings of passionate love. Interestingly, sexual *activity* (mean weekly number of "sexual events" in which partners engaged) is not associated with amount or depth of passionate love (Regan 2000). Pamela Regan, Elizabeth Koca, and Teresa Whitlock asked 120 college undergraduates to list all features that they considered *prototypical* of being in (passionate) love. Respondents generated a list that included 119 features, and sexual desire was the second most frequently mentioned feature (listed by 65.8% of the sample). Kissing (10%), touching or holding (17.5%), and sexual activity (25%) were mentioned far less often. Nevertheless, gender differences do exist, especially in terms of more casual relationships. (See Chapter 6 for a further discussion of sexuality.)

Besides sexual intimacy, there are many other ways we show intimacy and love. Some such displays occur openly, in public, as we say or do things that show others that we are a couple. Holding hands, being out together alone, telling others that we are a couple, and meeting our partner's parents are examples of public displays of affection and couple status (Vaquera and Kao 2005). More privately, we may exchange presents, tell each other how we feel (saying that we love each other), and just think about ourselves as a couple. Finally, the physical acts, from kissing to touching under clothes or with no clothes on, touching each other's genitals, and having sexual intercourse, are all "intimate displays" (Vaquera and Kao 2005).

Using data drawn from the National Longitudinal Study of Adolescent Health, with its large, nationally representative sample of high school students, Elizabeth Vaquera and Grace Kao examined how displays of affection varied between intraracial and interracial couples. Noting that interracial relationships are still a small percentage of all couple relationships in the

Understanding Yourself

"What Kind of Touching Makes You Feel Loved?"

For those of you who are or have been in romantic love relationships, which of the following makes or made you feel most loved, understood, or satisfied with your relationship? Is or was it the amount of hugs? Gentle massages or backrubs? Is or was it cuddling with or holding your partner? Being kissed on the face? Being kissed on the lips? Being caressed? What about simply holding hands? This is the subject Andrew Gulledge, Michelle Gulledge, and Robert Stahmann sought to explore in their study of 295 college students at Brigham Young University. How similar or different are your answers to their findings?

Gulledge, Gulledge, and Stahmann hypothesized that individuals who were more physically affectionate with their romantic partners would be more satisfied with their relationships and generally happier than those who were less physically affectionate. They looked specifically at the seven types of physical affection mentioned previously, asking respondents to rank each of the seven from most to least in terms of the following dimensions: favorite, frequent, intimate, and expressive of love. Before you read any further, try ranking them yourself, from most (1) to least (7), thinking about a current, former, or even anticipated or imagined relationship.

Think about your partner. Would his or her answers likely be the same? Now, consider how much you think each of the forms of physical affection affects whether you and your partner are (were, would be) satisfied with the relationship and with each other. In other words, are certain forms of physical affection more strongly associated with relationship happiness or satisfaction?

Finally, answer each of the following by indicating with a score of 1 to 7 (1 = Strongly disagree, 7 = Strongly agree) how you respond to each of the following statements. Where low rankings on the prior list indicated more favorite, or more intimate, and so on, low scores for the following items indicate strength of disagreement and high scores indicate how strongly you agree with the statement.

Item (PA = Physical affection)	Reply (1–7)
PA is important in achieving happiness or satisfaction in romantic relationships.	_____
There is less conflict in romantic relationships when partners give each other PA.	_____
PA is a good way of showing romantic love for another.	_____
I feel more loved by my romantic partner when he or she gives me PA.	_____
I feel more understood by my romantic partner when he or she gives me PA.	_____

Gulledge, Gulledge, and Stahmann found the following:

1. There were both similarities and differences in men's and women's rankings of favorite, frequent, intimate, and expressiveness of love associated with physical affection types. The rankings by gender are as follows:

Form of Affection	Favorite	Frequent	Intimate	Expressive of Love
1. Backrubs or massages				
2. Caressing or stroking				
3. Holding hands				
4. Cuddling or holding				
5. Kissing on the lips				
6. Kissing on the face				
7. Hugging				

	Favorite		Frequent		Intimate		Expressive of Love	
	Male	Female	Male	Female	Male	Female	Male	Female
(Most)	Kissing on lips	Cuddling	Cuddling	Holding hands	Kissing on lips	Kissing on lips	Kissing on lips	Kissing on lips
	Cuddling	Kissing on lips	Hugging	Cuddling	Cuddling/ holding	Cuddling/ holding	Cuddling/ holding	Cuddling/ holding
	Hugging	Hugging	Kissing on lips	Hugging	Caressing/ stroking	Caressing/ stroking	Caressing	Kissing on face
	Backrubs	Holding hands	Holding hands	Kissing on lips	Kissing on face	Kissing on face	Kissing on face	Caressing
	Caressing	Kissing on face	Caressing	Kissing on face	Backrubs	Backrubs	Hugging	Holding
	Kissing on face	Backrubs	Kissing on face	Caressing	Hugging	Holding hands	Holding hands	Hugging
(least)	Holding hands	Caressing	Backrubs	Backrubs	Holding hands	Hugging	Backrubs	Backrubs

Both men and women favor kissing on lips and cuddling (as 1 and 2 or 2 and 1), but women favor holding hands significantly more than men do. Men favor giving backrubs more than women do. Fairly consistent agreement characterizes the rankings for most intimate, and slightly less but still consistent rankings are found between women and men in terms of how expressive of love each kind of physical affection is.

2. All types of physical affection except holding hands and caressing/ stroking are significantly correlated with satisfaction with relationship or partner. Most highly correlated with relationship satisfaction is the amount of backrubs a couple gives to each other. Gulledge, Gulledge, and Stahmann suggest this may be because of the more "selfless" nature of the display, and the fact that they take more energy for a sustained period, suggesting determination and dedication. Also worth noting, conflict was more easily resolved with increasing amounts of kissing on the lips, cuddling or holding, and hugging.

3. Respondents most strongly agree (mean = 6.01) that they feel more loved, and more understood (mean = 5.01), when receiving physical affection. They further believe strongly that physical affection is a good way to show romantic love (mean = 5.97) and is important to achieve happiness or satisfaction in a relationship (mean = 6.05).

Some qualification on these findings is necessary. Gulledge, Gulledge, and Stahmann note that the absence of sexual activity among the rated acts of physical affection may limit the findings, as can the absence of other nonsexual, even nonphysical acts (gazing, talking together, saying I love you, and so on). Can you think of still other things that the researchers may have left out?

The sample is also a potential limiting factor. The researchers note that cultural differences in the meaning of some acts make the findings more limited. The sample is further limited in that it consists of college students, most of whom are members of the Church of Jesus Christ of Latter-day Saints, a conservative religious body that frowns on premarital sex and warns against excessive displays of physical affection. A more diverse sample may therefore generate different results.

Still, research such as this demonstrates how much we use physical means to display and convey intimacy and love.

SOURCE: Gulledge, Gulledge, and Stahmann 2003, 233–242.

United States and therefore may not be as openly or enthusiastically accepted as intraracial couple relationships, Vaquera and Kao look at whether the potential "stigmatizing" of such relationships may lead to different ways of behaving as a couple. Other research has determined that interracial couples limit their exposure, hoping in part to avoid negative reactions, rejection, and pressure. Vaquera and Kao found that in terms of both public and private displays, interracial couples display lower levels of affection. However, when it comes to intimate displays of affection, no difference was found between interracial and intraracial couples. This is consistent with the attention to stigma, because only in settings that involve no others are there no differences between interracial and intraracial couples.

Vaquera and Kao also report differences in the displays of affection across racial groups (that is, among intraracial couples of different racial backgrounds). They determined that compared with Caucasians, African American couples displayed less public affection but more intimate affection. Hispanics, Asians, and Native Americans display lower levels of intimate affection than do African American or Caucasian couples. All minority couples displayed less public affection than white couples did. In terms of "private displays," no statistically significant differences were found among racial groups (Vaquera and Kao 2005).

Gender, Love, and Sexuality

For both women and men, sexual desire—but not sexual activity—is associated with passionate love (Regan 2000). Yet gender differences have been observed in the relationship between love and sex. Men and women who are not in an established relationship have different expectations. Men are more likely than women to more easily separate sex from affection, whereas women attach greater importance to relationships as the "context" for sexual expression (Laumann et al. 1994; Diamond 2004). Lisa Diamond suggests that there are a number of possible reasons for this gender difference. First, men are more likely than women to first experience sexual arousal "in the solitary context of masturbation," whereas women are more likely to experience sexual arousal for the first time within a heterosexual relationship. Second, as shown in the next chapter, women and men have been differently socialized about the legitimacy of sexual expression. Women have been expected and encouraged to restrict sexual desire and activity to intimate relationships in

which they find themselves. Men have been raised with more "license" regarding casual sexual relationships. Finally, Diamond notes that biological factors may partly explain the gender difference. Specifically, certain neurochemicals, such as oxytocin, that mediate bonding also mediate sexual behavior. Much as oxytocin might be associated with caregiving, it is also released in greater amounts in women than in men during sexual activity. Oxytocin is also associated with orgasmic intensity (Diamond 2004).

Sexual Orientation and Love

Love is equally important for heterosexuals, gay men, lesbians, and bisexuals (Patterson 2000; Aron and Aron 1991; Keller and Rosen 1988; Kurdek 1988; Peplau and Cochran 1988). Given that men, in general, are more likely than women to separate love and sex, it is unsurprising that gay men are especially likely to make this separation. Although gay men value love, they also tend to value sex as an end in itself. Furthermore, they place less emphasis on sexual exclusiveness in their relationships (Patterson 2000). Researchers suggest, however, that heterosexual males are not very different from gay males in terms of their acceptance of casual sex. Lesbians and heterosexual couples tend to be more supportive than gay men of monogamy and sexual fidelity. This is probably because of gender more than sexual orientation; heterosexual males would be as likely as gay males to engage in casual sex if women were equally interested. Women, however, are not as interested in casual sex; as a result, heterosexual men do not have as many willing partners available as do gay men (Foa et al. 1987; Symons 1979).

For lesbians, gay men, and bisexuals, love has special significance in the formation and acceptance of their identities. Although significant numbers of women and men have had sexual experiences with members of the same sex or both sexes, relatively few identify themselves as lesbian or gay. An important element in solidifying such an identity is loving someone of the same sex. Love signifies a commitment to being gay or lesbian by unifying the emotional and physical dimensions of a person's sexuality (Troiden 1988). For the gay man or lesbian, it marks the beginning of sexual wholeness and acceptance. Some researchers believe that the ability to love someone of the same sex, rather than having sex with him or her, is the critical element that distinguishes being gay or lesbian from being heterosexual (Money 1980).

Sex without Love, Love without Sex

How exactly are love and sex linked? Is love necessary for sex? *Must* romantic love have a sexual component? Most of us might assume that love and sex *should be* connected, but our assumption is based mostly on our values and therefore cannot be answered by reference to empirical or statistical data. What we can address empirically is the extent to which this assumption is shared. Pamela Regan (2000) has determined that the link between sex and love is really a link between sexual *desire* and love, *not* between sexual *activity* and love. Sexual desire is assumed to be a basic "distinguishing feature" of passionate love, whereas it is understood that sexual activity may take place in or outside of a love relationship. In research with a sample of heterosexual women and men in relationships, Regan found that the respondents who felt the greatest desire for their partners also reported the greatest amount of love but that sexual activity levels were unrelated to the amount of love respondents felt.

To address the sex–love connection from the other direction, as in the question of the possibility of love without sex, Lisa Diamond notes that "it seems that individuals are capable of developing intense, enduring, preoccupying affections for one another regardless of either partner's sexual attractiveness or arousal" (2004, 116). She uses the examples of prepubertal children who describe intense romantic infatuations without having experienced the hormonal changes necessary for true sexual desire and of individuals who fall in love with partners of the "wrong gender" (such as heterosexuals falling in love with partners of the same gender and lesbians or gay men who fall in love with partners of the opposite sex). Although, as Diamond indicates, some may suspect that such relationships reflect suppressed sexual feelings, analysis of written reports of those involved in such situations suggests that they more genuinely reflect the presence of love without sexual desire.

Still, the normative expectations clearly suggest connections between sexual desire and romantic love. First, they prescribe that sex within a romantic love relationship is more acceptable and more legitimate than sex outside of a relationship context. Second, they convey the expectation that sexual longing and desire are part of loving another.

To believe that sex does not require love as a justification, argues John Crosby, does not deny the significance of love and affection in sexual relations. Love and affection are important and desirable for enduring relationships. They are simply not necessary, Crosby believes, for affairs in which erotic pleasure is the central feature (Crosby 1985).

Ironically, although sex without love may violate social norms, it is a less threatening form of infidelity. As you will see before this chapter's end, even those who accept their partners' having sex outside the relationship find it especially difficult to accept their partners' having a meaningful affair. As Philip Blumstein and Pepper Schwartz put it, "They believe that two intense romantic relationships cannot coexist and that one would have to go" (Blumstein and Schwartz 1983).

Love, Marriage, and Social Class

Gender is only one variable related to how we experience love and how love is associated with marriage. In many ways, our romantic view of love-based marriage represents a middle-class version of marriage. Among upper-class families, there is a greater urgency in assuring that our children marry the "right kind" because considerable wealth and social position may be at stake. Furthermore, upper-class families have more ability to exercise such control by the threat of withholding inheritance from the maverick child who dares act without consideration of parental preference (Goode 1982). Among the working class, marriage was often entered as a means to escape economic instability and parental authority and to be seen as an adult (Rubin 1976, 1992). This may now be less true, as working-class marriages have taken on more characteristics of the middle-class ideal (for example, expecting more sharing and communication) (Rubin 1995). Still, the economic circumstances that define someone's life may induce different ways of linking love and marriage.

But What *Is* This "Crazy Little Thing Called Love"?

Despite centuries of discussion, debate, and complaint by philosophers and lovers, no one has succeeded in finding a single definition of love on which all can agree. Ironically, such discussions seem to engender conflict and disagreement rather than love and harmony.

Because of the unending confusion surrounding definitions of love, some researchers wonder whether such definitions are even possible (Myers and Shurts 2002). In the everyday world, however, most of us seem to have something in mind and agree on what we mean when we tell someone we love him or her. We may not so much have formal definitions of love, as we do **prototypes** of love (that is, models of what we mean by love) stored in the backs of our minds. Some researchers suggest that instead of looking for formal definitions of love, it is more important that we examine people's prototypes; that is, we consider what people mean by the concept of love when they use it. When we say "I love you," we are referring to our prototype of love rather than its definition. If we find ourselves thinking about our partners all the time, feeling happy when we are with them and sad (or less happy) when we are apart, and spending all our available time together, we compare these thoughts, feelings, and behaviors to our mental models or prototypes of love (Regan 2003). If our experiences match the different characteristics of love, we then define ourselves as in love. By thinking in terms of prototypes, we can study how people actually use the word *love* in real life and how the meanings they associate with love help define the progress of their intimate relationships.

To discover people's prototypes, researcher Beverly Fehr (1988) asked 172 respondents to rate the central features of love and commitment. In order, the 12 central attributes of love they listed are as follows:

- Trust
- Caring
- Honesty
- Friendship
- Respect
- Concern for the other's well-being
- Loyalty
- Commitment
- Acceptance of the other the way he or she is
- Supportiveness
- Wanting to be with the other
- Interest in the other

There are many other characteristics identified as features of love (euphoria, thinking about the other all the time, butterflies in the stomach, and so on). These, however, tend to be peripheral. As relationships progress, the central aspects of love become more characteristic of the relationship than the peripheral ones. According to Fehr (1988), the central features "act as

true barometers of a move toward increased love in a relationship." Similarly, violations of central features of love are considered more serious than violations of peripheral ones. A loss of caring, trust, honesty, or respect threatens love, whereas the disappearance of butterflies in the stomach does not.

Love is also expressed behaviorally in several ways, with the expression of love often overlapping thoughts of love:

- Verbally expressing affection, such as saying "I love you"
- Self-disclosing, such as revealing intimate personal facts
- Giving nonmaterial evidence, such as offering emotional and moral support in times of need and showing respect for the other's opinion
- Expressing nonverbal feelings such as happiness, contentment, and security when the other is present
- Giving material evidence, such as providing gifts or small favors or doing more than the other's share of something
- Physically expressing love, such as by hugging, kissing, and making love
- Tolerating and accepting the other's idiosyncrasies, peculiar routines, or annoying habits, such as forgetting to put the cap on the toothpaste

These behavioral expressions of love are consistent with the prototypical characteristics of love. In addition, research supports the belief that people "walk on air" when they are in love. Researchers have found that those in love view the world more positively than those who are not in love (Hendrick and Hendrick 1988).

Although little research exists on ethnicity and attitudes and behaviors associated with love, one study of Mexican American college students suggests that they share many of the same attitudes and behaviors described previously (Castaneda 1993). Both females and males valued communication or sharing, trust, mutual respect, shared values and attitudes, and honesty. Data from white, middle-class adults indicate that men and women are quite similar in their love attitudes across adulthood (Montgomery and Sorell 1997).

Studying and Measuring Love

A review of the research on love finds a number of definitions, which are tied to a variety of research instruments that have been developed to measure love.

Jane Myers and W. Matthew Shurts (2002) reviewed the instruments that researchers have developed and that other researchers and/or clinicians might use, ultimately identifying nine different instruments. We look briefly here at the four most frequently used instruments and the definitions of love that they contain.

Hendrick and Hendrick's Love Attitude Scale

Hendrick and Hendrick's Love Attitude Scale is a 42-item instrument based on and designed to measure John Lee's six styles of love (Lee 1973, 1988):

- **Eros.** Romantic or passionate love
- **Ludus.** Playful or game-playing love
- **Storge.** Companionate or friendship love

These first three are "primary" styles, which can be combined to generate the following secondary styles:

- **Mania.** A combination of ludus and eros, mania is obsessive love, characterized by an intense love–hate relationship.
- **Agape.** A combination of eros and storge, agape is altruistic love.
- **Pragma.** A combination of storge and ludus, pragma is a practical, pragmatic style of love.

The six basic types can be described in greater detail:

- *Eros.* Erotic lovers delight in the tactile, the sensual, the immediate; they are attracted to beauty (although beauty may be in the eye of the beholder). They love the lines of the body, its feel and touch. They are fascinated by every detail of their beloved. Their love burns brightly but soon flickers and dies.
- *Ludus.* For ludic lovers, love is a game, something to play at rather than to become deeply involved in. Love is ultimately ludicrous. Love is for fun; encounters are casual, carefree, and often careless. "Nothing serious" is the motto of ludic lovers.
- *Storge.* Storge (pronounced *STOR-gay*) is the love between companions. It is, writes Lee, "love without fever, tumult, or folly, a peaceful and enchanting affection." It usually begins as friendship and then gradually deepens into love. If the love ends, it also occurs gradually, and the couple often becomes friends once again.
- *Mania.* The word *mania* comes from the Greek word for madness. For manic lovers, nights are marked by sleeplessness and days by pain and anx-

iety. The slightest sign of affection brings ecstasy briefly, only to have it disappear. Satisfactions last but a moment before they must be renewed. Manic love is roller-coaster love.

- *Agape.* Agape (pronounced *ah-GA-pay*) is love that is chaste, patient, selfless, and undemanding; it does not expect to be reciprocated. Agape emphasizes nurturing and caring as their own rewards. It is the love of monastics, missionaries, and saints more than that of worldly couples.
- *Pragma.* Pragmatic lovers are primarily logical in their approach toward looking for someone who meets their needs. They look for a partner who has background, education, personality, religion, and interests compatible with their own. If they meet a person who meets their criteria, erotic or manic feelings may develop. But, as Samuel Butler warned, "Logic is like the sword—those who appeal to it shall perish by it."

These styles, Lee cautions, are relationship styles, not individual styles. The style of love may change as the relationship changes or when individuals enter different relationships. In addition to these pure forms, there are mixtures of the basic types: storgic–eros, ludic–eros, and storgic–ludus. According to Lee, a person must thus find a partner who shares the same style and definition of love to have a mutually satisfying love affair. The more different two people are in their styles of love, the less likely it is that they will understand each other's love.

Love styles are also linked to gender and ethnicity (Hendrick and Hendrick 1986). Research indicates that heterosexual and gay men have similar attitudes toward eros, mania, ludus, and storge and that gay male relationships have multiple emotional dimensions (Adler, Hendrick, and Hendrick 1989). As to cultural differences, different styles tend to characterize Asians, African Americans, Latinos, and Caucasians. Asian Americans have a more pragmatic style of love than do Latinos, African Americans, or Caucasians, and they place a high value on affection, trust, and friendship (pragma and storge). Latinos often score higher on the ludic characteristics (Regan 2003).

Hatfield and Sprecher's Passionate Love Scale

Hatfield and Sprecher's Passionate Love Scale is based on a 30-item instrument measuring how much passionate love a relationship has. Hatfield and Sprecher divide love into two types, passionate and companionate. As shown earlier in the "Exploring Diversity" box, passionate love is "an intense longing for union with another" and is familiar to us because it most fits our ideas of being in love (Kim and Hatfield 2004). Passionate love can be seen through cognitive, emotional, and behavioral indicators. Companionate love refers more to the warm and tender affection we feel for close others. It is milder, less intense, and produces less of the extreme highs and lows people experience from passionate love (Kim and Hatfield 2004).

Rubin's Love Scale

In 1970, Zick Rubin developed a 13-item love scale to study what he called romantic love. Seeing love as an attitude one person has toward another that moves them to "think, feel and behave in certain ways toward the other person," Rubin (1973) found that there were four feelings identifying love:

- Caring for the other; that is, wanting to help him or her
- Needing the other; that is, having a strong desire to be in the other's presence and to be cared for by the other
- Trusting the other; that is, mutually exchanging confidences
- Tolerating the other; that is, accepting his or her faults

Of these, caring appears to be the most important, followed by needing, trusting, and tolerating. **Rubin's Love Scale** was designed to measure and assess three core elements of romantic love: affiliated and dependent need, predisposition to help, and exclusiveness and absorption.

Sternberg's Triangular Love Scale

Sternberg's Triangular Love Scale is based on his **triangular theory of love.** According to the theory, love is composed of three elements that can be visualized as the points of a triangle: intimacy, passion, and decision or commitment. The intimacy component refers to the warm, close feelings of bonding you experience when you love someone. It includes such things as giving and receiving emotional support to and from your partner, being able to communicate with your partner about intimate things, being able to understand each other, and valuing your partner's presence in your life. The passion component refers to the elements of romance, attraction, and sexuality in a relationship. These may be fueled by the desire to increase self-esteem, to be sexually active or fulfilled, to affiliate with others, to dominate, or to subordinate. The decision or commitment component consists of two parts, one short term and one long term. The short-term part refers to your decision that you love someone. You may or may not make the decision consciously, but it usually occurs before you decide to make a commitment to that person. The commitment represents the long-term aspect; it is the maintenance of love, but a decision to love someone does not necessarily entail a commitment to maintaining that love.

Each of these components can be enlarged or diminished in the course of a love relationship, and their changes will affect the quality of the relationship. They can also be combined in different ways in different relationships or even at different times in the same love relationship. Each combination offers a different type of love—for example, romantic love, infatuation, empty love, and liking. According to Robert Sternberg (1988), the intimacy, passion, and decision or commitment can be combined in eight ways, with these combinations forming the basis for classifying love:

- Liking (intimacy only)
- Romantic love (intimacy and passion)
- Infatuation (passion only)
- Fatuous love (passion and commitment)
- Empty love (decision or commitment only)
- Companionate love (intimacy and commitment)
- Consummate love (intimacy, passion, and commitment)
- Nonlove (absence of intimacy, passion, and commitment)

These types represent extremes that probably few of us experience. Not many of us, for example, experience infatuation in its purest form, in which there is no intimacy. The categories are nevertheless useful for examining love (except for empty love, which is not really love):

- *Liking.* Liking represents the intimacy component alone. It forms the basis for close friendships but is neither passionate nor committed. As such, liking is often an enduring kind of love. Boyfriends and girlfriends may come and go, but good friends remain.

- *Romantic love.* Romantic love combines intimacy and passion. It is similar to liking, but it is more intense as a result of physical or emotional attraction. It may begin with an immediate union of the two components—with friendship that intensifies with passion or with passion that develops intimacy. Although commitment is not an essential element of romantic love, it may develop.

- *Infatuation.* Infatuation is, like love at first sight, the kind of love that idealizes its object, rarely seeing the other as a "real" person with flaws. Marked by sudden passion and a high degree of physical and emotional arousal, it tends to be obsessive and consuming. The person has no time, energy, or desire for anything or anyone but the beloved (or thoughts of him or her). To the dismay of the infatuated individual, infatuations are usually asymmetrical: The passion (or obsession) is rarely returned equally, and the greater the asymmetry, the greater the distress in the relationship.

- *Fatuous love.* Fatuous, or deceptive, love is whirlwind love; it begins the day a couple meets and quickly results in cohabitation or engagement and then marriage. It goes so fast we hardly know what has happened. Often, nothing did happen that will permit the relationship to endure. As Sternberg (1988) observes, "It is fatuous in the sense that a commitment is made on the basis of passion without the stabilizing element of intimate involvement—which takes time to develop." Passion fades soon enough, and all that remains is commitment. But commitment that has had relatively little time to deepen is a poor foundation on which to build an enduring relationship. With neither passion nor intimacy, the commitment wanes.

- *Companionate love.* Companionate love is essential to a committed relationship. It often begins as romantic love, but as the passion diminishes and the intimacy increases, it is transformed. Some couples are satisfied with such love; others are not. Those who are dissatisfied in companionate love relationships may seek extra relational affairs to maintain passion in their lives. They may also end the relationship to seek a new romantic relationship in the hope that it will remain romantic.

- *Consummate love.* Consummate love is born when intimacy, passion, and commitment combine to form their unique constellation. It is the kind of love we dream about but do not expect in all our love relationships. Many of us can achieve it, but it is difficult to sustain over time. To sustain it, we must nourish its different components, each of which is subject to the stress of time.

- *Nonlove.* Nonlove can take many forms, such as attachment for financial reasons, fear, or fulfillment of neurotic needs.

The shape of the love triangle depends on the intensity of the love and the balance of the parts. Intense love relationships create triangles with greater area; such triangles occupy more of our lives. Just as love relationships can be balanced or unbalanced, so can love triangles. The balance determines the shape of the triangle (see Figure 5.1). A relationship in which the intimacy, passion, and commitment components are equal forms an equilateral triangle. But if the components are not equal, unbalanced triangles form. The size and shape of a person's triangle give a good pictorial sense of how that person feels about another. The greater the match between each person's triangle in a relationship, the more likely each is to experience satisfaction in the relationship.

These four instruments are the most widely used. They have adequate reliability (measurement consistency) and validity (fit between instrument and concepts). They are accessible, short, and easy to use and interpret (Myers and Shurts 2002). They are not perfect, however; no research instrument is. Because they have been generated from samples of college students, the norms defining different types of love, or the expected extent of different components, may not fit noncollege populations. Similarly, it is questionable how well the standards for interpreting scores and applying concepts pertain in other cultures or to gay and lesbian relationships.

Love and Attachment

The **attachment theory of love** maintains that the degree and quality of attachments we experience in early life influence our later relationships. It has been increasingly used to study personal relationships, including love. It examines love as a form of attachment that finds its roots in infancy (Hazan and Shaver 1987; Shaver, Hazan, and Bradshaw 1988). Phillip Shaver and his associates (1988) suggest that "all important love relationships—especially the first ones with parents and later ones with lovers and spouses—are attachments." On the basis of infant–caregiver work by John Bowlby (1969, 1973, 1980), some researchers suggest

Figure 5.1 ■ The Triangles of Love

The passion, intimacy, and decision or commitment components of love can be combined in a variety of ways to form different shaped triangles. The shape of a love triangle may change over time. In addition, the greater the intensity of love we experience, the greater will be a love triangle in area. The greater a given component of love, the further the point from the center of the triangle. Triangle A reflects a balanced love, in which intimacy, passion, and commitment are equally intense. Triangle B illustrates infatuation (passion only). C reflects empty love (containing commitment or decision only). D is romantic love (intimacy and passion). E is companionate love (containing intimacy and commitment). The five triangles reflect five different kinds of love, as a result their triangles are differently shaped.

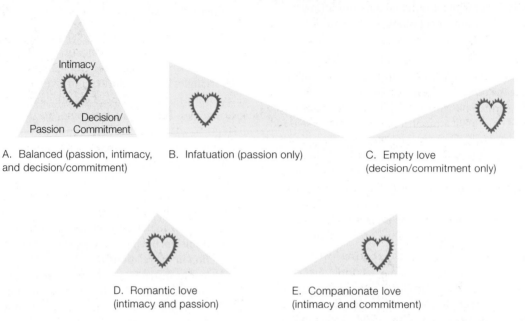

A. Balanced (passion, intimacy, and decision/commitment)

B. Infatuation (passion only)

C. Empty love (decision/commitment only)

D. Romantic love (intimacy and passion)

E. Companionate love (intimacy and commitment)

numerous similarities between attachment and romantic love (Downey, Bonica, and Rincon 1999; Bringle and Bagby 1992; Shaver et al. 1988).

These include the following:

Attachment	Love
Attachment formation and quality depend on attachment object's responsiveness, interest, and reciprocation.	Feelings of love are related to lover's feelings.
When attachment object is present, infant is happier.	When lover is present, person feels happier.
Infant shares toys, discoveries, and objects with attachment object.	Lovers share experiences and goods and give gifts.
Infant coos, talks baby talk, "sings."	Lovers coo, talk baby talk, and sing.
There are feelings of oneness with attachment object.	There are feelings of oneness with lover.

According to research by Geraldine Downey (1996), rejection by parents of their children's needs can lead to the development of **rejection sensitivity,** or the tendency to anticipate and overreact to rejection. Individuals who develop rejection sensitivity seek to avoid rejection by their partners and closely monitor, even overanalyze, the relationship dynamics for signs of potential rejection. As Pamela Regan (2003) notes, even "minimal or ambiguous" rejection cues may lead to feelings of rejection and to anger, jealousy, and despondency. Rejection-sensitive people tend to be less satisfied with their relationships and more likely to see them end.

Based on studies conducted by Mary Ainsworth and colleagues (1978, cited in Shaver et al. 1988) there are three styles of infant attachment: (1) secure, (2) anxious or ambivalent, and (3) avoidant. In *secure attachment*, the infant feels secure when the mother is out of sight. He or she is confident that the mother will offer protection and care. In *anxious or ambivalent attachment*, the infant shows separation anxiety when the mother leaves. He or she feels insecure when the mother is not present. In *avoidant attachment*, the infant senses the mother's detachment and rejection when he or she desires close bodily contact. The infant shows avoidance behaviors with the mother as a means of defense. In Ainsworth's study, 66% of the infants were secure, 19% were anxious or ambivalent, and 21% were avoidant.

Some researchers (Feeney and Noller 1990; Shaver et al. 1988) believe that the styles of attachment developed during infancy continue through adulthood. Others, however, question the validity of applying infant research to adults, as well as the stability of attachment styles throughout life (Hendrick and Hendrick 1994). Still others found a significant association between attachment styles and relationship satisfaction (Brennan and Shaver 1995).

Secure Adults

Secure adults find it relatively easy to get close to others. They are comfortable depending on others and having others depend on them. They believe they are worthy of love and support and expect to receive them in their relationships (Regan 2003). They generally do not worry about being abandoned or having someone get too close to them. More than avoidant and anxious or ambivalent adults, they feel that others generally like them; they believe that people are generally well intentioned and good hearted. In contrast to others, secure adults are less likely to believe in media images of love and more likely to believe that romantic love can last. Their love experiences tend to be happy, friendly, and trusting. They are more likely to accept and support their partners. Reportedly, compared to others, secure adults find greater satisfaction and commitment in their relationships (Pistol, Clark, and Tubbs 1995).

Anxious or Ambivalent Adults

Anxious or ambivalent adults feel that others do not or will not get as close as they themselves want. They worry that their partners do not really love them or that they will leave them. They feel unworthy of love and need approval from others (Regan 2003). They also want to merge completely with the other person, which sometimes scares that person away. More than others, anxious or ambivalent adults believe that it is easy to fall in love. Their experiences in love are often obsessive and marked by a desire for union, high degrees of sexual attraction and jealousy, and emotional highs and lows.

Avoidant Adults

Avoidant adults feel discomfort in being close to others; they are distrustful and fearful of becoming dependent (Bartholomew 1990). Thus, to avoid the pain they expect to come from eventual rejection, they maintain distance and avoid intimacy (Regan 2003). More than others, they believe that romance seldom lasts but that at times it can be as intense as it was at the beginning. Their partners tend to want more closeness than they do. Avoidant lovers fear intimacy and experience emotional highs and lows and jealousy.

In adulthood, the attachment styles developed in infancy combine with sexual desire and caring behaviors to give rise to romantic love. Comparing across these three types of attachment styles indicates that women and men with secure attachment styles tend to be the preferred type of romantic partner by women and men alike. They also tend to find more satisfaction in their relationships, experience more happiness, hold more positive views of their partners, and display fewer negative emotions (Regan 2003).

Love and Commitment

We expect our romantic partner to be there for us through "thick and thin." When we enter marriage, we pledge our love, "for better for worse, for richer, for poorer, in sickness and in health, till death do us part." In other words, we expect that, along with loving us, our partners will be committed to us and to our relationship. Although we generally make commitments to a relationship because we love someone, love alone is not sufficient to make a commitment last. Our commitments seem to be affected by several factors that can strengthen or weaken the relationship. Ira Reiss (1980a) believes that there are three important factors in commitment to a relationship:

1. *The balance of costs and benefits.* Whether we like it or not, humans have a tendency to look at romantic and marital relationships from a cost–benefit perspective. Most of the time, when we are satisfied, we are unaware that we judge our relationships in this manner. But as shown in our discussion of social exchange theory in Chapter 2, when there is stress or conflict we might ask ourselves, "Just what am I getting out of this relationship?" Then we add up the pluses and minuses. If the result is on the plus side, we are encouraged to continue the relationship; if the result is negative, we are more likely to discontinue it.

2. *Normative inputs.* Normative inputs for relationships are the values that you and your partner hold about love, relationships, marriage, and family. These values can either sustain or detract from a

commitment. How do you feel about a love commitment? A marital commitment? Do you believe that marriage is for life? Does the presence of children affect your beliefs about commitment? What are the values that your friends, family, and religion hold regarding your type of relationship?

3. *Structural constraints.* The structure of a relationship will add to or detract from commitment. Depending on the type of relationship—whether it is dating, living together, or marriage—different roles and expectations are structured. In marital relationships, there are partner roles (husband–wife) and economic roles (employed worker–homemaker). There may also be parental roles (mother–father).

These factors interact to increase or decrease the commitment.

Commitments are more likely to endure in marriage than in cohabiting or dating relationships, which tend to be relatively short lived. They are more likely to last in heterosexual relationships than in gay or lesbian relationships (Testa et al. 1987). Ethnicity may also be the greatest predictor of satisfaction and commitment to a friendship (deVries, Jacoby, and Davis 1996). The reason commitments tend to endure in marriage may or may not have anything to do with a couple being happy. Marital commitments may last because norms and structural constraints compensate for the lack of personal satisfaction.

For most people, love seems to include commitment and commitment seems to include love. Beverly Fehr (1988) found that if a person violated a central aspect of love, such as caring, that person was also seen as violating the couple's commitment. If a person violated a central aspect of commitment, such as loyalty, it called love into question.

Because of the overlap between love and commitment, we can mistakenly assume that someone who loves us is also committed to us. As one researcher points out: "Expressions of love can easily be confused with expressions of commitment. . . . Misunderstandings about a person's love versus commitment can be based on honest errors of communication, on failures of self-understanding" (Kelley 1983). Or a person can intentionally mislead the partner into believing that there is a greater commitment than there actually is. Even if a person is committed, it is not always clear what the commitment means: Is it a commitment to the person or to the relationship? Is it for a short time or a long time? Is it for better and for worse?

How Love Develops: Spinning Wheels and Winding Clocks

As shown earlier, one of the core beliefs that comprises the ideology of romantic love in the United States is the idea of love at first sight. This is often articulated by romantic partners in describing how they "just knew" they were meant for each other upon their first meeting, the first time they gazed at each other, or when they first heard the other laugh or speak. Increasingly, people believe they fell in love upon their first e-mail exchange. In fact, love develops through a process, beginning with first meeting but commencing through an intensification of the relationship and eventually a definition or interpretation of feelings as "love."

One of the more popular models depicting this process is Ira Reiss's **wheel theory of love** (Reiss 1960, 1980a). According to wheel theory, the development of love can be depicted as a spinning wheel, consisting of four spokes, each of which drives the others as the wheel spins forward. The four spokes are (1) rapport, (2) self-revelation, (3) mutual dependency, and (4) fulfillment of the need for intimacy.

- **Rapport.** When two people meet, they quickly sense if rapport exists between them. This rapport is a sense of ease, the feeling that they understand each other in some special way. We tend to feel rapport with those who share the same social and cultural background as ourselves. If one or both feel as though they have much in common, they are more likely to feel as though they can understand each other; they may even experience a comfort that makes them feel like they have known each other a long time (or before). However, if one person has only a grade-school education and the other a college education, it is not as likely that they will share many of the same values. If one person is upper class and the other is working class, their life experiences have probably been quite different. Such differences may make the building of rapport more challenging, although it is not impossible (Borland 1975).

- **Self-revelation.** Wheel theory posits that the greater the rapport we feel with someone, the more likely we are to feel relaxed and confident around them and to develop trust about the relationship. As a result, self-revelation—the disclosure of intimate feelings—is more likely to occur. We will reveal more about ourselves and more of a personal nature with greater confidence and trust. Furthermore, disclosure becomes mutual. Self-revelation may depend on more than the presence or absence of rapport. It may also depend on what is considered proper within our ethnic group or economic class. Certain groups have more of a tendency to be reserved about themselves. Others (for example, middle-class Americans) feel more comfortable in revealing intimate aspects of their lives and feelings.

- **Mutual dependency.** After two people feel rapport and begin revealing themselves to each other, they may become mutually dependent. Each needs the other to share pleasures, fears, and jokes, as well as sexual intimacies; each becomes the other's confidant. Each person develops ways of acting and being that cannot be fulfilled alone. Going for a walk is no longer something done alone; they walk together. Sleeping no longer takes place in a single bed but in a larger one with the partner. The two people form a couple.

  Here, too, social and cultural background is important. The forms of mutually dependent behavior that develop are influenced by each person's conception of the role of courtship. Interdependency may develop through dating, getting together, or living together. Premarital intercourse may or may not be acceptable.

- **Fulfillment of intimacy needs.** According to Reiss (1980a), we all have a basic need for intimacy—"the need for someone to love, the need for someone to confide in, and the need for sympathetic understanding." These needs are important for fulfilling our roles as a partner or parent. If we find that our needs for love and intimacy are met by our partner, rapport will deepen, setting the stage for more

self-revelation, increased mutual dependency, and greater fulfillment of our intimacy needs.

Reiss describes the relationship among the four processes, which culminate in intimacy, as follows:

By virtue of rapport, one reveals oneself and becomes dependent, and in the process of carrying out the relationship one fulfills certain basic intimacy needs. To the extent that these needs are fulfilled, one finds a love relationship developing. In fact, the initial rapport that a person feels on first meeting someone can be presumed to be a dim awareness of the potential intimacy need fulfillment of this other person for one's own needs. If one needs sympathy and support, and senses these qualities in a date, rapport will be felt more easily; one will reveal more and become more dependent, and if the hunch is right, and the person is sympathetic, one's intimacy needs will be fulfilled.

Reiss called his model the wheel theory of love and represented the four processes as spokes to emphasize this interdependence. Relationships, like wheels, can spin in reverse, as well as forward. In other words, we can "fall out of love," and the wheel theory addresses this phenomenon. A reduction in any one of the four spokes affects the development or maintenance of the love relationship (see Figure 5.2). If we feel less comfort (that is, rapport) with the other, we may reveal fewer thoughts or feelings, feel less dependent on the other for a sense of happiness or contentment, and seek and fulfill our intimacy needs elsewhere. This seems to approximate what happens through the process of divorce or the ending of intimate relationships as well (Vaughan 1986). If a couple habitually argues, the arguments will affect the partners' mutual dependency and their need for intimacy; this in turn will weaken their rapport. Thus, the model can depict falling in *or* out of love. The "+" and "−" in Figure 5.2 indicate the directions in which the processes can increase or decrease love. The outer ring on the diagram, "sociocultural background," produces the next ring, "role conceptions." All four processes are influenced by role conceptions, which define what a person should expect and do in a love relationship.

To capture that as relationships persist they tend to deepen—we grow closer and our connection "tightens"—Dolores Borland (1975) suggested thinking not so much in terms of a wheel but rather a "clockspring." The "most intimate aspects of the 'real self'" are at the

Figure 5.2 ■ Graphical Representation of Reiss's Wheel Theory of Love

According to this theory, the development of intimacy is most likely to take place between those who share the same sociocultural background and role conceptions. Intimacy develops from a feeling of rapport, which leads to self-revelation; self-revelation leads to mutual dependency, which in turn may lead to intimacy need fulfillment.

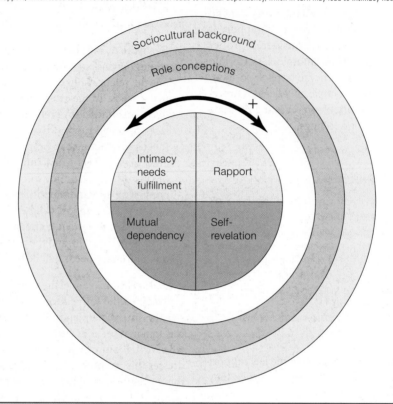

center of the clockspring. As rapport leads to self-revelation, revelation leads to mutual dependency, and mutual dependency leads us to seek and find satisfaction of our need for intimacy in our partner, we wind closer to a relationship with the "real inner self of the other person" (Borland 1975). A clockspring representation can also depict the depth of a relationship (by how much of our "real self" including our vulnerabilities and sensitivities we expose to others), as well as the difficulty and time it will take to "unwind" a relationship. Borland's "clockspring," depicted in Figure 5.3, is meant mostly as an aid in teaching about and better understanding the basic elements put forth by Reiss. Such elements are important if we are to fully understand how intimate relationships begin, develop, persist, and/or end.

Although mostly a model of how love develops, Reiss's wheel theory has been used to examine variations in patterns of marriage and family life in different societies (Haavio-Mannila and Rannik 1987).

Comparing marital relationships in then socialist Estonia with marriages in Finland, Elina Haavio-Mannila and Erkki Rannik (1987) suggest that Finnish couples more often experience a feeling of rapport

Figure 5.3 ■ The Clockspring Variation on Reiss's Wheel Theory

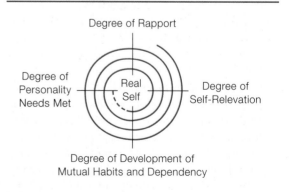

SOURCE: Borland 1975, 289–292.

than do Estonian couples, Estonian men engage in more self-revelation than Finnish men (the women in the two countries did not differ), the two countries were similar in their levels of mutual dependency (with wives more dependent on their husbands than husbands were on their wives), and Estonian respondents reporting that they receive more social support and greater satisfaction of their needs for intimacy than did couples in Finland.

Unrequited Love

As most of us know from painful experience, love is not always returned. We may reassure ourselves that, as Alfred, Lord Tennyson wrote 150 years ago, "'Tis better to have loved and lost / Than never to have loved at all." Too often, however, such words sound like a rationalization.

Who among us does not sometimes think that it is better never to have loved? **Unrequited love**—love that is not returned—is a common experience.

Several researchers (Baumeister, Wotman, and Stillwell 1993) accurately captured some of the feelings associated with unrequited love in the title of their research article: "Unrequited Love: On Heartbreak, Anger, Guilt, Scriptlessness, and Humiliation." They found that unrequited love was distressing for both the would-be lover and the rejecting partner. Would-be lovers felt both positive and intensely negative feelings about their unlucky attempt at a relationship. Nearly half of them (44) reported that the unreciprocated love caused them pain, suffering, jealousy, and anger. Almost a quarter of them (22%) experienced fears about rejection. However, positive feelings were more common than negative feelings. More than half looked back on the experience positively (Regan 2003). The rejecters, however, felt uniformly negative about the experience. Unlike the rejecters, the would-be lovers felt that the attraction was mutual, that they had been led on, and that the rejection had never been clearly communicated. Rejecters, by contrast, felt that they had not led the other person on; moreover, they felt guilty about hurting him or her. Nevertheless, many found the other person's persistence intrusive and annoying; they wished the other would have simply gotten the hint and gone away. Approximately half (51%) felt annoyed by the unwanted attention, 61% felt badly about having to reject the other, and 70% felt a range of negative emotions such as frustration, and resentment

■ *Unrequited love, when one's love is not reciprocated, is a painful experience.*

(Regan 2003). Whereas rejecters saw would-be lovers as self-deceptive and unreasonable, would-be lovers saw their rejecters as inconsistent and mysterious.

Unrequited love presents a paradox: If the goal of loving someone is an intimate relationship, why should we continue to love a person with whom we could not have such a relationship? Arthur Aron and his colleagues addressed this question in a study of almost 500 college students (Aron et al. 1989). The researchers found three different attachment styles underlying the experience of unrequited love:

■ *The Cyrano style.* Named after Cyrano de Bergerac, the seventeenth-century French poet and musketeer, whose love for Roxanne was so great that it was irrelevant that she loved someone else, this refers to the desire to have a romantic relationship with a specific person regardless of how hopeless the love is. The benefits of loving someone are considered so great that it does not matter how likely the love is to be returned. is.

■ *The Giselle style.* The misperception that a relationship is more likely to develop than it actually is. This might occur if we misread the other's cues, such as in mistakenly believing that friendliness is a sign of love. This style is named after Giselle, the tragic ballet heroine who was misled into believing that her love was reciprocated.

- *The Don Quixote style.* The general desire to be in love, regardless of whom we love. Here, the benefits of being in love—are more important than actually being in a relationship. This style is named after Cervantes's Don Quixote, whose love for the common Dulcinea was motivated by his need to dedicate knightly deeds to a lady love.

Using attachment theory, the researchers found that some people were predisposed to be Cyranos, others Giselles, and still others Don Quixotes. Anxious or ambivalent adults tended to be Cyranos, avoidant adults often were Don Quixotes, and secure adults were likely to be Giselles. Those who were anxious or ambivalent were most likely to experience unrequited love; those who were secure were least likely to experience such love. Avoidant adults experienced the greatest desire to be in love in general; yet they had the least probability of being in a specific relationship. Anxious or ambivalent adults showed the greatest desire for a specific relationship; they also had the least desire to be in love in general.

Reflections

Have you experienced unrequited love? How did it differ from requited love? Do you have a "style" of unrequited love? Have you been the object of someone's unrequited love? How did you handle it?

Stalking as Extreme Unrequited Love

When unrequited love is joined by obsessive thinking, the stage is set for what has come to be known as **stalking** or *obsessive relational intrusion* (Regan 2003).

The Bureau of Justice Assistance defines stalking as "non-consensual communication and/or harassment of another person" (in Spitz 2003). The National Violence Against Women (NVAW) survey defines it as "repeated (two or more) occasions of visual or physical proximity, nonconsensual communication, or verbal, written or implied threats that would cause fear in a reasonable person" (McFarlane, Campbell, and Watson 2002). In such instances, one person pursues another seeking to initiate or maintain an intimate relationship that the victim does not desire. The pursuer may send unwanted letters or gifts, make phone calls, vandalize property, steal mail, spread gossip

about, and visit, watch, or follow the target of his or her affection. Mary-Ann Leitz Spitz (2003) notes other disturbing examples, including stealing underwear; going through the victim's garbage; hurting, stealing, or killing pets; and obtaining items or services in the victim's name.

Although the more extreme forms are less common, perhaps as many as 30% of victims report such behaviors (Regan 2003). Consequences including assault and homicide have also been reported. In the NVAW survey, 81% of women who were stalked by former husbands or cohabiting partners also were assaulted by the stalker. As other research corroborates, stalkers with past intimate relationships with their victims, especially sexually intimate relationships, are most likely to be violent (McFarlane, Campbell, and Watson 2002).

The consequences experienced by the targets of various forms of relational stalking may include anger, self-blame, curtailed lifestyle, distrust of others, and physical symptoms including illness. Kathleen Basile and colleagues (2004) report that stalking, like physical, sexual, and psychological abuse, is significantly related to experiencing symptoms of post-traumatic stress disorder. (See the "Issues & Insights" box for more on the experience of stalking victims.)

Targets may try a variety of strategies to deal with the unwanted attention. Avoidance (ignoring, not responding, not accepting gifts, and so on) is common. Other strategies include direct confrontation, retaliation, and seeking of formal protection. These may not achieve the desired outcome of lessening or stopping the behavior. As Pamela Regan notes, avoidance strategies may be too ambiguous and therefore misunderstood by pursuers. Direct confrontation may actually give the pursuer what she or he is seeking, more contact. Both retaliation and the use of formal protection may serve to anger not stop the pursuer (Regan 2003). It is disturbing but important to note that stalking appears to be increasing; however, as with other forms of intimate violence or abuse the trend may be more an artifact of improved reporting and record keeping than the result of a real increase in frequency of the behaviors (Spitz 2003).

Stalking, like other forms of intimate violence or abuse, seems to be about issues of power and control (Brewster 2003). Brewster's sample of stalking victims reported that their stalkers were trying to control them, using whatever manner of control they could—social, emotional, financial, psychological, and the threat or use of physical violence.

Obsession appears to be at the center of stalking, although Spitz also differentiates simple and love obsessional stalking from **erotomania.** Stalkers suffering from erotomania, suffer from delusions in which they believe that they are involved in relationships with their victims. **Simple obsessional stalking,** where real relationships exist or existed between stalker and victim, is the most common form of stalking. It typically emerges after a relationship between stalker and victim ends, including but not necessarily involving a sexually intimate relationship. Simple obsessional stalking is used to punish the person who ended the relationship or to try to force him or her back into the relationship. This is the type most likely to result in violence toward the victim (Spitz 2003). **Love-obsessional** stalkers are not psychotic, but they pursue targets with whom they have never been involved in relationships (Spitz, 2003).

Either women or men can be victimized in such a way, and either gender can be stalkers, though research suggests many more women than men are victimized. Bonnie Fisher, Francis Cullen, and Michael Turner's thorough literature review and their own finding show the following range of estimated stalking victimization:

Year	Study	Sample	Prevalence
1996	Fremouw et al.	Female undergraduates in two psychology classes	31%
1997	Coleman	141 female students	29%
1998	Tjaden and Thoennes	8,000 women	8–12%
1999	Mustaine and Tewksbury	861 women in introductory sociology or criminal justice courses	10.5%
2000	Bjerregaard	512 women in randomly selected courses	25%
2000	Logan	84 women in a communications course	29%
2002	Fisher, Cullen, and Turner	4,446 college women	13.1%

Spitz (2003) reports research estimating that approximately 10 million Americans have been stalked and that 3% of all men and 8% of all women *will be* victims of stalking at some point in their lives. When men are stalking victims, their stalkers are more often colleagues and acquaintances. When women are stalked they are most often stalked by men with whom they have had romantic relationships. Judith McFarlane and colleagues report that 62% of female victims are stalked by a current or former intimate partner (38% by a current or former husband, 10% by a current or former cohabiting partner, and 14% by current or former boyfriends or dates) (McFarlane et al. 2002). As a general profile, stalkers are typically white males between 26 and 50, with at least high school educations. This makes them older and better educated than most convicted of other crimes (Spitz 2003). Victims tend to be never married or divorced women, on average 35 years old, with at least some college education (Spitz 2003).

The preceding descriptions show that there are both "milder" and more extreme forms of stalking behavior (Spitzberg and Cupach 2001; Regan 2003). It should be noted, however, that even the "milder" forms of stalking, such as repeatedly calling the victim and arguing, begging for another chance, or hanging up without speaking, are intrusive, unwanted, and disturbing to the victim. Such "lesser" forms of stalking should not be ignored, and we should not trivialize or dismiss any behaviors that cause victims discomfort and/or force them to alter their daily routines. At the extreme end, when stalkers follow and spy on their victims, leave them threatening notes on their cars, or—if there are children—threaten to harm children, there is a much greater risk of victims being physically injured or killed by their stalkers (McFarlane et al. 2002).

Jealousy: The Green-Eyed Monster

In addition to bringing us great joy, love relationships are often the source of painful insecurities and jealousy. What exactly is jealousy? As studied by researchers, **jealousy** can be defined as "a complex of behaviors, thoughts, and emotions resulting from the perception of harm or threat to the self and/or the romantic relationship by a real or potential rival relationship" (White and Mullen 1989). It is an aversive response that occurs because of a partner's real, imagined, or likely involvement with a third person (Bringle and Buunk 1985; Sharpsteen 1993). Jealousy sets the boundaries for what an individual or group feels are important relationships; others cannot trespass these limits into other emotional and/or sexual relationships without evoking jealousy.

Sometimes we may think that jealousy proves love and, by flirting with another person, may try to test our partner's interest or affection by attempting to make him or her jealous. If our date or partner becomes

©Jutta Klee/CORBIS

■ *Jealousy is not necessarily a sign of love.*

jealous, the jealousy is taken as a sign of love. But making jealousy a litmus test of love is dangerous, because jealousy and love are not necessarily related. Jealousy may be a more accurate yardstick for measuring insecurity or possessiveness than love (see Mullen 1993 for a discussion of changing cultural attitudes toward jealousy).

Social psychologists suggest that there are two types of jealousy: suspicious and reactive (Bringle and Buunk 1991). **Suspicious jealousy** is jealousy that occurs when there is either no reason to be suspicious or only ambiguous evidence to suspect that a partner is involved with another. **Reactive jealousy** is jealousy that occurs when a partner reveals a current, past, or anticipated relationship with another person.

Suspicious jealousy generally occurs when a relationship is in its early stages. The relationship is not firmly established, and the couple is unsure about its future. The smallest distraction, imagined slight, or inattention can be taken as evidence of interest in another person. Even without any evidence, a jealous person may worry ("Is my partner seeing someone else but not telling me?"). This person may engage in vigilance, watching the partner's every move ("I'd like to audit your marriage and family class"). He or she may snoop, unexpectedly appearing in the middle of the night to see if someone else is there ("I was just passing by and thought I'd say hello"). The partner may try to control the other's behavior ("If you go to your friend's party without me, we're through"). Suspicious jealousy may have both legitimate and nega-

tive functions in a relationship. Although it may be a reasonable response to circumstantial evidence and warn the partner what will happen if there are serious transgressions, if unfounded, it can be self-defeating.

Reactive jealousy occurs when one partner learns of the other's present, past, or anticipated sexual involvement with another. This usually provokes the most intense jealousy. If the affair occurred in the early part of the present relationship, the unknowing partner may feel that the primary relationship has been based on a lie. Trust is questioned. Every word and event must be reevaluated in light of this new knowledge: "If you slept with him when you said you were going to the library, did you also sleep with him when you said you were going to the Laundromat?" Or "How could you say you loved me when you were seeing her?" The damage can be irreparable.

As our lives become more and more intertwined, we become less and less independent and our commitment to each other grows stronger. For some, this loss of independence increases the fear of losing the partner, and indeed there is evidence that the strength of the commitment, the more we rely on the relationship for fulfillment of personal and interpersonal needs, the more threatened we will feel at the thought of losing our partner to a rival. Commitment alone will not evoke jealousy. We must perceive, rightly or wrongly, that our relationship is being threatened (Rydell, McConnell, and Bringle 2004).

Gender Differences in Jealousy

Both men and women are susceptible to jealous fears that their partner might be attracted to someone else because of dissatisfaction with the relationship, attractiveness of a rival, or the desire for sexual variety. Women feel especially vulnerable to losing their partner to a physically attractive rival, whereas in men jealousy is evoked more by a rival's status (Buunk and Dijkstra 2004). Furthermore, men and women become jealous about different matters. Men tend to experience more jealousy when they feel their partner is sexually involved with another man. Women, by contrast, tend to experience jealousy over intimacy issues (Buunk and Dijkstra 2004; Cramer et al. 2001, 2002). This gender difference has been found in research in the United States, as well as in China, Germany, Japan, Korea, the Netherlands, and Sweden (Cramer et al. 2001, 2002).

Psychologist Robert Cramer and colleagues asked a sample of undergraduate women and men to indicate which of two infidelities would distress or upset them more, by circling either alternative A or alternative B.

A. Imagining your partner forming a deep emotional attachment to another person.

B. Imagining your partner enjoying passionate sexual intercourse with another person.

Other respondents were asked to imagine their partners committing *both* infidelities: "Imagine your partner forming a deep emotional attachment to another person *and also* enjoying passionate sexual intercourse with that person." Participants were then asked to indicate which infidelity, assuming both had occurred, would distress or upset them more.

Results for both questions show a striking gender pattern (see Table 5.1).

Like much jealousy research, Cramer and colleagues use evolutionary theory to account for these gender differences. They suggest that emotional infidelity is more distressing for women than for men because, at least in theory, it threatens a romantic partner's commitment and, therefore, continued access to material resources and economic stability needed to assure the healthy growth and development of offspring. Men, on the other hand, are more distressed by sexual infidelity than women are because it decreases their "paternity certainty" through the loss of sexual exclusivity (Cramer et al. 2001, 2002).

Both men and women react to jealousy with a host of emotions. Betrayal, anger, rejection, hurt, distrust, anxiety, worry, suspicion, and sadness are all possible. The kind of emotional reaction appears to depend on the type of infidelity that provokes it. Following emotional infidelity, such feelings as anxiety, suspicion, worry, distrust, and threat are more common. Bram Buunk and Pieternel Dijkstra call this type of jealousy *suspicious* or *preventive* jealousy. Following sexual infidelity, jealousy was expressed more through anger, sadness, a sense of betrayal, hurt, and rejection. Buunk and Dijkstra label this *fait accompli* (after the fact) jealousy (Buunk and Dijkstra 2004). Further differentiating the genders, following emotional infidelity, jealousy was evoked in men by a rival's dominance and was experienced mostly as a sense of threat. Following sexual infidelity, men's jealousy was evoked by his rival's physical attractiveness, not his dominance, and was experienced as betrayal or anger. For women, after emotional infidelity a rival's physical attractiveness evoked a sense of threat, whereas after sexual infidelity women's jealousy responses were unaffected by any particular characteristics of her rival (Buunk and Dijkstra 2004).

Managing Jealousy

Jealousy can be unreasonable or a realistic reaction to genuine threats. Unreasonable jealousy can become a problem when it interferes with an individual's well-being or that of the relationship. Dealing with irrational suspicions can often be difficult, because such feelings touch deep recesses in ourselves. As noted earlier, jealousy is often related to personal feelings of insecurity and inadequacy. The source of such jealousy lies within a person, not within the relationship.

If we can work on the underlying causes of our insecurity, then we can deal effectively with our irrational jealousy. Excessively jealous people may need considerable reassurance, but at some point they must confront their own irrationality and insecurity. If they do not, they may emotionally imprison their partner. Their jealousy may destroy the very relationship they were desperately trying to preserve.

Managing jealousy requires the ability to communicate, the recognition by each partner of the feelings and motivations of the other, and a willingness to reciprocate and compromise (Ridley and Crowe 1992). If the jealousy is well founded, the partner may need to modify or end the relationship with the "third party" whose presence initiated the jealousy. Modifying the third-party relationship reduces the jealous response and, more important, symbolizes the partner's commitment to the primary relationship. If the partner is unwilling to do this—because of lack of commitment, unsatisfied personal needs, or other problems in the primary relationship—the relationship is likely to reach a crisis. In such cases, jealousy may be the agent for profound change.

Table 5.1 ■ Percentage of Women and Men Selecting Emotional or Sexual Infidelity as More Distressing

		Men	Women
Forced choice	Emotional	12.9%	54.5%
	Sexual	87.1%	45.5%
Assuming both, which is worse?	Emotional	13.3%	40.6%
	Sexual	86.7%	59.4%

Jealousy: The Psychological Dimension

Jealousy is a painful experience. It is an agonizing compound of hurt, anger, depression, fear, and doubt. We may feel less attractive and acceptable to our partner when we are jealous (Bush, Bush, and Jennings 1988). Jealous responses are most intense in committed or marital relationships because both partners assume "specialness." This specialness occurs because our intimate partner is different from everyone else: It is with him or her that we are most confiding, revealing, vulnerable, caring, and trusting. There is a sense of exclusiveness. To have sex outside the relationship violates that sense of exclusiveness because sex symbolizes "specialness." Words such as *unfaithfulness, cheating,* and *infidelity* reflect the sense that an unspoken pledge has been broken. This pledge is the normative expectation that serious relationships, whether dating or marital, will be sexually exclusive (Lieberman 1988).

Jealousy represents a boundary marker. It points out what the boundaries are in a particular relationship. It determines how, to what extent, and in what manner others can interact with members of the relationship. It also shows the limits within which the members of the relationship can interact with those outside the relationship. Culture prescribes the general boundaries of what evokes jealousy, but individuals adjust them to the dynamics of their own relationships.

Boundaries may vary, depending on the type of relationship, gender, sexual orientation, and ethnicity. Sexual exclusiveness is generally important in serious dating relationships and cohabitation; it is virtually mandatory in marriage (Blumstein and Schwartz 1983; Buunk and van Driel 1989; Hansen 1985; Lieberman 1988). Men are generally more restrictive toward their partners than women; heterosexuals are more restrictive than gay men and lesbians. Although we know little about jealousy and ethnicity, traditional Latinos and new Latino and Asian immigrants appear to be more restrictive than Anglos and African Americans (Mindel, Habenstein, and Wright 1988). Despite variations on where the boundary lines are drawn, jealousy guards those lines.

Although our culture sets down general marital boundaries, each couple evolves its own boundaries. For some, it is permissible to carve out an area of individual privacy. In some relationships, partners may have few or many friends of their own (of the same or other sex), activities, and interests apart from the couple. In others there are no separate spheres because

of jealousy or a lack of interest. But wherever a married couple draws its boundaries, each member understands where the line is drawn. The partners implicitly or explicitly know what behavior will evoke a jealous response (Bringle and Buunk 1991). For some, it is having lunch with a member of the other sex (or same sex, if they are gay or lesbian); for others, it is having dinner; for still others, it is having dinner and seeing a movie. It is often disingenuous for a married partner to say that he or she didn't know that a particular action (a flirtatious suggestion, a lingering touch, or dinner with someone else) would provoke a jealous response.

It's important to understand jealousy for several reasons. First, jealousy is a painful emotion filled with anger and hurt. Its churning can turn us inside out and make us feel out of control. If we can understand jealousy, especially when it is irrational, then we can eliminate some of its pain. Second, jealousy can help cement or destroy a relationship. It helps maintain a relationship by guarding its exclusiveness, but in its irrational or extreme forms, it can destroy a relationship by its insistent demands and attempts at control. We need to understand when and how jealousy is functional and when and how it is not. Third, jealousy is often linked to violence (Follingstad et al. 1990; Laner 1990; Riggs 1993). It is a factor in precipitating violence or emotional abuse in dating relationships among both high school and college students; among marital partners it is often used by abusive partners to justify their violence (Adams 1990). Rather than being directed at a rival, jealous aggression is often used against a partner (Paul and Galloway 1994).

The Transformation of Love: From Passion to Intimacy

Intense, passionate love does not last forever at the same high level. Instead, it fades or transforms itself into a more enduring love based on intimacy.

The Instability of Passionate Love

Ultimately, romantic love may be transformed or replaced by a quieter, more lasting love. Those in secure companionate love relationships, according to one study, experience the highest levels of satisfaction; they are much more satisfied than those in traditional romantic relationships (Hecht, Marston, and Larkey 1994).

The Passage of Time: Changes in Intimacy, Passion, and Commitment

According to researcher Robert Sternberg (1988), time affects our levels of intimacy, passion, and commitment.

INTIMACY OVER TIME. When we first meet someone, intimacy increases rapidly as we make critical discoveries about each other, ranging from our innermost thoughts of life and death to our preference for strawberry or chocolate ice cream. As the relationship continues, the rate of growth decreases and then levels off. After the growth levels off, the partners may no longer consciously feel as close to each other. This may be because they are beginning to drift apart, or it may be because they are becoming intimate at a different, less conscious, deeper level. This kind of intimacy is not easily observed. It is a latent intimacy that nevertheless is forging stronger, more enduring bonds between the partners.

PASSION OVER TIME. Passion is subject to habituation. What was once thrilling—whether love, sex, or roller coasters—becomes less so the more we get used to it. Once we become habituated, more time with a person (or more sex or more roller-coaster rides) does not increase our arousal or satisfaction.

If the person leaves, however, we experience withdrawal symptoms (fatigue, depression, anxiety), just as if we were addicted. In becoming habituated, we have also become dependent. We fall beneath the emotional baseline we were at when we met our partner. Over time, however, we begin to return to that original level.

COMMITMENT OVER TIME. Unlike intimacy and passion, time does not necessarily diminish, erode, or alter commitments. Our commitment is most affected by how successful our relationship is. Even initially, commitment grows more slowly than intimacy or passion. As the relationship becomes long term, the growth of commitment levels off. Our commitment will remain high as long as we judge the relationship to be successful. If the relationship begins to deteriorate, after a time the commitment will probably decrease. Eventually, it may disappear and an alternative relationship may be sought.

Disappearance of Romance as Crisis

The disappearance (or transformation) of passionate love is often experienced as a crisis in a relationship. A study of college students (Berscheid 1983) found that half would seek divorce if passion disappeared from their marriage. But intensity of feeling does not necessarily measure depth of love. Intensity, like the excitement of toboggan runs, diminishes over time. It is then that we begin to discover if the love we experience for each other is one that will endure.

Our search for enduring love is complicated by our contradictory needs. Elaine Hatfield and William Walster (1981) offer this observation:

> What we really want is the impossible—a perfect mixture of security and danger. We want someone who understands and cares for us, someone who will be around through thick and thin, until we are old. At the same time, we long for sexual excitement, novelty, and danger. The individual who offers just the right combination of both ultimately wins our love. The problem, of course, is that, in time, we get more and more security—and less and less excitement—than we bargained for.

The disappearance of passionate love, however, enables individuals to refocus their relationship. They are given the opportunity to move from an intense one-on-one togetherness that excludes others to a togetherness that includes family, friends, and external goals and projects. They can look outward on the world together.

The Reemergence of Romantic Love

Contrary to what pessimists believe, many people find that they can have both love and romance and that the rewards of intimacy include romance.

Romantic love may be highest during the early part of marriage and decline as stresses from childrearing and work intrude on the relationship. Most studies suggest that marital satisfaction proceeds along a U-shaped curve, with highest satisfaction in the early and late periods. Romantic love may be affected by the same stresses as general marital satisfaction. Romantic love begins to increase as children leave home. In later life, romantic love may play an important role in alleviating the stresses of retirement and illness.

New research on the differences in love attitudes across family life stages reveals some unexpected and perhaps encouraging news for older romantics. Marilyn Montgomery and Gwendolyn Sorell (1997) write:

> The love attitudes endorsed by the broad age-range sample contradicts notions that romantic, passionate love is the privilege of youth and young relationships, functioning to bring partners together. Instead, individuals throughout the life-stages of marriage consistently endorse the love attitudes involving passion, romance, friendship, and self-giving love, and these results indicate that any popularization of young single adulthood as the enviable passionate idea is erroneous.

So it is that, among those whose marriages survive, passion and romance do not necessarily decline over time.

Intimate Love: Commitment, Caring, and Self-Disclosure

Perhaps one of the most profound questions we can ask about love is how to make it stay. The key to making love stay seems to be not in love's passionate intensity but in the transformation of that intensity into intimate love. Intimate love is based on commitment, caring, and self-disclosure.

Commitment

Commitment is an important component of intimate love because it is a "determination to continue" a relationship or marriage in the face of bad times, as well as good (Reiss 1980a). It is based on conscious choice rather than on feelings, which, by their very nature, are transitory. Commitment is a promise of a shared future, a promise to be together come what may.

Commitment has become an important concept in recent years. We seem to be as much in search of commitment as we are in search of love or marriage. We speak of "making a commitment" to someone or to a relationship. (Among singles, commitment is sometimes referred to as "the C-word.") A committed relationship has become almost a stage of courtship, somewhere between dating and being engaged or living together.

■ *Being physically limited does not inhibit love and sexuality any more than being able-bodied guarantees them.*

© Jean Higgins/Envision

What we expect and experience of love varies across the life span of a relationship. The romantic mystique that defines the early stages of committed relationships in our youth may be difficult to sustain across many years together. It also may become less definitive of the kind of lifelong relationship implied by "till death do us part." Consider the following story of love's "final days," poignantly told by journalist Mike Harden. It captures what is meant when we exchange our vows and promise to love each other forever.

In the End, Real Love Means in Sickness and in Health

When Frank Steger pushed himself into an upright position in the hospital bed, the heart monitor's fluid cursive line disintegrated into an erratic scribble.

"I told the doctor," he said, peeking at the edge of the curtain to make sure wife, Mary, was not within earshot, "that I felt like I was drowning. He said, 'This is how it happens with congestive heart disease.' I told him I'd rather he throw me off the roof instead."

Mary returned to the room, drawing a chair to his bedside.

"Thirsty," he complained.

She lifted the straw to his lips as he pulled the oxygen mask aside.

The medicine was making him sick. She fetched the basin, wrapped a firm arm around his spasm-wracked shoulders, mopped the sweat from his forehead.

In sickness and in health, I thought. They were supposed to be preparing for a Florida vacation, not holding on to one another in the cardiac care unit at Mount Carmel East Hospital.

"Help me sit up," he whispered hoarsely.

In the end, love comes down to this; not Gable's devilish first appraisal of Leigh, not Lancaster and Kerr rolling in the surf. But, "Help me sit up."

A late December rain spattered against the pane. Christmas had come and gone in the half-darkened room, a blur of canned carols punctuated by beeps and buzzes, lit by the winking light on the intravenous monitor.

"Merry Christmas," the cardiologist hailed, parting the curtains.

Christmas had always been a festive time for them. Standing rib roast, all the trimmings. Lift the glasses to the new year. To your health and the health of all who sleep beneath your roof.

When breath came harder, he slept sitting up in the chair next to the bed. By then, the body had turned against itself, the mutinous kidneys loosing their slow poison on the weakened heart.

Mary paused in the waiting room to remove her street shoes and put on her slippers. She did not want to wake him now that sleep was such a rationed luxury. Soundlessly, she slipped into the chair next to his.

In the end, love is not the smoldering glance across the dance floor, the clink of crystal, a leisurely picnic spread upon summer's clover. It is the squeeze of a hand. I'm here. I'll be here, no matter how long the fight, even when what you want most is to close your eyes and be done with it all. Water? You need water? Here. Drink. Let me straighten your pillow.

"Help me into bed," he said, he who had once been warrior triumphant in the business world. He was tough, demanding, but never as much on others as himself. If you gave him your best, no one could hurt you. If you gave him less, no one could hide you. He was never accused of being a yes man. She had been beside him when the future was golden, beside him when health sent his career into eclipse.

Mary. Faithful Mary.

"I'm thirsty," he said.

"Here," she said, "let me get you something."

Along the road they had once traveled so often to visit family, the hearse wound its way past stubbled fields, shuttered roadside markets. The minister, clutching his Bible against his chest as though it alone was sufficient cloak against the wind whipping across Pickaway County, passed final benediction:

"Ashes to ashes, dust to dust."

He stopped to pick up his hat as the funeral director placed the folded flag in Mary's lap.

When all is said and done, love is not rapture and fire. It is a hand steadier than one's own squeezing harder than a heartbeat. Wine changes back to water. Roses no longer come with love messages, but best wishes for a quick recovery. Endearment is exhibited by what once might have been considered insignificant kindnesses, but which, in the end, become the tenderest of ministrations.

On the day after the funeral, trying to busy herself with chores that could easily wait, she plopped the laundry basket in front of her granddaughter. The child tugged out the end of the sheet her Frank had always held when they did the wash. When the child brought the folded end to meet the corners her grandmother held, she kissed her playfully, just as he had once done.

"I'm thirsty, Grandma."

"Here, let me get you something."

SOURCE: Reprinted with permission, from *The Columbus Dispatch.* Mike Harden, "In the End, Real Love Means in Sickness and in Health."

Caring

Caring is placing another's needs before your own. As such, caring requires treating your partner as valued for simply being himself or herself. It requires what the philosopher Martin Buber called an I–thou relationship. Buber described two fundamental ways of relating to people: I–thou and I–it. In an I–thou relationship, each person is treated as a thou—that is, as a person whose life is valued as an end in itself. In an I–it relationship, each person is treated as an It; a person has worth only as someone who can be used. When a person is treated as a thou, his or her humanity and uniqueness are paramount.

Self-Disclosure

When we self-disclose, we reveal ourselves—our hopes, our fears, our everyday thoughts—to others. Self-disclosure deepens others' understanding of us. It also deepens our own understanding, because we discover unknown aspects as we open ourselves to others. Without self-disclosure, we remain opaque and hidden. If others love us, such love leaves us with anxiety. Are we loved for ourselves or for the image we present to the world?

Together, commitment, caring, and self-disclosure help transform love. But in the final analysis, perhaps the most important means of sustaining love is our words and actions. Caring words and deeds provide the setting for maintaining and expanding love (Byrne and Murnen 1988).

■ As we age the dynamics that characterize our intimate relationships change even when the relationships themselves endure.

Although we increasingly understand the dynamics and varied components of love, the experience of love itself remains ineffable, the subject of poetry rather than scholarship. A journal article is not a love poem, and romantics should not forget that love exists in the everyday world. Researchers have helped us increasingly understand love in the light of day—its nature, its development, its varied aspects—so that we may better be able to enjoy it in the moonlight.

Summary

- Love is of major significance in American society. Popular culture prominently features romantic love themes in songs, books, and films. Families are formed on the basis of love.

- Humans have a basic need for intimacy or closeness with others. Intimacy consists of affection, personal validation, trust, and self-disclosure.

- Intimate relationships such as love and friendship offer numerous emotional, psychological, and health benefits.

- We can see the differences between friendship and love by contrasting the qualities we seek in friends versus lovers.

- In the twentieth century, love became a more central theme in our search for a mate and our expectations for marriage.

- Americans have an ideology of romanticism, in which love is seen as blind, irrational, uncontrollable, and likely to strike at first sight. In addition, it is believed that there is a "one and only" for each of us.

- In more individualistic societies like the United States, a high value is placed on passionate love. In more collectivist societies, individual happiness is subordinate to the well-being of the group (including especially the family) and companionate love is more highly valued (and romantic love is devalued and frowned upon).

- Cultural expectations surrounding friendship and love in the United States define such intimacy in more feminine ways, involving heavy emphasis on self-disclosure.

- There are consistent gender differences in experiences and expectations surrounding friendship, love, and intimacy.

- Many factors help account for gender differences in styles of intimacy, including gender socialization, early childhood experiences of being mothered, the kinds of role models we have, and evolutionary influences on reproductive strategies.

- Sexual intimacy is an expected part of romantic or passionate love relationships. We also display love through other forms of physical contact, some of which differ between interracial and intraracial couples.

- Males and females do not attach the same meanings to how sexual expression fits within love relationships.

- Heterosexuals, gay men, and lesbians all value meaningful loving relationships.

- *Prototypes* of love and commitment are models of how people define these two ideas in everyday life. The central aspects of the love prototype include trust, caring, honesty, friendship, respect, and concern for the other; central aspects of the commitment prototype include loyalty, responsibility, living up to our word, faithfulness, and trust.

- Of the many ways in which love has been studied and measured, four are more common. These are Hendrick and Hendrick's Love Attitude Scale, Hatfield and Sprecher's Passionate Love Scale, Rubin's Love Scale, and Sternberg's Triangular Love Scale.

- Commitment is affected by the balance of costs to benefits, normative inputs, and structural constraints.

- The *wheel theory of love* emphasizes the interdependence of four processes: (1) rapport, (2) self-revelation, (3) mutual dependency, and (4) fulfillment of intimacy needs.

- According to John Lee, there are six basic styles of love: *eros, ludus, storge, mania, agape,* and *pragma.*

- The *triangular theory of love* views love as consisting of three components: (1) intimacy, (2) passion, and (3) decision or commitment.

- The *attachment theory of love* views love as being similar in nature to the attachments we form as infants. The attachment (love) styles of both infants and adults are secure, anxious or ambivalent, and avoidant.

- *Unrequited love* is a common experience. Occasionally, unrequited love is expressed through obsessive relational intrusion, or *stalking*. Most stalkers are male and most victims are female.

- *Jealousy* is an aversive response that occurs because of a partner's real, imagined, or likely involvement with a third person. Jealousy acts as a boundary marker for relationships.

- Time affects romantic relationships. The rapid growth of intimacy tends to level off, and we become habituated to passion. Commitment tends to increase, provided that the relationship is judged to be rewarding.

- Romantic love tends to diminish. It may either end or be replaced by intimate love. Many individuals experience the disappearance of romantic love as a crisis. Intimate love is based on commitment, caring, and self-disclosure.

Key Terms

agape 167

attachment theory of love 169

commitment 155

companionate love 156

companionate marriage 154

conjugal family 155

co-rumination 157

deep friendship 161

eros 167

erotomania 177

expressive versus instrumental 155

feminization of love 155

fulfillment of intimacy needs 173

Hatfield and Sprecher's Passionate Love Scale 167

Hendrick and Hendrick's Loved Attitude Scale 167

homogamy 154

intimacy 151

jealousy 177

love-obsessional stalking 177

ludus 167

mania 167

mutual dependency 173

passionate love 156

peer marriage 161

pragma 167

prototypes 166

rapport 172

reactive jealousy 178

rejection sensitivity 170

Resources on the Internet

Companion Website for This Book

http://www.thomsonedu.com/sociology/strong

Gain an even better understanding of this chapter by going to the companion website for additional study resources. Take advantage of the Pre- and Post-Test quizzing tool, which is designed to help you grasp difficult concepts by referring you back to review specific pages in the chapter for questions you answer incorrectly. Use the flash cards to master key terms and check out the many other study aids you'll find there. Visit the Marriage and Family Resource Center on the site. You'll also find special features such as access to Info-Trac© College Edition (a database that allows you access to more than 18 million full-length articles from 5,000 periodicals and journals), as well as GSS Data and Census information to help you with your research projects and papers.

CHAPTER 6

Understanding Sex and Sexualities

Outline

Are the following statements TRUE or FALSE?

You may be surprised by the answers (see answer key on the bottom of this page).

T	F	1	Unlike most human behavior, sexual behavior is instinctive.
T	F	2	A significant number of women require manual or oral stimulation of the clitoris to experience orgasm.
T	F	3	It is normal for children to engage in sexual experimentation with other children of both sexes.
T	F	4	Kissing is the most common and most accepted sexual activity.
T	F	5	A decline in the frequency of intercourse almost always indicates problems in the marital relationship.
T	F	6	Most married women and men have had an extramarital sexual relationship.
T	F	7	Bisexuality is more widely accepted than male homosexuality or lesbianism.
T	F	8	Latinos are generally less permissive about sex than African Americans or Anglos.
T	F	9	Because of their knowledge, college students rarely put themselves at risk for HIV and AIDS.
T	F	10	Condoms are not very effective as contraceptive devices.

Answer Key for What Do You Think

1 False, see p. 190;   2 True, see p. 216;   3 True, see p. 198;   4 True, see p. 215;   5 False, see p. 220; 6 False, see p. 218;   7 False, see p. 209;   8 True, see p. 218;   9 False, see p. 226;   10 False, see p.218.

It is now time to consider sex. For many of you, that must seem like a silly statement. Quite apart from this book, we often consider, think about, or take steps to pursue—or avoid—sexual encounters. Our popular culture is heavily sexualized. Advertising, in particular, uses sexual innuendo and image to sell us any number of products. Furthermore, being sexual is an essential part of being human. Through our sexuality we are able to connect with others on the most intimate levels, revealing ourselves and creating deep bonds and relationships. Sexuality is a source of great pleasure and profound satisfaction. It is the typical means by which we reproduce, transforming ourselves into mothers and fathers. Paradoxically, sexuality also can be a source of guilt and confusion, a pathway to infection, and a means of exploitation and aggression. Examining the multiple aspects of sexuality helps us understand our sexuality and that of others. It provides the basis for enriching our relationships.

In this chapter, we offer an overview of sexuality and sexual issues, especially as they are interconnected with relationships, marriages, and family life. We begin by considering the sources of our sexual learning and proceed through sexual development and expression in young, middle, and later adulthood, including the gay, lesbian, or bisexual identity process. We consider the shifts in sexual scripts from traditional to modern and the social control of sexuality. When we examine sexual behavior, we cover the range of activities and relationships in which people engage. Ultimately, we look at nonconsensual sexual relations, sexual problems, and dysfunctions; birth control; sexually transmissible diseases, human immunodeficiency virus (HIV), and acquired immunodeficiency syndrome (AIDS); and sexual responsibility. We hope that this chapter will help you make sexuality a positive element in your life and relationships.

Gender, Sexuality, and Sexual Scripts

Linking Chapter 4 with the present one, our gender roles are critical in learning sexuality, telling us what sexual behavior is appropriate, legitimate, and acceptable for each gender. Organizing and directing our sexual impulses are culturally shared sexual scripts, which we learn and act out. A **sexual script** consists of expectations of how to behave sexually as a female or male and as a heterosexual, lesbian, or gay male. Like a sexual road map or blueprint offering us general directions, a sexual script enables each individual to organize sexual situations and interpret emotions and sensations as sexually meaningful (Hynie et al. 1998). It may be more important than our own experiences in guiding our actions. Over time, we may modify or change our scripts, but we will not throw them away.

The scripts we are "given" for sexual behavior tend to be traditional. They are most powerful during adolescence, when we are first learning to be sexual. Gradually, as we gain experience, we modify and change our sexual scripts. As children and adolescents, we learn our sexual scripts primarily from our parents, peers, and the media. As we get older, interactions with our partners become increasingly important. In adolescence, both middle-class Caucasians and middle-class African Americans appear to share similar values and attitudes about sex and male–female relationships (Howard 1988).

Traditional Male Sexual Scripts

In traditional sexual scripts, men are perceived to be sexually aggressive and their sexual response, once set in motion, is thought to be difficult to control (Denov 2003). Traditional male sexual scripts also portray sex as "recreational," or pleasure centered for men (Hynie et al. 1998). Therapist Bernie Zilbergeld (1993) has identified the following assumptions in the male sexual script:

- *Men should not have (or at least should not express) certain feelings.* Men should be assertive, confident, and aggressive. Tenderness and compassion are not masculine feelings, and doubts should be kept to oneself.

- *A man always wants sex and is always ready for it.* It doesn't matter what else is going on; a man wants sex. He is always able to become erect.

- *Performance is the thing that counts.* Sex is something to be achieved, more about orgasm than intimacy.

- *The man is in charge.* As in other things, the man is the leader; he initiates sex and gives the woman her orgasm. A "real man" knows what women like, he doesn't need to be told.

- *All physical contact leads to sex.* Men are basically sexual "machines" any physical contact, including touching, is a sign for, or step toward, sex. There is no physical pleasure except sexual pleasure.

- *Sex equals intercourse.* All erotic contact leads to or is intended to lead to sexual intercourse. **Foreplay** is just that: warming up, getting your partner excited for penetration. Kissing, hugging, erotic touching, and oral sex are only preliminaries to intercourse.

- *Sexual intercourse always leads to orgasm.* The orgasm is the proof of the pudding; the more orgasms, the better the sex. If a woman does not have an orgasm, the male feels that he is a failure because he was not good enough to give her pleasure. If she requires clitoral stimulation to have an orgasm, she is considered to have a problem.

Researchers who study sexual stereotypes observe that men's sexual identity may depend heavily on a capricious physiological event: getting and maintaining an erection. They note that the following traits are associated with the traditional male role: (1) sexual competence, (2) ability to give partners orgasms, (3) sexual desire, (4) prolonged erection, (5) being a good lover, (6) fertility, (7) reliable erection, and (8) heterosexuality (Riseden and Hort 1992). Common to all these beliefs, sex is seen as a performance in which men are both the directors and the principal actors.

Traditional Female Sexual Scripts

Contrary to male sexual scripts, traditional female sexual scripts focus on feelings more than sex, on love more than passion. In the traditional female sexual script, sex is relational, a way of "expressing or achieving emotional and psychological intimacy *within certain prescribed relationships*" (Hynie et al. 1998, emphases added). It includes the following assumptions (Barbach 1982):

- *Sex is both good and bad.* What makes sex good or bad is whether it occurs in marriage or a committed relationship as opposed to a casual or uncommitted relationship. When not sanctioned by love or marriage, sexually active women risk their reputations.

- *Girls don't want to know about their bodies "down there."* Girls are taught not to look at, touch, and explore their genitals. As a result, women may know little about their genitals. They are often concerned

Although much of what we look at within this chapter can be generalized across boundaries of race or ethnicity, we need to recognize that ethnicity often shapes our sexual identities and experiences. Using the case of Asian Americans, we can illustrate ways in which culture can affect sexuality.

We recognize that "Asian American" actually encompasses as many as "thirty separate and distinct groups (Chan 2004)—such as Chinese, Japanese, Filipino, Vietnamese—the following sketch pertains mostly to East Asians.

Among Asians, sexuality in not as openly discussed as among Americans, even among one's closest friends. Among East Asian cultures, such as China and Japan, there is minimal open, public discussion. Asian cultures strongly emphasize modesty and sexual restraint. Sexual expression before or outside of marriage is mostly seen as inappropriate. It is especially expected that individuals will withhold expression of anything that might shame or embarrass their families. Occasionally (mis)construed as Asian *asexuality,* sexuality is a valued and normal but private part of life.

When their behavior or demeanor appears to be too sexual in nature, young Asians can expect to receive "strong and direct messages" from parents, including "expressions of disappointment and a strong sense of shame" (Chan 2004, 109).

Behaviorally Speaking

Asian American teens and young adults are significantly less sexually active than Caucasian, African American, or Hispanic young adults; 47% of a sample of single, hetero-sexual,18- to 25-year-old Asian American college students were sexually active compared to 72% of whites, 84% of blacks, and 59% of Hispanics (Chan 2004). A study of Los Angeles County high schoolers found that Asian Americans were far more likely to be virgins (73%) than Caucasian (50%), Latino (43%), and African American (28%) high school students. Among Canadians, Asian Canadians hold "more conservative sexual attitudes" and possess less knowledge about sex than their non-Asian peers.

Compared to other ethnic groups, Asian American high school and college students report the "oldest 'best' age" for first intercourse and have "significantly later normative and personal sexual timetables" for all sorts of sexual behavior (Okazaki 2004). Indeed, Asian American males had the highest median age at first sex (18.1) among a sample of Los Angeles–area young people. A study of Southern California college students found that more than 50% of the Asian men and 60% of the women were virgins. Among Caucasians, only 25% of men and less than 30% of women were virgins (Okazaki 2004).

Sexuality-Related Issues

Asian Americans, especially recent immigrants, are more likely to believe that Pap tests and gynecological care, more generally, should wait until after marriage and that they are too invasive and inappropriate beforehand (Okazaki 2004).

Consequences of such practices suggest that the greater sexual modesty of Asian Americans comes at a price. Compared to Caucasian women, Asian American women tend to be diagnosed with more advanced cases of breast and cervical cancer. Asian American men also show the same health service utilization pattern. A study by the National Asian Women's Health Organization of more than 800 Asian American men age 18–65 found that 89% had *never received reproductive or sexual health-care services* (Okazaki 2004). Other possible consequences of conservative sexual attitudes include Asian women's "extreme reluctance" to discuss or report sexual assault or abuse. Also, Asian American women are more likely than Caucasians to have been socialized to believe they have fewer sexual rights than their husbands, to believe rape myths, and to have negative attitudes toward rape victims. On the positive side, however, data indicate that Asian American women are less often victims of sexual assault and Asian American men are less often perpetrators of sexual assault (Okazaki 2004).

SOURCES: Okazaki 2004, 159–169; Chan 2004, 106–112.

about vaginal odors, which may make them uncomfortable about cunnilingus (oral sex).

- *Sex is for men.* Men are supposed to want sex; women are supposed to want love. Women are supposed to be sexually passive, waiting to be aroused.

Sex is not supposed to be a pleasurable activity as an end in itself; it is something performed by women for men.

- *Men should know what women want.* Men are supposed to know what women want, even if women don't tell them. To keep her image of sexual innocence, and remain pure, a woman does not tell a man what she wants.

- *Women shouldn't talk about sex.* Many women cannot talk about sex easily because they are not expected to have strong sexual feelings. Some women may know their partners "well enough" to have sex with them but not well enough to communicate their needs to them.

- *Women should look like beautiful models.* The ideal woman is unrealistically depicted with slender hips, firm, full breasts, and no fat; they are always young, have no pimples, wrinkles, or gray hair. Many women become self-conscious, worrying that they are too fat, too plain, or too old. They often feel awkward without clothes on to hide their imagined flaws.

- *Women are nurturers.* Women are supposed to give themselves, their bodies, their pleasures. Men are supposed to receive. His needs come first—his desire, his orgasm, his enjoyment.

- *There is only one right way to experience orgasm.* Women often "learn" that the only right way to experience orgasm is from penile stimulation during sexual intercourse. But there are many ways to reach orgasm: through oral sex; manual stimulation before, during, or after intercourse; masturbation; and so on. Women who rarely or never have an orgasm during heterosexual intercourse may be deprived by not sexually expressing themselves in other ways.

Sexual scripts can affect the ways in which we look at a range of sexual and sexually related behaviors. For example, by portraying males as active, sexually aggressive initiators whose sexual response cannot be easily controlled or constrained and females as sexually passive, innocent, and even nonsexual, traditional scripts ignore the possibilities of sexually reluctant males or sexually coercive females. This, in turn, obscures the phenomenon of female sexual offending, especially with male victims, and leaves police, victims, and helping professionals in the dark about what may be a more common phenomenon than is understood. Although rates of female sexual offending on male victims are quite low in the United States, Canada, and the United Kingdom, the 1%–8% of cases don't easily fit within the widely shared traditional male and female sexual scripts. In addition to the female sexual script, gender role norms suggest that females are nurturing and nonaggressive. As a consequence, author-

ities may not recognize and respond and victims are more reluctant to report victimization by a female out of fear that their claims may be met by disbelief, ridicule, or trivializing. Given the belief that women "don't do things like that" victims may decide to forego reporting altogether (Denov 2003).

Contemporary Sexual Scripts

As gender roles have changed, so have sexual scripts. To a degree, traditional sexual scripts have been replaced by more liberal and egalitarian ones. Sexual attitudes and behaviors have become increasingly liberal for both Caucasian and African American males and females, but African American attitudes and behaviors have been and continue to be somewhat more liberal than those of Caucasians (Wyatt et al. 1988). We know less about how Latino sexuality and Asian American sexuality have changed, as there is less research on the sexual scripts, values, and behaviors in those cultures.

Many women have explicitly rejected the more traditional scripts, especially the good girl–bad girl dichotomy and the belief that "nice" girls don't enjoy sex (Moffatt 1989). College-age women, as well as older, professional women who are single, are among those most likely to reject the old images (Davidson and Darling 1988).

Contemporary sexual scripts include the following elements for both sexes (Gagnon and Simon 1987; Rubin 1990):

- Sexual expression is positive.

- Sexual activities are a mutual exchange of erotic pleasure.

- Sexuality is equally involving of both partners, and the partners are equally responsible.

- Legitimate sexual activities are not limited to sexual intercourse but also include masturbation and oral–genital sex.

- Sexual activities may be initiated by either partner.

- Both partners have a right to experience orgasm, whether through intercourse, oral–genital sex, or manual stimulation.

- Nonmarital sex is acceptable within a relationship context.

- Gay, lesbian, and bisexual orientations and relationships are increasingly open and accepted or tolerated, especially on college campuses and in large cities.

Contemporary scripts both give greater recognition to female sexuality and are relationship-centered rather than male-centered. As we said earlier, traditional scripts have been replaced *to a degree*. Women who have several concurrent sexual partners or casual sexual relationships, for example, are still more likely to be regarded as more promiscuous than are men in similar circumstances (Williams and Jacoby 1989). The "suppression of female sexuality" is often carried out by women, whether through maternal influence or the judgments of female peer groups (Miracle, Miracle, and Baumeister 2003).

A 1999 study of 165 young, heterosexual women showed both the liberalization of attitudes and the continuation of a **sexual double standard.** On the liberalization side, 99% of the university sample "strongly agreed" or "agreed" that women can enjoy sex as much as men do. Of the women, 69% disagreed or strongly disagreed with the statement that "women are less interested in sex than men are." However, 95% of the women believed that there continues to be a double standard, wherein it is less accepted for a woman to have many sexual partners than it is for a man. In addition, 93% "probably" or "definitely agreed" that women who have many partners are more harshly judged than men with many partners, 49% indicated that women were labeled and penalized, but 48% stated that men were encouraged to have and rewarded for having many partners.

Robin Milhausen and Edward Herold (1999) concluded that women "overwhelmingly" perceive that there is still a double standard in society, even though they claim not to support it. Furthermore, women's own attitudes were influenced by their experiences; the more sexual partners a woman had had, the more accepting she was likely to be of men and women who have had many partners.

Psychosexual Development in Young Adulthood

At each period in our psychosexual development, we are presented with different challenges. In purely physical terms, adolescents are sexually mature (or close to it) but they are still learning their gender and social roles; they may also be struggling to understand the meaning of their sexual feelings for others and their sexual orientation. During young adulthood—from the late teens through mid-30s—many of the same tasks continue and new ones are added.

How Do We Know What We Know?

Before we examine the developmental tasks of young adulthood, let's look at some sources of our sexual learning. Children and adolescents are subjected to gender specific messages about sexuality, as well as both subtle and explicit socialization into heterosexuality (and away from homosexuality and bisexuality). Sources of influence include parents, peers, media, and school.

Parental Influence

Children learn a great deal about sexuality from their parents. Often, however, they learn not because their parents set out to teach them but because they are avid observers of their parents' behavior. Rather than openly and actively pursuing sexual topics, in many families sexuality remains "hidden" (Roberts 1983). When silence surrounds sexuality, it suggests that one of the most important dimensions of life is off limits, bad to talk about and dangerous to think about.

Parents convey sexual attitudes to their children in a number of ways. What parents say or do, for example, to children who touch their "private parts" or try

Ebby May/Getty Images/Taxi

■ *Parents, especially mothers, are important sources of information and advice about sexuality. Although both sons and daughters speak more to mothers than to fathers about sexual issues, most parent–child sexual communication is really between mothers and daughters.*

to touch either their mother's or some other woman's breasts conveys meanings about sex to a child. Parents who overreact to children's sexual curiosity may create a sense that sex is wrong. On the other hand, parents who acknowledge sexuality rather than ignoring or condemning it, help children develop positive body images, comfort with sexual matters, and higher self-esteem (Miracle, Miracle, and Baumeister 2003).

As young people enter adolescence, they are especially concerned about their own sexuality, but they may be too embarrassed or distrustful to ask their parents directly about these "secret" matters. Furthermore, many parents are ambivalent about their children's developing sexual nature. They are often fearful that their children (daughters especially) will become sexually active if they have too much information. They tend to indulge in wishful thinking: "I'm sure Jessica's not really interested in boys yet;" "I know Jason would never do anything like that." As a result, they may put off talking seriously with their children about sex, waiting for the "right time," or they may bring up the subject once, say their piece, breathe a sigh of relief, and never mention it again. Sociologist John Gagnon calls this the "inoculation" theory of sex education: "Once is enough" (cited in Roberts 1983). But children may need frequent "boosters" where sexual knowledge is concerned.

Because parents assume that their children are (or will be) heterosexual, they may avoid—intentionally or merely without thinking it relevant—discussion about sexual orientation. In so doing, they leave their children less aware of homosexuality, which, itself, becomes invisible. In addition, even simple comments such as "When you grow up and get married someday . . ." assumes that the child is or will become heterosexual (Shibley-Hyde and Jaffe 2000).

Research is somewhat mixed about the nature and consequences of parent–child communication about sexuality. Some research suggests that "early, clear communication" between parents and their teenage children leads to lower levels of teen sexual activity, and, for those who become sexually active, to greater understanding and use of safe-sex practices (Lehr, Demi, DiIorio, and Facteau 2005; Leland and Barth 1993; DiIorio, Kelly, and Hockenberry-Eaton 1999). In addition, research cited by Lehr and colleagues indicates that mothers who discuss sex-related issues influence their children's later protective sexual behaviors, including condom use (Dittus, Jaccard, and Gordon 1999; Miller, Levin, Whitaker, and Xu 1998). Other researchers report (Newcomer and Udry 1985;

O'Sullivan et al. 1999) little to no effect of mother–child sexual communication on subsequent teen sexual behavior, and one study suggests that it is associated with greater involvement in sexual activity, although it may be as much a consequence as a cause (Paulson and Somers 2000). Existing research indicates that most parent–child discussion about sex is really mother–daughter discussion about sex, that sons receive much less parental insight and information than daughters do, and that what information sons do receive they tend to learn from their mothers not their fathers (Lehr, Demi, DiIorio and Facteau 2005).

Although parental norms and beliefs are generally influential, they do not always have the strong desired effect on an adolescent's decision to become sexually active, especially in comparison with peer influence. A lack of family rules and structure are related to more permissive sexual attitudes and premarital sex among adolescents (Forste and Heaton 1988; Hovell et al. 1994), whereas a strong bond with parents appears to lessen teens' dependence on the approval of their peers and to lessen the need for interpersonal bonding that may lead to sexual relationships (DiBlasio and Benda 1992).

Peer Influence

Adolescents garner a wealth of information, as well as much *misinformation,* from one another about sex. They often put pressure on one another to carry out traditional gender roles. Boys encourage other boys to become and be sexually active even if the others are unprepared or uninterested. Those who are pressured must camouflage their inexperience with bravado, which increases misinformation; they cannot reveal sexual ignorance. Even though many teenagers find their earliest sexual experiences less than satisfying, many still seem to feel a great deal of pressure to conform, which may mean becoming or continuing to be sexually active. For many young people, virginity may be experienced as a stigma, whereas virginity loss is seen either as a way to shed the stigma (more true for males) or merely as part of growing up (Carpenter 2002).

Encouragingly, four large national probability samples from the Youth Risk Behavior Survey indicate that the 1990s saw a shift in adolescent sexual activity toward greater responsibility and restraint. According to data collected between 1991 and 1997, there was an 11% increase in the "incidence of virgin adolescents." This shift occurred mostly among males and among blacks and whites (but not Hispanics). Still, it represents a "significant reversal" from the 1970s and 1980s (Christopher and Sprecher 2000).

Other signs that adolescent sexual activity has declined can be seen in the findings of a Health and Human Services study, indicating that from 1995 to 2002 fewer teenage boys and younger teenage girls were sexually active. Among never-married females 15 to 17 years old, the percentage who ever had sexual intercourse declined from 38% in 1995 to 30% in 2002. For females 18 to19, there was virtually no change over this same time period. Among males there were more dramatic drops. From 1995 to 2002, the percentage of 15- to 17-year-olds reporting ever having had intercourse dropped from 43% to 31%. Among 18- to 19-year-old males, the decline was from 75% in 1995 to 64% in 2002. Clearly, more teens are delaying their initial experience of sexual intercourse, thus creating a somewhat different peer culture. We will look further at the trend in sexual involvement in a later section of this chapter.

Media Influence

The media profoundly affect our sexual attitudes (Wolf and Kielwasser 1991; McMahon 1990). Writing almost two decades ago, anthropologist Michael Moffatt noted that although about a third of the students he studied at Rutgers University mentioned the effect of college and college friends on their sexual development and another third mentioned their parents and religious values, the bigger and major influence on their sexuality was contemporary American pop culture, including "movies, popular music, advertising . . . TV . . . *Playboy, Penthouse, Cosmopolitan, Playgirl,* and so on; Harlequins and other pulp romances (females only); the occasional piece of real literature; sex education and popular psychology . . .; classic softcore and hard-core pornographic movies, books, and (recently) videocassettes" (Moffatt, 1989). To these we can add DVDs, which now virtually have replaced videocassettes, and—even more significantly—the Internet, with its numerous sexually oriented websites and weblogs (see the "Popular Culture" box later in this chapter).

Whereas the effects of certain media—notably television—on attitudes about sex have received more scholarly attention, the effects of others, such as magazines, have received less (Kim and Ward 2004). Thus, even though women's magazines are "replete with sexual content," including "frank advice about sex," they remain somewhat understudied and therefore inadequately understood as a source of sexual learning.

Existing research suggests that the more "teen-focused magazines" expose young women to a contradictory message that encourages them be *sexually provocative* in their demeanor and dress but discourages them from being *sexually active.* Encouraged to devote much time and effort toward making themselves physically appealing to boys and to presenting themselves as sexual objects, at the same time girls are discouraged from and warned about pursuing sexual relationships. Males are negatively portrayed as "either emotionally inept . . . or as sexual predators" (Kim and Ward, 2004), neither of which are flattering to males or encouraging for females entering the world of heterosexual relationships.

More "adult-focused magazines," such as *Cosmopolitan,* convey a different message. Sexually aggressive women are portrayed positively, almost in the same way as a stereotypical male is portrayed. College-age women who more frequently read magazines such as *Cosmopolitan* were less likely to perceive sex as risky or dangerous and "more likely to view sex as a fun, casual activity and to be supportive of women taking charge in their sexual relationships" (Kim and Ward 2004).

SEX ON THE NET. From its inception in 1983 to the present, the Internet has revolutionized the way we live. By 2000, there were a reported 1 billion web pages. Just

■ *Popular magazines such as* Seventeen *or* Cosmopolitan *are part of the sexual socialization that many young women in the United States experience. Although both are widely read, they convey different messages about sexuality.*

Popular Culture   "Blogging" about Sex

One of the most popular innovations to emerge in our increasingly wired world is the **weblog,** or "blog." First appearing in the early 1990s, weblogs now number in the millions and have become resources for people with shared interests, people seeking specialized information, and people who wished to produce online journals. As defined by Blogger.com:

> A blog is a personal diary. A daily pulpit. A collaborative space. A political soapbox. A breaking-news outlet. A collection of links. Your own private thoughts. Memos to the world. Your blog is whatever you want it to be. There are millions of them, in all shapes and sizes, and there are no real rules.

Along with Blogger.com and its 1.1 million registered users, other popular blog sites include Blogwise.com, Globe of Blogs, Blogs.com (acquired by Yahoo), and even the popular but controversial Myspace.com. Describing itself as "a place for friends," Myspace.com also claims more than 61 million blog posts (not 61 million different bloggers) for bloggers to choose from.

Blogs can be accessed by people who are seeking others of similar backgrounds, interests, or points of view or by the less determined and more casual "surfer" on the net. In this way, blogs offer interested individuals a "place" where they can interact and yet retain a sense of anonymity.

Some sexually oriented blogs are really venues for posting, viewing, and/or sharing nude photos, sexually explicit videos, and/or pornographic images. Others are personal sexual "diaries," in which individuals share, either for a closed membership or for the anonymous public, their sexual fantasies, thoughts, and/or supposed exploits (anyone can claim to be or have done whatever they wish to have others think). They are a kind of "anonymous exhibitionism."

There are also more informational, sexually oriented blogs, where we can learn about various sexual issues. For young people, adolescents and young adults, the existence of the weblog community offers an expanded, although invisible and anonymous, network from whom they can learn about—and with whom they can talk about—sex and sexuality. For example, there are blogs by gay, lesbian, and bisexual bloggers (such as http://www.comingout.blogspot.com). There are blogs by "swingers" and "polyamorists" and blogs filled with more informational or educational resources.

There has been only limited research so far, and little "hard data" on the number of blogs with sexual content or themes or the number of users or readers. Clearly, however, the existence of blogs broadens the opportunities for online sexual communication and sexual learning. It also introduces a problem for parents of minors, because there are many blogs that are sexually explicit and quite graphic. Although parents can put "blocks" on their computers (to prevent children from accessing objectionable content), in the absence of such blocks, all children need do is claim to be 18 (or in some instances 21) and they can enter a world filled with explicit language, graphic images, and highly sexualized content. With the sense of "community" that sometimes surfaces among bloggers or within websites, these resources can act like peers do in the process of sexual learning yet at other times, and for other users, may be more like media content.

3 years later, a Google search reviewed more than 3 billion websites (Griffin-Shelley 2003). Beginning in the 1990s, researchers looked with increasingly critical eyes at "sex and the net." Topics such as "addiction" to pornography; exploitation and entrapment of children; sexual harassment; and deviant pornography are among the issues and concerns that emerged. A 1996 study of six patients in a sex offender program stirred much concern with assertions about the potential addictiveness of Internet sex.

Aside from the rare but disturbing victimization that occurs, such material significantly broadens young people's access to sexual knowledge and material and removes it further from parental control.

As we get older, parents, peers, and the media eventually become less important in our sexual learning. As we experience interpersonal sexuality, our sexual partners become the most important source of modifying traditional sexual scripts. In relationships, men and women learn that the sexual scripts and models they learned from parents, peers, and the media won't necessarily work in the real world. They adjust their attitudes and behaviors in everyday interactions. If they are married, sexual expectations and interactions become important factors in their sexuality.

Sexuality in Adolescence and Young Adulthood

Sexual Developmental Tasks

Several tasks challenge young adults as they develop their sexuality:

- *Establishing a sexual orientation.* Children and adolescents may engage in sexual experimentation such as playing doctor, kissing, and fondling members of both sexes without such activities being associated with sexual orientation. By young adulthood a heterosexual, gay, or lesbian orientation emerges. Most young adults develop a heterosexual orientation. Others find themselves attracted to members of the same sex and begin to develop a gay, lesbian, or bisexual identity.

- *Integrating love and sex.* As we move into adulthood, we need to develop ways of uniting sex and love instead of polarizing them as opposites.

- *Forging intimacy and commitment.* Young adulthood is characterized by increasing sexual experience. Through dating, cohabitation, and courtship, we gain knowledge of ourselves and others as potential partners. As relationships become more meaningful and intimate, sexuality can be a means of enhancing intimacy and self-disclosure, as well as a means of obtaining physical pleasure.

- *Making fertility or childbearing decisions.* Childbearing is socially discouraged during adolescence, but fertility issues become critical, if unacknowledged, for single young adults. If sexually active, how important is it for them to prevent or defer pregnancy? What will they do if the woman unintentionally becomes pregnant?

- *Developing a sexual philosophy.* As we move from adolescence to adulthood, we reevaluate our moral standards, using our personal principles of right and wrong and of caring and responsibility. We develop a philosophical perspective to give coherence to our sexual attitudes, behaviors, beliefs, and values. Sexuality must be placed within the larger framework of our lives and relationships, integrating our personal, religious, spiritual, or humanistic values with our sexuality. (Gilligan 1982; Kohlberg 1969).

Adolescent Sexual Behavior

If we take a long view of changes in teenage sexual behavior, today's teenagers are more tolerant of premarital sex and more likely to engage in it than teenagers were, say, 30 or 50 years ago. Comparing members of the graduating classes of 1950, 1975, and 2000 from the same northeastern high school, Sandy Caron and Eilean Moskey found a steady decline in negative attitudes about premarital sex and an increase in the percentages who reported having had sexual intercourse while in their teens. Furthermore, graduates of the class of 1950 were not only much less likely to have had intercourse but also less likely to have had more than one sexual partner, used birth control, or even talked about sex with their parents. Of the respondents from the class of 1950, 25% were sexually active while in high school, compared with 65% of the class of 1975 and 69% of the class of 2000. Similarly, where 76% of sexually active members of the class of 2000 "always used birth control," 68% of the sexually active members of the class of 1950 *never* used birth control (Caron and Moskey 2002).

Yet, as noted earlier and revealed in what follows, teenage and young adult sexuality have undergone some pronounced changes in more recent years. For example, between 1993 and 2001, an estimated 2.5 million adolescents took public virginity pledges, promising to abstain from sexual intercourse until they married. As sponsored and organized by the Southern Baptist Church, such pledges did tend to delay first intercourse, often for a long time (Bearman and Bruckner 2001).

A more comprehensive picture of some recent trends can be seen in the following data on adolescent sexual behavior:

- Through the 1990s, the percentage of teenagers reporting having sexual intercourse dropped 5.7% and the teen pregnancy rate was down 14%. According to sociologists Pepper Schwartz and Barbara Risman, there were gender and race differences in the changes in sexual expression. The number of high school boys—but not girls—under 18 who remained virgins dramatically increased. The rate of sexual activity of black females was sharply reduced, and among Caucasian and Hispanic females it remained generally stable (http://www2.asanet.org/media/cntrisman.html).

- Cumulatively, looking at ninth- through twelfth-grade girls and boys, 47% report having had sexual intercourse, a decline of 6% since 1993.

- In 2003, 33% of ninth graders and 62% of twelfth graders reported that they had ever had sexual intercourse.

- The median age at first sexual intercourse was 16.9 for males and 17.4 for females.

- In 2003, 66% of teens were "sexually abstinent," meaning that they had refrained from sexual intercourse for at least 3 months.

- The percentage of teens who first had sexual intercourse before having turned 14 years of age declined from 8% of girls and 11% of boys in 1995 to 6% of girls and 8% of boys in 2002.

- In 2003, among those teens who were sexually active, 98% reported using at least some form of contraception, with condoms (94%) and birth control pills (61%) being the most commonly used methods (Kaiser Family Foundation 2005).

Nearly one in ten 15–17 year olds had been physically forced to have intercourse. Among those who were sexually active, 24% reported having done something sexually that they didn't really want to, and 21% had participated in oral sex to avoid sexual intercourse. Racial differences surface here; Leslie Houts reports that black females (at 13%) were most likely to report their first sexual intercourse as "unwanted," Hispanics were least likely to describe their initial experience of intercourse as forced (4%), with whites (6%) slightly more likely to respond that way (Houts 2005).

The earlier the age at which we first experience sexual intercourse, the less likely it is that the experience was wanted. The relationship context also makes a difference in desirability. The more committed the relationship in which it occurs (such as in engagement or going steady), the more likely the first sexual encounter is to have been "wanted" (Houts 2005).

Emily Impett and Letitia Peplau report that more than a third of college men and more than half of college women report having consented to unwanted sex, and between 21% and 32% of college women said they engaged in unwanted sex out of fear that their partners would leave them. They further link attachment style to whether and why some college women consent to unwanted sex and others don't. "Anxiously attached" women, because they feared their partners would leave them, were more willing to conceive of engaging in unwanted sex. They also expressed more of a desire to avoid conflict and a concern for preventing their partners from losing interest than more securely attached women reported (Impett and Peplau 2002).

Virginity and Its Loss

What makes one a virgin? This fairly simple and straightforward-sounding question is a little more complicated. Is virginity more broadly the preservation of "innocence" through the lack of sexual experience, or is it more narrowly the lack of sexual intercourse experience? Most people agree that we maintain **virginity** as long as we refrain from sexual intercourse. But occasionally we hear people speak of "technical virginity," to refer to people who have had a variety of sexual experiences but have not had sexual intercourse. Such individuals are hardly sexually naïve and lack some other connotations associated with the concept of virginity (innocence and purity, for example). Data indicate that a "very significant proportion of teens has had experience with oral sex, even if they haven't had sexual intercourse, *and may think of themselves as virgins*" (Lewin 2005, emphasis added). Research findings from the National Survey of Family Growth released by the National Center for Health Statistics reveal that "more than half of all teenagers aged 15 to 19 have engaged in oral sex—including nearly a quarter of those who have never had intercourse" (Lewin 2005).

The proportion of teenagers who have had oral sex was slightly higher than the proportion that has had intercourse: 55% of the boys and 54% of the girls reported having given or received oral sex, and 53% of the girls and 49% of the boys reported having had sexual intercourse. Other research, especially research looking into virginity loss, reports that 35% of virgins, defined as people who have never engaged in vaginal intercourse, have—nonetheless—engaged in one or more other forms of heterosexual sexual activity (for example, oral sex, anal sex, or mutual masturbation).

Sociologist Laura Carpenter, author of *Virginity Lost: An Intimate Portrait of First Sexual Experiences* (2005), acknowledges that losing virginity has different meaning for males and females (see the "Issue & Insights" box later in this chapter). For starters, at least a quarter of women, age 15–24, reported in the National Survey of Family Growth that, although "voluntary," their first experience of heterosexual intercourse was "not really wanted," rating it low (1 to 4) on a 10-point scale of "wantedness" (Houts 2005).

Young women and men not only attach different meaning but also experience virginity loss in different ways. Women are more likely to be worried about negative outcomes of their first experience of intercourse.

Researchers looking into the meanings people attach to the loss of their virginity and how they negotiate the transition from virgin to nonvirgin find that virginity has different meanings for women of different ages (Carpenter 2002; Houts 2005). As Leslie Houts says, "the meaning of being a virgin at 14 is very different than at age 24, or at age 34" (2005, 1,097). At younger ages virginity may be culturally expected, at a somewhat later age it may be respected and celebrated, and at "too old an age" it may be viewed with curiosity or suspicion. In each of these scenarios, the meaning of virginity loss is different; in none of these scenarios is the physical pleasure of sexual intercourse manifest (Houts 2005).

Laura Carpenter's research with 61 women and men suggests that people draw upon three themes to make sense of their lost virginity: virginity

as a gift, virginity as a stigma, and virginity loss as part of the transition to adulthood. Although many individuals indicated more than one of the following categories, the following pattern of response was revealed:

- *Virginity as a gift.* Half of Carpenter's informants recalled that at some point in their lives they had thought of virginity as a gift they were giving to someone, ideally to someone they loved, and to which the recipient would give enhanced love and commitment in return.

- *Virginity as stigma.* More than a third of Carpenter's sample saw their virginity as something to hide, and something they wished to shed as soon as possible ("at the first available opportunity, often with relatively casual partners, such as friends or strangers"). The sexual double standard of even contemporary sexual scripts made it easier for women both to hide, and to shed, their virgin status.

- *Virginity loss as part of growing up.* More than half of Carpenter's

interviewees thought that the loss of virginity was inevitable and desirable, "just another experience" in the process of becoming an adult, with minimal gender differences in the interpretation of the experience. Where gender did surface prominently was in how much physical pleasure or enjoyment was experienced with the loss of virginity. For a majority in this group, including three-fourths of the women and three-fifths of the men, virginity loss was not physically enjoyable.

Sexual orientation also colored people's interpretations of their loss of virginity. Gay men and lesbians were more likely than heterosexuals to have seen the loss of virginity as a step in the process of growing up (73% versus 46%). Heterosexual women and men were more likely to have perceived virginity as a gift than were gays or lesbians (54% versus 31%). Interestingly, among gay men, lesbians, bisexuals, and heterosexuals who shared an interpretive framework, experience of virginity loss was quite similar.

In addition, women are more worried about pregnancy, more likely to be nervous, more likely to be in pain, and less likely to experience orgasm. They are also more likely to experience postcoital guilt and express with regret the wish that they had waited.

Converging Patterns for Women and Men

As recently as the 1980s, young women were more likely to value virginity and to contemplate its loss primarily within committed romantic relationships and men welcomed opportunities for casual sex and expressed disdain for virginity. Research in the 1990s revealed increasing similarities between women and men. More young men than before were expressing pride and happiness about being virgins. Growing numbers of young women were perceiving virginity

in neither a positive nor a negative light, with a minority eagerly anticipating "getting it over with." By the 1990s, gender differences in the age at which one first engages in intercourse had all but disappeared. By 1999, age at first vaginal sex was between 16 and 17 for both females and males (Carpenter 2002).

Gay, Lesbian, and Bisexual Identities

In contemporary America, people are generally classified as **heterosexual** (sexually attracted to members of the other gender), **homosexual** (sexually attracted to members of the same gender), or **bisexual** (attracted to both genders). Although today we may automatically accept these categories, such acceptance has not always been the case and the categories do not necessarily reflect reality. As late as the nineteenth century, there was no concept of "homosexuality." Both the label *homosexual* and the label *heterosexual* first appeared in print in the United States in a medical journal in 1892 (Katz 2004). At still other times, both *homosexual* and *heterosexual* were terms referring to sexual perversions, with heterosexuals being people with sexual inclinations toward both sexes and toward "abnormal methods of gratification" (Katz 2004).

The familiar threefold categorization of sexual orientation used today may not accurately depict the range that exists in our sexual orientations—who we are attracted to, who we have relations with, who we fantasize about, the type of lifestyle we live, and how we identify ourselves. On any of these items we may be *exclusively* oriented toward the other sex or our sex, *mostly* drawn to the other sex or our sex, or oriented to both sexes *about equally*. In addition, the interaction of numerous factors—social, biological, and personal—leads to the unconscious formation of sexual orientation. The two most important components of sexual orientation are the gender of our sexual partner and whether we label ourselves heterosexual, gay, lesbian, or bisexual. Finally, our sexual orientation may change over time. Thus, what was true of past relationships or attractions may not fit with the present or may differ from what we envision for our future (Klein 1990; Miracle, Miracle, and Baumeister 2003).

Because *homosexual* carries negative connotations and obscures the differences between what women and men experience, we refer to **gay men** and **lesbians.** In addition, replacing the term *homosexual* may help us see individuals as whole people; sexuality is not the only significant aspect of the lives of gay men, lesbians, bisexuals, or heterosexuals. Love, commitment, desire, caring, work, possibly children, religious devotion, passion, politics, loss, and hope are also, if not more, important.

At different times, especially in the past, those with lesbian or gay orientations have been called sinful, sick, perverse, or deviant, reflecting traditional religious, medical, and psychoanalytic approaches. Contemporary thinking in sociology and psychology has rejected these older approaches as biased and unscientific and has focused, instead, on how women and men come to identify themselves as lesbian or gay, how they interact among themselves, and what effect society has on them (Heyl 1989). As noted sociologist Howard Becker (1963) has pointed out, "Deviant behavior is

behavior that people so label." Deviance is created by social groups that make rules whose violation results in violators being labeled deviant and treated as outsiders. Lesbian and gay behavior, then, is deviant only insofar as it is called deviant.

How does one "become" gay, lesbian, bisexual, or even heterosexual, for that matter? Such a question is neither easily answered nor inconsequential. If sexual orientation is biologically based, discrimination against gay men, lesbians, or bisexual women and men is especially unjustified. It becomes no different than discriminating against someone because of their age, their gender, or their race, all statuses over which we exercise no control.

Research on the self-identification process suggests that we can divide the gay, lesbian, and bisexual population into two groups of people: one group comprising men and women who say that they knew, from a much earlier age while growing up, that they were "different" from others. The second group grew up "never questioning the suitability of a heterosexual identity" until later in their lives, such as college age or middle age. Most men and many of the women in this latter group attribute this delayed identification to denial. However, many women (but not many men) reject the idea that they were driven by uncontrollable or irresistible desires, saying, instead, that they "chose" to become involved with a same-sex partner and that their choice was a political one, associated with their particular feminist politics. For others, it was a choice motivated out of the desire for more equal, more intimate relationships than they believed they could have with men (Butler 2005).

If sexual orientation is chosen, given the cultural, social, and legal changes that have occurred in recent decades, we might expect an increase in the percentage of the population that engages in same-sex sexual relationships. Indeed, data bear this out. Data from the General Social Survey indicate that between 1988–1990 and 1996–1998 the percentage of American men reporting having had a same-sex sexual partner the previous year more than doubled, from 1.7% to 3.9%. Among women a similar pattern held, as the percentage of women having had a same-sex sexual partner rose from 0.7% to 2.7%. Amy Butler (2005) extended the time frame for analysis and incorporated data through the 2002 General Social Survey. Her findings are shown in Table 6.1.

Butler reminds us that these increases may be interpreted in a few ways. Rather than an increase in the percentage of women and men who *have* same-sex

Table 6.1 ■ Sex of Sex Partner, 1988–2002 (in percentages)

	1988	1990	1992	1994	1996	1998	2000	2002
Males								
Same sex	2.4	2.0	2.5	2.6	3.7	4.1	3.8	2.9
Opposite sex only	82.3	88.4	84.1	83.6	82.9	80.6	80.5	82.5
No partner	10.9	6.6	10.9	10.9	8.6	11.3	11.7	12.5
No answer	4.4	3.0	2.4	2.8	4.9	4.0	4.0	2.1
Females								
Same sex	0.2	0.8	1.2	1.0	2.5	2.6	3.3	3.5
Opposite sex only	83.5	85.1	81.7	80.3	79.7	80	78.3	78.3
No partner	12.8	10.8	13.6	14.0	13.1	14.6	15.5	16.4
No answer	3.6	3.3	3.4	3.7	4.8	2.7	3.0	1.8

attractions or desires, they may reflect increases in the *willingness to act* on desires that otherwise would have been suppressed, ignored, or denied. Another alternative to consider is that the changing or liberalized climate may simply encourage people to report more honestly what they are doing sexually.

The actual percentage of the population that is lesbian, gay, or bisexual is not known. Among women, about 13% have had orgasms with other women, but only 1% to 3% identify themselves as lesbians (Fay et al. 1989; Kinsey, Pomeroy, and Martin 1948, Kinsey et al. 1953; Marmor 1980c). Among males, including adolescents, as many as 20% to 37% have had orgasms with other males, according to Alfred Kinsey's studies. Of these, 10% were predominantly gay for at least 3 years; 4% were exclusively gay throughout their entire lives (Kinsey et al. 1948). A review of studies on male same-sex behavior between 1970 and 1990 estimated that a minimum of 5% to 7% of adult men had had sexual contact with other men in adulthood. Based on their review, the researchers suggested that about 4.5% of men are exclusively gay (Rogers and Turner 1991). A large-scale study of 3,300 men age 20 to 39 reported that 2% had engaged in same-sex sexual activities and 1% considered themselves gay (Billy et al. 1993). In 1994, the National Health and Social Life Study found that, of the participants, 2.8% of men and 1.4% of women described themselves as homosexual or bisexual, although approximately 6% of men and 4% of women said they had had a sexual experience with someone of the same sex at least once since puberty (Laumann et al. 1994).

What can we make of the differences among studies? In part, the variances may be explained by different methodologies, interviewing techniques, sampling,

or definitions of homosexuality. Furthermore, sexuality is more than simply sexual behaviors; it also includes attraction and desire. One can be a virgin or celibate and still be gay or heterosexual. Finally, sexuality is varied and changes over time; its expression at one time is not necessarily its expression at another.

Identifying Oneself as Gay or Lesbian

Many researchers believe that a person's sexual interest or direction as heterosexual, gay, or lesbian is established by age 4 or 5 (Marmor 1980a, 1980b). But identifying oneself as lesbian or gay takes considerable time and includes several phases, usually beginning in late childhood or early adolescence (Blumenfeld and Raymond 1989; Troiden 1988). **Homoeroticism**—erotic attraction to members of the same gender—almost always precedes gay or lesbian activity by several years.

We noted in Chapter 4 that people commonly, although incorrectly, assume that a person's masculinity or femininity reveals their sexual preference. Further complicating the connection, or lack thereof, between gender and sexual orientation are the retrospective accounts, more often revealed by gay men than by lesbians, of "being different" in childhood, of not fitting in with or desiring to conform to gender appropriate behavior. However, many heterosexuals remember their childhoods in similar ways; 60% or more of heterosexual women recall being tomboys, enjoying male activities and play, and engaging in gender nonconforming behavior (Gottshalk 2003). Research has shown that more heterosexual women than gay men enjoyed stereotypical masculine play and activities as children. Also, gay men and lesbians who piece together

Understanding Yourself

The Social Control of Sexuality and Your Sexual Scripts

Our discussion of sexual scripts illustrates some ways in which society influences our sexual attitudes and behavior. This is not unique to the United States; every society regulates and controls the who, what, when, where, and why of sexuality. Your script will change over time, depending on your age, sexual experience, and interaction with intimate partners and others. Examine some questions you are likely to encounter.

Who

Society tells you to have sex with people who are unrelated, around your age, and of the other sex (heterosexual). Less acceptable is having sex with yourself (masturbation), with members of the same sex (gay or lesbian sexuality), and with relatives (incest). In most societies, extramarital relationships are prohibited. Examine the "whos" in your sexual script, then think about the following questions:

- With whom do/would you engage in sexual behaviors?
- How do your choices reflect homogamy and heterogamy?
- What social factors influence your choice?
- Does your autoerotic behavior change if you are in a relationship? How? Why?

What

Society classifies sexual acts as good or bad, moral or immoral, and appropriate or inappropriate. Although these designations may seem absolute, they are culturally relative.

- What sexual acts are part of your sexual script?
- How are they regarded by society?

- How important is the level of commitment in a relationship in determining your sexual behaviors?
- What level of commitment do you need for kissing? Petting? Sexual intercourse?
- What occurs if you and your partner have different sexual scripts for engaging in various sexual behaviors?

When

You might make love when your parents are out of the house or, if a parent yourself, when your children are asleep. Usually, such timing is related to privacy, but it may also be related to the age at which sexual activity is expected to start and stop, how often people are expected to engage in sexual relations, and when in a relationship sex should begin. Finally, it may pertain to times when sex is considered appropriate or inappropriate. Some societies frown upon a woman engaging in sex during her menstrual flow, for a period after the birth of a child, or while nursing (Miracle, Miracle, and Baumeister 2003).

- When do you engage in sexual activities?
- Are the times related to privacy?
- When did you experience your first erotic kiss?
- At what age did you first have sexual intercourse? If you have not had intercourse, at what age do you think it would be appropriate?
- How was the timing for your first intercourse determined, or how will it be determined?
- What influences (friends, parents, religion) are brought to bear on the age-timing of sexual activities?

Where

Where do sexual activities occur with society's approval? In our society, they usually occur in the bedroom, where a closed door signifies privacy. For adolescents, automobiles, fields, beaches, and motels may be identified as locations for sex; churches, classrooms, and front yards usually are not. "Where" may also extend to where it might be considered appropriate or inappropriate to discuss sex or to expose parts of your body.

- Where do you think the acceptable places to be sexual are?
- What makes them acceptable for you?
- Have you ever had conflicts with partners about the "wheres" of sex? Why?

Why

There are many reasons for having sex: procreation, love, passion, revenge, intimacy, exploitation, fun, pleasure, relaxation, boredom, achievement, relief from loneliness, exertion of power, and on and on. Some of these reasons are approved by society; others are not. Some we conceal; others we do not.

- What are your reasons for sexual activities?
- Do you have different reasons for different activities, such as masturbation, oral sex, and sexual intercourse?
- Do the reasons change with different partners? With the same partner?
- Which reasons are approved by society, and which are disapproved?
- Which reasons do you make known, and which do you conceal? Why?

retrospective accounts do so from an adult vantage point *as gay or lesbian.* Thus, they may "read" their life story in such a way as to make it fit their adult sexuality (Gottshalk 2003).

Stages in Acquiring a Lesbian or Gay Identity

The first stage in acquiring a lesbian or gay identity is often marked by fear, confusion, and denial, and the perception that desires mark the person as different from others. The person may find it difficult to label the emotional and physical desires for the same sex. Adolescents especially fear their family's discovery of their homoerotic feelings. In the second stage, if these feelings recur often enough, the person recognizes the attraction, love, and desire as homoerotic. The third stage includes the person's self-definition as lesbian or gay. This may take a considerable struggle, because it entails accepting a label that society generally calls deviant. Questions then arise about whether to tell parents or friends, whether to hide the identity ("to be in the closet") or make the identity known ("to come out of the closet").

Some gay men and lesbians may go through two additional stages. One stage is to enter the gay subculture. A gay person may begin acquiring exclusively gay friends, going to gay bars and clubs, or joining gay activist groups. In the gay world, gay and lesbian identities incorporate a way of being in which sexual orientation is a major part of the identity as a person. Pat Califia (quoted in Weeks 1985) explains the process: "Knowing I was a lesbian transformed the way I saw, heard, perceived the whole world. I became aware of a network of sensations and reactions that I had ignored all my life."

The final stage begins with a person's first lesbian or gay affair. This marks the commitment to unifying sexuality and affection. Sex and love are no longer separated. Most lesbians and gay men have had such affairs, despite the stereotypes of anonymous gay sex.

Coming Out

Being lesbian or gay is often associated with a total lifestyle and way of thinking. In making the gay or lesbian orientation a lifestyle, **coming out**—publicly acknowledging one's gayness—has become especially important as an affirmation of sexuality. Coming out may jeopardize many relationships, but it is also an important means of self-validation. By publicly acknowledging a gay or lesbian orientation, a person be-

■ *Two significant factors in identifying sexual orientation are (1) the gender of one's partner, and (2) the label one gives oneself (lesbian, gay, bisexual, or heterosexual).*

gins to reject the stigma and condemnation associated with it. Generally, coming out occurs in stages, first involving family members, especially the mother and siblings and later the father. Coming out to the family often creates a crisis, but generally the family accepts the situation and gradually adjusts (Holtzen and Agresti 1990). Religious beliefs, prejudice, and misinformation about gay and lesbian sexuality, however, often interfere with a positive parental response, initially making adjustment difficult (Borhek 1988; Cramer and Roach 1987). After the family, friends may be told and, in fewer cases, employers and coworkers.

Gay men and women are often "out" to varying degrees. Some may be out to no one, some to their lovers, others to close friends and lovers but not to their families, employers, associates, or fellow students. Still others may be out to everyone. Because of fear of reprisal, dismissal, or public reaction, lesbian and gay schoolteachers, police officers, members of the military, politicians, and members of other such professions are rarely out to their employers, coworkers, or the public.

Outing refers to the practice of publicly identifying "closeted" gays or lesbians. Some claim that outing is politically justified, rationalizing that if gays and lesbians stay quiet about their sexual orientation, negative stereotypes about homosexuals remain unchallenged. They reason that as heterosexuals discover that some of their friends and family members, or even public figures that they are familiar with and respect, are gay or lesbian, they may modify their attitudes about homosexuality in a more accepting direction (Miracle, Miracle, and Baumeister 2003).

Gay and Lesbian Relationships versus Heterosexual Relationships

In reviewing the existing research comparing same-sex and heterosexual relationships, the literature is mixed as to how similar or different they are from each other. In many ways, same-sex couples want, experience, and struggle with many of the same things as heterosexual couples. Comparative research has indicated that gay men, lesbians, and heterosexuals report the same levels of relationship satisfaction, attraction and love for their partners, and relationship adjustment. Following couples over an 18-month period, among couples who had been together 10 years or more, only modest differences have been found in their rate of breaking up: 6% among lesbians, 4% among gay couples, and 4% of married couples. Among those together 2 years or less there were some differences in that rate: 22% for lesbian couples, 16% for gay male couples, and 4% for married couples (over the same period, 17% of heterosexual cohabitants together less than 2 years broke up).

Gay, lesbian, and heterosexual couples struggle over the same sorts of issues: money, housework, power, and abuse. When relationships end—because of breakup or death—they suffer similarly. However, gay and lesbian couples often lack the supportiveness of family, friends, and others that married heterosexual couples can mostly take for granted. Thus, when relationship issues arise, conflict occurs, and losses result, gay men and lesbians may not receive the encouragement, support, advice, and sympathy that heterosexuals receive (Peplau, Veniegas, and Campbell 2004).

Major areas of difference have been identified in the importance attached to gender and gender role behavior (as expected, greater among heterosexuals than in gay or lesbian relationships), the presence or absence of role models for healthy relationships and for resolution of difficulties (scarcer for gay and lesbian couples), and in sexual behavior. Sexual exclusivity is lower among gay male couples than among heterosexual or lesbian couples. Sexual behavior is less fre-

quent among lesbian couples than among gay male or heterosexual couples, although nongenital or nonsexual affection (for example, cuddling, kissing, and hugging) is reportedly more common. Monogamy and romantic love are more important to lesbians and to heterosexual women than to men in heterosexual or gay relationships (Spitalnik and McNair 2005). Lesbians and gay men also have fewer barriers than do heterosexuals to ending their relationships once troubles surface. This makes it unlikely that lesbians and gay men will live in long-term, dissatisfying, "miserable and deteriorating" relationships, but more gay and lesbian relationships than heterosexual relationships will end that could have been saved or improved with patience and effort. In addition, gay male and lesbian couples must deal with disagreements about how much they wish to disclose their sexuality to others. Such disagreements may lead a more open partner to pressure a less open partner with the threat of disclosure or leave the more open partner feeling as though the less open partner is less committed to the relationship (Peplau, Veniegas, and Campbell 2004).

Antigay Prejudice and Discrimination

Antigay prejudice is a strong dislike, fear, or hatred of lesbians and gay men because of their homosexuality. **Homophobia** is an irrational or phobic fear of gay men and lesbians. Not all antigay feelings are phobic in the clinical sense of being excessive and irrational. They may be unreasonable or biased. (Nevertheless, they may be within the norms of a biased culture.) Because prejudice may not be clinically phobic, the less clinical term, *antigay prejudice,* may be more appropriate (Haaga 1991).

Antigay prejudice justifies discrimination and violence based on sexual orientation. In his classic work on prejudice, Gordon Allport (1958) states that social prejudice is acted out in three stages: (1) offensive language, (2) discrimination, and (3) violence. Gay men and lesbians experience each stage. They are called *faggot, dyke, queer,* and *homo.* They are discriminated against in terms of housing, equal employment opportunities, insurance, adoption, parental rights, family acceptance, and so on, and they are the victims of violence known as *gay bashing* or *queer bashing.*

Such negative attitudes and hostile behaviors often exist among college students. One study of college freshmen found that 50% felt that homosexual behavior was wrong and that gay men were disgusting.

Real Families   Memoirs of a Sissy

The following autobiographical excerpts from writer Tommi Avicolli's longer essay "He Defies You Still: The Memoirs of a Sissy" describe treatment he received as a young boy and later teen in school and reveal the depth of painful consequences of the harassment he faced.

Scene One

A homeroom in a Catholic high school in South Philadelphia. The boy sits quietly in the first aisle, third desk, reading a book. He does not look up, not even for a moment. He is hoping no one will remember he is sitting there. He wishes he were invisible. The teacher is not yet in the classroom so the other boys are talking and laughing loudly.

Suddenly, a voice from beside him:

"Hey; you're a faggot, ain't you?"

The boy does not answer. He goes on reading his book, or rather pre-

tending. . . . It is impossible to actually read the book now.

"Hey, I'm talking to you!"

The boy still does not look up. He is so scared his heart is thumping madly; it feels like it is leaping out of his chest and into his throat. But he can't look up.

"Faggot, I'm talking to you!"

. . . Suddenly, a sharpened pencil point is thrust into the boy's arm. He jolts, shaking off the pencil, aware that there is blood seeping from the wound.

"What did you do that for?" he asks timidly.

"Cause I hate faggots," the other boy says laughing. Some other boys begin to laugh, too. A symphony of laughter. The boy feels as if he's going to cry. But he must not cry. . . . So he holds back the tears and tries to read the book again. . . .

When the teacher arrives a few minutes later, the class quiets down. The boy does not tell the teacher what has happened. He spits on the wound to clean it, dabbing it with a tissue until the bleeding stops. For weeks he fears some dreadful infection from the lead in the pencil point.

Scene Two

The boy is walking home from school. A group of boys (two, maybe three, he is not certain) grab him from behind, drag him into an alley and beat him up. When he gets home, he races up to his room, refusing dinner ("I don't feel well," he tells his mother through the locked door) and spends the night alone in the dark wishing he would die. . . .

These are not fictitious accounts—I *was* that boy. Having been branded a sissy by neighborhood children because I preferred a jump rope to baseball and dolls to playing soldiers, I was often taunted with "hey sissy" or "hey faggot" or "yoo hoo, honey" (in a mocking voice) when I left the house.

To avoid harassment, I spent many summers alone in my room. I went out on rainy days when the street was empty.

. . . I came to like being alone. I didn't need anyone. . . . Contact with others meant pain. Alone, I was protected. I began writing poems, then short stories. There was no rea-

---

And 30% said they would prefer to not go to school with gays and lesbians (D'Augelli and Rose 1990). Antigay prejudice can even extend to heterosexuals who voluntarily choose to room with a lesbian or gay man. They are assumed to have "homosexual tendencies" and to have many of the negative stereotypical traits of gay men and lesbians, such as poor mental health (Sigelman et al. 1991).

Typically, males harbor more prejudice and express more negative attitudes toward gays and lesbians than women do. As they move through adolescence toward young adulthood, male prejudice increases where female prejudice diminishes. Antigay prejudice is experienced in many different ways. Nearly all (98%) first-year college students in Anthony D'Augelli and M. L. Rose's study (1990) reported hearing disparag-

ing remarks on campus about gays and lesbians. In terms of victimization accounts, a nationwide survey of 15- to 21-year-old gay males, lesbians, and bisexuals revealed disturbing evidence pointing to wide ranging forms of mistreatment (see Figure 6.1).

Antigay prejudice adversely affects heterosexuals, too, by doing the following:

- Creating fear and hatred, aversive emotions that cause distress and anxiety

- Alienating heterosexuals from gay family members, friends, neighbors, and coworkers (Holtzen and Agresti 1990)

- Limiting expression of a range of behaviors and feelings, such as hugging or being emotionally intimate, with same-sex friends for fear that such in-son to go outside anymore. I had a world of my own. . . .

Scene 4....

High school religion class. Someone has a copy of *Playboy*. Father N. is not in the room yet. . . . Someone taps the boy roughly on the shoulder. He turns. A finger points to the centerfold model, pink fleshy body, thin and sleek. . . . The other asks, mocking voice, "Hey, does she turn you on?"

The boy smiles, nodding meekly; turns away.

The other jabs him harder on the shoulder, "Hey, whatsamatter, don't you like girls?"

Laughter . . . unbearable din of laughter. . . . The laughter seems to go on forever. . . .

What did being a sissy really mean? It was a way of walking (from the hips rather than the shoulders); . . . of talking (often with a lisp or a high pitched voice); . . . of relating to others (gently, not wanting to fight, or hurt anyone's feelings). It was being intelligent . . . getting good grades. It means not being interested in sports, not playing football in the street after school; not

discussing teams and scores and play-offs. And it involved not showing fervent interest in girls, not talking about scoring . . . not concealing naked women in your history books, or porno books in your locker.

On the other hand, anyone could be a "faggot." It was a catch-all. If you did something that didn't conform to what was acceptable behavior of the group . . . if you didn't get along with the "in" crowd, you were a faggot. It was the most commonly used put-down. It kept guys in line . . . The word had power. It toppled the male ego . . . violated the image he projected. He was tough. Without feeling. Faggot cut through all this. It made him vulnerable. Feminine. And feminine was the worst thing he could possibly be. Girls were fine for [sex], but no boy in his right mind wanted to be like them. A boy was the opposite of a girl. He was not feminine . . . not feeling . . . not weak.

Scene Five

. . . Realizing I was gay was not an easy task. Although I knew I was attracted to boys by the time I was

about eleven, I didn't connect this attraction to homosexuality. I was not queer. Not I. I was merely appreciating a boy's good looks, his fine features, his proportions. It didn't matter that I didn't appreciate a girl's looks in the same way. There was no twitching in my thighs when I gazed upon a beautiful girl. But I wasn't queer.

I resisted that label—queer—for the longest time. Even when everything pointed to it, I refused to see it. I was certainly not queer. Not I.

Epilogue

The boy marching down the Parkway. Hundreds of queers. Signs proclaiming gay pride. Speakers. Tables with literature from gay groups. A miracle, he is thinking. Tears are coming loose now. Someone hugs him.

> You could not control
> The sissy in me
> Nor could you exorcise him
> Nor electrocute him
> You declared him illegal illegitimate
> Insane and immature
> But he defies you still.

SOURCE: Avicolli 1985, 4–5.

---

timacy may be "homosexual" (Britton 1990; Garnets et al. 1990)

- Leading to exaggerated displays of masculinity by heterosexual men trying to prove they are not gay (Mosher and Tomkins 1988)

Education and positive social interactions appear to be important vehicles for changing attitudes and reducing hostility. Some research reveals increased tolerance following human sexuality courses (Stevenson 1990). Negative attitudes about homosexuality may also be reduced by arranging positive interactions between heterosexuals and gay men or lesbians, especially in settings of equal status, common goals, cooperation, and a moderate degree of intimacy. Such interactions may occur when family members or close

friends come out. Other interactions may emphasize common group membership (religious, social, ethnic, or political, for example) on a one-to-one basis.

Bisexuality

As we noted earlier, bisexuals are individuals attracted to members of both genders. Asked what their bisexual identities meant to them, most of Paula Rust's respondents said it meant that they had "the potential to be sexually, emotionally and/or romantically attracted to members of both sexes or genders" (Rust 2004). For many it is the capacity or potential, not necessarily the actual experience that makes them identify themselves as bisexual. For some, bisexuality is

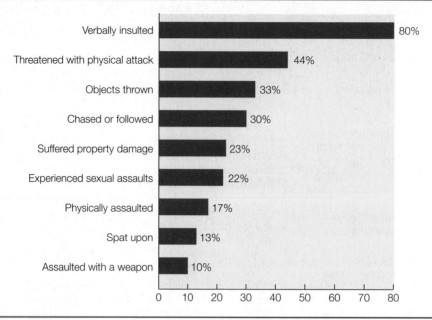

Category	Percentage
Verbally insulted	80%
Threatened with physical attack	44%
Objects thrown	33%
Chased or followed	30%
Suffered property damage	23%
Experienced sexual assaults	22%
Physically assaulted	17%
Spat upon	13%
Assaulted with a weapon	10%

SOURCE: Hershberger and D'Augelli 1995.

expressed in alternating relationships with women and men. Others have equal sexual relationships with women and men (for example, "I have a girlfriend, with whom I have sex often and am very attracted to, but am still attracted to men, with whom I also have sexual relations") (Rust 2004, 218). Still others base their self-definitions more on feelings than on any actual relationships, past or present (for example, "over

99% of my sexual interactions have been heterosexual. But I fantasize about women a great deal and enjoyed the one-on-one encounter I had") (Rust 2004).

Becoming bisexual requires the rejection of two recognized categories of sexual identity: heterosexual and homosexual. In a nationwide study by Samuel and Cynthia Janus (1993), about 5% of men and 3% of women identified themselves as bisexual. Data from the comprehensive survey of sexual behavior in the United States, the National Health and Social Life Survey, reveal a smaller percentage (less than 1%) who self-identified themselves as bisexual. If we look at reports of "sexual attraction," 3.9% of men and 4.1% of women report themselves attracted to "mostly the opposite gender," both genders, or "mostly the same gender" (Laumann et al. 1994).

Because it is only since the 1980s that bisexuality has become more visible and bisexuals more politicized and organized, it shouldn't be surprising that research on the "bisexual experience" is less abundant than the similar literature on gays and lesbians (Herek 2002). What research we have indicates that, like gays and lesbians, bisexuals are often the targets of hostility and harassment. Herek, Gillis, and Cogan report that 15% of bisexual women and 27% of bisexual men in their sample had experienced a property or violent crime. These rates are similar to what was reported

© Liss Steve/CORBIS/Sygma

■ *The hate-based killing of Matthew Shepherd inspired memorial demonstrations and raised awareness about the extent of homophobia in the United States.*

by lesbians (19%) and gay men (28%). A Kaiser Family Foundation survey of 405 lesbians, gay men, and bisexuals found that among bisexuals, because of their sexual orientation, 60% had experienced some form of discrimination, 52% had suffered verbal abuse, and 26% felt that they were not accepted by their families. These percentages were lower than the comparable percentages for lesbians and gay men, suggesting that bisexuality may be somewhat less stigmatized than homosexuality (Herek 2002).

More negative attitudes toward bisexual men and women seem to be associated with certain individual characteristics, such as frequent attendance at religious services, a conservative political ideology, and having had minimal prior contact with bisexual men or women. These same factors are associated with heterosexual attitudes toward gay men and lesbians (Herek 2002).

Because they might be perceived as rejecting both heterosexuality and homosexuality, bisexuals can also be stigmatized by gay men and lesbians who might view bisexuals as "fence-sitters" not willing to admit their homosexuality or as people simply "playing" with their orientation (Herek 2002). Thus, bisexuality may not be taken seriously by either group. Loraine Hutchins and Lani Kaahumanu (1991) believe that bisexuality arouses hostility because it "challenges current assumptions about the immutability of people's orientations and society's supposed divisions into discrete groups."

Gregory Herek (2002) looked at the attitudes of heterosexuals toward bisexuals by having his sample of more than 1,270 men and women rate them on a "feeling thermometer." He asked them to rate on a scale of 0–100 a number of different groups, including religious groups (Protestants, Catholics, and Jews); gay men, lesbians, and bisexuals; racial, ethnic, and national groups (including blacks, Mexican Americans, Puerto Ricans, whites, and Haitians); pro-life and pro-choice groups; people with AIDS; and people who inject illegal drugs. Higher numbers reflect "warmer" feelings. If the respondents felt "neither warm nor cold," they were instructed to rate a group with a 50. His results are shown in Table 6.2.

Bisexual women and men received similar but lower ratings than the average "feeling scores" for lesbian and gay men. The only group to receive "colder" ratings than bisexual women (45.8) and men (43.4) was illegal drug users (21.0). More than 400 people rated illegal drug users with the "coldest" possible score of zero. The two groups who had the next highest number of zero ratings were gay men (134) and bisexual men (140). Ratings for lesbians were more favorable than ratings for gay men, and among bisexuals, women received "warmer" scores than did bisexual men (Herek 2002).

Table 6.2 ■ "Feeling Thermometer" Measuring Attitudes toward Gays, Lesbians, Bisexuals, and Other Groups

Thermometer Target	Average Rating	# of Extreme Scores	
		# Coldest (rating of 0)	# Warmest (rating of 100)
Whites	70.4	1	223
Catholics	67.7	2	230
Blacks	66.8	4	190
Protestants	66.5	3	210
Mexican Americans	64.9	5	167
Jews	64.8	3	167
Puerto Ricans	63.5	7	162
Haitians	60.5	8	143
Pro-lifers	56.3	54	146
People with AIDS	55.6	48	96
Pro-choice people	53.3	116	117
Lesbians	47.5	116	57
Gay men	46.1	134	63
Bisexual women	45.8	116	57
Bisexual men	43.4	140	54
People who inject illegal drugs	21.0	414	19

Becoming Bisexual

There is considerable HIV and AIDS research on bisexual behavior among men who identify themselves as heterosexual, but compared with research on gay men and lesbians there is much less research on bisexuality. In 1994, the first model of bisexual identity formation was developed (Weinberg, Williams, and Pryor 1994). According to this model, bisexual women and men go through four stages in developing their identity:

1. *Initial confusion.* This may last years. People may be distressed by being sexually attracted to both sexes, may believe that their attraction to the same sex means an end to their heterosexuality, or may be disturbed by their inability to categorize their feelings as either heterosexual or homosexual.

2. *Finding and applying the bisexual label.* For many, discovering there is such a thing as bisexuality is a turning point. Some find that their first heterosexual or same-sex experience permits them to view sex with both sexes as pleasurable; others learn of the term *bisexuality* from friends and are able to apply it to themselves.

3. *Settling into the identity.* At this stage, bisexuals begin to feel at home with and accept the bisexual label.

4. *Continued uncertainty.* Bisexuals don't have a community or social environment that reaffirms their identity. Despite being settled in, many feel persistent pressure from gay men and lesbians to relabel themselves as homosexual and to engage exclusively in same-sex activities.

Sexuality in Adulthood

Psychosexual development and change does not end in young adulthood. It continues throughout our lives. In middle age and old age, our lives, bodies, sexuality, relationships, and environment continue to change. New tasks and new satisfactions arise to replace or supplement older ones.

Developmental Tasks in Middle Adulthood

In the middle adult years, some tasks of psychosexual development begun, but only partly completed, or deferred in young adulthood (for example, issues sur-rounding intimacy or childbearing) may continue. Because of separation or divorce, we may find ourselves facing the same intimacy and commitment tasks at age 40 that we thought we completed 15 years earlier (Cate and Lloyd 1992). But life does not stand still; it moves steadily forward, whether we're ready or not. Other developmental issues appear, including the following:

- *Redefining sex in marital or other long-term relationships.* In new relationships, sex is often passionate, intense, and may be the central focus. But in long-term marital or cohabiting relationships, the passionate intensity associated with sex is often eroded by habituation, competing parental and work obligations, fatigue, and unresolved conflicts. Sex may need to be redefined as a form of intimacy and caring. Individuals may also need to decide how to deal with the possibility, reality, and meaning of extramarital or extrarelational affairs.

- *Reevaluating sexuality.* Single men and women may need to weigh the costs and benefits of sex in casual or lightly committed relationships. In long-term relationships, sexuality often becomes less central to relationship satisfaction. Nonsexual elements, such as communication, intimacy, and shared interests and activities, become increasingly important to relationships. Women who have deferred their childbearing begin to reappraise their decision: Should they remain childfree, "race" against their biological clocks, or adopt a child?

- *Accepting the biological aging process.* As we age, our skin wrinkles, our flesh sags, our hair grays (or falls out), our vision blurs—and we become in the eyes of society less attractive and less sexual. By our 40s, our physiological responses have begun to slow noticeably. By our 50s, society begins to "neuter" us, especially if we are women who have gone through menopause. The challenges of aging are to accept its biological mandate and to reject the stereotypes associated with it.

Sexuality and Middle Age

Men and women view and experience aging differently. As men approach their 50s, they fear the loss of their sexual capacity but not their attractiveness; for women the reverse is true. As both age, purely psychological stimuli, such as fantasies, become less effective for arousal. Physical stimulation remains effective, however.

Among American women, sexual responsiveness continues to grow from adolescence until it reaches its peak in the late 30s or early 40s; it is usually maintained near the same level into the 60s and beyond. Data from both the United States and elsewhere have yielded inconsistent research findings on women's sexuality at midlife. Some studies suggest that rates of sexual intercourse, levels of sexual interest, frequency of orgasm, extent of sexual fantasizing, vaginal lubrication, and satisfaction with a partner all decline in midlife. Others show no decline in sexual interest, responsiveness, or "functioning." About the only thing that can be safely concluded is that considerable variability occurs in midlife women's sexuality.

Having emotional and psychological needs met (feeling attractive, appreciated, independent, understood, and productive) is related both to feeling attractive and to satisfaction with one's sex life. Frequency of intercourse and orgasm and finding sex pleasant, enjoyable, and satisfying are associated with higher levels of marital adjustment and contentment, although it is not clear whether marital quality causes or follows sexual satisfaction (Fraser, Maticka-Tyndale, and Smylie 2004).

Data from the United States, Great Britain, and France indicate age differences that may be the result of cohort differences (based on differences in sexual socialization and changing cultural attitudes) or possible effects of aging (Table 6.3).

In addition to age differences in whether and how often women report having engaged in sexual intercourse, data from the 1994 Sex in America survey reveal information about sexual problems or dysfunctions for women of different ages (see Table 6.4). We can see the effect of aging on women's sexuality in a number of reported dysfunctions.

As these data reveal, relative to other ages, high levels of orgasmic difficulty, lack of pleasure and interest, and trouble with vaginal lubrication are reported by 55- to 59-year-olds. Keep in mind that these data are from "sexually active women" and as such may even understate the effect of aging, as more women may become sexually inactive as they move through their 40s and into their 50s. Sexually inactive women and any problems they have that might cause them to refrain from sex are not represented in this data (Fraser, Maticka-Tyndale, and Smylie 2004).

Around the age of 50, the average American woman begins **menopause,** which is marked by a cessation of the menstrual cycle and an end to fertility. Menopause is not a sudden event. Usually, for several years preceding menopause, the menstrual cycle becomes increasingly irregular. Menopause does not end interest in sexual activities. The decrease in estrogen, however, may cause thinning and dryness of the vaginal walls, which makes intercourse painful. The use of vaginal lubricants will remedy much of the problem.

There is no male equivalent to menopause. Male fertility slowly declines, but men in their 80s are often fertile. Men's physical responsiveness is greatest in late adolescence or early adulthood; beginning in men's 20s, responsiveness begins to slow imperceptibly. Changes in male sexual responsiveness become apparent only when men are in their 40s and 50s. As a man ages, achieving erection requires more stimulation and time and the erection may not be as firm. In a subsequent section we examine some drugs used for erectile dysfunctions. For now, however, the point is that because of physical changes, "middle-aged couples may be misled into thinking that this change heralds a sexual decline as an accompaniment to aging" (Katchadourian 1987).

Table 6.3 ■ Women Reporting Intercourse in Past Year

	Percentage in Past Year			Mean Frequency in Past Month	
Age	U.S.	France	Great Britain	U.S.	France
35–44	87%	96%	92%	5.8	8.1
45–54	82%	90%	78%	5.0	6.1
55–59	59%	66%	NA	3.5	4.0

NA means data not available.
SOURCE: Fraser, Maticka-Tyndale, and Smylie, 2004.

Table 6.4 ■ Percentage of Sexually Active U.S. Women Reporting Various Sexual Dysfunctions

Age	Pain	Not Pleasurable	Unable to Orgasm	Lack of Interest	Trouble Lubricating
35–39	13.0	18.3	26.9	37.6	18.1
40–44	12.0	15.7	20.8	36.0	15.9
45–49	10.3	15.4	18.8	33.7	22.6
50–54	7.4	15.3	20.2	30.2	21.4
55–59	8.7	16.4	21.8	37.0	24.8

SOURCE: Laumann et al., 1994.

Table 6.5 ■ Age and Sexual Desire for Men and Women

| | Women | | Men | |
Age	% Low Desire	% High Desire	% Low Desire	% High Desire
60–64	23.26	13.85	18.29	4.89
65–69	26.92	10.26	21.13	5.63
70–74	46.05	7.90	38.00	2.00
75–79	49.12	5.26	27.08	2.08
80–84	85.29	2.94	50.0	3.85
85–89	73.0	0.0	50.0	0.0
90–94	100.0	0.0	NA	NA

SOURCE: DeLamater and Sill 2005.
NA means data not available.

Psychosexual Development in Later Adulthood

As we leave middle age, new tasks confront us, especially dealing with the process of aging itself. Our health and the presence or absence of a partner are key aspects of this time in our lives.

Developmental Tasks in Later Adulthood

Many of the psychosexual tasks older Americans must undertake are directly related to the aging process:

■ *Changing sexuality.* As physical abilities change with age, sexual responses change as well. A 70-year-old person, although still sexual, is not sexual in the same manner as an 18-year-old. Sexuality tends to be more diffuse, less genital, and less insistent. Chronic illness and increasing frailty understandably result in diminished sexual activity and desire (see Table 6.5).

These considerations contribute to the ongoing evolution of the individual's sexual philosophy.

■ *Loss of partner.* One of the most critical life events is the loss of a partner. After age 60, there is a significant increase in spousal deaths. As having a partner is the single most important factor determining an older person's sexual interactions, the death of a partner signals a dramatic change in the survivor's sexual interactions.

The developmental tasks of later adulthood are accomplished within the context of continuing aging. Their resolution helps prepare us for acceptance of our own eventual mortality.

Adult Sexual Behavior

In this section we examine various sexual behaviors. For a discussion of sexual structure and the sexual response cycle, see Appendix A on the book website.

■ *Sexuality among the aged tends to be sensual and affectionate. Older couples may experience an intimacy forged by years of shared joys and sorrows that is as intense as the passion of young love.*

Keri Pickett

Autoeroticism

Autoeroticism consists of sexual activities such as sexual fantasies, masturbation, and erotic dreams that involve only the self. Autoeroticism is one of our earliest and most universal, yet also less accepted, expressions of sexual stirrings. By condemning it, our culture sets the stage for the development of deeply negative inhibitory attitudes toward sexuality.

Sexual Fantasies

Erotic fantasizing is probably the most universal of all sexual behaviors, but because they may touch on feelings or desires considered personally or socially unacceptable, typically they are not widely discussed. Although fantasies are normal and serve certain functions (such as escape or rehearsal for later sexual behavior), they may also interfere with an individual's self-image, causing a loss of self-esteem, as well as confusion.

Various studies report that between 60% and 90% of respondents fantasize during sex—the percentage depending on gender, age, and ethnicity (Miracle, Miracle, and Baumeister 2003; Knafo and Jaffe 1984; Price and Miller 1984). A large-scale study (Michael et al. 1994) found that 54% of the men and 19% of the women thought about sex daily.

Women and men have sexual fantasies, although their fantasies often differ. Can you tell the gender of the individuals who supplied the following fantasies?

- "A tropical island. I've always dreamed about making love in a crystal blue sea, with a waterfall in the background, then moving on shore to a white, sandy beach."
- "It's eveningtime [sic], the sun is setting, I'm on a tropical island, a light breeze is blowing into my balcony doors and the curtains [white] are fluttering lightly in the wind. The room is spacious and there is white everywhere, even the bed. There are flowers of all kinds and the light fragrance fills the room."
- "Ménage à trois."
- "Have sex on the beach."

If you guessed that the first two fantasies are from women and the third and fourth are from men, you guessed correctly. These are real examples that Michael Kimmel and Rebecca Plante received from undergraduates at three New York colleges or universities.

Men's and women's fantasies contained similarities (for example, in the acts they described), but the differences were more striking: women's fantasies were longer and more vivid, using more emotional and sensual imagery, especially in describing the setting; men more often fantasized about *doing something sexual to someone,* whereas women's fantasies were often more passive and gentler, of *having something sexual done to them;* and women's fantasies tended to have more emotional and romantic content (47% of women described their fantasy partners as boyfriends or husbands; only 15% of men depicted their fantasy partners as "significant others"). Women's fantasies were also often romantic stories of love and affection, men's had less romance and less emotional language or context.

Masturbation

Masturbation is the manual stimulation of one's genitals. Individuals masturbate by rubbing, caressing, or otherwise stimulating their genitals to bring themselves sexual pleasure. Masturbation is an important means of learning about our bodies. Girls, boys, women, and men may masturbate during particular periods or throughout their entire lives. An analysis of research articles on gender roles and sexual behavior found that the greatest male–female difference was in masturbation (Oliver and Hyde 1993). Males had significantly more masturbatory experience than females.

By the end of adolescence, virtually all males and about two-thirds of females have masturbated to orgasm (Knox and Schacht 1992; Lopresto, Sherman, and Sherman 1985). Masturbation continues after adolescence. Gender differences, however, continue to be significant (Atwood and Gagnon 1987; Leitenberg, Detzer, and Srebnik 1993).

Although the rate is significantly lower for those who are married, many people, especially men, continue to masturbate even after they marry. There are many reasons for continuing the activity during marriage: masturbation is a pleasurable form of sexual excitement; a spouse may be away or unwilling to engage in sex; sexual intercourse may not be satisfying; the partners may fear sexual inadequacy; one partner may want to act out fantasies. In marital conflict, masturbation may act as a distancing device, with the masturbating spouse choosing masturbation over sexual intercourse as a means of emotional protection (Betchen 1991).

Cohabitation has a different effect than marriage on frequency of masturbation. Many cohabiting men masturbate often, despite the presence or availability of a sexual partner. Thus, social factors other than the presence of a partner affect masturbation. In citing reasons for why they masturbate, only a third of women and men list an unavailable partner (Laumann et al. 1994).

Matter of Fact

A study of college students in a human sexuality class found that 87% of the men and 58% of the women had masturbated (Knox and Schacht 1992). In a larger study, among adults of all ages, 63% of the men and 42% of the women had masturbated in the previous year (Michael et al. 1994).

Interpersonal Sexuality

We often think that sex is sexual intercourse and that sexual interactions end with orgasm (usually the male's). But sex is not limited to sexual intercourse. Heterosexuals engage in a variety of sexual activities, which may include erotic touching, kissing, and oral and anal sex. Except for sexual intercourse, gay and lesbian couples engage in sexual activities similar to those experienced by heterosexuals.

Touching

Because touching, like desire, does not in itself lead to orgasm, it has largely been ignored as a sexual behavior. Sex researchers William Masters and Virginia Johnson (1970) suggest a form of touching they call **pleasuring**—nongenital touching and caressing. Neither partner tries to stimulate the other sexually; the partners simply explore each other. Such pleasuring gives each a sense of his or her own responses; it also allows each to discover what the other likes or dislikes. We can't assume we know what any particular individual likes because there is too much variation among people. Pleasuring opens the door to communication; couples discover that the entire body is erogenous, rather than just the genitals.

As we enter old age, touching becomes increasingly significant as a primary form of erotic expression. Touching in all its myriad forms—ranging from holding hands to caressing, massaging to hugging, walking with arms around each other to fondling—becomes the touchstone of eroticism for the elderly. One study found touching to be the primary form of erotic expression for married couples more than 80 years old (Bretschneider and McCoy 1988).

Kissing

Kissing as a sexual activity is probably the most common and acceptable of all premarital sexual activities, occurring in more than 90% of all cultures (Jurich and Polson 1985; Fisher 1992, cited in Miracle, Miracle, and Baumeister 2003). The tender lover's kiss symbolizes love, and the erotic lover's kiss simultaneously represents passion. Both men and women in one study regarded kissing as a romantic act, a symbol of affection and attraction (Tucker, Marvin, and Vivian 1991). A cross-cultural study of jealousy found that kissing is also associated with a couple's boundary maintenance: In each culture studied, kissing a person other than the partner evoked jealousy (Buunk and Hupka 1987).

The lips and mouth are highly sensitive to touch. Kisses discover, explore, and excite the body. They also involve the senses of taste and smell, which are especially important because they activate unconscious memories and associations. Often we are aroused by familiar smells associated with particular sexual memories: a person's body smells, perhaps, or perfumes associated with erotic experiences. In some cultures—among the Borneans, for example—the word *kiss* literally translates as "smell." Among traditional Eskimos and Maoris there is no mouth kissing, only the nuzzling that facilitates smelling.

Kissing is probably the most acceptable premarital sexual activity.

Although kissing may appear innocent, it is in many ways the height of intimacy. The adolescent's first kiss is often regarded as a milestone, a rite of passage, the beginning of adult sexuality (Alapack 1991). Philip Blumstein and Pepper Schwartz (1983) report that many of their respondents found it unimaginable to engage in sexual intercourse without kissing. They found that those who have a minimal (or nonexistent) amount of kissing feel distant from their partners but engage in coitus nevertheless as a physical release.

The amount of kissing differs according to orientation. Lesbian couples tend to engage in more kissing than heterosexual couples, and gay male couples kiss less than heterosexual couples. As many as 95% of lesbian couples, 80% of heterosexual couples, and 71% of gay couples engage in kissing whenever they have sexual relations (Blumstein and Schwartz 1983).

Oral-Genital Sex

In recent years, oral sex has become part of our sexual scripts. It is engaged in by heterosexuals, gay men, and lesbians. The two types of oral–genital sex are cunnilingus and fellatio. **Cunnilingus** is the erotic stimulation of a woman's vulva by her partner's mouth and tongue. **Fellatio** is the oral stimulation of a man's penis by his partner's sucking and licking. Cunnilingus and fellatio may be performed singly or simultaneously. Oral sex is an increasingly common part of adolescent and young adult sexual development, as we noted earlier. It is also an important and healthy aspect of adults' sexual selves (Wilson and Medora 1990).

Although oral–genital sex is increasingly accepted by Caucasian middle-class Americans, it remains less permissible and less commonly practiced among certain ethnic groups. African Americans and Latinos, have lower rates of oral genital sex than do Caucasians (Wilson 1986; Wyatt and Lyons-Rowe 1990, Laumann et al. 1994). Although less is known about older Asian Americans and Asian immigrants, college-age Asian Americans appear to accept oral–genital sex to the same degree as middle-class Caucasians (Cochran, Mays, and Leung 1991).

Among both sexes, the same percentages report receiving and performing oral sex (Laumann et al. 1994). A study of university students of both sexes found that oral sex was regarded as an egalitarian, mutual practice (Moffatt

1989). Students felt less guilty about it than about sexual intercourse because oral sex was not "going all the way."

Sexual Intercourse

Sexual intercourse or **coitus**—the insertion of the penis into the vagina and subsequent stimulation—is a complex interaction. As with many other types of activities, the anticipation of reward triggers a pattern of behavior. The reward may not necessarily be orgasm, however, because the meaning of sexual intercourse varies considerably at different times for different people. There are many motivations for sexual intercourse; sexual pleasure is only one. Other motivations include showing love, having children, gaining power, ending an argument, demonstrating commitment, seeking revenge, proving masculinity or femininity, or degrading someone (including oneself).

Although sexual intercourse is important for most sexually involved couples, its significance is different for men and women. More than any other heterosexual sexual activity, sexual intercourse involves equal participation by both partners. Ideally, both partners equally and simultaneously give and receive. Many women report that this sense of sharing during intercourse is important to them.

Men tend to be more consistently orgasmic than women in sexual intercourse. Part of the reason may be that the clitoris often does not receive sufficient stimulation from penile thrusting alone to permit orgasm. Many women need manual stimulation during intercourse to be orgasmic. They may also need to be more assertive. A woman can manually stimulate herself or be stimulated by her partner before, during, or after intercourse. But to do so, she has to assert her own sexual needs and move from the idea that sex is centered around male orgasm.

Matter of Fact

According to a scientific, nationwide study of adults of all ages, about one-third of Americans have sexual intercourse twice a week, one-third a few times a month, and one-third a few times a year or not at all. Married couples are more likely to engage in coitus than singles; married women are more likely to be orgasmic. About 40% of married couples and 25% of singles report having coitus twice a week (Michael et al. 1994).

Anal Eroticism

Sexual activities involving the anus are known as **anal eroticism.** The male's insertion of his erect penis into his partner's anus is known as **anal intercourse.** Both heterosexuals and gay men may participate in this activity. For heterosexual couples who engage in it, anal intercourse is generally an experiment or occasional activity rather than a common mode of sexual expression. About 10% of men and 9% of women report engaging in anal sex in the previous year (Michael et al. 1994), and one in four men and one in five women reported having ever experienced anal sex (Laumann et al. 1994).

Anal intercourse is less common than oral sex but remains an important ingredient in the sexual satisfaction of many gay men (Blumstein and Schwartz 1983). From a health perspective, anal intercourse is the riskiest form of sexual interaction and the most prevalent sexual means of transmitting the HIV among both gay men and heterosexuals. Because the delicate rectal tissues are easily torn, HIV (carried within semen) can enter the bloodstream. (HIV will be discussed later in the chapter.)

Sexual Enhancement

Sexual behavior cannot be isolated from our personal feelings and relationships. Sometimes dissatisfaction arises because the relationship itself is unsatisfactory, other times the relationship itself is good but the erotic fire needs to be lit or rekindled. Such relationships may grow through **sexual enhancement**—improving the quality of a sexual relationship—which, according to noted sex therapist Bernie Zilbergeld (1992), consists of the following:

- Accurate information about sexuality, especially your own and your partner's.
- An orientation toward sex based on pleasure (including arousal, fun, love, and lust) rather than on performance and orgasm.
- Being involved in a relationship that allows each person's sexuality to flourish.
- An ability to communicate verbally and nonverbally about sex, feelings, and relationships.
- Being equally assertive and sensitive about your own sexual needs and those of your partner.
- Accepting, understanding, and appreciating differences between partners.

Being aware of our sexual needs is often critical to enhancing our sexuality. Gender-role stereotypes and negative learning about sexuality often cause us to lose sight of our sexual needs. Zilbergeld (1993) suggests that to fully enjoy our sexuality, we need to explore our "conditions for good sex," those things that make us "more relaxed, more comfortable, more confident, more excited, more open to your experience."

Different individuals report different conditions for good sex. More common conditions include the following:

- *Feeling intimate with your partner.* Emotional distance can take the heart out of sex.

- *Feeling sexually capable.* Generally this relates to an absence of anxieties about sexual performance

- *Feeling trust.* Both partners may need to know they are emotionally safe with the other and confident that they will not be judged, ridiculed, or talked about.

- *Feeling aroused.* A person does not need to be sexual unless he or she is sexually aroused or excited. Simply because your partner wants to be sexual does not mean that you have to be.

- *Feeling physically and mentally alert.* Both partners should not feel particularly tired, ill, stressed, preoccupied, or under the influence of excessive alcohol or drugs.

- *Feeling positive about the environment and situation.* A person may need privacy, to be in a place where he or she feels protected from intrusion.

©LWA-Dann Tardif/CORBIS

■ *Common conditions for a satisfying sexual relationship include feelings of intimacy, capability, trust, arousal, alertness, and positiveness about the environment and situation.*

Sexual Expression and Relationships

Sexuality exists in various relationship contexts that may influence our feelings and activities. These include nonmarital, marital, and extramarital contexts.

Nonmarital Sexuality

Nonmarital sex encompasses sexual activities, especially sexual intercourse, that take place outside of marriage. We use the term *nonmarital sex* rather than *premarital sex* to describe sexual behavior among unmarried adults in general. When we use the term **premarital sex** we are referring to never-married adults under the age of 30. There are several reasons to make premarital sex a subcategory of nonmarital sex. First, because increasing numbers of never-married adults are over 30, "premarital sex" does not adequately describe the nature of their sexual activities. Second, at least 10% of adult Americans will never marry; it is misleading to describe their sexual activities as "premarital." Third, many adults are divorced, separated, or widowed; 30% of divorced women and men will never remarry. Fourth, between 3% and 10% of the population is lesbian or gay, and gay and lesbian sexual relationships cannot be categorized as "premarital" until gays and lesbians are given the right to marry.

Sexuality in Dating Relationships

Over the last several decades, there has been a remarkable increase in the acceptance of premarital sexual intercourse, a decline in the numbers of people who believe that premarital sex is "always wrong," and an increase in the percentages who feel it is "not wrong at all." This trend has been interpreted as a shift toward "moral neutrality" regarding intercourse before marriage (Christopher and Sprecher 2000).

For adolescents and young adults, the combination of effective birth control methods, changing gender roles that permit females to be sexual, and delayed marriages have played a major part in the rise of premarital sex. For middle-aged and older adults, increasing divorce rates and longer life expectancy have created an enormous pool of once-married men and women who engage in nonmarital sex. Only **extramarital sex**—sexual interactions that take place outside the marital relationship—continues to be consistently frowned upon.

The increased legitimacy of sex outside of marriage has transformed both dating and marriage. Sexual intercourse has become an acceptable part of the dating process for many couples, whereas only petting was acceptable before. As a consequence, many people no longer feel that they need to marry to express their sexuality in a relationship (Scanzoni et al. 1989).

There appears to be a general expectation among students that they will engage in sexual intercourse sometime during their college careers. Although college students expect sexual involvement to occur within an emotional or loving relationship (Robinson et al. 1991), this emotional connection may be relatively transitory.

FACTORS LEADING TO PREMARITAL SEXUAL INVOLVEMENT. Examining the sexual decision making process closely, researcher Susan Sprecher (1989) identifies individual, relationship, and environmental factors affecting the decision to have premarital intercourse:

- *Individual factors.* Those with more premarital sexual experience, with more liberal sexual attitudes, and who do not feel high levels of guilt about sexuality are more likely to engage in sex, as are those who value erotic pleasure. Men tend to initiate sexual activity more than women, but both women and men use similar tactics to initiate sex (implying commitment, increasing attention, and displaying "status cues"). There is a gender difference in "compliance" with partner-initiated sex, such that women are more likely than men to comply, and they do it to maintain their relationships (Christopher and Sprecher 2000).
- *Relationship factors.* Two of the most important factors determining sexual activity in a relationship are the level of intimacy and the length of time the couple has been together. Even those with less permissive sexual attitudes accept sexual involvement if the relationship is emotionally intimate and long standing. Less committed individuals are less likely to make their relationships sexual. Finally, people in relationships in which power is shared equally are more likely to be sexually involved than those in inequitable relationships.
- *Environmental factors.* The opportunity for sex may be precluded by the presence of parents, friends,

roommates, or children (Tanfer and Cubbins 1992). The cultural environment, too, affects premarital sex. The values of parents or peers may encourage or discourage sexual involvement. A person's ethnic group also affects premarital involvement. Generally, African Americans are more permissive than Caucasians, and Latinos are less permissive than non-Latinos (Baldwin, Whitely, and Baldwin 1992). Furthermore, a person's subculture—such as the university or church environment or the gay and lesbian community—influences sexual decision making.

INITIATING A SEXUAL RELATIONSHIP. After we meet someone, we weigh each other's attitudes, values, and philosophy to see if we are compatible. If the relationship continues in a romantic vein, we may include physical intimacy. To signal the transition from nonphysical to physical intimacy, one of us must make the first move, marking the transition from a potentially sexual relationship to an actual one.

According to traditional gender-role patterns, as described earlier, males make the first move to initiate sexual intimacy, whether it is kissing, petting, or engaging in sexual intercourse (O'Sullivan and Byers 1992). Initial sexual involvement can occur as early as the first meeting or much later as part of a well-established relationship. Some people become sexually involved immediately ("lust at first sight"), but most being their sexual involvement in the context of an ongoing relationship. Even in one-night stands or short-term affairs, more couples knew each other at least a year before engaging in sex than knew each other just for a couple or few days (Miracle, Miracle, and Baumeister 2003).

DIRECTING SEXUAL ACTIVITY. As we begin a sexual involvement, we have several tasks to accomplish:

1. *We need to practice safe sex.* Ideally, we need information about our partners' sexual history and whether he or she practices safe sex, including the use of condoms. Unlike much of our sexual communication, which is nonverbal or ambiguous, we need to use direct verbal discussion in practicing safe sex.

2. *Unless we are intending a pregnancy, we need to discuss birth control.* Condoms alone are only moderately effective as contraception, although they help prevent the spread of sexually transmitted diseases. To be more effective, they must be used with contraceptive foam or jellies or with other devices.

3. *We need to communicate about what we like and need sexually.* What kind of foreplay or afterplay do we like? Do we like to be orally or manually stimulated during intercourse? What does each partner need to be orgasmic? Many of our needs and desires can be communicated nonverbally by our movements or other physical cues. But if our partner does not pick up our nonverbal signals, we need to discuss them directly and clearly to avoid ambiguity.

Sexuality in Cohabiting Relationships

As shown in Chapter 8, cohabitation has become a widespread phenomenon in American culture. In contrast to married men and women, cohabitants have sexual intercourse more often, are more egalitarian in initiating sexual activities, and are more likely to be involved in sexual activities outside their relationship (Waite and Gallagher 2001; Blumstein and Schwartz 1983). The higher frequency of intercourse, however, may be because of the "honeymoon" effect: Cohabitants may be in the early stages of their relationship, the stages when sexual frequency is highest. The differences in frequency of extrarelational sex may result from a combination of two factors: Norms of sexual fidelity may be weaker in cohabiting relationships, and men and women who cohabit tend to conform less to conventional norms.

Sexuality in Gay and Lesbian Relationships

Because of their socialization as males, gay men are likely to initiate sexual activity earlier and more often in the relationship than are lesbians. This is largely because both partners are free to initiate sex and because men are not expected to refuse sex, as women are (Isensee 1990). Lesbians may feel uncomfortable initiating sex because women are not socialized to do so.

In both gay and lesbian relationships, the more emotionally expressive partner is likely to initiate sexual interaction. The partner who talks more about feelings and who spontaneously gives the partner hugs or kisses is the one who more often begins sexual activity.

One of the major differences between heterosexuals and gay men and lesbians is in how they handle extrarelational sex. In the gay and lesbian culture, sexual exclusivity is more negotiable and not necessarily equated with commitment or fidelity among gay men, although it often is among lesbians (Renzetti and Curran 1995). As a result of these differing norms, gay men and lesbians must decide early in the relationship whether they will be sexually exclusive (Isensee 1990). If they choose to have a nonexclusive relationship, they need to discuss how outside sexual interests will be handled. They need to decide whether to tell each other, whether to have affairs with friends, what degree of emotional involvement will be acceptable, and how to deal with jealousy.

Marital Sexuality

When people marry, they discover that their sexual life is different than it was before marriage. Sex is now morally and socially sanctioned. It is in marriage that most heterosexual interactions take place, yet as a culture we seem ambivalent about marital sex. On the one hand, marriage is the only relationship in which sexuality is fully legitimized. On the other hand, marital sex is an endless source of humor and ridicule: "Marital sex? What's that?" On television, more sexual encounters portrayed are between unmarried than married couples. An early 1990s study found four times as much extramarital sex depicted as marital sex (Hanson and Knopes 1993).

Sexual Interactions

A variety of large-scale studies report consistent findings in regard to how often married couples engage in sexual intercourse and in how sexual frequency changes over the course of a marriage. Married couples report engaging in sexual relations about once or twice a week, or about six to seven times a month (Christopher and Sprecher 2003).

Sexual intercourse tends to diminish in frequency the longer a couple is married. For newly married couples, the average rate of sexual intercourse is about three times a week. Data from more than 13,000 respondents in the National Survey of Families and Households reported that couples under the age of 24 had sex on average 11.7 times per month (or approximately three times per week). (Call, Sprecher, and Schwartz 1995, cited in Christopher and Sprecher 2000). As couples get older, sexual frequency drops. In early middle age, married couples make love an average of 1.5 to 2 times a week. After age 50, the rate is about once a week or less. Among couples 75 and older, the frequency is a little less than once a month (Christopher and Sprecher 2000).

This decreased frequency, however, does not necessarily mean that sex is no longer important or that the marriage is unsatisfactory. For dual-worker families and families with children, fatigue and lack of private time may be the most significant factors in the decline of frequency (Olds 1985). Couples also report "being accustomed" to each other. In addition, activities and interests other than sex engage them. The decline in interest and frequency of sex may begin within the first 2 years of marriage (Christopher and Sprecher 2000).

Bringing New Meanings to Sex

Sex within marriage is significantly different from premarital sex in at least three ways: it is expected to be monogamous; procreation is a legitimate goal; and such sex takes place in the everyday world. These differences present each person with important tasks.

MONOGAMY. One of the most significant factors shaping marital sexuality is the expectation of monogamy. Before marriage or following divorce a person may have various sexual partners, but within marriage all sexual interactions are expected to take place between the spouses. Approximately 90% of Americans believe extramarital sexual relations are "always" or "almost always" wrong (Miracle, Miracle, and Baumeister 2003; Christopher and Sprecher 2000). This expectation of monogamy lasts a lifetime; a person marrying at 20 commits to 40 to 60 years of sex with the same person. Within a monogamous relationship, each partner must decide how to handle fantasies, desires, and opportunities for extramarital sexuality. Do you tell your spouse that you have fantasies about other people? Do you have an extramarital relationship? If you do, do you tell your spouse? How do you handle sexual conflicts or difficulties with your partner?

SOCIALLY SANCTIONED REPRODUCTION. Sex also takes on a procreative meaning within marriage. In most segments of society, marriage remains the more socially approved setting for having children. In marriage, partners are confronted with one of the most crucial decisions they will make: the task of deciding whether and when to have children. Having children will profoundly alter a couple's relationship. If they decide to have a child, lovemaking may change from simply an erotic activity to an intentionally reproductive act as well.

CHANGED SEXUAL CONTEXT. The sexual context changes with marriage. Because married life takes place in a day-to-day living situation, sex must also be expressed in the day-to-day world. Sexual intercourse must be arranged around working hours and at times when the children are at school or asleep. One or the other partner may be tired, frustrated, or angry.

Two examples from interviews one of this book's authors did illustrate this quite vividly. In the first, a 33-year-old father of one contrasted where he and his wife prioritized sex:

It's more important to me than to my wife . . . My wife always says, "I can't just have sex like you. Everything's gotta be . . . you know, the dishes gotta be washed, the place has gotta be cleaned. I got a thousand things on my mind." I say, "Yeah, well I got a thousand things on my mind too, but the first thing is sex!" They [women] can't do that.

However, in a second example, a 30-year-old husband describes life before and after marriage:

You don't think of [this] when you're single. You go out with the guys, you work all day, then you go out and play basketball for a couple of hours, afterward you go out, have a couple of beers, come home exhausted, and just plop into bed. Nobody's there to complain. Do the same thing when you're married, and you come home and your wife says, "Hi sweetheart. How about tonight?" You say, "Aaaaahhhhh . . . I'm really exhausted, honey, please. . . ." And she says, "But that's what you said last night." . . . After we got married, the honeymoon came and went fine, but then you get into your routine. And I'm not one of those guys who can handle that every night. When I go to bed I like to go to sleep. (Cohen 1986)

In marriage, some emotions associated with premarital sex may disappear. For many, the passion of romantic love, especially as experienced in the earliest period of a relationship, eventually disappears as well, to be replaced with a love based on intimacy, caring, and commitment. The relationship rests more on qualities that we earlier identified as *companionate love*. Still, as humorist Garrison Keillor (1994) reminds us, even within the changed context marital sex can be intensely gratifying:

Despite jobs and careers that eat away at their evenings and weekends and nasty whiny children who dog their footsteps and despite the need to fix meals and vacuum the carpet and pay bills, [married] couples still manage to encounter each other regularly in a lustful, inquisitive way and throw their clothes in the corner and do thrilling things in the dark and cry out and breathe hard and afterward lie sweaty together feeling *extreme pleasure.*

Relationship Infidelity and Extramarital Sexuality

As we noted, a fundamental assumption in our culture is that marriages are sexually and emotionally monogamous. This assumption is not unique to the United States. Eric Eric Widmer, Judith Treas, and Robert Newcomb (1998) undertook comparative research using a 24-country sample of more than 33,000 respondents and found strong and widespread disapproval of extramarital sex, although people in different countries varied some in their levels of disapproval, with some being more tolerant than the majority (for example, those in Russia, Bulgaria, and the Czech Republic). Within the United States, nearly 80% of Americans believe extramarital sex is "always wrong" (Blow and Hartnett 2005).

How Much Infidelity and Extramarital Sex Is There?

Although we sometimes overstate the amount of "cheating" that goes on, it is neither an isolated phenomenon nor restricted to married couples. As we reported previously, there is more nonmonogamy among cohabiting than among married couples and among gay male couples than among lesbians or heterosexual couples. There are widely varying estimates of how prevalent extramarital sex is in the United States. Adrian Blow and Kelley Hartnett (2005) cite a number of studies, with varying estimates:

- Of General Social Survey respondents in the 1991–1996 surveys, 13% admitted to having sex outside of their marriages.

- Of respondents, 1.5% acknowledged having had sex with someone other than their spouse or partner in the previous 12 months.

- In one survey, 25% of married men and 15% of married women said that at some point in their marriages they had sex with someone other than their spouses (Blow and Hartnett 2005).

- Findings from a study of 2,598 men and women ages 18–59 who had ever been married or lived with a partner suggested that 11% had been unfaithful; among those married only once, 16% acknowledged having had extramarital sex.

- Finally, when asked about behavior over the prior 12 months, 5% of the 2,010 respondents who had been married over that span of time admitted to infidelity (Olenick 2000).

Attempting to bring these disparate findings to some conclusion, Blow and Hartnett suggest the following (emphasis added)

> We can conclude that *over the course of married, heterosexual relationships* in the United States, [extramarital] sex occurs in less than 25% of committed relationships, and more men than women appear to be engaging in infidelity. . . . From studies of other countries, it appears that rates of infidelity are higher or lower in some places and that gender differences vary considerably.

Findings on extramarital sex from the widely hailed Sex in America survey are shown in Figure 6.2.

Types of Infidelity

We tend to think of extramarital involvements as being sexual, but they may actually assume several forms (Moore-Hirschl et al. 1995; Thompson 1993). They may be (1) sexual but not emotional, (2) sexual and emotional, or (3) emotional but not sexual (Thompson 1984). Less is known about extramarital relationships in which the couple is emotionally but not sexually involved.

People who engage in extramarital affairs have a number of different motivations, and these affairs satisfy a number of different needs (Adler 1996; Moultrup 1990).

Characteristics of Extramarital Sex

Most extramarital sexual involvements are sporadic. Most extramarital sex is not a love affair; it is generally more sexual than emotional. Affairs that are both emotional and sexual appear to detract more from the marital relationship than do affairs that are only sexual or only emotional (Thompson 1984). More women than men consider their affairs emotional; almost twice as many men as women consider their affairs only

Figure 6.2 ■ Lifetime Incidence of Infidelity by Gender and Age

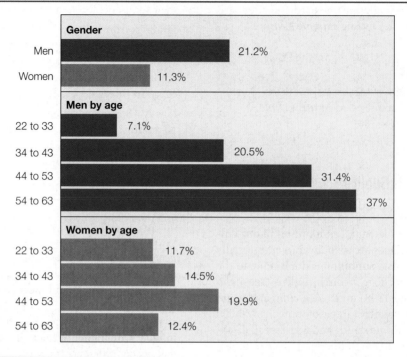

SOURCE: National Opinion Research Center, 1994.

sexual. About equal percentages of men and women are involved in affairs that they view as both sexual and emotional. Research suggests that men are more bothered by the sexual nature of a partner's infidelity where women are disturbed more by the emotional aspect (Christopher and Sprecher 2000).

An emotionally significant extramarital affair creates a complex system of relationships among the three individuals (Moultrup 1990). Long-lasting affairs can form a second but secret "marriage." In some ways, these relationships resemble polygamy, in which the outside person is a "junior" partner with limited access to the other. The involved partners, who know their system is triadic, must try to meet each other's needs for time, affection, intimacy, and sex while taking the uninvolved partner into consideration. Such extramarital systems are stressful and demanding. Most people find great difficulty in sustaining them. If both people involved in the affair are married, the dynamics become even more complex.

Are Gay Male Relationships Sexually Open?

Although the outbreak of the AIDS crisis in the 1980s and 1990s affected behavior and changed how accepting of nonmonogamy gay men were, research conducted both before and after the onset of the epidemic show that a proportion of gay men maintain relationships in which both partners agree to be nonexclusive (LaSala 2004). Among heterosexual couples, lesbian couples, and gay male couples, gay male couples have been and continue to be more likely to accept and experience sexual nonmonogamy. Furthermore, in comparing monogamous and non-monogamous (that is, "faithful" and "unfaithful") gay male couples, research often finds no differences in relationship adjustment or satisfaction (LaSala 2004). Some clinicians have gone as far as to deem those who condemn nonmonogamy as dysfunctional "hetero-centrist," meaning that they are applying standards that may pertain to heterosexual relationships too broadly. Given that males think about and act differently regarding sexual relationships, we might assume that gay male couples would display these tendencies even more than heterosexual couples (each of whom has a female partner) and certainly more than lesbian couples (who obviously have no male partners). Michael LaSala reminds us that research has established that, compared to women, men are more likely to separate sex and love; to engage in sexual relationships in the absence of emotional involvement; to engage in sexual relations within even "casual relationships;" and to consider having sex with strangers.

Just as a portion of gay men maintain nonexclusive relationships, many gay couples construct relationship boundaries that proscribe (prohibit) sex with others. Like heterosexual and lesbian couples, such men come to see infidelity as a breach of trust. In LaSala's sample of 121 gay male couples, 60% described their "relationship agreements" as assuming monogamy (40% were in sexually open relationships). However, among these 73 couples, 33 had breached this expectation and broken their monogamous agreement. It was this latter group that had the lowest scores on satisfaction, expression of affection toward one's partner, and relationship adjustment. Interestingly, however, when those who engaged in nonmonogamous sex in the prior 12 months were removed from the analysis, there appeared to be no real difference between those whose monogamous expectations had been upheld and those whose expectations had been violated.

Sexual Problems and Dysfunctions

Many of us who are sexually active may experience sexual difficulties or problems. Recurring problems that cause distress to the individual or his or her partner are known as **sexual dysfunctions.** Although some sexual dysfunctions are physical in origin, many are psychological. Some dysfunctions have immediate causes, others originate in conflict within the self, and still others are rooted in a particular sexual relationship.

Both men and women may suffer from hypoactive (low or inhibited) sexual desire (Hawton, Catalan, and Fagg 1991). Other dysfunctions experienced by women are orgasmic dysfunction (the inability to attain orgasm), arousal difficulties (the inability to become erotically stimulated), and **dyspareunia** (painful intercourse). The most common dysfunctions among men include **erectile dysfunction** (the inability to achieve or maintain an erection), **premature ejaculation** (the inability to delay ejaculation after penetration), and delayed orgasm (difficulty in ejaculating) (Spector and Carey 1990). Figure 6.3 shows the percentage of heterosexual adults in the general U.S. population who reported experiencing sexual problems during the previous year in response to a recent survey (Laumann et al. 1994).

Origins of Sexual Problems

Physical Causes

It is generally believed that between 10% and 20% of sexual dysfunctions are structural in nature. Physical problems may be *partial* causes in another 10% or 15% (Kaplan 1983; LoPiccolo 1991). Various illnesses may have an adverse effect on a person's sexuality (Wise, Epstein, and Ross 1992). Alcohol and some prescription drugs, such as medication for hypertension, may affect sexual responsiveness (Buffum 1992; "Drugs" 1992).

Among women, diabetes, hormone deficiencies, and neurological disorders, as well as alcohol and alcoholism, can cause orgasmic difficulties. Painful intercourse may be caused by an obstructed or thick hymen, clitoral adhesions, a constrictive clitoral hood, or a weak pubococcygeus muscle. Coital pain caused by inadequate lubrication and thinning vaginal walls often occurs as a result of decreased estrogen associated with menopause. Lubricants or hormone replacement therapy often resolve the difficulties.

Among males, diabetes and alcoholism are the two leading physical causes of erectile dysfunctions; atherosclerosis is another important factor (LoPiccolo 1991; Roenrich and Kinder 1991). Smoking may also contribute to sexual difficulties (Rosen et al. 1991).

Psychological or Relationship Causes

Two of the most prominent causes of sexual dysfunctions are performance anxiety and conflicts within the self. **Performance anxiety**—the fear of failure—is probably the most important immediate cause of erectile dysfunctions and, to a lesser extent, of orgasmic dysfunctions in women (H. Kaplan 1979). If a man does not become erect, anxiety is a fairly common response. Some men experience their first erectile problem when a partner initiates or demands sexual intercourse. Women are permitted to say no, but many men have not learned that they too may say no to sex. Women suffer similar anxieties, but they tend to center around orgasmic abilities rather than the ability to have intercourse. If a woman is unable to experience orgasm, a cycle of fear may arise, preventing future orgasms. A related source of anxiety is an excessive need to please one's partner.

Conflicts within the self are guilt feelings about one's sexuality or sexual relationships. Guilt and emotional

Figure 6.3 ■ Heterosexual Sexual Dysfunctions in a Nonclinical Sample

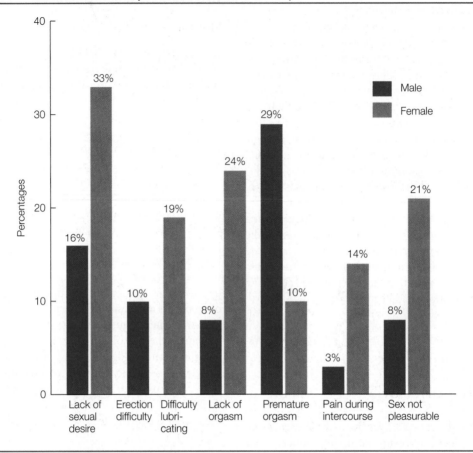

SOURCE: Adapted from Laumann et al. 1994, 370–371.

conflict do not usually eliminate a person's sexual drive; rather, they inhibit the drive and alienate the person from his or her sexuality. Such inner conflicts often are deeply rooted. Among gay men and lesbians, concerns about sexual orientation may be an important cause of such conflicts (George and Behrendt 1987). The relationship itself, rather than either individual, sometimes can be the source of sexual problems. Disappointment, anger, or hostility may become integral parts of a deteriorating or unhappy relationship. Such factors affect sexual interactions, because sex can become a barometer for the whole relationship.

Relationship discord can affect our sexuality in several ways, such as through poor communication that inhibits our ability to express our needs and desires; power struggles in which sexuality becomes a tool in struggles for control; and sexual sabotage where part-

ners ask for sex at the wrong time, put pressure on each other, and frustrate or criticize each other's sexual desires and fantasies. People most often do this unconsciously (Kaplan 1979).

Sex between Unequals, Sex between Equals

Sociologist Pepper Schwartz (1994) identified a number of sexual problems that plague traditional marriages because of the gender hierarchy and absence of empathy that characterize such marriages. Partners are "too distant, too different, and too inequitable" to enjoy complete sexual fulfillment. Sexual problems among traditional couples include the following:

- *Failure of timing.* This results when one person is more in charge of the couple's sexual relationship and his or her needs define when the couple has sexual relations.

- *Failure of intimacy.* If traditional couples lack the same depth of intimacy (that is, sharing and communication) that Schwartz contends more egalitarian relationships possess, it is apparent in their sexual relationship. According to Schwartz, this can prevent them from finding complete fulfillment in their sexual relationship.

- *Failure of sexual empathy.* Some couples fail to realize that what one finds pleasing the other may not. This is particularly true in the most traditional marriages, where "men and women have little experience of each other's lives." They may show little respect for each other's sexual needs and refuse to make the effort to learn what each other wants (Schwartz 1994).

- *Failure of reciprocity.* Inequality outside the bedroom can spill into the bedroom. Often the woman, but potentially either partner, feels as if she gives more than she receives. There is less mutual massage than desired, or she feels that she is touched less or receives less oral sex than she gives or performs.

- *Failure of overromanticization.* Women in more traditional relationships may possess overly romanticized expectations of sexual relations. These are often beyond what most "ordinary" men can live up to.

Schwartz notes that peer marriages, relationships built on deep friendship and commitments to fairness, sharing, and equality, avoid these particular sexual problems. What they may suffer from, instead, is a decline in sexual intensity. Some of this results just from habituation. More specific to peer couples are other problems that can diminish sexual excitement, most notably an inability to transform themselves from their everyday identities based on sameness and openness to erotic identities based on "principles of opposites and mystery" (Schwartz 1994). Thus, the same things that differentiate peer relationships from their more common and less equal counterparts may make it hard for peer couples to sustain sexual energy. These problems are not insurmountable, but they do require special effort on the part of peer couples to create a separate and special sexual environment removed from more mundane life matters.

Resolving Sexual Problems

Sexual problems can be embarrassing and emotionally upsetting. Perhaps the first step in dealing with a sexual problem is to turn to immediate resources. Talking about the problem with one's partner, finding out what he or she thinks, discussing specific strategies that might be useful, and simply communicating feelings and thoughts can sometimes resolve the difficulty. One can also go outside the relationship, seeking friends with whom to safely share feelings and anxieties, asking whether they have had similar experiences, and learning how they handled them. Sexual problems can become self-fulfilling, because couples may focus so much on the difficulties that they are having that additional pressure is placed on sexual performance. Thus, keeping perspective—and often a sense of humor—may be quite helpful.

Aside from one's circle of friends and intimates, there are ever-increasing, additional resources on which one can draw. A growing number of self-help books dealing with sexuality and relationship issues line the shelves in bookstores and libraries. There are also numerous websites one can access and consult. For example, Yahoo searches for sites dealing with erectile dysfunction, premature ejaculation, and female sexual dysfunction, generated 9,680,000, 4,890,000, and 3,130,000 hits, respectively. Not all of these will be sites offering help or advice. Some may be pornographic, and others may carry exaggerated claims designed to sell products, but many websites offer information compiled or overseen by medical, psychiatric, psychological, nursing, or educational specialists.

Cumulatively, partners, friends, websites, and books may provide information and grant individuals needed "permission" to engage in sexual exploration and discovery by making such inquiries normal. From these sources we may learn that our sexual issues, problems, fantasies, and behaviors are not unique. Such methods are most effective when the dysfunctions arise from a lack of knowledge or mild sexual anxieties.

If, despite conversation with one's partner, consultation with friends, and/or reading books, magazines, or other resources one remains unable to resolve his or her sexual difficulties, seeking professional assistance is the logical next step. It is important to realize that seeking such assistance does not signal personal weakness or failure. Rather, it demonstrates an ability to reach out and a willingness to change. It is a sign of caring for one's partner, one's relationship, and oneself.

For those whose problems stem mostly from psychological or relationship causes, therapists can help deal with sexual problems on several levels. Some focus directly on the problem, such as lack of orgasm, and suggest behavioral exercises, such as pleasuring and masturbation, to develop an orgasmic response. Others focus on the couple relationship as the source of difficulty. If the relationship improves, they believe that sexual responsiveness will also improve. Still others work with the individual to help develop insight into the origins of the problem to overcome it. Therapy can also take place in a group setting. Group therapy may be particularly valuable for providing partners with an open, safe forum in which they can discuss their sexual feelings and experience and discover commonalities with others.

A relatively new development for men who suffer from sexual problems is medication. In March 1998, the Food and Drug Administration approved **Viagra,** the first oral treatment for male impotence. With as many as 50% of men, 40 and older, suffering from at least occasional and mild impotence, Viagra quickly became an economic and cultural phenomenon. In just its first year of availability, Viagra had sales of $1 billion, propelling its manufacturer, Pfizer, to the second spot among the world's largest drug companies. In 2002, Viagra had sales in excess of $1.7 billion. Optimistic forecasts predicting continued growth and sales success turned out to be exaggerated, but Viagra definitely has made its mark on the economy, society, and culture.

There is still no equivalently successful prescription drug for women suffering from orgasmic difficulties or other sexual dysfunctions.

Issues Resulting from Sexual Involvement

Birth Control

Most of us think of sexuality in terms of love, passionate embraces, and entwined bodies. Sex involves all of these, but what we so often forget (unless we are worried) is that sex is also a means of reproduction. Whether we like to think about it or not, many of us

(or our partners) are vulnerable to unintended pregnancies. Not thinking about pregnancy does not prevent it; indeed, not thinking about it may increase the likelihood of its occurring. Unless we practice **abstinence,** refraining from sexual intercourse, we need to think about unintended pregnancies and then take the necessary steps to prevent them.

Sexually Transmitted Diseases, HIV, and AIDS

Americans are in the middle of the worst epidemic of sexually transmitted diseases (STDs) in our history. There are an estimated 15 million new cases of sexually transmitted infections in the United States each year, the highest rate of infection of any industrialized nation in the world (Miracle, Miracle, and Baumeister 2003). College students are as vulnerable as anyone else. Untreated, chlamydia and gonorrhea can lead to pelvic inflammatory disease (PID) in women, a major cause of infertility.

Still, most of us would wince if asked on a first date, "Do you have chlamydia, gonorrhea, herpes, syphilis, HIV, or any other sexually transmissible disease that I should know about?" However, given the risks of contracting an STD and the consequences associated with infection, it is a question whose answers you should know before you become sexually involved. Just because a person is "nice" or good looking, or available and willing, is no guarantee that he or she does not have one of the STDs discussed later in this chapter. You could be infected through such sexual contact as sexual intercourse, oral sex, or anal sex. Unfortunately, no one can tell by a person's looks, intelligence, or demeanor whether he or she has contracted an STD. The costs of becoming sexually involved with a person without knowing about the presence of any of these diseases are potentially steep.

Principal STDs

The most prevalent STDs in the United States are chlamydia, gonorrhea, genital warts, genital herpes, syphilis, hepatitis, and HIV and AIDS. Conditions that may be sexually transmitted include urethritis (in both women and men) and vaginitis and PID (in women). Table 6.6 briefly describes the symptoms, exposure intervals, treatments, and other information regarding the principal STDs.

HIV and AIDS

The **human immunodeficiency virus (HIV)** is the virus that causes **acquired immunodeficiency syndrome (AIDS)**. The disease is so termed because of its characteristics:

acquired—because people are not born with it

immunodeficiency—because the disease relates to the body's immune system, which is lacking in immunity

syndrome—because the symptoms occur as a group

Overall, the effects of HIV have been devastating, with the worst effects happening not in the United States but in other parts of the world. According to a report from the U.S. Department of Health and Human Services, National Institutes of Health, and National Institute of Allergies and Infectious Diseases, nearly 25 million people have died from HIV or AIDS worldwide, including nearly 0.5 million Americans. Furthermore, an estimated 40 million people worldwide are living with the disease, including 1 million or more Americans. Each year there are an estimated 40,000 new infections, with men representing 70% of the cases (National Institute of Allergies and Infectious Diseases 2005). The Centers for Disease Control (2003) estimates that approximately 56% of the infections in the United States were transmitted through male–male or heterosexual sexual contact.

HIV and AIDS cases have hit African Americans and Latinos especially hard, with each group infected at disproportionate rates. Despite being just 12% of the population of the United States, blacks make up more than half of all new HIV infections. Perhaps even more striking, AIDS is *the No. 1 cause of death* among African American men of all ages (National Institute of Allergies and Infectious Diseases 2005).

Although AIDS was initially discovered in gay men and was thought of early on as a "gay disease" or the "gay plague," sexually transmitted cases among heterosexuals increased at a rate at least comparable to that among gay men. This was partly because of how successful the gay community was in incorporating safer-sex practices, especially during the 1980s and 1990s. Unfortunately, there are some troubling signs that HIV and AIDS infections are again increasing among gay men, and not just among young men and/or minority men, at rates that Spencer Cox, of the AIDS Community Research Initiative of America, calls "alarming." Citing "dramatic increases in risky sex (and other kinds of risky behavior) among older, white, and

relatively affluent gay men in major cities—traditionally the group for whom prevention efforts were most effective," Cox asserts that this new wave of infections will be difficult to combat, especially since it appears to be tied in with other lifestyle choices, such as crystal methamphetamine use. Crystal methamphetamine is a highly addictive drug and associated with more sexual risk taking, such as engaging in unprotected anal intercourse and having numerous casual partners. Evidence from cities such as Chicago, New York, and San Francisco reveal that between 15% and 20% of gay men report using methamphetamines (Cox 2006).

Without discounting or diminishing the devastation that the gay community suffered from AIDS and HIV, or signs of resurging rates of infection, it is important to keep in mind that heterosexuals and bisexuals are also at risk and become infected. Virtually all adults in the United States are or will soon be related to, personally know, work with, or go to school with people infected with HIV or will know others whose friends, relatives, or associates test HIV-positive.

As of April 2006, there is still no surefire vaccine to prevent HIV, nor is there a cure for those who are or become infected. Significant strides have been made in fighting the disease, suppressing its symptoms, and prolonging life for those who are infected. In addition, between 1996 and 2001, AIDS death rates were reduced by 80% and postdiagnosis survival had doubled in length. Those diagnosed after 1998 could expect to live 9 to 10 years longer than those who were diagnosed during the mid-1980s (Fallon 2005).

In addition to new treatments that can lengthen the life span of an AIDS-infected person as much as 15 years (Fallon 2005), we have considerable knowledge about the nature of the virus and how to reduce the likelihood of infection:

- *HIV attacks the body's immune system.* HIV is carried in the blood, semen, and vaginal secretions of infected people. A person may be HIV-positive (infected with HIV) for years before developing AIDS symptoms.

- *HIV is transmitted only in certain clearly defined circumstances.* It is transmitted through the exchange of blood (as by shared needles or transfusions of contaminated blood), through sexual contact involving semen or vaginal secretions, and from an infected woman to her fetus through the placenta. Infected mothers may also transmit the infection during delivery or by nursing (Miracle, Miracle, and Baumeister 2003).

Table 6.6 ■ Principal Sexually Transmitted Diseases

STD and Infecting Organism	Time from Exposure to Occurrence	Symptoms	Medical Treatment	Comments
Chlamydia (*Chlamydia trachomatis*)	7–21 days	Women: 80% asymptomatic; others may have vaginal discharge or pain with urination. Men: 30%–50% asymptomatic; others may have discharge from penis, burning urination, pain and swelling in testicles, or persistent low fever.	Doxycycline, tetracycline, erythromycin	If untreated, may lead to pelvic inflammatory disease (PID) and subsequent infertility in women.
Gonorrhea (*Neisseria gonorrhoeae*)	2–21 days	Women: 50%–80% asymptomatic; others may have symptoms similar to chlamydia. Men: itching, burning, or pain with urination; discharge from penis ("drip").	Penicillin, tetracycline, or other antibiotics	If untreated, may lead to PID and subsequent infertility in women.
Genital warts (*Human papilloma virus*)	1–6 months (usually within 3 months)	Variously appearing bumps (smooth, flat, round, clustered, fingerlike, white, pink, brown, and so on) on genitals, usually penis, anus, vulva, vagina, or cervix.	Surgical removal by freezing, cutting, or laser therapy. Chemical treatment with podophyllin (80% of warts eventually reappear)	Virus remains in the body after warts are removed.
Genital herpes (*Herpes simplex virus*)	3–20 days	Small, itchy bumps on genitals, becoming blisters that may rupture, forming painful sores; possibly swollen lymph nodes; flulike symptoms with first outbreak.	No cure although acyclovir may relieve symptoms. Nonmedical treatments may help relieve symptoms.	Virus remains in the body, and outbreaks of contagious sores may recur. Many people have no symptoms after the first outbreak.
Syphilis (*Treponema pallidum*)	Stage 1: 1–12 weeks Stage 2: 6 weeks to 6 months after chancre appears	Stage 1: Red, painless sore (chancre) at bacteria's point of entry. Stage 2: Skin rash over body, including palms of hands and soles of feet.	Penicillin or other antibiotics	Easily cured, but untreated syphilis can lead to ulcers of internal organs and eyes, heart disease, neurological disorders, and insanity.

SOURCE: Strong and DeVault 1997.

■ *All those with HIV (whether or not they have AIDS symptoms) are HIV carriers.* They may infect others through unsafe sexual activity or by sharing needles; if they are pregnant, they may infect the fetus.

■ *Heterosexuals, bisexuals, gay men, and lesbians are all susceptible to the sexual transmission of HIV.* Sexual transmission accounts for 68% of AIDS and 57% of HIV infections among men through 2002. Male–male sexual contact is attributed to 55% of all AIDS cases and 47% of all non-AIDS HIV infections among men. The rate of heterosexual HIV transmission is rising faster than the rate of gay transmission. Among women, heterosexual contact accounts for 42% of HIV and AIDS cases.

■ *There is a definable progression of HIV infection and a range of illnesses associated with AIDS.* HIV attacks the immune system. AIDS symptoms occur as opportunistic diseases—diseases that the body normally resists—infect the individual. The most common opportunistic diseases are pneumocystis carinii pneumonia and Kaposi's sarcoma, a skin cancer. It is an opportunistic disease rather than HIV that kills the person with AIDS.

■ *The presence of HIV can be detected through various kinds of antibody testing.*

Anonymous testing is available at many college health centers and community health agencies. HIV antibodies develop between 1 and 6 months after infection. Antibody testing should take place 1 month after possible exposure to the virus and, if the results are negative, again 6 months later. If the antibody is

Table 6.6 ■ Continued

STD and Infecting Organism	Time from Exposure to Occurrence	Symptoms	Medical Treatment	Comments
Hepatitis (*hepatitis A or B virus*)	1–4 months	Fatigue, diarrhea, nausea, abdominal pain, jaundice, and darkened urine due to impaired liver function.	No medical treatment available; rest and fluids are prescribed until the disease runs its course.	Hepatitis B is more commonly spread through sexual contact and can be prevented by vaccination.
Urethritis (*various organisms*)	1–3 weeks	Painful and/or frequent urination; discharge from penis. Women may be asymptomatic.	Penicillin, tetracycline, or erythromycin, depending on organism	Laboratory testing is important to determine appropriate treatment.
Vaginitis (*Gardnerella vaginalis, Trichomonas vaginalis, or Candida albicans*) (women only)	2–21 days	Intense itching of vagina and/or vulva; unusual discharge with foul or fishy odor; painful intercourse. Men who carry organisms may be asymptomatic.	Depends on organism; oral medications include metronidazole and clindamycin, and vaginal medications include clotrimazole and miconazole	Not always acquired sexually. Other causes include stress, oral contraceptives, pregnancy, tight pants or underwear, antibiotics, douching, and dietary imbalance.
HIV infection and AIDS (*human immunodeficiency virus*)	Several months to several years	Possible flulike symptoms but often no symptoms during early phase. Variety of later symptoms including weight loss, persistent fever, night sweats, diarrhea, swollen lymph nodes, bruiselike rash, and persistent cough.	No cure available, although many symptoms can be treated with medications and antiviral drugs may strengthen the immune system. Good health practices can delay or reduce the severity of symptoms.	Cannot be self-diagnosed; a blood test must be performed to determine the presence of the virus.
Pelvic inflammatory disease (PID) (*women only*)	Several weeks or months after exposure to chlamydia or gonorrhea (if untreated)	Low abdominal pain; bleeding between menstrual periods; persistent low fever.	Penicillin or other antibiotics; surgery	Caused by untreated chlamydia or gonorrhea; may lead to chronic problems such as arthritis and infertility.

present, the test will be positive. That means that the person has been infected with HIV and an active virus is present. The presence of HIV does not mean, however, that the person necessarily will develop AIDS symptoms in the near future; symptoms generally occur 7 to 10 years after the initial infection.

Protecting Yourself and Others

As with avoiding unintended pregnancies, the safest practice to avoid STDs is *abstinence,* foregoing sexual relations. There is no chance of contracting STDs, although HIV infection can and does occur through nonsexual transmission (for example, intravenous drug use with shared needles). If you are sexually active,

however, the key to protecting yourself and others is to talk with your partner about STDs in an open, nonjudgmental way and to use condoms. The best way of finding out whether your partner has an STD is by asking. If you feel nervous about broaching the subject, you can rehearse talking about it. It may be sufficient to ask in a lighthearted manner, "Are you as healthy as you look?" or because many people are uncomfortable asking about STDs, you can open the topic by revealing your anxiety: "This is a little difficult for me to talk about because I like you and I'm embarrassed, but I'd like to know whether you have herpes, or HIV, or whatever." If *you* have an STD, you can say, "Look, I like you, but we can't make love right now because I have a chlamydial infection and I don't want you to get it."

Remember, however, that not every person with an STD knows she or he is infected. Women with chlamydia and gonorrhea, for example, generally don't exhibit symptoms. Both men and women infected with HIV may not show any symptoms for years, although they are capable of spreading the infection through sexual contact. If you are or are planning to be sexually active but don't know whether your partner has an STD, use a condom. Even if you don't discuss STDs, condoms are simple and easy to use without much discussion. Both men and women can carry them. A woman can take a condom from her purse and give it to her partner. If he doesn't want to use it, she can say, "No condom, no sex."

Sexual Responsibility

Because we have so many sexual choices today, we need to be sexually responsible. Sexual responsibility includes the following:

- *Disclosure of intentions.* Each person needs to reveal to the other whether a sexual involvement indicates love, commitment, recreation, and so on.
- *Freely and mutually agreed-upon sexual activities.* Each individual has the right to refuse any or all sexual activities without the need to justify his or her feelings. There can be no physical or emotional coercion.
- *Use of mutually agreed-upon contraception in sexual intercourse if pregnancy is not intended.* Sexual partners are equally responsible for preventing an unintended pregnancy in a mutually agreed-upon manner.
- *Use of "safer sex" practices.* Each person is responsible for practicing safer sex unless both have been monogamous with each other for at least 5 years or have recently tested negative for HIV. Safer sex practices do not transmit semen, vaginal secretions, or blood during sexual activities and guard against STDs, especially HIV and AIDS.
- *Disclosure of infection from or exposure to STDs.* Each person must inform his or her partner about personal exposure to an STD because of the serious health consequences, such as infertility or AIDS,

that may follow untreated infections. Infected individuals must refrain from behaviors—such as sexual intercourse, oral–genital sex, and anal intercourse—that may infect their partner. To help ensure that STDs are not transmitted, a condom should be used.

- *Acceptance of the consequences of sexual behavior.* Each person needs to be aware of and accept the possible consequences of his or her sexual activities. These consequences can include emotional changes, pregnancy, abortion, and STDs.

Responsibility in many of these areas is facilitated when sex takes place within the context of an ongoing relationship. In that sense, sexual responsibility is a matter of values. Is responsible sex possible outside an established relationship? Are you able to act in a sexually responsible way? Sexual responsibility also leads to the question of the purpose of sex in your life. Is it for intimacy, erotic pleasure, reproduction, or other purposes?

As we consider the human life cycle from birth to death, we cannot help but be struck by how profoundly sexuality weaves its way through our lives. From the moment we are born, we are rich in sexual and erotic potential, which begins to take shape in our sexual experimentations of childhood. As children, we are still unformed, but the world around us haphazardly helps give shape to our sexuality. In adolescence, our education continues as a mixture of learning and yearning. But as we enter adulthood, with greater experience and understanding, we undertake to develop a mature sexuality: we establish our sexual orientation as heterosexual, gay, lesbian, or bisexual; we integrate love and sexuality; we forge intimate connections and make commitments; we make decisions regarding our fertility and sexual health; we develop a coherent sexual philosophy. Then, in our middle years, we redefine sex in our intimate relationships, accept our aging, and reevaluate our sexual philosophy. Finally, as we become elderly, we reinterpret the meaning of sexuality in accordance with the erotic capabilities of our bodies. We come to terms with the possible loss of our partner and our own end. In all these stages, sexuality weaves its bright and dark threads through our lives.

Summary

- Our sexual behavior is influenced by *sexual scripts:* the acts, rules, stereotyped interaction patterns, and expectations associated with male and female sexual expression. These provide general guidelines of what is expected from or accepted of us.

- Traditional female sexual scripts include the following ideas: Sex is both good and bad (depending on the context); men should know what women want; and there is only one right way to experience an orgasm.

- Traditional male sexual scripts include the following: Men should not have (or at least should not express) certain feelings; the man is in charge; and all physical contact leads to sex.

- Contemporary sexual scripts are more egalitarian, consisting of the following beliefs: Sexual expression is positive; sexuality involves both partners equally and both partners are equally responsible; and legitimate sexual activities include masturbation and oral–genital sex.

- Even with the emergence of the contemporary sexual script, evidence suggests that there is a *sexual double standard,* in which different sexual behaviors are accepted and expected of men and women.

- There are several tasks that we must undertake in developing our sexuality as young adults, including (1) integrating love and sex, (2) forging intimacy and commitment, (3) making fertility or childbearing decisions, (4) establishing a sexual orientation, and (5) developing a sexual philosophy.

- We learn about sexuality from multiple sources: parents, peers, the mass media, and increasingly the Internet. Most sexual socialization by parents is from mothers, and daughters have more sexual communication with mothers than sons do.

- Even amid a longer-term trend toward more open acceptance of nonmarital sexual behavior, a recent trend seems to point toward a decline in sexual activity among teenagers.

- Between 1% and 10% of American men are *gay,* and between 1% and 3% of American women are *lesbian* at one time or another in their lives. Identifying oneself as gay or lesbian occurs in stages. Although some gay men and lesbians assert that they "always knew" they were different from their het-

- erosexual counterparts, some—especially among lesbians—report that their sexuality was partly a political choice.

- Gay men and lesbians maintain intimate relationships that have much in common with heterosexual relationships though they lack comparable social support, and legal rights and protections.

- Lesbian, gay, and bisexual individuals may be subject to prejudice or hostility, including verbal abuse, discrimination, or violence. Attitudes toward bisexuals may be even harsher than attitudes toward gay men and lesbians.

- *Bisexuals* are attracted to members of both genders. In developing a bisexual identity, men and women go through several stages: (1) initial confusion, (2) finding and applying the bisexual label, (3) settling into the identity, and (4) continued uncertainty. Bisexuals don't have a community or social environment that reaffirms their identity.

- Developmental tasks in middle adulthood include (1) redefining sex in marital or other long-term relationships, (2) reevaluating one's sexuality, and (3) accepting the biological aging process.

- The main determinants of sexual activity in old age are health and the availability of a partner.

- Chronic illnesses; medications; declining levels of testosterone in men and estrogens in women; negative attitudes and cultural stereotypes; and loss of, absence of, or monotony with a partner all contribute to the reduction in sexual desire for aging women and men.

- *Autoeroticism* consists of sexual activities that involve only the self. It includes sexual fantasies, masturbation, and erotic dreams.

- Gender and race differences in masturbation have been identified. More men than women masturbate, and men masturbate more often. In marriage, men masturbate to supplement their sexual activities, whereas women tend to masturbate as a substitute for such activities. European American women report higher rates of masturbation than either African American or Hispanic women.

- The most common and acceptable of all premarital sexual activities is kissing, which occurs in more than 90% of all cultures.

- *Oral–genital sex,* which includes *cunnilingus* and *fellatio,* is practiced by heterosexuals, gay men, and lesbians. Data indicate increasing rates of oral sex among teenagers and young adults.

- *Sexual intercourse (coitus)* is the insertion of the penis into the vagina and the stimulation that follows.

- *Anal eroticism* is practiced by both heterosexuals and gay men. From a health perspective, *anal intercourse* is dangerous because it is the most common means of sexually transmitting HIV.

- *Sexual enhancement* is based on accurate information about sexuality, developing communication skills, fostering positive attitudes, and increasing self-awareness.

- *Nonmarital sex* includes all sexual activities, especially sexual intercourse, that take place outside of marriage. *Premarital sex* has gained in acceptability.

- Marital sex tends to decline in frequency over time, but this does not necessarily signify marital deterioration.

- *Extramarital sex* is widely condemned, although people in some countries are more tolerant than those in others. Race, residential location, gender, and frequency of thinking about sex are associated with rates of infidelity.

- Extrarelational sex occurs among heterosexual cohabitants, gay male couples, and lesbian couples at higher rates than among married couples.

- **Nonconsensual sexual behaviors** range from exhibitionism and voyeurism to rape, sexual harassment, and child sexual abuse.

- *Sexual dysfunctions* (such as orgasmic or arousal difficulties in women or erectile dysfunction or premature ejaculation in men) are recurring problems in giving and receiving erotic satisfaction that may be physiological or psychological in origin.

- Therapeutic and medicinal assistance is available for people experiencing sexual dysfunctions.

- *Sexually transmitted diseases (STDs),* especially chlamydia and gonorrhea, are epidemic. *Acquired immunodeficiency syndrome (AIDS)* is caused by the *human immunodeficiency virus (HIV),* which attacks the body's immune system. HIV is carried in the blood, semen, and vaginal fluid of infected people.

- The rate of infection and death because of HIV is far greater in other parts of the world (for example, Africa and Asia) than in the United States, where. most HIV infections are the result of either heterosexual or male–male sexual contact, and where Hispanics and African Americans have been particularly hard hit.

- If someone is sexually active, the keys to protection against STDs, including HIV and AIDS, are communication and condom use.

- Anyone sexually active should practice sexual responsibility: disclose any STD infections, engage only in mutually agreed upon activities, use mutually agreed upon methods of contraception, and engage in safer sex.

Key Terms

abstinence 226

acquired immunodeficiency syndrome (AIDS) 227

anal eroticism 216

anal intercourse 216

antigay prejudice 205

autoeroticism 213

bisexual 200

coitus 216

coming out 204

cunnilingus 215

dyspareunia 223

erectile dysfunction 223

extramarital sex 218

fellatio 215

foreplay 191

gay men 201

heterosexual 200

homoeroticism 202

homophobia 205

homosexual 200

human immunodeficiency virus (HIV) 227

lesbians 201

masturbation 214

menopause 211

nonconsensual sexual behavior 232

nonmarital sex 217

outing 204

performance anxiety 223

pleasuring 214

premarital sex 217

premature ejaculation 223

sexual double standard 194

sexual dysfunctions 223

sexual enhancement 216

sexual intercourse 216

sexual script 191

Viagra 226

virginity 199

weblog 197

RESOURCES ON THE INTERNET

Companion Website for This Book

http://www.thomsonedu.com/sociology/strong

Gain an even better understanding of this chapter by going to the companion website for additional study resources. Take advantage of the Pre- and Post-Test quizzing tool, which is designed to help you grasp difficult concepts by referring you back to review specific pages in the chapter for questions you answer incorrectly. Use the flash cards to master key terms and check out the many other study aids you'll find there. Visit the Marriage and Family Resource Center on the site. You'll also find special features such as access to Info-Trac[c] College Edition (a database that allows you access to more than 18 million full-length articles from 5,000 periodicals and journals), as well as GSS Data and Census information to help you with your research projects and papers.

CHAPTER 7

Communication, Power, and Conflict

Are the following statements **TRUE** or **FALSE**?
You may be surprised by the answers (see answer key on the bottom of this page).

T   F   **1** Conflict and intimacy go hand in hand in intimate relationships.

T   F   **2** Touching is one of the most significant means of communication.

T   F   **3** Always being pleasant and cheerful is the best way to avoid conflict and sustain intimacy.

T   F   **4** Studies suggest that those couples with the highest marital satisfaction tend to disclose more than those who are unsatisfied.

T   F   **5** Negative communication patterns before marriage are a poor predictor of marital communication because people change once they are married.

T   F   **6** Good communication is primarily the ability to offer excellent advice to your partner to help him or her change.

T   F   **7** Physical coercion is the method men use most often when disagreement arises between them and their partners.

T   F   **8** The party with the least interest in continuing a relationship generally has the power in it.

T   F   **9** Latinos and Asian Americans tend to rely on the nonverbal expression of intense feelings in contrast to direct verbal expressions.

T   F   **10** Wives tend to give more negative messages than husbands.

"**Y**our mother called again."

What a simple, ordinary statement that sounds like. It hardly seems like the kind of comment that would provoke an argument, nor does it appear particularly revealing about the tone or quality of a marriage or relationship. It sounds so routine, so "matter-of-fact" that we might overlook its significance and potential effect on married or coupled life.

Of course, we only have the four words; we don't know *how* they were said. What was the tone of voice? The cadence or rhythm of speech—was it, "Your mother called again," or "Your mother called. Again." Or, combining tone and cadence, "Your mother called. Again!" We also have no information about the non-verbal signs. What was the expression on the face of the speaker—say a wife to a husband—when the statement was made? Did she smile? Roll her eyes? Frown? All of these aspects of nonverbal communication help reveal more of the meaning and significance of such a statement. Clearly, even such a simple comment as this may have greater importance than the four words otherwise convey.

Finally, of even greater significance is how the other person responds to a statement such as this one. Whether she or he responds with "an irritable groan," a laugh (as if to say "what, again!"), or with a positive discussion of his or her mother tells us a lot. A non-response may tell us yet more. It may suggest indifference and lack of interest in talking with the partner. Exchanges surrounding statements such as this one, "mundane and fleeting" as they may appear to be, can build and, in the process, greatly affect the quality of a relationship, the amount and nature of conflict, and the feeling of closeness and romance (Driver and Gottman 2004).

Thinking about the kinds of relationships that are the focus of this book, what is it you most want or expect from marriages, families, and other intimate

Answer Key for What Do You Think

1 True, see p. 259;   2 True, see p. 240;   3 False, see p. 238;   4 True, see p. 251;   5 False, see p. 244; 6 False, see p. 254;   7 False, see p. 256;   8 True, see p. 257;   9 True, see p. 241;   10 True, see p. 246.

relationships? Chances are, if you list the many characteristics or qualities you desire in such relationships, somewhere on that list will be "communication." We want our loved ones to share their feelings and ideas with us and to understand the ideas or feelings that we voice to them. After all, as shown in the last chapter, that is how we expect to share intimacy. We want to be able to communicate effectively.

Chances are that "conflict" will not be included among desired relationship characteristics. After all, who wants to argue? We tend to see conflict as a negative to be avoided. Yet, conflict is as much a feature of intimate relationships as are love and affection. As long as we value, care about, and live with others, we will experience occasions when we disagree. An absence of conflict is not only unrealistic, it would be unhealthy. How we resolve our disagreements tells us much about the health of our relationships.

Both communication and conflict are inextricably connected to intimacy. When we speak of communication, we mean more than just the ability to relay information (for example, "Your mother called"), discuss problems, and resolve conflicts. We also mean communication for its own sake: the pleasure of being in each other's company, the excitement of conversation, the exchange of touches and smiles, the loving silences. Through communication we disclose who we are, and from this self-disclosure, intimacy grows.

One of the most common complaints of married partners, especially unhappy partners, is that they don't communicate. But it is impossible not to communicate—a cold look may communicate anger as effectively as a fierce outburst of words. What these unhappy partners mean by "not communicating" is that their communication is somehow driving them apart rather than bringing them together, feeding and creating conflict rather than resolving it. Communication patterns are strongly associated with marital satisfaction (Noller and Fitzpatrick 1991).

In this chapter, we explore patterns and problems in communication in marital and intimate relationships. We also examine the role of power in marital relationships, where it comes from, and how it is expressed. Finally, we look at the relationship between conflict and intimacy, exploring different types of conflict and approaches to conflict resolution. We look especially at three of the more common areas of relationship conflict: conflicts about sex, money, and housework.

Verbal and Nonverbal Communication

When we communicate face to face, the messages we send and receive contain both a verbal and a nonverbal component. **Verbal communication** expresses the *basic content* of the message, whereas **nonverbal communication** reflects more of the *relationship* part of the message. The relationship part conveys the attitude of the speaker (friendly, neutral, or hostile) and indicates how the words are to be interpreted (as a joke, request, or command). To understand the full content of any message we need to understand both the verbal and nonverbal parts.

For a message to be most effective, both the verbal and the nonverbal components should be in agreement. If you are angry and say "I'm angry," and your facial expression and voice both show anger, the message is clear and convincing. But if you say "I'm angry" in a neutral tone of voice and a smile on your face, your message is ambiguous. More commonly, if you say "I'm not angry" but clench your teeth and use a controlled voice, your message is also unclear. Your tone and expression make your spoken message difficult to take at face value.

Nonverbal Communication

Whenever two or more people are together and aware of each other, it is impossible for them *not to communicate*. Even when you are not talking, you communicate by your silence (for example, an awkward silence, a hostile silence, or a tender silence). You communicate by the way you position your body and tilt your head, your facial expressions, your physical distance from the other person or people, and so on. Take a moment, right now, and look around you. If there are other people in your presence, how and what are they communicating nonverbally?

One of the problems with nonverbal communication, however, is the imprecision of its messages. Is a person frowning or squinting? Does the smile indicate friendliness or nervousness? A person may be in reflective silence, but we may interpret the silence as disapproval or distance. We may incorrectly infer meanings from expressions, eye contact, stance, and proximity that are other than what is intended. However, by acting on the meaning we read into nonverbal behavior, we give it more weight and make it of greater consequence than it initially might have been.

Functions of Nonverbal Communication

More than 20 years ago, an important study of nonverbal communication and marital interaction found that nonverbal communication has the following three important functions in marriage (Noller 1984): (1) conveying interpersonal attitudes, (2) expressing emotions, and (3) handling the ongoing interaction.

Conveying Interpersonal Attitudes

Nonverbal messages are used to convey attitudes. Gregory Bateson describes nonverbal communication as revealing "the nuances and intricacies of how two people are getting along" (quoted in Noller 1984). Holding hands can suggest intimacy; sitting on opposite sides of the couch can suggest distance. Not looking at each other in conversation can suggest awkwardness or lack of intimacy. Rolling eyes at another's statement conveys a negative attitude or reaction to what's being said or the person saying it, even if the eye-rolling culprit claims, "What? I didn't say *anything*."

Expressing Emotions

Our emotional states are expressed through our bodies. A depressed person walks slowly, head hanging; a happy person walks with a spring. Smiles, frowns, furrowed brows, tight jaws, tapping fingers—all express emotion. Expressing emotion is important because it lets our partner know how we are feeling so that he or she can respond appropriately. It also allows our partner to share our feelings, whether that means to laugh or weep with us. It is this feature of nonverbal communication that is most lacking from phone conversations and electronic communication. Without those emotional cues that we read and come to depend on, it is sometimes a challenge to know just what the person on the other end of the phone is "really saying."

Handling the Ongoing Interaction

Nonverbal communication helps us handle the ongoing interaction by indicating interest and attention. An intent look indicates our interest in the conversa-

tion; a yawn indicates boredom. Posture and eye contact are especially important. Are you leaning toward the person with interest or slumping back, thinking about something else? Do you look at the person who is talking, or are you distracted, glancing at other people as they walk by or watching the clock?

The Importance of Nonverbal Communication

According to psychologist John Gottman (1994), even seemingly simple acts, such as rolling one's eyes in response to a statement or complaint made by a spouse, can convey **contempt,** a feeling that the target of the expression is undesirable. Contempt can be displayed verbally as well through such things as insults, sarcasm, and mockery. Along with contempt, there are three other negative behaviors that indicate particularly troubled and vulnerable relationships. These others are criticism (especially when it is overly harsh), defensiveness, and stonewalling or avoiding. Together, these four behaviors made up Gottman's "four horsemen of the apocalypse," spelling potential for eventual divorce (Gottman 1994). Eventually, Gottman added a fifth—belligerence. Gottman suggested that these are all warning signs of serious risk of eventual divorce (Gottman 1994; Gottman et al. 1998). Conversely, couples who communicate with affection and interest and who maintain humor amid conflict can use such a *positive affect* to diffuse potentially threatening conflict (Gottman et al. 1998).

As you think about Gottman's danger signs, consider how easily they can be expressed and conveyed via nonverbal communication, as well as by things we say to each other. For example, failing to make eye contact is a way of avoiding or stonewalling. The common gesture of raising your hands in front of yourself and "pushing at the air" communicates defensiveness to those you are interacting with; it is as if you were saying "back off." In fact, nonverbally, you *are* saying just that.

Proximity, Eye Contact, and Touch

Three forms of nonverbal communication that have clear importance are proximity, eye contact, and touch.

Proximity

Nearness, in terms of physical space, time, and so on, is referred to as **proximity.** Where we sit or stand in relationship to another person can signify levels of intimacy or the type of relationship. Many of our words conveying emotion relate to proximity, such as feeling "distant" or "close," or being "moved" by someone. We also "make the first move," "move in" on someone else's partner, or "move in together."

In a social situation, the face-to-face distances between people when starting a conversation are clues to how the individuals wish to define the relationship. All cultures have an intermediate distance in face-to-face interactions that are considered neutral. In most cultures, decreasing the distance signifies an invitation to greater intimacy or a threat. Moving away denotes the desire to terminate the interaction. When you stand at an intermediate distance from someone at a party, you send the message that intimacy is not encouraged. If you want to move closer, however, you risk the chance of rejection. Therefore, you must exchange cues, such as laughter or small talk, before moving closer to avoid facing direct rejection. If the person moves farther away during this exchange or, worse, leaves altogether ("Excuse me, I think I see a friend . . ."), he or she is signaling disinterest. But if the person moves closer, there is the "proposal" for greater intimacy. As relationships develop, couples also engage in close gazing into each other's eyes, holding hands, and walking with arms around each other—all of which require close proximity.

But because of cultural differences, there can be misunderstandings. The neutral distance for Latinos, for example, is much closer than for Anglos, who may misinterpret the distance as close (too close for comfort). In social settings, this can lead to problems. As Carlos Sluzki (1982) points out, "A person raised in a non-Latino culture will define as seductive behavior the same behavior that a person raised in a Latin culture defines as socially neutral." Because of the miscue, the Anglo may withdraw or flirt, depending on his or her feelings. If the Anglo flirts, the Latino may respond to what he or she believes is the other's initiation. In addition, the neutral responses of people in cultures that have greater intermediate distances and less overt touching, such as Asian American culture, may be misinterpreted negatively by people with other cultural backgrounds.

Eye Contact

Much can be discovered about a relationship by watching whether, how, and how long people look at each other. Making eye contact with another person, if only for a split second longer than usual, is a signal of in-

terest. Brief and extended glances, in fact, play a significant role in women's expression of initial interest (Moore 1985). (The word *flirting* is derived from the Old English word *fliting*, which means "darting back and forth," as so often occurs when someone flirts with his or her eyes.) When you can't take your eyes off another person, you probably have a strong attraction to him or her. You can often distinguish people in love by their prolonged looking into each other's eyes. In addition to eye contact, dilated pupils may be an indication of sexual interest (or poor lighting).

Research suggests that the amount of eye contact between a couple having a conversation can distinguish between those who have high levels of conflict and those who don't. Those with the greatest degree of agreement have the greatest eye contact with each other (Beier and Sternberg 1977). Those in conflict tend to avoid eye contact (unless it is a daggerlike stare). As with proximity, however, the level of eye contact may differ by culture.

Reflections

Think about your nonverbal communication. In instances where you and another person had significant eye contact, what did the eye contact mean? As you think about touch, what are the different kinds of touch you do? What meanings do you ascribe to the touch you give and the touch you receive?

Touch

A review of the research on touch finds it to be extremely important in human development, health, and sexuality (Hatfield 1994). It is the most basic of all senses; it contains receptors for pleasure and pain, hot and cold, rough and smooth. "Skin is the 'mother sense' and out of it, all the other senses have been derived," writes anthropologist Ashley Montagu (1986). Touch is a life-giving force for infants. If babies are not touched, they may fail to thrive and may even die. We hold hands with small children and those we love. Many of our words for emotion are derived from words referring to physical contact: *attraction, attachment,* and *feeling.* When we are emotionally moved by someone or something, we speak of being "touched."

But touch can also be a violation. A stranger or acquaintance may touch you in a way that is too familiar. Your date or partner may touch you in a manner you don't like or want. Some sexual harassment consists of unwelcome touching.

Touching is a universal part of social interaction, but it varies in both frequency and meaning across cultures and between women and men (Dibiase and Gunnoe 2004). Often, touch has been taken to reflect social dominance. Based largely and initially on research by Nancy Henley in which men were found to touch women more than women touched men, the generalization was drawn that touch is a privilege that higher-status, more socially dominant individuals enjoy over lower-status, more subordinate others.

Extending the issue of touch beyond gender to incorporate social class, professional status, and cultural differences, Henley demonstrated that higher-status individuals were more likely to touch others than to be touched by lower-status individuals. This generalization was further modified some by research that revealed that when individuals were of close but different statuses, the lower-status person often strategically used touch as a means of "making a connection" with the higher-status person. Status differences also determined the *type of touch;* lower-status individuals were more likely to initiate handshakes, and higher-status individuals were more likely to initiate somewhat more intimate touch such as placing a hand on another's shoulder (Dibiase and Gunnoe 2004).

Others have refined the relationship between gender and touch further, showing that women and men use different types of touch (with men touching women more with their hands and women touching men more with other forms of touch) and that gender differences in touch varied by age: among people under 30 years of age, men touched women significantly more than women touched men. This pattern does not appear to hold among older people or among married couples (Dibiase and Gunnoe 2004).

What about culture? Differences surface in a number of interesting ways. For example, people in colder climates use relatively larger distance, and hence relatively less physical contact, when they communicate, whereas people in warmer climates prefer closer distances. Latin Americans are comfortable at a closer range (have smaller personal space zones) than Northern Americans. Middle Eastern, Latin American, and southern European cultures can be considered "high-contact cultures," where people interact at closer distances and touch each other more in social conversations than people from noncontact cultures, such as those of northern Europe, the United States, and Asia (Dibiase and Gunnoe 2004). In so-called high-contact cultures, the kind of touch used in greetings is more intimate, often consisting of hugging or kissing, whereas a firm but more distant handshake is an accepted greeting in noncontact cultures.

Comparing women and men in the United States, Italy, and the Czech Republic, Rosemarie Dibiase and Jaime Gunnoe found that gender differences in touch varied across the three cultures. Although men engaged in more "hand touch" than women and women engaged in more "nonhand touch" in all three cultures, the extent of gender difference was not the same in the three countries observed. Dibiase and Gunnoe report that "only in the Czech Republic did men touch women with their hands significantly more than women touched men with their hands . . . (and) . . . there was a tendency for women to do more nonhand touching than men did. However, there were not significant differences between men and women living in the United States, and there were only trends toward differences in Italy." Only in the Czech Republic were the gender differences in nonhand touching significant (Dibiase and Gunnoe 2004).

Touch can signify more than dominance; it often is a way to convey intimacy, immediacy, and emotional closeness. Touch may well be the most intimate form of nonverbal communication. One researcher (Thayer 1986) writes, "If intimacy is proximity, then nothing comes closer than touch, the most intimate knowledge of another." Touching seems to go "hand in hand" with self-disclosure. Those who touch seem to self-disclose more; touch seems to be an important factor in prompting others to talk more about themselves (Heslin and Alper 1983; Norton 1983).

The amount of contact, from almost imperceptible touches to "hanging all over" each other, helps differentiate lovers from strangers. How and where a person is touched can suggest friendship, intimacy, love, or sexual interest.

Sexual behavior relies above almost all else on touch: the touching of self and others and the touching of hands, faces, chests, arms, necks, legs, and genitals. Sexual behavior is skin contact. In sexual interactions, touch takes precedence over sight, as we close our eyes to caress, kiss, and make love. We shut our eyes to focus better on the sensations aroused by touch; we shut out visual distractions to intensify the tactile experience of sexuality.

The ability to interpret nonverbal communication correctly appears to be an important ingredient in successful relationships. The statement, "What's wrong? *I can tell something is bothering you,*" reveals the ability to read nonverbal clues, such as body language or facial expressions. This ability is especially important

■ *We convey feelings via a variety of nonverbal means—proximity, touch, and eye contact.*

in ethnic groups and cultures that rely heavily on nonverbal expression of feelings, such as Latino and Asian American cultures. Although the value placed on nonverbal expression may vary among groups and cultures, the ability to communicate and understand nonverbally remains important in all cultures. A comparative study of Chinese and American romantic relationships, for example, found that shared nonverbal meanings were important for the success of relationships in both cultures (Gao 1991).

Gender Differences in Communication

The idea that women and men communicate differently has been the subject of much research and writing (Rubin 1983; Tannen 1990; Gray 1993), including best sellers bemoaning our lack of understanding and inabilities to communicate with each other. Gender differences surface whether we examine nonverbal or verbal communication, and they become especially pronounced in cross-sex interaction.

Compared with men's nonverbal communication patterns, women smile more; express a wider range of emotions through their facial expressions; occupy, claim, and control less space; and maintain more eye contact with others with whom they are interacting (Borisoff and Merrill 1985; Lindsey 1997). In their use of language and their styles of speaking, further differences emerge (Lakoff 1975; Tannen 1990; Lindsey 1997). Women use more qualifiers (for example, "It's *sort of* cold out"), use more tag questions ("It's sort of cold out, *don't you think?*"), use a wider variety of intensifiers ("It was *awfully* nice out yesterday; now it's sort of cold out, don't you think?"), and speak in more polite and less insistent tones. Male speech contains fewer words for such things as color, texture, food, relationships, and feelings, but men use more and harsher profanity (Lindsey 1997). In cross-sex interaction, men talk more and interrupt women more than women interrupt men. In same-gender conversation, men disclose less personal information and restrict themselves to safer topics, such as sports, politics, or work (Lindsey 1997).

The male styles of both verbal and nonverbal communication fit more with positions of dominance, women's with positions of subordination. At the same time, women's style is one of cooperation and consensus; thus, it is also situationally appropriate and advantageous to relationship building and maintenance (Tannen 1990; Lindsey 1997). In light of these facts, researchers differ in their interpretations of these gender patterns: those who see women's style as artifacts of subordination versus those who see gender patterns as reflecting difference.

Communication Patterns in Marriage

Communication occupies an important place in marriage. When couples have communication problems they often fear that their marriages are seriously flawed. As shown in a subsequent section, one of the most common complaints of couples seeking therapy is about their communication problems (Burleson and Denton 1997).

Popular Culture

Buying into Mars versus Venus: Popularizing Gender Differences in Intimate Communication

Linguist Deborah Tannen tells this story:

A married couple was in a car when the wife turned to her husband and asked, "Would you like to stop for a coffee?"

"No, thanks," he answered truthfully. So they didn't stop.

The result? The wife, who had indeed wanted to stop, became annoyed because she felt her preference had not been considered. The husband, seeing his wife was angry, became frustrated. Why didn't she just say what she wanted?

To some of you, that story may be familiar, perhaps like one you encountered or witnessed yourself. Furthermore, it reflects a basic reality of communication and, more generally, of gender differences. Remember the discussion in Chapter 4, about the *gender lenses* that fundamentally shape our thinking about gender. Chief among these is *gender polarization*, the idea that there are basic and unavoidable differences between the genders, an idea. If women are expressive, men must be stoic. If men are aggressive and competitive, women must be passive and cooperative. Although there are real gender differences, we exaggerate many of them and fabricate still others.

Communication, especially between spouses or heterosexual intimate partners, is one area in which the idea of the genders as opposites is deeply believed and widely accepted. As we shall see, there *are* gendered styles of both verbal and nonverbal communication, as well as differences in how women and men approach, handle, and attempt to resolve conflict. These differences are not inevitable, although some research indicates that biological differences may help account for certain aspects of communication differences (especially regarding conflict). Furthermore, these differences are categorical ones; there are people whose style of communicating or approach to resolving conflict is more like the "opposite sex." In addition, gender differences are affected by culture and by the specific circumstances in which couples find themselves. Our point for now, is not whether differences exist or how wide they are but how widely we have accepted, even embraced, the notion that women's and men's communication patterns are so different.

One indicator of the extent of popular belief in gender-polarized communication can be found in the appeal of popular and "self-help" books that have addressed this divide. In 1990, Tannen published the hardcover edition of her book, *You Just Don't Understand: Women and Men in Conversation*. The book struck a nerve with readers, spending 8 months as the No. 1 best-selling book on the *New York Times* best seller list and remaining on the list for almost 4 years. Tannen is an accomplished scholar; she has a doctorate degree in linguistics, has authored more than 100 articles and books, and is a faculty member at Georgetown University. *You Just Don't Understand* brought Tannen's scholarly expertise on gender differences in communication to a popular audience. Its appeal was broad and international, as it was a best seller in a number of other countries, including Canada, England, Germany, Brazil, and Holland.

Tannen's thesis was straightforward: because men and women have such different needs and styles of communication, it is almost as though they are from different cultures, struggling to communicate despite speaking different languages. Communication across such differences invites frustration, misunderstanding, and conflict. Tannen (1990) located these communication differences in early socialization:

Little girls create and maintain friendships by exchanging secrets. Women regard conversation as the cornerstone of friendship. A woman expects her husband to be a new and improved version of a best friend. What is important is not the individual subjects that are discussed but the sense of closeness, of a life shared, that emerges when people tell their thoughts, feelings, and impressions. But . . . men don't know what kind of talk women want, and they don't miss it when it isn't there.

Tannen (1990) also raised the possibility that when women feel as though men "aren't really listening to them," they are—they just happen to *listen differently:*

The impression of not listening results from misalignments in the mechanics of conversation . . . the tendency of men to face away can give women the impression they aren't listening even when they are.

Additionally, where females may talk at length about a single topic, males jump from topic to topic, ". . . (a) habit that gives women the impression men aren't listening, especially if they switch to a topic about themselves."

Tannen went on to suggest other reasons for communication-related misunderstandings.

- Men don't make as much "listener noise" as women ("uh-huh," "yeah," and so on), even when they are paying attention. Expecting such reassuring signs of attentiveness, women may misinterpret men's silent attention as

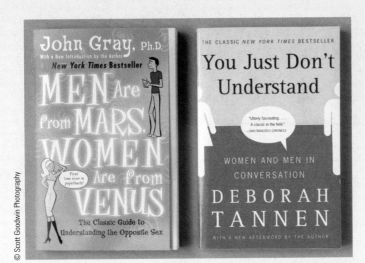

■ *Deborah Tannen's,* You Just Don't Understand..., *and John Gray's* Men Are from Marks, Women Are from Venus, *have sold millions of copies in the United States and worldwide. Each brings the issue of gender differences in communication to a popular audience.*

not paying attention. Meanwhile, men may interpret women's listener noise as overreaction or impatience and perceive women's tendencies to "overlap," (finish each other's sentences in anticipation of what the other is about to say) as intrusive interruption.

■ Women and men expect different things from conversation. When women talk to each other, they expect agreement and support. Many men perceive their "conversational duty" to be to represent the other side of an argument. To women, this is heard and, more importantly, felt as disloyalty. Women do want to see other points of view, but they don't want to feel directly challenged.

■ Women engage in what Tannen calls **rapport talk,** men in **report talk.** To men, language is a means to convey information. To women, talk is a way to build and sustain relationships, and conversations

are occasions to seek and give support and to reach consensus.

■ Women see conversation as a ritual means of establishing and sustaining intimacy. "If Jane tells a problem and June says she has a similar one, they walk away feeling closer to each other." In a relationship like marriage you can share your feelings and thoughts and still be loved. Women's greatest fear is being pushed away. Therefore, they may mistake men's ritual challenges for real attack. However, in men's experience talk maintains independence and status. They are on guard to protect themselves from being put down and pushed around.

Given these differences in how we communicate and what we expect from others, problems are nearly inevitable. But knowing the origin and understanding the motivations behind gender differences, we can come to an understanding of and

begin to try to fix communication problems in marriage. If we come to see communication differences as "cross-cultural" rather than as "right and wrong," or as difference rather than deficiency, it is easier to alter our behavior and our expectations of the other.

Tannen followed the success of *You Just Don't Understand* with books on gendered communication in the workplace, family communication between adults, and most recently communication between mothers and daughters. Her books have raised awareness and shaped the way we think about dynamics of family communication, especially across such divides as gender and generation.

Further evidence of how deeply accepted and widely embraced the idea of gender polarization is within the United States can be seen in the phenomenal success enjoyed by John Gray's *Men are From Mars, Women are From Venus,* and its many follow-ups. Most of Gray's books build off of the same clever idea first introduced in the 1992 best seller *Men are From Mars, Women are From Venus,* where Gray takes the issue of gender differences in a distinctive metaphorical direction. Instead of cross-cultural counterparts, women and men are portrayed as inhabitants of different planets, and—hence—as worlds apart.

As to Gray's Martian and Venutian ways of communicating, they clearly are meant to reflect observed patterns more typical of men and women. For example, when women/Venutians complain about something, they want to be heard and understood. When men/Martians voice feelings, they want action and solutions. Offering men "understanding" or women "solutions" will generate frustration and lead to problems.

Gray and Tannen have different approaches and backgrounds. Gray

Continues

INTERSTELLAR TRAVEL

http://www.cartoonstock.com

"I'm going to Venus. He's going to Mars."

lacks the academic background and scholarly approach that Tannen brought to *You Just Don't Understand.* As *Time* magazine reported in 1997, where Tannen's book has numerous footnotes to source material on gender and communication, Gray offers 1-800 numbers from which readers can order *Mars and Venus* products (including audio and videotapes, a CD-ROM, Mars and Venus vacations) (Gleick 1997).

Both Tannen and Gray have been the recipients of criticism, accused of overgeneralization (Shweder 1994); failure to look at larger social, cultural, or political contexts; or one-sidedness (Tannen accused by some of being "anti-male" and by others of being too soft on men; Gray labeled by some a misogynist with a sexist biases in his characterization of communication). In Gray's case, there have also been repeated questions to his claims about his academic credentials and training. He has also been accused of being, in his own words, a "watered-down version of Deborah Tannen" (Gleick 1997). But despite any criticism either may have received, they have been extremely influential in shaping how many people think about and act in relationships

There has been an explosion of research on premarital and marital communication in the last decade. Researchers are finding significant correlations between the nature of communication and satisfaction, as well as finding differences in male versus female communication patterns in marriage.

Premarital Communication Patterns and Marital Satisfaction

"Drop dead, you creep!" is hardly the thing someone would want to say when trying to resolve a disagreement in a dating relationship. But it may be an important clue as to whether such a couple should marry. Many couples who communicate poorly before marriage are likely to continue the same way after marriage, and the result can be disastrous for future marital happiness. Researchers have found that how well a couple communicates before marriage can be an important predictor of later marital satisfaction (Cate and Lloyd 1992). If communication is poor before marriage, it is not likely to significantly improve after marriage—at least not without a good deal of effort and help.

For example, *self-disclosure*—the revelation of our own deeply personal information—before or soon after marriage is related to relationship satisfaction later (see Chapter 5). In one study (Surra, Arizzi, and Asmussen 1988), men and women were interviewed shortly after marriage and 4 years later. The researchers found that self-disclosure was an important factor for increasing each other's commitment later. Talking about your deepest feelings and revealing yourself to your partner builds bonds of trust that help cement a marriage.

Whether a couple's interactions are basically negative or positive can also predict later marital satisfaction. In a notable experiment by John Markham (1979), 14 premarital couples were evaluated using "table talk," sitting around a table and simply engaging in conversation. Each couple talked about various topics. Using an electronic device, each partner electronically recorded whether the message was positive or negative. Markham found that the negativity or positivity of the couple's communication pattern barely affected their marital satisfaction during their first year. This protective quality of the first year is known as the **honeymoon effect**—which means you can say almost anything during the first year and it will not seriously affect marriage (Huston, McHale, and Crouter 1986). But after the first year, couples with negative premarital communication patterns were less satisfied than those with positive communication patterns. A later study (Julien, Markman, and Lindahl 1989) found that those premarital couples who responded more to each other's positive communication than to each other's negative communication were more satisfied in marriage 4 years later.

Cohabitation and Later Marital Communication

As shown in the next chapter, researchers have revealed a cohabitation effect on marriage. Specifically, couples who live together before marrying are more likely to separate and divorce than couples who don't live together before marriage. That may seem counterintuitive. Wouldn't couples who live together first find it easier to adjust to marriage? Doesn't cohabitation weed out the unsuccessful matches before marriage? In Chapter 8, we consider the range of explanations for this cohabitation effect. Here, we simply look at how communication patterns might contribute to later marital failure.

Among the possible explanations for the cohabitation effect, Catherine Cohan and Stacey Kleinbaum (2002) hypothesized that spouses who live together before marrying display more negative problem solving and support behavior compared with their counterparts who marry without first living together. Why would cohabitation lead to poorer marital communication? Cohan and Kleinbaum suggest three possible reasons:

1. Couples who live together come from backgrounds that may predispose them to poorer communication abilities. Compared with couples who don't cohabit, cohabitants tend to be younger, less religious, and more likely to come from divorced homes. Cohan and Kleinbaum point out that this translates into them being less mature, less traditional, and less likely to have had good parental role models for effective communication.

2. People who cohabit may be more accepting of divorce and less committed to marriage. Thus, they may expend less effort or energy developing good marital communication skills because they are less sure they will stay married.

3. Cohabitation is associated with factors such as alcohol use, infidelity, and lower marital satisfaction, which in turn are correlated with less effective communication.

In studying 92 couples who were in their first 2 years of marriage, Cohan and Kleinbaum found that premarital cohabitation was associated with poorer marital communication. Couples with one or more cohabitation experiences displayed poorer, more divisive, and more destructive communication behaviors than did couples with no prior cohabitation experience (Cohan and Kleinbaum 2002).

Marital Communication Patterns and Satisfaction

Researchers have found a number of patterns that distinguish the communication patterns in satisfied and dissatisfied marriages (Gottman 1995; Hendrick 1981; Noller and Fitzpatrick 1991; Schaap, Buunk, and Kerkstra 1988). Couples in satisfied marriages tend to have the following characteristics:

- Willingness to accept conflict but to engage in conflict in nondestructive ways.

- Less frequent conflict and less time spent in conflict. Both satisfied and unsatisfied couples, however, experience conflicts about the same topics, especially about communication, sex, and personality characteristics.

- The ability to disclose or reveal private thoughts and feelings, especially positive ones, to a partner. Dissatisfied spouses tend to disclose mostly negative thoughts to their partners.

- Expression by both partners of equal levels of affection, such as tenderness, words of love, and touch.

- More time spent talking, discussing personal topics, and expressing feelings in positive ways.

- The ability to encode (send) verbal and nonverbal messages accurately and to decode (understand) such messages accurately. This is especially important for husbands. Unhappy partners may actually decode the messages of strangers more accurately than those from their partners.

■ *Touch is one of our primary means of communication. It conveys intimacy, immediacy, and emotional closeness.*

Gender Differences in Partner Communication

In addition to overall gender differences in communication noted earlier, researchers have identified several gender differences in how spouses communicate (Klinetob and Smith 1996; Noller and Fitzpatrick 1991; Thompson and Walker 1989).

First, wives tend to *send clearer messages* to their husbands than their husbands send to them. Wives are often more sensitive and responsive to their husbands' messages, both during conversation and during conflict. They are more likely to reply to either positive messages ("You look great") or negative messages ("You look awful") than are their husbands, who may not reply.

Second, wives tend to *give more positive or negative messages;* they tend to smile or laugh when they send messages, and they send fewer clearly neutral messages. Husbands' neutral responses make it more difficult for wives to decode what their partners are trying to say. If a wife asks her husband if they should go to dinner or see a movie and he gives a neutral response, such as, "Whatever," does he really not care, or is he pretending he doesn't care to avoid possible conflict?

Third, although communication differences in arguments between husbands and wives are usually small, they nevertheless follow a typical pattern. Wives tend to *set the emotional tone* of an argument. They escalate conflict with negative verbal and nonverbal messages ("Don't give me that!") or deescalate arguments by setting an atmosphere of agreement ("I understand your feelings"). Husbands' inputs are less important in setting the climate for resolving or escalating conflicts. Wives tend to *use emotional appeals and threats more than husbands,* who tend to reason, seek conciliation, and find ways to postpone or end an argument. A wife is more likely to ask, "Don't you love me?" whereas a husband is more likely to say, "Be reasonable."

A prominent type of marital communication is referred to as **demand–withdraw communication**—a pattern in which one spouse makes an effort to engage the other spouse in a discussion of some issue of importance. The spouse raising the issue may criticize, complain, or suggest a need for change in his or her spouse's behavior. The other spouse, in response to such overtures, withdraws by either leaving the discussion, failing to reply, or changing the subject (Klinetob and Smith 1996).

In seeking change, the person making the demand is in a potentially vulnerable position and has less power than the person withdrawing from the interaction. The latter can choose to change or not. By withdrawing, she or he maintains the status quo. Withdrawal has other consequences. It keeps the conflict from escalating but may curtail needed communication and prevent necessary relationship adjustment (Sagrestano, Heavey, and Christensen 1999).

The demand–withdraw pattern has been found by researchers to be associated with gender. In 60% of couples, wives "demand" and husbands "withdraw." In 30% of couples, these roles are reversed. In the remaining 10%, spouses demand and withdraw about equally (Klinetob and Smith 1996). Researchers have considered a variety of explanations for the more common gender differences in demanding and withdrawing, including psychological, biological, and structural factors (Christensen and Heavey 1990). Research conducted by Nadya Klinetob and David Smith suggests that the demand and withdraw roles vary according to whose issue is being discussed: "During discussions of a wife-generated topic, she was the demander and her husband withdrew. During discussions of a husband-generated topic, he demanded and she withdrew" (1996, 954). They further suggest that because marriage relationships often favor husbands, husbands will be less likely to bring up issues for discussion, because the relationship *as is* is more acceptable to them. On the other hand, because wives may be more discontented with aspects of the relationship and bring them up for discussion, they more often occupy the "demand" position (Klinetob and Smith 1996).

Although there are certainly socialization influences behind these gender differences, biologically based gender differences may also come into play. Men and women may have different physiological responses to conflict, and these may help produce the familiar male withdrawal that is part of the female demand–male withdraw pattern of communication. With greater tolerance for physiological arousal, women can maintain the kinds of high levels of engagement that conflict contains. John Gottman and Robert Levenson (1992) reported that compared to women, men show different physiological reactions—more rapid heart beat, quickened respiration, release of higher level of epinephrine in their endocrine systems—to disagreements. To men, this arousal is highly unpleasant; thus, they act to avoid it by withdrawing from the conflict. Withdrawal may be a means of avoiding these reactions (Gottman and Levenson 1992; Levenson Carstensen, and Gottman 1994).

Although the demand–withdraw pattern is fairly common, it is not a particularly healthy style of communication and conflict resolution. It is associated with less marital satisfaction and higher likelihood of relationship failure (Regan 2003). It also may be a predictor of violence within the couple relationship, especially among couples with high levels of husband demand–wife withdraw. Such couples are more likely to experience violence than are couples who have low levels of this pattern. Conversely, the more common wife demand–husband withdraw pattern may have the consequence of preventing conflict from escalating into violence (Sagrestano, Heavey, and Christensen 1999). Although both patterns were associated with wives' verbal aggression, and with husbands' verbal aggression and violence, only husband demand–wife withdraw interaction was significantly related to women's use of violence (Sagrestano, Heavey, and Christensen 1999).

Sexual Communication

To have a satisfying sexual relationship, a couple must be able to communicate effectively with each other about expectations, needs, attitudes, and preferences (Regan 2003). Both the frequency with which couples engage in sexual relations and the quality of their involvement depend on such communication.

Among heterosexuals, in both married and cohabiting relationships, women and men often follow sexual scripts that leave the initiation of sex (that is, the communication of desire and interest) to men, with women then in a position of accepting or refusing men's overtures. Reviewing the literature on sexual communication, Pamela Regan observes that regardless of who takes the role of initiating, the efforts are usually met with positive responses. Both attempts to initiate and positive responses are rarely communicated explicitly and verbally (Regan 2003, 84):

> A person who desires sexual activity might turn on the radio to a romantic soft rock station, pour his or her partner a glass of wine, and glance suggestively in the direction of the bedroom. The partner . . . might smile, put down his or her book, and engage in other nonverbal behaviors that continue the sexual interaction without explicitly acknowledging acceptance.

Interestingly, lack of interest or refusal of sexual initiations is communicated directly and verbally (for example, "Not tonight, I have a lot of work to do"). By framing refusal in terms of some kind of account, the refusing partner allows the rejected partner to save face (Regan 2003).

Effective sexual communication may be difficult, but it is important if couples hope to construct and keep mutually satisfying sexual relationships. We must trust our partner enough to express our feelings about sexual needs, desires, and dislikes, and we must be able to hear the same from our partner without feeling judgmental or defensive (Regan 2003).

Problems in Communication

Studies suggest that poor communication skills precede the onset of marital problems (Gottman 1994; Markman 1981; Markman et al. 1987). Even family violence has been seen by some as the consequence of deficiencies in the ability to communicate (Burleson and Denton 1997).

Although we cannot *not* communicate, we can enhance the quality of our communication so that we can understand each other and enhance our relationships. We can learn to communicate constructively rather than destructively. What follows, we hope, will help you develop good communication skills so that your relationships are mutually rewarding.

Topic-Related Difficulty

Some communication problems are topic dependent more than individual or relationship based. By that we mean that some topics are more difficult for couples to talk about. As Keith Sanford (2003, 98) states, "it would seem easier to resolve a disagreement about what to do on a Friday night than a disagreement about whether one spouse is having an affair." If some topics are more difficult to discuss than others, couples are likely to display poorer communication when discussing those topics (98):

> If a couple is coping with a highly difficult, unresolved topic (for example, insults) . . . they might be likely to use poor communication in all their conflicts, whether the specific topic being discussed is easy or difficult for most couples.

In an attempt to determine the difficulty of different topics, Sanford gave a sample of 12 licensed Ph.D. psychologists a list of topics and asked them to provide their best guess as to how difficult each topic is for couples to discuss and resolve (from 1 = Extremely easy to 5 = Extremely difficult). The list consisted of 24 topics, generated from a sample of 37 couples who were asked to identify two unresolved issues in their relationships. The 10 topics to which the psychologists assigned the highest "difficulty scores" are listed in Table 7.1.

Other familiar relationship trouble spots and their assigned ratings include childrearing issues (3.42), finances (3.42), lack of listening (3.08), household tasks (2.33), and not showing sufficient appreciation (2.25). Interestingly, as determinants of communication behavior during attempts at problem solving, the difficulty of a topic showed only a small to negligible effect. Thus, although the scores demonstrate differences in the sensitivity contained in different marital issues, these differences do not, themselves, appear to determine how couples communicate about them (Sanford 2003).

Communication Styles in Miscommunication

Virginia Satir noted in *Peoplemaking* (1972), her classic work on family communication, that people can be classified according to four styles of *miscommunication:*

Table 7.1 ■ Ten Topics That Are Most Difficult for Couples to Discuss

Topic	Difficulty Score*
Relationship doubts (possibility of divorce)	4.58
Disrespectful behavior (lying, rudeness)	4.50
Extramarital intimacy boundary issues (use of pornography, jealousy)	4.42
Excessive or inappropriate display of anger (yelling, attacking)	4.25
Sexual interaction	4.17
Lack of communication (refusal to talk)	4.00
In-laws and extended family	3.83
Confusing, erratic, emotional behavior	3.75
Criticism	3.58
Poor communication skills (being unclear or hard to understand)	3.46

*1 = Extremely easy; 5 = Extremely difficult.

- *Placaters.* Always agreeable, placaters are passive, speak in an ingratiating manner, and act helpless. If a partner wants to make love when a placater does not, the placater will not refuse because that might cause a scene. No one knows what placaters really want or feel—and they themselves often do not know.

- *Blamers.* Acting superior, blamers are tense, often angry, and gesture by pointing. Inside, they feel weak and want to hide this from everyone (including themselves). If a blamer runs short of money, the partner is the one who spent it; if a child is conceived by accident, the partner should have used contraception. The blamer does not listen and always tries to escape responsibility.

- *Computers.* Correct and reasonable, computers show only printouts, not feelings (which they consider dangerous). "If one takes careful note of my increasing heartbeat," a computer may tonelessly say, "one must be forced to come to the conclusion that I'm angry." The partner who is interfacing, also a computer, does not change expression and replies, "That's interesting."

- *Distractors.* Acting frenetic and seldom saying anything relevant, distractors flit about in word and deed. Inside, they feel lonely and out of place. In difficult situations, distractors light cigarettes and

How partners express and handle conflict verbally, as well as nonverbally, says much about the direction in which the relationship is heading.

talk about school, politics, business—anything to avoid discussing relevant feelings. If a partner wants to discuss something serious, a distracter changes the subject.

Why People Might Communicate Ineffectively

We can learn to communicate, but it is not always easy. Traditional male gender roles, for example, work against the idea of expressing feelings. This role calls for men to be strong and silent, to ride off into the sunset alone. If men talk, they talk about things—cars, politics, sports, work, money—but not about feelings. Also, both men and women may have personal reasons for not expressing their feelings. They may have strong feelings of inadequacy: "If you really knew what I was like, you wouldn't like me." They may feel ashamed of, or guilty about, their feelings: "Sometimes I feel attracted to other people, and it makes me feel guilty because I should only be attracted to you." They may feel vulnerable: "If I told you my real feelings, you might hurt me." They may be frightened of their feelings: "If I expressed my anger, it would destroy you." Finally, people may not communicate because they are fearful that their feelings and desires will create conflict: "If I told you how I felt, you would get angry."

Obstacles to Self-Awareness

Before we can communicate with others, we must first know how we feel. Although feelings are valuable guides for actions, we often place obstacles in the way of expressing them. First, we suppress "unacceptable" feelings, especially feelings such as anger, hurt, and jealousy. After a while, we may not even consciously experience them. Second, we deny our feelings. If we are feeling hurt and our partner looks at our pained expression and asks us what we're feeling, we may reply, "Nothing." We may actually feel nothing because we have anesthetized our feelings. Third, we project our feelings. Instead of recognizing that we are jealous, we may accuse our partner of being jealous; instead of feeling hurt, we may say our partner is hurt.

Becoming aware of ourselves requires us to become aware of our feelings. Perhaps the first step toward this self-awareness is realizing that feelings are simply emotional states—they are neither good nor bad in themselves. As feelings, however, they need to be felt,

Reflections

Do you find that any of the stated styles of miscommunication characterize your own communication patterns? Your partner's style? What happens if you and your partner have similar styles? Different styles?

© Tony Freeman/PhotoEdit

Exploring Diversity Ethnicity and Communication

Different ethnic groups within our culture have different language patterns that affect the way they communicate. African Americans, for instance, have distinct communication patterns (Hecht, Collier, and Ribeau 1993). Language and expressive patterns are characterized by emotional vitality, realness, and valuing direct experience, among other things (White and Parham 1990). Emotional vitality is expressed in the animated use of words. Realness refers to "telling it like it is" using concrete, nonabstract words. Direct experience is valued because "there is no substitute in the Black ethos for actual experience gained in the course of living" (White and Parham 1990). "Mother wit"—practical or experiential knowledge—may be valued over knowledge gained from books or lectures.

Latinos, especially traditional Latinos, assume that intimate feelings will not be discussed openly (Guerrero Pavich 1986). One researcher (Falicov 1982) writes this about Mexican Americans: "Ideally, there should be a certain formality in the relationship between spouses. No deep intimacy or intense conflict is expected. Respect, consideration, and curtailment of anger or hostility are highly valued." Confrontations are to be avoided; negative feelings are not to be expressed. As a consequence, nonverbal communication is especially important. Women are expected to read men's behavior for clues to their feelings and for discovering what is acceptable. Because confrontations are unacceptable, secrets are important. Secrets are shared between friends but not between partners.

Asian American ethnic groups are less individualistic than the dominant American culture. Whereas the dominant culture views the ideal individual as self-reliant and self-sufficient, Asian American subcultures are more relationally oriented. Researchers Steve Shon and Davis Ja (1982) note the following about Asian Americans:

They emphasize that individuals are the products of their relationship to nature and other people. Thus, heavy emphasis is placed on their relationship with other people, generally with the aim of maintaining harmony through proper conduct and attitudes.

Asian Americans are less verbal and expressive in their interactions than are both African Americans and Caucasians; instead, they rely to a greater degree on indirect and nonverbal communication, such as silence and the avoidance of eye contact as signs of respect (Del Carmen 1990). Because harmonious relationships are highly valued, Asian Americans tend to avoid direct confrontation if possible. Japanese Americans, for example, "value implicit, nonverbal intuitive communication over explicit, verbal, and rational exchange of information" (Del Carmen 1990). To avoid conflict, verbal communication is often indirect or ambiguous; it skirts around issues instead of confronting them. As a consequence, in interactions Asian Americans rely on the other person to interpret the meaning of a conversation or nonverbal clues.

---

whether they are warm or cold, pleasurable or painful. They do not necessarily need to be acted on or expressed. It is the acting out that holds the potential for problems or hurt.

Problems in Self-Disclosure

Self-disclosure creates the environment for mutual understanding (Derlega et al. 1993). We live much of our lives playing roles—as student and worker, husband or wife, son or daughter. We live and act these roles conventionally. They do not necessarily reflect our deepest selves. If we pretend that we are only these roles and ignore our deepest selves, we have taken the path toward loneliness and isolation. We may reach a point at which we no longer know who we are. In the process of revealing ourselves to others, we discover who we are. In the process of our sharing, others share themselves with us. Self-disclosure is reciprocal.

Keeping Closed

Having been taught to be strong, men may be more reluctant to express feelings of weakness or tenderness than women. Many women find it easier to disclose their feelings, perhaps because from earliest childhood they are more often encouraged to express them (Notarius and Johnson 1982).

If distinct differences exist, they can drive wedges between men and women. One sex does not understand the other. The differences may plague a marriage until neither partner knows what the other wants; sometimes partners don't even know what they want for themselves. Sometimes, what is missing is the intimacy that comes from self-disclosure. People live together, or are married, but they feel lonely. There is no contact, and the loneliest loneliness is to feel alone with someone with whom we want to feel close.

How Much Openness?

Can too much openness and honesty be harmful to a relationship? How much should intimates reveal to each other? Some studies suggest that less marital satisfaction results if partners have too little *or too much* disclosure; a happy medium offers security, stability, and safety. But a review of studies on the relationship between communication and marital satisfaction finds that a linear model of communication is more closely related to marital satisfaction than the too little–too much curvilinear model (Boland and Follingstad 1987). In the linear model of communication, the greater the self-disclosure, the greater the marital satisfaction, provided that the couple is highly committed to the relationship and willing to take the risks of high levels of intimacy. High self-disclosure can be a highly charged undertaking. Studies suggest that high levels of negativity are related to marital distress (Noller and Fitzpatrick 1991). It is not clear whether the negativity reflects the marital distress or causes it. Most

likely, the two interact and compound each other's effects.

Research by Brant Burleson and Wayne Denton suggests that the relationship between communication skill and marital success and satisfaction is "quite complex" (1997, 889). In a study of 60 couples, the researchers explored the importance of four communication skills in determining marital satisfaction:

- *Communication effectiveness:* producing messages that have their intended effect
- *Perceptual accuracy:* correctly understanding the intentions underlying a message
- *Predictive accuracy:* accurately anticipating the effect of the message on another
- *Interpersonal cognitive complexity:* the capacity to process social information

Prior research had indicated that each of the preceding skills were important in differentiating satisfied from dissatisfied couples or nondistressed from distressed couples. Based on their research, Burleson and Denton suggest that *communication skill* may not adequately explain levels of distress or dissatisfaction. The intentions and feelings being communicated were more important factors separating distressed from nondistressed couples. Spouses in distressed couples had "more negative intentions" toward each other. "The negative communication behaviors frequently

■ *A pivotal aspect of effective communication, self-disclosure is reciprocal.*

© Laurie DeVault Photography

observed in distressed spouses may result more from ill will than poor skill" (1997, 897). Burleson and Denton also observe that good communication skills can worsen marital relationships when spouses have "negative intentions toward one another" (1997, 900).

"Can I Trust You?"

When we talk about intimate relationships, among the words that most often pop up are *love* and *trust*. As shown in the discussion of love in Chapter 5, trust is an important part of love. But what is trust? **Trust** is the belief in the reliability and integrity of a person.

When a person says, "Trust me," he or she is asking for something that does not easily occur. For trust to develop, three conditions must exist (Book et al. 1980). First, a relationship has to exist and have the likelihood of continuing. We generally do not trust strangers or people we have just met, especially with information that makes us vulnerable, such as our sexual anxieties. We trust people with whom we have a significant relationship.

Second, we must believe we are able to predict how the person will behave. If we are married or in a committed relationship, we trust that our partner will not do something that will hurt us, such as having an affair. If we discover that our partner is involved in an affair, we often speak of our trust being violated or destroyed. If trust is destroyed in this case, it is because the *predictability* of sexual exclusiveness is no longer there.

Third, the person must have other acceptable options available to him or her. If we were marooned on a desert island alone with our partner, he or she would have no choice but to be sexually monogamous. But if a third person, who was sexually attractive to our partner, swam ashore a year later, then our partner would have an alternative. Our partner would then have a choice of being sexually exclusive or nonexclusive; his or her behavior would then be evidence of trustworthiness—or the lack of it.

Matter of Fact

The happiest couples are those who balance autonomy with intimacy and negotiate personal and couple boundaries through supportive communication (Scarf 1995).

Trust is critical to communication in close relationships for two reasons (Book et al. 1980). First, the degree to which you trust a person influences the way you are likely to interpret ambiguous or unexpected messages. If your partner says he or she wants to study alone tonight, you are likely to take the statement at face value if you have a high trust level. But if you have a low trust level, you may believe your partner is going to meet someone else while you are studying in the library. Second, the degree to which we trust someone influences the extent of our self-disclosure. Revealing our inner selves—which is vital to closeness—makes a person vulnerable and thus requires trust. A person will not self-disclose if he or she believes that the information may be misused—for example, by a partner who resorts to mocking behavior or revealing a secret.

Trust in personal relationships has both a behavioral and a motivational component (Book et al. 1980). The behavioral component refers to the probability that a person will act in a trustworthy manner. The motivational component refers to the reasons a person engages in trustworthy actions. Whereas the behavioral element is important in all types of relationships, the motivational element is important in close relationships. One has to be trustworthy for the "right" reasons. As long as you trust your mechanic to charge you fairly for rebuilding your car's engine, you don't care why he or she is trustworthy. But you do care why your partner is trustworthy. For example, you want your partner to be sexually exclusive to you because he or she loves you or is attracted to you. Being faithful because of duty or because your partner can't find anyone better is the wrong motivation. Disagreements about the motivational bases for trust are often a source of conflict. "I want you because you love me, not because you need me" or "You don't really love me; you're just saying that because you want sex" are typical examples of conflict about motivation.

The Importance of Feedback

Self-disclosure is reciprocal. If we self-disclose, we expect our partner to self-disclose as well. As we self-disclose, we build trust; as we withhold self-disclosure, we erode trust. To withhold ourselves is to imply that we don't trust the other person, and if we don't, he or she will not trust us.

A critical element in communication is **feedback,** the ongoing process in which participants and their messages create a given result and are subsequently modified by the result (see Figure 7.1). If someone

Figure 7.1 ■ **Communication Loop**

In successful communication, feedback between the sender and the receiver ensures that both understand (or are trying to understand) what is being communicated. For communication to be clear, the message and the intent behind the message must be congruent. Nonverbal and verbal components must also support the intended message. Verbal aspects of communication include not only language and word choice but also characteristics such as tone, volume, pitch, rate, and periods of silence.

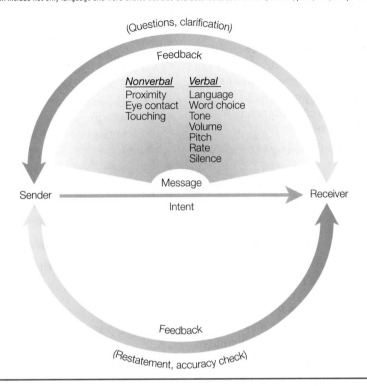

self-discloses to us, we need to respond to his or her self-disclosure. The purpose of feedback is to provide constructive information to increase self-awareness of the consequences of our behaviors toward each other.

If your partner discloses to you his or her doubts about your relationship, for example, you can respond in a number of ways:

- You can remain silent. Silence, however, is generally a negative response, perhaps as powerful as saying outright that you do not want your partner to self-disclose this type of information.

- You can respond angrily, which may convey the message to your partner that self-disclosing will lead to arguments rather than understanding and possible change.

- You can remain indifferent, responding neither negatively nor positively to your partner's self-disclosure.

- You can acknowledge your partner's feelings as being valid (rather than right or wrong) and dis-

close how you feel in response to his or her statement. This acknowledgment and response is constructive feedback. It may or may not remove your partner's doubts, but it is at least constructive in that it opens the possibility for change, whereas silence, anger, and indifference do not.

Some guidelines, developed by David Johnston for the Minnesota Peer Program, may help you engage in dialogue and feedback with your partner:

1. *Focus on "I" statements.* An "I" statement is a statement about your feelings: "I feel annoyed when you leave your dirty dishes on the living room floor." "You" statements tell another person how he or she is, feels, or thinks: "You are so irresponsible. You're always leaving your dirty dishes on the living room floor." "You" statements are often blaming or accusatory. Because "I" messages don't carry blame, the recipient is less likely to be defensive or resentful.

2. *Focus on behavior rather than the person.* If you focus on a person's behavior rather than on the person,

you are more likely to secure change. A person can change behaviors but not himself or herself. If you want your partner to wash his or her dirty dishes, say, "I would like you to wash your dirty dishes; it bothers me when I see them gathering mold on the living room floor." This statement focuses on behavior that can be changed. If you say, "You are such a slob; you never clean up after yourself," then you are attacking the person. He or she is likely to respond defensively: "I am not a slob. Talk about slobs, how about when you left your clothes lying in the bathroom for a week?"

3. *Focus on observations rather than inferences or judgments.* Focus your feedback on what you actually observe rather than on what you think the behavior means. "There is a towering pile of your dishes in the living room" is an observation. "You don't really care about how I feel because you are always leaving your dirty dishes around the house" is an inference that a partner's dirty dishes indicate a lack of regard. The inference moves the discussion from the dishes to the partner's caring. The question "What kind of person would leave dirty dishes for me to clean up?" implies a judgment: only a morally depraved person would leave dirty dishes around.

4. *Focus on observations based on a continuum.* Behaviors fall on a continuum. Your partner doesn't *always* do a particular thing. When you say that he or she does something sometimes or even most of the time, you are measuring behavior. If you say that your partner always does something, you are distorting reality. For example, there were probably times (however rare) when your partner picked up the dirty dishes. "Last week I picked up your dirty dishes three times" is a measured statement. "I always pick up your dirty dishes" is an exaggeration that will probably provoke a hostile response.

5. *Focus on sharing ideas or offering alternatives rather than giving advice.* No one likes being told what to do. Unsolicited advice often produces anger or resentment because advice implies that you know more about what a person needs to do than the other person does. Advice implies a lack of freedom or respect. By sharing ideas and offering alternatives, however, you give the other person the freedom to decide based on his or her own perceptions and goals. "You need to put away your dishes immediately after you are done with them" is advice. To offer alternatives, you might say, "Having to walk around your dirty dishes bothers me. What are the

alternatives other than my watching my step? Maybe you could put them away after you finish eating, clean them up before I get home, or eat in the kitchen. What do you think?"

6. *Focus the value of a response to the recipient.* If your partner says something that upsets you, your initial response may be to lash back. A cathartic response may make you feel better for the time being, but it may not be useful for your partner. If, for example, your partner admits lying to you, you can respond with rage and accusations, or you can express hurt and try to find out why he or she didn't tell you the truth.

7. *Focus on the amount the recipient can process.* Don't overload your partner with your response. Your partner's disclosure may touch deep, pent-up feelings in you, but he or she may not be able to comprehend all that you say. If you respond to your partner's revelation of doubts by listing all doubts you have ever experienced about yourself, your relationship, and relationships in general, you may overwhelm your partner.

8. *Focus on responding at an appropriate time and place.* Choose a time when you are not likely to be interrupted. Turn the television off and the phone answering machine on. Also, choose a time that is relatively stress free. Talking about something of great importance just before an exam or a business meeting is likely to sabotage any attempt at communication. Finally, choose a place that will provide privacy; don't start an important conversation if you are worried about people overhearing or interrupting you. A dormitory lounge during the soaps, Grand Central Station, a kitchen teeming with kids, or a car full of friends is an inappropriate place.

Mutual Affirmation

Good communication in an intimate relationship involves **mutual affirmation,** which includes three elements: (1) mutual acceptance, (2) liking each other, and (3) expressing liking in both words and actions. Mutual acceptance consists of people accepting each other as they are, not as they would like each other to be. People are who they are, and they are not likely to change in fundamental ways without a tremendous amount of personal effort, as well as a considerable passage of time. The belief that an insensitive partner will somehow magically become sensitive after marriage, for example, is an invitation to disappointment and divorce.

If you accept people as they are, you can like them for their unique qualities. Liking someone is somewhat different from being romantically involved. It is not rare for people to dislike those with whom they are romantically linked.

We also need to express our feelings of warmth, affection, and love. To a partner, unexpressed words, actions, thoughts, kindnesses, deeds, touches, caresses, and kisses can be the same as nonexistent or unfelt ones. "You know that I love you" without the expressions of love is a meaningless statement. A simple rule of thumb for communicating love is: If you love, show love.

Mutual affirmation entails our telling others that we like them for who they are, that we appreciate the little things, as well as the big things, that they do. Think about how often you say to your partner, your parents, or your children, "I like you," "I love you," "I appreciate your doing the dishes," or "I like your smile." Affirmations are often most common during dating or the early stages of marriage or living together. As you get to know a person better, you may begin noting things that annoy you or are different from you. Acceptance turns into negation and criticism: "You're selfish," "Stop bugging me," "You talk too much," or "Why don't you clean up after yourself?"

If you have a lot of negatives in your interactions, don't feel too bad. Many of our negations are habitual. When we were children, our parents may have been negating: "Don't leave the door open," "Why can't you get better grades?" "Stand straight and pull in your stomach." How often did they affirm? Once you become aware that negations are often automatic, you can change them. Because negative communication is a learned behavior, you can unlearn it. One way is to make the decision consciously to affirm what you like; too often we take the good for granted and feel compelled to point out only the bad.

Power, Conflict, and Intimacy

Although we may find it unusual to think about family life in these terms, day-to-day family life is highly politicized. By that we mean that the politics of family life—who has more power, who makes the decisions, who does what—can be every bit as complex and explosive as politics at the national level. Like other groups, families possess structures of power. As used here, **power** is the ability or potential ability to influence another person or group, to get people to do what you want them to do, whether they want to or not. Most of the time, we are not aware of the power aspects of our intimate relationships. We may even deny the existence of power differences because we want to believe that intimate relationships are based on love alone. Furthermore, the exercise of power is often subtle. When we think of power, we tend to think of coercion or force; as we show here, however, marital power takes many forms and is often experienced as neither coercion nor force. A final reason we are not always aware of power is that power is not constantly exercised. It comes into play only when an issue is important to both people and they have conflicting goals.

As a concept, power in marital and other couple relationships has been said to consist of power bases, processes, and outcomes. *Power bases* are the economic and personal assets (such as income, economic independence, commitment, and both physical and psychological aggression) that comprise the source of one partner's control over the other. *Power processes* are the "interactional techniques" or methods partners or spouses use to try to gain control over the relationship, the partner, or both, such as persuasion, problem solving, or demandingness. *Power outcomes* can be observed in such things as who has the final say and determines—or potentially could determine and control—the outcome of attempted decision making (Byrne, Carr, and Clark 2004; Sagrestano, Heavey, and Christensen 1999).

Power and Intimacy

The problem with power imbalances and the blatant use of power is the negative effects they have on intimacy. If partners are not equal, self-disclosure may be inhibited, especially if the powerful person believes his or her power will be lessened by sharing feelings (Glazer-Malbin 1975). Genuine intimacy appears to require equality in power relationships. Decision making in the happiest marriages seems to be based not on coercion or tit for tat but on caring, mutuality, and respect for each other. Women or men who feel vulnerable to their mates may withhold feelings or pretend to feel what they do not. Unequal power in marriage may encourage power politics. Each partner may struggle with the other to keep or gain power.

It is not easy to change unequal power relationships after they become embedded in the overall structure of a relationship; yet they can be changed. Talking, understanding, and negotiating are the best approaches. Still, in attempting changes, a person may risk estrangement or the breakup of a relationship. He or she must weigh the possible gains against the possible losses in deciding whether change is worth the risk.

Sources of Marital Power

Traditionally, husbands have held authority over their wives. In Christianity, the subordination of wives to their husbands has its basis in the New Testament. Paul (Colossians 3:18–19) states: "Wives, submit yourselves unto your husbands, as unto the Lord." Such teachings reflected the dominant themes of ancient Greece and Rome. Western society continued to support wifely subordination to husbands. English common law stated, "The husband and wife are as one and that one is the husband." A woman assumed her husband's identity, taking his last name on marriage and living in his house.

The U.S. courts formally institutionalized these power relationships. The law, for example, supports the traditional division of labor in many states, making the husband legally responsible for supporting the family and the wife legally responsible for maintaining the house and rearing the children. She is legally required to follow her husband if he moves; if she does not, she is considered to have deserted him. But if she moves and her husband refuses to move with her, she is also considered to have deserted him (Leonard and Elias 1990).

Legal and social support for the husband's control of the family has declined since the 1920s and especially since the 1960s. A more egalitarian standard for sharing power in families has taken much of its place. Especially through employment and wage earning, wives have gained more power in the family, increasing their influence in deciding such matters as family size and how money is spent.

The formal and legal structure of marriage may have made the male dominant, but the reality of marriage may be quite different. Sociologist Jessie Bernard (1982) makes an important distinction between authority and power in marriage. Authority is based in law, but power is based in personality. A strong, dominant woman is as likely to exercise power over a more passive man as vice versa, simply through the force of personality and temperament.

The relationship among gender, power, and violence is complex. Although some research suggests that men's violence is an expression of men's power over their wives (and of women's powerlessness), research also asserts that violence is more likely to be used by men with *less power*. Framed in this way, violence is a method through which men who lack power or have a need for power control their wives. Even the threat of violence can be an assertion of power, because it may intimidate women into complying with men's wishes even against their own (Sagrestano, Heavey, and Christensen 1999).

If we want to see how power works in marriage, we must look beneath gender stereotypes and avoid overgeneralizations. Women have considerable power in marriage, although they often feel that they have less than they actually do. They may fail to recognize the extent of their power; because cultural norms theoretically put power in the hands of their husbands, women may look at norms rather than at their own behavior. A woman may decide to work, even against her husband's wishes, and she may determine how to discipline the children. Yet she may feel that her husband holds the power in the relationship because he is supposed to be dominant. Similarly, husbands often believe that they have more power in a relationship than they actually do because they see only traditional norms and expectations.

Power is not a simple phenomenon. Researchers generally agree that family power is a dynamic, multidimensional process (Szinovacz 1987). Generally, no single individual is always the most powerful person in every aspect of the family. Nor is power always based on gender, age, or relationship. Power often shifts from person to person, depending on the issue.

According to J. P. French and Bertam Raven (1959), there are six types of marital power, each based on different beliefs or relationship dynamics:

1. *Coercive power* is based on the fear that one partner will punish the other. Coercion can be emotional or physical. A pattern of belittling, threatening, or being physical can intimidate and threaten another. This is the least common form of power, but it is used in partner rape or abuse.

2. *Reward power* is based on the belief that the other person will do something in return for agreement. If, for example, your partner attempts to understand your feelings about a specific issue, he or she may expect you to do the same.

3. *Expert power* is based on the belief that one partner has greater knowledge than the other. If you

believe that your partner has more wisdom about childrearing, for instance, you may defer the rewards, incentives, and discipline to him or her.

4. *Legitimate power* is based on acceptance of roles giving the other person the right to demand compliance. Gender roles are an important part of legitimacy because they give an aura to rights based on gender. Traditional gender roles legitimize male initiation in dating and female acceptance or refusal rights. Sociologists refer to legitimate power as *authority*.

5. *Referent power* is based on identifying with the partner and receiving satisfaction by acting similarly. If you have great respect for your partner's communication skills and his or her ability to actively listen, provide feedback, and disclose in an honest manner, you are more likely to model yourself after him or her.

6. *Informational power* is based on the partner's persuasive explanation. If, for example, your partner refuses to use a condom, you can provide information about the prevalence and danger of STDs and AIDS.

Explanations of Marital Power

Relative Love and Need Theory

Relative love and need theory explains power in terms of the individual's involvement and needs in the relationship. Each partner brings certain resources, feelings, and needs to a relationship. Each may be seen as exchanging love, companionship, money, help, and status with the other. What each gives and receives, however, may not be equal. One partner may be gaining more from the relationship than the other. The person gaining the most from the relationship is the most dependent. Constantina Safilios-Rothschild (1970) offers this observation:

> The relative degree to which the one spouse loves and needs the other may be the most crucial variable in explaining the total power structure. The spouse who has relatively less feeling for the other may be the one in the best position to control and manipulate all the "resources" that he has in his command to effectively influence the outcome of decisions.

Love itself is a major power resource in a relationship. Those who love equally are likely to share power equally (Safilios-Rothschild 1976). Such couples are likely to make decisions according to referent, expert, and legitimate power.

Principle of Least Interest

Akin to relative love and need as a way of looking at power is the **principle of least interest.** Sociologist Willard Waller (Waller and Hill 1951) coined this term to describe the curious (and often unpleasant) situation in which the partner with the least interest in continuing a relationship enjoys the most power in it. At its most extreme form, it is the stuff of melodrama. "I will do anything you want, Charles," Laura says pleadingly, throwing herself at his feet. "Just don't leave me." "Anything, Laura?" he replies with a leer. "Then give me the deed to your mother's house." Quarreling couples may unconsciously use the principle of least interest to their advantage. The less involved partner may threaten to leave as leverage in an argument: "All right, if you don't do it my way, I'm going." The threat may be extremely powerful in coercing a dependent partner. It may have little effect, however, if it comes from the dependent partner because he or she has too much to lose to be persuasive. Knowing this, the less involved partner can easily call the other's bluff.

Resource Theory of Power

In 1960, sociologists Robert Blood and Donald Wolfe studied the marital decision-making patterns as revealed by their sample of 900 wives. Using "final say" in decision making as an indicator of relative power, Blood and Wolfe inquired about a variety of decisions (for example, whether the wife should be employed, what type of car to buy, and where to live) and who "ultimately" decided what couples should do. They noted that men tended to have more of such decision-making power and attributed this to their being the sole or larger source of the financial resources on which couples depended. They further observed that as wives' share of resources increased, so did their roles in decision making (Blood and Wolfe 1960).

This **resource theory of power** has been met with both criticism and some empirical support. By focusing so narrowly on resources, the theory overlooks other sources of gendered power. Specifically, it fails to explain the power men continue to enjoy when they are outearned by their wives (Thompson and Walker 1989) or when they are househusbands and thus dependent on wives' incomes (Cohen and Durst 2000). The theory has also been criticized for equating power with decision making and for ignoring that power occasionally frees a spouse from having to make decisions. Although resources alone don't account for power, they may, with other factors, influence it, especially among heterosexual couples (Blumstein and Schwartz 1983; Schwartz 1994).

Rethinking Family Power: Feminist Contributions

Even though women have considerable power in marriages and families, it would be a serious mistake to overlook the inequalities between husbands and wives. As feminist scholars have pointed out, major aspects of contemporary marriage point to important areas in which women are clearly subordinate to men: the continued female responsibility for housework and childrearing, inequities in sexual gratification (sex is often over when the male has his orgasm), the extent of violence against women, and the sexual exploitation of children are examples.

Feminist scholars suggest several areas that require further consideration (Szinovacz 1987). First, they believe that too much emphasis has been placed on the marital relationship as the unit of analysis. Instead, they believe that researchers should explore the influence of society on power in marriage—specifically, the relationship between social structure and women's position in marriage. Researchers could examine, for example, the relationship of women's socioeconomic disadvantages, such as lower pay and fewer economic opportunities than men, to female power in marriage.

Second, these scholars argue that many of the decisions that researchers study are trivial or insignificant in measuring "real" family power. Researchers cannot conclude that marriages are becoming more egalitarian on the basis of joint decision making about such things as where a couple goes for vacation, whether to buy a new car or appliance, or which movie to see. The critical decisions that measure power are such issues as how housework is to be divided, who stays home with the children, and whose job or career takes precedence.

Some scholars suggest that we shift the focus from marital power to family power. Researcher Marion Kranichfeld (1987) calls for a rethinking of power in a family context. Even if women's marital power may not be equal to men's, a different picture of women in families may emerge if we examine power within the entire family structure, including power in relation to children. The family power literature has traditionally focused on marriage and marital decision making. Kranichfeld, however, feels that such a focus narrows our perception of women's power. Marriage is not family, she argues, and it is in the larger family matrix that women exert considerable power. Their power may not be the same as male power, which tends to be primarily economic, political, or religious. But if *power* is defined as the ability to change the behavior of others intentionally, "women in fact have a great deal of power, of a very fundamental and pervasive nature, so pervasive, in fact, that it is easily overlooked," according to Kranichfeld (1987). She further observes:

> Women's power is rooted in their role as nurturers and kinkeepers, and flows out of their capacity to support and direct the growth of others around them through their life course. Women's power may have low visibility from a nonfamily perspective, but women are the lynchpins of family cohesion and socialization.

Research on marital violence suggests that it is the level of absolute power that has consequence for couples. In relationships that are *either* male dominated *or* female dominated, we find the highest levels of violence. In relationships that are "power divided," there is less violence, and in egalitarian relationships we see the lowest levels of violence (Sagrestano, Heavey, and Christensen 1999).

The topic of "egalitarian relationships" is somewhat complicated by the question of whether such relationships truly are equal. Feminist scholarship has revealed that even among self-professed equal couples, power processes seem to favor men. Carmen Knudson-Martin and Anne Rankin Mahoney's 1998 study of equal couples—in which each spouse perceives the relationship to be characterized by mutual accommodation and attention and each spouse has the same ability to receive cooperation from the other in meeting needs or wants—is a case in point. Although couples described their relationships as equal and their roles as "nongender specific," men wielded more power than women. Wives made more concessions to fit their daily lives around their husbands' schedules than husbands did to fit their lives around the schedules of their wives. Women were also more likely than their husbands to report worrying about upsetting or offending their spouses, to do what their spouses wanted, and to attend to their spouses' needs (Fox and Murry 2000). It appears as if characterizing an unequal marriage as equal allows a couple to ignore real if covert power differences that might otherwise threaten their relationships (Fox and Murry 2000).

Intimacy and Conflict

Conflict between people who love each other seems to be a mystery. The coexistence of conflict and love has puzzled human beings for centuries. An ancient Sanskrit poem reflected this dichotomy:

> In the old days we both agreed
> That I was you and you were me.
> But now what has happened
> That makes you, you
> And me, me?

We expect love to unify us, but often times it doesn't. Two people don't really become one when they love each other, although at first they may have this feeling. It isn't that their love is an illusion, but their sense of ultimate oneness is. In reality, they retain their individual identities, needs, wants, and pasts while loving each other—and it is a paradox that the more intimate two people become, the more likely they may be to experience conflict. But it is not conflict itself that is dangerous to intimate relationships; it is the manner in which the conflict is handled. Conflict, itself is natural.

If this is understood, the meaning of conflict changes, and it will not necessarily represent a crisis in the relationship. David and Vera Mace (1979), prominent marriage counselors, observed that on the day of marriage, people have three kinds of raw material with which to work. First, there are things they have in common—the things they both like. Second, there are the ways in which they are different, but the differences are complementary. Third, unfortunately, there are the differences between them that are not complementary and that cause them to meet head on with a big bang. In every relationship between two people, there are a great many of those kinds of differences. So when they move closer to each other, those differences become disagreements.

The presence of conflict within a marriage or family doesn't automatically indicate that love is going or gone; it may mean quite the opposite. It is common and normal for couples to have disagreements or conflicts. The important factor is not *that* they have differences but *how* constructively or harmfully they resolve their differences. By using occasions of conflict to implement mutually acceptable behavior changes or decide that the differences between them are acceptable, couple relationships may grow as a product of their differences. Couples who resolve conflict with mutual satisfaction and who find ways to adapt to areas of conflict tend to be more satisfied with their relationships overall and are less likely to divorce.

Matter of Fact

When the communication patterns of newly married African Americans and Caucasians were examined, couples who believed in avoiding marital conflict were less happy 2 years later than those who confronted their problems (Crohan 1996).

Basic versus Nonbasic Conflicts

Relationships experience two types of conflict—basic and nonbasic—that have different effects on relationship quality and stability. Basic conflicts challenge the fundamental assumptions or rules of a relationship, leading to the possible end of the relationship. Nonbasic conflicts are more common and less consequential; couples learn to live with them.

Basic Conflicts

Basic conflicts revolve around carrying out marital roles and the functions of marriage and the family, such as providing companionship, working, and rearing children. It is assumed, for example, that a husband and a wife will have sexual relations with each other. But if one partner converts to a religious sect that forbids sexual interaction, a basic conflict is likely to occur because the other spouse considers sexual interaction part of the marital premise. No room for compromise exists in such a matter. If one partner cannot convince the other to change his or her belief, the conflict is likely to destroy the relationship. Similarly, despite recent changes in family roles, it is still expected that the husband will work to provide for the family. If he decides to quit work and not function as a provider, he is challenging a basic assumption of marriage. His partner is likely to feel that his behavior is unfair. Conflict ensues. If he does not return to work, his wife is likely to leave him.

Nonbasic Conflicts

Nonbasic conflicts do not strike at the heart of a relationship. The husband wants to change jobs and move to a different city, but the wife may not want to move. This may be a major conflict, but it is not a basic one. The husband is not unilaterally rejecting his role as a provider. If a couple disagrees about the frequency of sex, the conflict is serious but not basic because both agree on the desirability of sex in the relationship. In both cases, resolution is possible.

Experiencing and Managing Conflict

Differences and conflicts are part of any healthy relationship. If we handle conflicts in a healthy way, they can help solidify our relationships. But conflicts can go on and on, consuming the heart of a relationship, turning love and affection into bitterness and hatred. In the following section, we look at ways of resolving conflict in constructive rather than destructive ways. In this manner, we can use conflict as a way of building and deepening our relationships.

Dealing with Anger

Differences can lead to anger, and anger transforms differences into fights, creating tension, division, distrust, and fear. Most people have learned to handle anger by either venting or suppressing it. David and Vera Mace (1980) suggest that many couples go through a love–anger cycle. When a couple comes close to each other, they may experience conflict; then they recoil in horror, angry at each other because just at the moment they were feeling close their intimacy was destroyed. Each backs off; gradually they move closer again until another fight erupts, driving them apart. After a while, each learns to make a compromise between closeness and distance to avoid conflict. They learn what they can reveal about themselves and what they cannot.

Another way of dealing with anger is to suppress it. Suppressed anger is dangerous because it is always there, simmering beneath the surface. It leads to resentment, that brooding, low-level hostility that poisons both the individual and the relationship.

Anger can be dealt with in a third way; when conflict escalates into violence. Especially in a culture that cloaks families in privacy, surrounds people with beliefs that legitimize violence, and gives them the sense that they have a right to influence what their loved ones do, escalating anger can result in assault, injury, and even death. Given the relative power of men over women and adults over children, threats against one person's supposed advantage may provoke especially harsh reactions. We look closely at the causes, context, and consequences of family violence in Chapter 13.

Finally and most constructively, anger can be recognized as a symptom of something that needs to be changed. If we see anger as a symptom, we realize that what is important is not venting or suppressing the anger but finding its source and eliminating it. David and Vera Mace (1980) offer this suggestion:

> When your disagreements become overt conflict, the only thing to do is to take anger out of it, because when you are angry you cannot resolve a conflict. You cannot really hear the other person because you are just waiting to fire your shot. You cannot be understanding; you cannot be empathetic when you are angry. So you have to take the anger out, and then when you have taken the anger out, you are back again with a disagreement. The disagreement is still there, and it can cause another disagreement and more anger unless you clear it up. The way to take the anger out of disagreements is through negotiation.

Not all conflict is overt. Some conflict can go undetected by one of the partners. As such, it will have minimal effect on him or her and is not likely to lead to anger. In addition, not all "conflicts" (that is, of interest, goals, wishes, expectations, and so on) become *conflicts*. Spouses and partners can approach their differences in many ways short of overt conflict (Fincham and Beach 1999).

How Women and Men Handle Conflict

In keeping with observed gender differences in communication, research has identified differences in how men and women approach and manage conflict. As summarized by Rhonda Faulkner, Maureen Davey, and Adam Davey (2004), we can identify the following gender differences:

Real Families

"Did you bring me to this country for exploitation?"

Such is the plaintive appeal of 41-year-old Yong Ja Kim, a Korean immigrant, to her husband, Chun Ho Kim. What is it she is objecting to? In what way does she feel exploited? Sociologist Pyong Gap Min researched the consequences of immigration for marital relations among Korean immigrant couples. Existing research indicated that marital conflicts had emerged among Korean immigrants to the United States because of women's increased role in the economic support of families without concurrent changes in their husbands' gender attitudes or marital behavior. Min sought to delve more deeply into such conflicts.

Among Min's interviewees were Yong Ja Kim and Chun Ho Kim, husband and wife, who work together at their retail store 6 days a week from 9:30 a.m. to 6 p.m. Upon returning home, he watches Korean television programs and reads a Korean daily newspaper while she prepares dinner. Defensively, he retorts:

It makes no sense for her to accuse me of not helping her at home at all. In addition to house maintenance, I took care of garbage disposal more often than she and helped her with grocery shopping very often. I did neither of the chores in Korea.

To his wife, however, the comparison is not between what he did in Korea and what he does in the United States but between what *he does* and what *she does:*

I work in the store as many hours as you do, and I play an even more important role in our business than you. But you don't help me at home. It's never fair. My friends in Korea work full-time at home, but don't have to work outside. Here, I work too much both inside and outside the home.

Although conflicts such as this, in which wives contest an unequal division of household responsibility, are far from unique to Koreans or to immigrants, more generally, they take particular meaning and shape from the clash between the patriarchal Korean culture and the more egalitarian ideas espoused in the United States and from the discrepancy between men's status in Korea and their socioeconomic positions in the United States.

Culturally, there are noteworthy differences between the traditional status of husbands in Korea and the situations of most immigrant Korean men in the United States. Traditionally, Korean husbands were breadwinners and patriarchal heads of their families. Wives and children were expected to obey their husbands and fathers. Women were further expected to bear children and cater to their husbands and in-laws. Although the traditional South Korean family system has been "modified," it remains a patriarchal system, justified by Confucian ideology. As they have immigrated to the United States, Korean women's involvement in paid employment has increased "radically." In the process, traditional gender attitudes and male sense of self as patriarch and provider have been undermined.

Exacerbating the cultural transition are real economic adjustments. Min notes that with immigration to the United States, most Korean immigrant men encounter significant downward occupational mobility. This, in turn, results in further "status anxiety." They compensate by seeking ways to assert their authority in the household, only to find that their wives and children no longer grant them such status automatically. Min states that Mr. Kim "could not understand much and how fast his wife had changed her attitudes toward him since they had come to the United States. He did not remember her talking back to him in Korea."

She probably did not "talk back to him" in Korea. Min points out the marital conflicts and marital instability have increased alongside the increased economic role played by wives, the decreased economic status and power of their husbands, and women's pleas for greater male involvement in housework.

Min summarizes her research findings by noting that for Korean immigrant couples, the gulf between their gender-role behavior and their traditional gender attitudes may be greater than for many other ethnic groups. If so, and if such discrepancies are partly responsible for marital conflict, the situation for Korean immigrants may be harder than for other immigrant groups.

- Women are more likely than men are to initiate discussions of contested relationship issues.

- Where men have been found to be more likely to withdraw from negative marital interactions, women are more likely to pursue conversation or conflict.

- Typically, women are more aware of the emotional quality of and the events that occur in the relationship.

- In the course and processes of conflict management and resolution, men take on instrumental roles and women take on expressive roles. Men approach conflict resolution from a task-oriented stance, as in "problem solving"; women are more emotionally expressive as they pursue intimacy.

We need to bear in mind that the research designs used to study patterns of interaction in conflict management may have exaggerated the gender connection by commonly asking couples to engage in discussion of topics of greater salience to females than to males (for example, intimacy and childrearing practices). When researchers allowed members of couples to identify those areas in which they would like their partners to make changes and then had the couple hold two discussions, one for the topic that each person considered most important, gender patterns were more varied. Significantly more woman demand–man withdraw behavior occurred when couples addressed the woman's top issue, but there was significantly more man demand–woman withdraw behavior during discussions of issues most important to the man. Thus, it is crucial to avoid stereotyping gender patterns in partners' conflict styles; salience of the issue to each party also affects conflict behavior.

Conflict Resolution and Marital Satisfaction

Although we may perceive that "harmony" would guarantee "happiness," avoiding conflict is detrimental to relationships. However, *how couples manage conflict* is one of the most important determinants of their satisfaction and the well-being of their relationships (Greeff and DeBruyne 2000). Happy couples are not conflict free; instead, they tend to act in positive ways to resolve conflicts, such as changing behaviors (putting the cap on the toothpaste rather than denying responsibility) and presenting reasonable alternatives (purchasing toothpaste in a dispenser). Unhappy or distressed couples, in contrast, use more negative strategies in attempting to resolve conflicts (if the cap off the toothpaste bothers you, then *you* put it on).

Thus, we can talk of "constructive" and "destructive" conflict management (Greef and deBruyne 2000). Constructive conflict management is characterized by flexibility, a relationship rather than individual (self-interest) focus, an intention to learn from their dif-

ferences, and cooperation. Destructive conflict management consists of the following:

escalating spirals of manipulation, threat, and coercion

avoidance

retaliation

inflexibility

a competitive pattern of dominance and subordination

demeaning or insulting verbal and nonverbal communication

A study of happily and unhappily married couples found distinctive communication traits as these couples tried to resolve their conflicts (Ting-Toomey 1983). The communication behaviors of happily married couples displayed the following traits:

- *Summarizing.* Each person summarized what the other said: "Let me see if I can repeat the different points you were making."

- *Paraphrasing.* Each put what the other said into his or her own words: "What you are saying is that you feel bad when I don't acknowledge your feelings."

- *Validating.* Each affirmed the other's feelings: "I can understand how you feel."

- *Clarifying.* Each asked for further information to make sure that he or she understood what the other was saying: "Can you explain what you mean a little bit more to make sure that I understand you?"

In contrast, "distressed" or unhappily married couples displayed the following reciprocal patterns:

- *Confrontation.* Both partners confronted each other: "You're wrong!" "Not me, buddy. It's you who's wrong!"

- *Confrontation and defensiveness.* One partner confronted and the other defended: "You're wrong!" "I only did what I was supposed to do."

- *Complaining and defensiveness.* One partner complained and the other was defensive: "I work so hard each day to come home to this!" "This is the best I can do with no help."

- Overall, distressed couples use more negative and fewer positive statements. They become "locked in" to conflict. Thus, a major task for such couples is to find an effective or adaptive way out (Fincham and Beach 1999).

One of the strongest predictors of marital unhappiness and of the possibility of eventual divorce is whether couples engage in **hostile conflict.** Hostile conflict is a pattern of negative interaction wherein couples engage in frequent heated arguments, call each other names and insult each other, display an unwillingness to listen to each other, and lack emotional involvement with each other (Gottman 1994; Topham, Larson, and Holman 2005). Once such patterns become the normative pattern in a relationship, they are difficult to change.

What Determines How Couples Handle Conflict?

Many factors might affect how couples approach and attempt to manage the inevitable conflict that relationships contain. Among these, premarital variables, including carryover effects of upbringing, may be particularly influential. Glade Topham, Jeffrey Larson, and Thomas Holman (2005) suggest that such influence may be conscious or unconscious; may affect behaviors and patterns of interaction, as well as attitudes, beliefs, and self-esteem; and may remain even in the absence of contact with the family of origin.

Family of origin factors can be explained by social learning theory or attachment theory. Learning theory suggests that by observing parents and how they interact with each other we develop a **marital paradigm:** a set of images about how marriage ought to be done, "for better or worse" (Marks 1986). When, as children, we fail to experience a positive model of marriage, we may develop ineffective communication or conflict resolution skills. Attachment theory suggests that our attachment style influences the way conflict is expressed in relationships (Pistole 1989). Secure parent–child relationships lead us to be more self-confident and socially confident, more likely to view others as trustworthy and dependable, and more comfortable with and within relationships. Individuals who had insecure parent–child attachments are more demanding of support and attention, more dependent on others for self-validation, and more self-deprecating and emotionally hypersensitive (Topham, Larson, and Holman 2005).

In contrast to anxious or ambivalent and avoidant adults, secure adults are more satisfied in their relationships and use conflict strategies that focus on maintaining the relationship. Helping the relationship stay cohesive is more important than "winning" the battle. Secure adults are more likely to compromise than are anxious or ambivalent adults, and anxious or ambivalent adults are more likely than avoidant adults

to give in to their partners' wishes, whether they agree with them or not.

Although either husbands' or wives' family of origin experiences *could* negatively affect marital quality and conflict management, the influences are not equivalent. Wives' family of origin experiences—including the quality of relationships with their mothers, the quality of parental discipline they received, and the overall quality of their family environments—are more important than husbands' experiences in predicting hostile marital conflict (Topham, Larson, and Holman 2005).

There are two "analytically independent" dimensions of behavior in conflict situations: assertiveness and cooperativeness (Thomas 1976; Greeff and de Bruyne 2000). **Assertiveness** refers to attempts to satisfy our own concerns; **cooperativeness** speaks to attempts to satisfy concerns of others. With these two dimensions in mind, we can identify five conflict management styles:

- **Competing:** Behavior is assertive and uncooperative, associated with "forcing behavior and win–lose arguing." This style can lead to increased conflict, as well as to either or both spouses feeling powerless and resentful (Greeff and de Bruyne 2000).
- **Collaborating:** Behavior is assertive and cooperative; couples confront disagreements and engage in problem solving to uncover solutions. Collaborative conflict management may require relationships that are relatively equal in power and high in trust. Using this style then accentuates both the trust and the commitment couples feel.
- **Compromising:** This is an intermediate position in terms of both assertiveness and cooperativeness. Couples seek "middle ground" solutions.
- **Avoiding:** Behavior is unassertive and uncooperative, characterized by withdrawal and by refusing to take a position in disagreements.
- **Accommodating:** This style is unassertive and cooperative. One person attempts to soothe the other person and restore harmony.

Abraham Greeff and Tanya de Bruyne present these on a "conflict grid," depicting where each style falls on the axes of assertiveness and cooperativeness (see Figure 7.2).

Research has yielded inconsistent ("diverse") results about the relationship outcomes of each of these styles. Some studies favor one style—collaboration—

Figure 7.2 ■ Styles of Conflict Management

The "Conflict Grid" reveals the different combinations of assertiveness and cooperativeness that comprise the five styles of conflict management. For example, "competing" is a combination of a high degree of assertiveness and a low level of cooperativeness. An accommodating approach is low in assertiveness but high in cooperativeness.

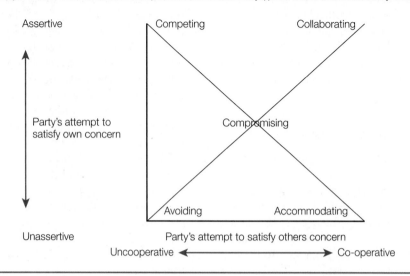

over all others as the only style displayed by satisfied couples. There is research suggesting that avoidance is dysfunctional and antisocial, and yet there is research that finds avoidance associated with satisfied, nondistressed couples. Still other research findings suggest that openly confronting conflict does not necessarily lead to higher marital quality. Finally, although some research suggests that when husbands and wives agree on how to manage conflict they have happier marriages, other findings indicate that discrepancies in spouses' beliefs about conflict are not predictive of how satisfied they are (Greeff and deBruyne 2000).

Greeff and de Bruyne point out that much literature on conflict management comes from studies of relatively young couples not long into marriage. They set out to examine the styles used by a sample of 57 Black South African couples married at least 10 years. Their findings reveal that the collaborating style led to the highest level of marital satisfaction for males and females, followed by the compromising style. Marital satisfaction was lowest when one or both spouses used the competing conflict management style. It was also low when one or both spouses used either an avoiding or an accommodating style.

They also considered how couples felt about their management of conflict. The collaborating and compromising styles were the ones with which couples expressed greatest satisfaction. Marriages where one or both spouses used a competitive approach to man-

aging conflict brought males great dissatisfaction with the conflict management in their relationships. Females were dissatisfied with conflict management when their husbands used a competitive approach but relatively satisfied when they, themselves, did. Both husbands and wives were also dissatisfied with the way conflict was managed in their marriages when either they or their spouses used a style of avoidance. Despite deliberately selecting older respondents, with longer-duration marriages, the patterns obtained were similar to what has been identified among younger couples. Like age, neither gender nor cultural background made much difference in which conflict management styles made people more satisfied with their marriages or their conflict management. Where gender did make a difference was in preferred style of conflict; males tended to use avoidance, compromise, and competition to manage conflict, whereas females showed a preference for accommodation, compromise, and avoidance (Greeff and de Bruyne 2000).

Conflict Resolution across Relationship Types

All couple relationships experience conflict. Using self-report and partner-report data, Lawrence Kurdek (1994) explored how conflicts were handled by 75 gay, 51 lesbian, 108 married nonparent, and 99 married parent couples. Essentially, the differences across couple type were less impressive than were the similarities.

■ *Conflict is an inevitable and normal part of being in a relationship. Rather than withdrawing from and avoiding conflict, we should use it as a way to build, strengthen, and deepen our relationships.*

© Laurie DeVault Photography

The four types of couples did not significantly differ in their level of ineffective arguing, and there were no noteworthy differences in their styles of conflict resolution as measured by the Conflict Resolution Styles Inventory (CRSI). The CRSI includes four styles of conflict resolution: (1) *positive problem solving* (including negotiation and compromise), (2) *conflict engagement* (such as personal attacks), (3) *withdrawal* (refusing to further discuss an issue), and (4) *compliance* (such as giving in). Ratings were obtained from both partners about themselves and the other partner. There was little indication that the frequency with which conflict resolution styles were used varied across couple type. As Kurdek (1994) notes, there is similarity in relationship dynamics across couple types.

Common Conflict Areas: Sex, Money, and Housework

Even if, as the Russian writer Leo Tolstoy suggested, every unhappy family is unhappy in its own way, marital conflicts still tend to center on certain recurring issues, especially communication, children, sex, money, personality differences, how to spend leisure time, in-laws, infidelity, and housekeeping. In this section, we focus on three areas: sex, money, and housework. Then we discuss general ways of resolving conflicts.

Fighting about Sex

Fighting and sex can be intertwined in several ways (Strong and DeVault 1997). A couple can have a specific disagreement about sex that leads to a fight. One person wants to have sexual intercourse and the other does not, so they fight. A couple can have an indirect fight about sex. The woman does not have an orgasm, and after intercourse, her partner rolls over and starts to snore. She lies in bed feeling angry and frustrated. In the morning she begins to fight with her partner over his not doing his share of the housework. The housework issue obscures why she is angry. Sex can also be used as a scapegoat for nonsexual problems. A husband is angry that his wife calls him a lousy provider. He takes it out on her sexually by calling her a lousy lover. They fight about their lovemaking rather than about the issue of his provider role. A couple can fight about the wrong sexual issue. A woman may berate her partner for being too quick during sex, but what she is really frustrated about is that he is not interested in oral sex with her. She, however, feels ambivalent about oral sex ("Maybe I smell bad"), so she cannot confront her partner with the real issue. Finally, a fight can be a cover-up. If a man feels sexually inadequate and does not want to have sex as often as his partner, he may pick a fight and make his partner so angry that the last thing she would want to do is to have sex with him.

Issues and Insights "What Are We Fighting About?"

Conflict is common. Living with other people introduces numerous points of potential disagreement. Not all disagreements are equally serious or carry equal risks for the health and future of the relationship. Certain problems, such as infidelity, men's jealousy, and reckless spending of money or poor money management are more significant predictors of eventual dissolution (Fincham and Beach 1999). Other problems, such as sexual disorders or substance abuse, may be beyond the couple's ability to resolve, requiring instead outside therapeutic assistance. When researchers surveyed therapists, seeking to identify the frequency, difficulty in treating, and severity of the effect of 29 problems couples might face, they found the following problems identified as the most frequent problems couples bring to therapy: unrealistic expectations, power struggles, communication problems, sexual problems, and conflict management difficulties. Problems deemed most difficult to treat included lack of loving feelings, alcoholism, extrarelational affairs, and power struggles (Whisman, Dixon, and Johnson 1997; Miller et al. 2003).

But not all couples need or come to therapy. Seeking to determine whether problems change over time in marriage, researchers have undertaken one of two research strategies. Either they have followed a sample of couples across a period of time (Storaasli and Markman 1990) or they

have compared couples of different marital durations, seeing whether their problem areas differ in intensity at the different points in which they find themselves in their marriages (Miller et al. 2003).

One longitudinal study by Raagnar Storaasli and Howard Markman followed 40 couples over a period: beginning before they married, shortly after they married, and after they had their first child. Some problem areas changed between premarriage and early marriage; jealousy and religious problems decreased and sexual problems worsened. Between early marriage and parenthood, communication and sexual problems increased. Overall, however, most problems remained unchanged, such as those having to do with relatives, friends, or money (Storaasli and Markman 1990; Miller et al. 2003).

Using a clinical sample of 160 couples married between 1 and 20 years, Richard Miller and colleagues (2003) sought to determine whether couples at different life cycle stages experience and seek help with different kinds of problems. Couples were asked to consider as problem areas: children, communication, housecleaning, gender-role issues, financial matters, sexual issues, spiritual matters, emotional intimacy, violence, commitment, values, parents-in-law, decision making, and commitment. Couples were asked to consider where each problem ranked in frequency, from "very often a problem" (5) to "never a problem" (1). Because it was a clinical sample, couples were also asked to consider from nine choices the problem that most brought them to therapy, including as possibilities communication,

violence, sexual issues, financial matters, emotional intimacy, separation or divorce concerns, extramarital affairs, commitment issues, or some other problem.

Problems with communication and financial matters were the most commonly reported. Also frequently mentioned were emotional intimacy, sexual issues, and decision making. Gender-role issues, values, violence, and spiritual issues were not common problems. These tendencies can be seen in Table 7.2, reflecting the percentage of spouses who listed a problem as either "Very often a problem" or "Often a problem."

As far as what problem area couples were most likely to identify as their "presenting problem," by far "communication problems" were most often mentioned by both males and females, *regardless of how long they were married.* Finally, as shown in Table 7.3, there were statistically significant gender differences for six problem areas.

According to Miller and colleagues, their findings indicate that problems experienced by couples are relatively stable as opposed to varying much over the life cycle. As to gender, they remind therapists that females generally perceive more problems than males within marital relationships. Somewhat consistent with the idea of "two marriages," males and females may indeed experience relationships and problems within those relationships differently (Storaasli and Markman 1990). Women's tendencies to report problem areas as more severe or frequent suggest "a complex picture of gender-related issues." Finally, regardless of how

---

In power struggles, sexuality can be used as a weapon, but this is generally a destructive tactic (Szinovacz 1987). A classic strategy for the weaker person in a relationship is to withhold something that the

more powerful one wants. In male–female struggles, this is often sex. By withholding sex, a woman gains a certain degree of power. A few men also use sex in its most violent form: They rape (including date rape and

Table 7.2 ■ Percentage Reporting Area Is Either "Very Often" or "Often" a Problem

Problem	Males			Females		
	< 3 years*	3–10 years	> 10 years	< 3 years	3–10 years	> 10 years
Communication	56.7%	63.8%	53.2%	62.9%	67.4%	66.6%
Financial matters	37.8	54.4	56.3	26.9	55.1	67.7
Decision making	27.0	34.4	25.0	34.2	42.7	48.4
Emotional intimacy	21.6$^+$	50.3	21.9	42.8	52.8	45.2
Sexual issues	21.6	34.1	28.2	37.2	38.2	29.0
Parent-in-law	27.0	24.2	19.4	28.5	31.5	22.6
Leisure activities	18.9$^+$	30.1	15.7	34.3	40.4	35.5
Dealing with children	18.2	22.8	28.1	26.9	35.6	29.1
Commitment	21.6	19.8	9.4	11.4	18.2	32.2
Housecleaning	13.5	25.6	18.8	17.1	28.1	29.0
Gender-role issues	10.8	13.5	0.0	14.3	16.8	9.7
Values	13.5	15.7	9.4	5.7	17.0	10.0
Violence	8.8	1.3	3.4	9.4	3.9	3.6
Spiritual matters	0.0	5.6	3.1	5.8	6.9	9.7

*Numbers represent duration of marriage.
$^+$Duration of marriage group differences for that gender significant at < 0.05.

Table 7.3 ■ Frequency of Reporting Areas

Problem	Males	Females
Dealing with children*	2.71	2.98
Emotional intimacy*	3.15	3.45
Sexual issues$^+$	2.90	3.08
Parents-in-law$^+$	2.62	2.84
Communication$^@$	3.70	4.00
Decision making$^+$	3.05	3.27

Range: (1) "never a problem" to (5) "very often a problem."
*Difference significant at $p < 0.01$
$^+$Difference significant at $p < 0.05$
$^@$Difference significant at $p < 0.001$

long a couple has been married, couples' therapists must be prepared to assess and treat problems dealing with communication, financial matters, sexual issues, decision-making skills, and emotional intimacy because such problem areas are consistent features of married life over which couples encounter difficulty.

It is worth pointing out that conflict is not only driven by "what" couples fight about but also by the wider social context in which relationships exist. Taking a broader view, we need to pay attention to the effects of negative life events, essentially nonmarital stressors, that may lead to more negative communication, poorer parenting, and lower satisfaction. Likewise, the amount of social support a couple enjoys outside the marriage may influence the direction and outcomes of conflict (Fincham and Beach 1999).

marital rape) to overpower and subordinate women. In rape, aggressive motivations displace sexual ones.

It is hard to tell during a fight if there are deeper causes than the one about which a couple is fighting.

Is a couple fighting because one wants to have sex now and the other doesn't? Or are there deeper reasons involving power, control, fear, or inadequacy? If they repeatedly fight about sexual issues without getting anywhere, the ostensible cause may not be the real one. If fighting does not clear the air and make intimacy possible again, they should look for other reasons for the fights. It may be useful for them to talk with each other about why the fights do not seem to accomplish anything. Also, it would be helpful if they step back and look at the circumstances of the fight; what patterns occur; and how each feels before, during, and after a fight.

Sexual tensions and strains arise because of these other conflicts that happen to play themselves out in the physical relationship. With a more "positive, respectful, affirming process of conflict resolution," partners may deepen the respect and admiration they feel for each other, develop a greater level of trust and of self-esteem in their relationship, and grow more confident that the relationship can withstand and grow through future conflict. These can create positive feelings and comfort with each other that facilitate sexual desire (Metz and Epstein 2002). Although the conflicts being resolved need not be sexual, positive and constructive relationship conflict resolution may provide affirmation of the love and intimacy two people share. This, along with the emotional relief that comes from resolving conflict may "directly or indirectly serve as a sexual aphrodisiac" (Metz and Epstein 2002). Thus, the intensity of pleasure supposedly accompanying "make-up sex" is another reminder of how conflict and its resolution can affect sex whether or not it is about sex.

Money Conflicts

An old Yiddish proverb addresses the problem of managing money quite well: "Husband and wife are the same flesh, but they have different purses." Money is a major source of marital conflict in families in the United States and abroad.

Intimates differ about spending money probably as much as, or more than, any other single issue.

WHY PEOPLE FIGHT ABOUT MONEY. Couples disagree or fight over money for a number of reasons. One of the most important has to do with power. Earning wages has traditionally given men power in families. A woman's work in the home has not been rewarded by wages. As a result, full-time homemakers have been placed in the position of having to depend on their husbands for money. In such an arrangement, if there are disagreements, the woman is at a disadvantage. If she is de-

ferred to, the old cliché "I make the money but she spends it" has a bitter ring to it. As women increased their participation in the workforce, however, power relations within families have shifted some. Studies indicate that women's influence in financial and other decisions increases if they are employed outside the home.

Another major source of monetary conflict is allocation of the family's income. Not only does this involve deciding who makes the decisions, but it also includes setting priorities. Is it more important to pay a past due bill or to buy a new television set to replace the broken one? Is a dishwasher a necessity or a luxury? Should money be put aside for long-range goals, or should immediate needs (perhaps those your partner calls "whims") be satisfied? Setting financial priorities plays on each person's values and temperament; it is affected by basic aspects of an individual's personality. A miser probably cannot be happily married to a spendthrift. Yet we know so little of our partner's attitudes toward money before marriage that a miser might well marry a spendthrift and not know it until too late.

Dating relationships are a poor indicator of how a couple will deal with money matters in marriage. Dating has clearly defined rules about money: Either the man pays, both pay separately, or they take turns paying. In dating situations, each partner is financially independent of the other. Money is not pooled, as it usually is in a committed partnership or marriage. Power issues do not necessarily enter spending decisions because each person has his or her own money. Differences can be smoothed out fairly easily. Both individuals are financially independent before marriage but financially interdependent after marriage. Even cohabitation may not be an accurate guide to how a couple would deal with money in marriage, as cohabitators generally do not pool all (or even part) of their income. It is the working out of financial interdependence in marriage that is often so difficult.

TALKING ABOUT MONEY. Talking about money matters is often difficult. People are secretive about money. It is considered poor taste to ask people how much money they make. Children often do not know how much money is earned in their families; sometimes spouses don't know either. One woman remarked that it is easier to talk with a partner about sexual issues than about money matters: "Money is the last taboo," she said. But, as with sex, our society is obsessed with money.

Why do we find it difficult to talk about money? There may be several reasons. First, we don't want to appear to be unromantic or selfish. If a couple is about to marry, a discussion of attitudes toward money may lead to disagreements, shattering the illusion of unity or selflessness. Second, gender roles make it difficult for women to express their feelings about money because women are traditionally supposed to defer to men in financial matters. Third, because men tend to make more money than women, women feel that their right to disagree about financial matters is limited. These feelings are especially prevalent if the woman is a homemaker and does not make a financial contribution, but they devalue her childcare and housework contributions.

Housework and Conflict

The division of responsibility for housework can be one of the most significant issues couples face, especially dual-earner couples (Kluwer, Heesink, and Van De Vliert 1997). It can become a source of tension and conflict within marriage (Hochschild 1989). Part of this is an understandable consequence of the inequality in each spouse's contribution; most men do not do much housework. Whether or not they are employed outside the home, and whether there are children in the home or not, wives bear the bulk of housework responsibility. A husband's lack of involvement can create resentment and affect the levels of both conflict and happiness in a marriage. Longitudinal research on married couples reveals that husbands whose wives perceived that the division of housework was unfair report higher levels of marital conflict over time (Faulkner, Davey, and Davey 2004). Similarly, in her acclaimed study of the division of housework among 50 dual-earner couples, Arlie Hochschild (1989) argued that men's level of sharing "the second shift" (that is, unpaid domestic work and childcare) influenced the levels of marital happiness couples enjoyed and their relative risk of divorce. This held true whether couples were traditional or egalitarian in their views of marriage.

In a study of 54 Dutch couples, Esther Kluwer, Jose Heesink, and Evert Van De Vliert (1996) found that conflict about household work was related to wives' dissatisfactions with their and their husbands' relative contributions and expenditures of time. They note that 72% of the wives preferred to do less than they actually did; that is, when they spent more time on house-

work than they preferred to, they were dissatisfied. They also tended to be dissatisfied if they perceived their husbands spending less time than they preferred them to spend on housework. In the study, 52% of the wives wished their husbands would do more housework than they actually did (Kluwer, Heesink, and Van De Vliert 1996).

How much each spouse contributes to the household is only the more observable aspect of the "politics of housework." In addition, couples must reach agreements about standards, schedules, and management of housework. Conflicts about standards are struggles over whose standards will predominate: Who decides whether things are "clean enough"? Similarly, disputes about schedules reflect whose time is more valuable and which partner works around the other's sense of priorities. Who waits for whom? Finally, arguments about who bears responsibility for organizing, initiating, or overseeing housework tasks are also disputes about who will have to ask the other for help, carry more responsibility in his or her head, and risk refusal from an uncooperative partner.

Thus, housework conflicts have both practical and symbolic dimensions. Practically, there are things that somehow must get done for households to run smoothly and families to function efficiently. Couples must decide who shall do them and how and when they should be done. On a more symbolic level, disputes over housework may be experienced as conflicts about the level of commitment each spouse feels toward the marriage. Because marriage symbolizes the union of two people who share their lives, work together, consult each other, and take each other's feelings and needs into consideration, resisting housework or doing it only under duress may be seen as a less-than-equal commitment. We look more in detail at the dynamics surrounding the division of housework in Chapter 12.

The absence of overt conflict over the allocation of tasks and time does not mean that there is no conflict. It means only that the conflict is not openly expressed. Wives in more traditional marriages are more likely than wives in egalitarian relationships to avoid conflict over housework even if they are dissatisfied with their domestic arrangements. They may withdraw from discussions of the division of labor as a way of avoiding the issue. Because egalitarian couples may engage in more open discussion and conflict over housework responsibilities, such conflict gives them more opportunity to establish a solution (Kluwer, Heesink, and Van De Vliert 1997).

Consequences of Conflict

Marital conflict has effects on a host of outcomes related to individual mental and physical health, family health, and child well-being. Frank Fincham and Steven Beach's thorough review (1999) of research on marital conflict showed the following outcomes.

Mental Health

There are links between experiencing marital conflict and suffering from depression, eating disorders, being physically and/or psychologically abusive of partners, and male alcohol problems (including excessive drinking, binge drinking, and alcoholism). There is less evidence connecting marital conflict to elevated levels of anxiety.

Physical Health

Marital conflict is associated with poorer overall physical health, as well as certain specific illnesses. These include cancer, heart disease, and chronic pain. The associations are stronger for wives than for husbands and may be the result of altered physiological functioning, including endocrine, cardiac, and immunological functioning, associated with the distress introduced by marital conflict.

Familial and Child Well-Being

Marital conflict may disrupt the entire family, especially if the conflict is frequent, intense, and unresolved. Marital conflict has been shown to be connected to poorer parenting, problematic parent–child attachments, and greater frequency and intensity of parent–child or sibling–sibling conflict. Consequences for children can be particularly harmful, "potentially profound," when the conflict centers on issues about the children and childrearing. The most destructive form of marital conflict appears to be when couples engage in attacking and withdrawing (hostility and detachment). In addition, when marriage is characterized by the absence of or low levels of warmth, mutuality, and harmony between parents, along with the presence of high levels of competitiveness and con-

flict, children develop more externalizing and peer problems (Katz and Woodin 2002). When parental marriages lack relationship cohesiveness, are devoid of playfulness and fun, and yet have high degree of conflict, children miss out on the warmth, intimacy, and security that healthy families can provide (Katz and Woodin 2002).

Research reveals numerous problematic effects of marital conflict on children including health problems, depression, anxiety, peer problems, conduct problems, and low self-esteem. When marital conflict is frequent, intense, and child centered, it has especially negative consequences for children.

How do children react to marital conflict? Research indicates that children are distressed by both verbal and physical conflict but reassured by healthy conflict resolution. Witnessing threats, personal insults, verbal and nonverbal hostility, physical aggressiveness between parents or by parents toward objects (for example, breaking or slamming things), defensiveness, and marital withdrawal all can give rise to "heightened negative emotionality" (Cummings, Goeke-Morey, and Papp 2003). Conversely, when parents engage in calm discussion, display affection and continued support even while engaged in conflict, children react positively.

Parents' displays of support, including providing validation to one another and affection during conflict, may reassure children that the marital relationship remains strong and loving even though parents disagree (Cummings, Goeke-Morey, and Papp 2003). However, the absence or failure of resolution causes anger, sadness, and distress. A frequently posed question, which we consider in Chapter 14, is whether the effects of conflict on children are worse than the effects of divorce.

Can Conflict Be Beneficial?

As we noted earlier, conflict is a normal and predictable part of living with other people, especially given the intensity of emotions that exist within marriage. Conflict, itself, is not necessarily damaging; there may be "reversal effects" of conflict, in which spouses' "conflict engagement" (especially that of husbands) predicts positive change in husbands' and wives' satisfaction with marriage. It appears as though some negative behavior—such as conflict—may be both healthy and necessary for long-term marital well-being. Too little conflict (suggestive of avoidance),

■ *Children react to parental conflict in a variety of ways, depending on how the parents handle themselves. Although children can be hurt by outward displays of anger and especially by witnessing violence, "healthy conflict management" may be beneficial for children to witness.*

like too much conflict, may lead to poorer outcomes. However, the outcome of conflict varies, along with the meaning and function of conflict behavior. It can as easily reflect engagement with a problem as it can suggest withdrawal from the problem (Christensen and Pasch 1993). Furthermore, it may be part of an effort to maintain the relationship or conversely indicate that one or both partners have given up on the relationship (Holmes and Murray 1996). Thus, as Frank Fincham and Steven Beach (1999) suggest, "we have to identify the circumstances in which conflict behaviors are likely to result in enhancement rather than deterioration of marital relationships" (1999, 54).

Resolving Conflicts

There are a number of ways to end conflicts and solve problems. You can give in, but unless you believe that the conflict ended fairly, you are likely to feel resentful. You can try to impose your will through the use of power, force, or the threat of force, but using power to end conflict leaves your partner with the bitter taste of injustice. Less productive conflict resolution strategies include *coercion* (threats, blame, and sarcasm), *manipulation* (attempting to make your partner feel guilty), and *avoidance* (Regan 2003).

More positive strategies for resolving conflict, include *supporting your partner* (through active listening, compromise, or agreement), *assertion* (clearly stating your position and keeping the conversation on topic), and *reason* (the use of rational argument and the consideration of alternatives) (Regan 2003). Finally, you can end the conflict through negotiation. In negotiation, both partners sit down and work out their differences until they come to a mutually acceptable agreement (see Figure 7.3). Conflicts can be solved through negotiation in three primary ways: agreement as a gift, bargaining, and coexistence.

Agreement as a Gift

If you and your partner disagree on an issue, you can freely agree with your partner as a gift. If you want to go to the Caribbean for a vacation and your partner wants to go backpacking in Alaska, you can freely agree to go to Alaska. An agreement as a gift is different from giving in. When you give in, you do something you don't want to do. When you agree without coercion or threats, the agreement is a gift of love, given freely without resentment. As in all exchanges of gifts, there will be reciprocation. Your partner will be more likely to give you a gift of agreement. This gift of agreement is based on referent power, discussed earlier.

Figure 7.3 ■ Family Problem-Solving Loop

Most family problem solving occurs in the ebb and flow of daily family events. Although family dynamics and transition take various forms, it is interesting to note the types that might have relevance for family issues.

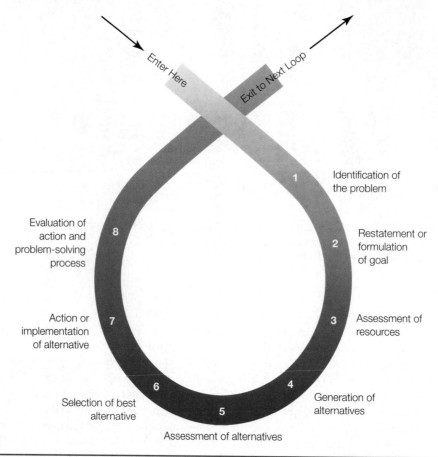

SOURCE: Kieren, Maguire, and Hurlbut 1996, 442–455. Copyright © 1996 by the National Council on Family Relations. Used by permission.

Bargaining

Bargaining in relationships—the process of making compromises—is different from bargaining in the marketplace or in politics. In relationships, you want what is best for the relationship, the most equitable deal for both you and your partner, not just the best deal for yourself. During the bargaining process, you need to trust your partner to do the same. In a marriage, both partners need to win. The result of conflict in a marriage should be to solidify the relationship, not to make one partner the winner and the other the loser. To achieve your end by exercising coercive power or withholding love, affection, or sex is a destructive form of bargaining. If you get what you want, how will that affect your partner and the relationship? Will your partner feel you are being unfair and become resentful? A solution has to be fair to both, or it won't enhance the relationship.

Coexistence

Although unresolved conflict may, over time, wear away at marital quality, sometimes differences simply can't be resolved. In such instances, they may need to be lived with. If a relationship is sound, often differences can be absorbed without undermining the basic ties. All too often we regard a difference as a threat rather than as the unique expression of two personalities. Rather than being driven mad by the cap left off the toothpaste, perhaps we can learn to live with it.

Helping Yourself May Mean Getting Help

Despite good intentions and communication skills, we may not be able to resolve our relationship problems on our own. Accepting the need for professional assistance may be a significant first step toward reconciliation and change. Experts advise counseling when communication is hostile, conflict goes unresolved, individuals cannot resolve their differences, and/or a partner is thinking about leaving.

Marriage and partners counseling are professional services whose purpose is to assist individuals, couples, and families gain insight into their motivations and actions within the context of a relationship while providing tools and support to make positive changes. A skilled counselor offers objective, expert, and discreet help. Much of what counselors do is crisis or intervention oriented.

It may be more valuable and perhaps more effective to take a preventive approach and explore dynamics and behaviors before they cause more significant problems. This may occur at any point in relationships: during the engagement, before an anticipated pregnancy, or at the departure of a last child.

Each state has its own degree and qualifications for marriage counselors. The American Association for Marital and Family Therapy (AAMFT) is one association that provides proof of education and special training in marriage and family therapy. Graduate education from an accredited program in social work, psychology, psychiatry, or human development is coupled with a license in that field

ensures both education and training, as well as offering the consumer recourse if questionable or unethical practices occur. This recourse is, however, only available if the practitioner holds a valid license issued by the state in which he or she practices. Mental health workers belong to any one of several professions:

- *Psychiatrists* are licensed medical doctors who, in addition to completing at least 6 years of postbaccalaureate medical and psychological training, can prescribe medication.
- *Clinical psychologists* have usually completed a Ph.D., which requires at least 6 years of postbaccalaureate course work. A license requires additional training and the passing of state boards.
- *Marriage and family counselors* typically have a master's degree and additional training to be eligible for state board exams.
- *Social workers* have master's degrees requiring at least 2 years of graduate study plus additional training to be eligible for state board exams.
- *Pastoral counselors* are clergy who have special training in addition to their religious studies.

Financial considerations may be one consideration when selecting which one of the preceding to see. Typically, the more training a professional has, the more he or she will charge for services.

A therapist can be found through a referral from a physician, school counselor, family, friend, clergy, or the state department of mental health. In any case, it is important to meet personally with the counselor to decide if he or she is right for you. Besides inquiring about his or her basic professional qualifications, it is important to feel comfortable with this person, to decide whether your value and belief systems are compatible, and to assess his or her psychological orientation. Shopping for the right counselor may be as important a decision as deciding to enter counseling in the first place.

Marriage or partnership counseling has a variety of approaches: Individual counseling focuses on one partner at a time; joint marital counseling involves both people in the relationship; and family systems therapy includes as many family members as possible. Regardless of the approach, all share the premise that to be effective, those involved should be willing to cooperate. Additional logistical questions, such as the number and frequency of sessions, depend on the type of therapy.

At any time during the therapeutic process, you have the right to stop or change therapists. Before you do, however, ask yourself whether your discomfort is personal or has to do with the techniques or personality of the therapist. Discuss this issue with the therapist before making a change. Finally, if you believe that your therapy is not benefiting you, change therapists.

Forgiveness

Related to the issues of conflict and its resolution is the topic of **forgiveness.** Conceptualized as a reduction in negative feelings and an increase in positive

feelings toward a "transgressor" after a transgression, an attitude of good will toward someone who has done us harm, and showing compassion and foregoing resentment toward someone who has caused us pain, research has determined that forgiveness has long-term physical and mental health benefits for the person forgiving. Forgiveness is associated with enhanced self-esteem, positive feelings toward the transgressor, and reduced levels of negative emotions such as anger, grief, revenge, and depression. In a relationship context, forgiveness has been defined as "the tendency to forgive partner transgressions over time and across situations" (Fincham and Beach 2002).

Forgiveness has been found to be a crucial element of married life. It is an important aspect of efforts to restore trust and relationship harmony after a transgression. Most "forgiveness narratives" mention motivations such as a partner's well-being, restoration of the relationship, and love (Fincham and Beach 2002). Forgiveness has been shown to resolve existing difficulties and prevent future ones. It also enhances marital quality, as can be seen in the positive association between forgiveness and marital satisfaction and longevity (Kachadourian, Fincham, and Davila 2004).

Research has identified both personal and relationship qualities associated with the ability or tendency to forgive. Qualities such as agreeableness, religiosity, humility, emotional stability, and empathy are associated with forgiveness. Pride and narcissism are associated with decreased tendencies to forgive. Individuals who are more accommodating within their relationships, more securely attached, and have more positive models of self and others are also more likely to be forgiving toward partners who have committed transgressions.

Not all relationship transgressions are equivalent. The ability to forgive relatively minor transgressions doesn't automatically guarantee forgiveness of more major transgressions. Wives who display tendencies to forgive seem able to do so in both minor and major transgressions. For husbands, on the other hand, tendencies to forgive apply more to major transgressions. It appears as though men may not consider minor transgressions important enough to warrant either receiving apologies or granting forgiveness (Kachadourian, Fincham, and Davila 2004).

If we can't talk about what we like and what we want, there is a good chance that we won't get either. Communication is the basis for good relationships. Communication and intimacy are reciprocal: Communication creates intimacy, and intimacy in turn helps create good communication.

If we fail to communicate, we are likely to turn our relationships into empty facades, with each person acting a role rather than revealing his or her deepest self. But communication is learned behavior. If we have learned *how not to* communicate, we can learn *how to* communicate. Communication will allow us to maintain and expand ourselves and our relationships.

Summary

- Communication includes both verbal and nonverbal communication. For the meaning of communication to be clear, verbal and nonverbal messages must agree.

- The functions of nonverbal communication are to convey interpersonal attitudes, express emotions, and handle the ongoing interaction. Much nonverbal communication, such as levels of touching, varies across cultures and between women and men.

- Nonverbal communication patterns can reveal whether a relationship is healthy or troubled.

- How well a couple communicates before marriage can be an important predictor of later marital satisfaction. *Self-disclosure* before marriage is related to relationship satisfaction later.

- Research indicates that happily married couples engage in less frequent and less destructive conflict, disclose more of their thoughts and feelings, and more accurately and effectively communicate.

- In marital communication, wives send clearer, less ambiguous messages; send more positive, more negative, and fewer neutral, messages; and take more active roles in arguments than husbands do.

- *Demand–withdraw communication* is common among married couples. One partner, more often the wife, will raise an issue for discussion, and the other partner, more likely the husband, will withdraw from the conversation instead of attempting to communicate.

- Satisfying sexual relationships require effective sexual communication.

- Some topics are more highly charged and more sensitive to discuss.

- Virginia Satir placed people into four categories based on their style of miscommunication: (1) placaters, (2) blamers, (3) computers, and (4) distractors. Placaters are passive, helpless, and always agreeable; blamers act superior, are often angry, do not listen, and try to escape responsibility; computers are correct, reasonable, and expressionless; and distractors are frenetic and tend to change the subject.

- Barriers to communication include the traditional male gender role; personal reasons, such as feelings of inadequacy; the fear of conflict; and an absence of self awareness.

- Some research reflects a curvilinear relationship between self-disclosure and marital satisfaction: both low and high levels of self-disclosure associated with lower levels of marital satisfaction. Other research supports a more linear model: high levels of self-disclosure result in higher levels of marital satisfaction.

- *Trust* is the belief in the reliability and integrity of a person. Self-disclosure requires trust. How much you trust a person influences the way you are likely to interpret ambiguous or unexpected messages from him or her.

- *Feedback* is the ongoing process in which participants and their messages create a given result and are subsequently modified by the result.

- The basis of good communication in a relationship is *mutual affirmation*. Mutual affirmation includes mutual acceptance, mutual liking, and expressing liking in words and actions.

- *Power* is the ability or potential ability to influence another person or group. There are six types of marital power: coercive, reward, expert, legitimate, referent, and informational.

- Self-described equal (or egalitarian) couples often still reveal power differences and inequalities that more often favor men.

- Conflict is natural in intimate relationships. *Basic conflicts* challenge fundamental rules; *nonbasic conflicts* do not threaten basic assumptions and may be negotiable. *Situational conflicts* are based on specific issues. *Personality conflicts* are unrealistic conflicts, potentially stemming from fundamental personality differences.

- People usually handle anger in relationships by suppressing or venting it. When anger arises, it is useful to think of it as a signal that change is necessary.

- Among heterosexual couples, women have greater awareness of the emotional quality of the relationship and are more likely to initiate discussion of contested issues. Men are more likely to approach conflict from a task oriented stance or to withdraw.

- Hostile conflict, characterized by frequent heated arguments, name calling, an unwillingness to listen to each other, is a particularly strong predictor of eventual divorce.

- Premarital variables help determine how we handle conflict. From observations of parental interaction, we develop a *marital paradigm*—a set of images of how marriage should be, "for better or worse." Conflict management may also be affected by our attachment style and by the wider social context in which relationships exist.

- There are two dimensions of behavior in conflict situations—*assertiveness* (attempting to satisfy our own concerns) and *cooperativeness* (attempting to satisfy the other person's concerns), which can be differently combined to form five styles of conflict management: *competing* (assertive and uncooperative), *collaborating* (assertive and cooperative), *compromising* (intermediate in both assertiveness and cooperativeness), *avoiding* (unassertive and uncooperative), and *accommodating* (unassertive and cooperative).

- Major sources of conflict include sex, money, and housework.

- Conflict can have effects on the mental and physical health of spouses or partners, the health of the relationship, and the well-being of the children. Especially when conflict is intense, frequent, and centers on issues related to the children, it is likely to negatively affect children. Seeing parents constructively engage in calm discussion and display affection and continued support even while engaged in conflict is beneficial for children.

- Happily married couples use certain techniques to resolve conflict, including summarizing, paraphrasing, validating, and clarifying. Unhappy couples use confrontation, confrontation and defensiveness, and complaining and defensiveness.

- Conflict resolution may be achieved through negotiation in three ways: agreement as a freely given gift, bargaining, and coexistence.

- *Forgiveness* is an important part of efforts to restore trust and rebuild relationship harmony. It is positively associated with both relationship satisfaction and stability (that is, longevity).

Key Terms

accommodating 263

assertiveness 263

avoiding 263

basic conflicts 259

collaborating 263

competing 263

compromising 263

contempt 238

cooperativeness 263

demand–withdraw
communication 246

feedback 252

forgiveness 273

honeymoon effect 245

hostile conflict 263

marital paradigm 263

mutual affirmation 254

nonbasic conflicts 260

nonverbal
communication 237

power 255

principle of least
interest 257

proximity 238

rapport talk 243

relative love and need
theory 257

report talk 243

resource theory
of power 257

trust 252

verbal
communication 237

Resources on the Internet

Companion Website for This Book

http://www.thomsonedu.com/sociology/strong

Gain an even better understanding of this chapter by going to the companion website for additional study resources. Take advantage of the Pre- and Post-Test quizzing tool, which is designed to help you grasp difficult concepts by referring you back to review specific pages in the chapter for questions you answer incorrectly. Use the flash cards to master key terms and check out the many other study aids you'll find there. Visit the Marriage and Family Resource Center on the site. You'll also find special features such as access to Info-Trac© College Edition (a database that allows you access to more than 18 million full-length articles from 5,000 periodicals and journals), as well as GSS Data and Census information to help you with your research projects and papers.

CHAPTER 8

©Scott Barrow

Singlehood, Pairing, and Cohabitation

Outline

What Do YOU Think?

Are the following statements TRUE or FALSE?
You may be surprised by the answers (see answer key on the following page).

T F **1** Looking for a mate can be compared to shopping for goods in a market.

T F **2** Generally, the most important factor in judging someone at the first meeting is how he or she looks.

T F **3** There is a significant shortage of single eligible African American men, which makes marriage less likely for African American women.

T F **4** If a woman asks a man on a first date, it is generally a sign that she wants to have sex with him.

T F **5** The lesbian subculture values being single and unattached more than being involved in a stable relationship.

T F **6** Singles, compared to their married peers, tend to depend more on their parents.

T F **7** An important dating problem that men cite is their own shyness.

T F **8** Cohabitation has become part of the courtship process among many young adults.

T F **9** Compared to married couples, cohabiting couples have a more accepting attitude toward infidelity.

T F **10** Previously married cohabitants are more likely than never-married cohabitants to view living together as a test of marital compatibility.

Do you know what this is?

Real Life Juliet Seeks Romeo: I have been searching the world over, looking for my true love. I am a friendly, ambitious, compassionate, hardworking female, who enjoys music, dancing, travel, and the beach. Looking for someone who wants to share a movie, dinner, a laugh, and maybe a lifetime. I know you're out there somewhere.

Of course, we all know this is a personal ad, one of the many found each day in newspapers and magazines, or on multiple sites on the Internet. Along with dating services, computer matchmakers, and singles clubs, such personal ads represent some more recent ways in which Americans go about trying to find their "one and only." In recent years, even reality television programs have been added into the mix, pushing such attempts into previously uncharted water. On February 15, 2000, the Fox Network aired *Who Wants to Marry a Multi-Millionaire?* Many wondered, what could be next? Now we know: two *Joe Millionaires*, *Who Wants to Marry My Dad?*, *Bachelor*, *Married by America*, and so on. Each of these reality shows has tried to capitalize on our age-old fascination with how people get together.

There is considerable social science interest, too, in understanding how people find their spouses or partners. In addition, researchers have studied *who* we choose and *why* we choose those particular individuals. In this chapter, we not only look at the general rules by which we choose partners but also examine dating, romantic relationships, and cohabitation. Not everyone is actively looking for a relationship or intending to ultimately marry. Thus, we look, too, in this chapter at the growth in the unmarried population and at the singles world.

Over the last several decades, many aspects of pairing, such as the legitimacy of premarital intercourse and cohabitation, have changed considerably, radically affecting marriage. Today, large segments of American society accept and approve of both premarital sex and cohabitation. Marriage has lost its exclusiveness as the only legitimate relationship in which people can have sex and share their everyday lives. Increasing numbers of Americans experience both premarital sex and cohabitation in their lives. These issues, too, are examined in this chapter.

Choosing Partners

How do we choose the people we date, live with, or marry? Your initial response might be, "Simple. We fall in love!" Although love is the major criterion used to select a spouse, and most people who marry would say they are doing so out of love, many factors operate alongside and upon love.

In theory, most of us are free to select as partners those people with whom we fall in love, but other factors enter the process and our choices become somewhat limited by rules of mate selection. Once you understand some principles of mate selection in our culture, without ever having met a friend's new boyfriend or girlfriend, you can deduce many things about him or her. For example, if a female friend at college has a new boyfriend, you would be safe in guessing that he is about the same age or a little older, probably taller, and a college student. Furthermore, he is probably about as physically attractive as your friend (if not, their relationship may not last); his parents probably are of the same ethnic group and social class as hers; and he is probably about as intelligent as your friend. If a male friend has a new girlfriend, many of the same things apply, except that she is probably the same age or younger and shorter than he is. Some relationships will depart from such conventions, and many will have one or two characteristics on which the partners differ (or differ more), but you will probably be correct in most instances. These are not so much guesses as deductions based on the principle of homogamy, discussed later in this chapter.

The Marketplace of Relationships

The process of choosing partners is affected by bargaining and exchange. We select each other in a kind of **marketplace of relationships.** We use the notion of a "marketplace" to convey that, as in a commercial marketplace, when we form relationships we enter

Answer Key for What Do You Think

1 False, see p. 280; 2 False, see p. 281; 3 True, see p.283; 4 True, see p. 296; 5 False, see p. 306; 6 False, see p. 305; 7 False, see p.297; 8 True, see p. 308; 9 True, see p. 313; 10 True, see p. 310.

exchange relationships, much as when we exchange goods.

Unlike a real marketplace, however, the "relationship marketplace" is more of a process, not a place, in which *we* are the goods exchanged. Each of us has certain resources—such as socioeconomic status, looks, and personality—that determine our marketability. As Matthijs Kalmijn (1998) puts it, "Potential spouses are evaluated on the basis of the resources they have to offer, and individuals compete with each other for the spouse they want most by offering their own resources in return." We bargain with the resources we possess. We size ourselves up and rank ourselves as a good deal, an average package, or something to be "remaindered"; we do the same with potential dates and, ultimately, mates. Our "exchanges" are more often between equally valuable goods. In other words, we tend to seek people about *as attractive* or *as intelligent* as ourselves.

Physical Attractiveness: The Halo Effect, Rating, and Dating

The Halo Effect

Pretend for a moment that you are at a party, unattached. You notice that someone is standing next to you as you reach for some chips or a drink. He or she says hello. In that moment, you have to decide whether to engage him or her in conversation. On what basis do you make that decision? Is it looks, personality, style, sensitivity, intelligence, or something else?

Most people consciously or unconsciously base this decision on appearance. If you decide to talk to the person, you probably formed a positive opinion about how she or he looked. In other words, he or she looked "cute," looked like a "fun person," gave a "good first impression," or seemed "interesting." Physical attractiveness is particularly important during the initial meeting and early stages of a relationship.

> ### Reflections
> How important are looks to you? Think back. Have you ever mistakenly judged someone by his or her looks? How did you discover your error? How did you feel?

Most people would deny that they are attracted to others *just* because of their looks. However, we tend to infer qualities based on looks. This inference is called the **halo effect**—the assumption that good-looking people possess more desirable social characteristics than unattractive people. In a well-known experiment (Dion et al. 1972), students were shown pictures of attractive people and asked to describe what they thought these people were like. Attractive men and women were assumed to be more sensitive, sexually responsive, poised, and outgoing than others; they were assumed to be more exciting and to have better characters than "ordinary" people. Furthermore, attractive people are preferred as friends, candidates, and prospective employees, and they receive more leniency when defendants in court (Ruane and Cerulo, 2004). Research indicates that overall, the differences between perceptions of attractive and average people are minimal. It is when attractive and average people are compared to those considered to be unattractive that there are pronounced differences, with those perceived as unattractive being rated more negatively (Hatfield and Sprecher 1986).

The Rating and Dating Game

In more casual relationships, the physical attractiveness of a romantic partner is especially important. Elaine Hatfield and Susan Sprecher (1986) suggest three reasons people come to prefer attractive people over unattractive ones. First, there is an "aesthetic appeal," a simple preference for beauty. Second, there is the "glow of beauty," in which we assume that good-looking people are more sensitive, modest, self-confident, sexual, and so on. Third, there is the deflected "status" we achieve by dating attractive people.

Research has demonstrated that good-looking companions increase our status. In one study, men were asked their first impressions of a man seen alone, arm-in-arm with a beautiful woman, and arm-in-arm with an unattractive woman. The man made the best impression with the beautiful woman. He ranked higher alone than with an unattractive woman. In contrast to men, women do not necessarily rank as high when seen with a handsome man. A study in which married couples were evaluated found that it made no difference to a woman's ranking if she was unattractive but had a strikingly handsome husband. If an unattractive man had a strikingly beautiful wife, it was assumed that he had something to offer other than looks, such as fame or fortune.

Trade-Offs

As we mix and meet people, we don't necessarily gravitate to the *most attractive* person in the room, but rather to those about as attractive as ourselves. Sizing up someone at a party or dance, a man may say, "I'd have no chance with her; she's too good-looking for me." Even if people are allowed to specify the qualities they want in a date, they are hesitant to select anyone notably different from themselves in social desirability.

We also tend to choose people who are our equals in terms of intelligence, education, and so on (Hatfield and Walster 1981). However, if two people are different in looks or intelligence, usually the individuals make a trade-off in which a lower-ranked trait is exchanged for a higher-ranked trait. A woman who values status, for example, may accept a lower level of physical attractiveness in a man if he is wealthy or powerful.

Are Looks Important to Everyone?

For all of us more ordinary-looking people, it will come as a relief to know that looks aren't everything. Looks are most important to certain types or groups of people and in certain situations or locations (for example, in classes, at parties, and in bars, where people do not interact with one another extensively on a day-to-day basis). Looks are less important to those in ongoing relationships and to those older than young adults. Those who interact regularly—as in working together—put less importance on looks (Hatfield and Sprecher 1986). In adolescence, the need to conform and the impact of peer pressure make looks especially important as we may feel pressured to go out with handsome men and beautiful women.

Men tend to care more about how their partners look than do women (Buunk et al. 2002; Regan 2003). This may be attributed to the disparity of economic and social power. Because men tend to have more assets (such as income and status) than women, they can afford to be less concerned with their potential partner's assets and can choose partners in terms of their attractiveness. Because women lack the earning power and assets of men, they may have to be more practical and choose a partner who can offer security and status. Unsurprisingly, then, women are more likely than men to emphasize the importance of socioeconomic factors (Regan 2003).

Most research on attractiveness has been done on first impressions or early dating. At lower levels of relationship involvement, physical attractiveness is more important. As relationship involvement increases, status and personality become more important, appearance less. For long-term relationships (for example, marriage) women and men prefer mates about as

■ *People tend to choose partners who are about as attractive as themselves.*

attractive as themselves. For short-term, less involved relationships, both men and women prefer more attractive mates. Bram Buunk and colleagues (2002) interpret this pattern to reflect potential costs of having as a long-term partner someone to whom others are strongly attracted.

Researchers are finding, however, that attractiveness is not *un*important in established relationships. Most people expect looks to become less important as a relationship matures, but Philip Blumstein and Pepper Schwartz (1983) found that the happiest people in cohabiting and married relationships thought of their partners as attractive. People who found their partners attractive had the best sex lives. Physical attractiveness continues to be important throughout marriage. It is, however, joined by other qualities, and these other attributes are deemed more important.

Bargains and Exchanges

Likening relationships to markets or choosing partners to an exchange may not seem romantic, but both are deeply rooted in marriage and family customs. In some cultures, for example, arranged marriages take place only after extended bargaining between families. The woman is expected to bring a dowry in the form of property (such as pigs, goats, clothing, utensils, or land) or money, or a woman's family may demand a bride-price if the culture places a premium on women's productivity. Traces of the exchange basis of marriage still exist in our culture in the traditional marriage ceremony when the bride's parents pay the wedding costs and "give away" their daughter.

Gender Roles

Traditionally, relationship exchanges have been based on gender. Men used their status, economic power, and role as protector in a trade-off for women's physical attractiveness and nurturing, childbearing, and housekeeping abilities; women, in return, gained status and economic security in the exchange.

The terms of bargaining have changed some, however. As women enter careers and become economically independent, achieving their own occupational status and economic independence, what do they ask from men in the marriage exchange? Clearly, many women expect men to bring more expressive, affective, and companionable resources into marriage. An independent woman does not have to "settle" for a man who brings little more to the relationship than a paycheck; she wants a man who is a partner, not simply a provider.

But even today, a woman's bargaining position may not be as strong as a man's. Women earn only about three-fourths of what men earn, are still significantly underrepresented in many professions, and have seen many of the things women traditionally used to bargain with in the marital exchange—such as children, housekeeping services, and sexuality—become devalued or increasingly available outside of relationships. Children are not the economic assets they once were. A man does not have to rely on a woman to cook for him, sex is often accessible in the singles world, and someone can be paid to do the laundry and clean the apartment.

Women are further disadvantaged by the **double standard of aging.** Physical attractiveness is a key bargaining element in the marital marketplace, but the older a woman gets, the less attractive she is considered. For women, youth and beauty are linked in most cultures. Furthermore, as women get older, their field of potential eligible partners declines because men tend to choose younger women as mates.

The Marriage Squeeze and Mating Gradient

An important factor affecting the marriage market is the ratio of men to women. Researchers Marcia Guttentag and Paul Secord (1983) argue that whenever there is a shortage of women in society, marriage

and monogamy are valued; when there is an excess of women, marriage and monogamy are devalued. The scarcer sex is able to weight the rules in its favor. It gains bargaining power in the marriage marketplace.

The **marriage squeeze** refers to the gender imbalance reflected in the ratio of available unmarried women and men. Because of this imbalance, members of one gender tend to be "squeezed" out of the marriage market. The marriage squeeze is distorted, however, if we look at overall figures of men and women without distinguishing between age and ethnicity. Overall, there are significantly more unmarried women than men: 87 single men for every 100 single women (U.S. Census Bureau 2003). This figure, however, is somewhat deceptive. From ages 18 to 44, the prime years for marriage, there are significantly more unmarried men than women, reversing the overall marriage squeeze. Combining widowed, divorced, and never-married people, in 2002 there were 113 unmarried men, aged 18 to 44, for every 100 unmarried women (U.S. Census Bureau 2003). Thus, women in this age group have greater bargaining power and are able to demand marriage and monogamy. But once ethnicity is taken into consideration, the many African American women of all ages are "squeezed out" of the marriage market. With eligible males scarcer, African American men have greater bargaining power and are less likely to marry because of more attractive alternatives (see Figure 8.1).

"All the good ones are taken" is a common complaint of women in their mid-30s and beyond, even if there are still more men than women in that age bracket. The reason for this is the **mating gradient,** the tendency for women to marry men of higher status. Sociologist Jessie Bernard (1982) comes to this conclusion:

In our society, the husband is assigned a superior status. It helps if he actually is superior in ways— in height, for example, or age or education or occupation—for such superiority, however slight, makes it easier for both partners to conform to the structural imperatives. The [woman] wants to be able to "look up" to her husband, and he, of course, wants her to. The result is a situation known sociologically as the marriage gradient.

Although we tend to marry those with the same socioeconomic status and cultural background, men tend to marry women slightly below them in age, education, and so on. The marriage gradient puts high-status women at a disadvantage in the marriage marketplace. Bernard continues:

> The result is that there is no one for the men at the bottom to marry, no one to look up to them. Conversely, there is no one for the women at the top to look up to; there are no men superior to them. . . . The never-married men . . . tend to be "bottom-of-the-barrel" and the women . . . "cream-of-the-crop."

The Field of Eligibles

The men and women we date, live with, or marry usually come from the **field of eligibles**—that is, those whom our culture approves of as appropriate potential partners. The field of eligibles is defined by two principles: **endogamy** (marriage within a particular group) and **exogamy** (marriage outside a particular group).

Endogamy

People usually marry others from within their same large group—such as the nationality, ethnic group, or socioeconomic status with which they identify—because they share common assumptions, experiences, and understandings. Endogamy strengthens group structure. If people already have ties as friends, neighbors, work associates, or fellow church members, a marriage between such acquaintances solidifies group ties.

To take an extreme example, it is easier for two Americans to understand each other than it is for an American and a Fula tribesperson from Africa. Americans are monogamous and urban, whereas the Fula are polygamous wandering herders. But another, darker force may lie beneath endogamy: the fear and distrust of outsiders, those who are different from ourselves. Both the need for commonality and the distrust of outsiders urge people to marry individuals like themselves.

Exogamy

The principle of exogamy requires us to marry outside certain groups—specifically, outside our own family (however defined) and outside our sex. Exogamy is enforced by taboos deeply embedded within our psychological makeup. The violation of these taboos may cause a deep sense of guilt. A marriage between a man

---

Figure 8.1 ■ Ratio of Unmarried Men to Unmarried Women by Age and Ethnicity, 2002

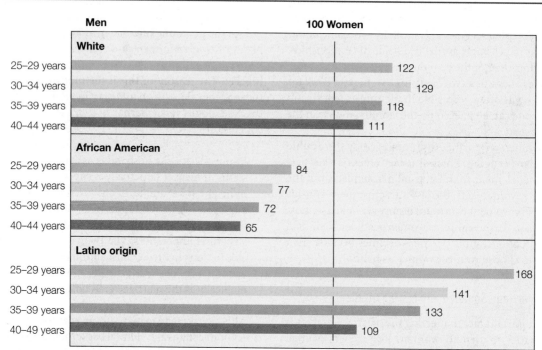

SOURCE: *Current Population Reports 2002*, unpublished Table 7.

and his mother, sister, daughter, aunt, niece, grandmother, or granddaughter is considered incestuous; women are forbidden to marry their corresponding male relatives. Beyond these blood relations, however, the definition of incestuous relations changes. One society defines marriages between cousins as incestuous, whereas another may encourage such marriages.

Some states prohibit marriages between stepbrothers and stepsisters, as well as cousins; others do not. In Chapter 9 we will further consider the laws that specify who can and can't marry. In general, there has been a growing tendency toward allowing individuals choice of partners without state interference. For example, in 1966, the U.S. Supreme Court ruled that laws prohibiting marriage between individuals of different races were unconstitutional (*Loving v. Virginia*). Massachusetts now allows same sex couples to marry and some other states provide same-sex couples the protection and rights heterosexuals receive when they marry. Denial of legal marriage rights and its many protections and benefits is otherwise unconstitutional.

Homogamy

Endogamy and exogamy interact to *limit* the field of eligibles. The field is further limited by society's encouragement of **homogamy,** the tendency to choose a mate whose personal or group characteristics are similar to ours. (See this chapter's "Understanding Yourself" box, on page 295, which discusses Internet personals, computer dating, and homogamy.) This is also known as *positive assortative mating* (Blackwell 1998). **Heterogamy** refers to the tendency to choose a mate whose personal or group characteristics differ from our own. The strongest pressures are toward homogamy. We may make homogamous choices regarding any number of characteristics, including age and race, but also such characteristics as height (Blackwell 1998). As a result, our choices of partners tend to follow certain patterns. These homogamous considerations generally apply to heterosexuals, gay men, and lesbians alike in their choice of partners.

The most important elements of homogamy are race and ethnicity, religion, socioeconomic status, age, and personality characteristics. These elements are strongest in first marriages and weaker in second and subsequent marriages (Glick 1988). They also strongly influence our choice of sexual partners, because our sexual partners are often potential marriage partners (Michael et al. 1994).

RACE AND ETHNICITY. Most marriages are between members of the same race. Of the nearly 55 million married couples in the United States in 2000, 98% of them consisted of husbands and wives of the same race. Nearly 6% of marriages in 2000 were between people from different racial backgrounds. Interestingly, as the overall phenomenon of interracial marriage has been increasing since the 1980s, it has especially increased among highly educated people (Harris and Ono 2005). Although most often taken to mean *black–white marriage,* such marital pairings are only approximately 25% of all racial intermarriages. This is the pairing most likely to be the target of hostility and prejudice (Leslie and Letiecq 2004).

Racial intermarriage varies greatly among different cities and regions in the United States. David Harris and Hiromi Ono assert that without taking into consideration "local" marriage markets, we can't completely and accurately understand racial marriage patterns.

Although the United States is 75.1% white, 12.3% black, 3.6% Asian, and 12.5% Latino, the racial composition of major cities exhibits substantial deviations from the national pattern, and many cities differ from one another in important ways. For example, whites are 45% of the population in Philadelphia but only 12% in Detroit. Asians are at least 25% of the population in San Jose, San Francisco, and Honolulu but no more than 2% of the population in Phoenix, San Antonio, and Detroit (Harris and Ono 2005, 238).

Harris and Ono contend that by failing to take into account the reality of local marriage markets and assuming, instead, a single national marriage market, projected levels of relative likelihood of racial homogamy are exaggerated between 19% (between whites and blacks) and 53% (between whites and Latinos). Where there is greater opportunity to find spouses of the same race, rates of homogamy are higher and intermarriage is less. On the other hand, racial and ethnic heterogeneity are associated with higher levels of intermarriage (Kalmijn 1998).

Of the more than 1 million interracial couples, one-fourth were marriages between blacks and whites (Fields and Casper 2001). By 1993, 12% of all new marriages involving African Americans were interracial. This is nearly double the percentage in 1980 (6.6%), and four times the percentage from 1970 (2.6%) (Besharov and Sullivan 1996). It is suggested that the reasons both black groom–white bride and black bride–white groom are increasing is the rise of a black middle class, making African American men and women more attractive to middle-class whites.

Black women still face obstacles to marriage of any kind; they are more than twice as likely to have children born out of wedlock. About 1.2% of marriages consist of one partner who is white and one from an Asian, Native American, or other nonwhite group (Fields and Casper 2001).

Based on their study of 76 black–white intermarriages (52 black male–white female couples and 24 black female–white male couples), Leigh Leslie and Bethany Letiecq (2004) suggest that success in black–white intermarriages may depend upon the degree to which the partners possess pride in their race or culture without diminishing other races. This appears to be especially true for the black spouse in such marriages and seems to influence the quality of married life well into the marriage.

Those who had resolved issues of racial identity and developed a strong black identity while showing racial tolerance and appreciation of other races, more positively evaluated their marriage, felt less ambivalent about it, and/or worked harder to maintain it. However, those who had more negative assessments of one culture or the other experienced lower marital quality (Leslie and Letiecq 2004, 570).

To an extent, this challenges the idea that, for interracial couples, race becomes irrelevant or unimportant (Leslie and Letiecq 2004). How they think and feel about race is of major significance in the quality of their marital experience. Unexpectedly, social support only "modestly" predicted marital quality. This could be a byproduct of the relative prevalence and acceptance of interracial marriage in the area where the research was done, the relatively comfortable economic circumstances of the couples studied, or evidence that interracial couples have learned to survive, if not thrive, even in the absence of social support (Leslie and Letiecq 2004).

A qualitative study of 19 individuals who were involved in interracial relationships uncovered a range of harassment and hostility to which they develop a number of management strategies (Datzman and Brooks Gardner 2000). These include ignoring the harassment, limiting the settings where they would be seen as a couple to those they knew would be supportive or to staying home altogether, having others with them who are more supportive, and directly confronting any harassment. Especially when such harassment is new, the emotional impact might include shock and surprise, numbness, sadness and shame, and ultimately resentment or anger. Eventually, the anger might be replaced by pity felt toward the harasser or harassers.

Bob Pool/Getty Images/Photographer's Choice

■ *Interracial relationships are increasing but are still relatively uncommon. Such couples often find themselves the recipients of negative reactions.*

The degree of intermarriage between ethnic groups is of concern to some members of these groups because it affects the rate of assimilation and continued ethnic identity (Stevens and Schoen 1988). Almost half of all Japanese Americans marry outside their ethnic group (Takagi 1994). More than half of all Native Americans are married to non–Native Americans (Yellowbird and Snipp 1994). For both Japanese Americans and Native Americans, intermarriage leads to profound questions about their continued existence as distinct ethnic groups in the twenty-first century. Among European ethnic groups in this country, such as Italians, Poles, Germans, and Irish, only one in four marries within the ethnic group. The ethnic identity of these groups has decreased considerably since the beginning of this century. Interestingly, Louisiana Cajuns have high rates of ethnic homogamy, especially for a group of their size and considering the length of time they have been in the United States. Among married Cajun women, more than 75% were married to Cajun men; among Cajun men, more than 70% were also homogamous (Bankston III and Henry 1999).

Matthijs Kalmijn points out that marrying *outside* of the group is not the same for all ethnic groups. For example, when Latinos marry "out," they are more likely marrying Latinos of a different cultural origin than they are white, European Americans. Asians, on the other hand, are much less likely to marry Asians of a different background and more likely when "marrying out" to marry whites (Kalmijn 1998). Kalmijn further indicates that the highest rates of homogamy are among blacks. The lowest rates are among European ethnic groups and among American Indians. Hispanics and Asians have intermediate homogamy rates (Kalmijn 1998).

RELIGION. Until the late 1960s, religion was a significant factor in marital choice. Today, most religions still oppose interreligious marriage because they believe it weakens individual commitment to the faith. Nonetheless, interreligious dating and marriage have been increasing. Almost half of all Catholics marry outside their faith (Maloney 1986). Tracking changes over a quarter century, research found that where in the early 1960s only 6% of Jews chose non-Jewish partners, by the late 1980s almost 40% of Jews were marrying non-Jewish spouses (Mindel, Haberstein, and Wright 1988). Intermarriages between Jews and Gentiles have continued to increase, as have marriages between Catholics and Protestants (Kalmijn 1998).

Data drawn from a study of 105 never-married undergraduates enrolled in courtship and marriage courses at a large southeastern university reveal a relatively small role played by religion in considerations of mate choice (Knox, Zussman, and Daniels 2002). Specifically, only one in five (22%) respondents agreed that, "I will only marry someone of the same religious background." Gender surfaced as an influence on attitudes, because females were more likely than males (27% to 15%) to agree that they would only marry someone of the same religion. Females were also more likely than males (20% to 15%) to believe that they would be disappointing their parents by dating outside their faith. Finally, Knox and colleagues determined that there were religious differences in the importance attached to religious homogamy, with Baptists being more likely than Methodists or Catholics to oppose marrying outside their faith, as evident in their belief that such marriages are at greater risk of divorce (Knox et al. 2002).

Those who marry from different religious backgrounds do have greater risk of divorce than those from similar backgrounds (Bumpass, Martin, and Sweet 1991; Lehrer and Chiswick 1993; Sander 1993). Jews who intermarry are twice as likely to divorce as those who marry homogamously (Chintz and Brown 2001). Apparently, being of different faiths is not the only consideration. It seems that the *larger* the "religious distance," or disparity between two people's backgrounds, the more likely they are to characterize their marriage as "unhappy" (Ortega, Whitt, and William 1988). In a study of Jewish marriages, what matters more in predicting the amount of conflict and instability is the extent of agreement or disagreement on Jewish issues, not what self-reported labels people use to identify themselves (Chintz and Brown 2001).

Religious groups tend to discourage interfaith marriages, believing that such marriages, in addition to weakening individual beliefs, lead to children being reared in a different faith or to secularization of the family. Such fears, however, may be overstated. Among Catholics who marry Protestants, for example, there seems to be little secularization by those who feel themselves to be religious (Petersen 1986). Some who are from different religious backgrounds, however, do convert to their spouses' religions.

SOCIOECONOMIC STATUS. Most people marry others of their own socioeconomic status and of the same or similar educational background. Even if a person marries outside his or her ethnic, religious, or age group, the selected spouse will probably be from the same socioeconomic level. Furthermore, some ethnic or racial homogamy may be increased because of tendencies toward socioeconomic homogamy (Bankston III and Henry 1999). Of the various dimensions of socioeconomic status (family background, education, and occupation), the weakest appears to be between spouses' class origins (correlation of about 0.30). The correlation between husbands and wives occupational statuses is stronger (around 0.40). However, the strongest correlation is between spouses' educational backgrounds (approximately 0.55). This holds true in the United States, as well as most other countries. In the United States, educational homogamy has "strongly" increased (Kalmijn 1998).

Socioeconomic homogamy results from the combination of *choice-shaping factors*, such as shared values, tastes, goals, and expectations, and *opportunity-determining factors*, such as residential neighborhood, school, and/or occupation. In addition, control is exerted by affluent families to ensure that their children marry at the "right" level.

Not everyone marries homogamously. Men more than women marry below their socioeconomic level (**hypogamy**); women more often "marry up" (a practice known as **hypergamy**). When class *intermarriage* occurs, it is rarely a case of spouses from opposite extremes (that is, paupers and princesses). Both the upper and the lower levels of the class spectrum appear more "closed" than the middle levels (Kalmijn 1998).

Looking at the education component of socioeconomic position a little more closely, we find that the biggest barrier is the one separating college graduates from those with lower levels of education. Occupationally, the divide is between those in white-collar and those in blue-collar occupations. It appears as though the cultural status, not the economic status, of occupations is a more important factor in determining compatibility and attractiveness for marriage (Kalmijn 1998).

AGE. Reflecting the data in Chapter 3 on trends in age at marriage, Americans have long tended to marry those of roughly the same or similar ages. Typically, the man is slightly older than the woman. Age is important because we view ourselves as members of a generation, and each generation's experience of life leads to different values and expectations. Furthermore, different developmental and life tasks confront us at different ages. A 20-year-old woman wants something different from marriage and from life than a 60-year-old man does. By marrying people of similar ages, we often ensure congruence for developmental tasks. The gap between grooms' and brides' ages has narrowed in recent years, as the ages at which both men and women enter marriage have climbed.

Research suggests that the importance individuals place on age varies *by age* differently for men than for women. As men age they prefer women progressively younger than themselves. Women, on the other hand, prefer for their partners to be about the same age (ranging from slightly younger through slightly older) up to 10 years older than themselves. This does not appear to vary much, even as women age. Generally, women prefer men slightly older than themselves as spouses (Buunk et al. 2002).

Interesting data from the United States and Australia reveal that the same age preferences that exist among heterosexuals exist among homosexual men and women—men prefer younger partners, women prefer partners of about the same age. This tendency first surfaces among older—middle-aged—gay men (Over and Phillips, 1997).

Despite the popular beliefs that we will be more compatible with partners similar to us in age and—

conversely—relationships with partners much older than ourselves will be plagued by problems of incompatibility, research suggests otherwise. A study by David Knox and Tim Britton of 97 female students and faculty involved with partners between 10 and 25 years older than themselves concluded that couples in "age discrepant" relationships were, indeed, happy. In the study, 80% indicated either agreement (40%) or strong agreement (40%) with the statement: "I am happy in my current relationship." Only 4% disagreed. Furthermore, more than 60% stated that if their current relationship ends, they would enter another age-discrepant relationship. They identified the following benefits of such relationships: maturity (mentioned by 58% of the women), financial security (58%), dependability (51%), and higher status (28%). Each of the relationship problems in Table 8.1 was identified by at least 25% of the women.

Also of note, only 25% of the women stated that their relationships had the support of their friends or parents. Fathers were most disapproving, with more than 40% identified as not being in support of the relationship (Knox and Britton 1997).

Marital and Family History

An interesting application of the concepts of homogamy and heterogamy (intermarriage) can be found with regard to marital history. Essentially, never-married people are more likely to marry other never-married people than they are to "intermarry" by marrying divorced people (Ono 2005). Hiromi Ono questions whether this is a "by-product" of other homogamous patterns (such as age, socioeconomic status, or parenthood status) or a deliberate choice that individuals make to marry someone of similar marital history. A divorced person may believe that only another divorced person will similarly understand and have experience with the lingering ties to prior marriages.

Table 8.1 ■ Problems Identified by Women in Age-Discrepant Relationships

Problem	Percentage Reporting
Money	39
In-laws	33
Recreation	33
Children	25

Conversely, never-married individuals who marry divorced partners may find that they have to deal with lower amounts of resources because of the continued demands of former spouses and the needs of children of former marriages. This, in turn, may give rise to jealousy and impede the development of needed levels of trust (Ono 2005). Ono determined that **marital history homogamy** occurs more as a result of deliberate choices than as a byproduct of other statuses. Ono also reasonably speculated that parental status, like marital history, operates in a similar fashion. Parents make lifestyle concessions to their parenting responsibilities that nonparents don't have to make. Where children and their needs become priorities for parents, nonparents can maintain other priorities.

The structure of an individual's family of origin also turns out to be a factor in the process of mate selection. Children of divorced parents often marry other children of divorced parents. Research by Nicholas Wolfinger suggests that coming from a divorced home increases by 58% the likelihood of choosing another child of divorce as a spouse. Although homogamy often is associated with a greater chance for marital happiness and stability, family structure homogamy may be a noteworthy exception because marriages in which both spouses are children of divorce face greater odds of marital failure. Marriages in which either spouse comes from a divorced family are twice as likely to fail as those in which neither spouse is a child of divorce. When both spouses are from divorced homes, their marriages face three times the likelihood of failure as marriages between two children of intact parental marriages (Wolfinger 2003).

Residential Propinquity

An additional homogamous factor is based on the principle of **residential propinquity**—the tendency we have to select partners (for relationships and for marriages) from a geographically limited locale. Put differently, the likelihood of marriage decreases as the distance between two people's residences increases. The obvious explanation behind this is one of opportunity. In most instances, to start dating or get together with someone you have to first meet. Our chances of meeting are greater when our daily activities (shopping, commuting, eating out, and so forth) overlap.

Although it is easy to trivialize this tendency as too obvious to be meaningful, consider the implications it has for the other patterns of homogamy. American communities are often segregated by class, race, or both.

In some towns, they may even have religious splits (for example, the Catholic side and the Protestant side of town or a Jewish neighborhood). Public schools, being neighborhood based, further the tendency for us to associate with others like ourselves. Thus, the types of people we are most likely to come into contact with and with whom we might develop intimate relationships or eventually marry are a lot like ourselves. Meeting at school promotes age, educational, and social class homogamy (Kalmijn and Flap 2001).

Thus, within a society somewhat residentially segregated by race or social class, residential propinquity may explain some other homogamous tendencies by how it limits our opportunity. But the story is more complicated than just where we live. After all, unmarried people do not just wander around a region looking for a spouse; they spend most of their lives in small and functional places, such as neighborhoods, schools, workplaces, bars, and clubs. Such local marriage markets are often socially segregated, which is why they are important for explaining marriage patterns. In the sociological literature, three local markets have been considered most often: the school, the neighborhood, and the workplace. Of these three, schools are considered the most efficient markets because they are homogeneous with respect to age and heterogeneous with respect to sex (Kalmijn 1998, 403).

At the same time that the opportunities to meet others like ourselves are so much greater than the opportunities to meet people unlike ourselves, the cultural beliefs that homogamous marriages are better or more likely to be stable might reinforce people's tendencies to "look locally," where they are more likely to be surrounded by people like themselves.

Reflections

Keeping heterogamy and homogamy in mind, think about those who are or have been your romantic or marital partners. In what respects have your partners shared the same racial, ethnic, religious, socioeconomic, age, and personality characteristics with you? In what respects have they not? Have shared or differing characteristics affected your relationships? How?

Understanding Homogamy and Intermarriage

Factors in the choice of partner interact with one another. Ethnicity and socioeconomic status, for example, are often closely related because of discrimination.

Many African Americans and Latinos are working class and are not as well educated as Caucasians. Caucasians generally tend to be better off economically and are usually better educated. Thus, a marriage that is endogamous in terms of ethnicity is also likely to be endogamous in terms of education and socioeconomic status.

Matthijs Kalmijn (1998) suggests that "in general, marriage patterns arise from three social forces: the preferences of individuals for resources in a partner, the influence of the social group, and the constraints of the marriage market." It appears that all three of these combine to produce the tendencies toward homogamy and the patterns of mate choice we observe, but it is difficult to determine the relative strength of the factors or what is "most influential" in shaping mate selection practices. What we can say with more certainty is that the presence of both opportunity constraints and outside influence (or "interference") makes it unwise to conclude that homogamy automatically reflects hostility or animosity toward others unlike oneself. It may not even illustrate an outright preference for people like oneself.

In addition to questioning causes, we might ask about consequences. Are homogamous relationships "better" or "stronger" relationships? Data on intermarriages by religion, race, and/or class are inconsistent on this question. Some studies reveal greater difficulties in non-homogamous relationships and higher likelihood of divorce among those who intermarry. Others fail to substantiate the negative outcome (Eshelman 1997). The most consistent findings are related to those risks associated with religious intermarriages, although these risks are not great.

There are three possible explanations as to why heterogamous marriages might be less stable than homogamous marriages (Udry 1974):

1. Heterogamous couples may have considerably different values, attitudes, and behaviors, which may create a lack of understanding and promote conflict.

2. Heterogamous marriages may lack approval from parents, relatives, and friends. Couples are then cut off from important sources of support during crises.

3. Heterogamous couples are probably less conventional and therefore less likely to continue an unhappy marriage for the sake of appearances.

Still other "consequences" of homogamy, especially by social class, education, or race, can be identified.

Hiromi Ono (2005, 304) contends that—especially with regard to race, education, and social class patterns but also with reference to marital history—homogamy has the potential to widen social inequality. What about consequences of intermarriage? Matthijs Kalmijn argues that intermarriage potentially has the following effects:

■ Intermarriage can decrease the importance of cultural differences because the children of mixed marriages are less likely to identify themselves with a single group. Even when mixed couples socialize children into the culture of a single group, the children are less likely to identify with that group when intermarriage in society is common.

■ Through intermarrying, individuals may question and lose negative attitudes they have toward other groups. Spouses and their wider networks (of kin and friends) gain the opportunity to get to know people "different" from themselves and question any biases and stereotypes they previously held.

Theories and Stages of Mate Selection

Say that you meet someone who fits all the criteria of homogamy: same ethnic group, religion, socioeconomic background, age, and personality traits—the person your parents always dreamed you'd marry. Unfortunately, you can't stand this person. Homogamy by itself doesn't work. A range of theories has been suggested to address the question of why we select particular individuals. Do "opposites attract"? Do "birds of a feather flock together"? Do we unconsciously select people like our parents? What is more important: finding someone who seems to think as we do about things, or finding someone whose behavior fits what we expect in a partner?

Each of the preceding questions illustrates an existing theory of mate selection. The commonsense notion that "opposites attract" is in keeping with **complementary needs theory,** the belief that people select as spouses those whose needs are different. Thus, an assertive person who has difficulty compromising will be drawn to a less outgoing and highly adaptable person. The notion that "birds of a feather flock together" is more in keeping with theories such as **value theory** or **role theory,** in which gratification follows from finding someone who feels and/or thinks like we do. Having someone who shares our view of what's important in life or who acts in ways that we desire in a partner validates us, and this sense of validation leads to an intensification of what we feel toward that other person. **Parental image theory** suggests that we seek partners similar to our opposite-sex parent. Some versions of parental image theory draw on Freudian concepts such as the Oedipus complex, whereas others point toward the lasting impressions made by our parents (Eshelman 1997; Murstein 1986).

Bernard Murstein developed a social exchange based, sequential theory known as **stimulus–value–role theory** to depict what happens between that "magic moment" with its mysterious chemistry of attraction and the decision to maintain a long-term relationship such as marriage. Murstein's theory identifies three stages of romantic relationships. At each stage, if the exchange seems equitable, the two will progress to the next stage and ultimately remain together (Murstein 1986). In the *stimulus* stage, each person is drawn or attracted to the other before actual interaction. This attraction can be physical, mental, or social. During the stimulus stage, with little other information on which to evaluate the other person, we make potentially superficial decisions. This is especially evident during first encounters.

In the next stage, the *value* stage, partners weigh each other's basic values seeking compatiblilty. Each person considers the other's philosophy of life, politics, sexual values, religious beliefs, and so on. Wherever they agree, it is a plus for the relationship. However, if they disagree—for example on religion—it is a minus for the relationship. Each person adds or subtracts the pluses and minuses along value lines. Based on the outcome, the couple will either disengage or go on to the next stage. Values are usually determined between the second and seventh meetings.

Eventually, in the *role* stage, each person analyzes the other's behaviors, or how the person fulfills his or her roles as lover, companion, friend, worker—and potential husband or wife, mother or father. Are the person's behaviors consistent with marital roles? Is he or she emotionally stable? This aspect is evaluated in the eighth and subsequent encounters.

Although the stimulus–value–role theory has been one of the more prominent theories explaining relationship development, some scholars have criticized it, especially regarding the question of whether we actually test the degree of "fit" between us and our partners. We might underestimate the importance of certain issues or, conversely, be focused more extensively on others. For example, religious fundamentalists and goddess worshippers may sometimes believe that they are compatible. They may not discuss religion; instead, they might focus on the "incredible" physical attraction in their relationship. They may believe that religion is not that important, only to discover after they are married that it is important.

Dating and Romantic Relationships

As increasing numbers of people delay marriage, never marry, or seek to remarry after divorce or widowhood, romantic relationships will, according to Catherine Surra (1991), "take different shapes at different points in time, as they move in and out of marriage, friendship, romance, cohabitation, and so on." As a result, researchers are shifting from the traditional emphasis on mate selection toward the study of the formation and development of romantic relationships, such as the dynamics of heterosexual dating, cohabitation, post-divorce relationships, and gay and lesbian relationships. The field of personal relationships is developing a broad focus that explores relationship dynamics (Duck 1994; Kelley et al. 1983; Perlman and Duck 1987).

Beginning a Relationship: Seeing, Meeting, and Dating

Although the general rules of mate selection are important in the abstract, they do not tell us how relationships begin. The actual process of beginning a relationship is discussed in the sections that follow.

Seeing

On a typical day, we may see dozens, hundreds, or even thousands of men and women. But seeing isn't enough; we must become aware of someone for a relationship

---

Reflections

What are some settings in which you "see" people? How do the settings affect the strategies you use to meet others? How do you move from meeting to "going out" with someone? What are your feelings at each stage of seeing, meeting, and dating?

to begin. It may only take a second from the moment of noticing to meeting, or it may take days, weeks, or months. Sometimes "noticing" occurs between two people simultaneously, other times it may take considerable time, and sometimes it never happens.

The setting in which you see someone can facilitate or discourage meeting each other (Murstein 1976, 1987). **Closed fields,** such as small classes or seminars, dormitories, parties, and small workplaces, are characterized by a small number of people who are likely to interact whether they are attracted or not. In such settings, you are likely to "see" and interact simultaneously. In contrast, **open fields,** such as beaches, shopping malls, bars, amusement parks, and large university campuses, are characterized by large numbers of people who do not ordinarily interact.

Meeting

How is a meeting initiated? Among heterosexuals, does the man initiate it? On the surface, the answer appears to be yes, but in reality, the woman often "covertly initiates . . . by sending nonverbal signals of availability and interest" (Metts and Cupach 1989). A woman will glance at a man once or twice and catch his eye; she may smile or flip her hair. If the man moves into her physical space, the woman then relies on nodding, leaning close, smiling, or laughing (Moore 1985).

Regardless of who initiates contact, a variety of verbal and nonverbal signals are used to convey attraction and interest to a potential partner. Smiling, moving closer to, gazing at, laughing, and displaying "positive facial expressions" are all gestures to convey interest or "flirt" (Regan 2003). Touch is also an important element in flirting, whether the touch consists of lightly touching the arm or hand or the face or hair of the target of interest or rubbing fingers across the other's arm (Regan 2003).

If a man believes a woman is interested, he often initiates a conversation using an opening line. The opening line tests the woman's interest and availability. You have probably used or heard an array of opening lines. According to women, the most effective are innocuous, such as "I feel a little embarrassed, but I'd like to meet you" or "Are you a student here?" The least effective are sexual come-ons, such as "You really turn me on. Do you want to have sex?" Women, more than men, prefer direct but innocuous opening lines over cute, flippant ones, such as "What's a good-looking babe like you doing in a college like this?"

A recent Web search for "pickup lines" identified more than *2 million* sites. There were sites specializing in math pickup lines, *Dr. Seuss* pickup lines, "Christian" pickup lines, "Jewish" pickup lines, and gothic pickup lines, as well as "cheesy," humorous, and bad pickup lines. There were lines for women to use with men, men to use with women, men to use with men, and women to use with women. The following list is a sampling of some opening lines men or women have used (or tried) to initiate contact: To many of us, these lines seem corny, shallow, and unlikely to generate the kind of impression that might lead to forming a relationship. Nevertheless, readers of this text may well spot one or a few that they have heard (or used).

"You must be tired, because you've been running through my mind all day."

"You know, if you held up eleven roses in front of a mirror, you would be looking at twelve of the most beautiful things in the world."

"Do you have a quarter? I promised my mother I'd call her when I met the girl (guy) of my dreams."

"Did they just turn on a fan in here or was that you blowing me away?"

"If I had a nickel for every time I met someone as beautiful as you, I'd have a nickel."

Much as they are more likely than women to use "a line," men are more likely to initiate a meeting directly, whereas women are more likely to wait for the other person to introduce himself or herself or to be introduced by a friend (Berger 1987). About a third or half of all relationships rely on introductions (Sprecher and McKinney 1993). An introduction has the advantage of a kind of prescreening, as the mutual acquaintance may believe that both may hit it off. Parties are the most common settings in which young adults meet, followed by classes, work, bars, clubs, sports settings, or events centered on hobbies, such as hiking (Marwell et al. 1982; Shostak 1987; Simenauer and Carroll 1982).

The Internet continues to gain popularity as a major way for people to "meet" a potential partner. Online, people can introduce themselves in fantasy-like images. A growing number of people first "meet" in cyberspace, find common interests, and form relationships that develop and intensify before they ever actually meet. Eleven percent of all internet using adults in the United States state that they have gone to an internet dating site for the purpose of meeting a potential partner (Madden and Lenhart, 2006). This translates to an estimated sixteen million adults. Moreover, Mary Madden and Amanda Lenhart report:

- Nearly a third of adults (31%), an estimated 63 million people, know someone who has used a dating website.

- A quarter of American adults (26%), 53 million people, claim to know someone who has gone on a date that was initiated via an internet site

- Fifteen percent of American adults, 30 million people, claim to know someone who has had a long-term relationship or married someone who they met on the Internet.

Single men and women also rely on printed personal classified ads, where men tend to advertise themselves as "success objects" and women advertise themselves as "sex objects" (Davis 1990). Their ads typically reflect stereotypical gender roles. Men advertise for women who are attractive and deemphasize intellectual, work, and financial aspects. Women advertise for men who are employed, financially secure, intelligent, emotionally expressive, and interested in commitment. Men are twice as likely as women to place ads. Other alternative forms of meeting others include video dating services, introduction services, and 1-900 party-line phone services.

Single men and women often rely on their churches and church activities to meet other singles. Black churches are especially important for middle-class African Americans, as they have less chance of meeting other African Americans in integrated work and neighborhood settings. They also attend concerts, plays, film festivals, and other social gatherings oriented toward African Americans (Staples 1991).

For lesbians and gay men, the problem of meeting others is exacerbated because they cannot necessarily assume that the person in whom they are interested shares their orientation. Instead, they must rely on identifying cues, such as meeting at a gay or lesbian bar or events, wearing a gay or lesbian pride button, or being introduced by friends to others identified as being gay or lesbian (Tessina 1989). Once a like orientation is established, gay men and lesbians usually engage in nonverbal processes to express interest. Lesbians and gay men both tend to prefer innocuous opening lines. To prevent awkwardness, the opening line usually does not make an overt reference to orientation unless the other person is clearly lesbian or gay.

Dating

For many of us, asking someone out for the first time is not easy. Shyness, fear of rejection, and traditional gender roles that expect women to wait to be asked may fill us with anxiety and nervousness. (Sweaty palms and heart palpitations are not uncommon when asking someone out the first time.) Both men and women contribute, although sometimes differently, to initiating a first date. Men are more likely to ask directly for a date: "Want to go see a movie?" Women are often more indirect. They hint or "accidentally on purpose" run into the other person: "Oh, what a surprise to see you *here* studying for your marriage and family midterm!" Although women may initiate dates, they do so less often than do men (Berger 1987).

In addition, research indicates that both women and men believe that men *should* initiate first dates, that men display a greater willingness to do so, and that men have a higher frequency of actual "first moves." Interestingly, men also express a desire for women to more actively participate in initiating relationships, either by asking directly for a date or at least hinting. Men report that the most passive stance, in which women wait for men to ask or initiate, is less preferred (Regan 2003).

Costs and Benefits of Romantic Relationships

As anyone who has had a romantic relationship can attest, relationships bring positive and negative experiences. In other words, when asked, people identify both rewards (companionship, sexual gratification, feeling loved and loving another, intimacy, expertise in relationships, and enhanced self-esteem) and costs (loss of freedom to socialize or date, investment of time and effort, loss of identity, feeling worse about oneself, stress and worry about the health or durability of the relationship, and other nonsocial costs like lower grades) of romantic relationships (Sedikedes, Oliver, and Campbell 1994, cited in Regan 2003).

Males and females differ some in what costs and rewards they identify. More males than females identify sexual gratification as a benefit of romantic relationships, and women are more likely than men to identify the benefit of enhanced self-esteem. More women than men mention loss of identity, feeling worse about themselves, or growing too dependent on their partners as relationship costs. Males, on the other hand, stress perceived loss of freedom (to socialize or

■ *Dating is a source of pleasure, as well as problems. It is also the process through which most Americans find their spouses.*

date) and financial costs more than women do (Regan 2003).

Issues arise from the question of who initiates, who touches, and who terminates sexual advances. Norms have prescribed male leadership and dominance. Even though many people do not wish to have unequal sexual relationships, modes of expression and resistance and difficulty in changing communication patterns help maintain an edge of inequality and imbalance among women. For equality to occur, women need to determine what they wish to express and how they wish to keep those behaviors that give them strength.

Problems in Dating

Dating is often a source of both fun and intimacy, but a number of problems may be associated with it. Think about your romantic relationships. When a disagreement occurs, who generally wins? Does it depend on the issue? When one person wants to go to the movies and the other wants to go to the beach, where do you end up going? If one wants to engage in sexual activities and the other doesn't, what happens?

Consistent with material presented on communication patterns, the female demand–male withdraw pattern found in many marriages is also common among dating couples. Of the 108 subjects in dating relationships studied by David Vogel, Stephen Wester, and Martin Heesacker (1999), 51% reported having a female demand–male withdraw communication

pattern. Another 28% described their communication as male demand–female withdraw, and 21% had no pronounced pattern. The female demand–male withdraw pattern was more often the style used by couples engaged in "difficult discussions." David Vogel and colleagues suggest that either version of demand–withdraw may prove to be a problem for dating couples as far as their relationship satisfaction and cohesion are concerned (Vogel, Wester, and Heesacker 1999). They recommend reduction of the overall level of demand–withdraw behavior as an important step toward enhancing the quality of relationships.

Dating Scripts and Female and Male Differences

Divergent gender-role conceptions may complicate dating relationships. Often, the woman is more egalitarian and the man is more traditional. Another problem is who pays when going out on a date. Some women may fear that male acquaintances would be put off if they offered to pay their share. Other women who offer to pay, whether traditional or egalitarian, may find their gestures are expected by their dates. Some men who accept offers by their dates to pay might nonetheless insist on choosing where they go, whether the women want to go there or not. Still other men allow their dates to pay but not publicly.

Although both women and men have ideas about what behaviors are most likely of men and of women on first dates, their ideas don't always match. This can be seen in the following data from 103 women and

Understanding Yourself

The Science Behind Internet Personals

Many search engines and websites allow users to search through personal ads to find a suitable match. For example, Yahoo and America Online each have sections of "personals." Sites specifically designed for matchmaking and finding dates are abundant. A Web search for "personals and dating services" netted nearly 1 million sites. There are free sites and pay sites; sites for finding Christian partners or Jewish partners; sites for people in the military, single parents, "shy folks over 30," and nondrinkers; sites that specialize in interracial matches; and some that specialize in matching gay men, lesbians, or bisexuals with potential partners. The more popular sites include Match.com, DreamMates, Matchdoctor, AmericanSingles.com, and eHarmony.

According to research by James Houran, Rense Lange, Jason Rentfrow, and Karin Bruckner, published in the *North American Journal of Psychology* (2004, 508),

> Internet dating services represent a significant and growing segment of online services and the general personals and dating services. Market data for [2003] . . . alone reveals that Web services accounted for approximately 43 percent of the $991 million United States dating-service sector, which also includes print and radio personal ads and other offline operations. Consumers tripled their spending on Internet dating services between 2001 and 2002, and Jupiter Research expects online dating sites to record over $640 million by 2007. Some have estimated that as many as 22 percent of the 98 million singles in the U.S. in 2002 used online dating. As the industry segment grows, its advertising is becoming ubiquitous.

Between 2000 and 2003, the number of online advertisements for internet dating services increased six-fold. As the stigma historically associated with Internet dating is seemingly diminishing, these services are targeting and reaching their intended audiences with unprecedented success.

Although users can often peruse the ads and photos without supplying information, most sites ask users to register and answer a series of screening questions about basic characteristics (such as height and weight (or body type), education, religious preferences, political views, and smoking habits), their interests and what they enjoy doing (for example, movies, outdoor activities, and travel), and what characteristics they seek in their "ideal matches." Upon completing these initial screening questions, the database is then made accessible, and users can search through the ads and e-mail those who interest them.

Some sites do the matching themselves, after asking members to complete more extensive questionnaires or personality profiles. In compiling profiles, some sites seek detailed self-assessments with close to 100 personality characteristics, including warmth, intelligence, submissiveness, impulsiveness, perfectionism, and generosity. People rate themselves on such characteristics and indicate their importance in a potential partner.

They are then matched with someone with whom they are deemed "compatible." Many of the more popular sites, such as eHarmony, Match.com, and Perfectmatch.com, claim that their methods of assessing compatibility and matching people successfully are based on sound scientific research about relationships. Although "virtually none of these services provide acceptable substantiation for their claims," eHarmony has patented its methodology and compatibility test (Houran et al. 2004). Central to its strategy is the principle of homogamy, the idea that more alike partners are more likely to be successful. Although there is considerable research supporting the idea that homogamy is beneficial, so too is there research stressing *complementarity*, wherein compatibility stems from the "harmonizing" of differences between partners' personalities and skills (Houran et al. 2004).

As you think about the processes described here, to what extent do you feel as though your "ideal match" could be found on the basis of homogamy? Do you accept the idea that people ought to be paired with people like themselves? Furthermore, what do you think would be your chances of liking a person who was your "ideal match"? What other characteristics would be important to you?

Finally, as Houran and colleagues (2004, 511–512) point out, matchmaking and dating online may succeed or fail for different reasons than relationships that commence from in-person meetings. As they note, we don't yet know enough about the outcomes of relationships initiated on-line to be able to conclude that the same variables that influence "offline" relationships, similarly affect relationships that begin on-line.

How confident are you that you could find your "perfect match" or relationship harmony online?

103 men, all of whom were upper-division college students. They were asked about many possible "first-date behaviors" and to identify whether the behavior would typically be something that the man or the woman would do or whether it would be equally possible and likely to be done by both. The results identified 14 activities as "the man's," 8 as "the woman's," and 7 as something either or both are equally likely to do. Overall, men's and women's **dating scripts** define first dates in fairly traditional terms, with such activities as who asks the other out, decides on the plans for the date, and pays the bill expected of the man (Table 8.2). In addition, he is expected to call the woman on the day of the date, buy her flowers, and pick her up. He is also identified by both women and men as the more likely to make affectionate moves, initiate sexual contact, and take the other home (Laner and Ventrone 2000).

Mary Riege Laner and Nicole Ventrone's findings also indicate that women are slightly more egalitarian than men; almost twice as many women as men thought either gender could do the inviting or initiating, and 22% of women compared to only 9% of men thought either person could pick up the bill.

Table 8.2 ■ **Percentage of Women and Men Identifying First-Date Behaviors as "Men's," "Women's" or "Either or Both"**

Behavior	Men's Responses (%)			Women's Responses (%)		
	Man	Woman	Either or Both	Man	Woman	Either or Both
1. Ask someone for a date	83	2	16	68	1	29
2. Wait to be asked for a date	4	86	10	2	87	8
3. Decide on plans by yourself	71	3	17	52	9	26
4. Discuss plans with date	43	16	38	17	26	54
5. Talk to friends about date	11	29	60	1	53	44
6. Buy new clothes for date	3	69	22	0	80	17
7. Select/prepare clothes for date	7	31	61	1	41	57
8. Groom for date (shave or put on makeup)	6	9	84	1	4	94
9. Take extra time to prepare	5	45	48	2	53	43
10. Call date on day of date	53	10	22	47	15	23
11. Prepare car (get gas, etc.)	83	1	13	69	8	18
12. Prepare house/apartment	24	18	56	7	44	47
13. Get money; collect keys	63	5	30	44	1	52
14. Get flowers to bring to date	83	7	8	79	4	2
15. Wait for date to arrive	13	82	5	11	76	11
16. Pick up your date	84	7	8	81	4	14
17. Greet/introduce date to family	16	50	33	5	58	35
18. Go to dinner	13	10	75	6	5	87
19. Eat light	5	78	16	0	87	5
20. Make small talk	31	13	54	15	20	60
21. Pay the bill	91	0	8	77	0	21
22. Open doors for date	88	5	4	89	1	3
23. Go somewhere else (e.g., movie)	22	4	71	11	1	86
24. Pay the bill	88	6	5	67	6	22
25. Go to bathroom to primp	4	76	17	2	73	17
26. Go somewhere else (e.g., drinks)	21	11	59	10	12	73
27. Have a deeper conversation	16	43	37	3	50	38
28. Pay the bill	82	3	15	67	4	23
29. Make affectionate move (e.g., hug)	60	6	30	52	7	39
30. Make sexual move	75	2	12	67	2	15
31. Take date home/walk to door	90	2	7	88	0	7
32. Discuss possible second date	59	5	34	38	5	53
33. Thank date for a good time	9	18	72	4	30	65
34. Call a friend to discuss date	9	54	36	0	67	31

Equalitarianism scores, calculated by adding the "either or both" category for men and for women and dividing by the 34 items, were: Men = 31.85; Women = 35.85
SOURCE: Laner and Ventrone 2000, 488–500.

These dating scripts introduce potential problems for men and women. A woman who wants to see a man again faces a dilemma: how to encourage him to ask her out again without engaging in more sexual activity than she really wants. Meanwhile, research with over 300 college-age men and women found that the No. 1 dating problem cited by men was communicating with their dates (Knox and Wilson, cited in Knox 1991). Men often felt that they didn't know what to say, or they felt anxious about the conversation dragging. Communication may be a particularly critical problem for men because traditional gender roles do not encourage the development of intimacy and communication skills among males. A second problem, shared by almost identical numbers of men and women, was where to go. A third problem, named by 20% of the men but not mentioned by women, was shyness. Although men can take the initiative to ask for a date, they also face the possibility of rejection. For shy men, the fear of rejection is especially acute. A final problem—and, again, one not shared by women—was money, cited by 17% of the men. Men apparently accept the idea that they are the ones responsible for paying for a date.

Extrarelational Sex in Dating and Cohabiting Relationships

You don't have to be married to be unfaithful (Blumstein and Schwartz 1983; Hansen 1987; Laumann et al. 1994). Both cohabiting couples and couples in committed relationships usually expect sexual exclusiveness. But, like some married men and women who take vows of fidelity, they do not always remain exclusive. Philip Blumstein and Pepper Schwartz (1983) found that those involved in cohabiting relationships had similar rates of extrarelational involvement as did married couples, except that cohabiting males had somewhat fewer partners than husbands did. Gay men had more partners than did cohabiting and married men, and lesbians had fewer partners than any other group.

Large numbers of both men and women have sexual involvements outside dating relationships considered exclusive. One study of college students (Hansen 1987) indicated that more than 60% of the men and 40% of the women had been involved in erotic kissing outside a relationship; 35% of the men and 11% of the women had experienced sexual intercourse with someone else. Of those who knew of their partner's affair, a large majority felt that it had hurt their own relationship. When both partners had engaged in

affairs, each believed that their partner's affair had harmed the relationship more than their own had. Both men and women seem to be unable to acknowledge the negative effect of their own outside relationships. It is not known whether those who tend to have outside involvement in dating relationships are also more likely to have extramarital relationships after they marry.

Breaking Up

"Most passionate affairs end simply," Elaine Hatfield and G. William Walster (1981) noted. "The lovers find someone they love more." Love cools; it changes to indifference or hostility. Perhaps the relationship ends because one partner shows a side that the other partner decides is undesirable. Or couples disclose *too much*, revealing negative feelings or ideas that lead to unhappiness and the demise of the relationship (Regan 2003).

Relationships are also susceptible to outside influences. Perhaps, some new opportunity for greater fulfillment appears in someone else or in a return to a more autonomous and independent state. Even

© Skjold/The Image Works

■ *Because of the variety of problems that plague relationships, many couples break up. In the process of breaking up, both the initiator and the rejected partner suffer.*

In American society, the expectation is that through the process of dating singles find their eventual life partner. Dating, or whatever else it might be called, allows us to test out our suitability for each other, develop stronger and closer relationships, fall in love, and select our life partners. Marriage without love goes against the culture of romantic love and these established patterns of mate selection, and is often the subject of soap operas and whispered gossip: "He just married her for her money." "She married his family name." Although we might consider marriage without love an exceptional case, anthropologists tell us that in traditional cultures most people do not consider love the basis for their entry into marriage.

Marriage customs vary dramatically across cultures, and marriage means different things in different cultures. If we consider how marriages come about—how they are "arranged"— we find that it is usually not the bride and groom who have decided to marry, as is the case in our own society today. Typically, the elders have done the matchmaking, sometimes relying on intermediaries and matchmakers to locate suitable spouses for their children. These strategies are neither "old news" (that is, they are still practiced) nor entirely restricted to other countries. *New York Times* journalist Stephen Henderson tells the story of Rakhi Dhanoa and Ranjeet Purewal.

Quoting one of their friends, Erica Loomba, Henderson captures some of the motivation behind using others to arrange marriages: "Each wanted a love marriage . . . yet neither would dream of marrying someone who wasn't a Sikh." An immigration lawyer in New York whose parents emigrated from Punjab, India, 27-year-old Dhanoa decided that she wanted to marry someone of the same faith. "I began to appreciate that my religion is based on complete equality of the sexes," she said. At the same time, Purewal was beginning to think about finding a partner. His mother had approached Jasbir Hayre, a Sikh matchmaker, living nearby in New Jersey. She told Henderson, "Ranjeet's mother had approached me several times to keep a lookout for a girl." So, when it came time to throw a party for her own daughter, Hayre invited both Dhanoa and Purewal. Although he had firmly believed in choosing for himself, on the basis of love, like Dhanoa, Purewal came to feel as though there were important issues to take into account. "I was adamant that I'd marry whoever I wanted. . . . But seeing how different cultures treated their families, I realized the importance of making the right match." After 2 months of mostly covert dating, "their cover was blown, on a double date, [and] the matchmaker was quickly summoned to negotiate marital arrangements" (Henderson 2002).

The story of Dhanoa and Purewal illustrates a variation of a phenomenon common in many parts of the world. In most cultures, marriage matches do not result from individuals meeting and dating; instead, the parents of the bride and groom are charged with arranging the marriage of their children. In some cultures, mothers are the primary matchmakers, as in traditional Iroquois culture. In others, fathers have a dominant voice in arranging marriage, as in traditional Chinese society. In still other cultures, the pool of elders involved in matchmaking is more extensive, including grandparents, aunts, uncles, and even local political and religious authorities, such as tribal chiefs and clan leaders. In all of these instances, though, marriage is a major event in the life of two *families*—both the bride's and the groom's—as well as for the clan, tribe, and community to which each family belonged. As such, important matters must be taken into account before agreeing to any particular match. Families must know how a particular marriage affects the family as a whole.

With issues of this magnitude at stake, marriage could not be left to the young people. Sentimental feelings of love would certainly cloud their judgment. Marriage was not primarily a personal or intimate event focused on a young couple alone. The feelings and love between an individual bride and groom were subordinate to the greater interests and welfare of the family, clan, and community.

Among the Bedouin of northern Egypt, marriages are usually arranged between a young man and a young woman who belong to different camps, thus creating blind marriages. That is, the bride and groom typically have not met before their engagement and marriage. The practice of arranging blind marriages further enhances the control and authority of the older generation over the young couple. Without ever having met, two people can hardly be in love at the time of their marriage. The emotion, attraction, and commitment that we mean by the word *love* may in time develop between husband and wife. However, in Bedouin society, as in most traditional societies, love is neither a necessary nor an advisable condition in arranging a good marriage.

satisfying relationships may end under these circumstances (Regan 2003). Over the course of their lifetimes, most people will experience multiple relationships and endure numerous breakups (Tashiro and Frazier 2003).

Breaking up is typically painful because few relationships end by mutual consent. The extent of distress caused by breakups is revealed by research indicating that many people include breaking up among the "worst events" they can experience; they are also among the biggest risk factors for adolescent depression (Tashiro and Frazier 2003). For college students, breakups are more likely to occur during vacations or at the beginning or end of the school year. Such timing is related to changes in the person's daily living schedule and the greater likelihood of quickly meeting another potential partner.

Research indicates that relationships often begin to sour as one partner grows quietly dissatisfied (Duck 1994; Vaughan 1990). Steve Duck (1994) calls this the *intrapsychic* phase, Diane Vaughan (1990) talks about "keeping secrets." One partner decides that something is wrong with the relationship, considers the possibility of ending the relationship, weighs the likely outcomes associated with being out of the relationship, and begins to build an identity as a "single." All of this may happen before the other partner learns what has happened. By the time the "initiator" informs the partner, the partner is forced to play "catch-up," in that the initiator is a few steps ahead in the exiting process. This is further discussed in Chapter 14.

Breaking up is rarely easy, whether you are on the initiating or "receiving" end. As Pamela Regan (2003) summarizes, the more satisfied you are with your partner, the closer you feel to your partner; the more difficult you believe it will be to find another relationship, the harder it is to experience a breakup. Social support and self-esteem appear to be important factors in helping someone recover more quickly and completely (Regan 2003).

Also important are the **attributions** we make to account for the demise of a relationship. Attributions may be important factors in efforts to avoid such problems in later relationships, shielding us from experiencing the heartache that accompanies a breakup. Ty Tashiro and Patricia Frazier suggest that there are four such attributions:

- *Person.* Personal traits and characteristics are identified as causes of relationship failure ("if only I hadn't been so jealous").

- *Other.* Personal traits and characteristics of the partner are seen as the causes of relationship failure ("he or she was always so insensitive").

- *Relational.* The unique combination of person and other is perceived as the cause of the breakup ("we just wanted different things").

- *Environmental.* The social environment is identified as the cause of the breakup. It comprises many things, from familial pressure and disapproval of the relationship, to work pressures, to "alternative romantic partners."

According to Tashiro and Frazier, "relational" attributions are usually cited by those who construct accounts to explain why their relationships failed. These are followed by "other" attributions, "person" attributions, and "environmental" attributions. Although environmental attributions are quite uncommon, environmental factors weigh heavily on relationships. Ironically, environmental factors may be the "real cause" of a breakup incorrectly attributed to something else.

Attributions are also related to how distressing a breakup is felt to be. People who apply relational attributions are happier, more confident, and more socially active. "Other" attributions are associated with greater distress, including sadness, lack of self-confidence, and greater pessimism. Research on person attributions is mixed, with some showing that it is related to less and some suggesting it is associated with more distress.

Research demonstrates that, alongside pain and distress, breakups can induce positive changes that improve the quality of subsequent relationships you might enter. We might expect that the degree to which breakups are associated with positive rather than negative outcomes (for example, growth rather than distress) would depend on such things as whether or not we initiated the breakup (nonintiators suffer greater distress), our gender (females more than males experiencing more positive emotions such as "growth" following breakups), and/or our personality (people high in traits such as "agreeableness" respond to stressful situations more positively; people high in "neuroticism" are more likely to suffer from distress). Of these, gender differences occur in "stress-related growth," with women reporting more growth following breakups than men. In addition, people high in "agreeableness" reported more post-breakup growth (Tashiro and Frazier 2003).

Popular Culture

Chocolate Hearts, Roses, and . . . Breaking Up?
What about "Happy Valentine's Day"?

Every year on February 14, millions of Americans exchange tokens of love and affection. As the day approaches, post office branches fill with Hallmark cards. Florists take orders and send roses around the country. Chocolate hearts show up on store shelves. Diamonds and gold jewelry are bought for and given to those we love. Millions of dollars are spent in efforts to show and tell our "one and only" how much we love them and, collectively, the country celebrates love and romance in the name of Valentine's Day. You may be familiar with such rituals as both a giver and a recipient. You may be wondering what if anything is noteworthy about such rituals. One of the lesser known aspects of this holiday devoted to love is the effect it can have on ongoing love relationships. This effect was provocatively captured by exploratory research undertaken by Katherine Morse and Steven Neuberg in their study following the relationship outcomes for 245 undergraduate students (99 male and 146 female; mean age of 19.5 years) from the week before to the week after Valentine's Day. The average relationship across all research participants was 18 months, suggesting that these were meaningful relationships (2004, 525). The results may surprise you, especially if you fashion yourself a romantic at heart.

It seems that "Valentine's Day is harmful to many relationships" (509). Although at first this may seem hard to imagine, Morse and Neuberg remind us that, although limited, research has shown that holidays can affect behaviors. The best illustration of this is the effect major holidays (for example, Christmas, Thanksgiving, and New Year's Eve and Day) have on suicide rates: essentially they postpone such acts from before to after the holiday. And although there is research on the effect of such events as spring break on relationships (especially infidelity), there hasn't been much attention paid to how "recurring cultural events and holidays" might have serious relationship implications (510).

Morse and Neuberg predicted that during the 2-week period straddling Valentine's Day (from 1 week before the holiday to 1 week after the holiday) there would be more breakups than in comparison periods from other times of year, and indeed, Valentine's Day posed relationship hazards. The overall odds of breaking up were 5.49 times greater during the Valentine's Day period than during the comparison months (which did not differ from one another). They further determined that the effect of the holiday on breakups was the result of a *catalyst effect*. The holiday had no effect on breakups among high-quality or improving relationships but did affect breakups among those in moderately strong and weak relationships if they were encountering relationship downswings. Already

suffering from diminishing expectations and unfavorable comparisons to other relationships or potential partners, such relationships might be deemed not worth the effort and expenses associated with trying to successfully play out the Valentine's Day script, thus "making the option of relationship dissolution more attractive" (512). In the absence of Valentine's Day's romantic expectations and comparisons, a relationship might weather the storm of disappointing comparisons and unmet expectations for at least a time, even eventually shifting to a more gratifying and healthier state. But couples in a "down" state or heading downward as Valentine's Day approaches are more vulnerable to breaking up.

They suggest that the catalyst effect may in part be a favor to troubled relationships in that it facilitates a breakup that was likely anyway, and hence "saved these couples the psychological stress, wasted time, and wasted resources that result from perpetuating a doomed relationship" (524). However, they also admit that because many long-term relationships go through periods of ups and downs, it is

> "at least plausible that a good number of our couples might have otherwise survived the downward blip in relationship expectations and quality had it not been for the catalytic effects of Valentine's Day" (524).

SOURCE: Morse and Neuberg 2004, 509–527.

A study of 92 undergraduates (75% female; age range 18–35 with a mean age of 20 years) found that positive change, such as personal growth, following breakups was common. On average, respondents reported positive changes that they believed would strengthen their future relationships and the chances for success in those relationships. Using the same four categories of attributions discussed earlier, the most commonly reported positive changes were "person-related" (for example, "I learned not to overreact") followed by "environmental" (for example, improved family relationships, increased success in school), "relational" (for example, better communication) and "other" (remember, "other" refers to characteristics of the other with whom we have a relationship). Individuals who use environmental attributions were most likely to report both distress from a breakup and having experienced growth as a result of the breakup. Potentially, those who can explain the failure of their relationship in terms of changeable environmental factors are in a better position to learn from and implement such changes in future relationships (Tashiro and Frazier, 2003).

Breakups among Gay and Lesbian Couples

As we discussed in Chapter 6, there are both similarities and differences between same-sex and heterosexual couples. Couple relationships, especially those that entail sharing a household, encounter many of the same day-to-day issues (for example, housework, money management, and the effects of outsiders such as family and friends on relationships). Furthermore, all couples need to manage issues that we dealt with in the last chapter, such as communication and conflict management. Given these similarities, how do same-sex couples and heterosexual couples compare in terms of susceptibility to breaking up?

Same-sex couples are more likely to break up than are heterosexual couples (Wagner 2006). Citing research findings by Lawrence Kurdek, Cynthia Wagner reports that in comparisons of married heterosexuals, cohabiting heterosexuals, gay and lesbian couples, married heterosexuals had the lowest rate (4%) of breaking up within 18 months of getting together ("relationship dissolution"), and lesbian couples had the highest (18%). However, Kurdek argues that the cause of differences is more likely the result of *marriage* than of *sexuality*. All cohabiting couples had similar "dissolution rates," and all were significantly higher than the rate found among married heterosexuals. Marriage is more likely to be associated with cultural acceptance and social support. Furthermore, once married, it is more difficult to simply walk away or to separate simply and easily. Using comparative data from Sweden and Norway, Kurdek illustrates that state-sanctioned and recognized legal unions between gay men or lesbians lowered the rates at which such relationships broke up, even though they still did so at levels greater than among married heterosexuals (Wagner 2006). With only recent passage of Massachusetts' gay marriage law and Vermont's civil union legislation, we don't yet have enough data on whether breakups and dissolutions have occurred at levels similar to those among married heterosexuals.

There will still be differences between married heterosexuals and married gay or lesbian couples that are products of something other than sexual orientation. Married gay or lesbian couples will not likely benefit from the same levels of social support and acceptance as married heterosexuals. Even if their own intimate networks of family and friends are supportive (and in the case of families that is far from automatic), the wider society doesn't offer the same climate of acceptance and support to gay and lesbian relationships. Thus, differences in rates of dissolution may follow from different levels of acceptance and support versus hostility.

What becomes of relationships once couples breakup? How do former romantic partners relate to each other? Research on 298 individuals from same-sex and 272 individuals from heterosexual romantic relationships reveals some interesting similarities in "post-dissolution relationships" (Lannutti and Cameron 2002). Many gay men and lesbians, as well as many heterosexuals, report remaining (or becoming) friends with former partners, especially following the "let's just be friends" type of breakup. Those friendships are different, however, from friendships in which two people have no shared romantic past. In comparing characteristics of post-dissolution relationships, Pamela Lannutti and Kenzie Cameron found the following: Heterosexuals reported moderate amounts of satisfaction and emotional closeness and low levels of interpersonal contact and sexual intimacy with former partners. Gay and lesbian respondents revealed high levels of satisfaction, moderate levels of emotional intimacy and personal contact, and low levels of sexual intimacy in their post-dissolution relationships. For both same-sex and heterosexual former partners, post-dissolution relationships are different from intact or ongoing romantic relationships and consistently platonic friendships (Lannutti and Cameron 2002).

Some Recommendations about Breakups

Regardless of your gender or sexual orientation, if you *initiate* a breakup, thinking about the following may help:

- *Be sure that you want to break up.* If the relationship is unsatisfactory, it may be because conflicts or problems have been avoided or confronted in the wrong way. Instead conflicts may be a rich source of personal development if they are worked out. Sometimes people erroneously use the threat of breaking up as a way of saying, "I want the relationship to change."

■ *Acknowledge that your partner will be hurt.* There is nothing you can do to erase the pain your partner will feel; it is only natural. Not breaking up because you don't want to hurt your partner may be an excuse for not wanting to be honest with him or her or with yourself.

■ *Once you end the relationship, do not continue seeing your former partner as "friends" until considerable time has passed.* Being friends may be a subterfuge for continuing the relationship on terms wholly advantageous to you. It will only be painful for your former partner because he or she may be more involved in the relationship than you. It may be best to wait to become friends until your partner is involved with someone else (and by then, he or she may not care if you are friends or not).

■ *Don't change your mind.* Ambivalence after ending a relationship is not a sign that you made a wrong decision; neither is loneliness. Both indicate that the relationship was valuable for you.

If your partner breaks up with you, keep the following in mind:

■ *The pain and loneliness you feel are natural.* Despite their intensity, they will eventually pass. They are part of the grieving process that attends the loss of an important relationship, but they are not necessarily signs of love.

■ *You are a worthwhile person, whether you are with a partner or not.* Spend time with your friends; share your feelings with them. They care. Do things that you like; be kind to yourself.

■ *Keep a sense of humor.* It may help ease the pain. Repeat these clichés: No one ever died of love. (Except me.) There are other fish in the ocean. (Who wants a fish?)

Singlehood

A quick question: Do you know what the third week of September is? Not Labor Day, that's weeks earlier. Give up? According to a report by the U.S. Census Bureau, the third week of September is Unmarried and

Single Americans Week, a week in which we are supposed to recognize singles, celebrate the single lifestyle, and acknowledge the contributions single people make to society. First started as National Singles Week in 1982 in Ohio by the Buckeye Singles Council and taken over by the American Association for Single People in 2001, the weeklong "celebration" was renamed in recognition that many unmarried people are in relationships or are widowed and don't identify with the "single" label (http://www.census.gov/Press-Release/www/releases/archives/facts_for_features_special_editions/005384.html). In 2002, the association changed its name to Unmarried America. Although the designated week has been around for more than a quarter century, and is recognized by mayors, city councils, and governors in some 33 states, as of 2005 it had yet to be "legitimized" and incorporated into mainstream American culture, as indicated by both the absence of greeting cards for the occasion and the number of people (including the millions of unmarried people) unaware that the weeklong recognition exists (Coleman 2005).

Even a casual inspection of demographics in this country illustrates the increasing phenomenon of singlehood. The trend, which has taken root and grown substantially since 1960, includes divorced, widowed, and never-married individuals. Each year more adult Americans are single (Table 8.3).

According to a 2005 U.S. Census Bureau report, there are *100 million* unmarried and single Americans, comprising 44% of all U.S. residents age 15 and over. Of this population, 64% have never married, 22% are

Table 8.3 ■ Percentage of Population 15 and Older Who Are Unmarried

Year	Men	Women
1890	48%	45%
1900	47%	45%
1910	46%	43%
1920	42%	43%
1930	42%	41%
1940	40%	40%
1950	32%	34%
1960	30%	34%
1970	34%	39%
1980	37%	41%
1990	39%	43%
2000	42%	45%

Marital status data for 1890–1970 from U.S. Census Bureau 1989. Data for 1980–2000 from U.S. Census Bureau 2001.

divorced, and 14% are widowed. There are 49 million households headed by single men or women. There are more single women than men; a ratio of 87 men to every 100 women of the U.S. population, 18 and older, has never married. And 15% of the unmarried population, representing 14.9 million people, are 65 years old or older (U.S. Census Bureau 2005).

The percentage of unmarried Americans varies by race and ethnicity; 20.6% of non-Hispanic whites, 39.4% of African Americans, 28% of Hispanics, and 28.5% of Asian and Pacific Islanders had never married. Furthermore, an additional 10.1% of non-Hispanic whites, 11.6% of African Americans, 7.7% of Hispanics, and 4.6% of Asian and Pacific Islanders were divorced (Fields 2000). Thus, the population of singles is quite large.

The varieties of unmarried lifestyles in the United States are too numerous to fit under one "umbrella" and too complex to be understood within any one category. They include: never married, divorced,

young, old, single parents, gay men, lesbians, widows, widowers, and so on, and represent diverse living situations that affect how singleness is experienced. In research on the unmarried, however, those generally regarded as "single" are young or middle age, heterosexual, not living with someone, and working rather than attending school or college. Although there are numerous single lesbians and gay men, they have not traditionally been included as singles in such research.

Unmarried in America: An Increasing Minority

The growth in the percentage of *never-married adults,* from 20.3% in 1980 to 24% in 2000, has occurred across all population groups. In part, this increase (like the creation of National Unmarried and Single Americans Week) reflects a change in the way in which society views this way of life. Many singles appear to be postponing marriage to an age which makes better economic and social sense (U.S. Census Bureau 2000). The growing divorce rate is also contributing to the numbers of singles. In 2000, 8.8% of men and 10.8% of women 18 and over were divorced (Fields 2000). The proportion of widowed men and women has declined somewhat but remains similar to past numbers. Among older people, singlehood most often occurs because of the death of a spouse rather than by choice. Nevertheless, as society moves toward valuing individualism and choice, the numbers of singles will likely continue to grow. In many large cities in the United States, including Washington, D.C., Cincinnati; Seattle; St. Louis; Minneapolis and Fort Lauderdale, 40% or more of the population consists of singles living in their own households.

The increases in the numbers of single adults are the result of several factors:

■ Delayed marriage. With a median age at first marriage of 27.1 years for men and 25.3 years for women in 2003 (U.S. Census Bureau 2003), Americans are waiting longer than ever to first enter marriage. The longer they postpone marriage, the greater the likelihood of never marrying. As shown in Table 8.4, the percentage of never-married men and women of typical "marrying ages" dramatically increased between 1970 and 2000. It is estimated that between 8% and 9% of men and women now in their 20s will never marry.

■ *There has been a steady increase in the numbers of single, unmarried Americans, as a result of such factors as delaying marriage, increases in divorce, and more economic opportunities for women.*

Table 8.4 ■ Percentage of Never-Married Women and Men by Age, 1970–2000

Age Year	Male		Female	
	1970	2000	1970	2000
20–24	35.8	83.7	54.7	72.8
25–29	10.5	51.7	19.1	38.1
30–34	6.2	30.1	9.4	21.9
35–39	5.4	20.3	7.2	14.3
40–44	4.9	15.7	6.3	11.8

SOURCE: Fields 2000.

- Increasingly expanded educational, lifestyle, and employment options open to women. These reduce women's economic need to be married and expand their lifestyle options outside of marriage.

- Increased rates of divorce coupled with somewhat decreased likelihood of remarriage, especially among African Americans.

- More liberal social and sexual standards.

- Uneven ratio of unmarried men to unmarried women.

Relationships among the Unmarried

When intentionally single people form relationships within the singles world, both the man and the woman tend to remain highly independent. Singles work, and, thus, tend to be economically independent of each other. They may also be more emotionally independent because their energy may already be heavily invested in their work or careers. Their relationships consequently tend to emphasize autonomy and egalitarian roles. Single women work and tend to be more involved in their work, either from choice or from necessity, but the result is the same: they are accustomed to living on their own without being supported by a man. Early analysis of the various factors that draw people to singlehood or marriage identified the various "pushes" and "pulls" of each lifestyle. These are illustrated in Table 8.5.

The emphasis on independence and autonomy blends with an increasing emphasis on self-fulfillment, which, some critics argue, makes it difficult for some to make commitments. Commitment requires sacrifice and obligation, which may conflict with ideas of "being oneself." A person under obligation can't

Table 8.5 ■ Pushes and Pulls toward Marriage and Singlehood

Pushes/Pulls toward Marriage	Cultural norms
	Love and emotional security
	Loneliness
	Physical attraction and sex
	Parental pressure
	Desire for children
	Economic pressure
	Desire for extended family
	Social stigma of singlehood
	Economic security
	Fear of independence
	Peer example
	Media images
	Social status as "grown up"
	Guilt over singlehood
	Parental approval
Pushes/Pulls toward Singlehood	Fundamental problems in marriage
	Freedom to grow
	Stagnant relationship with spouse
	Self-sufficiency
	Feelings of isolation with spouse
	Expanded friendships
	Poor communication with spouse
	Mobility
	Unrealistic expectations of marriage
	Career opportunities
	Sexual problems
	Sexual exploration
	Media images

SOURCE: Adapted from Stein 1975.

necessarily do what he or she "wants" to do; instead, a person may have to do what "ought" to be done (Bellah et al. 1985).

According to Barbara Ehrenreich (1984), men are more likely to flee commitment because they need women less than women need men. They feel oppressed by their obligation to be the family breadwinner. Men can obtain many of the "services" provided by wives—such as cooking, cleaning, intimacy, and sex—outside marriage without being tied down by family demands and obligations. Thus, men may not have a strong incentive to commit, marry, or stay married.

Nevertheless, "flying solo at midlife" appears to be more problematic for men than for women (Marks 1996). Single women appear to have better psychological well-being than do single men. For those socialized during an era of traditional gender roles and family values with marriage as the norm, there seemed to be a degree of mental health risk associated with singlehood, especially for men.

Culture and the Individual versus Marriage

The tension between singlehood and marriage is diminishing as society increasingly recognizes singlehood as an option rather than a deviant lifestyle. The singles subculture is glorified in the mass media; the marriages portrayed on television are situation comedies or soap operas abounding in extramarital affairs. Yet many are rarely fully satisfied with being single and yearn for marriage. They are pulled toward the idea of marriage by their desires for intimacy, love, children, and sexual availability. They are also pushed toward marriage by parental pressure, loneliness, and fears of independence. At the same time, married people are pushed toward singlehood by the limitations they feel in married life. They are attracted to singlehood by the possibility of creating a new self, having new experiences, and achieving independence.

Types of Never-Married Singles

Much depends on whether a person is single by choice and whether he or she considers being single a temporary or permanent condition (Shostak 1987). If the person is voluntarily single, his or her sense of well-being is likely to be better than that of a person who is involuntarily single. Arthur Shostak (1981, 1987) divided singles into four types:

- *Ambivalents.* Ambivalents are usually younger men and women actively pursuing education, career goals, or "having a good time." Voluntarily single, they consider their singleness temporary. Though not actively seeking marital partners, they remain open to the idea of marriage. Some ambivalents are cohabitors.
- *Wishfuls.* Wishfuls are involuntarily and temporarily single, actively and consciously seeking marital partners.
- *Resolveds.* Resolved individuals regard themselves as permanently single. They include priests and nuns, as well as single parents who prefer rearing their children alone. Most, however, are "hard-core" singles who prefer to be single.

- *Regretfuls.* Regretful singles prefer to marry but are resigned to their "fate." A large number of these are well-educated, high-earning women over 40 who find a shortage of similar men as a result of the marriage gradient.

Singles may shift from one type to another at different times. All but the resolveds share an important characteristic: they want to move from a single status to a romantic couple status. "The vast majority of never-married adults," writes Shostak (1987), "work at securing and enjoying romance." Never-married singles share with married Americans "the high value they place on achieving intimacy and sharing love with a special one."

Singles: Myths and Realities

There are many long-standing myths about singles (Cargan and Melko 1982; Waehler 1996). Although first identified more than 20 years ago, notice how familiar these notions still sound:

- *Singles depend on their parents.* Few real differences exist between singles and marrieds in their perceptions of their parents (regarding warmth or openness) and differ only slightly in the amount and nature of parental conflicts.
- *Singles are self-centered.* Singles value friends more than do married people.
- *Singles are more involved in community service projects.*
- *Singles have more money.* Married couples are better off economically than singles, in part because both partners often worked.
- *Singles are happier.* Singles tend to believe that they are happier than marrieds, whereas marrieds believe that they are happier than singles. Single men exhibited more signs of stress than did single women.
- *Singles view singlehood as a lifetime alternative.* Most singles expected to be married within 5 years. They do not view singlehood as an alternative to marriage but as a transitional time in their lives.

Leonard Cargan and Matthew Melko also determined that the following statements characterize singlehood more accurately:

- *Singles don't easily fit into married society.* Singles tend to socialize with other singles. Married people think that if they invite singles to their home, they must match them with an appropriate single member of the other sex. Married people tend to think in terms of couples.

- *Singles have more time.* Compared with their married peers, singles are more likely to go out two or three times a week, and they have more choices and more opportunities for leisure activities.

- *Singles have more fun.* Singles more often engage in sports and physical activities and have more sexual partners than do marrieds.

- *Singles are lonely.* Singles, especially formerly married singles, tend to be lonelier than married people.

Gay and Lesbian Singlehood

In the late nineteenth century, groups of gay men and lesbians began congregating in their own clubs and bars. There, in relative safety, they could find acceptance and support, meet others, and socialize. By the 1960s, some neighborhoods in the largest cities (such as Christopher Street in New York and the Castro district in San Francisco) became identified with gay men and lesbians. These neighborhoods feature not only openly lesbian or gay bookstores, restaurants, coffee houses, and bars but also clothing stores, physicians, lawyers, hair salons—even driver's schools. They have gay churches, such as the Metropolitan Community Church, where gay men and lesbians worship freely; they have their own political organizations, newspapers, and magazines (such as *The Advocate*). They have family and childcare services oriented toward the needs of the gay and lesbian communities; they have gay and lesbian youth counseling programs.

In these neighborhoods, men and women are free to express their affection as openly as heterosexuals. They experience little discrimination or intolerance, and they are more involved in lesbian or gay social and political organizations. Recently, with increasing acceptance in some areas, many middle-class lesbians and gay men are moving to suburban areas. In the suburbs, however, they remain more discreet than in the larger cities (Lynch 1992).

The urban gay male subculture that emerged in the 1970s emphasized sexuality. Although relationships were important, sexual experiences and variety were more important (Weinberg and Williams 1974). This changed with the HIV and AIDS epidemic. Beginning in the 1980s, the gay subculture placed increased emphasis on the relationship context of sex (Carl 1986; Isensee 1990). Relational sex has become normative among large segments of the gay population (Levine 1992). Most gay men have sex within dating or love relationships. (Some AIDS organizations are giving classes on gay dating to encourage safe sex.) One researcher (Levine 1992) says of the men in his study: "The relational ethos fostered new erotic attitudes. Most men now perceived coupling, monogamy, and celibacy as healthy and socially acceptable."

Beginning in the 1950s and 1960s, young and working-class lesbians developed their own institutions, especially women's softball teams and exclusively female gay bars as places to socialize (Faderman 1991). During the late 1960s and 1970s, **lesbian separatists,** lesbians who wanted to create a separate "womyn's" culture distinct from heterosexuals *and* gay men, rose to prominence. They developed their own music, literature, and erotica; they had their own clubs and bars. But by the middle of the 1980s, according to Lillian Faderman (1991), the lesbian community underwent a "shift to moderation." The community became more diverse, including Latina, African American, Asian American, and older women. It has developed closer ties with the gay community. They now view gay men as sharing much with them because of the common prejudice directed against both groups.

In contrast to the gay male subculture, the lesbian community centers its activities on couples. Lesbian therapist JoAnn Loulan (1984) writes: "Being single is suspect. A single woman may be seen as a loser no one wants. Or there's the 'swinging single' no one trusts. The lesbian community is as guilty of these prejudices as the world at large."

Lesbians tend to value the emotional quality of relationships more than the sexual components. Lesbians usually form longer-lasting relationships than gay men (Tuller 1988). Lesbians' emphasis on emotions over sex and the enduring quality of their relationships reflects their socialization as women. Being female influences a lesbian more than being gay.

Cohabitation

Few changes in patterns of marriage and family relationships have been as dramatic as changes in cohabitation. What in the 1960s was rare and relegated to hushed whispers and secrets from families is now a common experience (King and Scott 2005).

Table 8.6 ■ Numbers of Individuals Cohabiting, by Age	
Age	Number of People Cohabiting
<30	3.6 million
30–39	2.6 million
40–49	1.7 million
> 50	1.2 million

SOURCE: King and Scott 2005; 2000 Census Public Use Microdata Samples.

The Rise of Cohabitation

Over the past 40 years, cohabitation has increased 10-fold. It has increased across all socioeconomic, age, and racial groups. For example, just between 1980 and 1990 the rate of cohabitation nearly doubled among unmarried people less than 40 years old; during the same decade the cohabitation rate tripled among those 60 years old and older (King and Scott 2005).

Looking at the percentage of women, 15–44, who, according to a 2002 Centers for Disease Control study, have cohabited shows how commonplace this lifestyle has become (see Figure 8.2). This is especially true for women between the ages 25 to 44.

Figure 8.2 ■ Percentage of Women, 15–44, Who Have Ever Cohabited

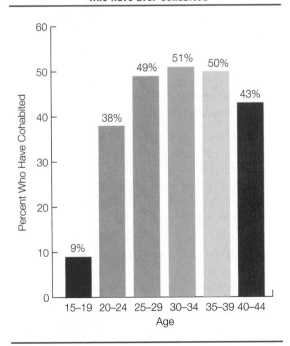

SOURCE: Centers for Disease Control and Prevention 2002.

As illustrated by both Table 8.6 and Figure 8.2, cohabitation appears no longer to be a moral issue but rather has increasingly become a family lifestyle. It also appears to be a lifestyle that is here to stay.

There are an estimated 5.5 million cohabiting couples in the United States, including 4.9 million heterosexual couples and nearly 600,000 gay and lesbian couples (U.S. Census Bureau 2002, Table 49). Forty years ago there were only approximately 400,000 such couples. Thus, we can see how steep an increase has occurred, especially since 1970 (see Figure 8.3).

In the United States, cohabiting couples still lack most of the rights that married couples enjoy, a topic we return to shortly. According to Judith Seltzer (2000), children of cohabiting couples may also be disadvantaged unless they have legally identified fathers. This situation differs greatly in many other parts of the world. In Sweden, for instance, the law treats unmarried cohabitants and married couples the same in such areas as taxes and housing. In many Latin American countries, cohabitation has a long and socially accepted history as a substitute for formal marriage (Seltzer 2000).

Cohabitation has increased, becoming not only more widespread but also more accepted in recent years for several reasons:

■ *The general climate regarding sexuality is more liberal than it was a generation ago.* Sexuality is more widely considered to be an important part of a person's life, whether or not he or she is married. Love rather than marriage is now widely regarded as making a sexual act moral.

■ *The meaning of marriage is changing.* Because of the dramatic increase in divorce for most of the last quarter of the twentieth century, marriage is no longer thought of as a necessarily permanent commitment. Permanence is increasingly replaced by serial monogamy—a succession of marriages, and the difference between marriage and living together is losing its sharpness.

Matter of Fact

Although cohabitation has increased for all educational groups and for Caucasians, Latinos, and African Americans, it is more common among those with lower levels of education and income (Seltzer 2000).

Figure 8.3 ■ Cohabitation: 1960 to 2001

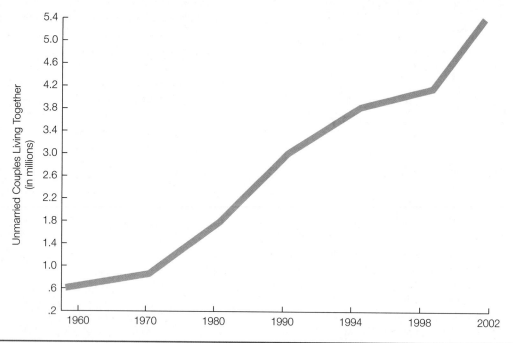

SOURCE: U.S. Bureau of the Census 2001.

- *Men and women are delaying marriage longer.* More than half of cohabiting couples eventually marry (Smock 2000). As long as children are not desired, living together offers advantages for many couples. When children are wanted, however, the couple usually marries.

Types of Cohabitation

There is no single reason to cohabit, just as there is no single type of person who cohabits or one type of cohabiting relationship. One typology differentiates among *substitutes or alternatives for marriage, precursors to marriage, trial marriages,* and *coresidential dating* (Casper and Bianch 2002; Phillips and Sweeney 2005). These can be distinguished by the expectations partners have for a married future, their perceptions of the stability of the relationship, and their general attitudes toward cohabiting relationships. In **trial marriages,** the motive for cohabiting is to assess whether partners have sufficient compatibility for marriage. They are undecided as to their likelihood of marriage and by cohabiting expect to assess their suitability. When the relationship is a *precursor to marriage,* there is an expectation that eventually the couple will marry.

In both of the other types (substitutes for marriage and coresidential dating), there is no expectation of marriage. As to the expected duration of the relationship, when either a substitute for marriage or precursor to marriage, couples expect to be together a long time. The coresidential dating situation is expected to last a short time. In the case of trial marriages, couples don't know whether and how long they will stay together (Heuveline and Timberlake 2004).

A second typology separates cohabitation into the following five types (Heuveline and Timberlake 2004):

- *Prelude to marriage.* Cohabitation is used as a "testing ground" for the relationship. Cohabitants in this type of situation would likely marry or break up before having children. The duration of this type is expected to be relatively short, and couples should transition into marriage.

- *Stage in the marriage process.* Unlike the prior type, couples may reverse the order of marriage and childbearing. They cohabit for somewhat longer periods, typically in response to opportunities that they can pursue "by briefly postponing marriage" (Heuveline and Timberlake 2004, 1,216). It is understood by both partners that they intend to eventually marry.

- *Alternative to singlehood.* Considering themselves too young to marry, and with no immediate intention to marry, such couples prefer living together to living separately. Having a commitment level more like a dating couple than that of a married couple, such relationships will be prone to separation and breaking up.

- *Alternative to marriage.* Couples choose living together over marriage but choose, as married couples would, to form their families. Greater acceptance of out-of-wedlock births and childrearing will increase the numbers of couples experiencing this type of cohabitation. Such couples would not likely transition into marriage but would likely build lasting relationships.

- *Indistinguishable from marriage.* Such couples are similar to the previous type but are more *indifferent* rather than *opposed* to marriage. As cohabitation becomes increasingly accepted and parenting receives support regardless of parents' marital status, couples lack incentive to formalize their relationships through marriage.

Consider one more typology of cohabitation, highlighting the factors couples consider in deciding to live together, the "tempo of relationship advancement" into cohabitation, and the language used, or story told, by couples in accounting for their cohabiting (Sassler, 2004). Sharon Sassler identifies six broad categories of reasons couples decide to cohabit: finances, convenience, housing situation, desire, response to family or parents, and as a trial, out of which she constructs a three category typology (2004, 498):

- *Accelerated cohabitants* decide to move in together quickly, typically before they had dated 6 months. Emphasizing the strength and intensity of their attraction and their connection, the fact that they were spending a lot of time together, and identifying finances and convenience as major reasons for their decision, they contend moving in together felt like "a natural process."

- *Tentative cohabitant* admitted to some uncertainty about moving in together. Together for 7 to 12 months before living together, they typically saw each other less often than the "accelerateds" did before moving in together (for example, 3 or maybe 4 nights a week) or had experienced disruptions in their relationships with one of the partners being gone for a period, which slowed their progression into cohabitation. They often

mentioned "unexpected changes in their residential situation" as a reason for their decision (2004, 500). Absent such a situation, they might not have moved in together when they did.

- *Purposeful delayers* were the most deliberate in the decision-making process. Their relationships progressed more gradually, taking more than a year before they decided to live together, and allowing them opportunity to discuss future plans and goals. They most often mentioned housing arrangements and finances as the reasons they moved in together.

Obviously, not all cohabitants desire, intend, or expect to marry. Although cultural attitudes and values—as well as ideas about singlehood, dating, marriage, and cohabitation—somewhat determine whether someone expects to marry, socioeconomic criteria are also of importance. Wendy Manning and Pamela Smock (2002) suggest that the percentage of cohabiting women who expect to marry their partners remained fairly stable from the late 1980s through the mid-1990s, with 74% of cohabitants expressing an expectation to marry. Those women with higher probability of expecting marriage are women who live with partners of high socioeconomic status. In addition, men's age and religiosity (strength of religious involvement and identification) make a difference in women's expectations of marriage. In terms of race, black women have lower probability of expecting marriage than either white or Hispanic women. Furthermore, despite relatively worse economic circumstances, Latinos have higher marriage rates than whites. Among cohabiting women, Latinos and whites have similar expectations regarding the likelihood that they will marry their partners (Manning and Smock 2002).

Matter of Fact

Children often turn cohabitation into marriage. Cohabiting couples in which the woman becomes pregnant have a greater likelihood of marrying than cohabiting couples where no pregnancy occurs. Also, cohabiting couples who already have children from previous relationships are more likely to marry than couples who don't have children (Seltzer 2000).

The meaning of cohabitation varies for different groups. For African Americans, cohabitation is more likely to be a substitute for marriage than a trial marriage, and blacks are more likely than whites to conceive, give birth, and raise children in a cohabiting household. Indirectly this implies that cohabitation is a more committed relationship for blacks than for whites, a more acceptable family status, and a more acceptable family form within which to rear children—even though blacks are no more likely than whites to say they approve of cohabitation (Phillips and Sweeney 2004).

The same appears to be true among Hispanics, where the idea of "consensual unions" outside of marriage goes back a long way in Latin America, especially among the economically disadvantaged. More than among whites, for Hispanics cohabitation is more likely to become an alternative to marriage. Again, we draw this conclusion from rates of nonmarital pregnancy and childbearing. Julie Phillips and Meghan Sweeney report that for Hispanics cohabitation "may be a particularly important context for *planned* childbearing" (2004, 299; emphasis added).

The most notable social effect of cohabitation is that it delays the age of marriage for those who live together. *Ideally,* cohabitation could encourage more stable marriages because the older people are at the time of marriage, the less likely they are to divorce. However, as we shall soon see, cohabition does *not* ensure more stable marriages.

Although there may be a number of advantages to cohabitation, there are also disadvantages. Parents may refuse to provide support for school as long as their child is living with someone, or they may not welcome their child's partner into their home. Cohabiting couples may also find that they cannot easily buy houses together, because banks may not count their income as joint; they also usually don't qualify for insurance benefits. If one partner has children, the other partner is usually not as involved with the children as he or she would be if they were married. Cohabiting couples may find themselves socially stigmatized if they have a child. Finally, cohabiting relationships generally don't last more than 2 years; couples either break up or marry.

Living together takes on a different quality among those who have been previously married. About 40% of cohabiting relationships have at least one previously married partner. Remarriage rates have dipped as postmarital cohabitation has increased (Seltzer 2000). Still, most remarriages are preceded by cohabitation (Ganong and Coleman 1994).

About a third of all cohabiting couples have children from their earlier relationships. As a result, the motivation in these relationships is often colored by painful marital memories and the presence of children (Bumpass and Sweet 1990). In these cases, men and women tend to be more cautious about making their commitments. Even though cohabiting couples are less likely to stay together compared to married couples, having children in the household somewhat stabilizes the couples (Wu 1995).

Common-Law Marriages and Domestic Partnerships

At one point, cohabiting couples would, after a short period of living together, enter what is known as **common-law marriage.** A couple who "lived as husband and wife and presented themselves as married," was considered to be married. Originating in English common law, as practiced in the United States common-law marriage was seen as a practical way to enable couples who wanted to be married but were too geographically removed from both an individual with the authority to marry them and a place where they could obtain a marriage license to marry (Willetts 2003). Common-law marriage became less necessary in the nineteenth century as the availability of officials who could perform marriage ceremonies grew. Although most states no longer allow or recognize common-law marriage, as recently as 2005, in the United States the following 11 states and the District of Columbia still did:

STATES WITH COMMON-LAW MARRIAGE

Alabama	Oklahoma
Colorado	Rhode Island
Iowa	South Carolina
Kansas	Texas
Montana	Utah
New Hampshire (for inheritance purposes only)	

If you happen to be reading this in one of those states and you meet the requirements described earlier, congratulations, you have just been pronounced married! Although we are being facetious, common-law marriage does unite into legal marriage two people who never sought and never obtained a marriage license. Once in such a marriage (Solot and Miller 2005),

> if you choose to end your relationship, you must get a divorce, even though you never had a wedding.

Legally, common law married couples must play by all the same rules as "regular" married couples. If you live in one of the common law states and don't want your relationship to become a common-law marriage, you must be clear that it is your intention not to marry.

In states that recognize common-law marriages, the amount of time a couple must live together before being considered married varies. What is essential is that they have presented themselves as if married, acting like they are married, telling people they are married, and doing the things married people do (including referring to each other as "husband" and "wife"). In states that don't recognize common-law marriages, no matter how long you live together or how married you act, you are not married.

Domestic partners—cohabiting heterosexual, lesbian, and gay couples in committed relationships—are gaining some legal rights. Domestic partnership laws, which grant some of the protection of marriage to cohabiting partners, are increasing the legitimacy of cohabitation. In some ways, domestic partnerships are alternative forms of cohabitation, with certain formal rights and protections. Civil unions are more like alternative versions of marriage (Willetts 2003).

In 1984, Berkeley, California, was the first U.S. city to enact a domestic partnership ordinance and extend it to both heterosexual and same-sex couples (Willetts 2003). In 1997, San Francisco extended health insurance and other benefits to their employees' domestic (which includes same-sex) partners. Individual employers, such as the Gap, Levi Strauss, and Walt Disney Company soon followed suit, introducing domestic partner policies, which have now become fairly commonplace in the private sector as well as in many local and state governments, colleges and universities. As of March, 2006, 49% of Fortune 500 (and 78% of Fortune 100) companies offered employees domestic partnership benefits, up from 25% of Fortune 500 companies in 2000.

A number of states—including California, Hawaii, Maine, Maryland, New Jersey and Alaska—have domestic partnership laws in place or pending. In Connecticut and Vermont, civil unions are available to same-sex couples. Additionally, some municipal governments provide domestic partner benefits even when their state governments do not. In New York City, for example, domestic partnership benefits extend to heterosexual or same sex couples, whereas the statewide laws are more narrowly framed, and available only to gay or lesbian couples. Furthermore, because heterosexual couples *could* marry whereas same-sex couples cannot, some domestic partnership protections, such as those provided by the state of New Jersey, are restricted to same-sex couples and to opposite sex couples in which one partner is at least 62 years old. Note that even in the absence of laws recognizing domestic partnerships, many employers offer benefits to domestic partners of their employees. As Marion Willetts (2003) details, thousands of private companies, along with hundreds of colleges and universities, provide employees' domestic partners with health benefits.

Domestic partners, whether heterosexual, gay, or lesbian, may still lack some of those legal rights and benefits that come automatically with marriage. Recalling only some of the rights and benefits noted in Chapter 1, these include the right to do the following:

- File joint tax returns
- Automatically make medical decisions if your partner is injured or incapacitated
- Automatically inherit your partner's property if he or she dies without a will
- Collect unemployment benefits if you quit your job to move with a partner who has obtained a new job
- Live in neighborhoods zoned "family only"
- Obtain residency status for a noncitizen partner to avoid deportation

Keep in mind that heterosexual domestic partnerships and same-sex domestic partnerships and civil unions frequently result from different motivations. Among heterosexuals, domestic partnership is a *deliberately chosen alternative to marriage*. This is illustrated in the "Real Families" box in this section. For at least some gay and lesbian couples, domestic partnerships or civil unions are the closest approximation to legal marriage available to them. Some same-sex couples would marry if marriage was an option.

Gay and Lesbian Cohabitation

The 2000 U.S. Census reported nearly 600,000 gay or lesbian couples living together. Other estimates put the number at more than 1.5 million same-sex cohabiting couples. The relationships of gay men and lesbians have been stereotyped as less committed than heterosexual couples because (1) lesbians and gay men cannot legally marry, (2) they may not emphasize sexual exclusiveness as strongly, and (3) heterosexuals

Real Families

Choosing Domestic Partnership

I get kind of upset when people say that a domestic partnership is an alternative to marriage. . . . That's not what it is. . . . It's a different approach to looking at partnerships in sort of a legal sense.

The preceding comment is from 26-year-old Marie as she talked to sociologist Marion Willetts about her 4-year-long licensed relationship. Willetts interviewed 22 other licensed heterosexual domestic cohabitants in the first study that attempted to uncover and document the motives for embarking on a domestic partnership instead of marriage. Although the rights and benefits bestowed by domestic partnership recognition could be obtained by marrying, some couples opt instead to enter licensed domestic partnerships. Typically, they must sign an affidavit declaring that they are not married to someone else and that they are not biologically or legally related to each other. They further pledge to be mutually responsible for each other's well-being and to report to authorities any change in their relationship—either marriage or dissolution (Willetts 2003, 939). Willetts notes that motives behind heterosexual couples' choice to *co-habit* rather than marry include economic benefits, as well as more personal and philosophical benefits, such as rejecting the assumptions that are part of legal marriage, not wanting the state to intervene in their relationships, and wanting to avoid past marital failures. But what about motives to license partnerships?

Economic benefits, including health insurance coverage, access to university-owned family housing or in-state tuition benefits, or access to family membership rates in outside organizations, was the motive most cited by Willetts's interviewees. For formerly married cohabitants, licensed partnerships allowed them to avoid reentering marriage yet obtain the protection and recognition of documentation. For others, such as 31-year-old Leslie, obtaining a domestic partnership license with her partner, Alan, was a means to obtain recognition in the eyes of others, that they had made a deep and meaningful commitment to each other, even in the absence of a wedding: "I guess [we wanted] to sort of be counted. There's [sic] relationships that mean a lot that aren't recognized by law and to sort of be counted in that count in the city."

When Willetts posed the question of why that mattered, Leslie continued:

It's difficult to be in a relationship where people are like, "Oh, aren't you married?" or "Are you not married?" . . . It's like an issue all the time. "Why aren't you married, you've been together for 10 years?" . . . so we were like, "We'll get a domestic partnership [to have some sort of documentation in response to these questions]." But it wasn't really something that meant a great deal to us. . . . It wasn't a big deal.

Although Leslie and Alan desired recognition, they wished to avoid too much interference, such as what accompanies a marriage license: "We didn't want to have any law interfere in our relationship, or we didn't feel we needed to have a legal stamp on our relationship."

Other respondents stressed wanting to avoid the trappings of the patriarchal institution they perceived marriage to be or wanting to demonstrate support for friends whose same-sex relationships were denied the right to marry. Licensed partnerships did not, however, give heterosexuals the same recognition and support with their families or friends that they would have had if they had married. Below, Marie comments on what her 4-year-long licensed partnership has lacked:

With a marriage license, there's that sort of social and economic and political legitimacy involved in it. . . . With our domestic partnership, nobody gave us any sort of crockery, nobody bought us a house, nobody sends us anniversary cards, and nobody sort of celebrated, or has celebrated that, you know, that special day [when she and her partner obtained their certificate].

Marie did not feel that legally defining licensed partners as though they were married was desirable: "Once the court says, 'Well, we're going to define this as marriage' . . . once you start having courts that intervene in using words that this is like a marriage, it takes away from, once again, the legitimacy of these other sorts of different types of families that can come about."

Willetts suggests that the wider implementation of civil union laws like those in Vermont may cause states and municipalities that already have domestic partnership ordinances to deem them no longer necessary and abandon them. Once same-sex couples can enter civil unions, and given that heterosexuals can legally marry, why continue to offer this other legal category?

misperceive love between gay couples and between lesbian couples as being somehow less "real" than love between heterosexual couples.

As we have already seen, numerous similarities exist between gay and heterosexual couples. Regardless of their sexual orientation, most people want a close, loving relationship with another person. For lesbians, gay men, and heterosexuals, intimate relationships provide love, romance, satisfaction, and security. There is one obvious difference, however. Heterosexual couples tend to adopt a gender divided model, whereas for same-sex couples these traditional gender divisions make no sense. Tasks are often divided pragmatically, according to considerations such as who likes cooking more (or dislikes it less) and work schedules (Marecek, Finn, and Cardell 1988). Most gay couples are dual-earner couples; furthermore, because gay and lesbian couples are the same gender, the economic discrepancies based on greater male earning power are absent. Although gay couples emphasize flexibility and egalitarianism, if there are differences in power they are attributed to personality or to dependency on the relationship (Peplau, Veniegas, and Campbell, 1996).

Letitia Peplau, Rosemary Veniegas, and Susan Miller Campbell (1996) describe gay and lesbian partners as maintaining a "friendship model" of relationships:

> In best friendships, partners are often of relatively similar age and share common interests, skills and resources . . . best friendships are usually similar in status and power.

With this model, tasks and chores are often shared, alternated, or done by the person who has more time. Usually, both members of the couple support themselves; rarely does one financially support the other.

Cohabitation and Marriage Compared

Different Commitments

A lesser level of commitment characterizes cohabiting couples when compared to married couples. When a couple lives together, their primary commitment is to each other, but it is a more transitory commitment. As long as they feel they love each other, they will stay together. In marriage, the couple makes a commitment not only to each other but to their marriage. Cohabitants are less committed to the certainty of a future together (Waite and Gallagher 2001; Forste and Tanfer 1996; Schwartz 1983). Hence, living together tends to

be a more temporary arrangement than marriage (Seltzer 2000; Teachman and Polonko 1990). Half of cohabiting relationships end within a year because the couple either marries or breaks up. Cohabiting couples are three times as likely as married couples (29% versus 9% for married couples) to break up within 2 years (Seltzer 2000). A man and woman who are living together may not work as hard to save their relationship. Less certain of a lifetime together, they live more autonomous lives. In marriage, spouses will do more to save their marriage, giving up dreams, work, ambitions, and extramarital relationships for marital success.

Unmarried couples are less likely than married couples to be encouraged to make sacrifices to save their relationships. Parents may even urge their children who are "living together" to split up rather than give up plans for school or a career. If a cohabiting couple encounters sexual difficulties, it is more likely that they will split up. It may be easier to abandon a problematic relationship than to change it. Among cohabitants who intend to marry, relationships are not significantly different from marriages. The intention to marry is highest among cohabiting couples with high incomes (Brown and Booth 1996).

Sex

There are differences in the sexual relationships and attitudes of cohabiting and married couples. Linda Waite and Maggie Gallagher (2001) suggest that married couples experience more fulfilling sexual relationships because of their long-term commitment to each other and their emphasis on *exclusivity*. Because they expect to remain together, married couples have more incentive to work on their sexual relationships and discover what most pleases their partners.

Cohabitants, however, have more frequent sexual relations. Whereas 43% of married men reported that they had sexual relations at least twice a week, 55% of cohabiting men said they had sex two or three times a week or more. Among married women, 39% said that they had sex at least twice a week, compared with 60% of never-married cohabiting women. Sex may also be more important in cohabiting relationships than in marriages. Waite and Gallagher (2001) go as far as calling it the "defining characteristic" of cohabitants' relationships.

Married couples are also more likely to be sexually monogamous. According to data from the National Sex Survey (see Chapter 6), 4% of married men said they had been unfaithful over the past 12 months; four times as many, 16%, of cohabitants reported infidelity. Among women, the equivalent comparison shows that 1% of married women compared with 8% of cohabiting women expressed having had sex outside of their relationship. Similar findings were obtained by Judith Treas and Deirdre Giesen even when they controlled for how permissive individuals were toward extramarital sex (2000, 59):

> This finding suggests that cohabitants' lower investments in their unions, not their unconventional values, accounted for their greater risk of infidelity.

Finances

Overall, cohabiting women and men have more precarious economic situations than married couples. The latter have higher personal earnings, higher household incomes, and are much less likely to live in poverty. There is also evidence that cohabitation carries an "economic premium" comparable to what accompanies marriage and that entering cohabiting relationships alleviates some financial distress, especially for Hispanic and African American women and their children (Avellar and Smock 2005). Unfortunately, as with the end of a marriage, when cohabiting relationships end there is considerable economic suffering, especially for women. Sarah Avellar and Pamela Smock

(2005) contend that where cohabiting men suffer modest effects when their relationships end, cohabiting women suffer "dramatic declines" in their standards of living. Men suffer declines of roughly 10% in their household income. For women, there is a more notable loss of household income (33%) and a striking spike in the level of poverty (nearly 30%) following breakups.

Cohabiting and married couples differ in whether and how they pool their money, typically a symbol of commitment (Waite and Gallagher 2001; Blumstein and Schwartz 1983). People generally assume that a married couple will pool their money, as it suggests a basic trust or commitment to the relationship and a willingness to sacrifice individual economic interests to the interests of the relationship. Among most cohabiting couples, money is not pooled. In fact, one of the reasons couples cohabit rather than marry is to maintain a sense of financial independence (Waite and Gallagher 2001; Blumstein and Schwartz 1983).

Finally, cohabitation brings financial benefits that result from our tax system and Social Security policies. When both partners earn approximately the same or similar amounts, by being legally single and filing their taxes as such they enjoy the benefit of larger standard deductions than they would if they were married. Regarding Social Security, some, especially elderly, men and women might decide to live together instead of

■ Heterosexuals, gay men, and lesbians cohabit. A significant difference between heterosexual and gay cohabitation is that many gay men and lesbians who would like to marry are prohibited by law from doing so.

© David Young-Wolff/PhotoEdit

marry because if they were to marry they would lose some of their Social Security benefits (Willetts 2003).

Children

The arrival of children tends to stabilize marriages, lowering the likelihood that couples will divorce. Young children and first-born children are especially associated with reductions in the likelihood of separation and divorce. Stepchildren have the opposite effect; their presence increases the risk of divorce (Manning 2004). How are cohabitors affected by the arrival and presence of children? Wendy Manning found that, for white women, conceiving a child while cohabiting promotes increased stability of the cohabiting relationship and increases the likelihood of marriage. Births during cohabitation do not seem to significantly affect—either positively or negatively—the cohabiting relationship. Such births reduce the likelihood of marriage for Latinas, but have no effect on the likelihood of marriage for either white or black women. However, cohabiting couples who give birth first but then marry face a greater risk of marital dissolution (Manning 2004).

Health

Marriage bestows health benefits on the married. Generally, married people live longer and healthier lives and suffer from fewer chronic or acute health concerns than the single, divorced, separated, or widowed. Some of this results from healthier lifestyles, evident in such things as lower rates of alcoholism and problem drinking and healthier body weights. Researchers have asked whether the advantages result from healthier people being more likely to marry than unhealthy or less healthy people—**the selection hypothesis**—or are part of the protection people receive from marriage itself. In addition, do the health benefits marriage bestows apply to cohabitants (Wu et al. 2003)? After analyzing Canadian data on the health status of 6,494 women and 5,368 men, Zheng Wu and colleagues conclude that married people have somewhat better general health than cohabitants, who, in turn, have better health than the separated and divorced, widowed, and never married. The difference between cohabitants and married people loses statistical significance once researchers control for other factors. By providing us with the social support of a loving partner, both marriage and cohabitation appear to "protect" the health of those in stable unions compared to those who lack such relationships. One thing to keep in mind: because cohabitation is typically of shorter duration, and more likely to fail or end, cohabitants are at a disadvantage compared to married people and may find that as their relationships end their health slides (Wu et al. 2003).

Relationship Quality and Mental Health

Research by Susan Brown and Alan Booth (1996) indicates that cohabiting couples have poorer relationship quality than do married couples, reporting lower levels of happiness with their relationships, more fighting, and more violence. However, these differences disappear or greatly diminish when we consider only cohabitants who have expressed the intention to marry (Brown and Booth 1996). Brown and Booth point out that the relationships experienced by those (> 75% of) cohabitants who plan to marry their partners *are not qualitatively different from marriage.*

Researchers have looked at the mental health characteristics of cohabitants as they compare to singles and married couples (Ross 1995; Horwitz and White 1998). Some report cohabitants to be like married people, experiencing similar levels of depression, with both being less depressed than those without partners (Ross 1995). However, Alan Horwitz and Helene White's research comparing rates of depression and alcohol problems among cohabiting, married, and single people found cohabitants to have higher rates of both depression and alcohol problems than married people. The mental health of cohabitants was more like single people than married ones. Furthermore, cohabiting men had the highest rate of alcohol problems of the three groups, suggesting that something about cohabitation (for example, unconventionality or financial pressures among those wishing to marry) may cause high rates of alcohol problems. Among married men, those who cohabited before marriage were no different in their level of alcohol problems than those who had not first cohabited (Horwitz and White 1998).

Work

Traditional marital roles call for the husband to work; it is left to the discretion of the couple whether the woman works. Contemporary families often cannot afford the luxury of a one-wage-earner household. Still, gender roles in marriage have emphasized men's economic provision as a major component of men's family responsibilities. In cohabiting relationships, the man is not expected to support his partner (Blumstein and Schwartz 1983). If the woman is not in school, she is expected to work. If she is in school, she is nevertheless expected to support herself.

Some married couples may fight about the wife going to work; such fights do not generally occur among cohabiting couples. With less certainty about the future of their relationships, cohabiting women may be less willing to restrict their outside employment or to spend time and energy on housework that could be spent on paid work.

Married couples often disagree about the division of household work. Both married and cohabiting women tend to do more of the domestic work than their male partners (Waite and Gallagher 2001; Seltzer 2000; Shelton and John 1993). But cohabiting women spend about 5–6 fewer hours on housework than do married women (Ciabatarri 2004). Cohabiting women who are not employed or who have children in the home tend to do more housework. Whether or not women intend to marry their partners does not significantly affect their time spent on housework. However, marital intentions loom large in influencing men's housework performance. Men who intend to marry *someone other than their current partner* (that is, intend to marry "someday" but not the woman they are living with) do 8 fewer hours of total housework and 4.4 fewer hours of core housework (house cleaning, cooking, laundry, shopping, and dishes) than men who definitely plan to marry their cohabiting partners. Cohabiting men with stronger commitments to their partners do more housework than men who are least committed to their relationships (Ciabatarri 2004).

Effect of Cohabitation on Marital Success

Although it may seem surprising and goes against the logic used by cohabiting couples who think that cohabitation helps prepare them for marriage, as we've noted before, such couples are more likely to divorce than those who do not live together before marriage (Bumpass and Lu 2000). In marriages that were previously cohabiting relationships, there are higher levels of disagreement and instability, lower levels of commitment, and greater likelihood of divorce.

The effect cohabitation has on subsequent marriage is not the same for all groups. Julie Phillips and Megan Sweeney report that for Caucasian women, 37% of those who cohabited before marriage saw their marriages end within 10 years compared to 28% who did not cohabit. The "cohabitation effect" is much smaller among African Americans and Hispanics. Among African American women, 51% of those who had cohabited before marriage and then married saw their marriages fail within 10 years compared to 48% of those who had never cohabited. Among Mexican American women, 32% who had cohabited experienced marital failure compared to 26% who hadn't cohabited. Among foreign-born Mexican Americans, there were more marital failures among women who *had not* cohabited than among those who had (Phillips and Sweeney 2004). Also of interest, cohabiting experience with only a subsequent spouse is still associated with risk of later marital failure for Caucasians but not for African Americans and Hispanics. Perhaps because cohabitation more often functions as a substitute for or precursor to marriage for blacks and Hispanics, they exercise more selectivity over their choice of partner than do whites. Phillips and Sweeney (2004) suggest that cohabitation among whites is more likely to consist of relationships between two people who begin their cohabitation less certain about their relationship.

Like race or ethnicity, age also matters. Older cohabitants are more likely than their younger counterparts to view their relationship as an alternative to marriage. Younger cohabitants more likely see their relationship as a prelude to marriage. Older cohabitants also report higher levels of relationship quality on numerous aspects of their relationships—fairness, having fewer disagreements, spending more time alone together, being less likely to argue heatedly, and being less likely to think that their relationship is in trouble or may end. Older cohabitants seem less negatively affected than younger cohabitants by the absence of plans to marry. Clearly, as Valarie King and Mindy Scott (2005, 283) suggest, "cohabiting relationships are indeed different for older and younger adults."

What is still unclear is what about cohabitation causes later marital difficulties. Is it the *types of people* who choose to live together before marrying or something about the *experience of living together* that causes problems later? Susan Brown and Alan Booth (1996) suggest that the characteristics of people who cohabit are more influential than the cohabiting experience itself. People who live together before marriage tend to be more liberal, more sexually experienced, and more independent than people who do not live together before marriage. They also tend to have slightly lower incomes and are slightly less religious than noncohabitants (Smock 2000).

At the same time, there is evidence that cohabitation itself may affect individual partners and their relationships. Compared with married couples, co-habiting partners tend to have more similar incomes and divide household tasks more equally. These arrangements may be harder to sustain once married, and strain or conflict may occur (Seltzer 2000).

As more people from different backgrounds enter cohabitation relationships, we will be better positioned to see whether the *experiences of cohabitation* or *characteristics of cohabitants* have greater effect on later marriage. As cohabitation grows in number and acceptability, its effects on marriage may also change. One thing we can suggest is that at least some poorly chosen relationships break up at the cohabitation stage. Thus, although it may not protect couples from later marital failure, it does show some high-risk couples that they were not meant for each other. This spares them the later experience of a divorce (Seltzer 2000).

As we have seen, whom we choose as a partner is a complex matter. Our choices are governed by rules of homogamy and exogamy as much as by the heart. But the process of dating or cohabiting helps us determine how well we fit with each other. Although these relationships may sometimes be viewed as a prelude to marriage, they are important in their own right. Whatever their outcome, these relationships provide a context for love and personal development.

Summary

- Aside from the popular emphasis on love, many factors shape our choices of partners and spouses, revealing the existence of rules of mate selection.

- The *marketplace of relationships* refers to the selection activities of men and women when sizing up someone as a potential date or mate. In this marketplace, each person has resources, such as social class, status, age, and physical attractiveness.

- In the marital exchange, women and men offer different resources.

- Initial impressions are heavily influenced by physical attractiveness. A *halo effect* surrounds attractive people, from which we infer that they have certain traits, such as warmth, kindness, sexiness, and strength.

- The *marriage squeeze* refers to the gender imbalance reflected in the ratio of available unmarried women to men. Overall, there are significantly more unmarried women than men. Marital choice is also affected by the *mating gradient*, the tendency for women to marry men of higher status.

- The *field of eligibles* consists of those of whom our culture approves as potential partners. It is limited by the principles of *endogamy* (marriage within a particular group) and *exogamy* (marriage outside a particular group), as well as by *homogamy* (the tendency to choose a mate whose individual or group characteristics are similar to our own).

- Interracial couples often receive negative nonverbal and verbal reactions from others.

- Similar gender-based age patterns in partner selection are evident among gay and lesbian couples and among heterosexual couples.

- Divorced people are more likely to select other divorced people as partners; adult children of divorced parents show a tendency to select other adult children of divorced parents.

- *Residential propinquity* refers to the tendency for partners to be selected from within a geographically limited locale.

- The patterns of mate selection and partner choice are affected by: preferences we form for certain types of people, reactions of and pressures from other people, and opportunities we have to interact and meet.

- The theories that attempt to explain mate selection include parental image, complementary needs, value, and role theories, and the three-stage *stimulus-value-role* theory.

- The setting in which you see someone can facilitate or discourage a meeting. A *closed field* allows you to see and interact simultaneously. An *open field*,

- characterized by large numbers of people who do not ordinarily interact, makes meeting more difficult.

- Women often covertly initiate meetings by sending nonverbal signals of availability and interest. Men then initiate conversation with an opening line.

- Power tends to be more equal in dating relationships than in marriage.

- *Dating scripts* prescribe certain behavior as expected of each gender. For women, problems in dating include sexual pressure, communication, and where to go on the date; for men, problems include communication, where to go, shyness, and money.

- Breakups are commonplace. In accounting for breakups, we attribute the cause to one of the following: our own or our partner's personal characteristics, characteristics within the relationship, and environmental influences.

- Gay and lesbian couples are more prone to breaking up. Gay and lesbian individuals report higher levels of satisfaction with post-breakup friendships with former partners than do heterosexuals.

- Due to delayed marriage, increased economic and educational opportunities and commitments for women, increased divorce, and liberal social and sexual standards, there has been a dramatic increase in the unmarried population (including both formerly married and never married).

- Relationships in the singles world tend to stress independence and autonomy. Singles may be classified into four categories: ambivalents, wishfuls, resolveds, and regretfuls, depending upon their desire and expectation to ever marry.

- *Domestic partnership* laws grant some legal rights to cohabiting couples, including gay and lesbian couples. Cohabitation has become increasingly accepted because of a more liberal sexual climate, the changed meaning of marriage, and delayed marriage.

- Reasons for and types of cohabitation vary. Cohabitation may be a substitute or alternative to marriage, a precursor to marriage, a trial marriage, or a convenient alternative to dating or to singlehood. The meaning and impact of cohabitation differs depending on the age and race of the partners.

- *Common-law marriage*—where couples who live together, present themselves as married, and are considered to be legally married—has gradually become less common. Only 11 states still recognize common-law marriage.

- Between 600,000 and 1.5 million gay men and lesbians cohabit. Whereas heterosexual cohabiting couples tend to adopt a traditional marriage model, lesbians and gay men use a "best friend" model that promotes equality in roles and power.

- Compared with marriage, cohabiting relationships are more transitory, have different commitments, lack economic pooling and social support. They also differ in sexual relationships, finances, health benefits, relationship quality, and household responsibilities.

- Cohabitants who later marry tend to be more prone to divorce, due to both selection factors (the type of people who cohabit) and experiential factors (consequences of cohabitation itself).

Key Terms

attributions 299

closed fields 292

common-law marriage 310

complementary needs theory 290

dating scripts 296

domestic partners 311

double standard of aging 283

endogamy 284

exogamy 284

field of eligibles 284

halo effect 281

heterogamy 285

homogamy 285

hypergamy 288

hypogamy 287

lesbian separatists 306

marital history homogamy 289

marketplace of relationships 280

marriage squeeze 283

mating gradient 283

open fields 292

parental image theory 291

residential propinquity 289

role theory 290

stimulus–value–role theory 291

trial marriages 308

value theory 290

Resources on the Internet

Companion Website for This Book

http://www.thomsonedu.com/sociology/strong

Gain an even better understanding of this chapter by going to the companion website for additional study resources. Take advantage of the Pre- and Post-Test quizzing tool, which is designed to help you grasp difficult concepts by referring you back to review specific pages in the chapter for questions you answer incorrectly. Use the flash cards to master key terms and check out the many other study aids you'll find there. Visit the Marriage and Family Resource Center on the site. You'll also find special features such as access to Info-Trac© College Edition (a database that allows you access to more than 18 million full-length articles from 5,000 periodicals and journals), as well as GSS Data and Census information to help you with your research projects and papers.

CHAPTER 9

Margaret Joanne Cotton Photography

Marriages in Societal and Individual Perspective

Outline

Are the following statements TRUE or FALSE?
You may be surprised by the answers (see answer key at the bottom of the page).

T F **1** More women than men tend to live with their parents.

T F **2** Couples who are unhappy before marriage significantly increase their happiness after marriage.

T F **3** Marriage, more than parenthood, radically affects a woman's life.

T F **4** The advent of children generally increases a couple's marital satisfaction.

T F **5** Age at marriage is a strong indicator of later marital success.

T F **6** In-law relationships tend to be characterized by low emotional intensity.

T F **7** Asian, Latino, and African American families are more likely than Caucasian families to take in extended family.

T F **8** The empty nest syndrome, characterized by maternal depression after the last child leaves home, is more a myth than a problem for American women.

T F **9** Most long-term marriages involve couples who are blissful and happily in love.

T F **10** The key to marital satisfaction in the later years is continued good health.

Answer Key for What Do You Think

1 False, see p. 350; 2 False, see p. 335; 3 False, see p. 336; 4 False, see p. 344; 5 True, see p. 335; 6 True, see p. 345; 7 True, see p. 352; 8 True, see p.359; 9 False, see p.354; 10 True, see p. 353.

I affirm the special bond and unique relationship that exists between us, and promise to keep it always alive. You are my partner in life and my one true love. I will cherish our union and love you more each day than I did the day before. I promise to support you in your goals, to honor, trust and respect you, today and for the rest of my life. I will laugh with you and cry with you, loving you faithfully through good times and bad, regardless of the obstacles we may face together. . . . I promise to always tell you what I feel about you, to leave no room for doubt and take nothing for granted. You will always know how much I love you and how beautiful you are to me. I promise to show you how much you have enriched my life and how much you allow me to feel that I never imagined ever again feeling for the rest of my life. I promise to stand with you always even in the darkest, hardest times and never let anyone hurt you. I will be beside you in whatever life throws our way and behind you, should you ever feel that what you need most is to know that there is someone there to catch you should you fall. I give you my hand and my heart. I promise to love you, comfort and encourage you, be open and honest with you, and stay with you from this day forward for as long as we both shall live.

As you no doubt recognize, those words are a version of wedding vows that, in some similar form or fashion, are exchanged between couples as they enter marriage. Some may add more religious language, some may be more or less traditional, some may be briefer and more concise, and others may be more personal and perhaps even playful. It is likely, however, that all will convey an intention to share life's ups and downs *together* for as long as both people live, as such is the essence and expectation of marriage.

Marriage is the foundation upon which American families are constructed. Although we recognize and value the ties connecting us with our wider families, marriage is the centerpiece of family life in the United States. In our nuclear family system, our relationships with our spouses are more important than our relationships with our extended families. In our lives as individuals, the person we marry is expected to be someone with whom we will share everything, a soul mate, and partner "for as long as we both shall live."

Yet the status and the direction of marriage in the United States are subjects of considerable ongoing controversy and debate. Although most Americans will, at some point, marry, fewer enter and stay in marriage today than did in the recent past. Is marriage less valued than it was in the past? As a society, are we less committed to marriage as a central life goal, and are those who marry less willing or able to work hard to make their marriages work? Is marriage "in trouble," destined to ever more gradual decline as more people divorce, remain single, live together, and have and raise their children without being married? Or, is marriage "doing just fine," having changed and adapted to, but having survived, ongoing societal and cultural changes? These "bigger picture" questions are addressed in the first part of this chapter.

In addition, we consider marriage from the vantage point of the married, by describing issues that confront couples as they enter marriage and as they attempt to craft and then share a lifetime together. Along the way we identify some factors that predict marital success, as well as some issues involved in the establishment of marital roles and boundaries. We describe the impact of children on marriages, especially the amount of satisfaction couples feel. We turn next to middle-aged marriages, examining families with young children and adolescents, families as launching centers of the young, and the process of reevaluating married life. Then we review later-life marriages, including connections with extended families, retirement, caregiving, and widowhood. Finally, we survey the different patterns and factors that characterize lasting marriages.

Marriage in Societal Context: The Marriage Debate

When it comes to marriage, these are confusing times. By this we mean that there is much difference of opinion over whether marriage is or isn't "endangered," whether it has or hasn't lost its appeal and its meaning as a major life goal to which people aspire and a relationship around which people build their lives. Even the marriage experts don't see eye to eye about what's going on and about whether people are turning from or continuing to value becoming and staying married. Consider, as illustration, the November 2004 issue of the *Journal of Marriage and the Family*, a special issue of one of the leading scholarly journals that focuses on family life. It contained a series of

articles and commentaries as part of a "Symposium on Marriage and Its Future." As article after article revealed, evidence can be marshaled on either side of what sociologist Paul Amato (2004) calls the **marriage debate**. As Amato points out, even "New professionals in the family field may find it curious that senior scholars can interpret the data on recent social trends in such strikingly different ways" (2004a, 960). Where some see marriage as "in decline," others portray it as dynamic, changing, and resilient. Just what is it that is so confusing?

Consider the following, clearly mixed, portrait of marriage in the United States:

- *Behaviorally,* almost three-fifths of adults in the United States are married. Another 17% are formerly married, being either widowed (6.6%) or separated or divorced (10.4%). Thus, three-fourths of adults are or have been married. In addition, 19% are never-married singles and nearly 6% are in cohabiting relationships (Centers for Disease Control and Prevention, National Center for Health Statistics).

- Over the quarter century from 1970 to 1996, the proportion of 25- to 29-year-olds who had never married *increased dramatically.* In 1970, 11% of 25- to 29-year-old females had never married; by 1996 the percentage had more than tripled, reaching 38%. Among men, the same period saw increases from 19% to 52% (Huston and Melz 2004). If not "foregoing" marriage, clearly people were "forestalling" it, as reflected in the unprecedented increases in the median age at which women and men enter their first marriages (27 years for men and 25 years for women in 2000) (Cherlin 2004).

- As shown in Chapter 3, cohabitation, births to unmarried mothers (either single or cohabiting), and divorces all increased over the last three decades of the twentieth century (though divorce decreased toward the end of the 1900s). Pessimistically, these might suggest that marriage had become less attractive, less essential as a prerequisite for having and raising children, and more fragile. None of these impressions are especially positive statements about the health and vitality of marriage (Huston and Melz 2004; Oropesa and Landale 2004).

- The preceding trends notwithstanding, experts estimate that nearly 90% of Americans will eventually marry.

- *Attitudinally,* marriage remains highly valued, even alongside increased acceptance of nonmarital lifestyles. Most young adults want to marry some-day and recognize that marriage brings benefits to their lives (Amato 2004b).

- Each year for more than a quarter of a century, around 80% of female high school seniors have expressed an expectation to marry someday. Among males, the percentage expecting to marry has increased during this period from 71% to 78% (Cherlin 2004).

- Marriage has been *and continues to be* seen as an "extremely important" part of life. Roughly 80% of young women and 70% of young men express such an attitude.

- Between 1980 and 2000, the norm of marriage as a life-long relationship received increased support (Amato 2004b).

- Marriage is not seen as essential even for those who wish to spend their lives with each other. Only 36% of U.S. adults disagree with the notion that "it is alright for a couple to live together without intending to get married" (Cherlin 2004).

Is There a Retreat from Marriage?

In the discussion of the status and vitality of marriage, we often hear that a retreat from marriage has taken place in the United States in recent decades. Just what does this mean, and how accurately does it represent marriage in America? R. S. Oropesa and Nancy Landale (2004) describe the **retreat from marriage** as "evident" in a number of recent and ongoing trends: "historic" delays in the age at which women and men first marry, nearly "unprecedented" proportions of the population never marrying, "dramatic" increases in cohabitation and nonmarital births, and continued high divorce rates. Robert Schoen and Yen-Hsin Alice Cheung (2006, 1) assert that marriage has actually "been in retreat for more than a generation," as fewer men and women "ever marry," and that the "U.S. withdrawal from marriage" persisted at least through 2003. The retreat from marriage appears to be associated with increases in employment of women, smaller gender wage gaps in earnings, wider inequality among men, and persistent economic inequality between racial groups (Schoen and Cheung 2006).

Economics and Demographics behind the Retreat from Marriage

Closer inspection of trends indicates that the retreat from marriage has not occurred among all social groups. Instead, both racial and economic differences can be identified. As shown earlier in Chapter 3, there are considerable differences in marital status for different racial, ethnic, and economic groups. Looking again, this time using data from the 1998, 2000, and 2002 *Current Population* surveys, you can see the following differences:

■ *Marriage patterns show significant race and ethnic differences in the likelihood of entering and remaining married.*

Table 9.1 ■ Marital Status by Ethnicity

Marital Status	White (%)	Hispanic (%)	Asian (%)	African American (%)
Married	57.4	50.9	57.4	34.0
Cohabiting	3.6	4.1	1.9	3.9
Widowed	6.8	3.5	4.1	6.6
Divorced	8.4	6.1	4.0	9.9
Separated	1.4	3.5	1.4	4.6
Never married	22.5	32.0	31.3	41.0

Table 9.2 ■ Percentage of Women Married, by Age, 1995

Age	18	20	25	30
White, non-Hispanic	8	26	63	81
Black, non-Hispanic	5	16	37	52
Hispanic	13	29	61	77
Asian	3	13	44	77
Total	8	25	59	76

SOURCE: Bramlett and Mosher 2002.

Where nearly three-fifths of Caucasians and Asians and half of Hispanics are married, only about a third of African Americans are married. Adding the widowed, divorced and separated to the portion married, nearly three-fourths of whites, two-thirds of Asians, and nearly two-thirds of Hispanics *are or have been* married, compared to more than half of African Americans (see Table 9.1).

Based on a National Center for Health Statistics report, *Cohabitation, Marriage, Divorce, and Remarriage in the United States* (Bramlett and Mosher 2002), we see differences in marital experiences for women of different racial backgrounds (Table 9.2).

Hispanic women are the most likely to marry young. By age 25 there is hardly any difference between Caucasians and Hispanics; more than three-fifths of women in both groups are married compared to less than half of Asian women and less than two-fifths of African Americans. By age 30, more than three-quarters of Caucasian, Asian, and Hispanic women are married as compared to just over half of African American women. Even by ages 35–39, only about two-thirds of African American women have married; a third are likely to never marry (Huston and Melz 2004).

The "Hispanic" and "Asian" categories reflect more diversity than can be addressed here. It is worth noting, however, that between 1970 and 2000 the percentage of Chinese American men and women who were married increased (from 50.7% to 63.4% among men and from 56% to 62.8% among women), as did the percentage of Japanese men who were married (from 57.4% to 59%). The percentage of Japanese women who were married decreased slightly (from 61.3% to 59.1%) during this same time period. Chinese men and women are more likely to be married and less likely to be divorced than are Japanese

American men and women. More generally, "marriage is still a strong institution for Chinese and Japanese Americans" (Ishii-Kuntz 2004). Among Hispanics, Mexican Americans and Cuban Americans "are generally more supportive of marriage than non-Hispanic whites" and tend to marry at similar levels (Oropesa and Landale 2004:906). Puerto Ricans, on the other hand, are considerably less likely to be married. They also display more acceptance of cohabitation, even without any plans to ever marry (Oropesa and Landale 2004).

One final example about race differences merits our attention. A three-state analysis of the percentage of white and black men and women marrying before age 50 in Virginia, North Carolina, and Wisconsin found the racial differences shown in Table 9.3 (Schoen and Cheng 2006).

In all three states, black women and men were considerably less likely to be married by age 50 than were white women and men. Clearly, three states do not reflect the whole of the United States, and states may differ in important ways in demographic or economic characteristics. Still, all three states have populations in excess of 4 million and represent both regional variation and variation in the proportion of the population

Table 9.3 ■ Percentage Marrying before Age 50

Virginia, North Carolina, and Wisconsin, circa 1990

Population Group	Virginia	North Carolina	Wisconsin
White men	89.2	82.9	86.8
Black men	85.7	67.6	69.2
White women	92.3	88.4	90.3
Black women	81.9	61.0	59.7

SOURCE: Schoen and Cheng 2006, 1–10.

■ *Despite data indicating that they are less likely to marry or expect to marry, African Americans express strong belief in the importance of marriage.*

that is African American (Wisconsin's 3–5% to North Carolina's 22%) (Schoen and Cheng 2006).

Some other racial differences to note: Although in general young people expect to marry someday, fewer young African Americans express an expectation to ever marry and report an older desired age at marriage than whites (Crissey 2005). African Americans who marry are more likely to divorce than Caucasians who marry. Divorced African Americans are less likely than divorced Caucasians to remarry. Blacks are also much more likely to bear children outside of marriage. Although a third of all children born in the United States are born to unmarried mothers, the race difference is pronounced: around a quarter of all births to Caucasian women compared to nearly 70% of births among African Americans are to unmarried mothers.

What about Class?

Within the shifts in marriage rates, there are notable socioeconomic differences. For example, although lifetime marriage rates among women have dropped by 5% in the United States, they have declined by 30% for women without a high school diploma (Gibson-Davis,

Edin, and McLanahan 2005). Among college-educated white women, the prospect of marrying has *grown greater,* whereas among those without college degrees it has decreased (Huston and Melz 2004). For both women and men, educational attainment is positively associated with the likelihood of marriage (Schoen and Cheng 2006). In addition, in the 1980s and 1990s, marriages among college-educated women became more stable than they had been in the previous decade; among women at the bottom of the educational distribution, marriage became less stable (Edin, Kefalas, and Reed 2004). In discussions of a retreat from marriage among Hispanics, R.S. Oropesa and Nancy Landale (2004) emphasize how limited economic opportunities may be major barriers or disincentives to marriage.

Look again at the data from Robert Schoen and Yen-Hsin Alice Cheng's study on marriage in Virginia, North Carolina, and Wisconsin, this time examining educational differences (Table 9.4).

As the data indicate, *for both men and women,* the percentages of people marrying by age 50 increased with education in all three states (except for the 13–15 years of education category, possibly indicating that starting but failing to complete college may make one less desirable as a marriage partner). When race and education are combined (not shown), the proportion of blacks with less than high school educations who are married by age 50 ranges from 38% to 65%. For blacks with less than 12 years of education, "never marrying was more likely than ever marrying" (Schoen and Cheng 2006, 9). For whites with college educations or more, the percentage who marry ranges from 89% to 96%.

Table 9.4 ■ Percentage Marrying, Divided According to Education

Population Group	Virginia	North Carolina	Wisconsin
By years of education			
Men, < 12 years	80.5	60.3	64.2
Men, 12 years	95.0	83.2	86.6
Men, 13–15 years	78.6	75.7	80.6
Men, 16 or > years	88.9	88.7	94.8
Women, < 12 years	81.0	55.0	63.8
Women, 12 years	97.0	85.4	91.3
Women, 13–15 years	82.3	75.2	81.5
Women, 16 or > years	92.3	90.8	95.4

Does Retreat from Marriage Suggest Rejection of Marriage?

Even if low socioeconomic status affects the likelihood of marriage, it may not signal an attitudinal rejection of marriage. Quite the contrary: Edin and colleagues (2004, 1,008) assert that "marriage has by no means lost its status as a cultural ideal among low-income and minority populations." Where only a minority of college graduates disapproved of cohabitation, two-thirds of high school dropouts disapproved or strongly disapproved of living together with no intention to marry. The difference is even more evident in the finding that after controlling for (comparing people of similar) race, age, marital status, presence of children, and religious attendance, individuals who hadn't completed high school were more than two times more likely to disapprove of cohabitation with no intention to marry than were college graduates (Edin et al. 2004).

Despite what the race data on marriage appear to suggest, African Americans remain "strong believers in the value of marriage" (Huston and Melz 2004). Some researchers have even found that unmarried blacks and Hispanics express greater interest in marrying than unmarried whites (Huston and Melz 2004). Overall, "research indicates very few significant racial or class differences in attitudes regarding the importance of marriage or aspirations toward marriage. . . . Even 70% of welfare recipients . . . say they expect to marry" (Gibson-Davis, Edin, and McLanahan 2005, 1,302).

Perhaps, then, a good portion of the "marriage retreat," at least that portion occurring among the most economically disadvantaged, is not really a *rejection of marriage*. Borrowing from the analysis done by Christina Gibson-Davis, Kathryn Edin, and Sara McLanahan, perhaps we should be asking ourselves, given their attitudes in favor of marriage and their expectations that they will someday marry, what keeps low-income unmarried parents from marrying? Interviews with a sample of low-income unmarried couples with children identified three barriers to marriage: financial concerns, concerns about the quality and durability of their relationships, and fear of divorce.

- *Financial concerns.* These concerns covered four aspects of financial matters: whether couples had the resources to "consistently make ends meet," whether they could exercise financial responsibility and wisely use what resources they possess, whether they could "work together toward long-term financial

goals," and whether they'd saved enough or had enough money for a "respectable wedding" (Gibson-Davis, Edin, and McLanahan 2005, 1,307).

- *Relationship quality.* Believing that marriage ought to be for life, that it is the "ultimate" relationship, couples want to make sure that their partners are suitable for marriage. One way they believe they can achieve this is by living together long enough to tell that their relationships are "up to the challenge" of marriage, that they can weather any storm, and that they have answered any doubts about whether they and their partners are ready and their relationships are strong enough for marriage.

- *Fear of and opposition to divorce.* Claiming not to believe in divorce as an option, and viewing marriage as somewhat "sacred," couples wait to marry until they fully believe that their relationships will last.

Expressed so well by Gibson-Davis, Edin, and McLanahan, what lies

at the heart of marital hesitancy is a deep respect for the institution of marriage. . . . The bar for marriage has grown higher for all Americans, making it increasingly difficult for those in the lower portions of the income distribution to meet the standards associated with marriage (2005, 1,311).

Religion and Marriage

Part of the supposed retreat from marriage consists of the delayed age at which women and men who marry are first entering marriage. Along with race and social class, religious affiliation is among the factors that may influence whether and when people choose to enter marriage. Religion has been shown to be associated with mate choice, childbearing and childrearing, the division of housework, domestic violence, marital quality, and divorce (Xu, Hudspeth, and Bartkowski 2005). Religious traditions and denominations differ in the kinds and degree of emphasis they place on marriage.

Importance of Marriage

Although Judeo-Christian religious groups tend to support marriage, uphold marriage and family as desirable and important lifestyles, and discourage both premarital and extramarital sex, there are differences among them, especially in the extent to which they support traditional gender roles and relationships and reject divorce, abortion, and homosexuality (Xu, Hudspeth, and Bartkowski 2005). Conservative Protestant denominations and Latter-day Saints (Mormons), articulate especially strong commitments to marriage, encouraging members to marry and stay married, by portraying marriage as "part of God's plan for self-development . . . in this life, as well as . . . long term spiritual salvation (Xu, Hudspeth, and Bartkowski 2005, 589–590)." Although, traditionally strongly promarriage, the Catholic Church has a "considerably less robust" promarriage orientation, as evidenced in the tendencies of American Catholics to move from the church's traditional teachings about marriage and toward a viewpoint that marital matters are subjects of individual choice. Liberal and moderate Protestants do not attach the same importance to marriage as do evangelical Protestants. Among Jews, we find greater emphasis on the importance of marriage and on more traditional gender roles in marriage among Orthodox Jews and considerably less encouragement to marry and bear children, as well as less emphasis on gender differences, among Reform Jews.

Timing of Marriage

Xiaohe Xu, Clark Hudspeth, and John Bartkowski found that women and men affiliated with moderate and conservative Protestant denominations and with the Mormon church are both more likely to marry and to marry young than those unaffiliated with a religious faith. Interestingly, they may face different consequences of early marriage. Baptists, among the most conservative Protestant denominations, have the highest divorce rate in the United States; Mormons are among those with the lowest likelihood of divorce.

Catholics and liberal Protestants also differ from the unaffiliated, but to a lesser extent. By emphasizing marriage as "the joining of two individuals with the goal of living a constructive, harmonious life" and "creating a good environment for rearing children" [(Xu, Hudspeth, and Bartkowski 2005, 589–590). Judaism, especially Reform Judaism, may encourage people to delay marriage. Indeed, Jews are more likely than Catholics, moderate and conservative Protestants, and Mormons to delay their entry into marriage. Jewish, liberal Protestant, and unaffiliated individuals were found to marry later (Xu, Hudspeth, and Bartkowski 2005, 589–590).

Between Decline and Resiliency

Perhaps the best way to understand what has happened and is happening to marriage is to use Andrew Cherlin's (2004) argument that marriage has been "deinstitutionalized." The **deinstitutionalization of marriage** refers to the "weakening of the social norms that define people's behavior in a social institution such as marriage" (848). As a result of wider social change, individuals can no longer rely on shared understandings of how to act in and toward marriage. Having undergone an earlier transformation from marriage as an institution to marriage as companionship, beginning in the 1960s the companionate marriage began to lose ground to a form of marriage Cherlin calls the **individualized marriage.** In individualized marriage, individual self-fulfillment and personal growth became the objectives people sought to satisfy through marriage. The companionate marriage had been the dominant form for more than half of the last century. Held together by love and friendship between spouses, not social obligations; characterized by egalitarian as opposed to the earlier patriarchal ideals for marriage; and allowing—indeed encouraging—spouses to focus on self-development and expression, the companionate marriage was by the 1950s the widely shared cultural ideal. More recently, partly as a product of "cultural upheavals of the 1960s and 1970s," the emphasis on the personal fulfillment and personal growth that is to come in marriage and the expectation that our spouses will be facilitators of such growth and sources of unprecedented support have given rise to the individualized marriage (Amato 2004a).

This is where the marriage debate centers. Some scholars see the changes and trends described here as worrisome because they undermine marriage as an institution that meets the needs of society. They believe that we have become too individualistic, too focused on personal happiness, and have less commitment to making our marriages work. Such attitudes help explain the increases in cohabitation, single parenthood, and divorce, as individuals pursue what they most want regardless of their effects on others. To proponents of this viewpoint, we need to enact policies to reinstitutionalize marriage, to restrict and decrease divorce, and to strengthen values such as marital commitment, obligation, and sacrifice.

Others put more emphasis on marriage as a relationship between two individuals and assert the value of such characteristics of contemporary marriage as self-development, freedom, and equality between spouses. Rejecting the idea that we have grown too individualistic or selfish, they also challenge the idea that ongoing trends should be seen with such negativity. Even the increase in divorce may be seen as an opportunity for happiness for adults and a means of escape for children from dysfunctional or dangerous environments.

As articulated by sociologist Paul Amato, neither the **marital decline perspective** nor the **marital resilience perspective** is consistently or uniformly supported by the variety of available data on marriage. As he says, "Recent social changes appear to have undermined marital unions in some respects and improved marital unions in other respects, with the current status of marriage lying somewhere between 'decline' and 'resilience'" (2004b, 101). Paul Amato, David Johnson, Alan Booth, and Stacy Rogers compared two national surveys of married women and men in the United States, one from 1980 the other from 2000. As expected, given some trends we have already discussed, the demographics of marriage had changed considerably; age at first marriage, the proportion of remarried individuals and couples marrying after first cohabiting, and the proportion of wives in the labor force and the share of household income that they contributed had all increased. Gender relations had changed in less traditional directions. Couples also became more religious and expressed greater support for the norm that marriage was for life.

Linking these sorts of changes to shifts in marital quality, data supported both the marital decline and the marital resilience perspectives. In other words, some changes were associated with declines in marital happiness and interaction and with increases in divorce proneness. Yet other changes were associated with improved marital quality. And the overall effect? Although the average level of marital interaction declined significantly (couples less likely to eat dinner together, go shopping together, visit friends together, and go out for recreation together), as Amato expresses (2004b, 101), "In general, these changes tended to offset one another, resulting in little net change in mean levels of happiness and divorce proneness in the U.S. population."

Who Can Marry?

Having looked at who is and isn't marrying (and at what ages or with what consequences they marry), we turn briefly to matters of legality. Not everyone can

AP Images/Mindi Sokoloski

■ Most states have passed laws declaring legal marriage to be available to heterosexuals only. Only Massachusetts allows gay and lesbian couples the legal right to marry.

marry the partner of their choice. In Chapter 1, we looked at some restrictions imposed on our marriage choices. As we noted then, who we are allowed to legally marry has undergone change and challenge over the past 150 years in the United States, over such issues as race and, more recently, over the question of marriage between two people of the same sex. As we remind you, no longer does any state prevent two people of different races from marrying, and *all but one state* restrict marriage to heterosexual couples.

What other criteria do state marriage laws specify regarding eligibility to marry? Each state enacts its own laws regulating marriage, leading to some discrepancies from state to state. Although some restrictions are uniform across all 50 states, others, such as those specifying minimum ages at which people can marry or addressing the question of cousin marriage, are more variable. As summarized by the Legal Information Institute of Cornell University Law School,

> The Supreme Court has held that states are permitted to reasonably regulate the institution by prescribing who is allowed to marry, and how the marriage can be dissolved. Entering into a marriage changes the legal status of both parties and gives both husband and wife new rights and obligations. One power that the states do not have, however, is that of prohibiting marriage in the absence of a valid reason.
>
> All states limit people to one living husband or wife at a time and will not issue marriage licenses to anyone with a living spouse. Once an individual is married, the person must be legally released from the relationship by either death, divorce, or annulment before he or she may remarry. Other limitations on individuals include age and close relationship. Limitations that some but not all states prescribe are: the requirements of blood tests, good mental capacity, and being of opposite sex.

Marriage between Blood Relatives

Nowhere in the United States is marriage allowed between parents and children, grandparents and grandchildren, brothers and sisters, uncles and nieces, and aunts and nephews. Perhaps this comes as no surprise to you, because such blood relations are clearly considered "too close" and marriage within such relationships is seen as incestuous and unacceptable. Some states disallow all "ancestor/descendant marriages," and a handful of states explicitly extend the prohibition to marriages between parents and children to parents and their adopted children.

The following example reflect the nature of such legal prohibitions or restrictions:

- New Jersey law uses language common to many other state marriage statutes:

  A man shall not marry any of his ancestors or descendants, or his sister, or the daughter of his brother or sister, or the sister of his father or mother, whether such collateral kindred be of the whole or half blood. A woman shall not marry any of her ancestors or descendants, or her brother, or the son of her brother or sister, or the brother of her father or mother, whether such collateral kindred be of the whole or half blood. A marriage in violation of any of the foregoing provisions shall be absolutely void.

You may have noticed from this statute that New Jersey allows first cousins to marry. Although many other state marriage statutes articulate similarly specific restrictions, some states, such as Ohio or Washington, more simply and generally prohibit marriage between relatives "closer than second cousins."

Some of you may find these laws surprising, thinking that first cousins can't marry, shouldn't marry, and—if they were to have children together—would face risks of passing genetic defects to their children. Although there are sociological and psychological arguments for the existence of incest restrictions, they tend to pertain mostly to the benefit of forcing people outside of their nuclear family of origin for a spouse. Furthermore, there is debate about the justification for prohibiting such marriages, common in many other parts of the world, including the Middle East, Europe, and South Asia. One genetics researcher estimates that as many as 20% of marriages worldwide are between first cousins (Willing 2002). As to the risk to offspring of such marriages, there is only a slightly elevated risk of such children inheriting recessive genetic disorders. Researchers "concluded that children of marriages between cousins inherited recessive genetic disorders, such as cystic fibrosis and Tay-Sachs disease, in 7% to 8% of cases. For the general population, the rate was 5%" (Willing 2002).

Age Restrictions

Although we have talked some and will talk more in later chapters about the effects of age at marriage on later marital success, here we simply note how state laws regulate and restrict marriage based on age. Throughout the United States, 48 of 50 states require both would-be spouses to be at least 18 years old to marry without parental consent. Two states set the age without parental consent higher: in Nebraska it is 19, and in Mississippi 21. Some states will waive the age requirement if the woman is pregnant, but in such instances she may need approval from a court. Many states allow couples to marry in their early to mid teens, providing they secure parental or court consent.

Number of Spouses

No state allows an individual to marry if she or he is already married. In other words, all 50 states consider monogamy the only legally accepted form of marriage. If a divorced or widowed man or woman wishes to re-marry, she or he must present evidence of the legal termination of the prior marriage or of the death of her or his former spouse

Gender of Spouses

In Chapter 1, we already considered the question of same-sex marriage. It is worth noting that many states have added to their state marriage laws explicit and emphatic declarations that same-sex marriage will not be recognized within the state, even if it is legally allowed elsewhere in the United States. A particularly emphatic example is illustrated in Chapter 3101 of Title 31 of the Ohio Revised Code on marriage:

> (1) Any marriage between persons of the same sex is against the strong public policy of this state. Any marriage between persons of the same sex shall have no legal force or effect in this state and, if attempted to be entered into in this state, is void ab initio and shall not be recognized by this state. (2) Any marriage entered into by persons of the same sex in any other jurisdiction shall be considered and treated in all respects as having no legal force or effect in this state and shall not be recognized by this state.

Most states have added similar amendments to their marriage laws, although typically in less extensive language. As of this writing, 43 of 50 states have either passed laws banning same-sex marriages or have had such laws approved by voters in ballot initiatives seeking to ban same-sex marriage (http://www .lambdalegal.org).

The Essence of Legal Marriage

Marriage creates a legal relationship between two people. As such, it imposes certain responsibilities and obligations but also bestows considerable rights and protections on spouses. As discussed in Chapter 1, marriage confers a wide range of benefits from tax breaks to rights to care for one another if hospitalized or to inherit. ("Marriage Rights and Benefits," http://www .nolo.com).

Marriage also imposes legal *responsibilities* and *obligations* on spouses, although these may not be spelled out. The "model marriage statute" is intended as a legislative device to provide "firmer guidance to courts and family law as a discipline about the nature and public purposes of marriage." (http://www .marriagedebate.com/ml_marriage/cat03-ml01.php) According to law professor Katherine Spaht, who drafted a "Model Marriage Obligations Statute," when they marry, husbands and wives owe each other mutual respect, fidelity, mutual support and assistance, and mutual commitment to and responsibility for the joint care of any children they have together (http://www .marriagedebate.com/ml_marriage/cat03-ml01.php):

> Respect requires each spouse to exhibit regard or esteem for the other. Fidelity is sexual faithfulness, precluding a spouse from sexual intercourse with another person. Support means economic resources sufficient to provide for not only the necessities of life, such as food, clothing, and shelter, but also the ordinary conveniences of life, including transportation and labor-saving devices. Assistance is cooperating in the accomplishment of tasks that support the spouses' life in common, including securing medical assistance for an ill or infirm spouse.

Fidelity, or sexual exclusivity, is described by Spaht as "the hallmark of marriage" and the essence that distinguishes marriage from "mere cohabitation." Cumulatively, the other designated obligations "embody well-understood community expectations as well as spousal expectations about appropriate marital behavior . . . [and] represent the principal core of a complex set of social norms that promote cooperation between spouses. . . . Other such norms include trust (incorporated within fidelity), reciprocity, and sharing (incorporated within respect, support, and assistance)."

However, most states do *not explicitly* define marriage responsibilities and obligations in statute, relying instead on common law understanding of marriage. Louisiana is a notable exception. According to Louisiana *Civil Code Art. 98. Mutual duties of married persons,* "Married persons owe each other fidelity, support, and assistance" (http://www.marriagedebate.com/ml_marriage/cat03-ml02.php).

Before we leave the topic of legal marriage, we ought to note that there is much ongoing disagreement and debate about what marriage does or ought to mean legally; whether legal marriage should or shouldn't be made available to same-sex, as well as heterosexual, couples; and whether its benefits and responsibilities ought to extend to unmarried couples. One way in which the debate has been framed, albeit by those from a more conservative perspective, is as a clash between two views of marriage: a conjugal view of marriage versus a close relationship model.

In a report prepared by the Council on Family Law, titled "The Future of Family Law: Law and the Marriage Crisis in North America," these are described as "dramatically different concepts of marriage and of the role of the state in making family law" (Council on Family Law 2005, 9). The **conjugal model of legal marriage** has at its core a view of marriage defined as "child centered," because it stresses the importance of "sustaining enduring bonds between women and men in order to give a baby its mother and father, to bond them to one another and to a baby" (13). A conjugal marriage is "a sexual union between a man and a woman who promise each other sexual fidelity, mutual caretaking, and the joint parenting of any children they may have" (7). On the other hand, the **close relationship model of legal marriage** sees marriage "as one in a universe of diverse close, private relationships, with intrinsic emotional, psychological, and sexual dimensions." From the conjugal model, only heterosexual legal marriage ought to be recognized in family law. In the close relationship model, the law ought to recognize and protect all relationships in which individuals share intimacy, commitment, interdependence, mutual support, and communication, regardless of whether partners are of the same or opposite sex and regardless of whether they legally marry or not.

Why Marry?

If you stopped each couple just before they exchanged their vows and asked, "Why are you doing this? Why are you getting married?" you would no doubt hear many different answers. You would also receive some baffled looks and possibly be pushed or shoved out of the way. More important for the moment, however, is the many reasons people can give for why they want to marry. You may recall that in the last chapter we identified some "pushes" and "pulls" that propel us either toward or from marriage (meaning from or toward cohabitation or singlehood). The greatest attraction of marriage is probably the love and intimacy

that we expect to find and share there. A nationally representative sample of 1,003 young adults (20–29 years old) demonstrated the extent to which our views about marriage and, perhaps, the appeal of marriage is rooted in the intimacy and love we hope to find there. More than 9 out of 10 never-married respondents endorsed the notion that "when you marry, you want your spouse to be your soul mate, first and foremost" (Whitehead and Popenoe, 2001, in Cherlin, 2006). In addition, more than 80% of women surveyed indicated that it was more important to "have a husband who can communicate his deepest feelings" than a husband who is financially successful (Cherlin 2004). Clearly, we are drawn to marriage in pursuit of a level of love and intimacy we believe may not be otherwise possible. As sociologist Paul Amato (2004b) puts it, we tend to see marriage as "the gold standard" for relationships.

Among the many reasons for marriage, we can easily recognize the role of possible economic and social pressures (that is, "pushes" toward marriage), as well as the strong desires to have and raise children, which, for many, seem to be best accomplished in marriage. As Amato (2004b) expresses, "most people will continue to see marriage as the best context for bearing and raising children" and, if they desire to become parents, will marry. Marriage may also symbolize that two people have reached a stage in their lives, as well as in their relationships, and that in it they have attained "a prestigious, comfortable, stable style of life" (Cherlin 2004, 857).

If the practical importance of marriage has diminished, if marriage can no longer be counted on to cement relationships, allowing spouses to confidently invest themselves in each other without fear, invest their time and energy in raising children together, and invest financially in acquiring such goods as cars and homes, the "symbolic significance" of marriage remains considerable and attractive. It has become less a marker of conformity as it has become more a marker of prestige (Cherlin 2004). This can be seen particularly well in the attitudes expressed by low-income, unmarried parents who continue to express a desire to someday marry. Although such women and men expressed economic incentives, more striking were their expectations of the kind of relationship marriage would offer: "a lifetime companion, a partner who will be their confidant and friend" (Edin, Kefalas, and Reed 2004, 1,012). As one woman articulated, "An understanding and loving man . . . that's what I'm looking for. It's like a fairy tale thing. . . .

I'm looking for that man who is totally devoted to me and is understanding, and has that undying love for me, you know?" (Edin, Kefalas, and Reed 2004, 1,012).

Benefits of Marriage

In what ways does being married benefit the women and men who marry? In comparing cohabitation and marriage in the last chapter, we looked at some advantages people obtain from marriage. We remind you here that marriage confers benefits in economic well-being, health, and happiness. Marriage provides clear economic benefits, and married couples are better off financially than those living in all other types of households (Hirschl, Altobelli, and Rank 2003). Marriage both reduces the risk of poverty and increases the probability of affluence. Defining *affluence* as living in a household that earns 10 times the poverty level, Thomas Hirschl, Joyce Altobelli, and Mark Rank conclude that married-couple households are more likely to attain affluence than those living outside of marriage. Women, in particular, face much greater likelihood of attaining affluence in marriage than outside of marriage (Hirschl, Altobelli, and Rank 2003).

Married people "enjoy better mental and physical health," lower risk of mortality, and lesser likelihood of alcohol problems, obesity, or both than the unmarried, although cohabitants experience similar benefits (Wu and Hart 2002). The Centers for Disease Control concluded that married women and men are less likely to smoke, drink heavily, or be physically inactive and are less likely to suffer from headaches and serious psychological distress. When marriages end, women suffer increased depression and men suffer poorer physical and mental health. Although married people generally report themselves as happier than unmarried people, this effect holds only for those whose marriages are satisfying or happy.

While marriage improves and protects men's physical and mental health, it appears to mostly just improve women's mental health. For men, marriage may have health benefits that are mostly the result of the social and emotional support men receive from wives and the control women exercise over their husbands' lifestyles and health-related behaviors. For women, health benefits of marriage may be more the byproducts of their increased economic well-being (Wu and Hart 2002).

© Scott Roper/CORBIS

■ *Rocky, high conflict courtships do not typically become smooth and harmonious marriages.*

Is It Marriage?

In considering the benefits that seem to accompany marriage, researchers have been somewhat divided as to whether these benefits truly followed marriage or were instead reflections of differences in the types of people who do and don't marry. Sometimes phrased as a difference between *selection* into marriage and *protection* afforded by marriage, it raises the question of whether there is something unique and beneficial about being married or whether those who marry are somehow unique compared to those who don't marry. In research on health and well being, selection is typically not the major factor, accounting instead for "only a small proportion of the variance in mental and physical health" (Wu and Hart 2002, 421).

For example, research into the effect of marriage and "union formation" on depression looked to differentiate between marriage effects and differences in the types of people who marry and those who don't.

The researchers concluded that marrying was associated with "substantively meaningful reduction" in rates of depression and that there was no indication that marriage was selective of less depressed people (Lamb, Lee, and DeMaris 2003).

Although we have painted these as alternatives—as *either* selection *or* protection—the two are not mutually exclusive. It is possible that both operate simultaneously. Thus, although healthier and more stable individuals may be more attractive as marriage partners, thus bringing better mental health with them into marriage, a good marriage also has a healthful and stabilizing effect on those who marry.

Experiencing Marriage: A Developmental Approach

Have you ever looked closely at a family photo album, say one that belonged to a parent or grandparent? If you have, you know that these albums are fascinating representations of the dynamics inherent in all families. If you get the chance, study one of your family's old albums closely. Typically, you'll find photos of now deceased relatives, which means you can "meet" ancestors that you never got to know in person. Many find it especially interesting to look at wedding photographs of parents or grandparents from years ago, pictures that capture in that instant the excitement and hope that they carried with them as they embarked on a shared married life. Eventually, there are baby pictures, where you may find these same spouses now new parents. As you turn the pages and study the photos, you can see other changes as children grow and parents age.

Understanding the basic truth conveyed by such visual images will enable you to better appreciate and understand the material to which we now turn. Marriages and families are dynamic. They are always changing to meet new situations, new emotions, new commitments, and new responsibilities.

The same can be said of individuals. Our individual identities, our sense of who we are, change as we mature. At different points in our lives, we are confronted with different developmental tasks, such as acquiring trust and becoming intimate. Our growth as humans depends on the way we perform these tasks. Psychologist Erik Erikson (1963) offered one of the most influential models describing human development (see Figure 9.1). In it, he depicted a life

Figure 9.1 ■ Erikson's Stages of the Life Cycle

Infancy: Trust versus mistrust. In the first year of life, children are wholly dependent on their parenting figures for survival. They learn to trust by having their needs satisfied and by being loved, held, and caressed. Without loving care, an infant may develop a mistrusting attitude toward others and toward life in general.

Toddler: Autonomy versus shame and doubt. Between ages 1 and 3, children learn to walk and talk and begin toilet training. They need to develop a sense of independence and mastery over their environment and themselves.

Early childhood: Initiative versus guilt. Ages 4 to 5 are years of increasing independence. The family must allow the child to develop initiative yet direct the child's energy. The child must not be made to feel guilty about his or her desire to explore the world.

School age: Industry versus inferiority. Between ages 6 and 11, children begin to learn that their activities pay off and that they can be creative. The family needs to encourage the child's sense of accomplishment. Failing to do so may lead to feelings of inferiority in the child.

Adolescence: Identity versus role confusion. The years of puberty, between ages 12 and 18, may be a time of turmoil, as well as discovery and growth. Adolescents try new roles as they make the transition to adulthood. To make a successful transition, they need to develop goals, a philosophy of life, and a sense of self. The family needs to be supportive as the adolescent tentatively explores adulthood. If the adolescent fails to establish a firm identity, he or she may drift without purpose.

Young adulthood: Intimacy versus isolation. In young adulthood, the adolescent leaves home and begins to establish intimate ties with other people through cohabitation, marriage, or other important intimate relationships. A young adult who does not make other intimate connections may be condemned to isolation and loneliness.

Adulthood: Generativity versus self-absorption. Generativity is the bearing of offspring, productiveness, or creativity. In adulthood, the individual establishes his or her own family and finds satisfaction in family relationships. It is a time of creativity. Work becomes important as a creative act, perhaps as important as family or an alternative to family. The failure to be generative may lead to self-centeredness and an attitude of "what's in it for me" toward life.

Maturity: Integrity versus despair. In old age, the individual looks back on life to understand its meaning—to assess what has been accomplished and to gauge the meaning of relationships. Those who can make a positive judgment have a feeling of wholeness about their lives. The alternative is despair.

cycle with eight developmental stages, each of which confronts us with an important developmental task to accomplish. Each stage intimately involves the family. As we enter young adulthood, these stages may also involve marriage or other intimate relationships (Nichols and Pace-Nichols 1993).

Throughout our life cycles, our goals and concerns change. Education and family-related goals, such as marriage and having children, are dominant among young adults. Among middle-aged adults, goals shift to concern about children's lives and about property, such as buying or maintaining homes. Among the elderly, health, retirement, leisure activities, and interest in the world predominate (Nurmi 1992).

As discussed in Chapter 2, some family scholars focus attention on how marriages and families predictably change across time. At various stages, the family has different developmental tasks to perform, and much is related to the presence and development of children. Families are often organized around child-rearing responsibilities and marriage relationships often become absorbed in these tasks.

We can use such insights to examine marriage. Spousal roles are different for couples with and without children, and they are different for parents of toddlers compared to parents of teens. Individual members and the family as a unit undergo changes that are better understood by locating the family in a developmental context. For example, couples who are parents of adolescents wrestle with the process of granting their children greater autonomy and independence. Meanwhile, a teenage daughter or son has an individual task of trying to develop a satisfactory identity. Simultaneously, an older sibling may be struggling with intimacy issues as a younger one develops "industry." Parents may struggle with issues of generativity while grandparents confront issues of integrity. A life course emphasis highlights the common experiences families have in the course of their shared lives.

In the Beginning

The marriage process may begin informally with cohabitation or more formally with engagement. Marriage ends with divorce, or continues legally but in a radically altered form with the death of a partner. When we enter marriage, we may find that the reality of marriage requires us to be more flexible than we had anticipated. We need flexibility to meet our needs,

our partners' needs, and the needs of the marriage. We may have periods of great happiness and great sorrow within marriage. We may find boredom, intensity, frustration, and fulfillment. Some of these may occur because of our marriage; others may occur despite it. But as we shall see, marriage encompasses constantly evolving changes and possibilities.

Again, Americans are waiting longer to marry today than in previous generations. Whatever the reasons, increasing age at time of marriage probably results in young adults beginning marriage with more maturity, independence, work experience, and education. Potentially, these are important assets to bring into marriage.

Predicting Marital Success

The period before marriage is especially important because couples learn about each other—and themselves. Courtship sets the stage for marriage. Many of the elements important for successful marriages, such as the ability to communicate in a positive manner and to compromise and resolve conflicts, develop during courtship. They are often apparent long before a decision to marry has been made (Cate and Lloyd 1992). Couples who are unhappy before marriage are more likely to be unhappy after marriage as well (Olson and DeFrain 1994).

Ted Huston and Heidi Melz (2004) describe three "prototypical courtship experiences," each of which has different likely consequences for couples who marry. Of critical importance in differentiating these courtships are personality characteristics of partners, which affect "both the dynamics of their courtships and the success of their marriages" (952). Some qualities, such as warmheartedness or an even temper, are important determinants of whether people create happy and stable marriages. Other qualities, such as being less stubborn, less independent minded, and more conscientious, are important factors in determining whether couples stay married. These personality characteristics are associated with the three courtships and marital outcomes that Huston and Melz identify as follows:

- *Rocky and turbulent courtships.* Such courtships are characterized by periods of upset and anger, distress and jealousy over potential rivals, and uneasiness about placing love in "undeserving hands" (950). They are more typically experienced by "difficult" personalities, people who are exceedingly independent minded, who lack conscientiousness, and who have high anxiety. If men are excessively independent, they may make poor husbands and their marriages are likely to be "brittle." If men and women high in anxiety marry each other, their marriages tend to be unhappy but lasting marriages.

■ *Sweet and undramatic courtships.* Partners are people with "good hearts" who are helpful, sensitive to the needs of others, gentle, warm, and understanding. Good-hearted couples find enjoyment and pleasure in each other's company. Their marriages are more likely to be satisfying and enduring.

■ *Passionate courtships.* These are characterized by partners "plunging into love, having sex early in the relationship, and deciding to marry one another within a few months" (950). Such couples begin marriage as "star-crossed lovers" sharing far more affection than typical of even newly married couples, "but over the first two years, much of the sizzle fizzles" (950). They are also vulnerable to divorce.

Huston and Melz contend that we can tell "from the psychological make-up of partners and how their courtships unfolded, whether they would be delighted, distressed or divorced years later" (2004, 949). How couples reach marriage, as well as what types of personal traits they bring into marriage, are important. We are not all of equal quality "marriage material" (Huston and Melz 2004).

Whether marriage is an arena for growth or disenchantment depends on the individuals and the nature of their relationship. It is a dangerous myth that marriage will change a person for the better: An insensitive single person simply becomes an insensitive husband or wife. Undesirable traits tend to become magnified in marriage because we must live with them in close, unrelenting, and everyday proximity.

Family researchers have found numerous premarital factors to be important in predicting later marital happiness and satisfaction. Although they may not necessarily apply in all cases—and when we are in love, we believe we are the exceptions—they are worth thinking about. According to Rodney Cate and Sally Lloyd (1992), these premarital characteristics include background, personality, and relationship factors.

Background Factors

Age at marriage is important. Adolescent marriages (where either party is younger than 20) are especially likely to end in divorce. Young marriages may be more divorce prone because of the immaturity and impulsivity of the partners (Clements, Stanley, and Markman 2004). Marriage age seems to have less effect as once people are past adolescence. In other words, differences between those who marry in their mid- to late 20s and those who marry in their 30s are slight. Length of courtship is also related to marital happiness. The longer you date and are engaged to someone, the more likely you are to discover whether you are compatible with each other. But you can also date "too long." Those who have long, slow-to-commit, up-and-down relationships are likely to be less satisfied in marriage. They are also more likely to divorce. Such couples may torture themselves (and their friends) with the familiar dilemma of whether to split up or marry—and then marry, to their later regret.

Level of education seems to affect both marital adjustment and divorce. Education may give us additional resources, such as income, insight, or status, that contribute to our ability to carry out our marital roles. Similarly, level of religiousness is a factor in shaping marital outcomes; higher religiousness, especially by wives, is associated with greater probability of happy and stable marriages (Clements, Stanley, and Markman 2004). Childhood environment, such as attachment to family members, parents' marital happiness and marital outcomes, and low parent–child conflict, is associated with marital happiness. This is especially true for women: some studies indicate that the woman's relationship with her family of orientation is crucial to later marital happiness. It may spell trouble if the man is too close to his family of orientation. Most studies on childhood environment, however, are based on men and women who came of age before the 1960s. The social context of marriage has changed dramatically since then, with the rise of divorce, smaller families, and changing gender roles. Parental divorce may cause someone either to shy from marriage or to marry with the determination not to repeat the parents' mistakes. Once married, the likelihood of success is negatively affected by parental divorce. Parental divorce increases risks to married children; those who grew up in households where parents divorced are more likely to experience a divorce themselves.

Personality Factors

How does having a flexible personality affect marital success? How about a contentious personality? A giving one? An obnoxious one? As you can imagine, your partner's personality will affect your life, your relationship, and your marriage considerably. We bring with us into our marriages personality characteristics, attitudes and values, habits and preferences, and unique personal histories and early experiences. Such characteristics are relatively stable and likely exert influence on the quality and outcomes of our marriages (Bradbury and Karney 2004).

We do know, however, that opposites do not usually attract; instead, they repel. We choose partners who share similar personality characteristics because similarity allows greater communication, empathy, and understanding (Antill 1983; Buss 1984; Kurdek and Smith 1987; Lesnick-Oberstein and Cohen 1984). It may be that personality characteristics are most significant during courtship. It is then that those with undesirable or incompatible personalities are weeded out—or ought to be—at least in theory.

Researchers tend to focus more attention on relationship process and change than on personality. Personality seems fixed and unchanging. Nevertheless, it clearly affects marital processes. For example, a rigid personality may prevent negotiation and conflict resolution and a dominating personality may disrupt the give-and-take necessary to making a relationship work, whereas warmth, an even temperament, a forgiving and generous attitude toward ones spouse contribute to happy, stable marriages. In Ted Huston's longitudinal study, following couples from courtship through early marriage and into "whatever destinations they arrived at nearly 14 years after they were wed," there was notable stability to assessments of spouses' personalities made when couples were first married. These early assessments predicted how these couples "behaved and felt about their marriages *almost 14 years later*" (Huston and Melz 2004, 953, emphasis added). Thus, such attributes and characteristics matter greatly in shaping marital outcomes.

Relationship Factors

Besides personality characteristics, researchers have also examined other aspects of premarital interaction and relationships that might predict marital success.

Loving each other did not seem to have much impact on whether couples fought. Couples who had other partners simultaneously prior to marriage or who compared their partners with others had lower levels of marital satisfaction. Another study on communication and marital satisfaction examined the same couples after 1, 2.5, and 5.5 years of marriage (Markman 1981, 1984). During the first year, there was no relationship between communication and marital satisfaction, but after 2.5 and 5.5 years, the more negative the communication, the less satisfactory the marriage.

Not all research substantiates the "intrinsically appealing" idea that marital success or failure is determined by how spouses communicate and solve problems (Bradbury and Karney 2004). Problem-solving skills are important, but not as important as the emotional climate within which such skills are implemented. "If spouses have a reservoir of good will and they show their affection regularly, they are more likely to be able to work through their differences, to warm to each other's point of view, and to cope effectively with stress" (Huston and Melz 2004).

If couples can maintain humor, express "genuine enthusiasm for what the partner is saying," and convey their continued affection for each other, couples with low levels of problem-solving ability will experience similar outcomes (in terms of shifts in marital satisfaction) as couples more skilled at problem solving (Bradbury and Karney 2004).

The same holds for conflict. The absence of conflict does not automatically result in positive feelings of warmth or more affection, nor does the presence of conflict early in marriage spell doom for couples. Researchers suggest that negative interactions did not significantly affect the first year of marriage because of the *honeymoon effect*, the tendency of newlyweds to overlook problems (Huston, McHale, and Crouter 1986; see also Chapter 7). Failure to fulfill a partner's expectations about marital roles, such as intimacy and trust, predicted marital dissatisfaction (Kelley and Burgoon 1991).

David Olson and John DeFrain (1994) asserted that we could predict an engaged couple's eventual marital satisfaction based on their current relationship. The factors they find significant in reviewing the research literature include the ability to do the following:

- Communicate well with each other
- Resolve conflicts in a constructive way
- Develop realistic expectations about marriage

- Like each other as people
- Agree on religious and ethical issues
- Balance individual and couple leisure activities with each other

In addition, how each person's parents related to each other and to their daughter or son is an important predictor. It is in our families of orientation that we learn our earliest (and sometimes most powerful) lessons about intimacy and relationships (Larsen and Olson 1989).

Engagement, Cohabitation, and Weddings

The first stage of the **family life cycle** may begin with engagement or cohabitation followed by a wedding, the ceremony that represents the beginning of a marriage.

Engagement

Engagement is the culmination of the premarital dating process. Today, in contrast to the past, engagement has more significance as a ritual than as a binding commitment to be married. Engagement is losing even its ritualistic meaning, however, as more couples start out in the less formal patterns of "getting together" or living together. These couples are less likely to become formally engaged. Instead, they announce that they "plan to get married." Because it lacks the formality of engagement, "planning to get married" is also less socially binding.

Engagements typically average between 12 and 16 months (Carmody 1992). They perform several functions:

- Engagement signifies a commitment to marriage and helps define the goal of the relationship as marriage.

- Engagement prepares couples for marriage by requiring them to think about the realities of everyday married life: money, friendships, religion, in-laws, and so forth. They are expected to begin making serious plans about how they will live together as a married couple.

- Engagement is the beginning of kinship. The future marriage partner begins to be treated as a

Weddings carry multiple meanings, both about the individuals marrying and the nature of their commitment.

© Bill Aron/PhotoEdit

member of the family. He or she begins to become integrated into the family system.

- Engagement allows the prospective partners to strengthen themselves as a couple. The engaged pair begin to experience themselves as a social unit. They leave the youth or singles culture and prepare for the world of the married, a remarkably different world.

Men and women may need to deal with a number of social and psychological issues during engagement, including the following (Wright 1990):

- *Anxiety.* A general uneasiness that comes to the surface when you decide to marry.

- *Maturation and dependency needs.* Questions about whether you are mature enough to marry and to be interdependent.

- *Losses.* Regret over what you give up by marrying, such as the freedom to date and responsibility for only yourself.

- *Partner choice.* Worry about whether you're marrying the right person.

- *Gender-role conflict.* Disagreement over appropriate male and female roles.

- *Idealization and disillusionment.* The tendency to believe that your partner is "perfect" and to become disenchanted when she or he is discovered to be "merely" human.

- *Marital expectations.* Beliefs that the marriage will be blissful and conflict free and that your partner will be entirely understanding of your needs.

- *Self-knowledge.* An understanding of yourself, including your weaknesses as well as your strengths.

Cohabitation

The rise of cohabitation has led to a new chapter in the story of contemporary families (Glick 1989; Surra 1991). As shown in the last chapter, for some people cohabitation is an alternative way of *entering marriage.* More than half of first unions result from cohabitation (Seltzer 2000; London 1991). For still others, cohabitation is an alternative *to marrying.*

Although cohabiting couples may be living together before marriage, their relationship is not legally recognized until the wedding, nor is the relationship afforded the same social legitimacy. For example, most relatives do not consider cohabitants as kin. As discussed in Chapter 8, there is evidence that marriages that follow cohabitation have a higher divorce rate than do marriages that begin without cohabitation (DeMaris and Rao 1992; Hall and Zhao 1995). Cohabitation does, however, perform some of the same functions as engagement, such as preparing the couple for some realities of marriage and helping them think of themselves as a couple.

Weddings

Weddings are ancient rituals that symbolize a couple's commitment to each other. The word *wedding* is derived from the Anglo-Saxon *wedd*, meaning "pledge." It included a pledge to the bride's father to pay him in money, cattle, or horses for his daughter (Ackerman 1994; Chesser 1980). When the father received his pledge, he "gave the bride away." The exchanging of rings dates back to ancient Egypt and symbolizes trust, unity, and timelessness because a

ring has no beginning and no end. It is a powerful symbol. To return a ring or take it off in anger is a symbolic act. Not wearing a wedding ring may be a symbolic statement about a marriage. Another custom, carrying the bride over the threshold, was practiced in ancient Greece and Rome. It was a symbolic abduction growing out of the belief that a daughter would not willingly leave her father's house. The eating of cake is similarly ancient, representing the offerings made to household gods; the cake made the union sacred (Coulanges 1960). The African tradition of jumping the broomstick, carried to America by enslaved tribespeople, has been incorporated by many contemporary African Americans into their wedding ceremonies (Cole 1993).

The honeymoon tradition can be traced to a pagan custom for ensuring fertility: Each night after the marriage ceremony, until the moon completed a full cycle, the couple drank mead, honey wine. The honeymoon was literally a time of intoxication for the newly married man and woman. Flower girls originated in the Middle Ages; they carried wheat to symbolize fertility. Throughout the world, gifts are exchanged, special clothing is worn, and symbolically important objects are used or displayed in weddings (Werner et al. 1992).

Wedding ceremonies, celebrations, and rituals such as those described are rites of passage encompassing rites of separation (for example, the giving away of the bride), aggregation, and transition. It is especially noteworthy as a rite of transition, wherein it marks the passage from single to married status. The wedding may also reflect the degree to which both the bride and groom's "social circles" are part of the transition into marriage. As such, weddings vary. As Matthijs Kalmijn (2004, 583) describes, they range from highly public to highly private:

> At one extreme is the lavish public wedding ceremony of a member of the royal family; at the other extreme is the Las Vegas wedding in a quarter of an hour at a wedding chapel without a best man or bridesmaids, without announcements or invitations, and without the parents' consent. The former . . . is extremely social and public, the latter . . . is socially isolated and almost private.

Kalmijn further elaborates, noting that in celebrating their marriage with a wedding ceremony and party, the "bride and groom show their friends and relatives the kind of spouse they have chosen, and they show others that they have chosen to go through life as a married couple (584)."

Marriage is a major commitment, and entering marriage may provoke considerable anxiety and uncertainty. Is this person right for me? Do I really want to get and be married? What is married life going to be like? Will I be a good wife or husband? These are examples of the kinds of anxieties brides and grooms might feel as they approach marriage. Kalmijn suggests that "by creating an audience that is witness to their decision, the couple may increase the commitment they have toward each other and to their new role. By increasing commitment, the couple also reduces the uncertainty they may feel about their marriage" (584).

Andrew Cherlin suggests that where weddings had historically been celebrations of a kinship alliance between two kin groups and later a reflection of parental "approval and support" for their child's marriage, today's weddings are more a symbolic demonstration of "the partners' personal achievements and a stage in their self-development" (2004, 856). A wedding is, in part, a statement, as is the buying of a house. It says, "look at what I have achieved. Look at who I have become." Seen this way, we can understand why, despite the economic obstacles they face, low-income couples can honestly contend that a major barrier preventing them from marrying is insufficient money to have a "real wedding" (that is, a church wedding and reception party). "Going down to the courthouse" is not a real or sufficient wedding (Smock 2004). A big wedding means a couple "has achieved enough financial security to do more than live from paycheck to paycheck" (Cherlin 2004, 857). Both "the brides and grooms of middle America" and low-income, unmarried parents alike desire "big weddings," even if the nature of "big" varies between the two (Edin, Kefalas, and Reed 2004). This is all part of the deinstitutionalization of marriage raised earlier. Marriage and the wedding that signifies its beginning has become more of a symbol of individual achievement and development. If it is no longer the foundation of adult life, it still serves as a capstone (Cherlin 2004).

To other analysts, weddings are seen as mostly "occasions of consumption and celebrations of romance" (Cherlin 2004, 857). Indeed, weddings of today are big business. Not all couples, however, have formal church weddings. Civil weddings now account for almost one-third of all marriage ceremonies (Ravo 1991). Because of the expense, some couples opt for civil ceremonies, which sometimes cost no more than $30, in addition to the marriage license.

Whether a first, second, or subsequent marriage, a wedding symbolizes a profound life transition. Most significantly, the partners take on marital roles. For young men and women entering marriage for the first time, marriage signifies a major step into adulthood. Some apprehension felt by those planning to marry may be related to their taking on these important new roles and responsibilities. Many will have a child in the first year of marriage. Therefore, the wedding must be considered a major rite of passage. When they leave the wedding scene, the couple leave behind singlehood. Transformed, they are now responsible to each other as fully as they are to themselves and more than they are to their parents.

However, if we focus too much on the ceremonial aspect of marriage, we overlook two important points. First, marrying is a process that begins well before and continues after the couple exchanges their vows. Second, the legal or ceremonial aspect of marrying may not be the most profound part of the transition.

The Stations of Marriage

Past analyses of both divorce and remarriage have used the concept of the stations of marriage to represent the dynamic and the multidimensional nature of transitions out of and back into marriage (Bohannan 1970; Goetting 1982). Yet these analyses work equally well to depict the multidimensional, complex process of marrying. (See Chapter 14 for further discussion of Bohannan's stations of divorce.) A decade later, Ann Goetting applied this same framework, with the same "six stations," to depict the complexities of remarriage.

Both Bohannan and Goetting stressed that that marital transitions are thick with complexity. Applying their notions of "stations," we can say that marrying consists of the following:

- *Emotional marriage.* The experiences associated with falling in love and the intensification of an emotional connection between two people. In the love-based marriages forming our society, as people fall in love they may contemplate an eventual marriage.

- *Psychic marriage.* The change in identity from an autonomous individual to a partner in a couple. As this occurs, we may encounter shifts in priorities, sense of self, perceptions of social reality, and expectations for the future (Berger and Kellner 1970).

- *Community marriage.* The changes in social relationships and social network that accompany the shift in priorities and identity described earlier. It is a two-way process of redefining and being redefined by others. Friends may perceive themselves as no longer able to make the same claims or hold the same expectations about a formerly single or unattached friend. People may begin to refer to each partner only as a couple. In other words, *Matt and Jen* replaces *Matt* or *Jen*. As relationships become even more serious, the couple will be introduced to each other's family and may also find a partner being incorporated into their own family events. This certainly occurs as couples become engaged and proceed toward marriage. Once married, new spouses are unquestionably looked on differently *because they are married.* They may even find their single friends becoming less interesting to or interested in them.

- *Legal marriage.* The legal relationship that—as we have seen—provides a couple with a host of rights and responsibilities. Clearly, legal marriage also restricts the individual's right to marry again without first ending the current marriage. However, aside from these and restrictions on whom we may marry (which, granted, are not insignificant matters), there are few legal interventions into marriage as long as both parties remain content with their marriage. We may not notice changes in our daily relationship exclusively caused by this dimension of marriage.

- *Economic marriage.* The variety of economic changes that people experience when they marry. If both are employed, they now have more financial resources that need to be managed and allocated in ways that differ from their single days. Whether the decision they face is which overdue bill to pay or whether to buy a Lexus or an SUV, they will have to change the way they previously made economic decisions and decide as part of a couple. Typically, there are stylistic differences in spending or money management that require some compromise.

- *Coparental marriage.* The changes induced in marriage relationships by the arrival (birth, remarriage, or adoption) of children. Important in both Bohannan's and Goetting's analyses, coparental marriage is not part of becoming married per se. With regard to divorce, the coparental station includes attending to such issues as daily care and custody, financial support, and visitation. In the coparental remarriage, the primary issue is to establish stepparenting roles and relationships (see Chapter 15). As far as a "station of marriage," we might say that if one party has any children, both partners will need to establish routines and share responsibilities. If childless at marriage, the coparental station would refer to those issues that change married relationships once children arrive (see Chapter 11).

Although neither Bohannan nor Goetting described a seventh station, we might include a *domestic marriage*—all of the negotiating, dividing, managing, and performing of daily household chores. Couples must establish a working division of household labor. Even if they have cohabited before marriage, there is no guarantee that their "cohabiting division of labor" will be sustained in marriage.

By conceptualizing becoming married in these terms, we can state the following important points. We may indeed feel and function as married before being legally married. That in no way guarantees success in marriage, because the research on cohabitants who marry is fairly pessimistic. But it does mean that when people think about the process of marrying, if they think in terms of before versus after wedding (essentially the legal station), the transition may seem less sweeping than it is.

Becoming married transforms lives in all of the ways depicted here. However, because you will likely encounter at least the emotional, psychic, and community (or some of it) stations of marriage by the time you enter legal marriage, you have an opportunity to begin to remake your life for marriage without yet being married. Bear in mind, too, that couples may experience these stations in different sequences. Cohabitants may experience all of these stations of marriage before legally marrying. Marriages entered into because of pregnancy or as escape from a single lifestyle will encounter these dimensions in a different order than those who marry out of first dating and falling in love. What's useful, however, about the concept of stations is how it helps us appreciate how broadly and deeply marriage changes two people.

Early Marriage

Ted Huston and Heidi Melz (2004) contend that early in marriage, newly married couples are affectionate, very much in love, and relatively free of excessive conflict, a state that might be called "blissful harmony."

Within a year, this affectionate climate "melts" into a more genial partnership. As they point out, "One year into marriage, the average spouse says, 'I love you,' hugs and kisses their partner, makes their partner laugh, and has sexual intercourse about half as often as when they were newly wed" (951). Even though conflict is not necessarily more frequent or intense, when it occurs it is less likely to be embedded in the highly affectionate climate of new marriage. Thus, it may feel worse.

Huston and Melz also found that couples establish a "distinctive emotional climate" from the outset that does not change over the initial 2 years of marriage; they are either happy or unhappy. Thus, it is not the case that unhappy couples begin on a blissful happy note and see things fail; instead, "most unhappy yet stable marriages fall short of the romantic ideal" from the beginning. All couples, even happy ones, have their ups and downs. Happy couples, however, typically contain two people who are both warm and even tempered (952).

Establishing Marital Roles

The expectations that two people have about their own and their spouse's marital roles are based on gender roles and their own experience. There are four traditional assumptions about husband or wife responsibilities: (1) the husband is the head of the household, (2) the husband is responsible for supporting the family, (3) the wife is responsible for domestic work, and (4) the wife is responsible for childrearing. More than mere expectations, these assumptions reflect traditional legal marriage (Weitzman 1981).

The traditional assumptions about marital responsibilities do not necessarily reflect marital reality, however. For example, the husband traditionally may be regarded as head of the family, but power tends to be more shared, although perhaps not equally. In dual-earner families, both men and women contribute to the financial support of the family. Although responsibility for domestic work still tends to reside largely with women, men are gradually increasing their involvement in household labor, especially childcare. The mother is generally still responsible for childrearing, but fathers are participating more.

Marital Tasks

Newly married couples need to begin a number of marital tasks to build and strengthen their marriages. The failure to complete these tasks successfully may contribute to what researchers identify as the **duration-of-marriage effect**—the accumulation over time of various factors such as unresolved conflicts, poor communication, grievances, role overload, heavy work schedules, and childrearing responsibilities that might cause marital disenchantment (see the "Issues & Insights" box in this section that examines marital satisfaction). These tasks are primarily adjustment tasks and include the following:

- *Establishing marital and family roles.* Discuss marital-role expectations for self and partner; make appropriate adjustments to fit each other's needs and the needs of the marriage; discuss childbearing issues; and negotiate parental roles and responsibilities.

- *Providing emotional support for the partner.* Learn how to give and receive love and affection, support the other emotionally, and fulfill personal identity as both an individual and a partner.

- *Adjusting personal habits.* Adjust to each other's personal ways by enjoying, accepting, tolerating, or changing personal habits, tastes, and preferences, such as differing sleep patterns, levels of personal and household cleanliness, musical tastes, and spending habits.

- *Negotiating gender roles.* Adjust gender roles and tasks to reflect individual personalities, skills, needs, interests, and equity.

- *Making sexual adjustments.* Learn how to physically show affection and love, discover mutual pleasures and satisfactions, negotiate timing and activities, and decide on the use of birth control.

- *Establishing family and employment priorities.* Balance employment and family goals; recognize the importance of unpaid household labor as work; negotiate childcare responsibilities; decide on whose employment, if either, receives priority; and divide household responsibilities equitably.

- *Developing communication skills.* Share intimate feelings and ideas with each other; learn how to talk to each other about difficulties; share moments of joy and pain; establish communication rules; and learn how to negotiate differences to enhance the marriage.

- *Managing budgetary and financial matters.* Establish a mutually agreed-upon budget; make short-term and long-term financial goals, such as saving for vacations or home purchase; and establish rules for resolving money conflicts.

- *Establishing kin relationships.* Participate in extended family and manage boundaries between family of marriage and family of orientation.

- *Participating in the larger community.* Make friends, nurture friendships, meet neighbors, and become involved in community, school, church, or political activities.

As you can see, a newly married couple must undertake numerous tasks as their marriage takes form. Marriages take different shapes according to how different tasks are shared, divided, or resolved. It is no wonder that many newlyweds find marriage harder than they expected. But if the tasks are undertaken in a spirit of love and cooperation, they offer the potential for marital growth, richness, and connection (Whitbourne and Ebmeyer 1990). If the tasks are avoided or undertaken in a selfish or rigid manner, however, the result may be conflict and marital dissatisfaction.

Identity Bargaining

People carry around idealized pictures of marriage long before they meet their marriage partners. They have to adjust these preconceptions to the reality of the partner's personality and the circumstances of the marriage. The interactional process of role adjustment is called **identity bargaining** (Blumstein 1975). The process is critical to marriage. A study of African American and Caucasian newlyweds, for example, found that marital interactions that affirmed a person's identity predicted marital well-being (Oggins, Veroff, and Leber 1993). Mirra Komarovsky (1987) points out that a spouse has a "vital stake" in getting his or her partner to fulfill certain obligations: "Hardly any aspect of marriage is exempt from mutual instruction and pressures to change."

Identity bargaining is a three-step process. First, a person has to identify with the role he or she is performing. A man must feel that he is a husband, and a woman must feel that she is a wife. The wedding ceremony acts as a catalyst for role change from the single state to the married state.

Second, a person must be treated by the other as if he or she fulfills the role. The husband must treat his wife as a wife; the wife must treat her husband as a husband. The problem is that partners rarely agree on what constitutes the roles of husband and wife. This is especially true now as the traditional content of marital roles is changing.

Third, the two people must negotiate changes in each other's roles. A woman may have learned that she is supposed to defer to her husband, but if he makes an unfair demand, how can she do this? A man may believe that his wife is supposed to be receptive to him whenever he wishes to make love, but if she is not, how should he interpret her sexual needs? A woman may not like housework (who does?), but she may be expected to do it as part of her marital role. Does she then do all the housework, or does she ask her husband to share responsibility with her? A man believes he is supposed to be strong, but sometimes he feels weak. Does he reveal this to his wife?

Eventually, these adjustments must be made. At first, however, there may be confusion; both partners may feel inadequate because they are not fulfilling their role expectations. Although some may fear losing their identity in the give and take of identity bargaining, the opposite may be true: a sense of identity may grow in the process of establishing a relationship. In the process of forming a relationship, we discover ourselves. An intimate relationship requires us to define who we are.

Establishing Boundaries

When people marry, many still have strong ties to their parents. Until the wedding, their family of orientation has greater claim to their loyalties than their spouse-to-be. After marriage, the couple must negotiate a different relationship with their parents, siblings, and in-laws. Loyalties shift from their families of orientation to their newly formed family. The families of orientation must accept and support these breaks. Indeed, opening themselves to outsiders who have become in-laws places no small stress on families (Carter and McGoldrick 1989). However, many so-called in-law problems may actually be problems between the couple. It's easier to complain about a mother-in-law, for example, than it is to deal with troubling issues in your own relationship (Silverstein 1992).

The new family must establish its own boundaries. The couple should decide how much interaction with their families of orientation is desirable and how much influence these families may have. The addition of extended family can bring into contact people who are very different from one another in culture, life experiences, and values. There are often important ties to the parents that may prevent new families from achieving their needed independence. First is the tie of habit. Parents are used to being superordinate; children are

Issues and Insights Examining Marital Satisfaction

Because marriage and the family have moved to the center of people's lives as a source of personal satisfaction, we generally evaluate them according to how well they fulfill emotional needs (although such fulfillment is not the only measurement of satisfaction). Marital satisfaction influences not only how we feel about our marriages and our partners but also how we feel about ourselves. If we have a good marriage, we tend to feel happy and fulfilled (Glenn 1991).

Considering the various elements that make up or affect a marriage—from identity bargaining to economic status—it should not be surprising that marital satisfaction ebbs and flows. Studies consistently indicate that marital satisfaction changes over the family life cycle, following a U-shape or curvilinear curve (Finkel and Hansen 1992; Glenn 1991; Suitor 1991; but see Vaillant and Vaillant 1993). Satisfaction is highest during the initial stages and then begins to decline, but it rises again in the later years.

Decline in Marital Satisfaction

Why does marital satisfaction tend to decline soon after marriage?

Researchers have suggested two explanations the presence of children, and the effects of time on marital satisfaction.

Children and Marital Satisfaction

Traditionally, researchers have attributed decline in marital satisfaction to the arrival of the first child: Children take from time a couple spends together, are a source of stress, and cost money. When children begin leaving home, marital satisfaction begins to rise again.

This seems paradoxical since for many people, children are among the things they value most in their marriages. First, attributing the decline to children creates a single-cause fallacy—that is, it attributes a complex phenomenon to one factor when there are probably multiple causes. Second, the arrival of children at the same time that marital satisfaction declines may be coincidental, not causal. Other undetected factors may be at work.

Although many societal factors make childrearing a difficult and sometimes painful experience for some families, it is also important to note that children create parental roles and the family in its most traditional sense. For some, the marital relationship may be less than fulfilling with children present, but many couples may make a trade-off for fulfillment in their parental roles. In times of marital crisis, parental roles may be

the glue that holds the relationship together.

The Duration of Marriage Effect and Marital Satisfaction

More recently, researchers have looked for other factors that might explain decline in marital satisfaction. The most persuasive alternative is the duration-of-marriage effect.

The duration-of-marriage effect is most notable during the first stage of marriage rather than during the transition to parenthood that follows (White and Booth 1985). This early decline may reflect the replacement of unrealistic expectations with more realistic ones.

Social and Psychological Factors in Marital Satisfaction

Social factors such as income level are a significant factor. Lower income creates financial distress. If a is deeply in debt, how to allocate resources—for rent, repairing the car, or paying dental bills—becomes critical, sometimes involving conflict-filled decisions.

Psychological factors also affect marital satisfaction (London, Wakefield, and Lewak 1990). Although it was once believed that marital satisfaction depended on a partner fulfilling complementary needs and qualities (an introvert marrying an extrovert, for example), research has failed to substantiate this assertion. Instead,

used to being subordinate. The tie between mothers and daughters is especially strong; daughters often experience greater difficulty separating themselves from their mothers than do sons. The adult child may feel conflicting loyalties toward parents and spouse (Cohler and Geyer 1982). Much conflict occurs when a spouse feels that an in-law is exerting too much influence on a partner (for example, a mother-in-law insisting that her son visit every Sunday and the son accepting

despite the protests of his wife). If conflict occurs, husbands and wives often must put the needs of their spouses ahead of those of their parents.

Also, newly married couples often have little money or credit, and ask parents to loan money, cosign loans, or obtain credit. But financial dependence keeps the new family tied to the family of orientation. The parents may try to exert undue influence on their children because their money is being spent.

marital success seems to depend on partners being similar in their psychological makeup and personalities. Outgoing people are happier with outgoing partners; tidy people like tidy mates. Furthermore, a high self-concept (how a person perceives himself or herself), as well as how the spouse perceives the person, contributes to marital satisfaction. Finally, similarity in perception, such as "seeing" events, relationships, and values through the same lenses, may be critical in marital satisfaction (Deal, Wampler, and Halverson 1992).

Attitudes toward gender and marital roles may affect marital satisfaction. One study found that the discrepancy between how you expect your partner to behave and his or her actual behavior could predict marital satisfaction. Discrepancies in expectations were particularly significant in terms of intimacy, equality, trust, and dominance. Interestingly, discrepancies were more important in predicting dissatisfaction than was the fulfillment of expectations (Kelley and Burgoon 1991). This finding is not entirely surprising. We seem to take for granted that our partner will fulfill our expectations, so it may be an unpleasant surprise to discover that our spouse is not interested in (or lacks the ability for) intimacy or that he or she is untrustworthy.

Expressiveness seems to be an important quality in marital satisfaction (L. King 1993). Wives whose husbands discussed their relationships tended to be more satisfied with their marriages than other wives (Acitelli 1992).

Even though much of the literature points to declines in marital satisfaction over time, we must remember that not all marriages suffer a significant decline. Even when there is a decline in marriage satisfaction, that may be offset by other satisfactions, such as pleasure in parental roles or a sense of security.

It is important to understand that marital satisfaction fluctuates over time, battered by stress, enlarged by love. The couple continuously maneuvers through myriad tasks, roles, and activities—from sweeping floors to kissing each other—to give their marriages form. Children, who bring us both delight and frustration, constrain our lives as couples but challenge us as mothers and fathers and enrich our lives as a family. Trials and triumphs, laughter and tears punctuate the daily life of marriage. If we are committed to each other and to our marriage, work together in a spirit of flexibility and cooperation, find time to be alone together, and communicate with each other, we lay the groundwork for a rich and meaningful marriage.

The arrival and presence of children profoundly affect marital relationships

A review of research on in-laws found that in-law relationships generally had little emotional intensity (Goetting 1989). The relationship between married women and their mothers-in-law and mothers seems to change with the birth of a first child (Fischer 1983). Mother–daughter relationships seem to improve as the mother shifts some of her maternal role to the grandchild. In-laws give minimal direct support. Bonding between in-laws tends to be between women, and

if there is a divorce, divorced women are more likely than their ex-husbands to maintain supportive ties with former in-laws (Serovich, Price, and Chapman 1991).

The critical task is to form a family that is interdependent rather than independent or dependent. It is a delicate balancing act as parents and their adult children begin to make adjustments to the new marriage. We need to maintain bonds with our families of

CATHY *Cathy Guisewite*

Panel 1: YOU'RE MOVING HOME WITH YOUR MOTHER?? ARE YOU CRAZY, ALEX???

Panel 2: DO YOU WANT YOUR MOTHER DECIDING WHAT YOU EAT FOR DINNER?? DO YOU WANT HER PICKING OUT YOUR CLOTHES??

Panel 3: DO YOU WANT SOMEONE HOVERING OVER YOUR EVERY MOVE AND TREATING YOU LIKE A BABY??!! / WELL, SURE!

Panel 4: MEN ARE TIED TO THE APRON STRINGS. WOMEN ARE STRANGLED BY THEM.

■ *Cathy*

orientation and to participate in the extended family network, but we cannot let those bonds turn into chains.

Social Context and Social Stress

Even with all the attention paid to the dynamics of spousal relationships, marital success may rest largely on things that happen outside of and around the married couple (Bradbury and Karney 2004). Marriages are affected by the wider context in which we live, including "the situations, incidents, and chronic and acute circumstances that spouses and couples encounter," as well as the developmental transitions they undertake (Bradbury and Karney 2004). Changes in employment, the transition to parenthood, health concerns, friends, finances, in-laws, and work experiences, can all affect the quality of marriage relationships. As Thomas Bradbury and Benjamin Karney (2004, 872) express, "Theoretically identical marriages are unlikely to achieve identical outcomes if they are forced to contend with rather different circumstances."

Similarly, they contend that marriages that are "rather different" in their internal dynamics may reach similar outcomes in quality depending on whether the wider context is especially healthy or especially "toxic" (Bradbury and Karney 2004). From their research on married couples, Bradbury and Karney offer the following points to consider:

■ Marital quality was lower among couples experiencing higher average levels of stress.

■ Marital quality dropped more quickly among couples reporting high levels of chronic stress.

■ During times of elevated stress, more relationship problems were perceived and partner's negative behaviors were more often viewed as selfish, intentional, and blameworthy.

Incorporating research findings from other studies, they also offer the following especially supportive evidence of the importance of social context on marital interaction and quality:

■ Observational research found that because of greater job stress, blue-collar husbands were more likely than white-collar husbands to respond with negative affect to negative affect from their wives in problem-solving discussions.

■ Among married male air-traffic controllers, on high stress days in which they received support from their wives they expressed less anger and more emotional withdrawal.

■ Among a sample of more than 200 African American couples, those living in more distressed neighborhoods (as measured by a composite that included such things as income and the proportion of the neighborhood on public assistance, living in poverty, unemployed, and in single parent households) experienced less warmth and more overt hostility.

Cumulatively, findings such as these remind us that improving the quality of marriage may require us to attend to and "fix" contextual circumstances, even if it means "bypassing couples and lobbying for change in environments and conditions that impinge on marriages and families" (Bradbury and Karney 2004, 876).

Marital Commitments

How often do we hear the statement, "marriage is a (lifelong) commitment"? What does that mean? Is it something internal to an individual, a reflection of attitudes, values, and beliefs, or is it something external, the outcome of constraints that keep us within a relationship? Just what does the commitment to marriage entail?

Trying to sort out the meaning and experience of **marital commitment,** Michael Johnson identifies three major types of commitment, each of which operates within marriage:

- *Personal commitment.* In essence, this is "the extent to which one wants to stay in a relationship" (Johnson, Caughlin, and Huston 1999, 161). It is affected by how strongly we are attracted to a spouse, how attractive our relationship is, and how central the relationship is to our concept of self.

- *Moral commitment.* This is the feeling of being "morally obligated" to stay in a relationship, resulting from our sense of personal obligation ("I promised to stay forever and I will"), the values we have about the lifelong nature of marriage (a "relationship-type obligation"), and a desire to maintain consistency in how we act in important life matters ("I am not a quitter, I have never been a quitter, I won't quit now").

- *Structural commitment.* This is feeling constrained from leaving a relationship, even in the absence of a strong sense of personal or moral obligation. It consists of the awareness and assessment we make of alternatives, our sense of the reactions of others and the pressures they may put on us, the difficulty we perceive in ending and exiting from a relationship, and the feeling that we have made "irretrievable investments" into a relationship and leaving the relationship would mean we had wasted our time and lost opportunities all for nothing.

Personal commitment is more a product of love, satisfaction with the relationship, and the existence of a strong couple identity. Moral commitment is the product of our attitudes about divorce, our sense of a personal "contract" with our spouse, and the desire

AP Images/The Index-Journal, Shavonne Potts

■ *Marriage relationships continue to face new challenges and circumstances as couples age.*

for personal consistency. Finally, structural commitment a product of attractive alternatives, social pressures, fear of termination procedures, and the feeling of sacrifices we have made and can't recover. Johnson and colleagues contend that in our efforts to understand why marriages do or don't last, we tend to look mostly at personal commitment. We need to move beyond that narrower focus and look at how all three types are experienced and how each influences the outcome and experience of marriage (Johnson, Caughlin, and Huston 1999).

Marital Impact of Children

Typically, husbands and wives both work until their first child is born; about half of all working women leave the workplace for at least a short period to attend to childrearing responsibilities after the birth of the first child. The husband continues his job or career. Although the first child makes the husband a father, fatherhood generally does not visibly alter his relationship with his work. For example, it may redefine his motivation for work and the responsibility he feels to provide. Thus, even if he appears to continue at work relatively unaffected, he may be experiencing important changes.

The woman's life, however, changes more dramatically and visibly with motherhood. If she continues her outside employment, she is usually responsible for arranging childcare and juggling her employment responsibilities when her children are sick, and if her story is like that of most employed mothers, she continues to have primary responsibility for the household and children. If she withdraws from the workplace, her contacts during most of the day are with her children and possibly other mothers. This relative isolation requires her to make a considerable psychological adaptation in her transition to motherhood, leading in some cases to unhappiness or depression.

Typical struggles in families with young children concern childcare responsibilities and parental roles. The woman's partner may not understand her frustration or unhappiness because he sees her fulfilling her roles as wife and mother. She herself may not fully understand the reasons for her feelings. The partners may increasingly grow apart during this period. During the day they move in different worlds, the world of the workplace and the world of the home; during the night they cannot relate easily because they do not understand each other's experiences. Research suggests that men are often overwhelmed by the emotional intensity of this and other types of conflict (Gottman 1994). With all that accompanies the transition to parenthood (see the next two chapters), it is unsurprising that more frequent conflict and tension ensue and that couples often change the ways in which they handle or resolve conflict (Crohan 1996).

For adoptive families, the transition to parenthood may differ somewhat from that of biological families (Levy-Shiff, Goldschmidt, and Har-Even 1991). Adoptive parents report more positive expectations about having a child, as well as more positive experiences in their transition to parenthood. This may be explained partly by adoptive parents' ability to fulfill parental roles that they vigorously sought. For them, parenting is a more conscious decision than for many biological parents; for biological parents, a pregnancy sometimes just "happens." For adoptive parents to become parents, considerable effort and expense must be undertaken; they are less likely to question their decision to become parents.

Individual Changes

Around the time people are in their 30s, the marital situation changes substantially. If there are children, they have probably started school and the mother begins to have more freedom from childrearing responsibilities. She evaluates her past and decides on her future. Most women who left jobs to rear children return to the workplace well before their children reach adolescence. By working, women generally increase their marital power.

Husbands in this period may find that their jobs have already peaked; they can no longer look forward to promotions. They may feel stalled and become depressed as they look into the future, which they see as nothing more than the past repeated for 30 more years. However, their families may provide emotional satisfaction and fulfillment as a counterbalance to workplace disappointments.

Middle-Aged Marriages

Middle-aged marriages, in which couples are in their 40s and 50s, are typically families with adolescents and/or young adults leaving home (stages 6 and 7 of Erikson's life cycle). Some parents may continue to raise young children; others, especially if one partner is considerably younger than the other, may choose to start a new family.

Matter of Fact

Family values, such as support, communication, and respect, along with marital satisfaction, face their greatest challenge in families with adolescent children (Larson and Richards 1994).

Families with Young Children

Increasing dramatically since 1970 are the women over 35 who have chosen to postpone childbearing until they are emotionally or financially ready. In 2000, more than 546,000 babies were born to women over 35 and

Marital Satisfaction

An important question in studying marital satisfaction is how to measure it (Fincham and Bradbury 1987). One measure widely used is Graham Spanier's Dyadic Adjustment Scale, which we have included a sample of here. This scale is an example of the type of questionnaire scholars use as they examine marital adjustment. What are the advantages of a questionnaire such as this? What are the disadvantages?

Answer the questions that follow and then ask yourself if you think they can measure marital satisfaction. (*Hint:* You must first define what marital satisfaction is.) If you are involved in a relationship or marriage, you and your partner might be interested in answering the questions separately and comparing your answers. Do you have similar perceptions of your relationship? At the end of this course, answer the questions again without referring to your first set of answers. Then compare your responses. What do you infer from this comparison?

The Marital Satisfaction Survey

	Always agree	Almost always agree	Occasionally agree	Frequently disagree	Almost always disagree	Always disagree
1. Handling family finances	5	4	3	2	1	0
2. Matters of recreation	5	4	3	2	1	0
3. Religious matters	5	4	3	2	1	0
4. Demonstrations of affection	5	4	3	2	1	0
5. Friends	5	4	3	2	1	0
6. Sex relations	5	4	3	2	1	0

© Multi-Health Systems

94,000 to women over 40 (U.S. Census Bureau 2002, Table 68). Although there have always been older women having children, in the past these mothers were having their last child, not their first. Because most of these women have a higher education, job status, and income, they also experience a lower divorce rate, are more stable, and are often more attentive to their young.

Families with Adolescents

Adolescents require considerable family reorganization on the part of parents: They stay up late, play loud music, infringe on their parents' privacy, and leave a trail of empty pizza cartons, popcorn, dirty socks, and Big Gulp cups in their wake. As Betty Carter and Monica McGoldrick (1989) point out:

Families with adolescents must establish qualitatively different boundaries than families with younger children. . . . Parents can no longer maintain complete authority. Adolescents can and do open the family to a whole array of new values as they bring friends and new ideas into the family arena. Families that become derailed at this stage are frequently stuck at an earlier view of their children. They may try to control every aspect of their lives at a time when, developmentally, this is impossible to do successfully. Either the adolescent withdraws from the appropriate involvements for this developmental stage or the parents become increasingly frustrated with what they perceive as their own impotence.

Although the majority of teenagers do not cause "storm and stress" (Larson and Ham 1993), increased family conflict may occur as adolescents begin to assert their autonomy and independence. Conflicts over tidiness, study habits, communication, and lack of responsibility may emerge. Adolescents want rights and privileges but have difficulty accepting responsibility. Conflicts are often contained, however, if both parents and adolescents tacitly agree to avoid "flammable" topics, such as how the teenager spends time or money. Such tactics may be useful in maintaining family peace, but in the extreme they can backfire by decreasing family closeness and intimacy. Despite the growing pains accompanying adolescence, parental bonds generally remain strong (Gecas and Seff 1991).

Families as Launching Centers

Some couples may be happy or even grateful to see their children leave home, some experience difficulties with this exodus, and some continue to accommodate their adult children under the parental roof.

The Empty Nest

As children are "launched" from the family (or "ejected," as some parents wryly put it), the parental role becomes increasingly less important in daily life. The period following the child's exit is commonly known as the **empty nest.** Most parents make the transition reasonably well (Anderson 1988). Marital satisfaction generally begins to rise for the first time since the first stage of marriage (Glenn 1991). For some parents, however, the empty nest is seen as the end of the family. Children have been the focal point of much family happiness and pain, and now they are gone.

Traditionally, it has been asserted that the departure of the last child from home leads to an "empty nest syndrome" among women, characterized by depression and identity crisis. However, there is little evidence that the syndrome is widespread. Rather, it is a myth that reinforces the traditional view that women's primary identity is found in motherhood. Once deprived of their all-encompassing identity as mothers, the myth goes, women lose all sense of purpose. (In reality, mothers may be more likely to complain when faced with adult children who have not left home.)

The couple must now re-create their family minus their children. Their parental roles become less important and less stressful on a day-to-day basis (Anderson 1988). The husband and wife must rediscover themselves as man and woman. Some couples may divorce at this point if the children were the only reason the pair remained together. The outcome is more positive when parents have other more meaningful roles, such as school, work, or other activities, to turn to (Lamanna and Riedmann 1997).

The Not-So-Empty Nest: Adult Children at Home

Just how empty homes are after children reach age 18 is open to question. Census data revealed that in 2000 56% of 18- to 24-year-old males and 43% of 18- to 24-year-old females were living with one or both parents (Fields and Casper 2001). Some are not moving out before their mid-20s, and many are doing an extra rotation through their family home after a temporary or lengthy absence. This later group is sometimes referred to as the **boomerang generation.**

In a 1995 survey of first-time college freshmen, 19% said wanting to leave home was an important reason to go to school. A larger share (25%) were living at home while they attended school, according to University of California at Los Angeles' Annual American Freshman Study.

Hispanics are more likely than other young adults to take a traditional route of staying home until they marry. Blacks are less likely than whites or Hispanics to leave home before marriage. Although family income may influence nest leaving, ethnic or racial tradition seems to be more important in determining whether young adults will leave home (American Demographics 1996). Most, however, move from home when they marry.

Researchers note that there are important financial and emotional reasons for this trend (Mancini and Blieszner 1991). High unemployment, expensive housing, and poor wages are factors causing adult children to return home. High divorce rates, as well as personal problems, push adult children back to the parental home for social support and childcare, as well as cooking and laundry services.

> ### Reflections
>
> **Recall your family** of orientation when you were an adolescent. How did you and your parents deal with establishing new family boundaries and with issues of autonomy and independence? What was the process of "launching" like? Has it been completed? If you continue to live at home, what difficulties has it caused you and your parents?

Young adults at home are such a common phenomenon that one of the leading family life cycle scholars suggests an additional family stage: *adult children at home* (Aldous 1990). This new stage generally is not one that parents have anticipated. Almost half reported serious conflict with their children. For parents, the most frequently mentioned problems were the hours of their children's coming and going and their failure to share in cleaning and maintaining the house. Most wanted their children to be "up, gone, and on their own."

Reevaluation

Middle-aged people find that they must reevaluate relations with their children, who have become independent adults, and must incorporate new family members as in-laws. Some must also begin considering how to assist their own parents, who are becoming more dependent as they age.

Couples in middle age tend to reexamine their aims and goals (Steinberg and Silverberg 1987). On the average, husbands and wives have 13 more years of marriage without children than they used to, and during this time their partnership may become more harmonious or more strained. The man may decide to stay at home or not work as hard as before. The woman may commit herself more fully to her job or career, or she may remain at home, enjoying her new child-free leisure. Because the woman has probably returned to the workplace, wages and salary earned during this period may represent the highest amount the couple will earn.

Matter of Fact

Average life expectancy is 74.4 years for men and 79.8 years for women. By the time individuals reach 65, their life expectancy rises to 81.4 years for men and 84.4 years for women. If they reach 75, they can anticipate living a decade (men) or dozen (women) years more (National Center for Health Statistics, http://www.cdc.gov/nchs/data/hus/tables/2003/03hus027.pdf).

As people enter their 50s, they probably have advanced as far as they will ever advance in their work. They have accepted their own limits, but they also have an increased sense of their own mortality. They not only feel their bodies aging but also begin to see people their own age dying. Some continue to live as if they were ageless—exercising, working hard, and keep-

ing up or even increasing the pace of their activities. Others become more reflective, retreating from the world. Some may turn outward, renewing their contacts with friends, relatives, and especially their children and grandchildren.

Later-Life Marriages

Later-life marriages represent the last two stages (stages 7 and 8) of the family life cycle. In families with children, a later-life marriage is one in which the children have been launched and the partners are middle age or older. Later-life families tend to be significantly more satisfied than families at earlier stages in the family life cycle (Mathis and Tanner 1991). Compared with middle-aged couples, older couples showed less potential for conflict and greater potential for engaging in pleasurable activities together and separately, such as dancing, travel, or reading (Levenson, Carstensen, and Gottman 1993). Research in the 1990s showed that older people without children experienced about the same level of psychological well-being, instrumental support, and care as those who have children (Allen, Bleiszner, and Roberto 2000).

During this period, the three most important factors affecting middle-aged and older couples are health, retirement, and widowhood (Brubaker 1991). In addition, these women and men must often assume roles as caretakers of their own aging parents or adjust to adult children who have returned home. Later-middle-aged men and women tend to enjoy good health, are firmly established in their work, and have their highest discretionary spending power because their children are gone (Voydanoff 1987). As they age, however, they tend to cut back on their work commitments for both personal and health reasons.

As they enter old age, men and women are better off, on the average, than young Americans (Peterson 1991). Beliefs that the elderly are neglected and isolated tend to reflect myth more than reality (Woodward 1988). Over half of all people age 65 and older live in either the same house or in the same neighborhood as one of their adult children (Troll 1994). In addition, a national study of people over 65 found that 41% of those with children see or talk with them daily, 21% do so twice a week, and 20% do so weekly. Over half have children within 30 minutes' driving time (U.S. Census Bureau 1988).

The Intermittent Extended Family: Sharing and Caring

Although many later-life families contract in size as children are launched, pushed, or cajoled out of the nest, other families may expand as they come to the assistance of family members in need. Families are most likely to become an intermittent extended family during their later-life stage (Beck and Beck 1989). An **intermittent extended family** takes in other relatives during a time of need. Such a family "shares and cares" when younger or older relatives are in need or crisis: It helps daughters who are single mothers; a sick parent, aunt, or uncle; or an unemployed cousin. When the crisis passes, the dependent adult leaves, and the family resumes its usual structure.

The incidence of intermittent extended families tends to be linked to ethnicity. Using national population studies, researchers estimate that the families of almost two-thirds of African American women and one-third of Caucasian women were extended for at least some part of the time during their middle age (Beck and Beck 1989; Minkler and Roe 1993). Latina women are more likely than non-Latina women to form extended households (Tienda and Angel 1982). Asian American families are also more likely to live at some time in extended families. There are two reasons for the prevalence of extended families among certain ethnic groups. First, extended families are by cultural tradition more significant to African Americans, Latinos, and Asian Americans than to Caucasians. Second, ethnic families are more likely to be economically disadvantaged. They share households and pool resources as a practical way to overcome short-term difficulties. In addition, there is a higher rate of single parenthood among African Americans, which makes mothers and their children economically vulnerable. These women often turn to their families of orientation for emotional and economic support until they are able to get on their own feet.

The Sandwich Generation

A relatively new phenomena, now referred to as the **sandwich generation,** are those middle-aged (or older) individuals who are sandwiched between the simultaneous responsibilities of caring for both their dependent children and their aging parents. Given the number of baby boomers now in their middle years, coupled with the increased longevity among their par-

ents, we can anticipate that this type of dual care will become increasingly common. As many as 20% to 30% of workers over age 30 may find themselves involved in caregiving to their parents, and this percentage is expected to grow (Field and Minkler 1993). Daughters outnumber sons as caretakers by more than three to one (Allen, Blieszner, and Roberto 2000; Cox 1993), although among Asian Americans, the eldest son may be expected to be responsible for his elders (Kamo and Zhou 1994). When sons are caretakers, whether in families with only sons or with sons and daughters, it is often daughters-in-law or grandaughters who actually provide the care (Allen, Blieszner, and Roberto 2000).

As people live longer, their disabilities, dependency, and the number of their long-term chronic illnesses increases. Complicating this is the shrinking number of young workers, facilities, and resources to care for the old and frail. All of this puts additional pressure on families to provide support for their elders. Care traditionally handled by health-care professionals—injections, monitoring of medications, bathing, and physical therapy—is now often in the hands of family members.

The trend today, whenever possible, is for the dependent aged to be cared for in the home (Freedman 1993). Placing added demands on family members' time, energy, and emotional commitment often results in exhaustion, anger, and in some cases, violence. Most people, however, are amazingly adept at meeting the needs of both their parents and their children. It is going to be an increasing challenge for society to acknowledge this phenomenon and provide services and support to both the elderly and those who care for them.

Retirement

Retirement, like other life changes, has the potential for both satisfactions and problems. In a time of relative prosperity for the elderly, retirement is an event to which older couples generally look forward. One key to marital satisfaction in these later years is continued good health (Brubaker 1991).

Widowhood

Marriages are finite; they do not last forever. Eventually, every marriage is broken by divorce or death. Despite high divorce rates, most marriages end with death, not divorce. "Till death do us part" is a fact for most married people.

In 2000, 66.5% of those between ages 65 and 74 were married. Among those 75 years old and older, however, only 46% were married; 46% were widowed. Because women live about 7 years longer on average than men, most widowed people are women. Women over 65 years of age outnumber men by a ratio of roughly 1.5 to 1. By age 85, this ratio has increased to approximately 4 women to every 1 man (Carr 2004).

These demographic facts of life expectancy yield many more widows than widowers, thus creating for men "many more opportunities to date and remarry should they choose to" (Carr 2004, 1,052). Indeed, greater proportions of older men than older women are married. Among women from 65 to 74 years old, 56% were married, but only 31% over age 75 had a spouse. In contrast, among men 65 to 74 years old, 79.6% lived with their wives; among those over 75, 69.3% lived with a spouse (U.S. Census Bureau 2001, Table 51). Three out of four wives will become widows.

Widowhood is often associated with a significant decline in income, plunging the grieving spouse into financial crisis and hardship in the year or so following death. This is especially true for poorer families (Smith and Zick 1986). Feelings of well-being among both elderly men and elderly women are related to their financial situations. If the surviving spouse is financially secure, she or he does not have the added distress of a dramatic loss of income or wealth.

Recovering from the loss of a spouse is often difficult and prolonged. A woman may experience considerable disorientation and confusion from the loss of her role as a wife and companion. Having spent much of her life as part of a couple—having mutual friends, common interests, and shared goals—a widow suddenly finds herself alone. Whatever the nature of her marriage, she experiences grief, anger, distress, and loneliness. Physical health appears to be tied closely to the emotional stress of widowhood. Widowed men and women experience more health problems over the 14 months following their spouses' deaths than do those with spouses. Over time, however, widows appear to regain much of their physical and emotional health (Brubaker 1991).

One common response of widowed women and men is to glorify or "sanctify" their marriages and their deceased spouses. This is especially true shortly after a spouse's death. Oftentimes, the "newly bereaved" retrospectively construct and offer "unrealistically positive portrayals" of their marriages (Carr 2004). One

The loss of one's spouse confronts women and men with a variety of deep and painful losses. Although both women and men lose their chief source of emotional support, women typically have wider and deeper friendship networks to turn to for support.

way in which women and men differ in their reactions is that women who had close marriages may feel less open to seeking and forming a new relationship with another man, retaining the feeling that they are "still married" to their late husbands. Men who were in close marriages may be especially motivated to establish another marriage. Having experienced and grown dependent on the emotional support and intimacy of their marriages, they may have few other alternative sources for support to whom they can turn. Thus, they have greater incentive to form a new emotionally supportive marriage or partner relationship (Carr 2004).

Eventually widowed women and men must in some way adjust to the loss. Some remarry; 2% of older widows and 20% of older widowers remarry. Each year, 3 of every 1,000 widows and 17 of every 1,000 widowers marry (Carr 2004). Many others adjust by learning to enjoy their new freedom. Others believe that they are too old to date or remarry; still others cannot imagine living with someone other than their former husband. (Those who had good marriages think of remarrying more often than those who had poor marriages.) A large number of elderly men and women live together without remarrying. For many widows, widowhood lasts the rest of their lives.

For both widowed women and men, remarriage and repartnering may be desired—but for different reasons. Given the multiple benefits men gain from marriage and the supportive presence of a wife, men may desire remarriage more, especially those men who were socially and emotionally dependent on their wives. The emptiness left by their wife's death and absence may be too great to bear. In addition to the loss of their confidant and chief source of emotional support, they may have limited experience managing households, cooking, and cleaning and as a result suffer from poor nutrition and distress over the conditions in which they live (Carr 2004). Widows certainly suffer, too, although they may be beneficiaries of more practical help from their children and draw on emotional support from a wider and deeper network of friends.

Enduring Marriages

Examining marriages across the family life cycle is an important way of exploring the different tasks we must undertake at different times in our relationships. A number of those who have studied long-term marriages lasting 50 years or more have discovered several common patterns. Two researchers (Rowe and Lasswell, cited in Sweeney 1982) have divided relationships into three categories: (1) couples who are happily in love, (2) unhappy couples who continue marriage out of habit and fear, and (3) couples who are neither happy nor unhappy and accept the situation. Lasswell and Rowe found that approximately 20% of long-term marriages were very happy and 20% were very unhappy.

Another way to look at marriage is according to stability rather than satisfaction. In other words, which marriages last? What researchers find is what many of us already know: little correlation exists between happy marriages and stable ones. Many unhappily married couples stay together, and some happily married couples undergo a crisis and breakup. In general, however, the quality of the marital relationship appears to show continuity over the years. Much of the discrepancy between happiness and stability results because happiness or satisfaction is an evaluative judgment of a marriage relative to what we expected from marriage and what better alternatives are available. Stability results more from assessments of the costs and rewards of staying in or leaving a marriage. Unhappy marriages may be enduring ones because there are no better alternatives, because the costs of leaving exceed the costs of staying married, or both.

Long-term marriages are not immune to conflict. As Figure 9.2 illustrates, as many as one-fourth of middle-aged couples, and between 12% and 20% of older couples, acknowledge engaging in conflict over such issues as children, money, communication, recreation, sex, and in-laws. Surviving together does not require couples to eliminate or avoid conflict.

A study by Robert and Jeanette Lauer used a more modest definition of *long term* to look at marriages that last. Their study of 351 couples married at least 15 years (most were married a good deal longer) found the following to be the "most important ingredients" identified by men and women to explain their marital success: "my spouse is my best friend," "I like my spouse as a person," "marriage is a long-term commitment," "marriage is sacred," "we agree on aims and goals," and "my spouse has grown more interesting" (Lauer and Lauer 1986). The correlation between husbands' and wives' lists was over 0.90, a remarkable consensus across gender lines. Summing up their results, the Lauers specify four keys to long-term satisfying marriages:

1. Having a spouse who is a best friend and whom you like as a person

2. Believing in marriage as a long-term commitment and sacred institution

3. Consensus on such fundamentals as aims and goals and philosophy of life

4. Shared humor

When assessing marriages, keep in mind that there is considerable diversity in married life. Thus, attempts have been made to document some types of marriages that couples construct (Cuber and Harroff 1965; Wallerstein and Blakeslee 1995; Schwartz 1994). One popular typology details five types of marriage, each of which could either last "till death do us part" or end in divorce. Thus, these are not degrees of marital success but rather different kinds of marriage relationships (Cuber and Harroff 1965).

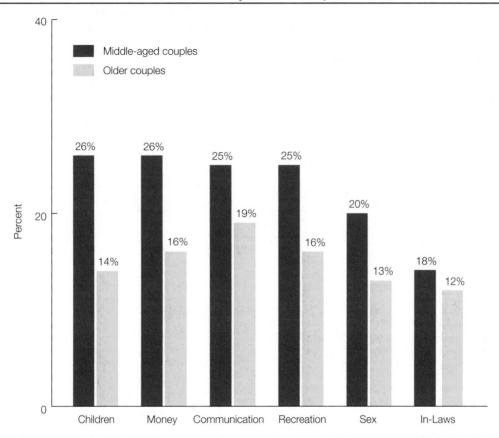

SOURCE: Levenson, Carstensen, and Gottman 1993, 307.

■ **Conflict-habituated marriages** are relationships in which tension, arguing, and conflict "permeate the relationship" (Cuber and Harroff 1965). One informant characterized his conflict-habituated marriage as a "long-running guerilla war" yet acknowledged that neither he nor his wife had ever thought of ending the marriage. It may well be that conflict is what holds these couples together. It is at least understood to be a basic characteristic of this type of marriage.

■ **Passive-congenial marriages** are relationships that begin without the emotional "spark" or intensity contained in our romantic idealizations of marriage. They may be marriages of convenience that satisfy practical needs in both spouses' lives. Couples in which both spouses have strong career commitments and value independence may construct a passive-congenial marriage to enjoy the

benefits of married life and especially parenthood. In some ways, these marriages are, and have been since their beginning, "emotional voids" (Cuber and Harroff 1965).

■ **Devitalized marriages** begin with high levels of emotional intensity that over time has dwindled. From the outside looking in, they may closely resemble passive-congenial relationships. What sets them apart is that they have a history of having been in a more intimate, sexually gratifying, emotional relationship that has become an emotional void. Obligation and resignation may hold such couples together, along with the lifestyle they have built and the history they have shared.

■ **Vital marriages** appeal more to our romantic notions of marriage because they begin and continue with high levels of emotional intensity. Such couples spend much of their time together and are "intensely bound together in important life matters" (Cuber and Harroff 1965). The relationship is the most valued aspect of their lives, and they allocate their time and attention based on such a priority. Conflict is not absent, but it is managed in such a manner as to make quick resolution likely.

- **Total marriages** are relationships in which characteristics of vital relationships are present and multiplied. In some ways they may be seen as multifaceted vital relationships where the "points of vital meshing" extended across more aspects of daily coupled life.

Differentiating between these five types, John Cuber and Peggy Harroff noted that the first three types were more common than the last two. As many as 80% of the relationships among their sample were of one of the first three types. Both vital and total marriages (what they called *intrinsic marriages*) were relatively rare. Again, we must remember that the researchers were not sorting relationships into "successful" versus "unsuccessful" or "good" versus "bad." Marriages of all five types were enduring marriages, and any of the five types could end in divorce, although the reasons for divorce would differ.

More recently, a seven-type typology was constructed by Yoav Lavee and David Olson (1993) from an analysis of the marriages of more than 8,000 couples voluntarily in marriage enrichment programs or marital therapy. Although such a sample may be more difficult to generalize, Lavee and Olson suggested that we could differentiate couples based on their satisfaction or dissatisfaction with nine areas of married life: personality issues, conflict resolution, communication, sexual intimacy, religious beliefs, financial management, leisure, parenting, and relationships with friends and family. Of their types, *vitalized couples* (9% of sample) reported themselves satisfied with all nine areas. At the opposite end, *devitalized couples* reported problems in all nine areas. Keeping in mind that the sample was drawn from either clinical or enrichment intervention, the devitalized were by far the most common type, representing 40% of their sample.

The remainder of the sample was relatively evenly divided across the other types: *balanced, harmonious, traditional, conflicted,* and *financially focused.* All types except the vitalized reported problems, although the areas and extent of problems differed across these types.

The financially focused (11%) had problems in all areas but financial matters. Traditional couples (10%) reported problems in their handling of conflict, communication, sexual intimacy, and parenting. The conflicted (14%) reported themselves generally satisfied with only their parenting, leisure activities, and religious beliefs. Even those couples designated as harmonious (8%) tended to have difficulties in areas such as religious beliefs, parenting, and relations with family and friends. Balanced couples (8%) were generally satisfied with all areas except financial matters.

For different reasons, we need to be cautious about generalizing too far from either Cuber and Harroff or Lavee and Olson. Nonetheless, in both typologies, 75% or more of the sample couples were in marriages that many would define as unattractive, seeming to be held together by something other than a deep emotional connection. In addition, both typologies should keep us from assuming that marriage has to be free of conflict to last. Most obviously, both typologies illustrate that marriages are not all alike. This is a simple and obvious but important point.

Throughout marriage, from the earliest most hopeful and optimistic beginning till death or divorce do us part, we are presented with opportunities for growth and change as we enter our roles as husbands or wives, become parents or stepparents, and still later become grandparents. Throughout all of these stages, marriage requires a deep commitment. As David and Vera Mace (1979) observe:

> Until two people, who are married, look into each other's eyes and make a solemn commitment to each other—that they will stop at nothing, that they will face any cost, any pain, any struggle, go out of their way so that they may learn and seek so that they may make their marriage a continuously growing experience—until two people have done that they are not in my judgment married.

As we have seen, marriages and families never remain the same. They change as we change; as we learn to give and take; as children enter and exit our lives; as we create new goals and visions for ourselves and our relationships. In our intimate relationships, we are offered the opportunity to discover ourselves.

As marriage continues to undergo changes, we are left to wonder about what the future holds. There is enough reason to believe, even in the face of the striking and sometimes troubling trends, that we have addressed that marriage will survive. We will not likely see a return to traditional marriages any more than we should expect a disappearance of marriage. If anything, we agree with Paul Amato's (2004b, 102) assessment that the future likely contains more of the same:

Alternatives to marriage will be accepted and widespread. People will continue to have sex prior to marriage, live together without being married, have children outside of marriage, avoid marriage altogether, and divorce if their marriages are flawed.

At the same time, most people will continue to view marriage as the "gold standard" for relationships . . . [and] to view marriage as the best context for bearing and rearing children. Helping more people to achieve healthy and stable marriages will require the efforts of marriage educators, counselors, therapists, and policy makers.

Summary

- Marriage is the foundation and centerpiece of the American family system.

- There is an ongoing *marriage debate* over the status and future of marriage. The two extreme positions in this debate are the *marital decline* and *marital resilience positions.*

- Behavioral indicators of a *retreat from marriage* include increasing percentages adults remaining unmarried, living together, having children outside of marriage, and divorcing. However, approximately 90% of Americans are expected to someday marry.

- The retreat from marriage varies considerably by race and economic status. African Americans are much less likely to marry, to stay married, and to have their children inside of a marriage than are other racial and ethnic groups. Socioeconomic factors are also important, as indicated by lower marriage rates among those with less education.

- Even those most likely to retreat from marrying continue to articulate support for and a desire to marry. Barriers to marriage for low-income, unmarried parents include financial concerns, concerns about relationship quality, and fear of divorce.

- There are religious differences in the importance of marriage and the push toward early marriage. Conservative Protestants and Latter-day Saints are most likely to marry young.

- The *deinstitutionalization of marriage* refers to weakening of the social norms that define people's behavior in a social institution such as marriage. In the move from companionate to *individualized marriage,* new emphases on personal self-fulfillment and freedom of choice become more important than marital commitment and obligation.

- Legal limits imposed on choice of marriage partner include gender, age, family relationship, and number of spouses.

- Marriage confers both rights and responsibilities onto married couples. Most states do not explicitly state what legal responsibilities are expected of married people. Benefits include tax benefits, government benefits, employment benefits, medical benefits, and housing and consumer benefits.

- Reasons to marry include both attractions of marriage and rejection of singlehood. Marital intimacy is the biggest attraction of marriage.

- Marriage provides various benefits to married people, including economic benefits, health benefits, and psychological benefits. Research supports both a selection effect (healthier and better adjusted people are more likely to marry) and a protection effect (marriage provides a range of protective resources enabling people to prosper).

- The eight developmental stages of the human life cycle described by Erik Erikson are (1) infancy: trust versus mistrust; (2) toddler: autonomy versus shame and doubt; (3) early childhood: initiative versus guilt; (4) school age: industry versus inferiority; (5) adolescence: identity versus role confusion; (6) young adulthood: intimacy versus isolation; (7) adulthood: generativity versus self-absorption; and (8) maturity: integrity versus despair. Each stage is intimately interconnected with family.

- The relationships that precede marriage often predict marital success because marital patterns emerge during these times. Premarital factors correlated with marital success include (1) background factors (age at marriage, length of courtship, level of education, and childhood environment), (2) personality factors, and (3) relationship factors (communication, self-disclosure, and interdependence).

- Engagement is the culmination of the formal dating pattern. It prepares the couple for marriage by involving them in discussions about the realities of everyday life, it involves family members with the couple, and it strengthens the couple as a social unit. Individuals must deal with key psychological issues, such as anxiety, maturation and dependency needs, losses, partner choice, gender-role conflict, idealization and disillusionment, marital expectations, and self-knowledge. Cohabitation serves many of the same functions as engagement.

- A wedding is an ancient ritual that symbolizes a couple's commitment to each other. About two-thirds are formal church weddings. The wedding marks a major transition in life as the man and woman take on marital roles. Marriage involves many powerful traditional role expectations, including assumptions that the husband is head of the household and is expected to support the family and that the wife is responsible for housework and childrearing.

- The process of marrying and becoming spouses consists of six dimensions of experience that can be classified as the stations of marriage: emotional, psychic, community, economic, legal, and parental. We should also recognize the domestic responsibilities that marriage introduces as another part of becoming married.

- Gender-role attitudes and behaviors contribute to marital roles. Women are more egalitarian than men in marital-role expectations, but both genders expect men to earn more money. Marital tasks include establishing marital and family roles, providing emotional support for the partner, adjusting personal habits, negotiating gender roles, making sexual adjustments, establishing family and employment priorities, developing communication skills, managing budgetary and financial matters, establishing kin relationships, and participating in the larger community.

- Couples undergo *identity bargaining* in adjusting to marital roles. This is a three-step process: (1) the person must identify with the role, (2) the person must be treated by the other as if he or she fulfills that role, and (3) both people must negotiate changes in each other's roles.

- Marital success is affected by the wider social context and the extent and kind of social stresses couples face.

- Marital commitments consist of personal commitments, moral commitments, and structural commitments. Personal commitment is a product of love, satisfaction with the relationship, and the existence of a strong couple identity; moral commitment is the product of our attitudes about divorce, our sense of a personal "contract" with our spouse, and the desire for personal consistency; and structural commitment a product of attractive alternatives, social pressures, fear of termination procedures, and the feeling of sacrifices we have made and can't recover.

- A critical task in early marriage is to establish boundaries separating the newly formed family from the couple's families of orientation. Ties to the families of orientation may include habits of subordination and economic dependency. In-law relationships tend to have little emotional intensity.

- In youthful marriages, about half of all working women leave the workforce to attend to childrearing responsibilities. Motherhood more radically alters a woman's life than fatherhood changes a man's life. Parental roles and childcare responsibilities need to be worked out.

- Middle-aged families must deal with issues of independence in regard to their adolescent children. Most women do not suffer from the *empty nest* syndrome. For many families, there is no empty nest because of the increasing presence of adult children in the home. As children leave home, parents reevaluate their relationship with each other and their life goals.

- In later-life marriages, usually no children are present. Marital satisfaction tends to be highest during this time. The most important factors affecting this life cycle stage are health, retirement, and widowhood. As a group, the aged have regular contact with their children, the lowest poverty level of any group, and good health through the early years of old age. Many families, especially among African Americans, Latinos, and Asian Americans, become *intermittent extended families* in which aging parents, adult children, or other relatives periodically live with them during times of need. This differs from the *sandwich generation,* which finds itself caring for children and aging parents at the same time.

- Long-term marriages may be divided into three categories: (1) couples who are happily in love, (2) unhappy couples who stay together out of habit or fear, and (3) couples who are neither happy nor unhappy. The percentage of couples who are happily in love is approximately 20%, the same percentage found for those who are unhappy.

- Some factors associated with long-term marriages are liking your spouse as a person, thinking of your spouse as your best friend, believing in marriage as a commitment, spousal agreement on life's goals, and a sense of humor.

- Marriages differ from one another. One popular typology contrasts five types of marriage: *conflict-habituated, devitalized, passive-congenial, vital,* and *total*. These reflect different conceptualizations and experiences of marriage, not different degrees of marital success.

Key Terms

boomerang generation 350

close relationship model of legal marriage 332

conflict-habituated marriages 355

conjugal model of legal marriage 332

deinstitutionalization of marriage 328

devitalized marriages 356

duration-of-marriage effect 342

empty nest 350

family life cycle 338

identity bargaining 343

individualized marriage 328

intermittent extended family 352

marital commitment 347

marital decline perspective 329

marital resilience perspective 329

marriage debate 323

passive-congenial marriages 356

retreat from marriage 324

sandwich generation 352

total marriages 356

vital marriages 356

Resources on the Internet
Companion Website for This Book

http://www.thomsonedu.com/sociology/strong

Gain an even better understanding of this chapter by going to the companion website for additional study resources. Take advantage of the Pre- and Post-Test quizzing tool, which is designed to help you grasp difficult concepts by referring you back to review specific pages in the chapter for questions you answer incorrectly. Use the flash cards to master key terms and check out the many other study aids you'll find there. Visit the Marriage and Family Resource Center on the site. You'll also find special features such as access to Info-Trac© College Edition (a database that allows you access to more than 18 million full-length articles from 5,000 periodicals and journals), as well as GSS Data and Census information to help you with your research projects and papers.

CHAPTER 10

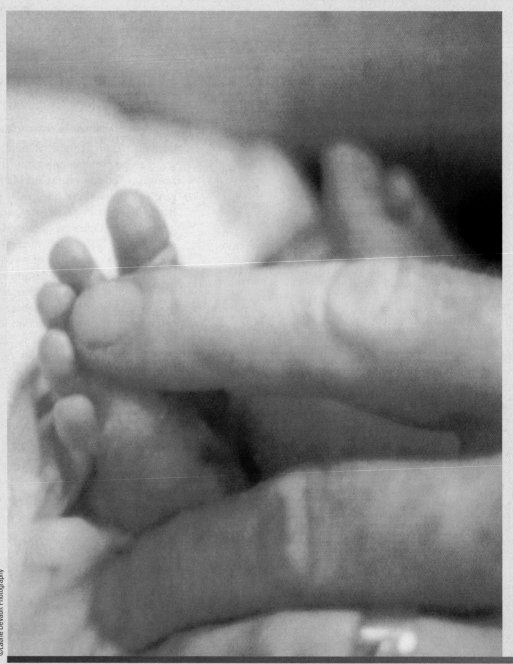

©Laurie DeVault Photography

Should We or Shouldn't We: Choosing Whether and How to Have Children

Outline

What Do YOU Think?

Are the following statements **TRUE** or **FALSE**?
You may be surprised by the answers (see answer key on the following page).

T F **1** The birthrate in the United States has risen steadily since 1990.

T F **2** It is estimated that a third of women who marry will forego having children.

T F **3** Abortions have declined over the past decade.

T F **4** It is usually unsafe for a woman to have sexual intercourse during the last two months of pregnancy.

T F **5** Miscarriage and stillbirth are major life events for parents.

T F **6** The rate of infant mortality in the United States is about what it is throughout the industrialized world.

T F **7** Adopted children tend to be poorer than children who live with their biological parents.

T F **8** Men and women both can suffer from "postpartum blues."

T F **9** There is often a decline in marital happiness following the transition to new parenthood.

T F **10** Stress is common among both biological and adoptive new parents.

I t's unbelievable. . . . There's really no way a non-parent can think like a parent. It's really knocked me for a loop. And in my wildest dreams, I never thought of it. . . . Something just creeps into your life and all of a sudden it dominates your life. It changes your relationship with everybody and everything, you question every value and every belief you ever had. And you say to yourself, "this is a miracle." It's like you take your life, open up a drawer, put it all in a drawer, and close the drawer.

These comments show a 33-year-old man's thoughtful reactions to becoming a first-time father. As he reflects on it, becoming a parent is life defining and life altering. He is not alone. Having and raising children introduce profound changes and impose labor-intensive responsibilities. As we examine over the next two chapters, parenthood changes how we see ourselves, how we live, what we think about, and how we feel. Simultaneously, parents experience changes in their social relationships and how they are viewed by others. These changes are neither minor nor temporary. Becoming a parent is as profound a life change as any other we make.

Not everyone decides to become a parent. With widespread availability of effective contraception and access to legal abortion, women and men can decide whether and when to have children. The bulk of this chapter focuses on the choices people make whether or not to have children and the range of factors that figure in to the decision-making process. We examine the characteristics of those who decide to or are forced to forego parenthood. But those who embark on parenthood face other choices. *How* should they become parents? For some, bearing a child is difficult or impossible, leading them to attempt to adopt or take advantage of the ever-expanding options presented by advances in reproductive technology. And *when* should they become parents? Is there an optimal time or age for entering motherhood or fatherhood? Throughout this chapter, we explore these choices.

Answer Key for What Do You Think

1 False, see p. 262; 2 False, see p. 363; 3 True, see p. 367; 4 False, see p. 369; 5 True, see p. 372; 6 False, see p. 373; 7 False, see p. 377; 8 True, see p. 380; 9 True, see p. 383; 10 True, see p. 380.

Our focus then shifts to how women and their partners experience pregnancy, the transition to parenthood, and the changes parenthood introduces into our lives. Chapter 11 then explores the meaning and special challenges confronting mothers and fathers in the United States today.

Fertility Patterns in the United States

There were more than 4.1 million births in the United States in 2004, up only 1% from 2003 (Hamilton et al. 2004). The **crude birthrate,** a statistic reflecting the number of births per every 1,000 people in the population, was 14.0 in 2004. This has varied little in recent years: it is down 1% from 2003, but the rate has hovered around the same rate for the last few years. In a somewhat longer view, the crude birthrate has declined some 17% since 1990.

In recent decades, the United States has also experienced a decrease in the **fertility rate,** the number of births annually per 1,000 women 15 to 44 years old, from 118 in 1960 to 66.3 in 2004 (National Center for Health Statistics 2003; Hamilton et al. 2004). This represents a more recent decline of 9% since 1990, but a slight increase from the last couple of years (64.8 in 2002, 66.1 in 2003). Finally, the **total fertility rate,** a more complicated statistic that estimates the number of births a hypothetical group of 1,000 women would have if they experience across their childbearing years the age-specific rates for a given year, indicates that there would be 2,048.5 births per 1,000 women, or 2 children per woman. This, too, reflects a decline (3%) since 1990.

Fertility and birthrates vary considerably according to social and demographic characteristics such as race, ethnicity, education, income, and marital status. Figure 10.1 shows variation by ethnicity. Within the Latino population, rates vary from a high among Mexican Americans to a low among Cuban Americans. Cultural, social, and economic factors play a significant part in influencing the number of children a family has. Because of a combination of higher fertility rates and continuing immigration patterns, Hispanics have become our nation's largest minority group, thereby surpassing African Americans.

Fertility rates also vary by education and income. Women with a high school education had the highest

Figure 10.1 ■ Fertility Rates Ethnicity, 2002

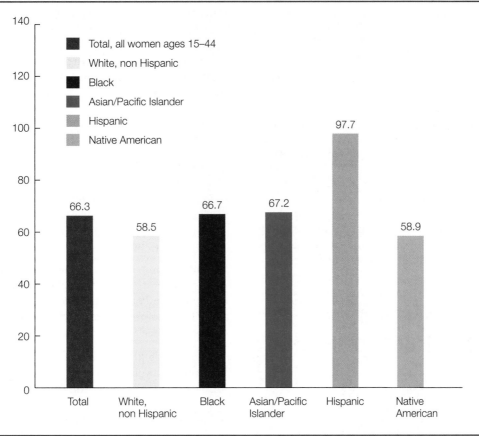

SOURCE: Hamilton, Martin, and Sutton 2003.

rate (67.4 per 1,000). College-educated women had the next highest rate (65.3), followed by women with graduate degrees (59.2) and women with less than high school (57.3). Income effects on childbearing are a little more complex. For females between the ages of 15 and 29, birthrates are highest among women with the lowest incomes (<$10,000) and lowest among women with high incomes (>$75,000). Among women 30 to 44, we see almost the opposite pattern. The birthrate is highest among women with high incomes ($50,000–$74,999) followed closely by women with family incomes more than $75,000; the lowest birthrates are found among women with family incomes between $10,000 and $29,999 (http://www.commissions.leg .state.mn.us/lcesw/, 1999).

Approximately 18% of American women between the ages of 40 and 44 have not had children. Among women of that same age who had ever married, 12% were reported to be childless (U.S. Census Bureau 2002). Among all women without children, most expect to have at least one child. Although the percentage who intend to forego parenting is difficult to estimate, it is likely no more than 10%.

Unmarried Parenthood

In 2004, a record number of unmarried women gave birth. Nearly 1.5 million nonmarital births occurred, an increase of 4% from the preceding year. These births represented almost 36% of all births, with increases occurring for women of all ages and races. As reflected in the data in Table 10.1, although there are increases among all of women there are also significant variations across ethnic and racial groups in the percentage of births to unmarried mothers.

There are prominent age differences as well in unmarried childbearing. More than four of five (82.6%) births to teenagers were outside of marriage. Among women 20–24 years old, more than half of the

Table 10.1 ■	Percentage of Births to Unmarried Mothers by Ethnic Origin, 2004	
Ethnic Origin of Mother	2004	2003
All Ethnic Groups	35.7	34.6
Non-Hispanic whites	24.5	23.6
Non-Hispanic blacks	69.2	68.5
American Indian	62.3	61.3
Asian or Pacific Islander	15.5	15.0
Hispanic	46.4	45.0

childbirths in 2004 were to unmarried women. Nearly 30% of the 25- to 29-year-old women who gave birth in 2004 were unmarried at the time of delivery.

"Maybe We Shouldn't": Foregoing Parenthood

What should we call couples who don't have children? In the past, the most common way to describe them was as *childless,* meaning simply that they had no children. However, the term *childless* conveys the sense that such women or couples were "less something" that they wanted or were *supposed to have.* This description no doubt describes the experiences of those women 15 to 44 years old who have an "impaired ability" to have children or those couples who seek help for infertility. Both are *involuntarily childless.* In 2002, roughly 2 million couples, 7% of married couples in the United States, reported not using contraception for 12 months without the woman becoming pregnant. Around 1.2 million women, 2% of the 62 million American women of reproductive age, sought help for infertility. An additional 6 million women reported having at some point in their lives received infertility services (Centers for Disease Control and Prevention 2005). We say more about such efforts and interventions in a later section.

In more recent years, the term *childless* has been joined by *child free.* In the United States, we have experienced a cultural and a demographic shift toward more voluntarily childless women and **child-free marriages**—couples who *expect and intend* to remain non-parents. The term *child free* suggests that those who do not choose to have children need no longer be

seen objects of sympathy, lacking something essential for personal and relationship fulfillment. The suffix -*free* suggests liberation from the bonds of a potentially oppressive condition (Callan 1985).

Using a U.S. Census Bureau report by Amara Bachu (1999), we can examine some trends and characteristics definitive of this trend. Looking at the last two decades of the twentieth century, Bachu notes the following:

■ Among 40- to 44-year-old women (an age by which most women woulkd have had their first child), the percentage without children nearly doubled between 1980 and 1998, from 10% to 19%.

■ Among married or previously married women, the percentage without children doubled between 1980 and 1998, from 7% to 14%.

■ Among never-married women, the percentage of 40- to 44-year-olds who had never had children *declined* from 79% in 1980 to 67% in 1998. This is a reflection of the increase in births to unmarried women noted earlier in this chapter.

The preceding sketch does not single out the child free from the involuntarily childless who are physically unable to have children. Kristin Park (2002) cites a variety of estimates from other research suggesting that between 6.6% and 9.3% of women within childbearing ages do not expect to have any children. She also suggests that as many as 25% of the childless population is truly child free because they are intentionally without children.

Who are the women who remain child free? Research indicates that compared to mothers, the child free are women with the highest levels of education, those employed in high status occupations such as managerial and professional occupations, and those in families with high levels of family income from dual-earner or dual-career marriages (Park 2002; Ambry 1992). They are also less religious, firstborn or only children, and less gender traditional (Park 2002). Hispanic women are less likely than black or white women to expect a childless future (Henslin 2000; U.S. Census Bureau 1998, Table 110). Bachu (1999) offered the following observation:

Childlessness among married couples today is no longer an uncommon situation. Compared to past decades, women are marrying and having their first birth much later in life. Among women in the child-bearing years, postponement of marriage and childbearing is viewed as pathway to a good job and economic independence. The cost of raising a child and the availability and affordability of childcare have further promoted childlessness among women.

What Rosemary Gillespie feels has changed over the past quarter century is the emergence of "a more radical rejection or push away from motherhood" (2003, 133). She asserts that an increasing number of women are resisting and rejecting the cultural expectations that automatically associate women with motherhood. Instead, she suggests, "modernity has given rise to wider possibilities for women" (2003, 134).

Couples usually have *some* idea that they will or will not have children before they marry. If the intent isn't clear from the start, or if one partner's mind changes, the couple may have serious problems ahead. Many studies of child-free marriages indicate a higher degree of marital adjustment or satisfaction than is found among couples with children. Given the time and energy required by childrearing, these findings are not particularly surprising. It has also been observed that divorce is more probable in child-free marriages, perhaps because child-free couples do not stay together "for the sake of the children," as do some other unhappily married couples.

Today, although greater in number than in the past, child-free women and couples may find themselves perceived as career oriented, materialistic, individualistic, or selfish, with child-free women more negatively perceived than child-free men. Of these stereotypes, only career orientated seems to accurately apply to the child free, especially the women (Park 2002). From in-depth interviews with 24 voluntarily child-free women and men, Park identified a variety of strategies they used to reduce the stigma attached to not wanting children (Park 2002). These strategies included the following:

- *Passing.* This involves pretending to intend someday to become parents.

- *Identity substitution.* This includes feigning an involuntary childless status, as well as letting other statuses (for example, as a voluntary single or an atheist) dominate a social identity.

- *Condemning the condemners.* In keeping with other instances of reaction to being labeled deviant, this reaction consists of suggesting that some people have children for the wrong, or for selfish, reasons or that they do so without thinking fully about the responsibility.

- *Asserting their "right" to self-fulfillment.* Park contends that this is a modern type of justification.

- *Claiming a biological "deficiency."* The individual lacks the desire or lacks the nurturing "instinct."

- *Redefining the situation.* This turns potential accusations around by showing how the lifestyle allows nurturing qualities to be used in other ways or allows the individual to be productive. Some also claim that their careers just don't allow for the inclusion of children.

An inspection of the preceding strategies shows that some are more defensive than others, suggesting the acceptance of pronatalist norms. Others are more proactive, redefining childlessness as something socially valuable (Park 2002).

Waiting a While: Parenthood Deferred

Although most women still begin their families while in their 20s, we can expect that the trend toward later parenthood will continue to grow, especially in middle- and upper-income groups. A number of factors contribute to this. More career and lifestyle options are available to single women today than in the past.

■ Many couples today (especially those in middle-and upper-income brackets) defer having children until they have established their own relationships and built their careers. These parents are usually quite satisfied with their choice.

Marriage and reproduction are no longer economic or social necessities. People may take longer to search out the "right" mate (even if it takes more than one marriage to do it), and they may wait for the "right" time to have children. Increasingly effective birth control (including safe, legal abortion) has also been a significant factor in the planned deferral of parenthood.

Besides giving parents a chance to complete education, build careers, and firmly establish their own relationship, delaying parenthood can be advantageous for other reasons. As shown, raising children is expensive. Waiting, delaying parenthood until economic position is more secure, makes good sense given the economic effect of parenthood. Older parents may also be more emotionally mature and thus more capable of dealing with parenting stresses (although age isn't necessarily indicative of emotional maturity).

Cost estimates that have tried to include both college expenses and estimated wages lost project that raising a "typical" child amounts to a 22-year investment of between $761,871 (lower-third income bracket) and $2.78 million dollars (upper-third income bracket). For middle-income-bracket families, the estimate is $1.45 million (Longman 1998).

Choosing When: Is There an Ideal Age at which to Have a Child?

Although we have briefly addressed the question of delayed or deferred parenthood, we should point out that delaying parenthood "too long," like having children "too young," carries risks and brings costs. For example, research on the health effects for mothers caused by their age at first birth reveal that both "unusually young" and "unusually old" mothers face health risks. Mothers who bear their first child in their teens face nearly twice the risk of anemia as women who have their first child between ages 30 and 35 (Mirowsky 2002).

But there are significant risks for pregnancy- and labor-related distress among older first-time mothers, too. For example, pregnancy-related hypertension rates are highest for mothers under 20 and over 40 years of age. Late first births and the care associated with infants take their physical toll on women. The kind of physical energy required to care for children tends to decline with age (Mirowsky 2002).

Both in the United States and elsewhere, women have increased their age at first motherhood. In the United States, the median age at first birth in 1972 was 22 years; in 1998 the median age had increased to 24.3 years. In Italy, the Netherlands, Denmark, and Spain, the typical age at first motherhood is near or beyond 30 years (Mirowsky 2002). Zheng Wu and Lindy MacNeill (2002) report that, in Canada, too, delaying childbearing has become increasingly popular. Probable factors to explain these widespread trends are the increasing age at which women and men marry, women's increased labor force participation and educational attainment, and the increasingly effective measures of reproductive control (Wu and MacNeill 2002).

Preventing and Controlling Conception

Effective pregnancy prevention or control is critical to both deferred or delayed parenthood and "child-free" lifestyles. According to a Kaiser Family Foundation Survey on men's role in contraception and pregnancy prevention, most women and men believe men should be more active participants in choosing methods of contraception. In pregnancy prevention, 66% of men and 70% of women endorse expanded roles for men; only 8% of men and 4% of women say men should play a smaller role (Ten Kate 1998). In addition, the survey found the following:

> ### Reflections
> If you don't have any children now, do you want to have them in the future? When? How many? What factors do you need to take into consideration when contemplating a family for yourself? Does your partner (if you have one) agree with you about having children?

Being Pregnant

Women and men who become parents enter a new phase of their lives. For those who bear their own children, this phase begins with pregnancy. From the moment it is discovered, a pregnancy affects people's feelings about themselves, their relationship with their partner, and the interrelationships of other family members.

In the United States in 2000, there were more than 6.4 million pregnancies, down 6% from the 1990 peak of 6.8 million but up from 6.3 million in 1999. Between 1990 and 2000, the pregnancy rate dropped for both married and unmarried women, about 8% and 12%, respectively ("Estimated Pregnancy Rates for the United States" 2004).

Although the 1990–2000 decline in pregnancy rates occurred across the board among all women under 30 years of age, the drop was steepest among teens of all racial groups. The teen pregnancy rate dropped 27% during this period, reaching a rate of 84.5 pregnancies per 1,000 women 15–19 years old—*the lowest recorded teen pregnancy rate since 1976* (National Center for Health Statistics 2004). Pregnancy rates remain highest for women in their 20s, with 20- to 24-year-old women having the highest rates, followed by 25- to 29-year-olds (National Center for Health Statistics 2004).

Of the more than 6 million pregnancies in the United States in 2000, 63% resulted in births, 20% in abortions, and 17% in stillbirths or miscarriages ("Estimated Pregnancy Rates for the United States" 2004). Since 1990, trends in birth, abortion, and fetal loss have all declined: live births by 9%, abortions by about 25%, and fetal losses by 4%.

Both marital status and race affect pregnancy outcomes. In 1999, 75% of pregnancies among married women resulted in a live birth; 7% resulted in an abortion. Meanwhile, about 50% of pregnancies of unmarried women resulted in live births; 40% ended in an abortion ("Revised Pregnancy Rates," 2004). Black women and white women report that they want about the same number of births, but black women experience more pregnancies. Among black women there is an average of 4.6 pregnancies per woman, compared with just 2.7 for white women. Black women's pregnancies are twice as likely to end in abortions as pregnancies among white and Hispanic women (National Center for Health Statistics 2000).

Emotional and Psychosocial Changes

A woman's feelings during pregnancy will vary dramatically according to who she is, how she feels about pregnancy and motherhood, whether the pregnancy was planned, whether she has a secure home situation, and many other factors. Her feelings may be ambivalent; they will probably change over the course of the pregnancy.

Planned versus Unplanned: Was It a Choice?

It is estimated that nearly a fourth of all pregnancies carried to full term are unplanned. If we assume that a much greater percentage of pregnancies that end in abortions were also unplanned, the true percentage of

■ *Both expectant parents may feel that the fetus is already a member of the family. They begin the attachment process well before birth.*

© Scott Barrow

all unplanned pregnancies is likely much higher than the 23% cited by researchers (Bouchard 2005). Comparing planned and unplanned pregnancies reveals the following:

■ Unplanned pregnancies present greater risks for problems associated with lack of readiness or preparedness for parenting on the part of pregnant women or expectant couples.

■ Babies born from unintended pregnancies are more likely to suffer both physical and social disadvantages.

■ Mothers who give birth to "unintended" babies are more likely to report experiencing psychological problems such as postpartum depression.

A woman's first pregnancy is especially important because it has traditionally symbolized the transition to maturity. Even as social norms change and it becomes more common and acceptable for women to defer childbirth until they have established a career or to choose not to have children, the significance of first pregnancy should not be underestimated. It is a major developmental milestone in the lives of mothers—and fathers (Marsiglio 1991; Notman and Lester 1988; Snarey et al. 1987).

A couple's relationship is likely to undergo changes during pregnancy. It can be a stressful time, especially if the pregnancy was unanticipated. Communication is particularly important at this time because each partner may have preconceived ideas about what the other is feeling. Both partners may have fears about the baby's well-being, the approaching birth, their ability to parent, and the ways in which the baby will affect their relationship. All of these concerns are normal. Sharing them, perhaps in the setting of a prenatal group, can deepen and strengthen the relationship (Kitzinger 1989). If the pregnant woman's partner is not supportive or if she does not have a partner, it is important that she find other sources of support—family, friends, women's groups—and that she not be reluctant to ask for help.

The first trimester (3 months) of pregnancy may be difficult physically and emotionally for the expectant mother. She may experience nausea, fatigue, and painful swelling of the breasts. She may also fear that she will miscarry or that the child will not be normal. Her sexuality may undergo changes, resulting in unfamiliar needs (for more, less, or differently expressed sexual love), which may in turn cause anxiety. (Sexuality during pregnancy is discussed later in this chapter.)

Education about the birth process, information about her body's functioning, and support from partner, friends, relatives, and health-care professionals are the best antidotes to her fear.

As illustrated in Table 10.2, not all pregnant women receive timely prenatal care or care commencing during the first trimester.

During the second trimester, most nausea and fatigue disappear and the pregnant woman can feel the fetus move within her. Worries about miscarriage will probably begin to diminish because the riskiest part of fetal development has passed. The pregnant woman may look and feel radiantly happy.

Some women, however, may be concerned about their increasing size; they may fear that they are becoming unattractive. A partner's attention and reassurance will ease this fear.

The third trimester may be the time of the greatest difficulties in daily living. The uterus, originally about the size of the woman's fist, has now enlarged to fill the pelvic cavity and is pushing up into the abdominal cavity, exerting increasing pressure on the other internal organs. Water retention (edema) is a fairly common problem during late pregnancy; it may cause swelling in the face, hands, ankles, and feet. It can often be controlled by reducing salt and refined carbohydrates (such as bleached flour and sugar) in the diet. If dietary changes do not help this condition, however, the woman should consult her physician.

Another problem is that the woman's physical abilities are limited by her size. She may be required by her employer to stop working at some point during her pregnancy. A family dependent on her income may suffer hardship. And the woman and her partner may become increasingly concerned about the upcoming birth.

Table 10.2 ■ Percentage of Mothers Beginning Prenatal Care in First Trimester, and Percentage with Late or No Prenatal Care		
	First Trimester	**Late or No Care**
All women	83.2	3.9
White (non-Hispanic)	88.5	3.2
African American	74.3	6.7
Hispanic	74.4	6.3

SOURCE: Hamilton, Martin, and Sutton 2003.

Some women experience periods of depression in the month preceding their delivery; they may feel physically awkward and sexually unattractive. Many, however, feel an exhilarating sense of excitement and anticipation marked by energetic bursts of industriousness. They feel that the fetus is a member of the family. Both parents may begin talking to the fetus and "playing" with it by patting and rubbing the expectant mother's belly.

Sexuality during Pregnancy

It is not unusual for a woman's sexual feelings and actions to change during pregnancy, although there is great variation among women in these expressions of sexuality. Some women feel beautiful, energetic, sensual, and interested in sex; others feel awkward and decidedly "unsexy." A woman's feelings may also fluctuate during this time. Some studies indicate a lessening of women's sexual interest during pregnancy and a corresponding decline in coital frequency. A study of 219 pregnant women found that although libido, intercourse, and orgasm declined, the frequency of oral and anal sex and masturbation remained at pre-pregnancy levels (Hart et al. 1991). Generally, however, by the third trimester of pregnancy, approximately 75% of first-time mothers report loss of desire; between 83% and 100% report reduced frequency of sexual intercourse (Judicubus and McCabe 2002).

Men may feel confusion or conflicts about sexual activity during this time. They, like many women, may have been conditioned to find the pregnant body unerotic. Or they may feel deep sexual attraction to their pregnant partner, yet fear their feelings are "strange" or unusual. They may also worry about hurting their partner or the baby.

Although there are no "rules" governing sexual behavior during pregnancy, a few basic precautions should be observed:

- If the woman has had a prior miscarriage, she should check with her health practitioner before having intercourse, masturbating, or engaging in other activities that might lead to orgasm. Powerful uterine contractions could possibly induce a spontaneous abortion in some women, especially during the first trimester.

- If there is bleeding from the vagina, the woman should refrain from sexual activity and consult her physician or midwife at once.

- If the insertion of the penis into the vagina causes pain that is not easily remedied by a change of position, the couple should refrain from intercourse.

- Pressure on the woman's abdomen should be avoided, especially in the final months of pregnancy.

- During oral sex, care should be taken not to blow air into the vagina, as there is a possibility of causing an embolism (an air bubble in the bloodstream).

- Late in pregnancy, an orgasm is likely to induce uterine contractions. Generally this is not considered harmful, but the pregnant woman may want to discuss it with her practitioner. (Occasionally, labor is begun when the waters break as the result of orgasmic contractions.)

A couple, especially during their first pregnancy, may be uncertain as to how to express their sexual feelings. The following guidelines may be helpful (Strong and DeVault 1997):

- Even during a normal pregnancy, sexual intercourse may be uncomfortable. The couple may want to try positions such as side by side or rear entry to avoid pressure on the woman's abdomen and to facilitate more shallow penetration.

- Even if intercourse is not comfortable for the woman, orgasm may still be intensely pleasurable. She may wish to consider masturbation (alone or with her partner) or cunnilingus.

- Both partners should remember that there are no rules about sexuality during pregnancy. This is a time for relaxing, enjoying the woman's changing body, talking a lot, touching each other, and experimenting with new ways—both sexual and non-sexual—of expressing affection.

Men and Pregnancy

Obviously, pregnancy is something men do not experience directly. It is the woman's body that carries the fetus and undergoes profound change along the way. For men, pregnancy is only accessible vicariously. Still, how men navigate the pregnancy process has consequences for their later conceptualization of and involvement in fathering (Marsiglio 1998).

During pregnancy, men experience changes in their sexual relations with their partners, especially in the amount and nature of fantasies, and alterations in their patterns of dreams. In their sexual fantasies, they reported feeling as if they were fertilizing, nurturing, or "feeding" their fetuses or their wives, thus revealing the connection they draw between pregnancy and sexuality (Marsiglio 1998).

Early in their partner's pregnancy, men's dreams occasionally take on qualities of mystery and awe, later shifting to dreams of being neglected or rejected by their partners (Marsiglio 1998). Men's anxieties during pregnancy cover a number of areas, including the health of both fetus and partner, whether they will be a good father, how fatherhood will affect their lives, and how well they will manage their economic responsibilities, especially given new expenses and reduced spousal income. Although a man's traditional role as father centered on providing, the concern over competence as a provider is not the major source of men's pregnancy anxieties. Men whose employment is unstable or whose incomes are insufficient will experience more provider anxiety than will men who simply take for granted that they can meet their financial responsibilities (Cohen 1993).

The roles men play in supporting their partners, participating in the preparation for parenthood, and at the birth also are significant. Not all men act in similar ways. Some may be relatively detached, others fully involved, and still others practical in their participation in the pregnancy (Marsiglio 1998). The way men act during pregnancy (reading material, attending prenatal classes, involving themselves in the birth process, and so on) may affect how they later relate with their newborns. Of particular note is the experience of witnessing the birth of their children, which reportedly opens men to a depth of emotional experience often otherwise absent from conventional cultural expressions of masculinity. Men are "feminized"; they speak poignantly, occasionally poetically, about what that experience was like or meant to them (Cohen 1987; Gerson 1993).

Contested Viewpoints on Childbirth

Women and couples planning the birth of a child have decisions to make in a variety of areas—birthplace, birth attendants, medications, preparedness classes, circumcision, and breastfeeding, to name but a few. The "childbirth market" is beginning to respond to consumer concerns, so it's important for prospective parents to fully understand their options.

The Critique against Medicalization: Hospital Birth

Through the last decades of the twentieth century, there was much criticism directed at what was seen as excessive and intrusive institutionalized control of women's birth experiences. The concept of the **medicalization of childbirth** depicts women receiving impersonal, assembly line–quality care during labor and delivery and lacking much input or control over their childbirth experiences. It is illustrated in the following comment from one woman describing her initial feelings in the hospital (Leifer 1990):

> When they put that tag around my wrist and put me into that hospital gown, I felt as if I had suddenly just become a number, a medical case. All of the excitement that I was feeling on the way over to the hospital began to fade away. It felt like I was waiting for an operation, not about to have my baby. I felt alone, totally alone, as if I had just become a body to be examined and not a real person.

During one of the most profound experiences of her life. a woman may find herself surrounded by strangers to whom birth is merely business as usual.

Central to the critique of medicalization is the idea of control: women have less say and control over the process than they should. This is illustrated in this second comment, this from another woman, describing the way she was treated (Leifer 1990):

> And then this resident gave me an internal [examination], and it was quite painful then. And I said: "Could you wait till the contraction is over?" And he said he had to do it now, and I was really upset because he didn't even say it nicely, he just said: "You'll have to get used to this, you'll have a lot of this before the baby comes."

The critique of medicalization also takes into account the environment and medical procedures used. Lighting, noise, routine use of monitoring devices, routine administering of enemas, rates of episiotomies (a surgical procedure to enlarge the vaginal opening by cutting through the perineum toward the anus), rates of Cesarean-section deliveries, use of forceps or vacuum suction to assist in pulling the fetus from the

Throughout the world, men envy and imitate both pregnancy and childbirth. In our culture, there are sympathetic pregnancies in which a man develops physical characteristics similar to those of his pregnant partner. If she has morning sickness, so does he; if her belly begins to swell, so does his. Also, men often use images of pregnancy and childbirth to describe their creative work. A man "conceives" an idea. He is in a "fertile" period in his artistic development.

Other cultures have the ritual of **couvade**. The word comes from the French *couver*, meaning "to hatch or brood." Among the Hopi, for example, a man is required to be careful not to hurt animals. If he does, his child may be born deformed.

The Huichol of Mexico traditionally practiced a ritual of couvade in which the husband squatted in the rafters of the house or the branches of a tree above his wife during labor. When the woman experienced a contraction, she pulled on ropes attached to his scrotum. In this way, the man shared the experience of childbirth.

The couvade is a dramatic symbol of the man's paternity and his "magical"

relation to the child. By pretending he is pregnant, he distracts evil spirits from harming his baby. Describing the magical effect of the couvade, Arthur and Libby Colman (1971) write: "The couvade phenomena have the important side effects of helping a husband play an important part in pregnancy and childbirth. . . . They help a man cope with the envy and competitiveness which he may feel at his wife's ability to perform such a fundamental and creative act."

In his activities to deceive the evil spirits, a man may also find a reasonable outlet for his own desire to take on something of the female role in life.

---

womb—all of these reflect what critics have suggested is society's increasing dependence on technology and medical control. Critics of medicalization contend that episiotomies, the use of forceps, and vacuum extraction are employed more for the convenience and control of the obstetrician than due to medical necessity. In general, critics recognize and value the potential lifesaving use of such interventions *when needed* but question procedures that seem to place medical convenience above women's interests or needs.

The Feminist Approach

The question of what women most want or need is central to a feminist critique of childbirth. Along with consumer advocates and government policymakers, feminists and activists in the "women's health movement," have raised concerns and objections about the medicalization of childbirth. For feminists, a woman's rights—"to be informed, fully conscious, and to experience childbirth as a 'natural' process" are paramount (Treichler 1990). Feminists question how much medical intervention and control are necessary to reduce risks associated with the "normal, natural physiological process" of childbirth. They assert that most pregnant women are essentially healthy and require

minimal medical management during the birth process.

Writing in 1990, Treichler spoke of a "crisis in childbirth" intensifying around such criticisms and the medical profession's defense. Many of the above criticisms have been heard and addressed by hospitals and medical practitioners. For example, most hospitals have responded to the need for family-centered childbirth. Fathers and other relatives or close friends typically participate today. "Birthing rooms," with softer lighting and more comfortable birthing chairs, are increasingly common. Some hospitals permit rooming-in (the baby stays with the mother rather than in the nursery) or a modified form of rooming-in.

What Mothers Say

With the preceding critique of medicalization in mind, we might predict women to express high levels of discontent with their experiences and treatment during pregnancy, while in labor and giving birth, and after they give birth. We see quite the opposite from data collected in the Listening to Mothers survey, conducted by Harris Interactive and the Maternity Center Association. Billed as "the first national U.S. survey of women's childbearing experiences," the report is based on telephone interviews and online surveys with a combined sample of 1,583 women who gave birth during the 2-year period between May 2000 and June 2002. If anything, the results suggest positive experiences and assessments of treatment and care received. Some key findings are as follows:

1. By far, most mothers felt "quite positive" about their birthing experiences:

   - 95% felt that they generally understood what was happening to them.
   - 93% felt comfortable asking questions.
   - 91% felt that they had received the necessary amount of attention.
   - 89% felt as involved as they desired in decision making about their deliveries.

2. Nearly all women (97%) reported giving birth in a hospital, and 80% were attended to by obstetricians. (10% used midwives, 4% were attended to by family physicians, and 5% by nurses or physician's assistants).

3. Qualitative assessments of the overall care and treatment women received from their physicians were quite positive. Approximately nine out of ten women reported that their doctor or midwife had been polite, supportive, and understanding. The biggest complaint was that physicians or midwives seemed "rushed."

4. Women reported experience with the following medical interventions during labor and delivery that included the following:

   - 93% reported being monitored by an electronic fetal monitor.
   - 86% were given an intravenous (IV) drip.
   - 63% were given epidural analgesias to relieve pain, 30% were given a narcotic pain reliever, and 5% were given general anesthesia.
   - 78% of the recipients of epidurals and 66% of those who were given general anesthesia reported those techniques to be "very helpful."
   - 61% reported trying to use breathing as a means of controlling or minimizing pain, and 30% reported using "mental strategies." Only about one in five women who used either of these two approaches considered them "very helpful."
   - One in four women had Cesarean sections. Only 11% of women had "assisted vaginal deliveries"

in which either forceps or vacuum extraction was used.

   - More than a third of women received episiotomies.

5. Nine out of ten women reported receiving "supportive care" or attention during labor and delivery from their spouses or partners. Medical personnel were also supportive; 83% of women reported that they received support from nurses, and 53% said they received support from their doctors.

Although critics may still question how much medical necessity accounts for the previously mentioned rates of pain medication, episiotomies, and fetal monitoring, it is hard to ignore the high rates of satisfaction expressed by women about their experiences giving birth.

Pregnancy Loss

One additional aspect of dependence on technology is feeling omnipotent and that we should be able to solve any problem. Thus, if something goes wrong with a birth—if a child is stillborn or has a disability, for example—we look for something or someone to blame. We have become unwilling to accept that some aspects of life and death are beyond human control.

The loss of a child through miscarriage, stillbirth, or death during early infancy is a devastating experience that has been largely ignored in our society. The statement "You can always have another one," although it may be meant as consolation, is particularly chilling to the ears of a grieving mother. In recent years, however, the medical community has begun to respond to the emotional needs of parents who have lost a pregnancy or an infant.

Spontaneous Abortion

Spontaneous abortion (miscarriage) is the most common form or type of pregnancy loss. Most occur during the first trimester, with only 3% of all intrauterine deaths occurring after 16 weeks of pregnancy (Layne 1997). The rate of miscarriage is lowest among women 20–24 years old and increases steadily to a high among women 35–39 years old. The estimated *rate* of pregnancy loss is nearly double among women of color compared to non-Hispanic white women, but white women suffer most miscarriages (Layne 1997).

About one out of four women is aware she has miscarried at least once (Beck 1988). Studies indicate that at least 60% of all miscarriages are caused by chromosomal abnormalities in the fetus (Adler 1986). Furthermore, as many as three-fourths of all fertilized eggs do not mature into viable fetuses (Beck 1988). One study found that 32% of implanting embryos miscarried (Wilcox et al. 1988).

Most miscarriages occur between the sixth and eighth weeks of pregnancy. Evidence is increasing that certain occupations involving exposure to chemicals or high levels of electromagnetism increase the likelihood of spontaneous abortions. Miscarriages may also occur because uterine abnormalities or hormonal levels are insufficient for maintaining the uterine lining.

Infant Mortality

The rate of **infant mortality** in the United States remains far higher than the rates in most of the developed world. The U.S. Public Health Service reported 6.8 deaths for every 1,000 live births in 2001 (National Center for Health Statistics 2002). Nevertheless, among developed nations, the United States does not fare well in low infant mortality. In 1999, the United States ranked 28th of 37 countries with populations of at least 1 million for which complete counts of live births and deaths were compiled. (This means that 27 countries had *lower* infant mortality rates than the United States.) In the same comparison in 1990, the United States ranked 11th (http://www.cdc.gov/nchs/data/hus/tables/2003/03hus025.pdf.). In comparison with many developing countries, the rate in the United States is quite low (UNICEF State of the World's Children 2001). Still, the U.S. rate is on par with the rate in Cuba, Malaysia, and Slovakia, all of which are far less wealthy than the United States (Ruane and Cerulo, 2004). Within the United States, the nation's capital has a higher infant mortality rate than any of the 50 states, at 12.5 per 1,000 live births (U.S. Census Bureau 2001, Table 104).

Looking at combined data, for 1995–2002, the U.S. infant mortality rate of 7.1 per 1,000 live births, represented more than 225,000 infant deaths during that period. The rate varied by race and ethnicity, from a low of 5.0 among Asians to a high of 13.9 among African Americans. The United States has targeted a goal rate for 2010 of 4.5, as well as an objective to eliminate racial and ethnic disparities (http://www.cdc.gov/mmwr/preview/mmwrhtml/mm5422a1.htm).

Of the thousands of American babies less than 1 year old who die each year, most are victims of the poverty that often results from racial or ethnic discrimination. Up to a third of these deaths could be prevented if mothers were given adequate health care (Scott 1990).

The United States is far behind many other countries in providing health care for children and pregnant women. In France, Sweden, and Japan, for example, all pregnant women are entitled to free prenatal care. Free health care and immunizations are also provided for infants and young children. Working Swedish mothers are guaranteed 1 year of paid maternal leave, and French families in need are paid regular government allowances (Scott 1990). One in six children born in the United States, is born to mothers who received no prenatal care through the first trimester of pregnancy. Almost one in eight children have no health insurance (Ruane and Cerulo, 2004).

Although many infants die of poverty-related conditions, others die from congenital problems (conditions appearing at birth) or from infectious diseases, accidents, or other causes. Sometimes, the causes of death are not apparent. Data from the Centers for Disease Control and the National Center for Health Statistics for 2001 attribute 2,234 infant deaths to **sudden infant death syndrome (SIDS),** a perplexing phenomenon wherein an apparently healthy infant dies suddenly while sleeping (http://www.cdc.gov/nchs/fastats/pdf/mortality/nvsr52_03t32.pdf).

A study from Australia identified four factors that appear to increase the chances of SIDS (Ponsonby et al. 1993): (1) a soft, fluffy mattress, (2) the baby being wrapped in a blanket, (3) the baby having a cold or other minor illness, and (4) allowing the baby to become too warm. Exposure to secondhand smoke also has been implicated (Klonoff-Cohen et al. 1995). It is also important that an infant not be placed to sleep on its stomach until it is strong enough to turn over ("Sleeping on Back Saves 1,500 Babies," 1996).

Coping with Loss

The depth of shock and grief felt by many who lose a child before or during birth is sometimes difficult to understand for those who have not had a similar experience (Layne 1997). What they may not realize is that most women form a deep attachment to their children even before birth. The loss of the child must be acknowledged and felt before psychological healing can take place. Instead, however, women typically find that friends, relatives, and coworkers want to pretend that "nothing has happened" (Layne 1997).

Equally problematic are the common reactions from medical personnel and midwives. Medical personnel, especially physicians, may perceive pregnancy loss as "medically unimportant" and as evidence of normal and natural processes at work. Midwives, who typically try to "demedicalize" pregnancy, tend to stress problems that result from overmedicalization. Thus, "nonmedically caused problems" (for example, naturally occurring spontaneous abortion) may remain beyond the domain of midwives. Finally, most "preparation for childbirth" literature and education glosses over or leaves out miscarriage. Thus, women who miscarry are likely to feel and be invisible to those involved in reproductive medicine and childbirth instruction and assistance (Layne 1997).

Women (and sometimes their partners) who lose a pregnancy or a young infant generally experience similar stages in their grieving process. Their feelings are influenced by many factors: supportiveness of the partner and other family members, reactions of social networks, life circumstances at the time of the loss, circumstances of the loss itself, whether other losses have been experienced, the prognosis for future childbearing, and the woman's unique personality. Physical exhaustion and, in the case of miscarriage, hormone imbalance often compound the emotional stress of the grieving mother.

The initial stage of grief is often one of shocked disbelief and numbness. This stage gives way to sadness, spells of crying, preoccupation with the loss, and perhaps loss of interest in the rest of the world. It is not unusual for parents to feel guilty, as if they had somehow caused the loss, although this is rarely the case. Anger (toward the physician, perhaps, or God) is also a common emotion.

Experiencing the pain of loss is part of the healing process (Vredevelt 1994). This process takes time—months, a year, perhaps more for some. Support groups and counseling are often helpful, especially if healing does not seem to be progressing or depression and physical symptoms do not appear to be diminishing. Planning the next pregnancy may be curative, too, although we must keep in mind that the body and spirit need some time to heal.

Giving Birth

Sociologist Karin Martin conducted intensive interviews with a small sample of first-time mothers. The twenty-six mostly white heterosexual women, ranging in age from 20 years to over 40, were interviewed within 3 months of having given birth. Instead of exploring the macro-level and institutional dimensions of childbirth, Martin wanted to know how women experienced childbirth and how their experiences were

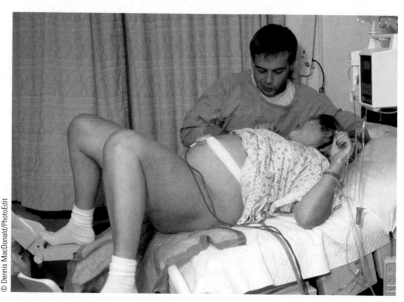

◾ *Family-centered childbirth allows fathers to participate alongside mothers in the birth process.*

© Dennis MacDonald/PhotoEdit

shaped by "internalized technologies of gender," those "aspects of the gender system that are in us, that become us" (2003, 56). These internalized ideas and practices help determine "how we think about and understand ourselves as men and women" (57).

Deep within us, even in "seemingly natural experiences like birth," are our culturally constructed gender identities (57). Even during childbirth, women are "doing gender," acting compliant, nice, and kind. Martin's informants recalled trying not to "bother" strangers in adjoining rooms, remembered trying hard to remain attentive during conversations, and described doing things that indicated they were putting the needs of others ahead of their own. Even though they had to impose on others (doctors, nurses, midwives, husbands, and so on) for things (backrubs, quiet, patience, information, and so on), they recalled feeling badly about doing so. They found it hard not to feel "rude" or "selfish" for making the demands and imposing on others.

Other ways in which women selflessly "acted like a girl" include allowing—indeed looking to—male partners to "describe, define and decide about their experiences during labor, even their bodily ones" (63). This included using their partners' experiences and views of the birth as definitive of what happened during birth. This was worsened by their inability to "see" the birth. Then, "at the height of labor's physical demands . . . just before an epidural . . . or when pushing the baby out," the women described "acting out," referring to themselves with words like "nasty," "crabby," and "out of control." Even though we might consider it understandable for women to yell, curse, or whine, in their own minds these reactions were neither understandable nor justified. They describe "feeling bad" when they "lost control," by which they mean when they were "not nice." They recall, both during and after giving birth, apologizing to their partners and medical providers. These apologies seem to validate Martin's contention that women were consciously trying to regulate their behavior, even amid the physically and emotionally demanding process of childbirth.

Martin suggests that the feminist critique of the medicalization of childbirth may be correct in highlighting how institutional control over birth shapes the experience. But it is only part of the story, but women's birth experiences are also regulated and controlled "from within," by internalized gender identities. Even when "given permission" to depart from gender expectations, to act in gender deviant ways, they found themselves at odds with such behavior. It was not "how they are" or "who they are" (Martin 2003).

Choosing How: Adoptive Families

Parenthood is not only entered biologically. Although adoption is being examined here as the traditionally acceptable alternative to pregnancy for infertile couples, it may also include the adoption of stepchildren in a remarriage, the adoption of a child by a relative, the adoption of adolescents, the adoption of two or more siblings, and the adoption of foster children who have been removed from their parental homes (Grotevant and Kohler, in Lamb 1999). Many people—married and single, with or without biological children—choose to adopt, not because they are unable to conceive or bear their own children but because they are ideologically committed to adoption. Perhaps they have concerns about overpopulation and the number of homeless children in the world. They may wish to provide families for older or disabled children. Thus, the population of adoptive families is diverse in terms of both motivation and circumstances.

Until recently, it has been difficult to say with certainty how common adoption is in the United States, given the relative absence of dependable or comprehensive data (Grotevant and Kohler, in Lamb 1999). The Center for Adoption Research and Policy estimated that more than 1 million children are currently in adoptive families and more than 5 million adults and children have been adopted (Grotevant and Kohler, in Lamb 1999). More recently, however, the U.S. Census Bureau undertook, for the first time, an effort to count and construct a profile of adoptive families (Kreider 2003). According to the census data, there are more than 2 million adopted children residing in 1.7 million households.

Table 10.3 shows other characteristics of households with adopted children. Nearly 2% of households with children have adopted children only. Another 1.8% have adopted and biological children together in the household, and 0.1% have adopted children with stepchildren or adopted, biological, and stepchildren together.

Census data further reveal the following:

- Adopted children are more likely to be female than male, which Kreider suggests results from both

Real Families Men and Childbirth

I remember thinking, "Nobody else cares." My wife was knocked out, everybody else in the room was taking care of my wife and the baby, and the baby was wet, cold, red, really an ugly looking thing. And yet, it was like your first puppy. No one else cares—no one else is going to take care of it at that moment. Truthfully, I was instantly bonded; it was like a marriage. She [my daughter] opened her eyes a little bit and I immediately began to relate to what she saw. . . . I was thrilled. I followed her after she opened her eyes and tried to imagine what she saw and what she might be seeing. . . . It was real exalting.

Mark, 37-year-old father of one

My wife had the shakes and couldn't hold the baby. I held her [my daughter] and sang her a lullaby. She was looking at my face, wasn't focusing, but I could see something going on. She could obviously hear too. I got to hold her for like fifteen minutes. It was all so exciting and incredible . . . and strange.

Bill, 36-year-old father of one

The preceding comments are the recollections of two fathers to having witnessed the births of their daughters. Told to sociologist Theodore Cohen, they reveal how deeply some men are moved by their involvement at birth. In the United States and many other countries, it is now commonplace for fathers to attend, witness, and often even actively assisting in the birth of their children. It may be so common that we forget how relatively recent it is for men to enjoy such access. In 1960, only about 15% of fathers attended the birth of their child in the delivery room. Although estimates vary, by the first years of the twenty-first century, between 75% and 80% of fathers were present at childbirth (*Washington Post* 2006).

We see the same trend elsewhere. In the United Kingdom, fathers are now in attendance at 80% of births (Johnson 2002). Similar trends have been observed in other European countries and in Canada. Attendance at birth offers fathers an opportunity to feel part of the birth process and to offer support to their partners. It may or may not lead to greater involvement in subsequent childrearing. Although some research speculates that it does, other research finds no such effect (Palkovitz 1985).

Among men in the United Kingdom, the most frequently cited motivations for attendance at birth are out of support for their partners, out of curiosity, or because of pressure. In the United States and Canada, there is a fourth reason: men often play the role of "coach," assisting their partners to implement what they have been taught in prenatal classes (Johnson 2002, 167):

Here it is the man's responsibility to help his partner practice the procedures learned in prenatal classes, requiring the acquisition of some knowledge and training.

Where once hospital practice and cultural expectations kept men out of the delivery room, now they are expected to be present. To illustrate this coercive element, in Martin Johnson's exploratory study of 53 British fathers, 57% of the men said they felt pressured to be there through labor and delivery. For example, "You don't get a choice, not really. It is assumed that you want to be there; I mean I did, but that is not the point. It's like not having a choice."

Finally, in Johnson's study, men's reactions to what they saw and experienced were both positive and negative.

On the negative side, 56% of the men identified as their most overwhelming memory the pain they witnessed their partners suffering. One man, Ben, claimed he felt as though he ought to be experiencing pain himself: "In a strange way, when she dug her nails into my hands, I wanted to embrace the pain; it was like my share."

On the positive side, and unsurprisingly, men were deeply moved by the birth and awed by their partners' strength and resilience.

Two of Johnson's informants, Ken and Bill, made the following comments:

For the first time in a long, long time, I had tears rolling down my face. (Ken)

When her head came out, I thought, I did this, she is half me, I have given the world a part of the future. (Bill)

---

desirability and availability. Specifically, "women in general express a preference for adopting girls, and single women more frequently have adopted girls than boys" (2003, 8). In addition, with regard to international adoptions, more female children are available for adoption from those countries that are "leading sources for adopted children" (8).

- Higher proportions of adopted children than biological children are African American and Asian; 16% were black (compared to 13% of biological

Table 10.3 ■ Households by Type of Children and Number of Adopted Children

	Number	Percent
Households by type of children	**45,490,049**	**100.0**
Adopted children only	816,678	1.8
Stepchildren only	1,485,201	3.3
Biological children only	40,657,816	89.4
Adopted and biological children	808,432	1.8
Adopted children and stepchildren	29,575	0.1
Biological children and stepchildren	1,659,924	3.6
Biological children, adopted children and stepchildren	32,423	0.1
Households with adopted children	**1,687,108**	**3.7**
One	1,383,149	3.0
Two	247,600	0.5
Three or more	56,359	0.1

SOURCE: http://www.census.gov/prod/2003pubs/censr-6.pdf

children) and 7.4% were Asian (more than double the 3.5% of biological children). A slightly smaller percentage of adopted children than biological children are Hispanic (14% versus 16%).

■ Economically, families with adopted children are somewhat better off than those without. Smaller percentages of adopted children than biological children are poor (11.8% vs. 16%). "Adoptive households" had higher median incomes than households with only biological children ($56,138 compared to the $48,200) and a third of adopted children as opposed to 27% of biological children lived in households with incomes of $75,000 or more.

■ Adopted children were more likely to have some disability than were biological children (15% versus 7% of boys, 9% versus 4% of girls). "Mental disabilities" were the most common disability, consisting of "difficulty learning, remembering, or concentrating."

■ More adopted children (78%) than biological children (74%) lived in two-parent households.

Of the 1.7 million households with adopted children, about 308,000 (18%) contained members of different races. This is twice the proportion found among the 43.8 million households with no adopted children, where 4.1 million had members of different races. Kreider notes that this is largely a result of the adoption of foreign-born children by U.S. residents.

The costs of adoption can be quite steep. The Child Welfare Information Gateway, provides the following estimates of adoption-specific costs. Such costs vary, depending on the type of adoption, the type of agency used, whether they adopt domestically or internationally, and so on. The costs from range from a low of $0 to $2,500 for foster care adoption, from $5,000 to $40,000 for domestic infant adoption, and from $7,000 to $30,000 for international adoption, plus travel costs.

If there is a need for a home study to determine the suitability of prospective adoptive parents, costs may exceed $3,000 (Child Welfare Information Gateway, http://www.childwelfare.gov/pubs/s_cost/s_costa.cfm).

Adoption laws vary widely from state to state; some prohibit private adoption, and other states have laws that are considered quite supportive of it.

With confidentiality no longer the norm, the trend is toward **open adoption** in which there is contact between the adoptive family and the birth parents (McRoy, Grotevant, and Ayers-Lopez 1994). This involvement can be either mediated (through an adoption agency) or direct, where the birth mother and adoptive family have contact with each other. Many adoption experts agree that some form of open adoption is usually in the best interests of both the child and the birth parents.

Adoptive families face unique problems and stresses. They may struggle with physical and emotional strains of infertility; endure uncertainty and disappointment as they wait for their child; and may spend all their savings and then some in the process. They often face insensitivity or prejudice. For example, an adopted child may be asked, "Who is your *real* mother?" or "Are you their *real* daughter?" Adoptive parents may be congratulated by well-meaning folks— "Oh, you're doing such a good thing!"—as though

Popular Culture   Covering Adoption

When you think about adoption, what comes to mind? Would you guess that most adoptions are successful? Do you consider adoptive families to be "real families"? What impressions of adoptees do you have? Researchers Susan Kline, Amanda Karel, and examined the role of broadcast journalism in shaping attitudes about adoption.

Looking at 292 adoption-related news stories that aired on morning news programs, news magazine programs, and evening news programs on either NBC, CBS, or ABC between 2001-2004, they found somewhat mixed coverage.

More than half (162 stories, 56%) of all adoption related stories contained "problematic depictions" of adoptees, such as portraying them as having emotional difficulties, health problems or engaging in antisocial behavior. In 121 stories (41%) adoptees were portrayed in socially desirable ways. In 34% of the stories (99), there were both positive and negative depictions, as in the case of an adoptee who had been "out of control," but bonded with adoptive parents. Stories with solely negative portrayals (63; 22%) were almost three times as common as stories with only positive depictions (22; 8%).

Almost half the stories contained nothing about biological parents. Of the 53% (156) that did feature birth parents, more negative than positive portrayals were found. In 68 stories (23%) only negative portrayals of birth parents were conveyed; conversely, 29 stories (10%) conveyed only positive images. Overall, more stories contained problematic depictions (44%; 127 stories) than contained positive portrayals (30%; 88 stories) of birth parents. The negative portrayals featured such issues as abandonment of one's children or food stamp fraud. Interestingly, the tone of the coverage shifted when considering adoptive parents.

Interactions in adoptive families were featured in 91% of the 292 stories. Positive depictions (for example, showing family cohesion, support, love, accomplishments) were somewhat more common than negative (for example, critical, aggressive, or abusive) ones (62% vs. 57%). Only 27% of the stories featured solely negative depictions of adoptive family interactions, while 41% portrayed such interactions in solely positive ways.

Kline, Karel and Chaterjee recommend that more news coverage focus on "solely positive portrayals of adoptees" and pay greater attention to birth parents' views of the adoption experience. They also warn practitioners who work with prospective adoptive parents to be aware of the following. Given the problematic tone of the portrayals of adoptees, the limited coverage of why birth parents place children for adoption, and the tendency for news stories to stigmatize or stereotype adoptees and birth parents, prospective parents may hold negative images, have limited knowledge of what motivates birth parents to place their children for adoption, and misunderstand adoption and adoptive family life. Thus, it is important to consider the slant of news coverage of adoption, as especially in the absence of direct interaction with adoptees and adoptive families, mass media images can help reduce stigma.

---

they had made a sacrifice of some kind in choosing to build a family in this way. Even grandparents may reject adopted grandchildren (at least initially), especially if the adoption is interracial. The idea that adoption is not quite "natural" is all too common in our society.

At the same time, adopted children may feel uniquely loved. Suzanne Arms recalling her son Joss's explanation to a friend, recounts, "When Joss was six, he was overheard explaining to a friend how special it was to be adopted. Apparently," she adds, "he made a good case for it, because when his friend got home, he told his mother he wanted to be adopted so he could be special too" (Arms 1990).

Reflections

Is the ability to create a child important to your sense of self-fulfillment? If you discovered that you were infertile, what do you think your responses would be? Would adoption be an option for you? Why or why not?

Becoming a Parent

The time immediately following birth is a critical period for family adjustment. No amount of reading, classes, and expert advice can prepare expectant parents for the real thing. The 3 months or so following child-birth (the "fourth trimester") constitute the **postpartum period.** This is a time of physical stabilization and emotional adjustment.

New mothers, who may well have lost most of their interest in sexual activity during the last weeks of pregnancy, will probably find themselves slowly returning to prepregnancy levels of desire and coital frequency. Some women may have difficulty reestablishing their sexual life because of fatigue, physiological problems such as continued vaginal bleeding, and worries about the infant (Reamy and White 1987).

Both relationship satisfaction and postpartum depression are important predictors of the levels of sexual desire and satisfaction and of changes in sexual frequency following childbirth. Enjoyment returns gradually. According to research by Margaret De Judicibus and Marita McCabe, at 2 weeks postpartum, few new mothers report sexual intercourse as pleasurable; by 12 weeks, two thirds of women say that sex is "mostly enjoyable." Even then, however, 40% complain of some difficulties. Relationship satisfaction is at its lowest at this point. Nearly six out of seven couples report reduced frequency of intercourse at 4 months postpartum. By 6 months postpartum, many women continue to report significantly lower levels of desire, sexual frequency, and sexual satisfaction when compared to the levels before conception. By this point, the quality of the mother role was also strongly associated with sexuality, as was fatigue. Over the first few postpartum months, there is evidence of reductions in both reported love for the partner and affection expressed between partners (De Judicibus and McCabe 2002).

De Judicibus and McCabe identify each of the following as factors associated with reduced sexual desire, decreased frequency of relations, and lower levels of satisfaction:

- Adjustment to changes in social roles during transition to parenthood.
- Declining marital satisfaction. This has been reported in many countries. After a first-month "honeymoon" period, the trend toward lower levels of satisfaction becomes stronger by the third postpartum month.
- Postpartum mood or postnatal depression; 35–40% of women report *some* depressive symptoms.
- Fatigue.

- Physical changes with birth of a child. These can result in *dyspareunia,* or painful intercourse, the symptoms of which may include "a burning, ripping, tearing, or aching sensation associated with penetration. The pain can be at the vaginal opening, deep in the pelvis, or anywhere between. It may also be felt throughout the entire pelvic area and the sexual organs and may occur only with deep thrusting" (http://www.healthscout.com/ency/1/474/main.html).
- Breastfeeding.
- The demands imposed by infants.

The postpartum period also may be a time of significant emotional upheaval. Even women who had easy and uneventful births may experience a period of "postpartum blues" characterized by alternating periods of crying, unpredictable mood changes, fatigue, irritability, and occasional mild confusion or lapses of memory. For some, this can be truly devastating, leaving them feeling as though they are "losing their minds" as they struggle with postpartum reactions that include psychosis, depression, panic disorder, and obsessive compulsive disorder (Layne 1997). A woman may have irregular sleep patterns because of the needs of her newborn, the discomfort of childbirth, or the strangeness of the hospital environment. Some mothers may feel lonely, isolated from their familiar world. Infants of women suffering postpartum depression also suffer, as postpartum depression interferes with mothers' abilities to respond to their newborns' needs and may lead to poor emotional and cognitive development (Layne 1997).

Many women blame themselves for their fluctuating moods. They may feel that they have lost control over their lives because of the dependency of their newborns.

Biological, psychological, and social factors are all involved in postpartum depression. Biologically, during the first several days following delivery, there is an abrupt fall in certain hormone levels. The physiological stress accompanying labor, dehydration, blood loss, and other physical factors contribute to lowering the woman's stamina. Psychologically, conflicts about her ability to mother, ambiguous feelings toward or rejection of her own mother, and communication problems with the infant or partner may contribute to the new mother's feelings of depression and helplessness. Finally, the social setting into which the child is born is important, especially if the infant represents a financial or emotional burden for the family. Post-partum counseling before discharge from the hospital can help couples gain perspective on their situation so that they will know what to expect and can evaluate their resources.

Although the postpartum blues are felt by many women, they may be especially problematic for young mothers. Donna Clemmens (2002) reports that as many as 48% of adolescent mothers suffer depressive symptoms, a rate more than 3.5 times that among adult mothers (13%). Clemmens identifies the following six themes as evident in depressive reactions of young mothers (19 years old or younger):

1. *Suddenly realizing motherhood.* Struck by the "sudden cold realization of being a mother, motherhood hits like a Nor'easter" (a severe storm that seems to come out of nowhere).

2. *Torn and pulled between two realities.* Mothers described being pulled and torn between the realities of new motherhood and being adolescents in school, sometimes leading to regrets over lives that "could have been."

3. *Constantly questioning and trying to explain the unexplainable.* Feeling depressed, participants had a difficult time explaining their depression. They reported feeling an emptiness, a state of "wanting to die," feelings that they couldn't shake but also feelings that they couldn't effectively explain to others.

4. *Feeling alone, betrayed, and abandoned by those that you need to love you.* This speaks to the feeling of abandonment by boyfriends and friends, leaving them feeling stressed and betrayed, as if they had nothing going for them.

5. *Everything is falling down on you and around you.* The sadness, anger, mood swings, fatigue, confusion, and crying symptomatic of depression felt like a heavy weight being carried around. Although they sometimes felt happy, their moods would drop for no apparent reason.

6. *You are changing and regrouping, seeing a different future.* Some mothers felt that having "survived the storm" they wanted to warn other teens about what early motherhood was really like. In this way they hoped to make something constructive of their hurt.

Clemmens notes that all the young women maintained warm feelings for and commitment to their children. They acknowledged regrets about having had sex so early and having become pregnant but not about having their children. There were also numerous statements about feeling stronger, more responsible, and reliable as a result of becoming mothers (Clemmens 2002).

Men, too, seem to get a form of postpartum blues. When infants arrive, many fathers do not feel prepared for their new parenting and financial responsibilities. Some men are overwhelmed by the changes that take place in their marital relationship. Fatherhood is a major transition for them, but their feelings are overlooked because most people turn their attention to the new mother.

The transition to parenthood can be made somewhat easier if the new parents understand in advance that a certain amount of fatigue and stress is inevitable. They need to ascertain what sources of support will be helpful to them, such as friends or family members who can help out with preparing meals or running errands. They also need to keep their lines of communication open—to let each other know when they are feeling overwhelmed or left out. It's also important that they plan time to be together, alone or with the baby—even if it means telling a well-meaning relative or friend they need time to themselves.

For many women and men, the arrival of a child is one of life's most important events, filling mothers and fathers alike with a deep sense of accomplishment. The experience itself is profound and totally involving. A father describes his wife (Kate) giving birth to their daughter (Colleen) (Armstrong and Feldman 1990):

> Toward the end, Kate had her arms around my neck. I was soothing her, stroking her, and holding her. I felt so close. I even whispered to her that I wanted to make love to her—It wasn't that I would have or meant to—it's just that I felt that bound up with her.

> Colleen was born while Kate was hanging from my neck. . . . I looked down and saw Mimi's [the midwife's] hands appearing and then, it seemed like all at once, the baby was in them. I had tears streaming down my face. I was laughing and crying at the same time. . . . Mimi handed her to me with all the goop on her and I never even thought about it. She was so pink. She opened her eyes for the first time in her life right there in my arms. I thought she was the most beautiful thing I had ever seen. There was something about that, holding her just the way she was. . . . I never felt anything like that in my life.

Taking on Parental Roles and Responsibilities

Even more than marriage, parenthood signifies adulthood—the final irreversible end of youthful roles. The irrevocable nature of parenthood may make the first-time parent doubtful and apprehensive, especially during the pregnancy. Despite the many months of pregnancy, the actual transition to parenthood happens in the instant of birth. Such an abrupt transition from a nonparent to a parent may create considerable stress. Parents take on parental roles literally overnight, and the job goes on without relief around the clock. Many parents express concern about their ability to meet all the responsibilities of childrearing.

There have been a number of important analyses of the transition to parenthood (Rossi 1968; LaRossa and LaRossa 1981; Cowan and Cowan 1992). An early and influential analysis of what parents experience as they enter the new social reality of parenthood was offered by Alice Rossi (1968). According to Rossi, entering parenthood is stressful because of the nature of the role of parent and the characteristics of the parental role transition.

Rossi singled out the following features of entering parenthood:

- *Irreversibility.* Unlike nearly any other role, once we enter parenthood we cannot easily leave without incurring significant social or legal repercussions. Even "deadbeat parents," who have left their children and ceased to support them, are still considered responsible for their children's welfare.

- *Lack of preparation.* There is almost nowhere and no way to practice parenting. Parenting books, childbirth classes, and babysitting experience, pale in comparison to the reality we face upon having children. Furthermore, little systematic effort is made to equip people with more realistic understanding or even practical skills to more effectively parent.

- *Idealization and romanticization.* Related to the lack of preparation are the expectations we have about what parenthood, which are often unrealistic and overly idealized. If and when reality turns out to be less than ideal, we become frustrated and disappointed.

- *Suddenness.* Despite what might be 8 months of awareness of impending parenthood, the actual transition is sudden. There is no opportunity for expectant parents to ease into the role; we go from nonparent to parent in the moment of childbirth and assume all of the role responsibilities with that same suddenness.

- *Role conflict.* The parental role affects all of the other roles we play, encroaching upon time spent with a spouse or partner and complicating paid employment.

Based on their research on new parents, Ralph and Maureen Mulligan LaRossa suggested that the major

Although becoming a parent is stressful, the role of mother or father is deeply fulfilling for many people.

© Laurie DeVault Photography

adjustment new parents face is *temporal*. Like a hospital or fire station, new parenthood is a **continuous coverage system;** infants must have someone available to care for them 24 hours a day, 7 days a week. When direct care is not needed (for example, during naps or nights when infants sleep), someone must at least be "on call" (a secondary level of accessibility), ready to move to more direct interaction. Finally, when at least two competent caregivers are present, one may move to a state of "downtime" (tertiary level), wherein one is free to pursue other activities without concern for the infant's needs. The LaRossas suggest that new parents (in a two-parent household) compete with each other and experience conflict over downtime. Much of the LaRossas' analysis follows from that, looking at who has more downtime (fathers), who does most of the primary parenting (mothers), and why. In answering the latter question, they note some wider cultural beliefs that value mother care, as well as some relationship and individual-level factors that press mothers toward more of the work associated with children and turn fathers into "helpers" and "playmates" (LaRossa and LaRossa 1981).

Still more recently, Carolyn and Phillip Cowan identified five domains in which new parents experience change as a result of the arrival of children (1992, 2000):

- *Identity and inner-life changes.* New parents discover that they no longer think of themselves the same way they did before their children were born. Their priorities and personal values also change. Issues that previously seemed remote, unimportant, or abstract become personal, meaningful, and real.
- *Shifts within the marital roles and relationship.* Parenthood alters how couples divide tasks or allocate responsibilities. Because they are also experiencing fatigue (from reduced sleep and more work), their relationship quality may diminish.
- *Shifts in intergenerational relationships.* Becoming parents alters—often improving and intensifying, sometimes straining—the relationship between new parents and *their* parents.
- *Changes in roles and relationships outside the family.* New parenthood, especially new motherhood, may force changes in other nonfamily roles and relationships, such as at work or in friendships. Although some of these changes may be temporary (for example, leaving work only for the length of a parental leave), they nonetheless compound other things to which new parents are adjusting.

- *New parenting roles and relationships.* New parenthood means that a couple must arrive at an agreeable division of childcare. New parents learn how difficult it is to maintain equal and/or equitable divisions of childcare. One parent may feel put upon or taken advantage of in the way the couple allocates their individual time and energy to childcare tasks.

The Cowans suggest that the difficulties associated with the parental transition are more difficult for contemporary parents because of some major features of the social climate in which they parent. First, contemporary parenthood is more discretionary or optional, making decisions about whether and when to have children subject to more discussion, negotiation, and potential dispute. Second, many new parents, especially middle-class parents, are relatively isolated, geographically, from their wider kin groups and other long-term social supports. Third, changes in women's roles have introduced more role conflict for new mothers and have increased women's need and legitimate demand for more sharing by their partners. Fourth, the social policies that address the needs of parents are weak to nonexistent. Fifth, there are few enviable or attractive role models for effective parenting; *Leave It to Beaver* families are unrealistic, and yet there is no equivalent cultural model of dual-earner parents to draw upon. If we cannot parent like our own parents did, who can we emulate? Sixth, today's families are supposed to fulfill all of our emotional needs. Parenting is stressful and requires mutual effort and sacrifice. But effort and sacrifice don't fit compatibly with individual emotional fulfillment.

Thus, the difficulties may become sources of resentment and estrangement (Cowan and Cowan 1992, 2000).

Reflections

If you have children, did you plan to have them? What considerations led you to have them? What adjustments have you had to make? How did your relationship with your partner change?

Stresses of New Parenthood

Many of the stresses felt by new parents closely reflect gender roles. Overall, mothers seem to experience greater stress than fathers. Although a couple may have an egalitarian marriage before the birth of the first child, the marriage usually becomes more traditional once a child is born. If the mother, in addition to the father, continues to work, or if the woman is single, she will have a dual role as both homemaker and provider. She will also probably have the responsibility for finding adequate childcare, and it will most likely be she who stays home to take care of a sick child. Multiple role demands are the greatest source of stress for mothers.

There are various other sources of parental stress. Fathers often describe severe stress associated with their work. Both mothers and fathers must be concerned about having enough money. Other sources of stress involve infant health and care, infant crying, interactions with the spouse (including sexual relations), interactions with other family members and friends, and general anxiety and depression (Harriman 1983; McKim 1987; Ventura 1987; Wilkie and Ames 1986).

Changes in marital quality and marital conflict were studied among a sample of Caucasian and African American spouses as they transitioned to parenthood (Crohan 1996). The results of this study showed a decline in marital happiness and more frequent conflicts among both Caucasian and African American spouses. Caucasian parents also reported higher marital ten-

sion and a greater likelihood to become quiet and withdrawn after the birth of their child. This increase in avoidance behaviors may be because of the limited time and energy that new parents have to devote to conflict resolution.

Although the first year of childrearing is bound to be stressful, the partners experience less stress if they (1) have already developed a strong relationship, (2) are open in their communication, (3) have agreed on family planning, and (4) originally had a strong desire for the child. Despite planning, the reality for most is that this is a stressful time. Accepting this fact while developing time management skills, patience with oneself, and a sense of humor can be most beneficial.

Having a child is unlike any other experience we undertake. The changes in our lives are wide ranging and irreversible, the potential rewards are great, and the sacrifices are many. Increasing numbers of women and couples are deciding to forego parenthood, largely to avoid its many and profound consequences. Most people, however, continue to decide to embark on the journey described in this chapter and take on the challenging tasks that we look at in Chapter 11.

■ *Becoming a parent introduces changes in intergenerational relationships.*

© Myrleen Ferguson Cate/PhotoEdit

Summary

- With wider availability of effective contraceptives and access to legal abortion, and with increases in delayed parenthood and *child-free marriage,* parenthood may now be considered more a matter of choice.

- There were more than 4 million births in the United States in 2004.

- Both fertility and birthrates vary across such social and demographic characteristics as race, ethnicity, education, income, and marital status.

- A record number of unmarried women gave birth in 2004. There were nearly 1.5 million nonmarital births, representing 36% of all 2004 births.

- Between 8% and 9% of women 15–44 have an "impaired ability" to have children. Between 7 and 8% of couples are involuntarily childless. Each year, more than 2 million couples seek assistance for problems of infertility.

- Estimates suggest that between 6.6% and 9.3% of women of childbearing age do not expect to have children.

- A common pattern leading couples to child-free lifestyles is to initially postpone having children for a definite period, when that time lapses to extend it indefinitely, and ultimately to perceive more advantages than disadvantages to remaining childless.

- Even with greater acceptance of voluntary childlessness, women and men who forego parenthood experience social pressure to justify or change their statuses and suffer from negative stereotypes.

- The pattern of delaying or deferring parenthood is increasingly common.

- Having children both at younger and at older ages exposes women to greater health risks.

- Between 1990 and 2000 the pregnancy rate declined for unmarried and married women. Pregnancy rates are highest among 20- to 24-year-old women, followed by 25- to 29-year-old women.

- Teen pregnancy rates declined through the 1990s and in the first years of this century.

- Of the more than 6 million pregnancies in a year, there were more than 4 million births (63% of pregnancies), 1.3 million abortions (20% of pregnancies), and 1 million stillbirths or miscarriages (17%).

- Nearly a quarter of pregnancies carried to full term are unplanned. Babies born from unplanned pregnancies face greater health and social risks, and their mothers have greater risks of postpartum depression.

- A couple's relationship is likely to undergo changes during pregnancy, especially if the pregnancy was unanticipated. Both partners may have fears about the baby's well-being, the birth, their ability to parent, and how the baby will affect their relationship.

- Although indirectly and vicariously, men are affected by a partner's pregnancy. Men's involvement in the pregnancy and birth process may affect their later parenting.

- Feelings about sexuality are likely to change during pregnancy for both women and men. Sexual activity is generally safe during pregnancy unless there is a prior history of miscarriage, bleeding, or pain.

- Critics have alleged that the *medicalization of childbirth*—making this natural process into a medical "problem"—has caused an overdependence on technology and an alienation of women from their bodies and feelings.

- Research with new mothers documented positive experiences and assessments of treatment and care received.

- Miscarriages are the most common form of pregnancy loss. Most occur early, during the first trimester; only 3% occur after the 16th week of pregnancy.

- *Infant mortality* rates in the United States are higher than in other industrialized nations. Loss of pregnancy or death of a young infant is a serious life event, although pregnancy loss is often met by silence.

- Birth may be an occasion where women "do gender" as they attempt to maintain the niceness and politeness of femininity despite the physical and emotional stress of childbirth.

- According to the U.S. Census Bureau the more than 2 million adopted children, 2.5% of all children living with a parent. They are more likely to be female, are economically better off than those living with biological parents and, racially, adopted children are more likely to be African American or Asian compared to children who live with biological parents.

- Adoptive families face unique problems and stresses; nevertheless, most report feeling greatly enriched.

- The transition to parenthood is unlike other role transitions. It is irreversible and sudden, and it comes with little preparation.

- Reduced sexual desire and depression during the *postpartum period* are among the potential problematic reactions to childbirth. Teenage mothers are much more likely than adult mothers to suffer from postpartum depression.

- Parental roles can create considerable and multiple stresses. Both mothers and fathers face multiple role demands (parent, spouse, and provider). Other sources of stress are associated with not having enough money; worries about infant care and health; and interactions with spouse, family, and friends. The *continuous coverage* that infants require also introduces stress and potential conflict into the lives of new parents.

Key Terms

child-free marriages 364

continuous coverage
system 382

couvade 371

crude birthrate 362

fertility rate 362

infant mortality 373

medicalization of
childbirth 370

open adoption 377

postpartum period 379

spontaneous
abortion 372

sudden infant death
syndrome (SIDS) 373

total fertility rate 362

RESOURCES ON THE INTERNET

Companion Website for This Book

http://www.thomsonedu.com/sociology/strong

Gain an even better understanding of this chapter by going to the companion website for additional study resources. Take advantage of the Pre- and Post-Test quizzing tool, which is designed to help you grasp difficult concepts by referring you back to review specific pages in the chapter for questions you answer incorrectly. Use the flash cards to master key terms and check out the many other study aids you'll find there. Visit the Marriage and Family Resource Center on the site. You'll also find special features such as access to InfoTrac© College Edition (a database that allows you access to more than 18 million full-length articles from 5,000 periodicals and journals), as well as GSS Data and Census information to help you with your research projects and papers.

CHAPTER 11

Experiencing Parenthood: Roles and Relationships of Parents and Children

Outline

T	F		
T	F	1	A maternal instinct has been proved to exist in humans.
T	F	2	Employed mothers earn less than women without children.
T	F	3	Egalitarian marriages usually remain so after the birth of the first child.
T	F	4	Behavior of fathers has changed more than cultural beliefs about fatherhood.
T	F	5	Studies consistently show that regular day care by nonfamily members is detrimental to intellectual and social development.
T	F	6	Children of higher-earning families are less likely to be cared for by parents only or by other relatives and are more likely to be cared for by nonrelatives.
T	F	7	Children raised by authoritarian parents tend to be less cheerful, more moody, and more vulnerable to stress.
T	F	8	Children of gay or lesbian parents are likely to be gay themselves.
T	F	9	In situations such as a parent's serious illness or death or a parental divorce, children may become caregivers for their parents.
T	F	10	Many parents follow the advice of "experts" even though it conflicts with their own opinions, ideas, or beliefs.

You might wonder, "What is it like to be a parent and to raise children?" Journalist and novelist Anna Quindlen (1988) expresses how deep and broad her responsibility for her children is in the following way:

> I am aghast to find myself in such a position of power over two other people [her sons]. Their father and I have them in thrall simply by having produced them. We have the power to make them feel good or bad about themselves, which is the greatest power in the world. Ours will not be the only influence, but it is the earliest, the most ubiquitous, and potentially the most pernicious. Lovers and friends may make them blossom and bleed, but they move on to other lovers and friends. We are the only parents they will ever have.

Economist Sylvia Ann Hewlett (1992, 122) adds this:

> Responsible parenthood involves the expenditure of a great deal of energy and effort. Done properly it is a noisy, exhausting, joyous business that uses up a chunk of one's best energy and taps into prime time. Well developing children dramatically limit personal freedom and seriously interfere with the pursuit of an ambitious career. When psychiatrist David Guttmann talks about the "routine unexamined heroism of parenting," he is describing the manifold ways dedicated parents "surrender their own claims to personal omnipotentiality" in the wake of childbirth, conceding these instead to the new child.

Being Parents

Over the last four decades or so, major changes in society have profoundly influenced parental roles. Parents today cannot necessarily look to their own parents as models. Most mothers and fathers of today's children have some things in common with mothers and fathers

Answer Key for What Do You Think

1 False, see p. 390;   2 True, see p. 416;   3 False, see p. 406;   4 False, see p. 393;   5 False, see p. 397; 6 True, see p. 397;   7 True, see p. 403;   8 False, see p. 409;   9 True, see p. 415;   10 True, see p. 401.

Ready for Parenthood? The Insider's Test

Unlike marrying—a process that unfolds gradually as two people form, maintain, and intensify a coupled relationship before entering marriage—parenthood is more sudden. There is less opportunity to experience "being a parent" before having a child to care for. Most parents discover the great extent to which they were either unprepared or incorrectly prepared. Here, we present anonymously written, jointly crafted, preparation-for-parenthood "tests," parts of which have been circulated and posted widely on the Internet. The excerpts from the tests are from the website nokidding.net.

So, are you ready to have children? Take these tests and see.

The mess test. With your hands, smear peanut butter and grape jelly on the sofa and curtains. Now rub your hands in the wet flower bed and smear them on the walls. Cover the stains with crayons. Pee on your carpets and cloth-covered furniture just for fun. . . .

The grocery store test. Borrow one or two small animals (goats are best) and take them with you as you shop at the grocery store. Without the aid of a leash, always keep them in sight and pay for everything they eat or damage.

The dressing test. Obtain one large, unhappy octopus. Stuff it into a small net bag with large holes, making sure that all tentacles stay inside. Time allowed: all morning. . . .

The night test. Fill a cloth bag with 8–12 pounds of sand. Soak it thoroughly in water. At 8 p.m. begin to waltz and hum with the bag until 9 p.m. Lay down your bag and set your alarm for 10 p.m. Get up, pick up your bag, and sing every song you have ever heard. Make up about a dozen more songs and sing these until 4 a.m. Set the alarm for 5 a.m. Get up and make breakfast. Keep this up for 5 years. Look cheerful and alert. . . .

The patience test. Always repeat everything you say at least five times. Always repeat everything you say at least *five times.* Always repeat everything you say at least *five times. Always* repeat everything you say at least *five times. Always* repeat everything you say AT LEAST *five times. AL-WAYS* REPEAT EVERYTHING YOU SAY AT LEAST FIVE TIMES. . . .

The finance test. Go to the nearest drugstore. Set your wallet on the counter. Ask the clerk to help him/herself. Now proceed to the nearest food store. Go to the office and arrange for your paycheck to be directly deposited to the store's account. Purchase a newspaper. Go home and read it peacefully for the last time in your life.

The final assignment. Find a couple who already has a small child. Lecture them on how they can improve their discipline, patience, tolerance, toilet training, and child's table manners. Emphasize to them that they should never allow their children to run rampant. Enjoy this experience. It will be the last time you have all the answers.

Obviously, the preceding is meant (and ideally received) with humor. We hope, too, that the logic underlying these "tests" made you think because behind this humor is the reality that we can't necessarily envision the profound changes that accompany parenthood. In fact, as a society we do very little to prepare people for what parenthood entails.

SOURCE: http://www.nokidding.net, "Humor Page."

---

throughout history, such as the desire for their children's well-being. But in many areas they have had to chart a new course. Here we briefly review motherhood and fatherhood, highlighting some major changes of the last quarter century that have transformed the meaning and experience of each.

Motherhood

To many, a chapter about parenting would be assumed to be about mothers and children, since "parenting" and "nurturing" are treated as though they are syn-

onymous with "mothering." Furthermore, many women see motherhood as their "destiny." Given the choice of becoming mothers or not (with "not" made possible and more controllable through birth control and abortion), most women would choose to become mothers at some point in their lives and they would make this choice for positive reasons. Some women make no conscious choice; they become mothers without weighing their decision or considering its effect on their own lives and the lives of their children and partners. The potential negative consequences of a nonreflective decision—bitterness, frustration, anger, or depression—may be great. Yet it is possible that a woman's nonreflective decision will turn out to be "right" and that she will experience unique personal fulfillment as a result.

Although researchers are unable to find any purely instinctual motives for having children among humans, they recognize many social motives impelling women to become mothers. When a woman becomes a mother, she may feel that her identity as an adult is confirmed. Having a child of her own proves her womanliness because, from her earliest years, she has been trained to assume the role of mother. The stories a girl has heard, the games she has played, the textbooks she has read, the religion she has been taught, the television she has watched—all have socialized her for the mother role. Jessie Bernard (1982), a pioneer in family studies, writes, "The pain and anguish resulting from deprivation of an acquired desire for children are as real as the pain and anguish resulting from an instinctive one." Whatever the reason, most women choose to experience motherhood.

Still, in the face of mixed messages from the wider culture, motherhood leaves many women feeling ambivalent. Author Liz Koch summarized some of that ambivalence:

> We fear we will lose ourselves if we stay with our infants. We resist surrendering even to our newborns for fear of being swallowed up. We hear and accept both the conflicting advice that bonding with our babies is vital, and the opposite undermining message that to be a good mother, we must get away as soon and as often as possible. We hear that if we mother our own babies full time, we will have nothing to offer society, our husbands, ourselves, even our children. We fear isolation, lack of self-esteem, feelings of entrapment, of emotional and financial dependency. We fear that we will be left behind— empty arms, empty home, empty women, when our children grow away. . . . The reality is that in many ways contemporary America does not honor mothering.

Also, when sociologists Deirdre Johnston and Debra Swanson (2003) were interviewed for The Mothers Movement Online (http://www.mothersmovement .org/), they were asked why they undertook a study of the depictions of mothers and motherhood in popular magazines. Their comments reveal continued cultural ambiguity about motherhood:

> As mothers ourselves, we experience the tensions of balancing work and family. We are enmeshed in the myths of motherhood that create cultural ideals about who is a "good mother" and who is not. On days that we went into the office, we felt guilty, crying as we left our young children at childcare. On other days, we stayed home, watching the clock, waiting for each painful minute to go by, calling a spouse at the office, waiting desperately for an adult to walk through the front door.

Popular culture certainly contributes to these mixed feelings. To uncover some wider cultural messages about motherhood, Johnston and Swanson examined the portrayals of mothers in five magazines targeted to mothers: *Good Housekeeping, Family Circle, Parent's Magazine, Working Mother,* and *Family Fun.* These portrayals put both at-home mothers and employed mothers on the defensive, in difficult, no-win situations—a point we return to shortly.

Koch further observed that the "job" of mother is devalued because it is associated with "menial tasks of housekeeper, cook, laundry maid," and so on. Although seeming to speak specifically about at-home mothers, Koch's plea for greater recognition of the contributions made by mothers is as relevant for mothers employed outside the home as those who work at home "full-time." As Koch (1987) articulates we need to better celebrate the "special state" of motherhood:

> Being mothers is truly immersing ourselves in a special state, a moment to moment state of being. It is difficult to look at our day and measure success quantitatively. The day is successful when we have shared moments, built special threads of communication, looked deeply into our children's eyes and felt our hearts open. . . . It is important that we see our job as vitally important to our own growth, to our community, to society, and to world peace. Building family ties, helping healthy, loved children grow to maturity is a worthwhile pursuit. . . . The transmission of values is a significant reason to raise our own children. We are there to answer their questions and to show children, through our example, what is truly important to us.

The idea of a maternal instinct reflects a belief that mothering comes naturally to women. For women who struggle with the new roles and responsibilities that motherhood brings, such an idea can be frustrating and can produce guilt. Add to this the assumption that mothers instinctively or intuitively "know" how to nurture children, the lack of confidence by both parties in a father's ability to parent, and the inherent ability of women to breastfeed, and we can quickly see the enormous pressures that can face new mothers more than new fathers.

Compounding the situation are those ambiguous cultural expectations alluded to earlier. "Too much" mothering? "Not enough" mothering? What do children really need, and what should mothers give and do? Women receive unclear, often contradictory, messages. Furthermore, the standards against which mothers are judged (and come to judge themselves) are often unrealistic and idealized, putting women in a situation of comparing themselves to a model to which it is difficult to "measure up." Sociologist Sharon Hays refers to our cultural expectations of mothers as the **ideology of intensive mothering.** This ideology portrays mothers as the essential caregivers, who should be child centered, guided by experts, and emotionally absorbed in the labor-intensive, and financially demanding task of childrearing. As a result, mothers "see the child as innocent, pure, and beyond market pricing. They put the child's needs first, and they invest much of their time, labor, emotion, intellect, and money in their children" (Hays 1996). In today's cultural climate, this view of motherhood contrasts with the business market ideology of efficiency, rationality, time saving, and profit.

The "intensive mothering" ideology confronts mothers and women who contemplate motherhood with cultural contradictions. Living up to its standards is difficult even for stay-at-home mothers. For women employed outside the home, the ideology can provoke self-doubt, guilt, and a sense of being judged by others. As Hays notes, there is almost no woman who can resolve this cultural no-win situation. Women who forgo childbearing may be perceived as "cold" and "unfulfilled." An employed woman with children may be told she is selfishly neglecting her children. If she scales back her workload but stays in a job, she may be "mommy tracked," put in a less demanding but also less important and less upwardly mobile position. Finally, at-home mothers, in meeting the intensive mothering mandates, will be seen by some as "useless" or "unproductive" (Hays 1996). In Deirdre Johnston and Debra Swanson's research, popular culture depicts at-home mothers as somewhat incompetent and yet underrepresents employed mothers, rendering them less visible models of motherhood. This occurs despite the fact that more than 60% of mothers are in the paid labor force (Johnston and Swanson 2003). Clearly, neither employed nor at-home mothers are well served by their portrayal in popular culture.

Motherhood affects women's employment experiences, as shown in Chapter 12. One notable way that women are affected is in their earnings. Estimates differ, but it is clear that women with children earn less than their counterparts without children (Budig and England 2001). Plus, regardless of their employment status, the responsibilities of parenthood continue to fall more heavily upon women than upon men, even as children age and move into their teens (Kurz, 2002).

Fatherhood

Beginning in the mid-1990s, a number of books appeared on fathers and fatherhood (Blankenhorn 1995; Coltrane 1996; Gerson 1993; Popenoe 1996; Hawkins and Dollahite 1997) as the depth and breadth of male involvement or absence in the lives of their children became a source of increasing societal concern (Eggebeen and Knoester 2001). In all the commentary and analysis, however, we are left with something short of a consensus about the state of fatherhood in America. This is even evident in the ambiguity of the idea of fathering. When we speak of mothering a child, everyone knows what we mean: a process that involves nurturing and caring for, the physical and emotional well-being of the child almost daily for at least 18 consecutive years. The popular meaning of *fathering* is quite different— impregnating the child's mother.

Nurturing behavior by a father toward his child has not typically been referred to as *fathering*. As used today, the term *parenting* is intended to describe the child-tending behaviors of both mothers and fathers (Atkinson and Blackwelder 1993).

As we have seen, the father's traditional roles of provider and protector are *instrumental*; they satisfy the family's economic and physical needs. The mother's role in the traditional model is *expressive*; she gives emotional and psychological support to her family. However, the lines between these roles are becoming increasingly blurred because of economic pressures, women's expanded involvement in so-called instrumental tasks, and new societal expectations and desires.

From a developmental viewpoint, the father's importance to the family derives not only from his roles as a breadwinner or as a representative of society, connecting his family and his culture, but also from his role as a developer of self-control and autonomy in his children. Research indicates that although mothers are inclined to view both sons and daughters as "simply children" and to apply similar standards to both sexes, fathers tend to be more closely involved with their sons than with their daughters (Morgan, Lye, and Condran 1988; Smith and Morgan 1994). This involvement generally involves sharing activities rather than sharing feelings or confidences (Cancian 1989; Starrels 1994). This may place a daughter at a disadvantage because she has less opportunity to develop instrumental attitudes and behaviors. It may also be disadvantageous to a son, because it can limit the development of his expressive patterns and interests (Gilbert et al. 1982; Starrels 1994).

In analyzing today's fathers and today's families, we find a diversity of opinion and a range of experiences of fathering. There is evidence indicating that fathers have become more emotionally connected to and involved in the lives of their children (Eggebeen and Knoester 2001). Some commentators point proudly to our embracement of this "new father" model against which many men now measure themselves (Lamb 1986, 1993). Feminist ideology is credited with being influential in shifting the emphasis to a more expressive model of fathering, but many men pursue more involved versions of fatherhood as part of their own quest for deeper relationships with their children (Griswold 1993; Daly 1993). When pressed, most men today compare themselves favorably with their own fathers in both the quality and the quantity of involvement they have with their children. The new "nurturant father," as Michael Lamb (1997) refers to him, is able to participate in virtually all parenting practices (except, of course, gestation and lactation) and experience all the emotional states that mothers experience. It is clear that fathers can feel a connection to their infants that men were often thought to lack (Doyle 1994). Furthermore, *father involvement* has been reconceptualized to include the many ways fathers are influential participants in their children's development. Fathering activities such as communicating, teaching, caregiving, protecting, and sharing affection, are viewed as beneficial to the development and well-being of both children and adults (Palkovitz 1997; Hawkins and Dollahite 1997).

Although this new standard of fatherhood has been widely hailed, it is unclear how much it reflects actual behavior (LaRossa 1988; Gerson 1993). As described by Ralph LaRossa, the **culture of fatherhood** has clearly changed in the directions described here; it is less clear how much the **conduct of fatherhood** has kept pace. Also, when we look at how *fathers compare to mothers*, fathers are neither as involved with nor as close to their children, including their teenaged

© Christine Mendes/Buena Vista Photography

■ *Fathers are increasingly involved in parenting roles—not just playing with their children but also changing their diapers, bathing, dressing, feeding, and comforting them.*

children, as mothers are (Kurz 2002). This an important reminder that reality may be different from rhetoric when it comes to what people actually do or believe they should do in their families.

Although the subject of much positive commentary today's "new fathers" have also faced criticism from both traditional and less traditional sources. Critics who embrace a more traditional perspective question the efficacy and desirability of a fatherhood that becomes *too much* like motherhood. They advocate more traditional models of men as fathers (Blankenhorn 1995). Still others focus more narrowly on the behaviors of the most irresponsible fathers, especially those who, after divorce, neither provide the expected financial support nor even maintain contact with their children. Instead, as shown in Chapter 14, such fathers simply disappear. Other negative expressions of fathering can be seen in data on child abuse. When we control for mothers' and fathers' different levels of responsibility and time spent in childcare, males are more often physically abusive to their children (Gelles 1998). These negative aspects of fathering tarnish the cultural celebration of the new nurturant father.

One way to resolve the apparent contradiction between positive and negative depictions of today's fathers is to recognize the two sides of contemporary fatherhood. Frank Furstenberg Jr. (1988) differentiates between "good dads" and "bad dads." This **bifurcation of fatherhood** results from the declining division of labor in the family, especially the decline of the male good-provider role. By rejecting this narrower notion of a father's primary role as provider, some men felt "freed" from their sense of duty toward their spouses and children (especially children of ex-spouses); yet other men found that this liberated them to construct expanded, more expressive versions of fathering.

When sociologist Kathleen Gerson (1993) interviewed 138 fathers, she uncovered an interesting diversity in men's perceptions of their family roles. Roughly a third of her sample was *traditional,* identifying themselves largely in terms providing financially for their families. Another near-third conceptualized their role in the family and as fathers in deeply nurturant ways. The final third avoided involvement in childrearing because of how it would impose on their freedom and autonomy. They either had no children or were estranged from their children because they had divorced or separated from their children's mothers.

As the preceding examples show, it is difficult and potentially risky to generalize too widely about today's fathers. Although more of today's fathers may aim to be more broadly involved with their children than what they perceive fathers to have been in the past, and although most may recognize father involvement as beneficial, many are confused. They are unsure of what is expected of them. Because models of highly involved fathers are relatively new, many fathers today "focus on being a model to their children to create for them a new set of standards for *who the father is*" (Daly 1993). The creation of a new role understandably can provoke both doubt and anxiety.

Women often can't identify with and may not understand why men don't automatically "know" what to do with and for children. Such stresses between mothers and fathers are common, according to a study by the Families and Work Institute (Levine 1997; Martin 1993). Although men are often willing to "help out" their wives, this can pose a problem. Women often wish their partners would take on an equal share of the work rather than simply "helping." In assessing fathers, we shouldn't neglect other important aspects of fathering. Fathers still see their roles as breadwinners as making important contributions because doing so provides financial resources for the family (Cohen 1993). As shown shortly, most fathers are not as involved as most mothers; still, most are emotionally involved with their children.

It is clear, however, that fathers and mothers are not held to the same parenting standards or expectations for involvement with their children. For example, a sample of college students was told of a hypothetical employed parent who showed a lack of involvement in caring for his or her child. They rated fathers and mothers lower for behavior described as "home but uninvolved" compared to "uninvolved because of business trips." However, mothers were rated even more negatively than fathers for the lack of involvement at home (Riggs 2005). The cultural stereotype is that mothers are *supposed to be involved.*

What Fatherhood Means to Men

Over the last 15 years or so, a number of fathers have written books to help guide their peers through the joys and perils of more involved fatherhood. Psychologist and writer Jerrold Shapiro (1993) says, "Whether men have been enticed or cajoled, the fact is that we're around our kids a lot more. And when you're around your kids, you get to like it." More than a matter of liking it or not liking it, we might wonder what the consequences of fatherhood are for men. David Eggebeen and Chris Knoester (2001) raised this question, looking at the experiences of 5,226 men age 19–65, comparing fathers and nonfathers, and examining different "versions" or "settings" of fatherhood: men living with their own (biological or adopted) dependent children, men living apart from their dependent children, men whose children are independent adults, and men who are stepfathers. Interested specifically in psychological and physical health, men's social connections, their intergenerational familial ties, and their work behavior, Eggebeen and Knoester found the following.

Generally, there were not big differences between fathers and nonfathers in psychological and health dimensions. In the other three areas—social, intergenerational/familial, and occupational, there were "clear and compelling differences between fathers and nonfathers," as well as interesting differences across fatherhood settings (390). Men living with dependent children were significantly less likely to participate in social activities with friends or leisure pursuits. Men without children, men who lived away from their children, and men who lived with stepchildren attended church much less often than men who lived with their own biological or adopted children.

Fathers who lived with their own biological or adopted children were more likely to have regular contact with aging parents and adult siblings than were men without children or men with stepchildren. Even fathers who lived apart from their children had more frequent contact with parents and siblings, suggesting that "fatherhood tightens intergenerational family ties" (389).

Overall, Eggebeen and Knoester believe evidence reveals that fatherhood, clearly and "unequivocally," has the power to "profoundly shape the lives of men" (Eggebeen and Knoester 2001, 390).

Who Takes Care of the Children?

Childcare responsibility varies according to the marital status of parents and their employment roles and schedules. In a two-parent family, care for children is more the responsibility of mothers than fathers (Yeung et al. 2001). When we examine data on actual involvement in tasks associated with childcare or time spent with children, mothers are more involved in such tasks than fathers. In making such comparisons, it is helpful to differentiate between **engagement** with children, or time spent in *direct interaction* with a child across any number of different activities, and **accessibility,** or *availability to a child,* when the parent is at the same location but not in direct interaction (Yeung et al. 2001). Even though fathers' proportional involvement with children has increased, it is estimated that fathers' engagement with children is less than 45% that of mothers' and their accessibility to children is less than 66% that of mothers' (Yeung et al. 2001).

Circumstances affect how much time fathers spend with children. One study, based on analyses of data from the Panel Study of Income Dynamics (Yeung, et al. 2001), noted that in two-parent homes a child's direct engagement with biological fathers ranged from an average of 1 hour and 13 minutes on weekdays to 2 hours and 29 minutes on weekends. The total time (engagement plus accessibility) these fathers h are involved with their children 12 years and younger is roughly 2.5 hours a day on weekdays and 6.5 hours a day on weekends.

Some interesting differences can be observed in fathers' time with children. For children who live with only their mothers (with or without a stepfather), the time spent with their biological fathers averaged 5 minutes a day on weekdays and 21 minutes a day on weekends. For children living with only their biological fathers (with or without a stepmother), the time spent with fathers averaged 64 minutes a day on weekdays and 90 minutes a day on weekends (Yeung et al. 2001).

Active Childcare

Active, hands-on childcare is more "in the hands" of mothers than fathers. Mothers take care of and think about their children more than fathers do (Walzer 1998). In most two-parent households, mothers' childcare responsibility and involvement greatly exceed fathers' involvement (Bird 1997; Aldous and Mulligan 2002). For every hour that fathers spend actively involved with their children, mothers spend between 3 and 5 hours (Bird 1997).

What do mothers and fathers do in the time they spend with children? Research from the 1960s through the 1980s suggested that fathers spend more time in interactive activities, such as play or helping with homework, whereas mothers spend time doing custodial childcare, such as feeding and cleaning (Yeung et al. 2001).

Also, fathers are more involved with sons than daughters, with younger children more than older children, and with firstborn more than with later born children (Pleck 1997, cited in Doherty, Kouneski, and Erickson 1998). Fathers are engaged with or accessible to their infants and toddlers an average of a little over 3 hours per day during the week. By ages 9–12, the combined (engagement and accessibility) weekday time between fathers and children declined to 2 hours and 15 minutes (Yeung et al. 2001). Fathers spend 18 minutes more per day in play and companionship activities with sons than with daughters during the week.

Research suggests that fathers who work more hours and who have prestigious but time-demanding

Exploring Diversity Beyond the Stereotypes of Young African American Fathers

Among the many familiar stereotypes that persist in the United States is that of the "dysfunctional and deviant young African American male" (Smith et al. 2005). The image of young African American fathers "as sexual predators likely to abandon their children and the child's mother," has "seeped into the nation's conscience . . . influencing public policy on public assistance and related issues" (Smith et al. 2005, 977). Yet there are men like 18-year-old Terrell Pough, named by *People* magazine as an "outstanding father" in a feature story in August 2005. Pough was described as a "rare breed of teenaged dads who are trying to raise children." A devoted father to his daughter, Diamond, who was not yet 2 years old, Pough juggled finishing high school, working, and caring for Diamond, of whom he had custody, when featured by the magazine. As he told the magazine, "She's what I work for, what I live for, why I wake up. . . . She's everything." Pough asserted his determination that "If something ever happens to me . . . no one can ever tell her that her dad didn't take care of her." Tragically, something did happen to Pough. He was shot to death while returning home from work November 17, 2005.

According to research by Carolyn Smith, Marvin Krohn, Rebekah Chu, and Oscar Best, although Pough may have been exceptional in his dedication and sacrifice, his commitment to his daughter may be more representative of young, single African American fathers than the negative stereotypes. Using data from the

Rochester Youth Development Study, a longitudinal study that followed 1,000 seventh- and eighth-grade adolescents over a number of years, Smith and colleagues focused on the experiences of 193 young fathers, 67.4% of whom were African Americans, 20.7% Hispanics, and 11.9% whites. Interested in the extent of a father's contact and involvement and the matter of financial support of his child or children, Smith and colleagues offered the following findings.

Approximately 33% of the African American fathers reported that they live with their child. Although the ethnic differences are not statistically significant, this percentage is higher than that of Hispanics (25.9%) but less than that of Caucasians (45.5%). Even among the nonresident fathers, 61.8% of the African American men reported "at least weekly" contact, an amount not widely different from that of Caucasians (67.7%) or Hispanics (54.3%). Only 11.4% of African American fathers reported "no contact," slightly more than the percentage among Caucasians (9.3%) but less than among Hispanics (15.5%).

Looking at the extent to which non-resident nonresident fathers pro-

vide financial support for their children, revealed the following patterns.

Although again not statistically significant (largely because of sample sizes), the data suggest that the levels of support provided by nonresident African American fathers was about the same as that of Hispanic fathers. Combining this finding with the data on contact reveals two important points: (1) African American fathers are more similar to than different from Hispanic fathers and, in terms of contact, not that different from Caucasians. In both the amount of contact with and financial support for their children, these nonresident fathers *do not fit* the racial stereotype.

Based on research findings such as these, we need to reconsider the stereotype of uninvolved and irresponsible young black fathers. Even when the majority of fathers were not living with their oldest child, many had regular contact and two-thirds provided some to all of the financial support as arranged. No doubt, there are still men who make and maintain no commitment to their children. However, they can be found among all races and are not the norm among men of any particular race.

SOURCE: Smith et al. 2005, 975–1,001.

Table 11.1 ■ Financial Support for Children Provided by Nonresident Fathers, by Race		
	African American	**Hispanic**	**Caucasian**
No support arranged or 0% paid	33.2%	36.4%	17.7%
1% to 99% of arranged support	12.6%	9.5%	—
100% of arranged support	54.2%	54.1%	82.3%

occupations tend to be less engaged in childrearing (NICHD Early Child Care Research Network 2000). On weekends, fathers become somewhat more equal caregivers, and their involvement is greater when mothers contribute a "substantial" portion of the family income (Yeung et al. 2001). Although fathers "help" mothers with the caregiving work and supervision involved in raising teenaged children, most fathers do less of such work than most mothers (Kurz 2002).

Mental Childcare

Responsibility for childcare doesn't only consist of what we *do* with and for our children. In her book *Thinking about the Baby: Gender and Transitions into Parenthood,* sociologist Susan Walzer (1998) examines the division of responsibility for infants in 25 two-parent households. Her focus is less on "who does what" with their children than on "who thinks what, and how often" about their children. Walzer identifies this "invisible" parenting as **mental labor**—the process of worrying about the baby, seeking and processing information about infants and their needs, and managing the division of infant care in the household (that is, seeking the "assistance" of their spouse).

Sociologist Demie Kurz reports similar kinds of mental labor among mothers of teenaged children. Fearful for their adolescents' safety, and especially fearful about the sexual vulnerability of their teenaged daughters, mothers worry (Kurz 2002). Thus, mothers continue to worry as children grow.

Key to understanding this mental labor at both the earliest and the late adolescent or young adult stages is that mothers feel responsible for and judged by what happens to their children in ways that most fathers do not.

Nonparental Childcare

Day Care and Supplemental Childcare

Discussions of who cares for children cannot begin and end just with parents. Supplementary childcare is a crucial issue for today's parents of young children. Given the prevalence of two-earner households (addressed more in Chapter 12) and single-parent households, many parents must look outside their homes for assistance in childrearing. In 2001, 63% of married women with children younger than 6 years were in the labor force. Also in the paid labor force were 70% of never-married mothers of preschool-age children and 76% of divorced widowed or separated women with preschool-age children (U.S. Census Bureau 2002, Table 570). In 2001, 58% of married mothers with husbands present and children under 1 year of age were employed outside the home (U.S. Census Bureau 2002, Table 571). The combination of trends in employment status, marital status, and childbearing has increased the need for outside caregivers.

Despite the clear need for quality childcare, the United States compares poorly to many European countries. Take, for example, France, where childcare is publicly funded as part of early education (Clawson and Gerstel 2002). Nearly all 3- to 5-year-olds are enrolled in full-day programs taught by well-paid teachers.

■ *Mothers do more mental labor, including worrying, involved with raising their young and teenaged children.*

© David Young Wolff/PhotoEdit

Based upon sample estimates from the National Center for Education Statistics, 77% of the more than 8 million 3- to 5-year-olds in the United States are in some form of nonparental childcare (U.S. Census Bureau 2002, Table 550). This varies by age of child, as 31% of 3-year-olds, compared with 18% of 4-year-olds, and 13.5% of 5-year-olds are in parental care only. Income also makes a difference, as children of higher-earning families are less likely to be cared for by parents or other relatives only and are more likely to be cared for by nonrelatives.

Most experts agree that the ideal environment for raising a child is in the home with the parents and family. Intimate daily parental care of infants for the first several months to a year is particularly important. Because this ideal is often not possible, the role of day care needs to be considered. Day care homes and centers, nursery schools, and preschools can relieve parents of some of their childrearing tasks and furnish them with some valuable time of their own. Among children in nonrelative care, about 7% are looked after in their own homes. Family day care enrolls about 27%, and centers about 66% (National Household Education Survey 2001, in Wrigley and Dreby 2005).

What is the effect of early outside childcare on children? The results of research are mixed. In evaluating such data, it is important to keep in mind the family's education, the personalities involved, and the family interests—key factors that play a part in which parents choose to return to work and which must return to work once a child is born (Crouter and McHale 1993). Furthermore, a child's personality, the child's age when the custodial parent reentered the workforce, the involvement of the other parent in the home, the quantity of time spent working or with the child, the nature of the work, and the quality of care all contribute to how childcare affects the child.

When mothers of infants enter the workforce, there is some evidence that these infants are at risk for insecure attachments between the ages of 12 and 18 months (Brooks 1996). They are also at risk for being considered noncompliant and aggressive between 3 and 8 years of age (Howes 1990). Other consequences, such as behavior problems, lowered cognitive performance, distractibility, and inability to focus attention, have been noted. These negative effects are not necessarily the consequence of being cared for by outside caregivers. Rather, they may be the result of *poor-quality* childcare. It has been noted that high-quality care, given by sensitive, responsive, and stimulating caregivers in a safe environment with low teacher-to-student ratio, can actually facilitate the development of positive social qualities, consideration, and independence (Field 1991). In school-age and adolescent children, maternal employment is associated with self-confidence and independence, especially for girls whose mothers become role models of competence (Hoffman 1979).

■ *As more women return to the workforce, a critical issue is the quality of the day care for their children. High-quality day care can facilitate the development of positive social qualities.*

National concern periodically is focused on day care by revelations of sexual abuse of children by their caregivers. Although these revelations have brought providers of childcare under close public scrutiny and have alerted parents to potential dangers, they have also produced a backlash within the childcare profession. Some caregivers are now reluctant to have physical contact with the children; male childcare workers feel especially constrained and may find their jobs at risk. However, children have a far greater likelihood of being sexually abused by a father, stepfather, or other relative than by a day care worker.

Parents with children in childcare should take some degree of comfort from the evidence demonstrating that children in outside, especially in organized, childcare facilities are safe. Overall, all types of childcare are safer than care within children's own families (Finkelhor and Ormrod 2001; Wrigley and Dreby 2005).

According to a 2005 study, between 1985 and 2003 more than 1,300 children died while in childcare (Wrigley and Dreby 2005). Of these, only 110 were in center care. The total number of fatalities that occurred in "home-based care" numbered 1,030: 270 in the child's home, 656 in the caregiver's home, and another 104 cases that occurred in private homes that were undesignated as to whose homes they were.

Of those infants who died from violence in home-based care settings, more than 90% of the acts were perpetrated by caregivers; more than 60% of the deaths were the result of shaking. What can parents do to ensure quality care for their young children? In addition to the obvious requirements of cleanliness, comfort, nutritious food, and a safe environment, parents should be familiar with the state licensure regulations for childcare. They should also check references and observe the caregivers with the child. Although the needs of young children differ from those of older ones, the American Academy of Child and Adolescent Psychiatry (1992) suggests that parents seek day care services that meet specific standards:

- More adults per child than older children require
- A lot of individual attention provided for each child
- Trained, experienced teachers who understand, praise, and enjoy children
- The same day care staff for a long period
- Opportunity for creative work, imaginative play, and physical activity
- Space to move indoors and out

- Enough teachers and assistants—ideally, at least 1 for every 5 (or fewer) children (studies have shown that 5 children with one caregiver is better than 20 children with four caregivers)
- An ample supply of drawing and coloring materials and toys, as well as equipment such as swings, wagons, and jungle gyms
- Small rather than large groups if possible

Finally, if the child shows persistent fear about leaving home, parents should discuss the problem with the childcare provider and their pediatrician.

As with a number of critical services in our society, those who most need supplementary childcare are those who can least afford it. The United States is one of the few industrialized nations without a comprehensive national day care policy. In fact, beginning in 1981, the federal government dramatically cut federal contributions to day care; many state governments followed suit.

School-Age Childcare and Self-Care

Although there are particularly acute needs when children are young, the need for childcare is not restricted to families of preschoolers. We need to pay attention to the circumstances confronting parents of children in middle school. A number of terms used to refer to caregiving for these older children, including *after-school, around school, out-of-school,* and *school-age care* (Polatnik 2002).

Many children express strong opposition to after-school programs, seeing them as geared toward "little kids", but they find activities such as sports or other recreational or artistic programs more appealing (Polatnik 2002). Unfortunately, even when the programs and activities are free or when the costs are affordable, they are neither consistent nor continuous enough to cover the whole time children are out of school before parents return from work. Many parents of these children feel pressed to allow them to stay home alone. Research indicates that approximately a third of 11- to 12-year-olds are in **self-care**—that is, care for themselves without supervision by an adult or older adolescent (Hochschild 1997; Polatnik 2002; Casper and Smith 2002).

Self-care increased through the 1980s and 1990s, and some estimates of children in self-care range as high as 7 million, including 0.5 million preschoolers (Hewlett and West 1998). In fact, self-care is rarely used for very young children. Lynn Casper and Kristin Smith (2002) report that 3% of 5- to 7-year-olds are in self-care. The percentage increases to 11% among

8- to 10-year-olds and jumps to 33% among 11- to 13-year-olds.

Self-care exists in families of all socioeconomic classes, although—contrary to stereotypes—higher income parents are more likely to allow their children to remain in self-care than are lower-income parents. After age 7 or 8, Caucasian children are more likely than either African Americans or Hispanics to be in self-care (Casper and Smith 2002).

Parents need to evaluate whether self-care is appropriate for their children. Ideally, parents and educators together would see to it that children develop such self-care skills as basic safety, time management, and other self-reliance skills, before being faced with actually having to care for themselves (Polatnik 2002, 745).

Raising Children: Theories of Socialization, Advice to Parents, and Styles of Parenting

Attitudes and beliefs about parenting flow from attitudes and beliefs about children and their development. Current attitudes about children still reflect the influence of a number of psychological theories concerning child socialization. Ultimately, as we will see, these theories have been influential in shaping some of the parenting advice offered by prominent authors in their childrearing advice books.

Psychological theories of human development give prime importance to the role of the mind, particularly the subconscious mind, which, according to psychoanalytic theory, motivates much of our behavior without our being consciously aware of the process. According to these theories, many aspects of our psychological makeup are inborn; our minds grow and develop with our bodies.

Psychological Theories

Psychoanalytic Theory

The emphasis by Sigmund Freud (1856–1939) on the importance of unconscious mental processes and on the stages of psychosexual development has greatly influenced modern psychology. Freud's **psychoanalytic theory** of personality development holds that we are driven by instinct to seek pleasure, especially sexual pleasure. This part of the personality, called the **id,** is kept in check by the **superego**—what we might call the conscience. The third component of personality, the rational **ego,** mediates between the demands of the id and the constraints of society. Freudian theory views the uninhibited id of the infant as gradually becoming controlled as the individual internalizes societal restraints. Too much restraint, however, leads to repression and the development of **neuroses**—psychological disorders characterized by anxiety, phobias, and so on.

Freud viewed the parents as the primary force responsible for the child's psychological development. He posited that between the ages of 4 and 6 years, the child identifies with the parent who is of the same sex. Not becoming like that parent was seen as a failure to reach maturity. Freud divided **psychosexual development** into five stages spanning the time from birth through adolescence: (1) oral, (2) anal, (3) phallic, (4) latency, and (5) genital (Table 11. 2).

Psychosocial Theory

Erik Erikson (1902–1994) based much of his work on psychoanalytic theory, but he emphasized the effects of society on the developing ego, creating a model that has come to be known as **psychosocial theory** (Erikson, 1963). Stressing parental and societal responsibilities in children's development, each of Erikson's life cycle stages (see Table 11.2 and Chapter 9) is centered on a specific emotional concern based on individual biological influences and external sociocultural expectations and actions.

Learning Theories

Learning theorists emphasize the aspects of behavior that are acquired rather than inborn or instinctual. Return to Chapter 4 to review behaviorism, which explains human behaviors entirely on the basis of what can be observed, and social learning theory, which emphasizes the role of cognition, or thinking, in learning.

Cognitive Development Theory

Beginning in the 1930s, Swiss psychologist Jean Piaget (1896–1980) began intensively observing and interviewing children, formulating what has become known as *cognitive development theory* (see Chapter 4). Piaget suggested that cognitive development occurs in discrete stages for all infants and children. These stages are linked to the development of the brain and the nervous system, and can be seen as building blocks, each of which must be completed before the next one can be put into place. In Piaget's view, children develop their cognitive abilities through interaction with the world and adaptation to their surroundings. Children adapt by **assimilation** (making new information compatible with their world understanding) and **accommodation** (adjusting their cognitive framework to incorporate new experiences) (Dworetsky, 1990). Piaget identified four stages of cognitive development: (1) sensorimotor, (2) preoperational, (3) concrete operational, and (4) formal operational (see Table 11.2).

The Developmental Systems Approach

Parents do not simply give birth to children and then "bring them up." According to the **developmental systems approach,** the growth and development of children takes place within a complex and changing family system that both influences and is influenced by the child. The family system is part of a number of larger systems (extended family, friends, health care, education, and local and national government, to name a few), all of which mutually interact. Many models or theories that use a developmental systems approach including Urie Bronfenbrenner's (1979) ecological model, discussed in Chapter 2.

Parent-Child Interactions

Children also are socializers in their own right. When an infant cries to be picked up and held, to have a diaper changed, or to be burped, or when he or she smiles when being played with, fed, or cuddled, the parents are being socialized. The child is creating strong bonds with the parents (see the discussion of attachment later in Chapter 5). Although the infant's actions are not at first consciously directed toward reinforcing parental behavior, they nevertheless have that effect. In this sense, even very young children can be viewed as participants in creating their own environment and in contributing to their further development (see Peterson and Rollins, 1987).

In the developmental systems model of family growth, social and psychological development are seen as lifelong processes, with each family member having a role in the development of the others. In terms of the eight developmental stages of the human life cycle described by Erikson, parents are generally at the seventh stage (generativity) during their children's growing years, and the children are probably anywhere from the first stage (trust) to the fifth (identity) or sixth (intimacy). The parents' need to establish their generativity is at least partly met by the child's need to be cared for and taught. The parents' approach to childrearing will inevitably be modified by the child's inherent nature or temperament.

Sibling Interactions

More than 80% of American children have one or more siblings. Siblings influence one another according to their particular needs and personalities. They are also significant agents for socialization. Although rivalry and aggression may appear to be the foundation of such interactions, young siblings at home spend a large percentage of their time actually playing together.

The quality of sibling interaction may have consequences for the child's later behavior (Newcombe,

Table 11.2 ■ Stages of Development: Freud, Piaget, and Erikson Compared

	Freud	Piaget	Erikson
Infancy	Oral	Sensorimotor	Trust versus mistrust
Toddler	Anal		Autonomy versus shame and doubt
Early childhood	Phallic	Preoperational	Initiative versus guilt
Late–middle childhood	Latency	Concrete operational	Industry versus inferiority
Adolescence	Genital	Formal operational	Identity versus confusion
Early adulthood			Intimacy versus isolation
Middle adulthood			Generativity versus stagnation
Late adulthood			Ego Integrity versus despair

1996). Close, affectionate sibling relationships contribute to the development of desirable characteristics such as social sensitivity, communication skills, cooperation, and understanding of social roles. Moreover, sibling relationships continue to be meaningful well into adulthood. As examined by Shelley Eriksen and Naomi Gerste (2002), adult siblings have perhaps the "most egalitarian" of all family relationships, and provide each other with a variety of supportive resources throughout their adult lives. Relationships between sisters or between brothers are often much like friendships, and sisters are especially close with each other.

Symbolic Interaction Theory

Symbolic interaction theory is the sociological theory that most applies to the process of socialization. The ways in which this theory explains partner relationships was discussed in Chapter 2; here, we focus on how the theory pertains to development.

Symbolic interactionists such as Charles Horton Cooley and George Herbert Mead stressed the processes through which we develop a *social self,* the sense of who we are and how we are perceived by those around us. To interactionists, the self is not with us at birth but emerges out of interactions with others. In Cooley's formulation, three key components comprise the **looking-glass self,** the self-concept that develops from our sense of how others view us. First, we imagine how others perceive us. Second, we draw conclusions about how others judge us. And third, based on these, we develop our ideas about ourselves (Henslin, 2000).

Mead emphasized that the self consists of both an active, spontaneous part (the "I") and a more passive, acted upon part (the "me"), in which we see ourselves as an object of other people's actions toward us (Henslin, 2000). This social self develops early in life and can be seen in the developing sophistication of children's play. Play forces children to see things from someone else's vantage point, what Mead called **taking the role of the other.** Mead noted that until about age 3 children really don't "play" but rather engage in imitative behavior. In the **play stage** (3 to 6 years old), children play at being specific individuals, often by dressing up. By the **game stage,** they have developed sufficient self-awareness to be able to simultaneously take into account multiple perspectives and anticipate how other players might act in a given situation.

In symbolic interactionist terms, family members, especially parents, are among the more "significant" significant others in influencing the opinions we form of ourselves. They are perhaps the purest example of what Cooley called *primary groups,* characterized by intimate, face-to-face interaction, and crucial in the development of our social selves.

From the Theoretical to the Practical: Expert Advice on Childrearing

About 150 years ago, Americans began turning to books rather than one another to learn how to act and live. They began to lose confidence in their abilities to make appropriate judgments about childrearing.

The vacuum that formed when traditional ways broke down under the effect of industrialization was filled by the so-called "experts" who dispensed their wisdom through books, radio, and TV. The old values and ways had been handed down from parents to child in an unending cycle, but with increasing mobility, this continuity between generations ceased and parents increasing turned to these experts for help.

Contemporary parents, too, are surrounded by expert advice, some of which may conflict with their own beliefs. If an expert's advice counters their understanding, parents should critically examine that advice, as well as evaluate their own beliefs.

Twentieth-century parenting was shaped by childrearing advice from such notable authorities as Benjamin Spock, T. Berry Brazelton, and Penelope Leach. These three authors sold well over 40 million copies of their books advising parents, especially mothers, as to the best ways to raise their children. Building on psychological theories of development, they stressed the importance of parents understanding their child's cognitive and emotional development.

So what do these experts advocate as effective parenting? Sharon Hays (1996) suggests that they all advocate the ideology of intensive mothering, discussed earlier in this chapter. Aside from the belief in the special nurturing capacities of mothers, this ideology contains the following assumptions about what children need from parents:

■ Raising children is and should be an emotionally absorbing experience characterized by affectionate nurture. Emotional attachment is essential for healthy development; parental unconditional love and loving nurture are seen as critical to the child, no less essential, Spock asserts, than "vitamins and calories" (Spock 1985, quoted in Hays 1996).

- It is the mother's job to respond to the needs and wants of her child. Parents should follow the cues given by their child, submit to the child's desires, and understand "what every baby knows" it needs from its parent (Brazelton 1987, quoted in Hays 1996). This requires knowledge of children's needs and developmental phases, as well as great parental sensitivity.

- Parents must develop sensitivity to the particular needs of their child. This includes, for example, recognizing the different meanings of the child's crying and understanding the unique and individual developmental pattern of the child.

- Physical punishment is frowned upon. Instead, setting limits, providing a good example of what parents expect from their child, and giving the child lots of love are preferred ways to convince the child to internalize and act upon parents' standards. Punishment consists of "carefully managed temporary withdrawal of loving attention," a labor-intensive, emotionally absorbing method of discipline. Once a child can question, parents are urged to reason with the child, negotiate, and discuss motives and alternative ways of acting. This strategy obviously involves more time and effort than spanking.

Contemporary Childrearing Strategies

One of the most challenging aspects of childrearing is knowing how to change, stop, encourage, or otherwise influence children's behavior. We can request, reason, command, cajole, compromise, yell, or threaten with physical punishment or the suspension of privileges; alternatively, we can just get down on our knees and beg. Some of these approaches may be appropriate at certain times; others clearly are never appropriate. The techniques of childrearing currently taught or endorsed by educators, psychologists, and others involved with child development differ somewhat in their emphasis but share most of the tenets that follow:

- *Respect.* Mutual respect between children and parents must be fostered for growth and change to occur. One important way to teach respect is through modeling—treating the child and others respectfully. Counselor Jane Nelsen (1987) writes, "Kindness is important in order to show respect for

the child. Firmness is important in order to show respect for ourselves and the situation."

- *Consistency and clarity.* Consistency is crucial in childrearing. Without it, children become hopelessly confused and parents become hopelessly frustrated. Patience and teamwork (maintaining a united front when there are two parents) on the parents' part help ensure consistency. Parents should beware of making promises or threats they won't be able to keep, and a child needs to know the rules and the consequences for breaking them.

- *Logical consequences.* One of the most effective ways to learn is by experiencing the logical consequences of our actions. Some of these consequences occur naturally—if you forget your umbrella on a rainy day, you are likely to get wet. Sometimes parents need to devise consequences appropriate to their child's misbehavior. Rudolph Dreikurs and Vicki Soltz (1964) distinguish between logical consequences and punishment. The "three R's" of logical consequences dictate that the solution must be *related* to the problem behavior, *respectful* (no humiliation), and *reasonable* (designed to teach, not to induce suffering).

- *Open communication.* The lines of communication between parents and children must be kept open. Numerous techniques exist for fostering communication. Among these are active listening and the use of "I" messages. In *active listening,* the parent verbally reflects the child's communications to confirm they have a mutual understanding. *"I" messages* are important because they impart facts without placing blame and are less likely to promote rebellion in children than are "you" messages. Also, regular weekly *family meetings* provide an opportunity to be together and air gripes, solve problems, and plan activities.

- *No physical punishment.* Many physicians, psychologists, and sociologists have become harsh and vocal critics of physical punishment. Both the American Psychological Association and the American Medical Association oppose physical punishment of children. Many sociologists, most notably scholars who study family violence, such as Murray Straus, oppose corporal punishment; they note that it is related to later aggressive behavior from children, including later perpetration of spousal violence (Straus and Yodanis 1996). Although such punishment is used widely (Straus and Yodanis estimate more than 90% of parents of toddlers use corporal punishment) and may "work" in the short run by stopping undesirable behavior, its long-range results—anger, resentment, fear, hatred, aggressiveness, family violence—may be extremely problematic (Dodson 1987; Straus and Yodanis 1996; McLoyd and Smith 2002). Besides, it often makes parents feel confused, miserable, and degraded right along with their children.

- *Behavior modification.* Effective types of discipline use some form of behavior modification. Rewards (hugs, stickers, or special activities) are given for good behavior, and privileges are taken away when misbehavior is involved. Good behavior can be kept track of on a simple chart listing one or several of the desired behaviors. Time-outs—sending the child to his or her room or to a "boring" place for a short time or until the misbehavior stops—are useful for particularly disruptive behaviors. They also give the parent an opportunity to cool off (Dodson 1987; see also Canter and Canter 1985).

Styles of Childrearing

Authoritarian, Permissive, Authoritative, and Uninvolved Parents

A parent's approach to training, teaching, nurturing, and helping a child will vary according to cultural influences, the parent's personality, the parent's basic attitude toward children and childrearing, and the role model that the parent presents to the child.

One popular formulation contrasts four basic styles of childrearing: authoritarian, permissive or indulgent, authoritative, and uninvolved (Baumrind 1971, 1983, 1991). *Style of parenting* refers to variations between parents in their efforts to socialize and control their child (Baumrind 1991). All four styles are part of the normal variation among parents. Thus, although research tends to identify one of the following as more effective than the others, none of them is abusive or deviant (Davis 1999).

Parents who practice **authoritarian childrearing** typically require absolute obedience. The parents' ability to maintain control is of primary importance. "Because I said so" is a typical response to a child's questioning of parental authority, and physical force may be used to ensure obedience. Working-class families tend to be more authoritarian than middle-class families. Diana Baumrind (1983) found that children

of authoritarian parents tend to be less cheerful than other children and correspondingly more moody, passively hostile, and vulnerable to stress.

Permissive or **indulgent childrearing** is a more popular style in middle-class families than in working-class families. The child's freedom of expression and autonomy are valued. Permissive parents rely on reasoning and explanations. Yet permissive parents may find themselves resorting to manipulation and justification. The child is free from external restraints but not from internal ones. The child is supposedly free because he or she conforms "willingly," but such freedom is not authentic. Although children of permissive parents are generally cheerful, they exhibit low levels of self-reliance and self-control (Baumrind 1983).

Parents who favor **authoritative childrearing** rely on positive reinforcement and infrequent use of punishment. They direct the child in a manner that shows awareness of his or her feelings and capabilities. Parents encourage the development of the child's autonomy within reasonable limits and foster an atmosphere of give-and-take in parent–child communication. Parental support is a crucial ingredient in child socialization. It is positively related to cognitive development, self-control, self-esteem, moral behavior, conformity to adult standards, and academic achievement (Gecas and Seff 1991). Control is exercised in conjunction with support by authoritative parents.

Finally, **uninvolved parenting** refers to parents who are neither responsive to their children's needs nor demanding of them in their behavioral expectations. Children and adolescents of uninvolved parents suffer consequences in each of the following areas or domains: social competence, academic performance, psychosocial development, and problem behavior (Davis 1999).

Much research points to the authoritative style as especially effective. Children raised by authoritative parents tend to approach novel or stressful situations with curiosity and show high levels of self-reliance, self-control, cheerfulness, and friendliness (Baumrind 1983).

Even bigger differences, however, are found between children of more involved parents as opposed to unengaged parents (Davis 1999).

Reflections

In your family, what childrearing attitudes (authoritarian, permissive, or authoritative) predominated? Do you think these attitudes influenced your development? If so, how? Which might (or do) you find useful in raising your child?

What Influences Child Development?

Although the relative effects of physiology and environment on human development are still often much debated by today's experts, it is clear that *both* nature and nurture play important roles in children's development. In addition to biological factors, important factors affecting early development include the formation of attachments (especially maternal) and individual temperamental differences.

Biological Factors

According to biological determinists, much of human behavior is guided by genetic makeup, physiological maturation, and neurological functioning. Psychologist Jerome Kagan (1984) presented a strong case for the role of biology in early development. He asserted that the growth of the central nervous system in infants and young children ensures that motor and cognitive abilities such as walking, talking, using symbols, and becoming self-aware will occur "as long as children are growing in any reasonably varied environment where minimal nutritional needs are met and [they] can exercise emerging abilities." Furthermore, according to Kagan, children are biologically equipped for understanding the meaning of right and wrong by the age of 2 years, but although biology may be responsible for the development of conscience, social factors can encourage its decline.

Individual Temperament

Most parents with more than one child will tell you of the differences between their children that were evident almost from the moment of birth. Even parents of an only child will recount how their child seemed to come with a personality. A child's unique temperament, such as "inhibited/restrained/watchful" or "uninhibited/energetic/spontaneous," influences the way in which he or she develops (Kagan 1984). Temperamental differences may be rooted in the biology of the brain (Kagan and Snidman 1991), but temperament is also developed by interaction with the environment. For example, a baby who is vigorous, strong, and outgoing will probably encourage her parents to reinforce the lively, extroverted, and spontaneous aspects of her personality. An infant who is shy, fearful, and cries easily, however, may inhibit them from interacting with him, thus causing him to become more shy and fearful. It is important for parents to understand "how they create the meaning of the child's individuality by their own temperaments, and their demands, attitudes, and evaluations," according to psychologists Richard and Jacqueline Lerner (Brooks 1994). Lerner and Lerner stress that if parents are sensitive to a child's unique temperament, they are better able to seek appropriate ways to influence the child's behavior.

What Do Children Need?

Parents often want to know what they can do to raise healthy children. Are there specific parental behaviors or amounts of behaviors (say 12 hugs, three smiles, a kiss, and a half hour of conversation) that all children need to grow up healthy? Of course not. Apart from saying that basic physical needs must be met (adequate food, shelter, clothing, and so on), along with some basic psychological ones, experts cannot give parents such detailed instructions.

Noted physician Melvin Konner (1991) lists the following needs for optimal child development—which, he writes, "parents, teachers, doctors, and child development experts with many different perspectives can fairly well agree on":

- Adequate prenatal nutrition and care
- Appropriate stimulation and care of newborns
- Formation of at least one close attachment during the first 5 years
- Support for the family "under pressure from an uncaring world," including childcare when a parent or parents must work
- Protection from illness
- Freedom from physical and sexual abuse
- Supportive friends, both adults and children
- Respect for the child's individuality and the presentation of appropriate challenges leading to competence
- Safe, nurturing, and challenging schooling
- Adolescence "free of pressure to grow up too fast, yet respectful of natural biological transformations"
- Protection from premature parenthood

In today's society, especially in the absence of adequate health care and schools in so many communities, it is difficult to see how even these minimal needs can all be met. Even when the necessary social supports are present, parents may find themselves confused, discouraged, or guilty because they do not live up to their own expectations of perfection.

Yet children have more resiliency and resourcefulness than we may ordinarily think. They can adapt to and overcome many difficult situations. A mother can lose her temper and scream at her child, and the child will most likely survive, especially if the mother later apologizes and shares her feelings with the child. A father can turn his child away with a grunt because he is too tired to listen, and the child will not necessarily grow up neurotic, especially if the father spends some "special time" with the child later.

Self-Esteem

High self-esteem—what Erik Erikson called "an optimal sense of identity"—is essential for growth in relationships, creativity, and productivity in the world at large. Low self-esteem is a disability that afflicts children (and the adults they grow up to be) with feelings of powerlessness, poor ability to cope, low tolerance for differences and difficulties, inability to accept responsibility, and impaired emotional responsiveness. Self-esteem has been shown to be more significant than intelligence in predicting scholastic performance. A study of 3,000 children found that adolescent girls had lower self-images, lower expectations from life, and less self-confidence than boys (Brown and Gilligan 1992). At age 9, most of the girls felt positive and confident, but by the time they entered high school, only 29% said they felt "happy" the way they were. The boys also lost some sense of self-worth, but not nearly as much as the girls.

Ethnicity was an important factor in this study. African-American girls reported a much higher rate of self-confidence in high school than did Caucasian or Latina girls. Two reasons were suggested for this discrepancy.

First, African-American girls often have strong female role models at home and in their communities; African-American women are more likely than others to have a full-time job and run a household. Second, many African-American parents specifically teach their children that "there is nothing wrong with them, only with the way the world treats them" (Daley 1991). According to researcher Carole Gilligan, their study "makes it impossible to say that what happens to girls is simply a matter of hormones. . . . [It] raises all kinds of issues about cultural contributions, and it raises questions about the role of the schools, both in the drop of self-esteem and in the potential for intervention" (quoted in Daley 1991).

Parents can foster high self-esteem in their children by (1) having high self-esteem themselves, (2) accepting their children as they are, (3) enforcing clearly defined limits, (4) respecting individuality within the limits that have been set, and (5) responding to their child with sincere thoughts and feelings.

It is also important to single out the child's behavior—not the whole child—for criticism (Kutner 1988). Children (and adults) can benefit from specific information about how well they've performed a task. "You did a lousy job" not only makes us feel bad but also gives us no useful information about what would constitute a good job.

Misusing the concept of self-esteem with superficial praise is probably the most common way parents mishandle the issue. Children notice when praise is insincere. If, for instance, Martha refuses to comb her hair, yet we continually tell her how good it looks, Martha quickly realizes that we either have low expectations or do not have a clue about hair care. Instead, parents can accomplish more by giving children timely, honest, specific feedback. For example, "I like the way you discussed Benjamin Franklin's inventions in your essay" is more effective than, "You're wonderful!" Each time parents treat their child like an intelligent, capable person, they increase the child's self-esteem.

Psychosexual Development in the Family Context

It is within the context of our overall growth, and perhaps central to it, that our sexual selves develop.

Within the family we learn how we "should" feel about our bodies—whether we should be ashamed, embarrassed, proud, or indifferent. Some families are comfortable with nudity in a variety of situations: swimming, bathing, sunbathing, dressing, or undressing. Others are comfortable with partial nudity from time to time: when sharing the bathroom, changing clothes, and so on. Still others are more modest and carefully guard their privacy. Most researchers and therapists suggest that all these styles can be compatible with the development of sexually well-adjusted children as long as some basic needs are met:

- The child's body (and nudity) is accepted and respected.
- The child is not punished or humiliated for seeing the parent naked, going to the toilet, or making love.
- The child's needs for privacy are respected.

Families also vary in the amount and type of physical contact. Some families hug and kiss, give back rubs, sit and lean on each other, and generally maintain a high degree of physical closeness. Some parents extend this closeness into their sleeping habits, allowing their infants and small children in their beds each night. (In many cultures, this is the rule rather than the exception.)

Other families limit their contact to hugs and tickles. Variations of this kind are normal. Concerning children's needs for physical contact, we can make the following generalization. First, all children (and adults) need a certain amount of freely given physical affection from those they love. Although there is no prescription for the right amount or form of such expression, its quantity and quality affect both children's emotional well-being and the emotional and sexual health of the adults they will become.

Second, children should be told, in a nonthreatening way, what kind of touching by adults is "good" and what kind is "bad." They need to feel that they are in charge of their own bodies, that parts of their bodies are private property, and that no adult has the right to touch them with sexual intent.

It is not necessary to frighten a child by going into great detail about the kinds of things that might happen. A better strategy is to instill a sense of self-worth and confidence in children so that they will not allow themselves to be victimized (Pogrebin 1983).

What Do Parents Need?

Although some needs of parents are met by their children, parents have other needs. Important needs of parents during the childrearing years are personal developmental needs (such as social contacts, privacy, and outside interests) and the need to maintain marital satisfaction. Yet so much is expected of parents that they often neglect these needs. Parents may feel varying degrees of guilt if their child is not happy or has some "defect", an unpleasant personality, or even a runny nose.

However, many forces affect a child's development and behavior. Accepting our limitations as parents (and as humans) and accepting our lives as they are (even if they haven't turned out exactly as planned) can help us cope with some of the many stresses of childrearing in an already stressful world. Contemporary parents need to guard against the "burnout syndrome" of emotional and physical overload. Parents' careers and children's school activities, organized sports, Scouts, and music, art, or dance lessons compete for the parents' energy and rob them of the unstructured (and energizing) time that should be spent with others, with their children, or simply alone.

The Effects of Parenthood on Marriage and Mental Health

Early research depicted the transition to parenthood as a crisis leading to a decline in marital quality and satisfaction. We now know, however, that the impact of parenthood is variable. Although marital satisfaction declines for many new parents, it also declines for couples without children during the early years of marriage. Thus, what may have appeared to be an effect of parenthood may just reflect the ebbs and flows of marital satisfaction (Helms-Erikson 2001). That doesn't mean that parenthood has no effect on marriage; indeed, it does, but its effects depend at least somewhat on when couples become parents and on how couples negotiate the new responsibilities. As Heather Helms-Erikson puts it, parenthood leaves "some couples faring better following the birth of their first child, others worse, and still others seemingly unchanged" (2001, 1,100).

New parents show more traditional divisions of duties and lower levels of companionship compared to couples without children, but marital discontent is by no means inevitable. Even these outcomes—traditionalization and declining marital quality—depend upon the circumstances under which they become parents. Couples who become parents "early" (that is, in their late teens or early 20s) are more likely to divide their household tasks on "traditional gender lines, with wives being responsible for the bulk of housework and childcare" and men becoming more involved only when pushed. Couples who become parents in their late 20s and 30s tend to display more "collaborative" divisions of roles, and fathers' involvement tends to be both more self-determined and reflect more liberal gender ideals (Helms-Erikson 2001, 1,101).

Mental health effects of parenthood have also been explored by researchers, but they come to different conclusions. Research suggests either: (1) parents and nonparents are similar to each other in their emotional well-being; (2) parents suffer "significantly more emotional distress" than nonparents.

Ranae Evenson and Robin Simon demonstrate that the picture is more complicated and cannot be summarized by a generalization about mental health outcomes. There are both positive and negative outcomes from parenthood; there is gratification, as well as an added sense of purpose and meaning to life, from being parents. But there are stresses and demands, especially when parents have young children, that may overshadow the benefits and undermine parents' mental health. Furthermore, the wider social and cultural context has reduced the significance, social value, and esteem attached to the parental role and left parents without the institutional supports that could make parenting less stressful (Evenson and Simon 2005).

Looking specifically at depression, Evenson and Simon compared childless adults with parents in different circumstances. After controlling for the effects of other demographic and social characteristics, compared to non-parents, parents reported *significantly higher* levels of depression. Contrary to their expectations, gender did not affect the relationship between parental status and depression. Among parents, those with minor and dependent children at home report *less, not more,* symptoms of depression than those with older children.

Embattled Parents and Societal Insensitivity to Raising Children

Even under ideal conditions, parenting is bound to be a difficult undertaking. Yet despite our cultural celebration of families and children, contemporary American society does little to ensure that families function effectively or that children are raised by involved and dedicated parents. Sylvia Hewlett and Cornel West (1998) note that in recent decades "public policy and private decision making have tilted heavily against the altruistic nonmarket activities that comprise the essence of parenting. In recent years, big business, government, and the wider culture have waged an undeclared and silent war against parents." Hewlett and West point to a number of examples of societal indifference to the needs of parents and children:

- *Economic issues.* Matters such as corporate downsizing, declining wages, and longer workweeks have led to more instability, impoverishment, and uncertainty, as well as less time between parents and children.

- *Popular culture.* Television programs, popular music, and movies undermine the efforts of parents through the parent bashing, violence, and sex to which they expose children.

- *Government insensitivity and neglect.* In such areas as housing and taxes, government policies have failed to support parents' efforts to raise their children.

- *Diminishment and devaluation of fathers.* Some social programs, especially in policies of poverty and divorce, have contributed to undermining the role of fathers in children's lives.

Combining these with alterations in household structure and increased economic vulnerability spells disaster for many fathers in their efforts to function effectively.

Diversity in Parent-Child Relationships

The diversity of family forms in our country creates a variety of parenting experiences, needs, and possibilities, as well as a range of parent–child relationships. The problems and strengths of single-parent and stepfamilies are discussed in more detail in Chapter 15 but will be touched upon here, along with the influences of ethnicity, sexuality (that is, lesbian and gay parenthood), and aging.

Effect of Parents' Marital Status

There is much research indicating that parental marital status affects children's upbringing and well-being. For example, comparisons of the experiences of children in married, "intact," two-parent households, where they reside with their biological parents, with those of children in single-parent households, remarried parent or stepparent households, and cohabiting parent households suggest that children living in families with their two married, biological parents fare best (Manning and Lamb 2003; Sun 2003). Reviewing the research literature, Yongmin Sun notes that compared to children in households with two biological parents, children in stepfamilies and single-parent families are more likely to have behavior and drug problems, show lower rates of graduation from high school, report lower levels of self-esteem, and perform worse on standardized tests (Sun 2003).

On a few measures, such as levels of delinquency and academic achievement, teens in married stepfamilies are somewhat advantaged compared to teens from cohabiting stepfamilies (that is, unmarried couples with one partner functioning as a stepparent) (Manning and Lamb 2003).

In accounting for the differences that surface between married and cohabiting stepfamilies, and among families with two biological parents, single-parent families, and stepfamilies, economic factors (for example, family income and parents' level of education) are especially important. Economic disadvantages faced by single mothers, as well as by stepfamilies, may explain why children in such households do less well (Sun 2003). The research is consistent in demonstrating that—whether because of economic advantage, social resources, amount and kind of parental attention and commitment, or some other factors—children who live with both of their biological parents benefit in a variety of ways when compared to peers in some "nontraditional" household structures (Manning and Lamb 2003; Sun 2003).

What about Nonparental Households?

Yet another way to see the effects of parents on children is to examine the experiences of children in households with *no* biological parents. In 1996, nearly 4% of all American children under 18 years of age—roughly 2.7 million children—lived in households with neither biological parent (Sun 2003). Three-quarters of children in **nonparental households** live with relatives, most with a grandparent.

Table 11.3 ■ Living Arrangements of Children in Households without Parents

Arrangement	Percentage of Children
Grandparents	47.9
Grandparents and other relatives	27.6
Nonrelative guardians	21.9
Other arrangements	2.7

SOURCE: Sun 2003, 894–909 (U.S. Census Bureau. *Detailed Living Arrangements of Children by Race And Hispanic Origin*, Table 1).

Sociologist Yongmin Sun reports that children 15–17 are twice as likely as children under 5 to live in one of these nonparental households. In addition to age, ethnicity makes a difference: 2.1% of Asian, 2.6% of Caucasian, 4.3% of Hispanic, and 7.9% of African American children live in a household without either biological parent (Sun 2003).

Generally, research has documented that children in nonparental households suffer when compared to children who live with at least one parent. Comparisons of children in foster care, albeit only one type of nonparental care, show negative effects in areas ranging from children's mental health, academic achievement, drug use, and behavioral problems (Sun 2003). Likewise, children in nonparental "kinship care" have been found to have poorer health, mental health, and school achievement than children in "parent present" families, whether single- or two-parent families (Sun 2003). Sun suggests that it is likely that the absence of mothers has the greatest impact. In accounting for the observed effects in nonparental households, Sun argues that the differences "are either completely or partially attributable to resource differences between these family structures." Key resources include income and parents' education, parents' expectations for their children's education, frequency of conversations between parents and children about school, involvement of parents with the schools and with other parents, and children's experiences of various cultural activities. No differences of note existed between kinship care and nonrelative care, and no differences were observed between girls and boys in how they fare in nonparental environments (Sun 2003).

Ethnicity and Parenting

There are other important differences among parents. A person's ethnicity is not necessarily fixed and unchanging. Researchers generally agree that ethnicity has both objective and subjective components. The objective component refers to ancestry, cultural heritage, and, to varying degrees, physical appearance. The subjective component refers to whether someone feels he or she is a member of a certain ethnic group. If a child has parents from different ethnic groups, ethnic identification becomes more complex. In such cases, the child may identify with both groups, only one group, or according to the situation—Latino when with Latino relatives and friends or Anglo when with Anglo friends and relatives, for example. However we choose to identify ourselves, our families are the key to the transmission of ethnic identification.

A child's ethnic background can affect how he or she is socialized. According to some researchers, minority families socialize their children to more highly value obligation, cooperation, and interdependence (Demo and Cox 2000). It has been suggested that Mexican American parents tend to value cooperation and family unity more than individualism and competition. Asian Americans and Latinos traditionally stress the authority of the father in the family. In both groups, parents command considerable respect from their children, even when the children become adults. Older siblings, especially brothers, have authority over younger siblings and are expected to set a good example (Becerra 1988; Tran 1988; Wong 1988). Many Asian Americans tend to discourage aggression in children and expect them to sacrifice their personal desires or interests out of loyalty to their elders and to family authority more generally (Demo and Cox 2000). In disciplining their children, Asian parents tend to rely on compliance based on the desire for love and respect.

African Americans, too, may have group-specific emphases in the ways they socialize their children. As reported in Chapters 3 and 4, African American parents tend to socialize their children into less rigid, more flexible gender roles. They reinforce certain traits, such as assertiveness and independence, in both their sons and their daughters. They also seek to promote such values as pride, closeness to other African Americans, and racial awareness (Demo and Cox 2000).

Groups with minority status in the United States may be different from one another in some key ways, but they also have much in common. Such groups often emphasize education as the means for the children to achieve success. Studies show that immigrant children tend to excel as students until they become acculturated and discover that it's not "cool." Minority groups are often dual-worker families, which means that the children may have considerable exposure to television while the parents are away from home. This may be viewed as a mixed blessing: on the one hand, television may help children who need to acquire English language skills; on the other, it can promote fear, violence, and negative stereotypes of women and minority-status groups. Some American children are raised with a strong positive sense of ethnic identification, however, that can also result in a sense of separateness is imposed by the greater society.

Discrimination and prejudice shape the lives of many American children. Parents of ethnic minority children may try to prepare their children for the harsh realities of life beyond the family and immediate community (Peterson 1985). According to Mary Kay DeGenova (1997), to reduce an environment of racism, it is important for us to identify the similarities among various cultures. These include people's hopes, aspirations, desire to survive, search for love, and need for family—to name just a few. Although superficially we may be dissimilar, the essence of being human is very much the same for all of us.

Gay and Lesbian Parents and Their Children

Researchers believe that the number of gay families is in the millions. They estimate between 2 million and 14 million children have at least one gay parent (Kantrowitz 1996; Stacey and Biblarz 2001). The high ends of these estimates include parents with adult children no longer in the home and use generous definitions of sexual orientation (including anyone with homoerotic desires). If we restrict the estimates to families with children 19 years or younger, there are anywhere from 1 million to 9 million children of lesbian or gay parents, representing between 1% to 12% of all children in this age group (Stacey and Biblarz 2001).

According to psychologist Charlotte Patterson, a leading authority on gay and lesbian parenting, the current research on the subject has some limitations. It has mostly focused on lesbian mothers, and on young children (pre-adolescent). Plus, it has been rare to have longitudinal studies in which researchers follow a sample of gay and lesbian parents and/or their children over time (Patterson 2005). These limitations aside, existing research fails to support the notion that children of lesbian mothers or gay fathers are negatively affected (Patterson 2000, 2005; Stacey and Biblarz 2001).

In fact, most gay or lesbian parents have been in heterosexual marriages (Patterson and Chan 1999). Concerns about gay and lesbian parents tend to center on questions about parenting abilities, fear of sexual abuse, and worry that the children will become gay or lesbian themselves. Research has failed to support such concerns or identify any significant negative outcomes for children. In fact, much research has failed to identify any meaningful differences between children of gay and heterosexual parents. Sociologists Judith Stacey and Timothy Biblarz's (2001) and psychologist Charlotte Patterson's (2000, 2005) reviews

■ *Families headed by lesbians or gay men generally experience the same joys and pains as those headed by heterosexuals, but they are also likely to face insensitivity or discrimination from society.*

© Paul Conklin/PhotoEdit

of existing research on the effect of parental sexual orientation on children finds that most research supports either a "no effects" or a "beneficial effects" interpretation.

In summarizing the research on children of gay and lesbian parents as they compare with children of heterosexual parents, Patterson notes that there are no significant differences in their gender identities, gender-role behaviors, self-concepts, moral judgment, intelligence, success with peer relations, behavioral problems, or successful relations with adults of both genders (Patterson 2000, 2005). Stacey and Biblarz suggest that there may be some defensiveness on the part of researchers, especially from those sympathetic to gay and lesbian parents. Aware of the social stigma and lack of support gay and lesbian families face, there may be a tendency to minimize differences. In so doing, some differences that might be strengths of gay and lesbian families may go underemphasized.

Fears about Gay and Lesbian Parenting

Heterosexual fears about the parenting abilities of lesbians and gay men are exaggerated and unnecessary. There are minimal differences between lesbians and heterosexual women in their "approaches to childrearing" or their mental health (Patterson 2005). No studies identify ways in which lesbian mothers or gay fathers are "unfit parents" or less fit than heterosexual parents.

Fears about gay parents' rejecting children of the other sex also are unfounded. Such fears reflect the popular misconception that being gay or lesbian is a rejection of members of the other sex. Many gay and lesbian parents go out of their way to make sure that their children have role models of both sexes (Kantrowitz 1996). Gay and lesbian parents also tend to say that they hope their children will develop heterosexual identities to be spared the pain of growing up gay in a homophobic society. Research finds children of gay males and lesbians to be well adjusted and no more likely to be gay as adults (Goleman 1992; Flaks et al. 1995; Kantrowitz 1996).

Ultimately, it is the quality of parenting and the harmony within the family—not the sexuality of the parents—that matters most to children. Like children of heterosexual parents, children whose gay or lesbian parents are in "warm and caring relationships," experiencing less stress and conflict, and receiving more support from partners (as well as from other family members) tend to fare better.

Consider, finally, the following account by Abigail Garner, author of *Families Like Mine: Children of Gay Parents Tell It Like It Is* (2005) and creator of the website, site FamiliesLikeMine.com:

When I was 5, my father came out as gay to his family and friends and moved in with another man. By the time I entered elementary school, I was learning about the cruelty of homophobia. "Faggot" was the favorite put-down among the boys in my class. I didn't know what it meant until my parents explained that it was a mean way of saying someone was gay. Since my classmates seemed to be so hostile about gay people, I decided I should keep quiet about my family.

People who knew me then are surprised by my outspokenness. "Can't you move on?" they ask. But I am driven to speak about my past because the consequences feel less risky now that I'm an adult. I no longer worry about people who might try to "protect" me from my father by taking me away from him. I don't have to wonder every time we go out: is this the time he gets "caught"? I remem-ber when I was about 8, I was walking down the street between my father and his partner and holding both of their hands. It felt dangerous, because by standing as a link between them I was "outing" them. What would happen if others realized my dad was gay? Would he lose his job? Get beaten up? Be declared an unfit parent?

While the threat of being separated from him was never real, I spent plenty of time worrying about it. Fortunately, my mother (who is heterosexual) made no attempt to limit my father's custody rights. If she had, she probably would have gained full custody. Our courts have a history of favoring straight parents over gay ones in custody battles.

My parents did their best to make me feel good about where I came from. They told me that even though they were divorced and my dad was gay, we were no less valid than any other family. But they could do nothing about the abundance of negative messages about homosexuality that I interpreted as direct attacks on my family.

Why did so many people—including TV evangelists and talk-show guests—think that my dad was such a terrible person? They didn't even know him. While my friends had monsters keeping them awake at night, I lost sleep over the anti-gay rhetoric spouted by right-wing politicians.

College marked a significant change in my life. The 1,500 miles between home and school gave me the distance I needed to figure out who I was, separate from my parents. I thought I had outgrown the label of "daughter from a gay family." Soon after I graduated, however, I connected with a group of teens with gay and lesbian parents while volunteering for a youth organization. When I realized how similar their stories were to mine, I was inspired to start talking openly about my own experiences.

When I do speak, many people assume I'm a lesbian. And for those who don't respect homosexuals, it's the only reason they need to dismiss my arguments for gay rights. Once I identify myself as straight, however, I'll watch their rigid, angry faces soften to ask me questions. I'll see the handful of college students in the audience who were rolling their eyes sit up and listen. It gives me hope that they'll hear my message: it wasn't having a gay father that made growing up a challenge, it was navigating a society that did not accept him and, by extension, me.

Summarizing the research on parenting by, and children of, gays and lesbians in a report for the American Psychological Association, Charlotte Patterson makes the following strong assertion:

"(T)here is no evidence to suggest that lesbian women or gay men are unfit to be parents. . . . Not a single study has found children of lesbian or gay parents to be disadvantaged in any significant respect relative to children of heterosexual parents. Indeed, the evidence to date suggests that home environments provided by lesbian and gay parents are as likely as those provided by heterosexual parents to support and enable children's psychosocial growth" (Patterson, 2005:15)

Parenting and Caregiving in Later Life

Parenting Adult Children

Many years ago, a Miami Beach couple reported their son missing (Treas and Bengtson 1987). Joseph Horowitz still doesn't understand why his mother became so upset. He wasn't "missing" from their home in Miami Beach: he had just decided to go north for the winter. Etta Horowitz, however, called authorities. Social worker Mike Weston finally located Joseph in Monticello, New York, where he was visiting friends. Etta, 102, and her husband, Solomon, 96, had feared harm had befallen their son Joseph, 75. As the Horowitz story reminds us, parenting does not end when children grow up.

By some measures, children are "growing up" later than at any time in the past. They lack the means to be financially independent and delaying entry into marriage, parenthood, and independent living, away from their families. In one study that compares 1960 census data to 2000 census data, researchers noted that there has been a significant decrease in the percentage of young adults who, by age 20 or 30, have completed all of the following five traditionally defined major adult transitions: leaving the parental home, completing their schooling, achieving financial independence (being in the labor force and/or—for women—being married and a mother), marrying, and becoming a parent. In 1960, more than three-fourths of women and two-thirds of men had reached all five of these markers by age 30, yet in 2000, less than half of women and less than a third of men had achieved all five of these (Furstenberg et al. 2004).

> More than at any time in recent history, parents are being called on to provide financial assistance (either college tuition, living expenses or other assistance) to their young adult children. Robert Schoeni and Karen Ross conservatively estimate that nearly one-quarter of the entire cost of raising children is incurred after they reach 17. Nearly two-thirds of young adults in their early 20s receive economic support from parents, while about 40 percent still receive some assistance in their late 20s (Furstenberg, et al, 2004).

Most parents with adult children still feel themselves to be parents even when their "children" are middle-aged. However, their parental role is considerably less important in their daily lives. They generally have some kind of regular contact with their adult children, usually by letters, phone calls, or e-mails; parents and adult children also visit each other fairly frequently and often celebrate holidays and birthdays together. Financially, they may make loans, give gifts, or pay bills for their children. Further assistance may come in the form of shopping, house care, and transportation and help in times of illness.

Parents tend to assist those whom they perceive to be in need, especially children who are single or divorced. Parents perceive their single children as being "needy" when they have not yet established themselves in occupational and family roles. These children may need financial assistance and may lack intimate ties; parents may provide both until the children are more firmly established. Parents often assist divorced children, especially if grandchildren are involved, by providing financial and emotional support. They may also provide childcare and housekeeping services.

Parents tend to be deeply affected by the circumstances in which their adult children find themselves. Adult children who seem well adjusted and who have fulfilled the expected life stages (becoming independent, starting a family, and so on) provide their aging parents with a vicarious gratification. On the other hand, adult children who have stress-related or chronic problems (for example, with alcohol) cause higher levels of parental depression (Allen, Blieszner, and Roberto 2000).

Some elderly parents never cease being parents because they provide home care for children who are severely limited either physically or mentally. Many elderly parents, like middle-aged parents, are taking on parental roles again as children return home for financial or emotional reasons. Although we don't know how elderly parents "parent," presumably they are less involved in traditional parenting roles.

Reflections

Think about your grandparents. How many are alive? What kind of relationship do you (or did you) have with them? What role do they (or did they) play in your life and your family's life?

Grandparenting

The image of the lonely, frail grandmother in a rocking chair needs to be discarded. Grandparents are often not old, nor are they lonely, and they are certainly not absent in contemporary American family life. Grandparents are "a very present aspect of family life, not only for young children but young adults as well," writes Gregory Kennedy (1990).

Grandparenting is expanding tremendously these days, creating new roles that relatively few Americans played a few generations back. Three-quarters of people aged 65 and older are grandparents (Aldous 1995). Grandparents play important emotional roles in American families; the majority appear to establish strong bonds with their grandchildren (Kennedy 1990; Strom et al. 1992–1993).

They help achieve family cohesiveness by conveying family history, stories, and customs. Grandparents influence grandchildren directly when they act as caretakers, playmates, and mentors. They influence indirectly when they provide psychological and material support to parents, who may consequently have more resources for parenting (Brooks 1996).

Grandparents seem to take on greater importance in single-parent and stepparent families and among certain ethnic groups (see Figures 11.1 and 11.2). They often act as a stabilizing force for their children and grandchildren when the families are divorcing and reforming as single-parent families or stepfamilies. The significance of grandparents appears to vary by family form (Kennedy and Kennedy 1993). When compared with children from intact families, children in single-parent families report greater closeness and active involvement with their grandparents; children in stepfamilies are even closer.

According to the 2000 Census, 5.8 million grandparents live in the same home as one of their grandchildren. In 42% of these 4.1 million households (some households have more than one grandparent), grandparents had primary caregiving responsibility for their

© Spencer Grant/PhotoEdit

■ *Grandparents are important to their grandchildren as caregivers, playmates, and mentors.*

grandchildren, age 18 or younger. Of these "grandparent caregivers," 39% had cared for their grandchildren for at least 5 years (Simmons and Dye 2003).

Grandparents, especially grandmothers, are often involved in the daily care of their grandchildren (see Figure 11.2). A recent study found that African Americans had twice the odds of becoming caregiving grandparents, partly reflecting the long tradition of caregiving that goes back to West African cultures. In

Figure 11.1 ■ **Percentage of Population, Age 30 Years or Older, Living with and Responsible for Grandchildren, 2000**

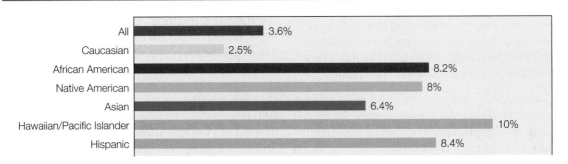

All — 3.6%
Caucasian — 2.5%
African American — 8.2%
Native American — 8%
Asian — 6.4%
Hawaiian/Pacific Islander — 10%
Hispanic — 8.4%

SOURCE: Simmons and Dye 2003.

Figure 11.2 ■ Percentage of Residential Grandparents Who Are Responsible for Grandchildren

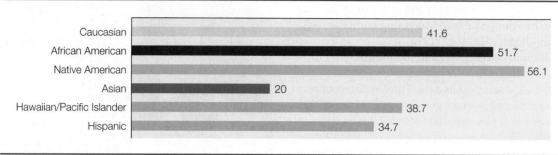

Caucasian	41.6
African American	51.7
Native American	56.1
Asian	20
Hawaiian/Pacific Islander	38.7
Hispanic	34.7

SOURCE: Simmons and Dye 2003.

the crack cocaine epidemic, grandmothers and great-grandmothers play critical roles in rearing the children of addicted parents (Minkler and Roe 1993).

Andrew Cherlin and Frank Furstenberg (1986) identified three distinct styles of grandparenting:

■ *Companionate.* Most grandparents perceive their relationships with their grandchildren as companionate. The relationships are marked by affection, companionship, and play. Because these grandparents tend to live relatively close to their grandchildren, they can have regular interaction with them. Companionate grandparents do not perceive themselves as rule makers or enforcers; they rarely assume parent-like authority.

■ *Remote.* Remote grandparents are not intimately involved in their grandchildren's lives. Their remoteness, however, is geographic rather than emotional. Geographic distance prevents the regular visits or interaction with their grandchildren that would bind the generations together more closely.

■ *Involved.* Involved grandparents are actively involved in what have come to be regarded as parenting activities: making and enforcing rules and disciplining children. Involved grandparents (most often grandmothers) tend to emerge in times of crisis—for example, when the mother is an unmarried adolescent or enters the workforce following divorce. Some involved grandparents may become overinvolved, however. They may cause confusion as the family tries to determine who is the real head of the family.

Single parenting and remarriage have made grandparenthood more painful and problematic for many grandparents. Stepfamilies have created step-grandparents, who are often confused about their grandparenting role. Are they really grandparents?

The grandparents whose sons or daughters do not have custody often express concern about their future grandparenting roles (Goetting 1990). Although research indicates that children in stepfamilies tend to do better if they continue to have contact with both sets of grandparents, it is not uncommon for the parents of the noncustodial parent to lose contact with their grandchildren (Bray and Berger 1990).

A variety of circumstances may lead to situations in which the grandparent role and the relationships with grandchildren are strained if not disrupted. Divorce and single parenthood may be the most prominent of such circumstances, but death of a spouse, distance, or estrangement between parents and children can all impede grandparent–grandchild relationships (Keith and Wacker 2002). Over the past 40 years, grandparent visitation statutes have been enacted in all 50 states and grandparents' visitation rights have been increased.

Generally, courts have not wanted to expand grandparents' rights at the expense of parents' rights, especially parents' rights to control the custody of their children (Keith and Wacker 2002).

Children Caring for Parents

Parent–child relationships do not flow in one direction. A common experience faced by many American families is the need to provide care for aging or ill parents. The idea of the *sandwich generation* (see Chapter 9) captures the experience of many adults, sandwiched between raising their own children and caring for their parents. However, there are circumstances that create **parentified children**—children forced to become caregivers for their parents well before adulthood (Boszormenyi and Spark 1973, quoted in Winton 2003). In situations of "parentification," children may be pressed into taking care of parents who have become chronically ill, chemically dependent, mentally ill, incapacitated after a divorce or widowhood, or socially isolated or incapacitated (Winton 2003).

Much of the psychological and sociological literature depicts parentified children as pathological or deviant. Psychologists may focus on how taking on caregiving responsibilities for a parent or parents while still a child or adolescent disrupts normal developmental processes. Sociologists tend to focus on the nonnormative nature of children being responsible for their parents. However, definitions of normative and nonnormative vary by culture. Among many populations other than white, middle-class, European Americans, parentification is expected and obligatory. Similarly, rather than pathological, parentification under certain circumstances may be beneficial for the development of certain personality traits, the maintenance of certain family relationships, and the acquisition of particular skills. Chester Winton (2003) suggests that parentification may be a normative part of childhood in many contemporary American families, where children *temporarily* take care of a parent (for example, after surgery or during an illness). This fits Gregory Jurkovic's continuum of caretaking roles, where parentification is normal and adaptive under certain conditions. *Destructive parentification* occurs when the circumstances become extreme and long-term and the responsibilities children carry are age-inappropriate (Jurkovic 1997, cited in Winton 2003).

Winton suggests the following as possible consequences of parentification:

- *Delayed entry into marriage.* If children have had to care for parents (or siblings) over a number of years, they may decide to delay taking on the caretaking that comes with marriage and choose, instead, to take time for themselves where they can concentrate on their own needs more than or instead of the needs of others.

- *Acquisition of certain personality characteristics.* Having played a parentified role over time might lead to the development of such traits or tendencies as the following:

  - Masochistic or self-defeating behavior because of having had to meet others' needs and suppress their own compulsive behavior, such as perfectionism

  - Feelings of excessive responsibility for others that make it difficult to say "no" to people, to set limits, or to concentrate on their own needs

- *Relationship and intimacy problems.* Parentified children may seek as adult partners people who they can be caretakers for—in other words, "dependent, needy people" who have emotional or physical disabilities or have been emotionally "wounded" by past experiences.

- *Career choices.* The "caretaker syndrome" associated with parentification may lead people to jobs where they can physically or emotionally take care of people, such as jobs in social work, medicine, nursing, teaching, or preschool childcare.

Caring for Aging Parents

Most elder care is provided by women, generally daughters or daughters-in-law (Mancini and Blieszner 1991). Psychologist Rita Ghatak estimates that "eighty percent of the time it's the female sibling who is taking most of the responsibility" (quoted in Rubin 1994). Elder caregiving seems to affect husbands and wives differently. Women report greater distress, greater decline in happiness, more hostility, less autonomy, and more depression from caregiving than do men (Fitting et al. 1986; Marks, Lambert, and Choi 2002). This may partly be because men approach their daily caregiving activities in a more detached, instrumental way. Another factor may be that women often are not only mothers but also workers; an infirm parent can sometimes be an overwhelming responsibility to an already burdened woman (Rubin 1994). Interestingly, when caring for a parent out of the household, women feel a *caregiver gain,* a greater sense of purpose in life than that felt by noncaregiving women (Marks, Lambert, and Choi 2002). Fortunately, most adult children participate in parental caregiving in some fashion when needed, whether it involves doing routine caregiving, providing backup, or giving limited or occasional care (Mancini and Blieszner 1991).

A study of 539 older participants found that although there are psychological benefits associated with intergenerational support, excessive support received from adult children may be harmful, eroding competence and imposing excessive demands (Silverstein, Chen, and Heller 1996). In balancing personal needs with those of families, it is important to define the level of care that is both appropriate and necessary.

Caregiver Conflicts

Even though elder care is often done with love, it can be the source of profound stress. Caregivers often experience conflicting feelings about caring for an elderly relative. The conflicts experienced by primary caregivers include the following (Springer and Brubaker 1984):

- Earlier unresolved antagonisms and conflicts
- The caregiver's inability to accept the relative's increasing dependence
- Conflicting loyalties between spousal or child-rearing responsibilities and caring for the elderly relative
- Resentment toward the older relative for disrupting family routines and patterns
- Resentment by the primary caregiver for lack of involvement by other family members
- Anger or hostility toward an elderly relative who tries to manipulate others
- Conflicts over money or inheritance

Coping Strategies

Caregiver education and training programs, self-help groups, caregiver services, and family therapy can provide assistance in dealing with the problems encountered by caregivers. In addition, elders receiving Medicaid may be eligible for respite care and homemaker or housework assistance. Because elder care involves complex emotions raised by issues of dependency, adult children and their parents often postpone discussions until a crisis occurs.

Summary

- Although today's mothers and fathers have many things in common with mothers and fathers throughout history, in many ways they have to chart a new course because both motherhood and fatherhood have changed.
- Many women find considerable satisfaction and fulfillment in motherhood. Although there is no concrete evidence of a biological maternal drive, it is clear that socialization for motherhood does exist.
- Whether employed outside the home or not, women who become mothers face high expectations and cultural contradictions. The *ideology of intensive mothering* portrays mothers but not fathers as essential caregivers and depicts childrearing as child centered, expert guided, emotionally absorbing, labor intensive, and financially expensive. At the same time, however, at-home mothers are often perceived negatively, as though they were unproductive. Such contradictions surface in popular culture, as well as in wider societal attitudes and beliefs.
- Employed mothers earn less than employed women without children. Married mothers pay a steeper wage price than never-married mothers.
- The role of the father in his children's development has been reexamined, and expectations of fathers have been redefined. The traditional instrumental roles are being supplemented, and perhaps supplanted, by expressive ones. This may be truer of our beliefs about fathers (the *culture of fatherhood*) than of fathers' real behavior (the *conduct of fathers*).
- There appear to be two extremes among contemporary fathers: Many men aspire for active, meaningful involvement with their children; others, especially divorced or never-married fathers, often maintain little contact with their children.
- Fatherhood affects many areas of men's lives. Fathers differ from nonfathers in their social activities, intergenerational family ties, and their occupational behavior.
- Most hands-on childcare is done by mothers. Fathers are less engaged with and accessible to their children than are mothers. When directly engaged with children, fathers more often play or assist in personal care activities. Age and gender of children, age and gender attitudes of fathers, and fathers' occupations and earnings all affect father involvement.

- Mothers also do more of the *mental labor* of child-care, including worrying about, gathering information, and managing fathers' involvement. Even as children enter adolescence, mothers do more of the mental labor, including monitoring where their children go, who they are with, and what they do.

- Supplementary childcare outside the home is a necessity for many families. Most children who receive outside care are in childcare centers. Overall, childcare is safe, and center-based care is safer than "family day care" or paid care by others in the child's home.

- The effect of childcare on children depends on the quality of care. The development and maintenance of quality day care programs should be a national priority.

- Increasing attention has been paid to school-age childcare. Many communities provide after-school care through the schools. A common alternative to such care is *self-care.*

- Children have a number of basic physical and psychological needs, including adequate prenatal care; formation of close attachments; protection from illness and abuse; and respect, education, and support from family, friends, and community. High self-esteem is essential for growth in relationships, creativity, and productivity. Parents can foster high self-esteem in their children by encouraging the development of a sense of connectedness, uniqueness, and power, and by providing models.

- *Psychosexual development* begins in infancy. Infants and children learn from their parents how they should feel about themselves as sexual beings.

- Parents differ in terms of their styles of parenting. Four styles are: *authoritarian, permissive* (or *indulgent*), *authoritative,* and *uninvolved.* Of these, most research portrays the authoritative as most effective.

- Today's parents often rely on expert advice. It needs to be tempered by confidence in their parenting abilities and in their children's strength and resourcefulness. Contemporary strategies for childrearing include the elements of mutual respect, consistency and clarity, logical consequences, open communication, and behavior modification in place of physical punishment.

- Parenthood has effects on marital relations and on mental health, especially depression rates, of par-ents. New parents tend to display more traditional role relationships, although this depends partly on the age at which they become parents. Across all parental statuses (married, single, step, custodial, and empty nest), parents appear to suffer more emotional distress than nonparents.

- Children's needs include adequate prenatal nutrition and care, appropriate stimulation and care of as newborns and infants, formation of at least one close attachment during the first 5 years of life, quality childcare when a parent or parents must work, protection from illness, freedom from physical and sexual abuse, supportive friends, safe and nurturing schools, and protection from premature parenthood.

- Overall, American society is not particularly supportive of the needs of parents and children. Economic, cultural, and political institutions have neglected to adopt policies that would allow parents and children deeper and more frequent contact with each other.

- Parents' marital status, ethnicity, and sexuality all influence parenting and child socialization.

- Children who live in households without any parents (either foster care or "kinship care" from other relatives) have lower academic performance, educational aspirations, and psychological well-being (self-esteem and locus of control) and greater likelihood of behavioral problems (for example, truancy and fighting) and cigarette smoking.

- Parents of ethnic minority status may try to give their children special skills for dealing with prejudice and discrimination.

- Most gay and lesbian parents are, or have been, married. Studies indicate that children of both lesbians and gay men fare best when the parents are secure in their sexual orientation.

- Parenting roles continue through old age. Older parents provide financial and emotional support to their children; they often take active roles in childcare and housekeeping for their daughters who are single parents. Divorced children and those with physical or mental limitations may continue living at home.

- Grandparenting is an important role for the middle-aged and aged; it provides them and their grandchildren with a sense of continuity. Grandparents often provide extensive childcare for grandchildren. Grandparenting can be divided into three styles: companionate, remote, and involved.

- In some instances, such as when parents are chronically ill, chemically dependent, mentally ill, or incapacitated after a divorce or widowhood, children become caregivers to their parents. Such *parentified children* may develop unique personality characteristics, experience problems in their intimate relationships, or develop and make career choices that are the results of having had to care for their parents.

- Family caregiving activities often begin when an aged parent becomes infirm or dependent. Conflicts that may arise involve previous unresolved problems, the caregiver's inability to accept the parent's dependence, conflicting loyalties, resentment, anger, and money or inheritance conflicts.

Key Terms

accessibility 394

assimilation accommodation 400

authoritarian childrearing 403

authoritative childrearing 403

bifurcation of fatherhood 393

conduct of fatherhood 392

culture of fatherhood 392

developmental systems approach 400

ego 399

engagement 394

game stage 401

id 399

ideology of intensive mothering 391

indulgent childrearing 403

looking-glass self 401

mental labor 396

neuroses 399

nonparental households 408

parentified children 414

permissive childrearing 403

play stage 401

psychoanalytic theory 399

psychosexual development 399

psychosocial theory 399

self-care 398

superego 399

taking the role of the other 401

uninvolved parenting 403

Resources on the Internet

Companion Website for This Book

http://www.thomsonedu.com/sociology/strong

Gain an even better understanding of this chapter by going to the companion website for additional study resources. Take advantage of the Pre- and Post-Test quizzing tool, which is designed to help you grasp difficult concepts by referring you back to review specific pages in the chapter for questions you answer incorrectly. Use the flash cards to master key terms and check out the many other study aids you'll find there. Visit the Marriage and Family Resource Center on the site. You'll also find special features such as access to InfoTrac© College Edition (a database that allows you access to more than 18 million full-length articles from 5,000 periodicals and journals), as well as GSS Data and Census information to help you with your research projects and papers.

CHAPTER 12

Marriage, Work, and Economics

Outline

Are the following statements **TRUE** or **FALSE**?
You may be surprised by the answers (see answer key on the bottom of this page).

T F   1 In contrast to single-worker couples, dual-career couples tend to divide household work almost evenly.

T F   2 More than 1 million American men are full-time homemakers with no outside employment.

T F   3 It is generally agreed by economists that welfare encourages poverty.

T F   4 Couples who work different shifts have more satisfying and stable marriages.

T F   5 Women in the United States currently make 90 cents for every dollar that men earn.

T F   6 Family economic well-being is a national priority.

T F   7 Many female welfare recipients are on welfare as a result of a change in their marital or family status.

T F   8 Most families are dual-earner families.

T F   9 Women tend to interrupt their work careers for family reasons far more often than do men.

T F   10 Married women tend to earn more and have higher-status jobs than single women.

Answer Key for What Do You Think

1 False, see p. 427; 2 True, see p. 441; 3 False, see p. 449; 4 False, see p. 440; 5 False, see p. 442; 6 False, see p. 451; 7 True, see p. 432; 8 True, see p. 433; 9 True, see p. 431; 10 False, see p. 426.

Imagine yourself at a party put on by your school's alumni association. As you float around the room, trying to meet and mingle with some people who graduated in recent years, you overhear the following exchanges among some of the other guests. Each snippet of conversation illustrates some unspoken assumptions people have about work and family. Can you recognize the assumptions and identify what is wrong in each exchange?

- *Exchange No. 1.* A trio of women is in a corner. "What do you do?" one of the women inquires politely while being introduced by a second woman to the third. "Nothing. I'm a housewife," the third responds. "Oh, that's . . . nice," the first woman replies, seeming to lose interest and turning toward the woman handling the introduction.

- *Exchange No. 2.* A bearded man is talking to a couple. "So, what do you two do?" the man asks. "I'm a doctor," the woman responds as she picks up her child, who is impatiently tugging on her. "And I'm an architect," her husband says while nursing their second child with a bottle.

Although they are subtle, we can observe the following assumptions being made and attitudes being displayed. In the first exchange, both women ignore that the woman who identified herself as a homemaker does, indeed, work. They also appear to devalue such unpaid work in comparison with paid work. In the second exchange, the woman identifies herself as a physician without acknowledging that she is also a parent. Her husband makes the same mistake. As husband and wife, father and mother, both the physician and the architect are unpaid family workers making important—but generally unrecognized—contributions to the family's economy.

Because it is unpaid, and perhaps because it is done mostly by women, family work is ignored and looked upon as inferior to paid work, regardless of how difficult, time consuming, creative, rewarding, and important it is for our lives and future as humans. This is not surprising, because in the United States employment takes precedence over family.

To understand the role of work in families, we may also need to rethink the meaning of *family*. We ordinarily think of families in terms of relationships and feelings—the family as an emotional unit. But families are also economic units that happen to be bound by emotional ties (Ross, Mirowsky, and Goldsteen

1991). Paid work and unpaid family work, as well as the economy itself, profoundly affect the way we live in and as families. Our most intimate relationships vary according to how we participate in, divide, and share paid work and family work (Voydanoff 1987).

Our paid work helps shape the quality of family life: it affects time, roles, incomes, spending, leisure, and even individual identities. Whatever time we have for one another, for fun, for our children, and even for sex is the time not taken up by paid work. Work regulates the family, and for most families, as in the past, a woman's work molds itself to her family, whereas a man's family molds itself to his work (Ross, Mirowsky, and Goldsteen 1991). We must constantly balance work roles and family roles. These facts are the focus of this chapter.

Workplace and Family Linkages

Time and Time Strains

Outside of sleeping, probably the single activity to which most employed men and women devote the most time is their jobs. Data suggest that, in contrast to declines throughout Europe, Americans are working more (Jacobs and Gerson 2004). Although European and American workers face similar "time dilemmas," the societal responses to these pressures have been vastly different. Jerry Jacobs and Kathleen Gerson (2004, 124) assert the following:

> Several European countries, especially those in Northern Europe, have made sustained, highly publicized, and well-organized efforts to reduce working time as a strategy for reducing unemployment, increasing family time, and reducing gender inequalities in the market and at home.

Conversely, ". . . the average American worker—including both part-timers and full-timers—puts in more hours per year on the job than the typical full-time worker in Europe" (Jacobs and Gerson, 2004, 127). The United States has the longest average work week and the highest percentages of men and women who work 50 hours per week or more. This is true of married women and men as well as unmarried, parents as well as people without children (Jacobs and Gerson, 2004, 125). The more we work, the less time we have for our families and leisure. Most of us know from experience that our work or even our studies affect our personal relationships.

It bears mentioning that although some categories of workers (for example, professional and managerial) have experienced an increase in the time demands upon them, others are underemployed and would prefer to work more (Jacobs and Gerson, 2004; Perry-Jenkins, Repetti, and Crouters, 2000). This **bifurcation of working time,** wherein some work longer and longer days and weeks while others work less hours than they need or want, is revealed by findings from the National Study of the Changing Workforce. Sixty percent of both men and women would prefer to work less; however, about one in five men (19.3%) and 18.5% of women would prefer to work more hours than they currently work (Jacobs and Gerson, 2004).

Whether we love, loathe, or merely learn to live with them, our jobs structure the time we can spend as families (Hochschild 1997). Time at work can create a feeling of **time strain,** in which individuals feel they do not have or spend enough time in certain roles and relationships. Kei Nomaguchi, Melissa Milkie, and Suzanne Bianchi (2005) found interesting gendered patterns in their investigation into the psychological effects of time strains:

- More fathers than mothers report feeling they do not have enough time with their children or their spouses. More mothers than fathers feel they have too little time for themselves.

- Life satisfaction is significantly reduced *for mothers but not for fathers* when they feel they have or spend "too little time with children."

- Feelings of time strains with a spouse are associated with significantly higher levels of distress *for women but not for men.*

- Feelings of insufficient time for oneself are associated with reduced levels of family and life satisfaction and with increased feelings of distress *for men but not for women.*

- Fathers articulate feeling strained for time with both their spouses and their children, but these feelings do not affect them as much psychologically as they do women.

Work and Family Spillover

In addition to the time we have available to our families, work affects home life in other ways. Common sense (as well as our own fatigue) suggests that our paid work has effects on other aspects of our lives. We can call this **work spillover**—the effect that work has on individuals and families, absorbing their time and energy and impinging on their psychological states. It links our home lives to our workplace (Small and Riley 1990). Work is as much a part of our marriages and home lives as love is. What happens at work—frustration or worry, a rude customer, an unreasonable boss—has the potential to affect our moods, perhaps making us irritable or depressed. Often, we take such moods home with us, affecting the emotional quality of our relationships.

Research demonstrates that work-induced energy depletion, fatigue, or, in more extreme cases, exhaustion can affect the quality of our family relationships. Fatigue and exhaustion can make us angry, anxious, less cheerful, and more likely to complain and can cause us to experience more difficulty interacting and communicating in positive ways. Yet according to one study, although both stress and exhaustion from work affect marital relationships, "stress is far more toxic" (Roberts and Levenson 2001,1,065). These researchers suggest that although common, job stress can seriously and negatively affect marital happiness, creating dynamics that unchecked may even contribute to divorce.

Scholars have increasingly looked at how and how often negative spillover affects us. Although negative work spillover occurs neither every day nor to everyone, it is accurate to consider it fairly commonplace (Roehling, Jarvis, and Swope 2005). This is revealed in the Figure 12.1, based upon data from the 1997 National Study of the Changing Workforce.

Such work–family tensions are greater for mothers and fathers than they are for employed women and men without children. Furthermore, the effects seem to be greater on mothers than on fathers, just as the differences between parents and nonparents are greater among women than among men. Jerry Jacobs and Kathleen Gerson note that children's ages make little difference in parents' experiences of work–family stress. Workplace stress often causes us to focus on our problems at work rather than on our families, even when we are home with our families. It can lead to fatigue, stomach ailments, and poorer health, as well as depression, anxiety, increased drug use, and problem drinking (Roehling, Jarvis, and Swope 2005; Crouter and Manke 1994).

Family-to-Work Spillover

As many employed parents can attest, the relationship between paid work and family life cuts both ways. The emotional climate in our homes can affect our morale and performance in our jobs. Positively, family can help alleviate some workplace stress. More research has focused on how the demands of our home lives may impinge on our concentration, energy, or availability at work (Jacobs and Gerson 2004).

Figure 12.1 ■ Work-to-Family Spillover

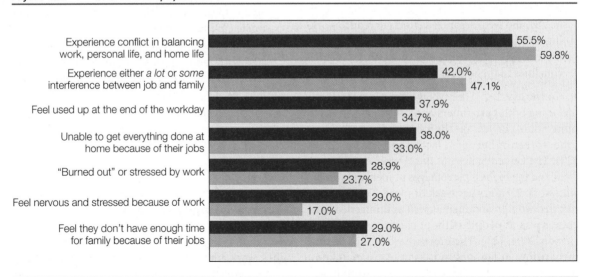

■ = Women   ■ = Men; (from Jacobs and Gerson, 2004:85)

Yet Jennifer Keene-Reid and John Reynolds (2005) argue that "workers who have control over their work schedules report feeling more successful at balancing work and family life." Furthermore, because family demands and needs can and do arise unexpectedly, the ability of employed parents to adjust their schedules accordingly is a useful and important family-friendly benefit.

Research indicates that because women, more often than men, face the intrusion of their family responsibilities into their work lives, they are forced to make more work-related adjustments because of family needs (Keene-Reid and Reynolds 2005). Additionally, higher levels of family to work spillover have been found among parents compared to non-parents (Roehling, Jarvis and Swope, 2005).

Meeting family demands such as assuming more household and childcare responsibility often comes with hidden or unanticipated work-related financial costs. Regardless of gender, those who carry responsibility for traditionally female housework chores are likely to suffer reduced wages. This is probably the result of having less effort and energy available to spend on paid work activities, although it may also reflect employer discrimination against those who perform female housework as a result of reduced effort and energy to devote to their employment (Noonan 2001).

Role Conflict, Role Strain, and Role Overload

Two-parent families in which both partners are employed face more severe work-related problems than do nonparents. Being an employed parent usually means performing three demanding roles simultaneously: worker, parent, and spouse or partner (Voydanoff and Donnelly 1989). In juggling these roles, we might experience role conflict, role strain, role overload, or a combination of these.

When the multiple social statuses or positions that we occupy (for example, spouse, parent, and worker) present us with competing, contradictory, or simultaneous role expectations, we experience **role conflict.** When the role demands attached to any particular status (for example, mother, husband, or employee) are contradictory or incompatible we experience **role strain.** Finally, when the various roles we play require us to do more than we can comfortably or adequately handle, or when we feel we have so much to do that we will never "catch up," or have enough time for ourselves we experience **role overload** ourselves (Crouter et al. 2001).

David Young-Wolff/PhotoEdit

■ *Work spillover and role strain affect many employed women and men, especially those who have children.*

In the specific case of family and paid work roles, when we feel torn between spending time with our spouses or children and finishing work-related tasks, we are experiencing role conflict. We cannot be in two places at once.

Men who see themselves as traditional providers may experience role strain when pressed into higher levels of housework or childcare. Employed wives exhausted by their combination of paid work, housework and childcare may also experience role strain and not enjoy sexual intimacy with their spouses.

There is some evidence suggesting that job stress has a "crossover effect" on a spouse or other family members. When one spouse feels a lot of pressure or overload at work, the other spouse may begin to feel depressed or overloaded as well. This may be especially true regarding a crossover effect of husbands' overload onto wives. Less clear is how much "crosses over" from parents to children or whether parent–child relationships are affected in similar ways as marriages. Crouter and colleagues found that both fathers' role overload and the amount of hours they worked affected the quality of their relationships with their adolescent children (2001). When fathers worked long hours but did not experience overload, their relationships with their adolescents do not seem to suffer. It appears that for fathers and children the combination of hours and overload have the greatest effect (Crouter et al. 2001).

Parental "availability" to children is affected by the levels of stress that parents experience. Particularly stressful days at work may be followed by parents being withdrawn at home. This may sometimes prove beneficial because, by withdrawing, less negative emotion is brought into the relationships (Perry-Jenkins, Repetti, and Crouter 2000).

Some research indicates that individuals with high self-esteem feel less role conflict than those with low self-esteem (Long and Martinez 1994). Women spend less time on housework if they are employed (Coltrane 2000; Greenstein 1996). But women with high self-esteem accept lower housekeeping standards as necessary and realistic adjustments to their multiple roles rather than as signs of inadequacy.

However, the more important sources of role conflict and overload are not within the person but rather *within the person's role responsibilities*. Men experience role conflict when trying to balance their family and work roles. Because men are expected to give priority to their jobs over their families, it is not easy for men to be as involved in their families as they may like. A study examining role conflict among men (O'Neil and Greenberger 1994; see also Marks 1994 and Greenberger 1994) found that men with the least role conflict fell into two groups. One group consisted of men who were highly committed to both work and family roles. They were determined to succeed at both. The other group consisted of men who put their family commitments above their job commitments. They were willing to work at less demanding or more flexible jobs, spend less time at work, and put their family needs first. In both instances, however, the men received strong encouragement and support from their spouses.

Married women employed full-time often prefer working fewer hours as a means of reducing role conflict (Warren and Johnson 1995). Some women work a shift different from that of their spouses or partners. Not surprisingly, because they have less role conflict, single women (including those who are divorced) are often more advanced in their careers than married women (Houseknecht, Vaughan, and Statham 1987). They are more likely to be employed full-time and have higher occupational status and incomes. They are also more highly represented in the professions and hold higher academic positions.

The various issues surrounding spillover, role conflict, role overload, and role strain vary depending upon the household structure and division of labor. Single-parent households with full-time working parents are easily susceptible to role overload and role conflict. Two-parent, dual-earner households also face versions of work-to-family spillover different from those of households with one provider and a partner at home full-time.

Comparing levels of expressed work–family interference from two large survey sources, the Quality of Employment Study in 1977 and the National Study of the Changing Workforce in 1997, Sarah Winslow (2005) offered the following conclusions about work–family conflict: Compared with respondents in 1977, respondents in 1997 reported greater difficulty balancing work and family. This was greatest among parents regardless of whether they were in dual-earner or single-earner households. Also, women and men reported similar levels of work–family interference.

Reflections

Much of the workplace–family linkage concept can be applied to the college environment. If you think of your student role as a work role and the college as the workplace, what types of work spillover do you experience in your personal or family life? If you are a homemaker or are employed (or both), what kinds of role strain do you experience?

The Familial Division of Labor

Families divide their labor in a number of ways. Some follow more traditional male–female patterns, most share wage earning, and a small number reverse roles. Even within a single family, there will likely be a number of divisions of labor over time, as the family members move through the various family life-cycle stages. How families allocate tasks and divide paid and unpaid work have a tremendous effect on how a family functions.

The Traditional Pattern

In what we often consider the "traditional" division of labor in the family, work roles are complementary: the husband is expected to work outside the home for wages, and the wife is expected to remain at home caring for children and maintaining the household. A man's family role is secondary to his provider role, whereas a woman's employment role is secondary to her family role (Blair 1993).

In the nineteenth century, industrialization transformed the face of America. It also transformed American families from self-sufficient farm families to wage-earning urban families. As factories began producing farm machinery such as harvesters, combines, and tractors, significantly fewer farm workers were needed. Workers migrated to the cities, where they found employment in the ever-expanding factories and businesses.

Because goods were now bought rather than made in the home, the family began to shift from being primarily a production unit to being a consumer and service-oriented unit. With this shift, a radically new division of labor arose in the family. Men began working outside the home in factories or offices for wages to purchase the family's necessities and other goods. Men became identified as the family's sole providers or "breadwinners." Their work began to be identified as "real" work and was given higher status than women's work because it was paid in wages.

Industrialization also created the housewife, the woman who remained at home attending to household duties and caring for children. With industrialization, because much of what the family needed had to be purchased with the husband's earnings, the wife's contribution in terms of unpaid work and services went unrecognized, much as it continues today (Ferree 1991).

In earlier times, the necessities of family-centered work gave marriage and family a strong center based on economic need. The emotional qualities of a marriage mattered little as long as the marriage produced an effective working partnership. Without its productive center, however, the family focused on the relationships between husband and wife and between parent and child. Affection, love, and emotion became the defining qualities of a good marriage.

This difference in primary roles between men and women in traditional households profoundly affects the most basic family tasks, such as who cleans the toilet, mops the floors, does the ironing, and washes the baby's diapers. Women—whether or not they are employed outside the home—remain primarily responsible for household tasks (Demo and Acock 1993). This is the family form that most fits the *two-person career* model (see Chapter 3). Women become the domestic and childrearing supports on whom families depend, freeing men to focus on wage earning and providing.

The division of family roles along stereotypical gender lines varies by race and class. It is more characteristic of Caucasian families than of African American families. African American women, for example, are less likely than Caucasian women to be exclusively responsible for household tasks. Latino and Asian families are more likely to be closer to the traditional than are African Americans or Caucasians (Rubin 1994).

Class differences are somewhat ambiguous. Among middle-class couples, greater ideological weight is given to sharing and fairness. Working-class couples, although less ideologically traditional than in the past, are still not as openly enthusiastic about more egalitarian divisions of labor. However, in terms of *who does what,* working-class families are more likely than middle-class families to piece together work-shift arrangements that allow parents to take turns caring for the children and working outside the home. Such arrangements may force couples to depart from tradition, even if they neither believe they should nor boast that they do (Rubin 1994).

Men's Family Work

The husband's role as provider is probably his most fundamental role in marriage. As Barbara Arrighi and David Maume (2000, 470) put it, "It is the activity in which they spend most of their time and depend on most for their identity." In the traditional equation, if the male is a good provider, he is a good husband and a good father (Bernard 1981). This core concept seems to endure despite trends toward more egalitarian and androgynous gender roles. A woman's marital satisfaction is often related to how well she perceives her husband as fulfilling his provider role (Blair 1993). It is not uncommon for women to complain of husbands who do not work to their full potential. They feel their husbands do not contribute their fair share to the family income.

Looking at marriages in which wives are "mutually dependent"—earning between 40–59% of the family income, such couples increased nearly 300% between 1970 and 2001. As many as 30% of dual-earner couples and 20% of all married couples fit such a pattern. In one-fourth of dual-earner couples, wives outearn husbands (an increase of 40% between 1987 and 2003). In 12% of dual-earner couples, wives earn *at least* 60% of the total income (Winslow-Bowe, 2006). Interestingly, neither pattern has a uniform effect on married life. Only when men have traditional gender attitudes despite finding themselves in nontraditional life situations and marriages do such income differentials negatively affect men (Brennan, Barnett, and Gareis 2001).

Men are traditionally expected to contribute to family work by providing household maintenance. Such maintenance consists primarily of repairs, light construction, mowing the lawn, and other activities consistent with instrumental male norms. (But, as one woman asked, how often do you have to repair the toaster or paint the porch?)

Men often also contribute to housework and childcare, although their contribution may not be notable in terms of the total amount of work to be done. Men tend to see their role in housekeeping or childcare as "helping" their partner, not as assuming equal responsibility for such work. Husbands become more equal partners in family work when they, their wives, or both have egalitarian views of family work or when such a role is pressed upon them by either circumstantial necessity or ultimatum (Hochschild 1989; Greenstein 1996). Men who believe they should act as traditional providers resist performing more housework or do so only reluctantly, whether their wives are employed outside the home or not. If both spouses share a traditional gender ideology (traditional beliefs about what each should contribute to paid and family work), men's low level of household participation is not problematic.

Women's Family Work

Although most women now earn salaries as paid employees, contributing more than 40% of family income in dual-earner households, neither traditional women nor their partners regard employment as a woman's fundamental role (Coontz 1997). For those with traditional gender ideologies, women are not duty-bound to provide; they are duty-bound to perform household tasks (Thompson and Walker 1991).

No matter what kind of work the woman does outside the home or how nontraditional she and her husband may consider themselves to be, there is seldom equality when it comes to housework. Women's family work is considerably more diverse than that of men, permeating every aspect of the family. It ranges from housekeeping to childcare, maintaining kin relationships to organizing recreation, socializing children to caring for aged parents and in-laws, and cooking to managing the family finances. Ironically, family work is often invisible to the women who do most of it (Brayfield 1992).

Sociologist Ann Oakley (1985) described four primary aspects of the **homemaker role:**

- Exclusive allocation to women, rather than to adults of both sexes
- Association with economic dependence
- Status as nonwork, which is distinct from "real," economically productive paid employment
- Primacy to women—that is, having priority over other women's roles

> ### Reflections
> List the tasks that make up family work in your family. What family work is given to women? To men? On what basis is family work divided? Is it equitable?

Most full-time housekeepers feel the same about housework: it is routine, unpleasant, unpaid, and unstimulating, but it provides a degree of autonomy. Full-time male houseworkers, however, do not as often call themselves *housekeepers* or *homemakers*. Instead, they identify themselves as retired, unemployed, laid off, or disabled (Bird and Ross 1993). Increasingly, they may call themselves *househusbands*, but they are less likely to do so than full-time female homemakers are to call themselves *housewives*. Many women find satisfaction in the homemaker role, even in housework. Young women, for example, may find increasing pleasure as they experience a sense of mastery over cooking, entertaining, or rearing happy children. If homemakers have formed a network among other women—such as friends, neighbors, or relatives—they may share many of their responsibilities. They discuss ideas and feelings and give one another support. They may share tasks, as well as problems.

■ Researchers Linda Thompson and Alexis Walker (1991) observe, "Family work is unseen and unacknowledged because it is private, unpaid, commonplace, done by women, and mingled with love and leisure."

© Laurie DeVault Photography

Women in the Labor Force

Women have always worked outside the home. Like many of today's families, early American families were **coprovider families**—families that were economic partnerships dependent on the efforts of both the husband and the wife. Although women may have lacked the economic rights that men enjoyed, they worked with or alongside men in the tasks necessary for family survival (Coontz 1997). Beginning in the early nineteenth century, "work" and "family work" were separated. Men were assigned the responsibility for the wage-earning labor that increasingly occurred away from the home in factories and other centralized workplaces.

Women stayed within the home, tending to household tasks and childrearing. But this gendered division of labor was never total. Single women have traditionally been members of the paid labor force. There have also been large numbers of employed mothers, especially among lower-income and working-class families, African Americans, and many other ethnic minorities. By the late 1970s, the employment rate of Caucasian women began to converge with that of African American women (Herring and Wilson-Sadberry 1993).

The most dramatic changes in women's labor force participation have occurred since 1960, resulting in the emergence of a family form in which both hus-bands and wives are employed outside the home. Although many viewed that family type as abnormal, in the 1980s married women's employment came to be seen as the norm. Recent research indicates that women's employment has positive rather than negative effects on marriages and families (Crosby 1991).

In 2002, more than 67 million women were employed in the civilian labor force. Women comprised 46.3% of the labor force, and 60% of adult women were employed. In comparison, 74% of adult men were employed (U.S. Census Bureau 2003, Tables 592, 596). African American women and Caucasian women had virtually the same rate of labor force participation (61.8% and 59.3%, respectively); Hispanic women were slightly less likely to be employed (57.6%).

Between 1960 and 2002, the percentage of married women in the labor force almost doubled—from 32% to about 61%; this compares to 77% of married men. During that same period, the number of employed married women between 25 and 34 years (the ages during which women are most likely to bear children) rose from 29% to 72%. More than 70% of married women with children were in the labor force in 2000, including 76.8% of those with children 6 to 17 years of age and 60.8% of those with children age 6 or younger (U.S. Census Bureau 2003, Table 597).

In 2002, there were more than 3 million single mothers in the labor force; of these, almost 60% had preschool-aged children (U.S. Census Bureau 2003, Table 598).

Why Women Enter the Labor Force

Four sets of factors influence a woman's decision to enter the labor force (Herring and Wilson-Sadberry 1993):

- *Financial factors.* To what extent is income significant? For unmarried women and single mothers, employment may be their only source of income. The income of married women may be primary or secondary to their husbands' incomes.

- *Social norms.* How accepting is the social environment for married women and mothers working at paid jobs? Does the woman's partner support her? If she has children, do her partner, friends, and family believe that working outside the home is acceptable? After the 1970s, social norms changed to make it more acceptable for white mothers to hold a job.

- *Self-fulfillment.* Does a job meet needs for autonomy, personal growth, and recognition? Is it challenging? Does it provide a change of pace?

- *Attitudes about employment and family.* Does the woman believe she can combine her family responsibilities with her job? Can she meet the demands of both? Does she believe that her partner and children can do well without her as a full-time homemaker?

Like men, women enter the labor force for largely financial reasons. According to Stephanie Coontz (1997), women's incomes keep approximately a third of dual-earner couples from falling into poverty. Economic pressures traditionally have been powerful influences on African American women. Among many married women and mothers, entry into the labor force or increased working hours are attempts to compensate for their husbands' loss in earning power because of inflation. In addition, the social status of the husband's employment often influences the level of employment chosen by the wife (Smits, Utee, and Lammers 1996).

Among the psychological reasons for employment are an increase in a woman's self-esteem and sense of control. A comparison between African American and Caucasian women found that personal preference was the primary employment motivation for about 42% of African American women and 46% of Caucasian women (Herring and Wilson-Sadberry 1993). Employed women are less depressed and anxious than nonemployed homemakers; they are also physically healthier (Gecas and Seff 1989; Ross, Mirowsky, and Goldsteen 1991). A 34-year-old Latina mother of three told social psychologist Lillian Rubin (1994) the following:

> I started to work because I had to. My husband got hurt on the job and the bills started piling up, so I had to do something. It starts as a necessity and it becomes something else.
>
> I didn't imagine how much I'd enjoy going to work in the morning. I mean, I love my kids and all that, but let's face it, being mom can get pretty stale. . . . Since I went to work I'm more interested in life, and life's more interested in me.
>
> I started as a part-time salesperson and now I'm assistant manager. One day I'll be manager. Sometimes I'm amazed at what I've accomplished; I had no idea I could do all this, be responsible for a whole business.

There are two reasons employment improves women's emotional and physical well-being (Ross, Mirowsky, and Goldsteen 1991). First, employment decreases economic hardship, alleviating stress and concern not only for the woman but also for other family members. A single parent's earnings may constitute her entire family's income.

Second, an employed woman receives greater domestic support from her partner. The more a woman earns relative to what her partner earns, the more likely he is to share housework and childcare.

■ *Women seek the same gratifications from paid work that men seek. These include but go beyond wages.*

How do you expect to divide household and employment responsibilities in marriage? More often than not, couples live together or marry without ever discussing basic issues about the division of labor in the home. Some think that things will "just work out." Others believe that they have an understanding, although they may discover later that they do not. Still others expect to follow the traditional division of labor. Often, however, one person's expectations conflict with the other's expectations.

The following questions cover important areas of understanding for a marriage contract. These issues should be worked out before marriage.

Although marriage contracts dividing responsibilities are not legally binding, they make explicit the assumptions that couples have about their relationships.

Answer these questions for yourself. If you are involved in a relationship, live with someone, or are married, answer them with your partner. Consider putting your answers down in writing.

- Which has the highest priority for you: marriage or your job? What will you do if one comes into conflict with the other? How will you resolve the conflict? What will you do if your job requires you to work 60 hours a week? Would you consider such hours to conflict with your marriage goals and responsibilities? What would your partner think? Do you believe that a man who works 60 hours a week shows care for his family? Why? What about a woman who works 60 hours a week?

- Whose job or career is considered the most important—yours or your partner's? Why? What would happen if both you and your partner were employed and you were offered the "perfect" job 500 miles away? How would the issue be decided? What effect do you think this would have on your marriage or relationship?

- How will household responsibilities be divided? Will one person be entirely, primarily, equally, secondarily, or not responsible for housework? How will this be decided? Does it matter whether a person is employed full-time as a salesclerk or a lawyer in deciding the amount of housework he or she should do? Who will take out the trash? Vacuum the floors? Clean the bathroom? How will it be decided who does these tasks?

- If you are both employed and then have a child, how will the birth of a child affect your employment? Will one person quit his or her job or career to care for the child? Who will that be? Why? If both of you are employed and a child is sick, who will remain home to care for the child? How will that be decided?

Women's Employment Patterns

The employment of women has generally followed a pattern that reflects their family and childcare responsibilities. Because of the family demands they face, women must consider the number of hours they can work, what time of day to work, and whether adequate childcare is available. Traditionally, women's employment rates dropped during their prime childbearing years, from 20 to 34 years. But this is no longer true; most women with children are in the labor force, regardless of age of child, marital status, and racial or ethnic affiliation.

Women no longer automatically leave the job market when they become mothers. Either they need the income or they are more committed to work roles than in previous generations (Coontz 1997). Among first-time mothers, more than half return to their jobs within 6 months of giving birth and two-thirds have returned by the time their child celebrates her or his first birthday. Looking only at women who worked during their pregnancies, only 20% stayed at home for the entire first year of motherhood. For those who returned to work for the same employer as before childbirth, 89% worked at least as many hours—if not more—than they had before they became mothers (Johnson and Downs 2005).

Because of family responsibilities, many employed women work part-time or work shifts other than the 9-to-5 workday. Furthermore, when family demands increase, wives, not husbands, are more likely to cut back their job commitments (Folk and Beller 1993). As a result of family commitments, women tend to interrupt their job and career lives more often than do men.

Although raising children may be among the most meaningful and fulfilling work anyone can do, as writer and journalist Ann Crittenden (2001) notes, it is seriously undervalued in the United States. As a result, mothers pay a price that punishes them socially and economically, just for caring for their children.

Crittenden shows that women incur a steep economic penalty for having invested themselves in raising their children. Among the more extreme aspects of the cost of motherhood is her assessment that a typical, college-educated mother in America loses around $1 million of lifetime earnings as a result of having had and raising a child. How? There are a variety of interconnected issues; a mother may have to forgo, for at least a time, some income she could have earned. She receives no Social Security for time in which she is not "employed" but is, instead, caring for and raising their children. She also cannot count on any other pensions to assist her in her "retirement years" and—if she divorces—cannot expect her contributions and sacrifices to count in her favor.

Sociologists Michelle Buddig and Paula England estimate that for the cohort of women currently in their childbearing years, mothers incur a "wage penalty" of approximately 7% per child that fathers do not suffer. In attempting to account for why mothers pay a 7% "price," they suggest that perhaps a third of the wage penalty results from motherhood leading to fewer years of continuous job experience and lost seniority. That leaves two-thirds of the motherhood penalty unaccounted for. They suggest that it may be the product of employer discrimination and the effects of motherhood on productivity (Budig and England 2001).

Crittenden also offers other examples of the problems women face when becoming mothers, including that a 30-year-old women without children may earn only 90% of men's wages, but a 30-year old woman with children earns only 70% of men's. The loss of income resulting from motherhood ("the mommy tax") may amount to as much as $1 million for college-educated American women. More than one-third of all divorced mothers have to go on welfare because child-support formulas don't factor in the cost of being the primary caregiver.

- Fathers are statistically less likely than mothers to spend money on their children's health and education.

- Eight states have laws protecting them from discrimination in the workplace.

Although these many aspects of "the mommy tax" are significant, Crittenden concludes that an even bigger price, perhaps the ultimate cost to women, is to not have children. A striking gender gap surfaced in a survey of 1,600 MBAs: although 70% of the males had children, only about 20% of the females did.

We recognize a potential danger in highlighting all of these statistics. One might conclude that, given the ways women are financially "punished" when they become mothers, perhaps women ought to rethink the desirability of becoming having children. Yet the issue is much more that changes should be made to lessen the price of motherhood. Crittenden makes more than a dozen recommendations for needed changes which could make a significant difference, such as extending paid parental leave, shortening the workweek, enacting divorce policies that would neither penalize mothers and children nor unduly reward either parent.

Crittenden notes that whatever changes like these could be put into practice would help move us away from the punishing and unfair ways that mothers have been made to suffer.

Researchers have found that a woman's decision to remain in the workforce or to withdraw from it during her childbearing and early childrearing years is critical for her later workforce activities. If a woman chooses to work at home caring for her children, she is less likely to be employed later. If she later returns to the workforce, she will probably earn substantially less than women who have remained in the workforce.

Dual-Earner Marriages

Since the 1970s, inflation, a dramatic decline in real wages, the flight of manufacturing, and the rise of a low-paying service economy have altered the economic landscape. These economic changes have reverberated through families, altering the division of household roles and responsibilities. Today, more than 60% of families with children under 18 years are two-earner families. This includes 66% of two-parent families with children 6 to 18 and 54% of families with children under 6 (U.S. Census Bureau 2003).

The sources of the dual-earner, or coprovider, household are many. Over the past 30 years, wages have *declined* for male high school and college graduates. Since 1973, men ages 25 to 34 have had their wages decline 25%. Sylvia Hewlett and Cornell West (1998) note that even during the economic expansion of the mid-1990s, men's wages dropped.

They point out that "32 percent of all men between 25 and 34 when working full-time now earn less than the amount necessary to keep a family of four above the poverty line" (1998). Among women, wages of high school graduates have also declined, but the drop was less because women started at lower wages. College-educated women saw their wages increase, although they remained well behind the wages paid to male college graduates (Vobejda 1994).

In 2001, the median income among families who depended on the wages of a male breadwinner was $50,926. Families in which both husbands and wives were employed had median incomes of $73,407. Families in which wives worked and husbands didn't had median incomes of $39,566 (U.S. Census Bureau 2003, Table 690).

Economic changes have led to a significant increase in dual-earner marriages. Most employed women are still segregated in low-paying, low-status, low-mobility jobs—secretaries, clerks, nurses, factory workers, waitresses, and so on. Rising prices and declining wages pushed most of them into the job market.

Employed mothers generally do not seek personal fulfillment in their work as much as they do additional family income. Their families remain their top priorities.

Dual-career families are a subcategory of dual-earner families. They differ from other dual-earner families insofar as both husband and wife have high-achievement orientations, a greater emphasis on gender equality, and a stronger desire to exercise their capabilities. Unfortunately, these couples may find it difficult to achieve both their professional and their family goals. Often they have to compromise one goal to achieve the other because the work world generally is not structured to meet the family needs of its employees, as Donna H. Berardo and colleagues (1987) point out:

The traditional "male" model of career involvement makes it extremely difficult for both spouses to pursue careers to the fullest extent possible, since men's success in careers has generally been made possible by their wives' assuming total responsibility for the family life, thus allowing them to experience the rewards of family life but exempting them from this competing set of responsibilities.

Typical Dual-Earners

We are increasingly seeing that marital satisfaction is tied to fair division of household labor (Blair 1993; Pina and Bengston 1993; Suitor 1991). A husband wielding a vacuum cleaner or cooking dinner while his partner takes off her shoes to relax a few moments after returning home from work is sometimes better than him presenting her a bouquet of flowers—it may show better than any material gift that he cares. In a world where both spouses are employed, dividing household work fairly may be a key to marital success (Hochschild 1989; Perry-Jenkins and Folk 1994; Suitor 1991).

Although we traditionally separate housework, such as mopping and cleaning, from childcare, in reality the two are inseparable (Thompson 1991). Although fathers have increased their participation in childcare some, they have made smaller increases in the frequency with which they swing a mop or scrub a toilet. If we continue to separate the two domains, men will take the more pleasant childcare tasks of playing with the baby or taking the children to the playground and women will take the more unpleasant duties of washing diapers, cleaning ovens, and ironing. Furthermore, someone must do behind-the-scenes dirty work for the more pleasant tasks to be performed. Alan Hawkins and Tomi-Ann Roberts (1992) note the following:

> Bathing a young child and feeding him/her a bottle before bedtime is preceded by scrubbing the bathroom and sterilizing the bottle. If fathers want to romp with their children on the living room carpet, it is important that they be willing to vacuum regularly. . . . Along with dressing their babies in the morning and putting them to bed at night comes willingness to launder jumper suits and crib sheets.

If we are to develop a more equitable division of domestic labor, we need to see housework and childcare as different aspects of the same thing: domestic labor that keeps the family running. (See Hawkins and Roberts 1992 and Hawkins 1994 for a description of a program to increase male involvement in household labor.)

Housework

Standards of housework have changed over the last few generations, as wryly noted by Barbara Ehrenreich (1993):

> Recall that not long ago, in our mother's day, the standards were cruel but clear: Every room should look like a motel room. The floors must be immaculate enough to double as plates, in case the guests prefer to eat doggie-style. The kitchen counters should be clean enough for emergency surgery, should the need at some time arise, and the walls should ideally be sterile.
>
> The alternative, we all learned in Home Economics, is the deadly scorn of the neighbors and probably the plague.

The engine of change was not the vacuum cleaner—which, in fact, seemed to increase hours spent in housework because it promised the possibility of immaculateness if its welder "simply" worked hard enough. What changed was that working women could no longer hold up the standards of their mothers—or

of household product advertisers. They now spend less time on housework. But Ehrenreich advises those who miss the good old days: "For any man or child who misses the pristine standards of yesteryear, there is a simple solution. Pitch in!"

Evidence indicates that although men do "pitch in," possibly more often than in the past, they are nowhere near sharing the burden of housework. As noted earlier, housework remains clearly unevenly divided between women and men. Scott Coltrane (2000) reports that the average married woman does more than three times the amount of routine housework as the average married man (32 hours versus 10 hours per week). This includes the most time-consuming chores such as cooking, cleaning, grocery shopping, laundry, and cleaning up after meals. Recall, too, the data in Chapter 2: looking across more than a dozen countries, 65.8% of the males and 72.7% of the females reported that routine housework is usually or always done by wives.

Other studies estimate that men do between 20% and 33% of all housework (Arrighi and Maume 2000; Baxter 1997). Mary Noonan (2001) estimates that women spend 25 hours a week to men's 7 hours on traditionally female household tasks (such as doing laundry and preparing children for school) and an additional 6 hours to men's 4 hours on "gender neutral" tasks (such as paying bills and "chauffeuring" family members). For more occasional tasks that comprise male household tasks, men perform 7 hours to

"My wife works, and I sit on the eggs. Want to make something of it?"

women's 2 hours. Totaling up all household tasks shows women with 33 hours a week to men's 19 hours.

As a result of the division of household tasks, employed married women have more to do, experience more stress, and have less leisure time than married men. They not only do more, they also "almost invariably" manage the housework that men do (Coltrane 2000). Even when at the office, many women, through planning and supervising, may be unable to escape entirely the burdens of their domestic responsibilities.

As indicated in Chapter 8, there are differences between cohabiting couples and married couples. One such difference is that cohabiting couples have more equitable divisions of household labor than do married couples (Baxter 2005). Cohabiting women also do significantly less housework than married women do (Shelton and John 1993). It seems that marriage, rather than living with a man, transforms a woman into a homemaker (Baxter 2005). Marriage seems to change the house from a space to keep clean to a home to care for.

Various factors seem to affect men's participation in housework. Men tend to contribute more to household tasks when they have fewer time demands from their jobs—that is, early in their employment careers and after retirement (Rexroad and Shehan 1987) or when they have jobs that demand fewer hours of actual time at work (Coltrane 2000; Arrighi and Maume 2000). They also participate more in housework when their hours and their wives' hours at work do not overlap (see the discussion of shift work later in this chapter). As their income rises, wives report more participation by their husbands in household tasks; increased income and job status motivate women to try to ensure their husbands share tasks. However, research by Julie Brines reviewed by Coltrane (2000) suggests that men who are economically dependent on their wives do less housework. Likewise, Barbara Arrighi and David Maume (2000) found that men whose wives earn the same or greater amounts of income may attempt to restore their masculinity by avoiding housework.

Other factors that appear to influence men's involvement in housework include the following:

- *Gender role attitudes.* Men who have more traditional gender role attitudes take on a smaller share of housework than do men who have egalitarian views.
- *Men's socialization experience and modeling of parents.* Although it does not seem to influence

women's participation in those same tasks, early parental division of labor acts as a strong predictor of men's involvement in the "female tasks" of housework (Cunningham 2001).

- *Men's status in the workplace.* Men who have their "masculinity challenged" at work reduce their involvement in housework as a way of avoiding feminine behavior (Arrighi and Maume 2000).
- *Men's age and generation.* Older men do less housework than younger men do. Arrighi and Maume speculate that this may be a reflection of generational change, with younger men having been socialized toward more participation than older men were.

Whether a couple has children or not is a factor affecting how much men participate in household labor. Even though the presence of young children increases women's and men's housework, it also skews the division of housework in even more traditional directions. Men tend to work more hours in their paid jobs and women tend to work fewer hours at paid work and more in the home. Women then end up with a larger share of housework than before the arrival of children.

One factor that may not be as strong a determinant as we might predict is the husband's **gender ideology**—what he believes he ought to do as a husband and how paid and unpaid work should be divided. As Arlie Hochschild's (1989) research showed, even traditional men can become more egalitarian if wives successfully use direct and indirect gender strategies. In some instances, repeated requests might be enough. In other cases, ultimatums may be necessary. Aside from these direct strategies, more indirect strategies—helplessness, withholding sexual intimacy, and so on—may work with husbands who otherwise would not do more.

Furthermore, necessity may create more male involvement. Wives with particularly demanding jobs or who work unusual hours (described later) may force their husbands to share more simply because they are not available (Gerson 1993; Rubin 1994). Women appeared to be more satisfied if their husbands shared traditional women's chores (such as laundry) rather than limiting their participation to traditional male tasks (such as mowing the lawn). African Americans are less likely to divide household tasks along gender lines than Caucasians.

We might assume that the stresses and inequalities of juggling paid work and domestic work undermine women's well-being, but research on consequences related to marital, mental health, and physical health tells a much different story. Analyzing more than a quarter century of General Social Survey data, sociologist Jason Schnittker finds that "women who are employed, regardless of the number of hours they work or how they combine work with family obligations, report better health than do those who are unemployed" (American Sociological Association 2004).

Women in dual-earner families appear to be mentally healthier than full-time housewives are (Crosby 1991). In juggling multiple roles, they suffer less depression, experience more variety, interact with a wider social circle, and have less dependency on their marital or familial roles to provide all of their needed gratification. These psychological benefits accrue despite the unequal division of labor.

Emotion Work

Although we might not typically think about them as "work," or include them in a discussion of "family work," there are other tasks that need to be performed to generate and maintain successful and satisfying marital relationships. Such tasks are often referred to as **emotion work** and include the following (Stevens, Kiger, and Riley 2001):

- Confiding innermost feelings
- Trying to bring our partner out of a bad mood
- Praising our partner
- Suggesting solutions to relationship problems
- Raising relationship problems for consideration and discussion
- Taking initiative to begin the process of "talking things over"
- Monitoring the relationship and sensing when our partner is disturbed about something

Although these might not cleanly fit your notion of "tasks," they may be experienced as work by those who feel unevenly burdened by them. According to research by Daphne Stevens, Gary Kiger, and Pamela Riley (2001), women do more of the emotion work in their relationships and report being less than satisfied with how these "responsibilities" are divided. This has important consequences, because both men's and women's satisfaction with the division of emotion work in their relationships was significantly and positively

associated with their marital satisfaction (Stevens, Kiger, and Riley 2001).

Childrearing Activities

As we examined in some detail in the last chapter, men increasingly believe that they should be more involved as fathers than men have been in the past. Yet the shift in attitudes has not been matched by changes in men's caregiving behavior. One study (Darling-Fisher and Tiedje 1990) found that the father's time involved in childcare is greatest when the mother is employed full-time (fathers become responsible for 30% of the care compared with mothers' 60%; the remaining 10% of care is presumably provided by other relatives, babysitters, or childcare providers). The father's involvement is less when the mother is employed part-time (fathers' 25% versus mothers' 75%) and least when she is a full-time homemaker (fathers' 20% versus mothers' 80%). At the other extreme, roughly 2 million fathers are the primary childcare providers while their wives are at work.

Generalizing from research on parental involvement in two-parent families, we find the following:

- Mothers spend from 3 to 5 hours of active involvement for every hour fathers spend, depending on whether the women are employed or not.
- Mothers' involvement is oriented toward practical daily activities, such as feeding, bathing, and dressing. Fathers' time is generally spent in play.
- Mothers are almost entirely responsible for childcare: planning, organizing, scheduling, supervising, and delegating.
- Women are the primary caretakers; men are the secondary.
- David Maume (2006) reports that when men become fathers, they work more hours of paid employment; new mothers reduce their hours of work.

Although mothers are increasingly employed outside the home, many fathers have yet to pick up the slack at home. Children especially suffer from the lack of parental time and energy when their fathers do not participate more. If children are to be given the emotional care and support they need to develop fully, their fathers must become significantly more involved (Hochschild 1989; Hewlett and West 1998).

Why It Matters: Consequences of the Division of Household Labor

Marital Power

An important consequence of women's employment is a shift in the decision-making patterns in a marriage. Although decision-making power in a family is not based solely on economic resources (personalities, for instance, play a large part), economics is a major factor. A number of studies suggest that employed wives exert greater power in the home than that exerted by non-employed wives (Blair 1991; Schwartz 1994). Marital decision-making power is greater among women employed full-time than among those employed part-time. Wives have the greatest power when they are employed in prestigious work, are committed to it, and have greater income than that of their husbands. Conversely, full-time housewives may find themselves taken for granted and, because of their economic dependency on their husbands, relatively powerless (Schwartz 1994).

Some researchers are puzzled about why many employed wives, if they do have more power, do not demand greater participation in household work on the part of their husbands. Joseph Pleck (1985) suggests several reasons for women's apparent reluctance to insist on their husbands' equal participation in housework. These include (1) cultural norms that housework is the woman's responsibility, (2) fears that demands for increased participation will lead to conflict, and (3) the belief that husbands are not competent.

Marital Satisfaction and Stability

How do patterns of employment and the division of family work affect marital satisfaction? Traditionally, this question was asked only of wives, not husbands; even then, it was rarely asked of African American wives, who had a significantly higher employment rate. In the past, married women's employment, especially maternal employment, was viewed as a problem. It was seen as taking from a woman's time, energy, and commitment for her children and family. In contrast, non-employment or unemployment was seen as a major problem for men. But as our discussion of issues surrounding paid work shows, it is possible that the husband's work may increase marital and family problems by preventing him from adequately fulfilling his role as a husband or father: he may be too tired, too busy, or never there. It is also possible that a mother's lack of employment may affect the family adversely: Her income may be needed to move the family out of poverty, and she may feel depressed from lack of stimulation (Menaghan and Parcel 1991).

How does a woman's employment affect marital satisfaction? There does not seem to be any straightforward answer when comparing dual-earner and single-earner families (Piotrkowski et al. 1987). This may be partly because there are trade-offs: a woman's income allows a family a higher standard of living, which compensates for the lack of status a man may feel for not being the "sole" provider. Whereas men may adjust (or have already adjusted) to giving up their sole-provider ideal, women find current arrangements less than satisfactory. After all, women are bringing home additional income but are still expected to do the overwhelming majority of household work. Role strain is a constant factor for women, and in general, women make greater adjustments than men make in dual-earner marriages.

Studies of the effect of women's employment on the likelihood of divorce are not conclusive, but they do suggest a relationship (Spitze 1991; White 1991). Many studies suggest that employed women are more likely to divorce. Employed women are less likely to conform to traditional gender roles, which potentially causes tension and conflict in the marriage. They are also more likely to be economically independent and do not have to tolerate unsatisfactory marriages for economic reasons. Other studies suggest that the only significant factor in employment and divorce is the number of hours the wife works. Hours worked may be important because full-time work for both partners makes it more difficult for spouses to share time together. Numerous hours may also contribute to role overload on the part of the wife (Greenstein 1990). African American women, however, are not more likely to divorce if they are employed. This may be because of their historically high employment levels and their husbands' traditional acceptance of such employment (Taylor et al. 1991).

Overall, despite an increased divorce rate, in recent years the overall effect of wives' employment on marital satisfaction has shifted from a negative effect to no effect or even a positive effect. The effect of a wife's full-time employment on a couple's marital satisfaction is affected by such variables as social class, the presence of children, and the husband's and wife's attitudes and commitment to her working. Thus, the more the wife is satisfied with her employment, the higher their marital satisfaction will be. Also, the higher the husband's approval of his wife's employment, the higher the marital satisfaction.

Data on the effects of the division of domestic labor on marital satisfaction indicate a relationship. Couples who share report themselves as happier and are less at risk of divorce than couples in which men do little of the family work. This appears to be true regardless of whether couples' gender ideologies are traditional, egalitarian, or transitional (somewhere between the other two) (Hochschild 1989). Also, the fewer hours women spend on household tasks, the more time they can spend in "status enhancement" activities and the greater their marital satisfaction (Stevens, Kiger, and Riley 2001). Daphne Stevens, Gary Kiger, and Pamela Riley report that marital satisfaction is affected by the way couples divide each of the three dimensions of domestic labor: domestic work, emotion work, and "status enhancement" work (helping a partner's career development by building goodwill with the partner's clients or coworkers, ensuring that the partner has the needed time to commit to work, and so on). Women felt more resentment and less marital satisfaction when they do the majority of both domestic and emotion work. Only among women with traditional gender ideologies did this differ. For them, marital satisfaction was positively influenced by their feeling that they have fulfilled their marital obligations. In the case of status enhancement work, the more of such tasks women do, the more satisfied they and their husbands report themselves to be with their marriages. Nevertheless, the division of emotion work was most related to marital satisfaction, and the performance of status enhancement activities was least related (Stevens, Kiger, and Riley 2001).

Atypical Dual-Earners: Shift Couples and Peer Marriages

There are some interesting lifestyle variations among dual-earner couples. Couples with these lifestyles differ from more common two-earner couples in one of two ways: they have constructed household arrangements in which the parents work opposite, mostly nonoverlapping shifts, and thus take turns working outside the home and caring for children, or they have consciously adopted a belief in equality and fairness into how they divide domestic responsibilities. As a result of either of these differences, such atypical couples show much higher rates of male participation in childcare and housework than among more typical dual-earners. Briefly consider each of these types.

Shift Couples

In 2001, nearly 15 million Americans worked hours other than the typical 9-to-5 or 8:30-to-4:30 daytime shifts. Harriet Presser, the leading authority on shift work and its consequences for individuals and families, notes that the proportion of Americans who work nonstandard schedules—evenings, nights, weekends, or on shifts that rotate—now exceeds 45%. Only 54.4% of Americans work Monday through Friday, on a fixed schedule, 5 days a week.

Presser (2003) identifies three macrolevel changes that have contributed to an increase in such work circumstances: changes in the economy, demographics, and technology:

- *Changes in the economy.* There has been a substantial increase in the service sector of the economy, which has a high prevalence of nonstandard schedules. Simultaneously, women's labor force participation doubled between 1975 and 2000, from one-third to two-thirds of all adult women.

- *Changes in demographics.* Both delayed age at marriage (by nearly 3 years between 1960 and 2000) and sizable increases in dual-earner couples have contributed to increased demand for entertainment and recreation at night and over weekends (Presser 2003, 4). In addition, as the U.S. population has aged, there has been a need for medical services available to people 24 hours a day, 7 days a week.

- *Changes in technology.* Computers, faxes, overnight mailing, and other communications technology have made round-the-clock offices a norm for many multinational corporations

Although such large-scale changes have expanded the opportunity to work atypical schedules, *why* do individuals choose to do so? More than 60% of individuals working nonstandard schedules identify job demands or constraints as the driving force behind their work schedules. These include such reasons as "they could not get any other job, the hours were mandated by the employer, or the nature of the job required non-standard hours" (Presser 2003, 20). Only among mothers of children under 5 years do we find as many as 43.8% identifying "caregiving" needs as a reason for their employment schedule. Looking specifically at childcare arrangements, 35% of mothers and 7.6% of fathers of children under 5 years identify "better childcare arrangements" as a main reason for their nonstandard shifts.

Couples in which one spouse works such a nonstandard shift and the other remains in a more typical shift are sometimes referred to as *opposite shift, split shift,* or simply *shift couples.* **Shift couples** structure their home and work lives into a turn-taking, alternating system of paid work and family work. When one is at work, the other is at home. When the at-work partner returns home, the at-home partner departs for work, giving them a kind of "hello, good-bye" lifestyle. Presser indicates that nearly 28% of dual-earner couples have at least one spouse working "other than a fixed day," and in only 2% of dual-earner couples were both spouses employed in the same nonstandard shift. Hence, about a fourth of all dual-earner couples are shift couples (Presser 2003).

When this lifestyle is the product of choice, shift couples may perceive it as a reasonable trade-off. Through it, they stress the importance of childrearing over the importance of marital relations. Spouses may not see each other much, but they strive to communicate frequently, even if doing so means notes on refrigerators, calls during breaks, or e-mail. Significantly, for the household to function, men are pressed to do a greater share of domestic work and especially childcare than among either traditional couples or more typical dual-earners (Presser 2003; Rubin 1994). If wives work second-shift (late afternoon through midnight) or third-shift (late night through morning) jobs, husbands must feed children dinner or breakfast, see that they do their homework, take baths, go to bed or get up for school, make lunches to take to school, and so on.

Aside from parental involvement, what does shift work do to family life? Much research is pessimistic about the effects of shift work. Harriet Presser reports that shift workers suffer more distress, greater dissatisfaction and higher risks of divorce. She also found that some shift combinations among dual earners increased men's participation in housework and childcare (Presser, 2003). Summarizing these findings, Blanche Grosswald (2004) notes that shift workers have been found to have lower levels of marital satisfaction, more disagreements,

marital and sexual difficulties, higher divorce rates, and more problematic relationships with their children. Grosswald observed that 69% of respondents to the Families and Work Institute's 1997 National Study of the Changing Workforce reported themselves to be satisfied with their family lives; however, among shift workers the results were as shown in Table 12.1.

There are some positive familial outcomes that result from shift work such as the abilities to take turns, to have a parent home with children when the other is at work, and to increase father–child closeness. Couples save money on childcare as well as reducing and reduce some of whatever stress parents might feel about outside caregivers. Additional economic benefits might include the opportunity to earn potentially higher wages and the flexibility to work a second job (Grosswald, 2004). However, Harriet Presser found that those who work nonstandard schedules are more likely to be economically disadvantaged than those who work more typical schedules,

Peer and Postgender Marriages

Among some dual-earner couples, there is explicit agreement that household tasks will be divided along principles of fairness. Many couples believe their family's division of labor is fair (Spitze 1991). Among couples who can afford household help, husbands may be excused from many household chores, such as cleaning and mopping. Because of their incomes, they are allowed to "hire" substitutes to do their share of housework (Perry-Jenkins and Folk 1994).

It is important to note that an equitable division is not the same as an equal division. Relatively few couples divide housework 50-50. For women, fair division

Table 12.1 ■ Family-Life Satisfaction of Workers

Shift	Percentage "Extremely" or "Very" Satisfied with Family Life
Day	71
Evening	56
Night	54
Rotating	63
Split	67
Flexible	74
Total (All Shifts)	69

of household work is more important than both spouses putting in an equal number of hours. There is no standard of fairness, however (Thompson 1991). Because most women work fewer hours than men spend in paid work, and because wives tend to work more hours in the home, some women believe that the household labor should be divided proportionately to hours worked outside the home. Other women believe that it is equitable for higher-earning husbands to have fewer household responsibilities. Still others believe that the traditional division of labor is equitable, wherein household work is women's work by definition.

Middle-class women are more likely to demand equity; equity is less important for working-class women, who are more traditional in their gender-role expectations (Perry-Jenkins and Folk 1994; Rubin 1994).

Peer marriages (or *postgender marriages,* to use Barbara Risman and Danette Johnson-Sumerford's term) take concerns for fairness and sharing to heart in how they structure each facet of their relationships. Rarer than shift couples, they, too, depart from the model of typical dual-earners described previously. Whereas shift arrangements may be the result of choice, necessity, or circumstance, peer relationships typically emerge from egalitarian values or conscious intent. Peer or postgender couples base their relationships on principles of deep friendship, fairness, and sharing. Hence, they monitor each other's level of commitment and involvement, maintain equally valued investments in their paid work, and share household tasks and childcare.

Research by Pepper Schwartz (1994) and Barbara Risman and Danette Johnson-Sumerford (1997) indicate that such relationships avoid many of the trappings that often accompany more traditional divisions of labor, including female powerlessness and resentment and male ingratitude and lack of respect. Furthermore, children receive attention and care from both parents, and men develop deeper relationships with their children than commonly found. Although such couples are rare, they show that the inequities in either the traditional or the more typical dual-earner household are not inevitabilities. Indeed, couples can—and some do—commit themselves to "doing it fairly" (Risman and Johnson-Sumerford 1997).

Coping in Dual-Earner Marriages

Dual-earner marriages are here to stay. They remain stressful today because society has not pursued ways to alleviate the work–family conflict.

The three greatest social needs in dual-earner marriages are (1) redefining gender roles to eliminate role overload for women, (2) providing adequate childcare facilities for working parents, and (3) restructuring the workplace to recognize the special needs of parents and families. Coping strategies include reorganizing the family system and reevaluating household expectations.

Husbands may do more housework. Children may take on more household tasks than before. Household standards—such as a meticulously clean house, elaborate meal preparation, and washing dishes after every meal—may be changed. Careful allocation of time and flexibility assist in coping. Dual-earner couples often hire outside help, especially for childcare, which is usually a major expense for most couples. One of the partners may reduce hours of employment, or both partners may work different shifts to facilitate childcare (but this usually reduces marital satisfaction) (White and Keith 1990).

The goal for most dual-earner families is to manage their family relationships and their paid work to achieve a reasonable balance that allows their families to thrive rather than merely survive. Achieving such balance will continue to be a struggle until society and the workplace adapt to the needs of dual-earner marriages and families.

Reflections

The chances are good that if you cohabit or marry, you will be in a dual-earner relationship. How will you balance your employment and relationship or family needs?

At-Home Fathers and Breadwinning Mothers

An additional departure from both the typical dual-earner and the traditional family is the family type in which spouses switch places or reverse roles. Although the term **role reversal** may be somewhat more familiar to us, it may be more accurate to suggest that what such spouses do is switch traditional places; husbands move into the domestic realm and provide housework and childcare, and wives support the family financially with outside paid work. Calling them *role reversed* implies that men do and experience what women traditionally experienced and that wives approach work and wage earning as husbands traditionally did. This appears not to be the case (Russell 1987; Cohen and Durst 2000).

Of the 23 million married-couple families with children under 15, in 2003, in 4.3% of them (1 million) fathers were home (U.S. Census Bureau, Current Population Survey, Annual Social and Economic Supplement 2003). The reasons men give for staying home do not typically identify "to care for home and family" (only 15.6% of the 1 million at-home men stated this as their reason). Most men are home because of disability (45%), unemployment (11%), retirement (10.7%), school (8.9%), or other reasons (8.7%), but they can and often do provide care for their children. In contrast, 88% of the 6.8 million mothers who are out of the labor force and home with children under age 15 cite "to care for home and family" as their reason (U.S. Census Bureau 2003).

What happens to such couples? Based on research conducted in the 1980s by Graeme Russell and recent research by Theodore Cohen and John Durst (2000), we can point to five areas in which couples experience some impact from having switched places:

- *Economic impact.* Couples live on less money but spend less on childcare. Hence, they may not suffer dramatic declines, especially if women's careers are enhanced and men's occupations were not high paying. Men gain an opportunity to take a "time-out," refocus, and try new career possibilities. They do, however, surrender the provider status and confront the reality of economic dependency. Interestingly, this dependency does not seem to have the same marital consequences for men as it does for at-home women.

- *Social impact.* Socially, men experience some isolation as they lose the primary source of social interaction—the workplace. In addition, couples may become the targets of curiosity, or even criticism, for their choices. Men, however, also receive supportive responses, especially from women. Women often receive envious reactions, especially from coworkers. In general, at-home fathers become visible in their domestic role in contrast to the invisibility that traditionally befalls housewives.

- *Marital impact.* This lifestyle leads to high levels of male involvement in housework and childcare. Although men don't take over everything to the same extent that housewives do, they are likely to share or do most domestic work. In addition, couple relationships change. Whereas Russell (1987) found

the changes to be negative, Cohen and Durst (2000) found high levels of communication, empathy, and appreciation among the couples they studied. In some ways, men who are home full-time, like traditional men before them, benefit from having wives. Wives, in particular, know what it takes to care for households and children. Full-time housewives often are married to men who lack such understanding and appreciation. Women are also aware that their spouses have taken risks and made sacrifices by staying home and support them in ways that breadwinning husbands probably don't support housewives.

- *Parental impact.* Perhaps the most noticeable area of impact is the enlarged relationship between fathers and children. Fathers get to know their children in ways that are not otherwise likely and may not even be possible. Children see fathers in nontraditional ways. Mothers maintain the same sorts of relationships as other employed mothers do with their children, but they have greater peace of mind. Children are not in day care, at the sitter, or home alone. They are home with dad.

- *Personal impact.* Being an at-home father changes the ways men look at their lives, resulting in a reshuffling of priorities and the construction of a new social identity. Breadwinning mothers may also enlarge their sense of themselves as providers, take advantage of the at-home resource, and make work a larger component of their own identities.

The increase in both actual involvement and social visibility of at-home fathers can be seen in a variety of ways and places. There are now a variety of websites (such as Daddyshome.com), a number of newsletters (such as *At-Home Dad,* which also has a website at http://www.athomedad.com), and an annual convention, which in 1999 drew more than 80 men from 20 different states, all catering to the needs and issues confronting at-home fathers (Marin 2000). There is good reason to think that the number of men with this lifestyle will increase in coming years, but it is difficult to know by how much.

Staging and Sequencing

What it means to be male and to be female is influenced not only by biology but also by the way in which families define those roles in their work and home life.

Role taking and role making are negotiated and renegotiated throughout the family life cycle and are influenced by changing patterns in society (Zvonkovic et al. 1996).

To reduce some of the complexity of the dual-earner lifestyle, many couples display a pattern of sequential work and family role staging. This pattern reflects the adjustments women try to make in balancing work and family demands. Many of women's choices about employment and careers are based on their plans for a family and whether and when they will want to work. The key event is first pregnancy.

Before pregnancy, most married women are employed. When they become pregnant, however, they begin leaving their jobs and careers to prepare for the transition to parenthood. By the last month of pregnancy, 80% have left the workforce. Within a year, two-thirds of these women have returned to employment. Those who return to employment are strongly motivated by economic considerations or need.

There are four common forms of sequential work and family patterns:

- *Conventional.* A woman quits her job after marriage or the birth of her first child and does not return.

- *Early interrupted.* A woman stops working early in her career to have children and resumes working later.

- *Later interrupted.* A woman first establishes her career, quits to have children (usually in her 30s), and then returns to work.

- *Unstable.* A woman goes back and forth between full-time paid employment and homemaking, usually according to economic need.

A major decision for a woman who chooses sequential work and family role staging is at what stage in her life to have children. Should she have them early or defer them until later? As with most things in life, there are pros and cons. Early parenthood allows women to have children with others in their age group; they are able to share feelings and common problems with their peers. It also enables them to defer or formulate career decisions. At the same time, however, if they have children early, they may increase economic pressures on their beginning families. They also have greater difficulty in reestablishing their careers.

Women who defer parenthood until they reach their middle career stage often are able to reduce the role conflict and economic pressures that accompany the new parent or early career stage of the traditional pattern. Such women, however, may not easily find other new mothers of the same age with whom to share their experiences. They may find the physical demands to be greater than anticipated.

Some may decide that they do not want children because motherhood would interfere with their careers.

Family Issues in the Workplace

Many workplace issues, such as economic discrimination against women, occupational stratification, adequate childcare, and an inflexible work environment, directly affect families. They are more than economic issues—they are also family issues.

Discrimination against Women

A woman's earnings significantly affect family well-being, regardless of whether the woman is the primary or secondary contributor to a dual-earner family or the sole provider in a single-parent family. Furthermore, as we have seen, women's family responsibilities significantly affect their earnings. Given the importance, however, of women's wage contributions to their families, we need to consider at least briefly economic discrimination against women and sexual harassment. By affecting women's employment status and experiences in their jobs, these become important family issues, as well as economic issues.

Economic Discrimination

The effects of economic discrimination can be devastating for women. In 1997, women in the United States made 74 cents for every dollar that men earned. By 2001, median earnings of men who worked full-

■ *In dual-earner families, interrole conflict is often high as parents try to balance family and work obligations.*

© Tom McCarthy/PhotoEdit

time, year-round were $38,884; for women, the median was $29,680, resulting in a **wage gap** wherein women earned 76% of what men did. Because of sizable differences in women's and men's wages, more women than men are condemned to poverty and federal assistance. Wage differentials are especially important to single women.

Women face considerable barriers in their access to well-paying, higher-status jobs (Bergen 1991). Although employment and pay discrimination are prohibited by Title VII of the 1964 Civil Rights Act, the law did not end the pay discrepancy between men and women.

Much of the earnings gap is the result of occupational differences, gender segregation, and women's tendency to interrupt their employment for family reasons and to take jobs that do not interfere extensively with their family lives. Earnings are about 30% to 50% higher in traditionally male occupations, such as truck driver or corporate executive, than in predominantly female or sexually integrated occupations, such as secretary or schoolteacher. The more an occupation is dominated by women, the less it pays.

Sexual Harassment

Sexual harassment is a mixture of sex and power, with power often functioning as the dominant element. Such harassment may be a way to "keep women in their place". **Sexual harassment** can be defined as two dis-

tinct types of harassment: (1) the abuse of power for sexual ends and (2) the creation of a hostile environment. In **abuse of power,** sexual harassment consists of unwelcome sexual advances, requests for sexual favors, or other verbal or physical conduct of a sexual nature as a condition of instruction or employment. Only a person with power over another can commit the first kind of harassment. In a **hostile environment,** someone acts in sexual ways to interfere with a person's performance by creating a hostile or offensive learning or work environment. Sexual harassment is illegal.

Some estimate that as many as half of employed women are harassed during their working years. Few women report their harassment (Koss et al. 1994). Nonetheless, sexual harassment can have a variety of serious consequences. Some people quit their jobs, others may be dismissed as part of their harassment. Victims also often report depression, anxiety, shame, humiliation, and anger (Paludi 1990).

Lack of Adequate Childcare

As we saw in the last chapter, even though mothers continue to enter the workforce in ever-increasing numbers, high-quality, affordable childcare remains an important but uncertain support. For many women, especially for those with younger children and for single mothers, the availability of childcare is critical to their employment.

Heather Boushey and Joseph Wright (2004) report that "over half of mothers of children under the age of six were employed—three-quarters of employed mothers worked 30 hours per week or more—and nearly all of this group—over 90%—reported using some kind of childcare." Approximately four-fifths of employed mothers use childcare arrangements for their preschool-age children (Boushey and Wright 2004).

For most employed mothers with children 5 to 14 years old, school attendance is their primary day-care solution. Women with preschool children, however, do not have that option; in-home care by a relative is their most important resource. As more mothers with preschool children become employed, families are struggling to find suitable childcare arrangements. This may involve constantly switching arrangements, depending on who or what is available and the age of the child or children (Atkinson 1994).

Women also often use multiple arrangements—the child's father, relatives living in or outside the household, day care, or a combination of these—before a child reaches school age. Of employed mothers, 30% have two childcare arrangements and 8% use three or more. In addition, 20% of working mothers use two day-care centers (Gullo 2000). For African American and Latina single mothers, living in an extended family in which they are likely to have other adults to care for their children is an especially important factor that allows them to find jobs (Rexroat 1990; Tienda and Glass 1985).

Frustration is one of the most common experiences in finding or maintaining day care. Changing family situations, such as unemployed fathers' finding work or grandparents' becoming ill or overburdened, may lead to these relatives being unable to care for the children. Family day-care homes and childcare centers often close because of low wages or lack of funding. Furthermore, As Heather Boushey and Joseph Wright (2004) show, childcare is expensive:

> On average, in 2001, a working mother using formal day care paid $92.30 per week per child, which adds up to an annual cost of $4,615 in 2002 dollars (this calculation assumes two weeks off for vacation—although many low-income mothers do not get vacations). . . . Nearly all mothers using formal or family day care paid for it and, in 2001, on average, this payment took up 9.0% of family income for formal day care and 7.4% for family day care. Working mothers are less likely to pay for relative care, but when they do, it can be a substantial burden: in 2001, on average, costs were $66.20 per week, or $3,310 for a 50-week year.

Other estimates suggest that costs may run between 10% and 35% of a family's budget (Children's Defense Fund 1998), depending on the family's socioeconomic status. The high cost of childcare is a major force that in the past kept mothers on welfare from working (Joesch 1991).

Parents who accept the home-as-haven belief—that the home provides love and nurturing—prefer to place their children in family day-care homes. They

■ *About 10% of children are regularly cared for by grandparents.*

© Laurie DeVault Photography

believe that a homelike atmosphere is more likely to exist in family day care than in preschools or children's centers, where greater emphasis is placed on education (Rapp and Lloyd 1989).

Effect on Employment and Educational Opportunities

The lack of childcare or inadequate childcare has the following consequences:

- It prevents many mothers from taking paid jobs.
- It keeps many women in part-time jobs, most often with low pay, few or no benefits, and little career mobility.
- It keeps many women in jobs for which they are overqualified and prevents them from seeking or taking job promotions or training necessary for advancement.
- It sometimes conflicts with women's ability to perform their work.
- It restricts women from participating in education programs.

For women, lack of childcare or inadequate childcare is one of the major barriers to equal employment opportunity. Many women who want to work are unable to find adequate childcare or to afford it.

Childcare issues may also play a significant role in women's choices concerning work schedules, especially among women who work part-time.

Reflections

Of the family economic issues discussed previously, have any affected you or your family? How? How were they handled?

Inflexible Work Environments and the Time Bind

In dual-worker families, the effects of the work environment stem from not just one workplace but two. Although some companies and unions are developing programs that are responsive to family situations (Crouter and Manke 1994), the workplace in general has failed to recognize that the family has been radi-

cally altered during the last 50 years. Most businesses are run as if every worker were male with a full-time wife at home to attend to his needs and those of his children. But the reality is that women make up a significant part of the workforce, and they do not have wives at home. Allowances are not made in the American workplace for flexibility in work schedules, day care, emergency time off to look after sick children, and so on. Many parents would reduce their work schedules to minimize work–family conflict. Unfortunately, many do not have that option.

Twenty years ago, Carol Mertensmeyer and Marilyn Coleman (1987) contended that our society provides little evidence that it esteems parenting. It appears that little has changed. This seems to be especially true in the workplace, where corporate needs are placed high above family needs. Mertensmeyer and Coleman suggest that family policymakers should encourage employers to be more responsive in providing parents with alternatives that alleviate forced choices that are incongruent with parents' values. For example, corporate-sponsored childcare may offset the conflict a mother feels because she is not at home with her child. Flextime and paid maternal and paternal leaves are additional benefits that employers could provide employees. These benefits would help parents fulfill self and family expectations and would give parents evidence that our nation views parenting as a valuable role.

Unfortunately, policies alone do not guarantee that employees will follow them. In her book *The Time Bind: When Work Becomes Home and Home Becomes Work,* Arlie Hochschild (1997) describes the official policies and corporate culture at a large corporation that she calls Amerco to protect its anonymity. At Amerco, workers could use a number of family-friendly time-enhancing policies, including job sharing, part-time work, parental leave, flextime, and "flexplace" (where workers could work from home). Despite the availability of such options, Hochschild notes that employees rarely used these opportunities.

Hochschild notes that Amerco employees are typical of employees at other large corporations. Citing a 1990 study of 188 Fortune 500 manufacturing companies, and reports that although companies tended to offer family-friendly policies, few employees used them. Of the companies, 88% offered part-time work, but only between 3% and 5% of their employees chose to work part-time. In addition, 45% of the companies offered flextime, but only 10% of employees used it.

Fewer companies offered job-sharing (6%) or work-at-home (3%) options, but among those that did 1% chose to share jobs and less than 3% used flexplace options. This lack of use is especially puzzling given that Amerco employees acknowledge not having enough family time.

In accounting for the lack of utilization of workplace policies, Hochschild considers and rejects a variety of explanations. Can employees afford to work fewer hours? Do they fear being laid off? Do employees even know about policies? Do they have insensitive and insincere supervisors?

These explanations have partial validity. Some hourly employees do fear potential layoffs or reduced wages. There are some supervisors who seem reluctant to embrace and resentful at having to accommodate family-friendly policies. But the biggest reason employees do not use potential family time-enhancing initiatives is because they do not want to. They would rather be at work.

In recent decades, with the dramatic changes in the division of labor and the growth of dual-earner families, home life has become more stressful and tightly scheduled. There is too much to do, too little time to do it, and not enough appreciation or recognition for what is done. On the other end of the work–family divide, many workplaces in the United States have implemented "humanistic management" policies designed to enhance worker morale and productivity and to reduce turnover. Thus, at work, people find social support, appreciation, and a sense of control and competence, which makes them feel better about themselves. In other words, for some, work has become homelike, and home often feels like a job (Hochschild 1997).

Because Hochschild studied only one company, it is hard to know how far we can generalize from her research. Clearly, as we have seen, employed American parents often feel they face a shortage of time to spend with their families. Other researchers have failed to support Hochschild's conclusions, at least to the same extent. For example, a study by Susan Brown and Alan Booth (2002), which uses the National Survey of Families and Households and is based on more than 1,500 dual-earner couples with children, indicates that Hochschild's findings may not be generalizable.

Job status seems to be an important determinant of whether individuals see their jobs as more satisfying than their home lives. Brown and Booth claim that this is true only among workers in positions of lower occupational status. Also, respondents who have high satisfaction with work and low satisfaction at home do not work significantly more hours at work. Only those who are satisfied with work, unsatisfied with home, and have adolescent children work more hours (Brown and Booth 2002).

Another study by K. Jill Kiecolt, based on General Social Survey data from 1973 to 1994, challenged several of Hochschild's conclusions. She argued, for example, that a "cultural reversal" in favor of work over home had not taken place, and employed parents with children under age 6 actually are more likely to find home rather than work to be a haven.

Even if Hochschild's findings are somewhat limited, her study is important for showing that policies are not deterministic (see also Blair-Loy and Wharton 2002). People must take advantage of policies. This suggests that people's values must be directed more toward home and family. Furthermore, cultural reinforcement for using family-friendly policies must be more widespread and reflected in company "cultures." If "time equals commitment" to a job, then work time can only be reduced at the risk of appearing undercommitted. By the same token, dual-earner family life must be made less stressful. One way in which this can occur is by men doing more of the "second shift" work discussed earlier, thereby reducing the overload and time drain that their wives more consistently feel.

Employees who feel supported by their employer with respect to their family responsibilities are less likely to experience work–family conflict. A model corporation would provide *and support* the use of family-oriented policies that would benefit both its employees and itself, such as flexible work schedules, job-sharing alternatives, extended maternity and/or paternity leaves and benefits, and childcare programs or subsidies. Such policies could increase employee satisfaction, morale, and commitment.

Living without Work: Unemployment and Families

Unemployment is a major source of stress for individuals, with its consequences spilling over into their families (Voydanoff 1991). Even employed workers suffer anxiety about possible job loss caused by economic restructuring and downsizing (Larson, Wilson, and Beley 1994). Job insecurity leads to uncertainty that affects the well-being of both worker and spouse. They feel anxious, depressed, and unappreciated. For some, the uncertainty before losing a job causes more emotional and physical upset than the actual job loss.

Economic Distress

Those aspects of economic life that are potential sources of stress for individuals and families make up **economic distress** (Voydanoff 1991). Major economic sources of stress include unemployment, poverty, and economic strain (such as financial concerns and worry, adjustments to changes in income, and feelings of economic insecurity).

In times of hardship, economic strain increases; the rates of infant mortality, alcoholism, family abuse, homicide, suicide, and admissions to psychiatric institutions and prisons also sharply increase. Patricia Voydanoff (1991), one of the leading researchers in family–economy interactions, notes the following:

> A minimum level of income and employment stability is necessary for family stability and cohesion. Without it, many are unable to form families through marriage and others find themselves subject to separation and divorce. In addition, those experiencing unemployment or income loss make other adjustments in family composition such as postponing childbearing, moving in with relatives, and having relatives or boarders join the household.

Furthermore, economic strain is related to lower levels of marital satisfaction as a result of financial conflict, the husband's psychological instability, and marital tensions.

The emotional and financial cost of unemployment to workers and their families is high. A common public policy assumption, however, is that unemployment is primarily an economic problem. Joblessness also seriously affects health and the family's well-being.

The families of the unemployed experience considerably more stress than that experienced by those of the employed. Mood and behavior changes cause stress and strain in family relations. As families adapt to unemployment, family roles and routines change. The family spends more time together, but wives often complain of their husbands' "getting in the way" and not contributing to household tasks. Wives may assume a greater role in family finances by seeking employment if they are not already employed. After the first few months of their husbands' unemployment,

wives of the unemployed begin to feel emotional strain, depression, anxiety, and sensitiveness in marital interactions. Children of the unemployed are likely to avoid social interactions and tend to be distrustful; they report more problems at home than do children in families with employed fathers. Families seem to achieve stable but sometimes dysfunctional patterns around new roles and responsibilities after 6 or 7 months. If unemployment persists beyond a year, dysfunctional families become highly vulnerable to marital separation and divorce; family violence may begin or increase at this time (Teachman, Call, and Carver 1994).

The types of families hardest hit by unemployment are single-parent families headed by women, African American and Latino families, and young families. Wage earners in African American, Latino, and female-headed, single-parent families tend to remain unemployed longer than other types of families. Because of discrimination and the resultant poverty, they may not have important education and employment skills. Young families with preschool children often lack the seniority, experience, and skills to regain employment quickly. Therefore, the largest toll in an economic downturn is paid by families in the early years of childbearing and childrearing.

Emotional Distress

Aside from the obvious economic effect of unemployment, job loss can have profound effects on how family members see each other and themselves. This in turn can alter the emotional climate of the family as much as lost wages alter the material conditions. Men are particularly affected by unemployment because wage earning is still a major way men satisfy their family responsibilities. Thus, when men fail as workers, they may feel they failed as husbands, fathers, and men (Rubin 1994; Newman 1988). As Lillian Rubin (1994) poignantly conveys in *Families on the Fault Line,* when men lose their jobs, "it's like you lose a part of yourself." Unemployed men may display a variety of psychological and relationship consequences, including emotional withdrawal, spousal abuse, marital distress, increased alcohol intake, and diminished self-identity (Rubin 1994). Katherine Newman (1998) suggests that when families suffer downward mobility as a result of male unemployment, relations between spouses or between fathers and children are likely to be strained. Although children and spouses may be initially supportive, their support may wear thin or run out if joblessness lasts and other resources are unavailable, thus preventing families from maintaining their previous economic lifestyle.

Women, too, suffer nonmaterial losses when they lose their jobs, but those losses are different in degree and kind from those that men are likely to suffer. Men have more of their self-identities, and especially their gendered identities, tied up in working; success at work comes to define successful masculinity (Arrighi and Maume 2001). Women have other acceptable ways of maintaining or achieving adult status (as mothers, for example). Thus, although both women and men will suffer from lost work relationships, lost gratification, and even lost structure and purpose to their day, women have not put as many of their "identity eggs" into the "work basket" as have most men.

Coping with Unemployment

Economic distress does not necessarily lead to family disruption. In the face of unemployment, some families experience increased closeness (Gnezda 1984). Families with serious problems, however, may disintegrate. Individuals and families use a number of coping resources and behaviors to deal with economic distress. Coping resources include an individual's psychological disposition, such as optimism; a strong sense of self-esteem; and a feeling of mastery. Family coping resources include a family system that encourages adaptation and cohesion in the face of problems and flexible family roles that encourage problem solving. In addition, social networks of friends and family may provide important support, such as financial assistance, understanding, and willingness to listen.

Several important coping behaviors assist families in economic distress caused by unemployment. These include the following:

- *Defining the meaning of the problem.* Unemployment means not only joblessness but also diminished self-esteem if the person feels the job loss was his or her fault. If a worker is unemployed because of layoffs or plant closings, the individual and family need to define the unemployment in terms of market failure, not personal failure.
- *Problem solving.* An unemployed person needs to attack the problem by beginning the search for another job, dealing with the consequences of unemployment (for example, by seeking unemployment

insurance and cutting expenses), or improving the situation (for example, by changing occupations or seeking job training or more schooling). Spouses and adolescents can assist by increasing their paid work efforts. Studies suggest that about a fifth of spouses or other family members find employment after a plant closing.

- *Managing emotions.* Individuals and families need to understand that stress may create roller-coaster emotions, anger, self-pity, and depression.

Family members need to talk with one another about their feelings; they need to support and encourage one another. They also need to seek individual or family counseling services to cope with problems before they get out of hand.

Poverty

Although poverty and unemployment may appear to be largely economic issues, as we saw in Chapter 3, the family and the economy are intimately connected to each other, and economic inequality directly affects the well-being of America's disadvantaged families. Poverty can drive families into homelessness. The poor have traditionally been isolated from the mainstream of American society (Goetz and Schmiege 1996). Poverty is consistently associated with marital and family stress, increased divorce rates, low birth weight and infant deaths, poor health, depression, lowered life expectancy, and feelings of hopelessness and despair. It is a major contributing factor to family dissolution.

Welfare Reform and Poor Families

Since the 1960s, when massive social programs known as the "war on poverty" cut the poverty rate almost in half, national priorities have shifted. In the last decade or so of the twentieth century, the war on poverty became a war on welfare—or, as some describe it, a war on the poor. Instead of viewing poverty as a structural feature of our society—caused by low wages, lack of opportunity, and discrimination—we increasingly blame the poor for their poverty (Aldous and Dumon 1991; Katz 1990). They are viewed as having become poor *because* they are "losers," "cheats," "lazy," "welfare queens," and "drug abusers"—people undeserving of assistance. Poverty is viewed as the result of individual character flaws—or even worse, as something inherently racial (Katz 1990).

Nearly 13 million people received Aid to Families with Dependent Children (AFDC) benefits in 1996 (U.S. Census Bureau 1998, Table 605). In addition, 27 million people received food stamps; their monthly value averaged $71. About 6.2 million children received free school breakfasts, and 7.2 million pregnant women, infants, and children under 2 years of age participated in supplemental food programs known as the Women, Infants, and Children (WIC) program (U.S. Census Bureau 1998).

There has been considerable antagonism toward welfare and welfare recipients. Much of the antiwelfare sentiment is based on stereotypes of welfare recipients, especially young unmarried mothers. (Whereas women receiving welfare are often described as "welfare queens," there are no equivalent "welfare kings.")

Joel Handler, a longtime welfare researcher (quoted in Herbert 1994), describes the stereotype of welfare recipients as "young women, without education, who are long-term dependents and whose dependency is passed on from generation to generation." He further notes: "The subtext is that these women are inner-city substance-abusing blacks spawning a criminal class." Furthermore, single mothers receiving welfare are stigmatized as incompetent and uncaring; some suggest that their children be placed in orphanages (Seeyle 1994). Conservative thinker Charles Murray, for example, believes most adolescent girls "don't know how to be good mothers. A great many of them have no business being mothers and their feelings don't count as much as the welfare of the child" (quoted in Waldman and Shackelford 1994).

Welfare became a central, emotional issue in 1990s politics. Many Americans who opposed welfare viewed it as violating the work ethic and destroying the traditional family. They believed that a person uses welfare as a way to avoid working and that welfare undermined the traditional family by "encouraging" women to become single mothers (Waldman and Shackelford 1994). They accused unmarried adolescent mothers

of becoming pregnant to collect welfare benefits. But it is doubtful that adolescents are thinking of welfare benefits as they contemplate premarital sex. In fact, part of the problem is that adolescents often don't make the connection between sex and pregnancy. Finally, studies indicate that government welfare policies had little to do with the rise of divorce, single-parent families, and births to single mothers (Aldous and Dumon 1991). Indeed, welfare benefits help stabilize families; those states with the most generous welfare benefits also have the lowest divorce rates (Zimmerman 1991).

Numerous approaches to welfare reform were considered on both the federal and the state levels. On August 22, 1996, President Bill Clinton signed into law the Welfare and Medicaid Reform Act of 1996, also known as the Personal Responsibility and Work Opportunity Act of 1996. This legislation, which became Public Law 104-193, was proclaimed as an effort to "end welfare as we know it." Proponents in Congress believed that welfare had created a climate of irresponsibility and family pathology and saw the reform as a way to prevent or dramatically reduce out-of-wedlock pregnancy, out-of-wedlock births, and single-parent families. The legislation replaced AFDC with **Temporary Assistance for Needy Families (TANF),** which sharply reduced the period during which someone could receive governmental assistance and imposed more restrictive expectations on what recipients were compelled to do to remain eligible for assistance. TANF programs "include mandatory work (public or private, subsidized or unsubsidized), education-, and job-related activities, including job training and job search, for the purpose of (1) providing such families with time-limited assistance to end their dependency on government benefits and achieve self-sufficiency; (2) preventing and reducing out-of-wedlock pregnancies, especially teenage ones; and (3) encouraging the formulation and maintenance of two-parent families" (Bill Summary, 104th Congress, 1996).

Beginning in October 1996, no family or individual was entitled to receive welfare help. Furthermore, recipients of TANF are limited to a maximum of 5 years, either consecutive or nonconsecutive, with exceptions allowed only for such misfortunes as battery or abuse victimization. The law requires that recipients be working within 2 years. The new legislation replaced AFDC entitlement with a block grant of federal funds given to states. States have the authority to decide how to provide assistance to eligible recipients, and the aid can be of some form other than money. Each state is required to operate a statewide welfare program and to provide certain social services (such as childcare or health care for employed mothers) but the specifics may vary within and between states. After a period of steady growth from the mid-1980s on, as a result of welfare reform, welfare rolls were sharply reduced. In Table 12.2, the figures for 1995 are "pre-reform," whereas the 2000 figures reflect the sharp reduction in welfare since the enactment of the 1996 reform act. As 2001 ended, the average number of monthly TANF cases was 57% lower than the number of AFDC cases pre-reform. The 5.4 million people receiving TANF was the lowest number to receive public assistance since 1961. In 2001, families on TANF received an average of $351 per month ($288 for one-child families, $362 for two-child families, $423 for three-child families, and $519 for families with four or more children). By September 2003, there had been still further reduction. There were just over 2.0 million families and nearly 4.9 million individuals receiving TANF assistance (U.S. Department of Health and Human Services, 2004).

Moderates and liberals stress the importance of education and work training to prepare welfare recipients for employment. They believe that affordable childcare should be made available for parents to work. Such solutions, however, entail spending public monies at a time many are demanding tax cuts and limits on spending. Moderates and liberals also criticize welfare programs that make children's welfare support dependent on their parents' reproductive or employment behavior (such as not having children if they are unmarried adolescents or finding employment, regardless of how low the pay). They point out that such programs penalize children if their parents "misbehave." Finally, they note that state bureaucracies may be as or more inefficient and unresponsive as the federal government. More important, states may not be equally willing to devote resources to helping welfare recipients out of poverty.

Other progressives argue that the problem was never welfare but poverty. People use welfare for the simple reason that they are poor. The best way to resolve welfare issues is by focusing on the poverty issues underlying it: low wages, unemployment, the high cost of housing, lack of affordable childcare, economic discrimination against women and ethnic groups, and a deteriorating education system.

No doubt our welfare system was in trouble, but punitive approaches that blame the poor for their poverty are not the only—and may not be the best—way to resolve the problem. Critics contend that more imaginative approaches are needed. To deal with childhood poverty, for example, we might use the approach used by all Western industrial nations (except ours): provide a minimum children's allowance. A children's allowance goes to all families and is based on the belief that a nation is responsible for the well-being of its children (Meyer, Phillips, and Maritato 1991). Because it is universal, no poor child is missed, nor is his or her family stigmatized as being "on welfare." When we examine our attempt to reform and revamp the welfare system, we can't help but wonder what effect the interplay between politics and economics will have on children. As the state creates jobs for parents, it must also pave the way to providing available and affordable childcare. But licensed day care is unlikely to meet the needs of the millions of welfare families and working poor who are mandated to work (Kilborn 1997). Furthermore, in cities such as New York, Chicago, and Boston, the cost of care for even one child may be almost equal to the earnings of a minimum-wage worker. This situation could encourage wider use of unqualified childcare providers or greater reliance on relatives.

Table 12.2 ■ Recipients of Aid to Families with Dependent Children (AFDC) and Temporary Assistance for Needy Families (TANF) 1975–2002

	1975	1980 (AFDC)	1985	1990	1995	2000 (TANF)	2002
Total recipients (in thousands)	11,165	10,597	10,812	11,460	13,652	5,778	5,066
Percentage of U.S. population	5.2	4.7	4.5	4.6	5.2	2.5	NA
Families receiving assistance (in thousands)	3,498	3,642	3,692	3,974	4,876	2,215	2,047

NA means data not available

One consequence of welfare reforms has been the "re-extension" of the family. As many single mothers enter the workforce, as is mandated by the new policies, it is often grandparents, especially grandmothers, who step into the childcare void they leave. The number of children in grandparental care has increased by 50% in the past decade. The new welfare policies may force it even higher. Critics of the reforms say the poor with legitimate reasons for parental unemployment may be caught without a safety net, especially if the economy were to go into a recession (Livernois 1997).

Welfare reform continues to be of acute concern. Evaluation of the legislative changes enacted in 1996 will continue for years, along with various experimental programs. For now, it appears that neither the costs critics feared nor the benefits proponents projected of moving mothers to work have come to pass (Morris 2002). The ongoing challenge remains the same: We must find ways for people to have adequate food and shelter in an environment that facilitates the development of life skills and assists parents to succeed in the labor force. At the same time, we must provide for the safety, care, and guidance of our children.

Reflections

Do you believe that welfare helps or hinders families? Have you, your family, or your friends received welfare assistance? If so, were its effects positive, negative, or both? Why?

Workplace and Family Policy

Family policy is a set of objectives concerning family well-being and the specific government measures designed to achieve those objectives. As we examine America's priorities, it is clear that we have an implicit family policy that directs our national goals. Given the host of issues raised in this chapter, we might argue that if families were truly the national priority we claim them to be, we would entertain and enact policy initiatives such as the following:

- Paid parental leave for pregnancy and sick children and paid personal days for child and family responsibilities
- Flexible work schedules for parents whenever possible and job-sharing alternatives

- Increased minimum wage so that workers can support their families
- Policies to ensure fair employment for all, regardless of ethnicity, gender, sexual orientation, or disability
- Pay equity between men and women for the same or comparable jobs and affirmative action programs for women and ethnic groups
- Corporate childcare programs or subsidies for families
- Individual and family counseling services and provision of flexible benefit programs

Once enacted, policies such as these must be supplemented by sincere cultural support for families and children. People must believe that if they commit themselves to their families they will not suffer unfair economic consequences. This is harder to convey and carry out than are most specific workplace policies.

Reflections

If you were to construct a coherent family policy that meets your needs and reflects your values, what would it be like? How would it compare to the preceding suggestions?

We also cannot help but compare the reality in the United States with that of other countries.

For example, the passage under President Clinton of the **Family and Medical Leave Act (FMLA) of 1993** finally gave unpaid, job-protected leave of up to 12 weeks for employees to care for an ill family member or take time off after childbirth.

However, because the leave is unpaid, many workers cannot afford to lose the income they would sacrifice for 3 months and therefore don't use it. Also, the stipulation this unpaid leave applies only to workplaces with 50 or more employees leaves as many as half of U.S. workers unprotected by the policy. This is in contrast to Europe and Canada, where paid maternity leave is common (Gornick and Meyers 2004). For example, in Finland, mothers receive 44 weeks with about 66% pay, resulting in an estimate of 29 weeks. In Canada, leaves are 50 weeks long at 55% of wages replaced, (resulting in an estimate of 28 weeks.) In Italy, "maternity leave is mandatory for the first five months after childbirth, and the benefit is 80% of the mother's earnings" (Henneck 2003).

Taking all of these into account, we concur with Gornick and Meyers' (2004) assessment of the harsh situation that exists for American families:

> The U.S. is the extreme case even among the English-speaking countries. Most American parents are left to design private solutions to the dilemma of supporting and caring for children. They are left to negotiate, often unsuccessfully, with their employers for paid family leave, reduced-hour options, and vacation time. Most American parents rely on private markets for childcare, especially during the first four years of their children's lives. They pay a substantial portion of their earnings for this care at a point in their careers when they may be least likely to have accumulated savings or to have advanced to high wage positions. And, ironically, they are often purchasing poor-quality care that may jeopardize their children's healthy development, while simultaneously impoverishing an overwhelmingly female childcare work force.

Our marriages and families are not simply emotional relationships—they are also work relationships in which we divide or share many household and childrearing tasks, ranging from changing diapers, washing dishes, cooking, and fixing running toilets and leaky faucets to planning a budget and paying the monthly bills. These household tasks are critical to maintaining the well-being of our families. They are also unpaid and insufficiently honored. In addition to household work and childrearing, there is our employment, the work we do for pay. Our jobs usually take us out of our homes from 20 to 80 hours a week. They are not only a source of income; they also help our self-esteem and provide status. They may be a source of work and family conflict as well.

Now that we have entered the twenty-first century, we need to rethink the relationship between our work and our families. Too often, household work, childrearing, and employment are sources of conflict within our relationships. We need to rethink how we divide household and childrearing tasks so that our relationships reflect greater mutuality. For many, poverty and chronic unemployment lead to distressed and unhappy families. We need to develop and support policies that help build strong families.

Summary

- Families are economic units bound together by emotional ties. Families are involved in two types of work: *paid work* at the workplace and unpaid *family work* in the household.

- Americans appear to be working more and facing family *time strains.* More fathers than mothers believe they do not have time for their families; more mothers than fathers report not having time for themselves.

- *Work spillover* is the effect that employment has on the time, energy, and psychological well-being of workers and their families at home. *Family-to-work spillover* is when the demands from home life reduce the time and energy available to succeed at work.

- *Role strain* refers to difficulties that individuals have in carrying out the multiple responsibilities attached to a role. *Role overload* occurs when the total prescribed activities for one or more roles are greater than an individual can handle. *Role conflict* occurs when roles conflict with one another.

- Evidence indicates that balancing work and family has become more difficult, especially for employed parents.

- The traditional division of familial labor is complementary: husbands work outside the home for wages and wives work inside the home without wages.

- There are four characteristics that define the *homemaker role:* (1) its exclusive allocation to women, (2) its association with economic dependence, (3) its status as nonwork, and (4) its priority over other roles for women.

- Women enter the workforce for economic reasons and to raise their self-esteem. Employed women tend to have better physical and emotional health than do nonemployed women.

- More than half of all married women are in dual-earner marriages. Husbands generally do not significantly increase their share of household duties when their wives are employed.

- Women do between 70% and 80% of daily housework and carry more responsibility for managing the division of housework. Women's household tasks tend to include the daily chores (such as cooking, shopping, cleaning, etc.) and childcare. Men's household tasks tend to be more occasional and often outdoors.

- Men's involvement in routine housework is affected by their gender role attitudes, their upbringing, their experiences and status at work, and their age.

- The division of paid and unpaid labor and the allocation of housework affect marital power, marital satisfaction, and marital stability (that is, the risk of divorce).

- Two contemporary arrangements are (1) *shift couples,* with spouses who work opposite shifts and alternate domestic and caregiver responsibilities and (2) households in which men stay home with children while women support the family financially.

- Nonstandard shift work has increased because of changes in the economy, demographic changes, and technological changes. It affects family experiences in both negative and positive ways.

- There are approximately 1 million fathers of children under 15 years who stay home full-time. In such households, we can identify marital, parental, economic, and social consequences that follow from this arrangement.

- Among the problems women encounter in the labor force are economic discrimination and *sexual harassment.* Families suffer from lack of adequate childcare and an inflexible work environment.

- Unemployment can cause both economic and emotional distress. Unemployment most often affects female-headed single-parent families, African American and Latino families, and young families.

- Welfare reforms have been enacted by the U.S. government. Stricter limits now exist in determining and maintaining eligibility.

- Family policy is a set of objectives concerning family well-being and the specific government measures designed to achieve those objectives.

Key Terms

abuse of power 443

bifurcation of working time 423

coprovider families 429

economic distress 447

emotion work 436

Family and Medical Leave Act of 1993 451

family policy 451

gender ideology 435

homemaker role 428

hostile environment 443

role conflict 425

role overload 425

role reversal 440

role strain 425

sexual harassment 443

shift couples 439

Temporary Assistance for Needy Families (TANF) 449

time strain 423

wage gap 443

work spillover 424

Resources on the Internet

Companion Website for This Book

http://www.thomsonedu.com/sociology/strong

Gain an even better understanding of this chapter by going to the companion website for additional study resources. Take advantage of the Pre- and Post-Test quizzing tool, which is designed to help you grasp difficult concepts by referring you back to review specific pages in the chapter for questions you answer incorrectly. Use the flash cards to master key terms and check out the many other study aids you'll find there. Visit the Marriage and Family Resource Center on the site. You'll also find special features such as access to Info-Trac© College Edition (a database that allows you access to more than 18 million full-length articles from 5,000 periodicals and journals), as well as GSS Data and Census information to help you with your research projects and papers.

CHAPTER 13

© Christina Mendes/Buena Vista Photography

Intimate Violence and Sexual Abuse

Outline

T F 1 Intimate relationships of any kind increase the likelihood of violence.

T F 2 Rape by an acquaintance, date, or partner is less likely than rape by a stranger.

T F 3 Male aggression is generally considered a desirable trait in our society.

T F 4 Studies of family violence have helped strengthen policies for dealing with domestic offenders.

T F 5 Physically abused children are often perceived by their parents as "different" from other children.

T F 6 Sibling violence is the most widespread form of family violence.

T F 7 More than 2 million elderly Americans are emotionally or physically abused by a family member.

T F 8 Deliberate fabrications of sexual abuse constitute nearly 25 percent of all reports.

T F 9 Most people who were sexually abused as children at least partially remember the abuse.

T F 10 Brother-sister incest is generally harmless.

Answer Key for What Do YouThink

1 True, see p. 467;   2 False, see p. 468;   3 True, see p. 459;   4 True, see p. 471;   5 True, see p. 474; 6 True, see p. 476;   7 True, see p. 477;   8 False, see p. 478;   9 True, see p. 481;   10 False, see p. 479.

ike most Americans, you might assume that when you lock your home at night, you are safe, protected from violence by locking out any would-be intruders. The sad reality is that many of us also *lock in* violence once we close and lock our doors to the outside world. It may seem a cruel irony, but the relationships we most value are also the relationships in which we are most violent. The people we love and live with are often the people most likely to hurt or assault us. It is an unhappy fact that intimacy or relatedness increases our likelihood of experiencing abuse, violence, sexual abuse, or even homicide.

Some widely publicized cases of domestic violence include such tragedies as Scott Peterson murdering his wife Laci and her unborn child, Andrea Yates drowning of her five children, the savage and fatal beating of 7 year-old Nixzmary Brown by her stepfather, and the Menendez brothers shooting of their parents after claims of years of sexual and emotional abuse. Although these cases are not typical of intimate violence and abuse, or representative of most homicides in the United States, they are a chilling reminder of the worst of violence among family members. Now, consider, too, the following:

- More than 8 million adults, 5.3 million women, and 3.2 million men experience some form of violence by an intimate partner—spouse, cohabiting partner, boyfriend, or girlfriend.

- Based on various studies, 30% to 40% of college students report violence in dating relationships.

- At least 1 million American children are physically abused by their parents each year.

- Almost 1 million parents are physically assaulted by their adolescents or younger children every year.

- Perhaps as many as two-thirds of teenagers commit an act of violence against a sibling.

- As many as 27% of American women and 16% of men have been the victims of childhood sexual abuse, much of it in their own families.

In addition, as many as 90% of American parents spank their children. Although clearly different from beatings, assaults, physical and sexual abuse, these, too, are violent acts and therefore merit attention and consideration in this chapter.

Think for a moment about who our society "permits" us to shove, hit, or kick. If we assault a stranger, push a coworker or employer, or spank or slap a fellow student or professor, we would run great risk of being arrested. It is with our intimates that we are "allowed" to do such things.

Those closest to us are the ones we are most likely to slap, punch, kick, bite, burn, stab, or shoot. And our intimates are the most likely to do these things to us (Gelles and Cornell 1990; Gelles and Straus 1988). Furthermore, living together provides people more opportunity to disagree, get angry at one another, and hurt one another. In effect, families and households can be very dangerous places.

To understand intimate violence and abuse, we need to consider a range of behaviors and examine the various factors—social, psychological, and cultural—that shed light on why it is that we often hurt the ones we most love. In this chapter, we look at violence between husbands and wives (including marital rape), between gay and lesbian partners, between dating partners (including acquaintance rape), and between siblings, as well as violence committed against children by parents and against parents by grown children. We look, too, at the various models researchers use in studying intimate violence, and we discuss the dynamics of battering relationships. We also discuss prevention and treatment strategies. In the last section of the chapter we discuss child sexual abuse—its forms, participants, and effects, as well as treatment and prevention strategies.

Intimate Violence and Abuse

In exploring the violent and abusive underside of families and intimate relationships, researchers have used different and changing terminology, trying to keep pace with increasing knowledge about the phenomenon (McHugh, Livingston, and Ford 2005). Many now use the terms **intimate partner violence** or **intimate partner abuse** to address the full scope of violence among intimate couples. Other forms of family violence, such as those between siblings or between parents and children, still most often fall under the broader umbrella term *family violence*. They will be addressed later in this chapter.

Researchers differentiate between violence and abuse. For the purpose of this book, we use the definition of **violence** offered by Richard Gelles and Claire Pedrick Cornell (1990): "an act carried out with the intention or perceived intention of causing physical pain or injury to another person." Abuse includes acts

such as neglect and emotional abuse, including verbal abuse, that are not violent. Thus, abuse is broader than family violence.

Violence may best be understood along a continuum, with "normal" and "routine" violence at one end and lethal violence at the other extreme (Gelles and Straus 1988). Thus, family violence ranges from spanking to homicide. We must look at the continuum as a whole to be concerned with "families who shoot and stab each other as well as those who spank and shove, . . . [as] one cannot be understood without considering the other" (Straus, Gelles, and Steinmetz 1980). In this chapter we focus most of our attention on physical violence and sexual abuse that occurs between intimate partners and between family members.

Types of Intimate Violence

Even when we narrow the discussion to violence in intimate couple relationships, we confront a range of behaviors that beg for some kind of differentiation. Michael Johnson and Kathleen Ferraro (2000) offer the following widely used typology of partner violence:

- **Common couple violence** (sometimes called **situational couple violence**) is violence that erupts during an argument when one partner strikes the other in the heat of the moment. Such violence is not part of a wider relationship pattern; it is as likely to come from a woman as a man or to be mutual. It rarely escalates, and it is less likely to lead to serious injury or fatality.

- **Intimate terrorism** occurs in relationships where one partner tries to dominate and control the other. Violent episodes that escalate, and emotional abuse are two common traits. Victims are left "demoralized and trapped" as their sense of self and their place in the world are greatly diminished by their partner's dominance. The violence in intimate terrorism is likely to recur, escalate, and lead to injury. It is also less likely to be mutual.

- **Violent resistance** encompasses what is often meant by "self-defensive" violence. It tends to be more commonly perpetrated by women than men and can signal that the victim is moving toward leaving the abusive partner.

- **Mutual violent control** refers to relationships in which both partners are violently trying to control each other and the relationship.

Distinctions such as these are important if we are to make sense of the data on who commits violence against a partner or spouse. Of the four types, common couple violence seems to be slightly more typical of men than of women, intimate terrorism is "essentially" perpetrated by men, and violent resistance is typically committed by women (Johnson and Ferraro 2000). Also this typology is useful because it differentiates motives and outcomes of violence. Not all intimate violence is an attempt to control a partner, and injuries and fatalities do not occur equally in all types. Other outcomes—economic, psychological, and health related—also differ by the type of violence.

Why Families Are Violent: Models of Family Violence

To better understand violence within the family, we must look at its place in the larger sociocultural environment. Cultural values and beliefs are important to keep in mind. Getting ahead at work, being assertive in relationships, and winning at sports are all culturally approved values. But does aggression necessarily lead to violence?

All families have their ups and downs, and all family members at times experience anger toward one another. But why does violence erupt more often and with more severe consequences in some families than in others? The principal models used in understanding family violence are discussed in the following sections.

Individualistic Explanations

An individualistic approach emphasizes how the abuser's violence is related to a personality disorder, mental or emotional illness, or alcohol or drug misuse (O'Leary 1993). The idea that people are violent because they are crazy or drunk is widely held (Gelles and Cornell 1990), although research indicates that fewer than 10% of family violence cases are attributable to psychiatric causes, and only about 25% of cases of wife abuse are associated with alcohol. Richard Gelles and Claire Pedrick Cornell suggest that this model is especially appealing to abusers because "if we

can persist in believing that violence and abuse are the products of aberrations or sickness, and, therefore, believe ourselves to be well, then our acts cannot be hurtful or abusive." But besides looking at the abuser, we must step back and look at the big picture—at the family and society that influence the abuser.

Ecological Model

The ecological model uses a systems perspective to explore child abuse. Psychologist James Garbarino (1982) suggests that cultural approval of physical punishment of children combines with lack of community support for the family to increase the risk violence within families. Under this model, a child who doesn't "match" well with the parents (such as a child with emotional or developmental disabilities) and a family that is under stress (from, for example, unemployment or poor health) and that has little community support (such as childcare or medical care) can be at increased risk for child abuse.

Feminist Model

The feminist model stresses the role of gender inequalities or cultural concepts of masculinity as causes of violence. Using a historical perspective, this approach holds that most social systems have traditionally placed women in a subordinate position to men, thus supporting male dominance even when that includes violence (Toews, Catlett, and McKenny 2005; Yllo 1993).

There is no doubt that violence against women and children, and indeed violence in general, has had an integral place in most societies throughout history. Feminist theory must be credited for advancing our understanding of domestic violence by insisting that the patriarchal roots of domestic relations be taken into account. However, the patriarchy model alone does not adequately explain the variations in degrees of violence among families in the same society (Yllo 1993). Women are sometimes violent toward their husbands and partners. More mothers are implicated in child abuse than fathers (although this has much to do with responsibility for and time with children). Finally, and most telling, rates of violence between lesbian partners may be as high as among heterosexual partners. Like heterosexual violence, when homosexual violence does occur it is more likely to be a recurrent feature of the relationship than a onetime event. Although it is clear that men's aggressiveness and even male violence are often met with cultural acceptance, not all forms of violence fit with the emphasis on patriarchy.

Social Structural and Social Learning Models

The social models are related to the ecological and feminist models in that they view violence as originating in the social structure.

First, the social *structural* model views family violence as arising from two main factors: (1) structural stress such as low income or illness, and (2) cultural norms such as the "spare the rod and spoil the child" ethic (Gelles and Cornell 1990). Groups with few resources, such as the poor, are seen to be at greater risk for family violence.

Second, the social *learning* model holds that people learn to be violent from society and their families (Ney 1992). The core premise is that children, especially boys, learn to become violent when they are a victim or witness to violence and abuse (Bevan and Higgins 2002). This is even more likely if the child experiences positive reinforcement for displaying violence. Although it is true that many perpetrators of family violence were abused as children, it is also true that many victims of childhood violence do not become violent parents. These theories do not account for this discrepancy. (See Egeland 1993 and Kaufman and Zigler 1993 for conflicting views on the significance of the intergenerational transmission of abuse.)

Resource Model

William Goode's (1971) resource theory can be applied to family violence. This model assumes that social systems are based on force or the threat of force. A person acquires power by mustering personal, social, and economic resources. Thus, according to Goode, the person with the most resources is the least likely to resort to overt force. Gelles and Cornell (1990) describe the typical situation: "A husband who wants to be the dominant person in the family but has little education, has a job low in prestige and income, and lacks interpersonal skills may choose to use violence to maintain the dominant position."

Exchange-Social Control Model

Richard Gelles (Gelles 1993b; Gelles and Cornell 1990) posits the two-part exchange–social control theory of family violence. The first part, exchange theory, holds that in our interactions, we constantly weigh the perceived rewards against the costs. When Gelles says that "people hit and abuse family members because they can," he is applying exchange theory.

The expectation is that "people will only use violence toward family members when the costs of being violent do not outweigh the rewards." The possible rewards of violence might be getting their own way, exerting superiority, working off anger or stress, or exacting revenge. Costs could include being hit back, being arrested, being jailed, losing social status, or dissolving the family. Three characteristics of families that may reduce those costs of violence, and thus reduce social control are the following:

- *Inequality.* Men are stronger than women and often have more economic power and social status. Adults are more powerful than children.

- *Private nature of the family.* People are reluctant to look outside the family for help, and outsiders (the police or neighbors, for example) may hesitate to intervene in private matters. The likelihood of family violence decreases as the number of nearby friends and relatives increases (Gelles and Cornell 1990).

- *"Real man" image.* In some American subcultures, aggressive male behavior brings approval.

A violent man may gain status among his peers for asserting his "authority."

The exchange-social control model is useful for looking at treatment and prevention strategies for family violence, discussed later in this chapter.

Each of these models has valuable insight to offer concerning a complex problem with no easy or single solution. Looking across the theories we see that several factors surface repeatedly.

Gender

Although there is female-on-male violence and female-on-female violence (discussed later), violence by males tends to be more extreme, often has different causes (power and control versus self-defense), and typically results in different consequences (in terms of both physical injuries and domination). Thus, gender matters a lot with family violence

Power

Central to many theories of intimate violence is the idea of power. Power is a central motive in much intimate violence, especially the long-term and extreme forms of spousal violence that Michael Johnson calls intimate terrorism. Also, powerlessness can be linked to violence when those who feel dominated and unable to legitimately assert their rights may turn to violence as a last resort.

Stress

As individuals are subjected to a variety of stresses (such as unemployment, underemployment, illness, pregnancy, work-related relocations, and difficult or disabled children) tensions among family members may rise. Stress-based explanations help account for the greater prevalence of violence among lower-income families and households facing unemployment, but stress alone cannot account for the breadth and depth of family violence (McCaghy, Capron, and Jamieson 2000; Straus, Gelles, and Steinmetz 1980). However, stress may raise the likelihood of violence, but it is not the cause. Somewhere, the individual must have learned that acting violently toward loved ones is appropriate and acceptable (Gelles and Straus 1988).

Intimacy

The heightened emotions and long-term commitments that characterize family relationships are qualities we value about those relationships. Those same qualities lead to a greater likelihood that we will have disagreements, that those disagreements will be more emotional. Furthermore, cultural beliefs promote the idea that we have the right to influence our loved one's behavior. Some abusive men explain that they assault their spouses "because they love them."

Also, as discussed in Chapter 3, we grant and expect privacy and even secrecy to family relationships. Even when family conflict is in a public setting, others are reluctant to intervene in such "domestic disputes." In some ways, our society thus legitimizes violence and force within families and then turns the other way when they occur.

Prevalence of Intimate Violence

It is difficult to know exactly how much violence there is in families and relationships in the United States. Part of the difficulty results from methodological limitations in the various data we gather. Depending on *how* we gather the information, estimates of *how much* there is and of *where it happens* will vary. You might think that there are "official statistics" we could use, such as arrest records or emergency room visits. Yet so much family violence is unreported that the official data incomplete (U.S. Bureau of Justice Statistics 1998). Plus, some people are better positioned to hide their abusive behavior from authorities and upper- and middle-class abusers may be given more credibility by police. People who can afford to use nonhospital medical resources (such as family doctors to treat injuries) may avoid suspicion since the incident won't show up in hospital records.

Data from domestic violence shelters are even more severely limited since most victims don't seek out a shelter. Also, most women who use shelters are from lower economic backgrounds (Cunradi, Caetano, and Schafer 2002). Thus, the information about shelter populations do not reflect the extent of the wider problem.

That leaves survey data. Many discussions of intimate violence rely on surveys of large random samples drawn from the wider U.S. population. Such studies include the National Family Violence Resurvey, the National Survey of Families and Households, the National Violence Against Women Survey, and the National Longitudinal Couples Survey. In addition, broader studies of crime and victimization such as the National Crime Victimization Survey, the FBI's Supplemental Homicide Reports, and the Study of Injured Victims of Violence, have been used to better estimate the prevalence of intimate violence and to understand the influence of social and economic factors (Field and Caetano 2005).

Of course, reports and estimates based on survey data are themselves prone to problems. In asking people to admit to family violence, researchers may receive underreports. Even in anonymous surveys, individuals may downplay their involvement in socially undesirable behavior. Nevertheless, the estimates from such large-scale, national surveys give us our best ideas of the frequency and spread of family violence. It is on such data that most estimates in this chapter are based.

Based on survey data from large, representative samples of heterosexual couples in the United States, approximately 12% of adult intimates experience some form of physical abuse from their partners; out of every 1,000 couples, 122 wives and 124 husbands are assaulted by their spouses (Renzetti and Curran 1999). Another national survey estimates nearly 9 million couples, one out of six marriages, experiencing some incident of violence every year (Gelles and Straus 1988; Newman 1999). The National Violence Against Women Survey found that 22% of women report physical assault from an intimate partner (Cherlin et al. 2004). Roughly one out of five couples in the general population report having experienced intimate partner violence according to 25 years of survey data summarized by Craig Field and Raul Caetano (2005).

Using multiple sources of data, the Bureau of Justice Statistics produced a report on violence between intimates (Bureau of Justice Statistics 1998). Key findings are as follows:

- There are an estimated 1 million rapes, sexual assaults, robberies, or assaults (simple or aggravated) between intimates each year.

- Approximately 85% of these incidents had female victims.

- 150,000 men were victims of violent crimes committed by an intimate.

- In 2000, there were nearly 1,700 murders attributed to spouses, ex-spouses, boyfriends, or girlfriends; 1 in every 11 homicides was a murder between intimate partners or ex-partners. Spousal homicides are down dramatically, however.

- Nearly 40% of violent incidents occur on weekends, and most occur in or around the victim's home.

- In 2000, 33% of female murder victims and 4% of male murder victims were killed by an intimate.

■ *Tension and conflict are normal features of family life but can escalate into violence under certain conditions.*

Women and Men as Victims and Perpetrators

"Battering", as used in the literature on family violence, includes slapping, punching, knocking down, choking, kicking, hitting with objects, threatening with weapons, stabbing, and shooting. Although the term *battering* does not specify the gender of the batterer, we most likely assume that the batterer is male and the victim is female. However, survey research has found that the number of women who report expressing violence toward their male partners is the *same as or greater than* the number of men who report expressing violence toward their female partners. This is true of research on spousal, cohabiting, and dating relationships.

However, it appears that most violence perpetrated by women on men (as well as most male-on-female partner violence) is of the more situational, routine, and relatively minor variety. It is not the sort of violence that typically leads to hospitals or shelters. Yet the less common and more extreme violence that escalates and causes serious injury or even death is usually committed by men against women (Johnson 1995).

Ignored or rejected by many researchers through the 1970's and 80's, or interpreted as signs of "self-defensive" or reactive violence by female victims, we now know that women use violence with male part-

ners about as often as men do with female partners (Frieze, 2005). One analysis of more than 80 studies of physical aggression between intimate partners found similar proportions of male and female violence (Archer, 2000, cited in Graham-Kevan and Archer, 2005).

However, we need to keep in mind that even when the rates of violence are similar for males and females, the motives and outcomes of male-on-female and female-on-male violence may not be. There is reason to suspect that women and men use violence for different reasons. As Maureen McHugh and colleagues (2005) assert, men's violence tends to be instrumental: they use violence to get what they want and to assert control and gain power over their partners. Women's motives include self-defense, retaliation, expression of anger, attention seeking, stress or frustration, jealousy, depression, and loss of self-control.

We also must remember that historically and culturally, women have unfortunately been considered "appropriate" victims of domestic violence (Gelles and Cornell 1990). Many mistakenly accept the misogynistic idea that women sometimes need to be "put in their place" by men, thus providing a disturbing cultural basis for the physical and sexual abuse of women. There is no comparable cultural justification for the physical or sexual abuse of men.

As far as outcomes are concerned, more female victims than male victims are injured from partner violence and their injuries tend to be more severe than those received by male victims. Even the same acts are not really the same: a slap that breaks the victim's jaw is not the same as a slap that reddens the victims face. In other words, men's slaps (or punches, shoves, kicks, and so on) are not identical to those of women (McHugh et al. 2005).

In violent relationships, a woman may not only suffer physical damage but also be seriously harmed emotionally by a constant sense of danger and the expectation of violence that weaves a "web of terror" about her (Edelson et al. 1985). Lenore Walker (1993) suggests that women who are repeatedly abused may develop a set of psychological symptoms similar to those of post-traumatic stress disorder (PTSD). She labels these symptoms *battered woman syndrome.*

Female Victims and Male Perpetrators

No one knows with certainty exactly how many women are victims of partner violence each year, but as shown earlier, the data we have paint a less-than-optimistic picture. Consider, too, these facts from the Bureau of Justice Statistics (1998, 2003):

- Of all violent crime experienced by women, 20% is from an intimate (spouse, ex-spouse, or boyfriend). In 2001, intimates accounted for 3% of nonfatal violence against men.

- In 1996, at least a third of women who experienced violence reported having been assaulted more than once within the 6 months before the survey; 12% were assaulted at least six times.

- Half of victims report an injury; one in five injured women seeks medical treatment.

- More than 55% of female victims call the police. Police typically respond in 10 minutes or less, although more than 40% of victims say police took 1 hour or more to arrive.

- Fortunately, trend data indicate that such violence may be declining. Between 1993 and 2001 intimate violence against women declined by nearly half. In that same time span, the rate against males dropped 42%.

Women of all races, ages, and socioeconomic statuses are victimized, although they are not victimized equally. Younger women, black women, lower-income women, and urban women are more frequent victims of partner violence. One out of every 50 women, ages 16 to 24, was a victim of intimate violence. This is the highest per capita rate of victimization. Black women

suffered higher rates of nonlethal violence than did white women. As income increased, the rate of female victimization decreased (Bureau of Justice Statistics 1998). Although no social class is immune to it, as shown later, marital violence is more likely to occur in low-income, low-status families (Gelles and Cornell 1990). (For an exception to this, see the "Exploring Diversity" box on upscale violence.)

Although early studies of battering relationships seemed to indicate a cluster of personality characteristics constituting a typical battered woman, more recent studies have not borne out this viewpoint. Factors such as low self-esteem or childhood experiences of violence do not appear to be necessarily associated with a woman being in an assaultive relationship (Hotaling and Sugarman 1990). Two characteristics, however, do appear to be highly correlated with wife assault. First, a number of studies have found that wife abuse is more common and more severe in families of lower socioeconomic status. However, this is partly due to the fact that higher income adults have greater privacy, and thus greater ability to conceal domestic violence (Fineman and Mykitiuk 1994). Second, marital conflict—and the inability to resolve conflict—is a factor in many battering relationships. Gerald Hotaling and David Sugarman (1990) found that common sources of conflict were the division of labor, the husband's heavy drinking, and the wife's superior educational level. These researchers concluded that it is not useful to focus "primarily on the victim in the assessment of risk to wife assault."

Characteristics of Male Perpetrators

A man who systematically inflicts violence on his wife or lover is likely to have some or all of the following traits (Edelson et al. 1985; Gelles and Cornell 1990; Goldstein and Rosenbaum 1985; Margolin, Sibner, and Gleberman 1988; Vaselle-Augenstein and Erlich 1992; Walker 1979, 1984):

- He believes the common myths about battering (see the "Understanding Yourself" box on page 461).

- He believes in the "traditional" home, family, and gender-role stereotypes.

- He has low self-esteem and may use violence as a means of demonstrating power or adequacy.

- He may be sadistic, pathologically jealous, or passive-aggressive.

- He may have a "Dr. Jekyll and Mr. Hyde" personality, being capable at times of great charm.

- He may use sex as an act of aggression.

- He believes in the moral rightness of his violent behavior (even though he may "accidentally" go too far).

Maureen McHugh and colleagues (2005) note that in addition to perpetrating violence, violent men are likely to be the target of violence, either in the present or in their past. In other words, they are either victims of mutual violence or have histories of being abused themselves. We often read or hear the mistaken notion that a major factor in predicting a man's violence is his childhood experience of violence in his family. According to research, a childhood troubled by parental violence accounts for only 1% of adult dating violence and approximately the same proportion of violence in marriage or marriage-like relationships (see review by Johnson and Ferraro 2000). Although it is true that sons of the *most* violent parents have a 1,000% greater rate of wife-beating than sons of nonviolent parents, the majority of these sons are not violent. A recent study noted that 80% of the sons of even the most violent parents were nonviolent for at least the past 12 months (Johnson and Ferraro 2000).

Female Perpetrators and Male Victims

The incidence and experiences of "battered husbands" are poorly understood. Although it is undoubtedly true that some men are injured in attacks by wives or lovers, most injured victims of severe intimate partner violence are women. Thus, we may not consider violence by women as significant as that committed by men (Straus 1993). Often, even if a woman attempts to inflict damage on a man in self-defense or retaliation, her chances of prevailing in hand-to-hand combat with a man are slim. A woman may be severely injured simply trying to defend herself. Remember, though, when we combine common couple violence and violent resistance, about the same rate of female-on-male acts of violence occur.

Suzanne Steinmetz (1987) suggests that some scholars "deemphasize the importance of women's use of violence." As such, there is a "conspiracy of silence [that] fails to recognize that family violence is never inconsequential." Sociologist Murray Straus (1993) offered four reasons for taking the study of female violence seriously:

- Assaulting a spouse—either a wife or a husband—is an "intrinsic moral wrong."

- Not doing so unintentionally validates cultural norms that condone a certain amount of violence between spouses.

- There is always the danger of escalation. A violent act—whether committed by a man or a woman—may lead to increased violence.

- Spousal assault is a model of violent behavior for children. Children are affected as strongly by viewing the violent behavior of their mothers as by viewing that of their fathers.

Furthermore, as Todd Migliaccio (2002) argues, if the experiences of abused women and abused men are similar, if they identify common themes and experiences, we will be better able to identify techniques abusers use *regardless of their sex or gender*. Indeed, from his exploratory interview study with a dozen male victims of female-on-male marital violence, he concluded that, indeed, common themes from past research on wife abuse can be employed to make sense of husband abuse, despite the size and strength differences between husbands and wives.

Class and Race

We often hear about how "democratic" intimate violence is, occurring among all groups, regardless of economic status, race, or sexual orientation. Indeed, there is truth to that statement: intimate partner violence *can* be found among all ethnic and economic groups; however, the amount of violence varies greatly.

Class

More than three decades of research demonstrates an association between socioeconomic status and partner violence. Consider the following sample findings from recent large, national surveys (Cunradi, Caetano, and Schafer, 2002):

- In the 1975 National Family Violence Survey, families classified as "low" income had more than four times the rate of wife assaults compared to those classified as "high" income: 16.4 per 100 compared to 3.5 per 100. The 1985 National Family Violence Survey found that even after controlling for alcohol use and beliefs about violence, blue-collar men abused their spouses at higher rates than white-collar men.

- Data from the 1987 National Survey of Families and Households found that those who had graduated from college were 30% less likely to report intimate partner violence than were high school graduates. Those who had not completed high school were 40% more likely than high school graduates to report intimate partner violence. Income also made a difference. Individuals with household incomes between $25,000 and $39,999 were 50–70% less likely to report experiencing partner violence than were those individuals with incomes less than $25,000.

- Data from the 1992 National Crime Victimization Survey found that among women, young women in low-income households were the most likely to experience partner violence.

- Using data from the 1995 National Alcohol Survey, Carol Cunradi, Raul Caetano, and John Schafer (2002) found that household income had the greatest influence on intimate partner violence, across racial and ethnic lines.

Although there are consistent and strong associations between low economic status and violence, research also reveals partner violence and abuse among high status couples as well (Weitzman, 2000). Their economic position may even create unique problems for women who are victimized.

Race

According to data from the National Family Violence Survey and the National Longitudinal Couples Survey, African Americans have higher rates of violence than either Caucasians or Hispanics and Hispanics have a higher rate than Caucasians. However, the difference between Caucasians and Hispanics tends to diminish if not disappear when we control for various demographic, familial, and social background variables (for example, history of violence between parents, violent victimization in childhood, alcohol problems, and drug use). Between whites and blacks, a significant difference remained in the experience of female-on-male partner violence, even after the demographic and social variables were controlled (Field and Caetano 2005). In research using data from the

National Alcohol Survey, African Americans reported double the rate of both types of violence—male-on-female (23% versus 11%) and female-on-male (30% versus 15%). Hispanics were between whites and blacks (17% for male-on-female, and 21% for female-on-male partner violence) (Caetano et al. 2000).

Marital Rape

One of the most serious, widespread, and overlooked forms of intimate violence, **marital rape** is a form of battering inflicted by husbands on wives, often as parts of a pattern of intimate terrorism.

Most legal definitions of **rape** include "unwanted sexual penetration, perpetrated by force, threat of harm, or when the victim [is] intoxicated" (Koss and Cook 1993). Rape may be perpetrated by males or females and against males or females; it may involve vaginal, oral, or anal penetration; and it may involve the insertion of objects other than the penis. Approximately 10% to 14% of wives have been forced by their husbands to have sex against their will (Yllo 1995).

Historically, marriage has been regarded as giving husbands unlimited sexual access to their wives. Beginning in the late 1970s, most states enacted legislation to make at least some forms of marital rape illegal. On July 5, 1993, marital rape became a crime in all 50 states. Throughout the United States, a husband can be prosecuted for raping his wife, although many states limit the conditions, such as requiring extraordinary violence. Less than half of the states offer full legal protection for wives (Muehlenhard et al. 1992). The precise definition of marital rape differs from state to state, however. In several states, wife rape is illegal only if the couple has separated.

Because of the sexual nature of marriage, marital rape has not been regarded as a serious form of assault, as Kersti Yllo (1995) explains:

A widely held assumption has been that an act of forced sex in the context of an ongoing relationship in which consensual sex occurs cannot be significant or traumatic. This assumption is flawed because it overlooks the core violation of rape that is coercion, violence and in the case of wife rape, the violation of trust.

Marital rape victims experience feelings of betrayal, anger, humiliation, and guilt. Following their rapes, many wives feel intense anger toward their husbands.

One woman recounted, "'So,' he says, 'You're my wife and you're gonna . . .' I just laid there thinking 'I hate him, I hate him so much.'" Another expressed the desire to resolve her humiliation and sense of "dirtiness" by taking a shower: "I tried to wash it away, but you can't. I felt like a sexual garbage can" (Finkelhor and Yllo 1985). Some feel guilt and blame themselves for not being better wives. Some develop negative self-images and view their lack of sexual desire as a reflection of their own inadequacies rather than as a consequence of abuse.

There still remains the problem of enforcing the laws. Many people discount rape in marriage as a "marital tiff" that has little to do with "real" rape (Yllo 1995). Many victims have difficulty acknowledging that their husbands' sexual violence is indeed rape. Caucasian females are more likely than African American females to identify sexual coercion in marriage as rape (Cahoon et al. 1995), and all too often judges seem sympathetic with the perpetrator than the victim, especially if he is intelligent, successful, and well educated.

There is also the "notion that the male breadwinner should be the beneficiary of some special immunity because of his family's dependence on him" (Russell 1990). Because of deeply entrenched attitudes and beliefs about what constitutes rape, and about marital and sexual relationships, it is estimated that two-thirds of sexual assault victims do not report the crime (U.S. Department of Justice 1997).

Violence in Gay and Lesbian Relationships

Until recently, little was known about violence in lesbian and gay relationships. One reason is that such relationships have not been given the same social status as those of heterosexuals. Also, long-term same-sex relationships are less common than long-term heterosexual relationships. Finally, many gays and lesbians are likely to be reluctant to identify their sexuality for fear of resulting stigma or mistreatment. However, understanding violence in same-sex relationships is important for at least two reasons: people are being victimized and their victimization is mostly invisible and unaddressed. Relationships between gay men or lesbians obviously lack the gender differences that otherwise reflect male dominance and female subordination. Clearly,

neither male dominance nor male socialization toward dominance, aggressiveness, or violence can account for physical abuse in lesbian relationships.

Recent research indicates that the rate of abuse in gay and lesbian relationships is comparable to that in heterosexual relationships. A recent estimate placed the range between 25% and 50% for lesbian couples (McClennen, Summers, and Daly 2002, in Frieze 2005). A study by Kimberly Balsam and Dawn Syzmanski found that of the 272 lesbian and bisexual women in their sample, 40% reported being violent and 44% reported being victims of violence within relationships with female partners (in Frieze 2005).

Furthermore, Claire Renzetti found that violence in same-sex relationships is rarely a one-time event; once violence occurs it is likely to reoccur. It also appears to be as serious as violence in heterosexual relationships, including physical, psychological, and/or financial abuse. Michael Johnson and Kathleen Ferraro (2000) note that intimate terrorism can be found among lesbian couples. One additional form of abuse, unique to same-sex couples, is the threat of "outing" (revealing another's gay orientation without consent). Threatening to out a partner to coworkers, employers, or family may be used as a form of psychological abuse in same-sex relationships.

For battered partners in same-sex relationships, there is often nowhere to go for support. Services for gay men and lesbians are often nonexistent or uninformed about the multifaceted issues that face such victims. Renzetti (1995) points out several policy issues that must be addressed among service providers and domestic violence agencies:

- Consider how homophobia inhibits gay and lesbian victims of abuse from self-identifying as such.

- Recognize that battered gay men and lesbians of color experience a triple jeopardy: as victims of domestic violence, as homosexuals, and as racial or ethnic minorities.

- Address the issue of gay men and lesbians as both batterers and victims who may seek services at the same time from the same agency.

Dating Violence and Date Rape

In the last two decades, researchers have become increasingly aware that violence and sexual assault can take place in all forms of intimate relationships.

Violence between intimates is not restricted to family members. Even casual or dating relationships can be marred by violence or rape.

Dating Violence and Abuse

The incidence of physical violence and emotional or verbal abuse in dating relationships, including those of teenagers, is alarming. Evidence suggests that it even exceeds the level of marital violence (Lloyd 1995). One study of relationships among college students found that of the sample of 572, 21% had engaged in "physically aggressive" behavior, acts that included throwing something at; pushing, grabbing, or hitting; slapping; kicking, biting, or punching; beating up; choking; and threatening to or using a gun or a knife on a partner. Verbal abuse was even more common: 80% acknowledged having been verbally abusive toward a dating partner in the previous 12 months. Verbal abuse consisted of such acts as insulting or swearing at a partner, sulking or refusing to talk with a partner, stomping out of the house or room, and saying or doing something to spite a partner (Shook et al. 2000). Although the males and females were similar in their verbally abusive behavior, females reported "significantly more use of physical aggression" against their partners than men did (Shook et al. 2000).

For both the females and males, the two variables most strongly associated with verbal aggressiveness were alcohol use 3 hours before the incident, and a childhood history of parent–child aggression (Shook et al. 2000).

In two studies of undergraduate couples (18–25 years old) in ongoing relationships, Jennifer Katz and colleagues found that a third to nearly half of the students were in relationships in which their partners had acted violently toward them. In both studies, rates at which men and women were victimized were similar, although men experienced higher levels of moderate violence (Katz, Kuffel, and Coblentz 2002).

Dating relationships among high school students are also prone to violence. Reviewing research from the 1980s and 1990s, Susan Jackson, Fiona Cram, and Fred Seymour found that rates of reported violence range from 12% to 59%. A 1997 observational study of high school couples reported that 51% of participating couples displayed some form of aggression, such as shoving or grabbing. In this same study, males were unilaterally violent in 4% of the cases, and females were unilaterally violent in 17%. Both were mutually violent in the remaining 30% (Capaldi and Crosby 1997, as summarized in Katz, Kuffel, and Coblentz 2002). Although some patterns are similar (for example, the gender symmetry), the issues involved in dating violence appear to be different than those generally involved in spousal violence. Whereas marital violence may erupt over domestic issues such as housekeeping and childrearing (Hotaling and Sugarman 1990), dating violence is far more likely to be precipitated by jealousy or rejection (Lloyd and Emery 1990; Makepeace 1989). For example, one young woman recounted the following incident (Lloyd and Emery 1990) of her boyfriend's furious treatment after seeing her chat with a group of male friends in front of the school. He was silent until they were home, then:

> He caught me on the jaw, and hit me up against the wall . . . He picked me up and threw me against the wall and then started yelling and screaming at me that he didn't want me talking to other guys.

Sally Lloyd and Beth Emery (1990) found that dating violence might also involve the man's use of alcohol or drugs, "unpredictable" reasons, and intense anger.

Although women and men may sustain dating violence at comparable levels, they do not appear to react similarly to it. As in the case of marital violence, women react with more distress than men do to relationship violence, even within mutually violent relationships (Katz, Kuffel, and Coblentz 2002). They also sustain more physical injuries from dating violence. More surprising is the finding that "partner violence generally is unrelated to decreased relationship satisfaction" (Katz, Kuffel, and Coblentz 2002, 250). One study cited by Jennifer Katz and colleagues found that more than 90% of adolescents in violent relationships described those relationships as "good" or "very good."

Many women leave a dating relationship after one violent incident; others stay through repeated episodes. Women who have "romantic" attitudes about jealousy and possessiveness and who have witnessed physical violence between their own parents may be more likely to stay in such relationships (Follingstad et al. 1992). Women with "modern" gender-role attitudes are more likely to leave than those with traditional attitudes (Flynn 1990). Women who leave violent partners cite the following factors in making the decision to break up: a series of broken promises that the man will end the violence, an improved self-image ("I deserve better"), escalation of the violence, and physical and emotional help from family and friends (Lloyd and Emery

1990). Apparently, counselors, physicians, and law enforcement agencies are not widely used by victims of dating violence.

Date Rape and Coercive Sex

Sexual intercourse with a dating partner that occurs against his or her will with force or the threat of force—often referred to as **date rape**—is the most common form of rape. Date rape is also known as **acquaintance rape.** One study found that women were more likely than men to define date rape as a crime. Disturbingly, date rape was considered less serious when the woman was African American (Foley et al. 1995).

Date rapes are usually not planned. Two researchers (Bechhofer and Parrot 1991) describe a typical date rape: He plans the evening with the intent of sex, but if the date does not progress as planned and his date does not comply, he becomes angry and takes what he feels is his right—sex. Afterward, the victim feels raped but the assailant believes that he has done nothing wrong. He may even ask the victim out on another date.

Alcohol or drugs are often involved. When both people are drinking, they are viewed as more sexual. Men who believe in rape myths are more likely to see drinking as a sign that females are sexually available (Abbey and Harnish 1995). In one study, 79% of women who were raped by their date had been drinking or taking drugs before the rape. In addition, 71% said their assailant had been drinking or taking drugs (Copenhaver and Grauerholz 1991). There are also high levels of alcohol and drug use among middle school and high school students who have unwanted sex (Rapkin and Rapkin 1991).

In recent years, certain "date-rape drugs," most often either gamma-hydroxybutyrate (GHB) or Rohypnol (flunitrazepam, popularly known as "roofies," "roofenol," "rochies," and other street names), have surfaced as major public safety concerns. Both drugs have sedative effects, especially when combined with alcohol. They may reduce inhibitions, and they affect memory. Both are used by some men to sedate and later victimize women, many of whom wake up unaware of where they are, how they got there, or what they have done. Samantha Reid, a 15-year-old, died as a result of drinking a soft drink that had been laced with GHB. Knowing only that the drink tasted funny, she died just hours later. Her friend, Melanie Sindone,

recovered after entering a coma that lasted less than a day. According to a *New York Times* article, the Drug Enforcement Agency estimates that between 1990 and 2000 there have been 65 deaths and have been 15 sexual assault cases involving 30 victims who had been given GHB. In Reid's death, three men were convicted of involuntary manslaughter, punishable by 15 years in prison (Bradsher 2000). In 2000, then President Bill Clinton signed into law the Hillory J. Farias and Samantha Reid Date-Rape Drug Prohibition Act of 2000, named for Reid and another teenage victim who died after unknowingly drinking a beverage mixed with GHB. It is a federal crime, punishable by up to 20 years in prison, to manufacture, distribute, or possess GHB (http://abcnews.go.com).

Incidence of Date Rape

Estimates of date rape vary considerably. If the definition is expanded to include attempted intercourse as a result of verbal pressure or the misuse of authority, then women's lifetime incidence increases significantly. When all types of unwanted sexual activity are included, ranging from kissing to sexual intercourse, half to three-quarters of college women report *sexual aggression* in dating (Cate and Lloyd 1992). There is also considerable *sexual coercion* in lesbian relationships and in relationships between gay men.

The National College Women Sexual Victimization Study surveyed more than 4,000 women during the 1996–1997 academic year. Asked about victimization just in the 7 months since school began in the fall, 1.7% of the women had been raped. Another 1.1% had experienced an attempted rape. Nine out of ten of these women knew their offenders.

Physical violence often goes hand in hand with sexual aggression. One researcher found, in a study of acquaintance rape victims, that three-fourths of the women sustained bruises, cuts, black eyes, and internal injuries. Some were knocked unconscious (Belknap 1989).

WHEN "NO" IS "NO." There is considerable confusion and argument about sexual consent. Much sexual communication is done nonverbally and ambiguously, as Charlene Muehlenhard and her colleagues (1992) note:

> Most sexual scripts do not involve verbal consent. One such script involves two people who are overcome with passion. Another such script involves a

male seducing a hesitant female, who, according to the sexual double standard, must not acknowledge her desire for sex lest she be labeled "loose" or "easy." Neither of these scripts involve explicit verbal consent from both people.

That we don't necessarily give verbal consent for sex indicates the importance of the nonverbal clues we do give off. However, as we saw in Chapter 6, nonverbal communication is imprecise. It can be misinterpreted easily if it is not reinforced verbally. For example, some men may even mistake a woman's friendliness for sexual interest (Johnson, Stockdale, and Saal 1991; Stockdale 1993). Others may misinterpret a woman's cuddling, kissing, and fondling as wishing to engage in sexual intercourse (Gillen and Muncher 1995; Muehlenhard 1988; Muehlenhard and Linton 1987). Our sexual scripts often assume "yes" unless a "no" is directly stated (Muehlenhard et al. 1992). This makes individuals "fair game" unless a person explicitly says "no."

The assumption of consent puts women at a disadvantage. First, because men traditionally initiate sex, men may feel it is legitimate to initiate sex whenever they desire without women explicitly consenting. Second, women's withdrawal can be considered insincere because consent is always assumed. Such thinking reinforces a common sexual script in which men initiate and women refuse so as not to appear promiscuous. In this script, the man continues believing that her refusal is token. One study found that almost 40% of the women had offered a "token no" at least once for such reasons as not wanting to appear "loose," uncertainty about how the partner feels, inappropriate surroundings, and game playing (Muehlenhard and Hollabaugh 1989; Muelhenhard and McCoy 1991).

MALE EXPERIENCES OF COERCIVE SEX. Rape is not the only form of unwanted sexual relations that are experienced between acquaintances or on dates. Nor are women the only ones who are subjected to unwanted sexual contact. A study of New Zealand high school students found that, like "emotional violence" and "physical violence," the 373 high school males and females reported high rates of coercive sexual contact. Defining such contact as unwanted kissing, hugging, French kissing (tongue kissing), genital contact ("being felt up"), and sex, as constituting sexual coercion, they found that more than three-fourths of their female subjects and two-thirds of their male respondents had

experienced one or more forms of such "sexual coercion" (Jackson, Cram, and Seymour, 2000). With the exception of "being felt up," similar percentages of male as and female respondents reported having experienced nonconsensual sexual activities.

AVOIDING DATE RAPE. To reduce the risk of date rape, women should consider the following points:

- When dating someone for the first time, go to a public place, such as a restaurant, movie, or sports event.

- Share expenses. A common scenario is a date expecting you to exchange sex for his paying for dinner, the movie, drinks, and so on (Muehlenhard and Schrag 1991; Muehlenhard et al. 1991).

- Avoid using drugs or alcohol if you do not want to be sexual with your date. Their use is associated with date rape (Abbey 1991).

- Avoid ambiguous verbal or nonverbal behavior. Examine your feelings about sex and decide early if you wish to have sex. Make sure your verbal and nonverbal messages are identical. If you only want to cuddle or kiss, tell your partner that those are your limits. Tell him that if you say "no," you mean "no." If necessary, reinforce your statement emphatically, both verbally ("No!") and physically (pushing him away) (Muehlenhard and Linton 1987).

- Be forceful and firm. Don't worry about being polite. Often men interpret passivity as permission and ignore or misunderstand "nice" or "polite" approaches (Hughes and Sandler 1987).

- If things get out of hand, be loud in protesting, leave, and go for help.

- Be careful about what you drink, who you accept drinks from, and where you place your unfinished drink if you put it down; be suspicious of any open drink that tastes funny (salty or flat). These strategies will help reduce the likelihood of having your drink laced with date-rape drugs.

Beyond these strategies and suggestions, however, is an important reality. As with avoidance of stranger rapes, you can do everything right and still be victimized. If you experience a sexual assault, rather than compound the trauma by blaming yourself and experiencing guilt, you should focus on doing what is necessary to restore your confidence and faith in future relationships.

When and Why Some Women Stay in Violent Relationships

Violence in relationships generally develops a continuing pattern of abuse over time. We know from systems theory that all relationships have some degree of mutual dependence, and battering relationships are certainly no different. Despite the mistreatment they receive, some women stay in or return to violent situations for many reasons. However, we need to be careful not to overstate the tendency for abuse victims to stay with their abusers. Johnson and Ferraro (2000) note, for example, "We need to watch our language; there is no good reason why a study in which two-thirds of the women have left the violent relationship is subtitled, 'How and why women stay' instead of 'How and why women leave.'" For the women who do stay in violent or abusive situations, their reasons include the following:

- *Economic dependence.* Even if a woman is financially secure, she may not perceive herself as being able to cope with economic matters. For low-income or poor families, the threat of losing the man's support—if he is incarcerated, for example— may be a real barrier against change.

- *Religious pressure.* She may feel that the teachings of her religion require her to keep the family together at all costs, to submit to her husband's will, and to try harder.

- *Children's need for a father.* She may believe that even a father who beats the mother is better than no father. If the abusing husband also assaults the children, the woman may be motivated to seek help (but this is not always the case).

- *Fear of being alone.* She may have no meaningful relationships outside her marriage. Her husband may have systematically cut off her ties to other family members, friends, and potential support sources. She has no one to go to for any real perspective on her situation. (See Nielsen, Endo, and Ellington 1992 for the relationship between social isolation and abuse.)

- *Belief in the American dream.* The woman may have accepted without question the myth of the perfect woman and happy household. Even though her existence belies this, she continues to believe that it is how it should (and can) be.

- *Pity.* She feels sorry for her husband and puts his needs ahead of her own. If she doesn't love him, who will?

- *Guilt and shame.* She feels that it is her own fault if her marriage isn't working. If she leaves, she believes, everyone will know she is a failure or her husband might kill himself.

- *Duty and responsibility.* She feels she must keep her marriage vows "till death us do part."

- *Fear for her life.* She believes she may be killed if she tries to escape.

- *Love.* She loves him; he loves her. On her husband's death, one elderly woman (a university professor) spoke of her 53 years in a battering relationship (Walker 1979): "We did everything together. . . . I loved him; you know, even when he was brutal and mean. . . . I'm sorry he's dead, although there were days when I wished he would die. . . . He was my best friend. . . . He beat me right up to the end. . . . It was a good life and I really do miss him."

- *Cultural reasons.* A woman from nonmainstream cultural backgrounds may face great obstacles to leaving a relationship. She may not speak English, may not know where to go for help, and may fear she will not be understood. She often fears that her husband will lose his job, retaliate against her, or take the children back to the country of origin (Donnelly 1993). Recent immigrants from Latin America, Asia, and South Asia may be especially fearful that their revelations will reflect badly on the family and community.

- *Nowhere else to go.* She may have no alternative place to live. Shelter space is limited and temporary. Relatives and friends may be unable or unwilling to house a woman who has left, especially if she brings children with her.

- *Learned helplessness.* Lenore Walker (1979, 1993) theorizes that a woman stays in a battering relationship as a result of **learned helplessness.** After being repeatedly battered, she develops a low self-concept and comes to feel that she cannot control the battering or the events that surround it. Through a process of behavioral reinforcement, she "learns" to become helpless and feels she has no control over the circumstances of her life.

Michael Johnson and Kathleen Ferraro's distinction between common couple violence and intimate terrorism is important to add here. Women subjected to situational violence are less likely to leave than victims of intimate terrorism. Victims of intimate terrorism leave their partners more often, most commonly seeking friends and relatives for help, and look for destinations that are safe and secret (Johnson and Leone 2005).

Reflections

In your family (including your extended family), has there been spousal violence? Have you experienced violence in a dating relationship? If so, what were the factors involved in causing it? In sustaining it? If you or your family have not been involved in such violence, what factors do you think have protected against it?

The Costs of Intimate Violence

The cumulative financial costs associated with intimate violence are considerable. Zink and Putnam report that add costs for direct medical and mental health services for victims of partner violence, rape, assault and stalking total in excess of *four billion dollars*. Add to these the millions of dollars worth of broken or stolen property and the wages lost to victims due to time out of work. The "bottom line" is indeed steep.

Then there are the nonfinancial costs. These include the actual health and mental health effects with which victims of violence must cope. DeMaris (2001) reports that thousands of women and men are treated in emergency rooms each year for injuries suffered in partner violence. Victims of intimate partner violence also suffer twice the rate of depression and four times the rate of posttraumatic stress disorder as non-victims (Zink and Putnam, 2005). According to the Centers for Disease Control and Prevention (2003) victims of severe intimate violence lose nearly 8 million days of paid work—the equivalent of more than 32,000 full-time jobs—and almost 5.6 million days of household productivity each year (2003).

Responding to Intimate Violence: Police Intervention, Shelters, and Abuser Programs

Professionals who deal with domestic violence have long debated the most appropriate strategy: control and deterrence versus compassion (Mederer and Gelles 1989). Both approaches have their place. Controlling measures such as arrest, prosecution, and imprisonment, as well as compassionate measures such as shelters, education, counseling, and support groups have been shown to be successful to varying degrees. Used together, these interventions may be quite effective. Helen Mederer and Richard Gelles (1989) suggest that controlling measures may be used to "motivate violent offenders to participate in treatment programs."

Battered Women and the Law

Early family violence studies and feminist pressure spurred a movement toward the implementation of stricter policies for dealing with domestic offenders. Once long ignored, in the last 10 to 15 years intimate violence has become a top concern for legislators and law enforcement agencies throughout the country (Wilson 1997). Today, many of the largest U.S. police forces have implemented **mandatory arrest** policies in which discretion is removed from police officers responding to a call about intimate violence. Under such policies, "if an officer finds probable cause that a crime occurred, he or she must arrest" (Goodman and Epstein 2005, 480). In addition, the adoption of **no-drop prosecution** policies compels prosecutors to proceed in the prosecution of an intimate violence case as long as evidence exists, regardless of a victim's expressed wishes (Goodman and Epstein 2005).

For police to play any effective role in combating intimate partner violence they must first *know of the violence*. According to a "Fact Sheet on Intimate Partner Violence" put out by the National Center for Injury Prevention and Control, only about a fifth of rapes or sexual assaults by a partner, a fourth of physical assaults, and half of the incidents of stalking directed toward women are reported (http://www.cdc.gov/ncipc/factsheets/ipvfacts.htm). The rate at which men report their victimization is even less.

Even when incidents are reported, we have reason to question how committed police officers are to becoming involved in domestic disputes. This has long been a complaint of women who are victimized and who find police reluctant to intervene, even under mandatory arrest policies. Male victims of female perpetrators find police are often dismissive of their concerns (Migliaccio 2002).

Aside from the sincerity of the commitment of criminal justice personnel, the innovations in policy have potentially mixed consequences. Lisa Goodman and Deborah Epstein (2005) use the following as examples to illustrate this:

> If a victim seeks to drop charges so that the father of her children can continue to work and provide financial support, a prosecutor is likely to refuse on the grounds that this would not serve the interests of the state. . . . No-drop policies also allow a district attorney little leeway in situations where a victim fears, realistically, that prosecution will provoke the batterer into retaliatory abuse against her; the district attorney may even subpoena the victim and force her to testify.

Abuser Programs

According to Richard Tolman (1995), "A comprehensive solution to violence against women in intimate relationships demands that perpetrators of abuse be held accountable for their behavior and that direct efforts be made with batterers to change their behavior." Treatment services for men who batter provide one important component of a coordinated response to domestic violence (see Gondolf 1993 for program and treatment issues). Psychotherapy, group discussion, stress management, or communication skills classes may be available through mental health agencies, women's crisis programs, or various self-help groups.

The extent to which attending batterers' groups changes the violent behavior of abusing men is difficult to measure (Gelles and Conte 1991). What has become apparent is the ineffectiveness of the "one size fits all" approach and the need to adopt a more sophisticated understanding of an individual's violent behaviors (Tolman 1995). Also, coordinated community

■ Battered women's shelters provide safe havens for women in abusive relationships. Shelters provide counseling and emotional support, as well as temporary lodging, meals, and other necessities for women and their children.

response that includes proactive police and criminal justice strategies, advocacy and services for battered women and their children, and responses by other community institutions that promote safety for battered women and sanctions for men who batter are necessary interventions (Tolman 1995).

As Michael Johnson and Janel Leone warn, failure to differentiate types of violence may also leave women who are victims of intimate terrorism vulnerable and endangered if they choose to use such interventions as couples counseling or mediation. The same strategies would be very appropriate for couples experiencing more situational common couple violence.

Child Abuse and Neglect

Child abuse was not recognized as a serious problem in the United States until the 1960s. At that time, C. H. Kempe and his colleagues (1962) coined the medical term **battered child syndrome** to describe the patterns of injuries commonly observed in physically abused children. The Children's Defense Fund (2005) reports the following:

- Every 30 seconds, a child is reported abused or neglected.
- Every 20 seconds, a child is arrested.
- Every 3 hours a child is killed by firearms.
- Every 5 hours a child commits suicide.
- Every 6 hours a child dies from abuse or neglect.

When we look at violence among children from a global perspective, we see an even larger shadow cast over our nation. A study by the Centers for Disease Control and Prevention (1997) found that nearly three out of four child slayings in the industrialized world occur in the United States. The statistics show that the epidemic of violence in recent years that has hit increasingly younger children is confined almost exclusively to the United States. The suicide rate alone for children 14 and under is double that of the rest of the industrialized world. No explanation for the huge gap between the rates of violent death for American children and those of other countries was given, although some experts speculate it is because of a growing number of children who are unsupervised or otherwise at risk. The low level of funding for social programs, sexism, racism, and epidemic rates of poverty among our young are other factors that continue to embarrass our nation. Parental violence is among the five leading causes of death for children between the ages of 1 and 18. About 1,300 children are killed by their parents or other close relatives each year (McCormick 1994).

As is true of partner relationships, children are subjected to other, nonphysical forms of mistreatment by parents. In examining the national prevalence of **psychological aggression** by parents, Murray Straus and Carolyn Field (2003) find that verbal attacks on children are so common as to be "just about universal." Based upon nearly 1,000 interviews with a nationally representative sample of households with at least one child under 18 years living at home, Straus and Field explore the prevalence of psychological aggression. They define psychological aggression as consisting of the following kinds of behaviors, with the latter three constituting "more severe" psychological aggression:

- Shouting, yelling, or screaming at one's child
- Threatening to spank or hit one's child but not actually doing it
- Swearing or cursing at one's child
- Threatening to send one's child away or kick him or her out of the house
- Calling one's child dumb or lazy, or making some other disparaging comment

Of the sample parents, 89% reported having committed at least one of the five kinds of psychological aggression and 33% reported at least one instance of the more severe forms. The prevalence of the various forms of psychological aggression are illustrated in Table 13.1.

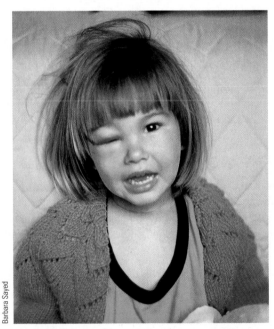

Barbara Sayed

■ *Children are the least protected members of our society. Much physical abuse is camouflaged as discipline or as the parent "losing" his or her temper.*

Table 13.1 ■ Prevalence of Psychological Aggression

Prevalence	Measure (% in last year)
Overall	88.6
Severe	33.4
Shouting, yelling, screaming	74.7
Threatening to spank	53.6
Swearing or cursing	24.3
Name-calling	17.5
Threatening to kick out of house	6.0

SOURCE: Straus and Field 2003.

Use of psychological aggression varies with the age of the child. A total of 43% of parents of infants reported using psychological aggression, and nearly 90% of parents of 2-year-olds use some form of psychological aggression. The percentage peaks at 98% at age 7, and as late as age 17 the rate still remains a high 90%.

Conversely, research on corporal punishment shows it declining with the age of the child; only 12% of parents of 17-year-olds report still using corporal punishment (Straus and Field 2003). However, more than 90% of toddlers in the United States are reportedly spanked (Straus and Field 2003). Most childrearing experts, currently advise that parents use alternative disciplinary measures.

Parents' ages matter, too. Younger parents (ages 18 to 29) reported the most frequent use of psychological aggression (22 times in past 12 months) compared to parents 30 to 39 (19 times in past 12 months), and parents over 40 (15 times in past 12 months). Aside from age differences, there was "a lack of demographic differences in use of psychological aggression; this means that nearly all parents, regardless of sociodemographic characteristics, used at least some psychological aggression as a disciplinary tactic" (Straus and Filed 2003, 805).

Families at Risk

Early research established that the following three sets of factors put families at risk for child abuse and neglect: (1) parental characteristics, (2) child characteristics, and (3) the family ecosystem—that is, the family system's interaction with the larger environment (Burgess and Youngblood 1987; Vasta 1982). The characteristics described in the next sections are likely to be present in abusive families (Straus, Gelles, and Steinmetz 1980; Turner and Avison 1985).

Parental Characteristics

Some or all of the following characteristics are likely to be present in parents who abuse their children:

- The abusing father was physically punished by his parents, and his father physically abused his mother.
- The parents believe in corporal discipline of children and wives.
- The marital relationship itself may not be valued by the parents. There may be spousal violence.

- The parents believe that the father should be the dominant authority figure.
- The parents have low self-esteem.
- The parents have unrealistic expectations for the child.
- There is persistent role reversal in which the parents use the child to gratify their own needs, rather than vice versa.
- The parents appear unconcerned about the seriousness of a child's injury, responding, "Oh well, accidents happen."

Child Characteristics

Who are the battered children? Are they any different from other children? Surprisingly, the answer is often "yes"; they are different in some way or at least are perceived to be so by their parents. Children who are abused are often labeled by their parents as "unsatisfactory," a term that may describe any of the following:

- A "normal" child who is the product of a difficult or unplanned pregnancy, is of the "wrong" sex, or is born outside of marriage
- An "abnormal" child, one who was premature or of low birth weight, possibly with congenital defects or illness
- A "difficult" child, one who shows such traits as fussiness or hyperactivity

Researchers note that all too often, a child's perceived difficulties are a result (rather than a cause) of abuse and neglect.

Family Ecosystem

As discussed earlier in this chapter, the community and the family's relation to it may be relevant to the existence of domestic violence. The following characteristics may be found in families that experience child abuse:

- The family experiences unemployment.
- The family is socially isolated, with few or no close contacts with relatives, friends, or groups.
- The family has a low level of income, which creates economic stress.
- The family lives in an unsafe neighborhood, which is characterized by higher-than-average levels of violence.

- The home is crowded, hazardous, dirty, or unhealthy.
- The family is a single-parent family in which the parent works and is consequently overstressed and overburdened.
- One or more family members have health problems.

Notice the clustering of such socioeconomic characteristics as unemployment, low income, neighborhood, and housing. This combination tells an important story. Like spousal or partner violence, the mistreatment of children can be found across the socioeconomic spectrum. But like spousal violence, it happens more often at the lower levels. As noted earlier, the culprit in these associations is most likely stress.

The likelihood of child abuse increases with family size. Parents of two children have a 50% higher abuse rate than do parents of a single child. The rate of abuse peaks at five children and declines thereafter. The overall child abuse rate by mothers has been found to be significantly higher than that by fathers. The responsibilities and tensions of mothering and the enforced closeness of mother and child are different and more demanding than those between father and child. They may lead to situations in which women are likely to abuse their children. But, as David Finkelhor (1983) and others have pointed out, if we "calculate [child] vulnerability to abuse as a function of the amount of time spent in contact with a potential abuser, . . . we . . . see that men and fathers are more likely to abuse."

Single parents—both mothers and fathers—are at especially high risk of abusing their children (Gelles 1989). According to Richard Gelles, "the high rate of abusive violence among single mothers appears to be a function of the poverty that characterizes mother-only families." He states that programs must be developed that are "aimed at ameliorating the devastating consequences of poverty among single parents." Single fathers, who show a higher abuse rate than single mothers, "need more than economic support to avoid using abusive violence toward their children."

Matter of Fact

American children are 12 times more likely to die by gunfire than their counterparts in the rest of the industrialized world (Meyer 1997).

Intervention

The goals of intervention in domestic violence are principally to protect the victims and to assist and strengthen their families. In dealing with child abuse, professionals and government agencies may be called on to provide medical care, counseling, and services such as day care, childcare education, telephone crisis lines, and temporary foster care.

Many of these services are costly, and many of those who require them cannot afford to pay. Our system does not currently provide the human and financial resources necessary to deal with these problems. The first step in treating child abuse is locating the children who are threatened. Mandatory reporting of suspected child abuse is now required of professionals such as teachers, doctors, and counselors in all 50 states. Reported incidents of child abuse have increased greatly during this time, but the actual number of incidents appears to have decreased. This is good news as far as it goes. Still, levels of violence against children remain unacceptably high, and not nearly enough resources are available to assist children. Child welfare workers are notoriously overburdened with cases, and adequate foster placement is often difficult to find (Gelles and Cornell 1990).

Society must address this tragedy of continued child abuse from a variety of levels:

- Parents must learn how to deal more positively and effectively with their children.
- Children need to be infused with self-esteem and taught skills to recognize and report abuse as soon as it occurs.
- Professionals working with children and families should be required to receive adequate training in child abuse and neglect and to be sensitive to cultural norms.
- Agencies should coordinate their efforts for preventing and investigating child abuse.
- Public awareness of child abuse needs to be created by methods such as posters and public service announcements.
- The workplace should promote educational programs to eliminate sexism, provide adequate childcare, and help reduce stress among its workforce.
- Government should support sex education and family life programs to help reduce the number of unwanted pregnancies.

- Criminal statutes should be developed and enforced to impose felony sentences on those who perpetuate child maltreatment.
- Research efforts concerning family violence and child maltreatment should be supported.

Reflections

If you became (or are) a parent, would you consider it violent to spank your child with an open hand on the buttocks if the child was disobedient? To slap your child across the face? Is it acceptable to spank your small child to teach him or her not to run into a busy street? To spank because you are angry?

Hidden Victims of Family Violence: Siblings, Parents, and the Elderly

Most studies of family violence have focused on violence between spouses and on parental violence toward children. There is, however, considerable violence between siblings, between teenage children and their parents, and between adult children and their aging parents. These are the "hidden victims" of family violence (Gelles and Cornell 1990).

Sibling Violence

More than a quarter century of research illustrates that violence between siblings is by far the most common form of family violence (Straus, Gelles, and Steinmetz 1980; Hoffman, Kiecolt, and Edwards 2005). Perhaps as many as three out of four children experience sibling violence every year. Although violence declines as children age, no less than two-thirds of teenagers annually commit an act of violence—pushing, slapping, throwing or hitting with an object, or something more severe—against a sibling. A recent study of 651 college undergraduates found that nearly 70% acknowledged having acted violently toward their closest-age sibling while seniors in high school. The violence most commonly consisted of hitting with a hand or object, pushing or shoving, and throwing things but often included slapping, punching, and pulling hair (Hoffman,

Kiecolt, and Edwards 2005). Most of this type of sibling interaction is simply taken for granted by our culture—"You know how kids are!"

The full scope and implications of sibling violence have not been rigorously explored. However, more than 25 years ago, Murray Straus, Richard Gelles, and Suzanne Steinmetz (1980) offered this observation, which remains just as relevant today:

> Conflicts and disputes between children in a family are an inevitable part of life. . . . But the use of physical force as a tactic for resolving their conflicts is by no means inevitable. . . . Human beings learn to be violent. It is possible to provide children with an environment in which nonviolent methods of solving conflicts can be learned. . . . If violence, like charity, begins at home, so does nonviolence.

Parents as Victims

Teenage Violence toward Parents

Most of us find it difficult to imagine children attacking their parents because it so profoundly violates our image of parent-child relations. Parents possess the authority and power in the family hierarchy. Furthermore, there is greater social disapproval of a child striking a parent than of a parent striking a child; it is the parent who has the "right" to hit. Although we know fairly little about adolescent violence against parents, scattered studies indicate that it is almost as prevalent as spousal violence.

Most children who attack parents are between the ages of 13 and 24. Sons are slightly more likely to be abusive than daughters; the rate of severe male violence tends to increase with age, whereas that of females decreases. Boys apparently take advantage of their increasing size and the cultural expectation of male aggression. Girls, in contrast, may become less violent because society views female aggression more negatively. Most researchers believe that mothers are the primary targets of violence and abuse because they may lack physical strength or social resources to defend themselves (Gelles and Cornell 1985).

Abuse of Elderly Parents

Of all the forms of hidden family violence, only the abuse of elderly parents by their grown children (or, in some cases, by their grandchildren) has received

Working the Front Line in the Fight against Child Abuse

An estimated 1,400 children a year are murdered by their parents or guardians. Some cases remain relatively unknown to the wider public, reported in small articles in mostly local newspapers if reported at all. Others become major news stories, the focus of not only local but also wider regional or even national attention. Both kinds of cases can be seen in the following list of cases that occurred over the past 20 years. The list includes Eli Creekmore, age 3, beaten to death by his father, in 1986; Elizabeth "Lisa" Steinberg, age 6, beaten to death by her adopted father, in 1987; Joseph

Wallace, age 3, hung by his mother, in 1993; Elisa Izquierdo, age 6, beaten to death by her mother, in 1995; Nadine Lockwood, age 4, intentionally starved to death by her mother in 1996; and James Pack, age 3, beaten to death by his father in 2003. In just a 3-month period, between late 2005 and early 2006, Sierra Roberts, age 7, Dahquay Gillians, age 16 months, and Joziah Bunch, age 1, died at the hands of their parents. Then there was Nixzmary Brown, age 7. As reported in the *New York Daily News,* Nixzmary had been "bound to a chair, tortured, sexually molested and starved for weeks before being killed by a savage blow to the head—even after child welfare authorities dismissed charges of abuse" (Dillon, Fenner, and Gendar 2006).

Her death in January 2006 drew widespread attention and considerable outrage at the system that is supposed to monitor and protect children.

This is but a partial list of child abuse homicides, selected because in each instance some agency or individuals in a position to intervene didn't—despite what in retrospect looked like clear and unambiguous evidence of severe abuse. Many of these cases were met by public outcry and led to changes in the policies used by the relevant protective agencies. Typically, the most extreme outrage is expressed at the parent perpetrators. Often there is also intense anger and blame directed at the agency or caseworkers who failed to rescue the child from his or her abusive, lethal surroundings.

considerable public attention. Elder mistreatment may be an act of commission (abuse) or omission (neglect) (Wolf 1995). It is estimated that approximately 500,000 elderly people are physically abused annually. An additional 2 million are thought to be emotionally abused or neglected. Although mandatory reporting of suspected cases of elder abuse is the law in 42 states and the District of Columbia, much abuse of the elderly goes unnoticed, unrecognized, and unreported (Wolf 1995). Elderly people are often confined to bed or a wheelchair, and many do not report their mistreatment out of fear of institutionalization or other reprisal. Although some research indicates that the abused elder may have been an abusing parent, more knowledge must be gained before we can draw firm conclusions about the causes of elder abuse (Egeland 1993; Kaufman and Zigler 1993; Ney 1992).

The most likely victims—in most cases, women—of elder abuse are suffering from physical or mental impairments, especially those with Alzheimer's disease. Their advanced age renders them dependent on their caregivers for many, if not all, of their daily needs. It may be their dependency that increases their likelihood

of being abused. Other research indicates that many abusers are financially dependent on their elderly parents; they may resort to violence out of feelings of powerlessness.

While researchers are sorting out the whys and wherefores of elder abuse, battered older people have a number of pressing needs. Karl Pillemer and Jill Suitor (1988) recommend the following services for elders and their caregiving families:

- Housing services, including temporary respite care to give caregivers a break and permanent housing (such as rest homes, group housing, and nursing homes)

- Health services, including home health care; adult day-care centers; and occupational, physical, and speech therapy

- Housekeeping services, including shopping and meal preparation

- Support services, such as visitor programs and recreation

- Guardianship and financial management

Reducing Family Violence

Based on the foregoing evidence, you may by now have concluded that the American family is well on its way to extinction as family members bash, thrash, cut, shoot, and otherwise wipe themselves out of existence. Statistically, the safest family homes are those with one or no children in which the husband and wife experience little life stress and in which decisions are made democratically. By this definition, most of us probably do not live in homes that are particularly safe. What can we do to protect ourselves (and our posterity) from ourselves?

Prevention strategies usually take one of two paths: (1) eliminating social stress or (2) strengthening families (Swift 1986). Family violence experts make the following general recommendations (Straus, Gelles, and Steinmetz 1980) (for specific prevention and treatment strategies, see Hampton et al. 1993):

- Reduce societal sources of stress, such as poverty, racism and inequality, unemployment, and inadequate health care.

- Eliminate sexism.

- Furnish adequate day care.

- Promote educational and employment opportunities equally for men and women.

- Promote sex education and family planning to prevent unplanned and unwanted pregnancies.

- Initiate prevention and early intervention efforts for young males before they become adult batterers.

- End social isolation. Explore means of establishing supportive networks that include relatives, friends, and community.

- Break the family cycle of violence. Eliminate corporal punishment and promote education about disciplinary alternatives. Support parent education classes to deal with inevitable parent-child conflict.

- Eliminate cultural norms that legitimize and glorify violence. Legislate gun control, eliminate capital punishment, and reduce media violence.

Child Sexual Abuse

Whether it is committed by relatives or nonrelatives, **child sexual abuse** is defined as any sexual interaction (including fondling, erotic kissing, or oral sex, as well as genital penetration) between an adult or older adolescent and a prepubertal child. It does not matter whether the child is perceived by the adult as freely engaging in the sexual activity. Because of the child's age, he or she cannot legally give consent; the activity can only be considered as self-serving to the adult.

Estimates of the incidence of child sexual abuse vary considerably. The first national survey found that 27% of the women and 16% of the men surveyed had experienced sexual abuse as children (Finkelhor et al. 1990). Most recently, Andrew Cherlin and colleagues report that available evidence indicates that each year "several million" children experience physical or sexual abuse and that data drawn from a review of 19 surveys that touched on sexual abuse suggest that 20% *or more* American women had been sexually abused as children (Cherlin et al. 2004). Although others estimate that perhaps as many as 25% of women and 10% of men have been sexually abused as children or teens, their abusers are different; those who abuse males are more likely to be nonfamily members. Although, overall, more perpetrators of child sex abuse are nonfamily, a higher percentage of those who abuse females are from within the family (Whealin 2006, http://www.ncptsd.va.gov/facts/specific/fs_male_sexual_assault.html).

For a variety of reasons, as the American Psychological Association (APA) reports, definitive statistics "are difficult to collect because of problems of underreporting and the lack of one definition of what constitutes such abuse." In lieu of specific statistics, the APA states that child sexual abuse is "not uncommon and is a serious problem in the United States" (http://www.apa.org/releases/sexabuse).

Different definitions of abuse, methodologies, samples, and interviewing techniques account for sometimes widely varied estimates. Fabricated reports of sexual abuse do occur, but deliberate fabrications constitute only 4% to 8% of all reports (Finkelhor 1995). Encouragingly, the Department of Justice reports that substantiated cases of child sexual abuse have declined, dropping by about a third between 1992 and 1998 (Cherlin et al. 2004).

Child sexual abuse is generally categorized in terms of kin relationship. **Extrafamilial sexual abuse** is conducted by nonrelated individuals. **Intrafamilial abuse** is conducted by related individuals, including steprelatives. The abuse may be pedophilic or nonpedophilic. **Pedophilia** is an intense, recurring sexual attraction to prepubescent children. Nonpedophilic sexual interactions with children are not motivated as much by sexual desire as by nonsexual motives, such as power or affection (Groth 1980).

The child's victimization may include force or the threat of force, pressure, or the taking advantage of trust or innocence. The most serious forms of sexual abuse include actual or attempted penile–vaginal penetration, fellatio, cunnilingus, and anilingus, with or without the use of force. Other serious forms range from forced digital penetration of the vagina to fondling of the breasts (unclothed) or simulated intercourse without force. The least traumatic sexual abuse ranges from kissing to intentional sexual touching of the clothed genitals, breasts, or other body parts with or without the use of force (Russell 1984).

Forms of Intrafamilial Child Sexual Abuse

The incest taboo, which is nearly universal in human societies, prohibits sexual activities between closely related individuals. **Incest** is generally defined as sexual intercourse between people too closely related to marry legally (usually interpreted to mean father–daughter, mother–son, or brother–sister). Sexual abuse in families can involve blood relatives (most commonly uncles and grandfathers) and steprelatives (most often stepfathers and stepbrothers). Grandfathers who abuse their granddaughters often sexually abused their own children as well. Step-granddaughters are at greater risk than are granddaughters (Margolin 1992).

Father–Daughter Sexual Abuse

There is general agreement that the most traumatic form of sexual victimization is father–daughter abuse, including that committed by stepfathers. Some factors contributing to the severity of reactions to father–daughter sexual relations include fathers being more likely to engage in penile–vaginal penetration than other relatives, fathers sexually abusing their daughters more frequently and being more likely to use force or violence.

In the past, many have discounted the seriousness of sexual abuse by a stepfather because incest is generally defined legally as sexual activity between two biologically related people. The emotional consequences are just as serious, however. Sexual abuse by

a stepfather still represents a violation of the basic parent–child relationship.

Brother–Sister Sexual Abuse

There are contrasting views concerning the consequences of brother–sister incest. Most researchers have tended to view it as harmless sex play or sexual exploration between mutually consenting siblings. The research, however, has generally failed to adequately distinguish between exploitative and nonexploitative brother–sister sexual activity. One resource (Niolon 2000, http://www.psychpage.com/family/library/sib _abuse.htm) defines brother–sister (or cousin) sexual interaction as *abuse*,

> when it is marked by a five year [age] difference; when the children are less than five years apart in age, the interaction is not deemed abusive unless force, coercion, injury, or penetration occurs. The criteria of force and/or coercion may be the most highly associated with negative outcomes, regardless of the specific sexual behavior (for example, kissing, fondling, simulated intercourse, or exhibition). Typically, the abuse begins when the victim is around six to seven years of age.

Diana Russell (1986) suggests that the idea that brother–sister incest is usually harmless and mutual may be a myth. Even more strongly, there are recent studies that assert that the circumstances, characteristics, and potential outcomes of brother–sister incest are as serious as, if not more than, those of father–daughter incest (Rudd and Herzberger 1999; Cyr et al. 2002).

In Russell's (1986) study, the average age difference between the brother (age 17.9 years) and the sister (age 10.7 years) was so great that the siblings could hardly be considered peers. The age difference represents a significant power difference. Furthermore, not all brother–sister sexual activity is "consenting"; considerable physical force may be involved. Russell writes:

> So strong is the myth of mutuality that many victims themselves internalize the discounting of their experiences, particularly if their brothers did not use force, if they themselves did not forcefully resist the abuse at the time, if they still continued to care about their brothers, or if they did not consider it abuse when it occurred. And sisters are even more likely than daughters to be seen as responsible for their own abuse.

Uncle-Niece Sexual Abuse

Both Alfred Kinsey (1953) and Diana Russell (1986) found the most common form of intrafamilial sexual abuse to involve uncles and nieces. Russell reported that almost 5% of the women in her study had been molested by their uncles, slightly more than the percentage abused by their fathers. The level of severity of the abuse was generally less in terms of the type of sexual acts and the use of force. Although such abuse does not take place within the nuclear family, many victims found it quite distressing. A quarter of the respondents indicated long-term emotional effects (Russell 1986).

Children at Risk

Not all children are equally at risk for sexual abuse. Although any child can be sexually abused, some groups of children are more likely to be victimized than others. A review of the literature (Finkelhor and Baron 1986) indicates that children at higher risk for sexual abuse are the following: female children, preadolescent children, children with absent or unavailable parents, children whose relationships with parents are poor, children whose parents are in conflict, children of single parents, and children who live with a stepfather. A variety of studies have found little or no association between sexual abuse and race and socioeconomic status (Finkelhor 1995).

Most sexually abused children are girls, but boys are also victims (Watkins and Bentovim 1992). David Finkelhor (1979) speculates that men tend to underreport sexual abuse because they experience greater shame; they feel that their masculinity has been undermined. Boys tend to be blamed more than girls for their victimization, especially if they did not forcibly resist: "A real boy would never let someone do that without fighting back" (Rogers and Terry 1984).

Most sexually abused children are between 8 and 12 years of age when the abuse first takes place. At higher risk appear to be children who have poor relationships with their parents (especially mothers) or whose parents are absent or unavailable and have high levels of marital conflict. A child in such a family may be less well supervised and, as a result, more vulnerable to manipulation and exploitation by an adult. Finally, children with stepfathers are at greater risk for sexual abuse. The higher risk may result from the weaker incest taboo in stepfamily relationships and because stepfathers have not built inhibitions resulting from parent–child bonding beginning from infancy. As a result, stepfathers may be more likely to view their stepdaughters sexually.

Effects of Child Sexual Abuse

There is extensive research indicating that potential "profound, long-term consequences for an adult's sexual behavior and intimate relationships" can result from child sexual abuse (Cherlin et al. 2004, 770). Among the numerous well-documented consequences of child sexual abuse are both initial and long-term consequences. Many abused children experience symptoms of PTSD (McLeer et al. 1992).

Initial Effects of Sexual Abuse

The initial consequences of sexual abuse—those occurring within the first 2 years—include these effects:

- *Emotional disturbances,* including fear, anger, hostility, guilt, and shame
- *Physical consequences,* including difficulty in sleeping, changes in eating patterns, and pregnancy
- *Sexual disturbances,* including significantly higher rates of open masturbation, sexual preoccupation, and exposure of the genitals (Hibbard and Hartman 1992)
- *Social disturbances,* including difficulties at school, truancy, running away from home, and early marriages among abused adolescents

Ethnicity appears to influence how a child responds to sexual abuse. For example, one study compared sexually abused Asian American children with a random sample of abused Caucasian, African American, and Latino children (Rao, Diclemente, and Pouton 1992). The researchers found that Asian American children suffered less sexually invasive forms of abuse. They tended to be more suicidal and to receive less support from their parents than did non-Asians. They were also less likely to express anger or to act out sexually. These different responses point to the importance of understanding the cultural context when treating ethnic victims of sexual abuse. (For a discussion of child sexual abuse histories among African American college students, see Priest 1992.)

Long-Term Effects of Sexual Abuse

Although the initial effects of child sexual abuse can subside to some extent, the abuse may leave lasting scars on the adult survivor (Beitchman et al. 1992).

These adults often have significantly higher incidences of psychological, physical, and sexual problems than the general population. Cherlin and colleagues (2004) list such outcomes as feelings of betrayal, lack of trust, feelings of powerlessness, low self-image, depression, and a lack of clear boundaries between self and others. Abuse as a child may predispose some women to early onset of sexual involvement, more involvement in sexually risky behavior, multiple partners, and sexually abusive dating relationships (Cherlin et al. 2004; Cate and Lloyd 1992). Cherlin and colleagues also identify the following:

- More frequent but less satisfying sexual encounters

- Greater anxiety and less pleasure from sex

- Behaviors such as using drugs and/or alcohol with sex that increase risk of sexually transmitted disease or HIV infection

- Engaging in sex soon after meeting a partner

Long-term problems include the following (Beitchman et al. 1992; Browne and Finkelhor 1986; Cherlin et al. 2004; Elliott and Briere 1992; Jeffrey and Jeffrey 1991; Wyatt, Gutherie, and Notgrass 1992; DeGroot, Kennedy, Rodin, and McVey 1992; Walker et al. 1992; Young 1992):

- Depression, the most frequently reported symptom of adults sexually abused as children

■ *This drawing was made by an adolescent who was impregnated by her father. According to psychologists, it expresses her inability to deal with body images, especially genitalia, and her rejection of her body's violation.*

- Self-destructive tendencies, including suicide attempts and thoughts of suicide

- Somatic disturbances and dissociation, including anxiety and nervousness, eating disorders (anorexia and bulimia), feelings of "spaciness," out-of-body experiences, and feelings that things are "unreal"

- Negative self-concept, including feelings of low self-esteem, isolation, and alienation

- Revictimization, in which women abused as children are more vulnerable to rape and marital violence

- Sexual problems, in which survivors find it difficult to relax and enjoy sexual activities or they avoid sexual relations and experience hypoactive (inhibited) sexual desire and lack of orgasm

- Interpersonal relationship difficulties, including lower relationship satisfaction, difficulties in relating to both sexes, parental conflict, problems in responding to their own children, and difficulty in trusting others

As Cherlin and colleagues (2004) point out, childhood sexual abuse victimization may affect the ability to maintain long-term intimate relationships in adulthood. "Overall, the relationship difficulties associated with childhood sexual abuse would seem to be more consistent with frequent, short-term unions than with long-term unions" (771).

CAN WE REMEMBER? In the past two decades, some adults have been accusing family members or others of abusing them as children. They say that they unconsciously repressed their traumatic childhood memories of abuse and only later, as adults, recalled them with the help of psychotherapy. These accusations have given rise to a fierce controversy about the nature of memories of abuse. A review of the research related to this topic was done by the American Psychological Association (1994) and the following conclusions were made:

- Most people who were sexually abused as children at least partially remember the abuse.

- Memories of sexual abuse that have been forgotten may later be remembered.

- False memories of events that never happened may occur.

- The process by which accurate or inaccurate recollections of childhood abuse are made is not well understood.

Because firm scientific conclusions cannot be made at this time, the debate is likely to continue.

Sexual Abuse Trauma

As we have seen, childhood sexual abuse has numerous initial and long-term consequences. Together, these consequences create a traumatic dynamic that affects the child's ability to deal with the world. David Finkelhor and Angela Browne (1986) suggest a model of sexual abuse that contains four components: traumatic sexualization, betrayal, powerlessness, and stigmatization. When these factors converge as a result of sexual abuse, they affect the child's cognitive and emotional orientation to the world. They create trauma by distorting a child's self-concept, worldview, and affective abilities. These consequences affect abuse survivors not only as children but also as adults.

TRAUMATIC SEXUALIZATION. The process in which a sexually abused child's sexuality develops inappropriately and the child becomes interpersonally dysfunctional is referred to as **traumatic sexualization.**

Finkelhor and Browne (1986) note the following: Sexually traumatized children learn inappropriate sexual behaviors (such as manipulating an adult's genitals for affection), are confused about their sexuality, and inappropriately associate certain emotions—such as loving and caring—with sexual activities.

As adults, sexual issues may become especially important. Survivors may suffer flashbacks, sexual dysfunctions, and negative feelings about their bodies. They may also be confused about sexual norms and standards. A fairly common confusion is the belief that sex may be traded for affection. Some women label themselves as "promiscuous," but this label may be more a result of their negative self-image than of their actual behavior. There seems to be a history of childhood sexual abuse among many prostitutes (Simons and Whitbeck 1991).

BETRAYAL. Children feel betrayed when they discover that someone on whom they have been dependent has manipulated, used, or harmed them. Children may also feel betrayed by other family members, especially mothers, for not protecting them from abuse. As adults, survivors may experience depression as a manifestation, in part, of extended grief over the loss of trusted figures. Distrust may manifest itself in hostility and anger or in social isolation and avoidance of intimate relationships. Anger may express a need for revenge or retaliation.

POWERLESSNESS. Children experience a basic kind of powerlessness when their bodies and personal spaces are invaded against their will. A child's powerlessness is reinforced as the abuse is repeated. In adulthood, powerlessness may be experienced as fear or anxiety; a person feels unable to control events. Adult survivors often believe that they have impaired coping abilities. This feeling of ineffectiveness may be related to the high incidence of depression and despair among survivors. Powerlessness may also be related to increased vulnerability or revictimization through rape or marital violence; survivors may feel unable to prevent subsequent victimization.

Other survivors, however, may attempt to cope with their earlier powerlessness by an excessive need to control or dominate others.

STIGMATIZATION. Ideas about being a bad person as well as feelings of guilt and shame about sexual abuse are transmitted to abused children and then internalized by them. Stigmatization is communicated in numerous ways. The abuser conveys it by blaming the child or, through secrecy, communicating a sense of shame. If the abuser pressures the child for secrecy, the child may also internalize feelings of shame and guilt. As adults, survivors may feel extreme guilt or shame about having been sexually abused. They also feel different from others because they mistakenly believe that they alone have been abused.

Treatment Programs

Child sexual abuse, especially father–daughter incest, is increasingly being treated through therapy programs working with the judicial system rather than through breaking up the family by removing the child or the offender (Nadelson and Sauzier 1986). Because the offender is often also the breadwinner, incarcerating him may greatly increase the family's emotional distress. The district attorney's office may work with clinicians in evaluating the existing threat to the child and deciding whether to prosecute, refer the offender to therapy, or both. The goal is not simply to punish the offender but to try to assist the victim and the family in coming to terms with the abuse.

Many of these clinical programs work on several levels at once: they treat the individual, the father–

■ *Children need to have someone, such as a teacher who they trust, in whom they can confide about their suffering.*

started to identify and help child or adult survivors of sexual abuse. (For an evaluation of commercially available materials for preventing child abuse, see Roberts et al. 1990.) Such prevention programs have been hindered, however, by three factors (Finkelhor 1986a, 1986b):

- The issue of sexual abuse is complicated by differing concepts of appropriate sexual behavior and partners, which are not easily understood by children.

- Sexual abuse, especially incest, is a difficult and scary topic for adults to discuss with children. Children who are frightened by what their parents tell them, however, may be less able to resist abuse than those who are given strategies of resistance.

- Sex education is controversial. Even where it is taught, instruction often does not go beyond physiology and reproduction. The topic of incest is especially opposed.

Child abuse prevention (CAP) programs typically aim at three audiences: children, parents, and professionals (especially teachers). The CAP programs stress that the child is not at fault when such abuse does occur.

They also try to give children possible courses of action if someone tries to sexually abuse them. In particular, children are taught that it's all right to say "no," and that it's important to tell someone they trust about what has happened—and to keep telling until they are believed (Gelles and Conte 1991).

daughter relationship, the mother–daughter relationship, and the family as a whole. They work on developing self-esteem and improving the family and marital relationships. If appropriate, they refer individuals to alcohol or drug abuse treatment programs.

A crucial ingredient in many treatment programs is individual and family attendance at self-help group meetings. These self-help groups are composed of incest survivors, offenders, mothers, and other family members. Self-help groups such as Parents United and Daughters and Sons United help the offender acknowledge his responsibility and understand the effect of the incest on everyone involved.

Preventing Sexual Abuse

The idea of preventing sexual abuse is relatively new (Berrick and Barth 1992). Prevention programs began about a decade ago, a few years after programs were

Reflections

Assume for a moment that a young child disclosed to you the fact that she was hurt by her father. What would you say to her? How would you feel? Whom would you tell?

Other programs focus on educating parents to warning signs of abusers. It is hoped that they will then educate their children. Such programs, however, need to be culturally sensitive, because Latinos and Asian Americans may be especially reticent about discussing these matters with their children (Ahn and Gilbert 1992).

CAP programs have also directed attention to professionals such as teachers, physicians, mental health professionals, and police officers. Because of their close contact with children, teachers are especially important. Professionals are encouraged to watch for signs of sexual abuse and to investigate children's reports of such abuse.

In recent years, both the American Medical Association (AMA) and the federal government have become more actively involved in fighting domestic violence. AMA guidelines advise doctors to question female patients routinely as to whether they have been attacked by their partners or forced to have sex. Physicians are also urged to investigate cases of injuries to women that are not well explained.

Obviously, the violence and abuse discussed in this chapter are complex phenomena. They are products of individual characteristics of perpetrators and victims, relationship dynamics, and certain social and cultural factors. Not every home becomes a center of violence and abuse, and most families are not embattled. We need to realize that those families and relationships that are violent or abusive are products of a blend of qualities and are affected on multiple levels. This understanding is important if we hope to reduce the prevalence of violence and abuse and if we care to help those who are most at risk or already victimized.

Summary

- Any form of intimacy or relatedness increases the likelihood of violence or abuse. *Violence* is defined as an act carried out with the intention or perceived intention of causing physical pain or injury to another person.

- Abuse and violence are separate, although certainly related and overlapping, phenomena. Not all abuse is violent, and some intimate violence is considered appropriate and not abusive.

- Violence ranges from routine to extreme, from *common couple violence,* which is typically less severe, to *intimate terrorism,* which is a more severe, most often male-on-female form of violence and abuse in which power and domination are key motives.

- *Violent resistance,* often considered under the idea of "self-defense" is more often used by women.

- Seven principal models are used to study sources of family violence: (1) individualistic explanations, which find the source of violence within the personality of the abuser; (2) the feminist model; (3) the social situational model; (4) the social learning model, (5) the resource model, and (6) the exchange–social control model. Three factors that may reduce social control are inequality of power in the family, the private nature of the family, and the "real man" image.

- Researchers have stressed the role played by gender, power and control, stress, and intimacy in explaining intimate violence.

- It is difficult to know exactly how much violence there is in intimate relationships. The use of official records and/or survey data give us underestimates of how much intimate violence there is in the United States.

- Wife battering is one of the most common and most underreported crimes in the United States. Two characteristics that correlate highly with wife assault are: low socioeconomic status and a high degree of marital conflict.

- **Gender symmetry** refers to the findings of similarity in both expressing and experiencing violence between the genders. Even The context and consequences of partner violence are not the same for men and women.

- Age, race, and social class all factor into domestic violence. Younger women, black women, and lower-income women experience more intimate violence than do other women.

- Research on abused husbands shows both similarities and differences with what research has revealed about male perpetrators and female victims.

- Although intimate violence can be found among all groups in society, it happens with greater frequency among lower-income individuals and among African Americans.

- *Marital rape* is a form of battering. Many people, including victims themselves, have difficulty acknowledging that forced sex in marriage is rape, just as it is outside of marriage.

- Violence among same-sex couples is similar to the levels of violence among heterosexuals. Because such relationships lack the social supports that

- heterosexual couples can draw upon the experience of victimization may be worse.

- The incidence of verbal abuse, physical violence, and coercive sex in dating relationships among high school and college students is alarming.

- Dating violence is often precipitated by jealousy or rejection. *Date rape* or *acquaintance rape* may not be recognized by either the assailant or the victim because they think that rape is something done by strangers.

- Dangerous date-rape drugs such as Rohypnol (flunitrazepam) and gamma hydroxybutyrate (GHB) are sometimes used by offenders to sedate and sexually victimize unsuspecting women, prompting the passage of date-rape drug prohibition laws.

- Reasons women may stay in, or return to, abusive relationships include economic dependency, religious pressure or beliefs, the perceived need for a father for the children, a sense of duty, fear, love, and reasons pertaining to their particular culture.

- Some women may also be paralyzed by *learned helplessness*

- Intimate violence generates high costs in terms of time lost at work, mental health, and medical expenses for injuries or trauma sustained.

- Domestic violence intervention can be based on either control or compassion. Arrest, prosecution, and imprisonment are examples of control; shelters and support groups (including abuser programs) are examples of compassionate intervention.

- Recent legal innovations such as *mandatory arrest* and *no-drop prosecution* have had mixed results. In some ways they raise the costs for victims of reporting the violence.

- At least 1 million children are physically abused and neglected by their parents each year in the United States. Most abuse cases are unreported. Parental violence is one of the five leading causes of childhood death.

- Families at risk for child abuse often have specific parental, child, and family ecosystem characteristics.

- Nearly 90% of parents acknowledged using some form of psychological aggression with at least one child during the prior 12-month period. Younger parents use such aggression more often, and as children move from infancy they are more often recipients of such behavior.

- Mandatory reporting of suspected child abuse may be helping to decrease the number of abused children in the United States. Early intervention and education also may help reduce abuse.

- The hidden victims of family violence include siblings (who have the highest rate of violent interaction), parents assaulted by their adolescent or youthful children, and elderly parents assaulted by their middle-aged children.

- Recommendations for reducing family violence include reducing sources of societal stress, such as poverty and racism; eliminating sexism; establishing supportive networks; breaking the family cycle of violence; and eliminating the legitimization and glorification of violence.

- *Incest* is defined as sexual intercourse between people too closely related to marry. Sexual victimization of children may include incest, but it can also involve other family members and other sexual activities. The most traumatic form of child abuse is probably father–daughter (or stepfather–stepdaughter) abuse.

- Children most at risk for sexual abuse include females, preadolescents, children with absent or unavailable parents, children with poor parental relationships, children with parents in conflict, and children living with a stepfather.

- *Child sexual abuse* has both initial and long-term effects. The survivors of sexual abuse often suffer from sexual abuse trauma, which is characterized by traumatic sexualization, betrayal, powerlessness, and stigmatization.

- Child sexual abuse offenders are increasingly being sent into treatment programs in an attempt to assist the incest survivor and family in coping with the crisis that incest creates. Self-help groups are important for many survivors of sexual abuse.

Key Terms

acquaintance rape 468

battered child syndrome 472

battering 462

child sexual abuse 478

common couple violence 457

date rape 468

extrafamilial sexual abuse 478

Resources on the Internet

Companion Website for This Book

http://www.thomsonedu.com/sociology/strong

Gain an even better understanding of this chapter by going to the companion website for additional study resources. Take advantage of the Pre- and Post-Test quizzing tool, which is designed to help you grasp difficult concepts by referring you back to review specific pages in the chapter for questions you answer incorrectly. Use the flash cards to master key terms and check out the many other study aids you'll find there. Visit the Marriage and Family Resource Center on the site. You'll also find special features such as access to Info-Trac© College Edition (a database that allows you access to more than 18 million full-length articles from 5,000 periodicals and journals), as well as GSS Data and Census information to help you with your research projects and papers.

CHAPTER 14

© Scott Barrow

Coming Apart: Separation and Divorce

Outline

Are the following statements **TRUE** or **FALSE**?
You may be surprised by the answers (see answer key on the bottom of this page).

T   F   1   More than 40% of couples who enter marriage are projected to end up divorcing within 7 years.

T   F   2   Divorce occurs as a single event in a person's life.

T   F   3   Americans have one of the highest marriage, divorce, and remarriage rates among industrialized nations.

T   F   4   The critical emotional event in a marital breakdown is separation rather than divorce.

T   F   5   Age at marriage is the best predictor of the likelihood of divorce.

T   F   6   Divorce is an important element of the contemporary American marriage system because it reinforces the significance of emotional fulfillment in marriage.

T   F   7   The higher an individual's employment status, income, and level of education, the greater the likelihood of divorce.

T   F   8   Many problems assumed to be caused by divorce are present before marital disruption.

T   F   9   Those whose parents are divorced have a significantly greater likelihood of divorcing themselves.

T   F   10   Marital conflict in an intact two-parent family is generally more harmful to children than living in a tranquil single-parent family or stepfamily.

A s one woman told sociologist Joseph Hopper (2001), there was nothing she and her husband could do:

> It's something that had to happen, and it wasn't something that either one of us really controlled. It was just an awful situation that we had to get out of, and I recognized it and he didn't.

A second person offered the following:

> I had wanted that forever—the white picket fence and the whole dream. But it didn't come true. But I was at least smart enough to realize it wasn't happening and no matter what I did it wasn't going to.

Are Americans pro-marriage? Are we too soft on divorce? Do we believe in the importance of marriage and the commitment we make when we exchange wedding vows? Or when we say "I do" are we really adding, perhaps not under our breath but in our heads, "at least for now"? Americans' feelings about marriage and divorce are paradoxical. Consider the following generalizations (Ganong and Coleman 1994; White 1991):

- Americans like marriage: they have one of the highest marriage rates in the industrialized world.

- Americans don't like marriage: they have one of the highest divorce rates in the world.

- Americans like marriage: they have one of the highest remarriage rates in the world.

What sense can we make out of being one of the most marrying, divorcing, and remarrying nations in the world? What does our high divorce rate tell us about how we feel about marriage? In this chapter, we hope to explain the paradox of high rates of marriage and divorce as we examine the divorce process, marital separation, divorce consequences, children and divorce, child custody, and divorce mediation. This exploration will help you better understand what parents, children, and families experience and how they

Answer Key for What Do You Think

1 True, see p. 492;   2 False, see p. 500;   3 True, see p. 490;   4 True, see p. 501;   5 True, see p. 496;   6 True, see p. 491;   7 False, see p. 493;   8 True, see p. 511;   9 True, see p. 498;   10 True, see p. 508.

cope with what increasingly has become part of our marriage system—divorce.

Some scholars suggest that divorce represents not a devaluation of marriage but an idealization of it. They reason that we would not divorce if we did not have so much hope about marriage fulfilling our various needs. According to Frank Furstenberg and Graham Spanier (1987), divorce may well be a critical part of our contemporary marriage system, which emphasizes emotional fulfillment and satisfaction.

Our high divorce rate also tells us that we may no longer believe in the permanence of marriage. Norval Glenn (1991) suggests that there is a "decline in the ideal of marital permanence and . . . in the expectation that marriages will last until one of the spouses dies." Instead, marriages disintegrate when love goes or a potentially better partner comes along. Divorce is a persistent fact of American marital and family life and one of the most important forces affecting and changing American lives today (Furstenberg and Cherlin 1991).

Before 1974, the view of marriage as lasting "till death do us part" reflected reality. However, a surge in divorce rates that began in the mid-1960s did not level off until the 1990s. In 1974, a watershed in American history was reached when more marriages ended by divorce than by death. Today approximately 50% of all new marriages are likely to end in divorce (U.S. Census Bureau 1996).

Divorce not only ends marriages and breaks up families, it also creates new forms from the old ones. It creates remarriages (which are different from first marriages). It gives birth to single-parent families and stepfamilies. Today about one out of every five American families is a single-parent family; more than half of all children will become stepchildren (U.S. Census Bureau 1996). Within the singles subculture is an immense pool of divorced men and women (most of whom are on their way to remarriage). Or consider the numbers of marriages that are truly remarriages for one or both spouses. As seen in Table 14.1, for 8.4% of currently married couples the marriage is a second marriage for *both wife and husband*. Nearly one in ten marriages in the United States consists of two people who have both been married before to other spouses.

The greatest concern that social scientists express about divorce is its effect on children (Aldous 1987; Wallerstein 1997; Wallerstein and Blakeslee 1989). But even in studies of the children of divorce, the research may be distorted by traditional assumptions about divorce being deviant (Amato 1991). For example, problems that children experience may be attributed to divorce rather than to other causes, such as personality traits. Although some effects are caused by the disruption of the family itself, others may be linked to the new social environment—most notably poverty and parental stress—into which children are thrust by their parents' divorce (McLanahan and Booth 1991; Raschke 1987). Some therapists suggest that we begin looking at those factors that help parents and children successfully adjust to divorce rather than focusing on risks, dysfunctions, and disasters (Abelsohn 1992).

Table 14.1 ■ Number of Times Married, for Those Currently Married[*]

Number of times wife has been married		Number of times husband has been married		
	Total	Married once	Married twice	Married three or more times
Number (in thousands)				
Total	57,728	44,965	10,274	2,489
Married once	45,389	40,288	4,421	681
Married twice	10,232	4,107	4,866	1,259
Married three or more times	2,106	571	987	549
Percentage of marriages				
Total	100.0	77.9	17.8	4.3
Married once	78.6	69.8	7.7	1.2
Married twice	17.7	7.1	8.4	2.2
Married three or more times	3.6	1.0	1.7	1.0

[*]This table includes only married people whose spouse is present.
SOURCE: Kreider 2005.

Measuring Divorce: How Do We Know How Much Divorce There Is?

How common is divorce and how likely is it to happen to us? The U.S. Census Bureau (2000) shows that there are nearly 20 million divorced people age 15 and older in the United States, representing more than 9% of the population. And many of you have probably heard the gloomy news that *one out of two marriages ends in divorce.* What exactly do those statistics mean and on what are they based? There are a variety of ways to measure and represent the prevalence of divorce in the United States. Look briefly at the most common measures.

Ratio Measure of Divorces to Marriages

The **ratio measure of divorce** is calculated by taking the number of divorces and the number of marriages in a given year and producing a ratio to represent how often divorce occurs relative to marriage. In 1998, for example, there were 1.13 million divorces and 2.24 million marriages—a ratio of 1 divorce for every 1.98 marriages. But recognize the difference between that statistic and a statement indicating that one of every two marriages *will end* in divorce. What the ratio measure truly reflects is the relative popularity or commonality of marriage and divorce.

Crude Divorce Rate

The **crude divorce rate** represents the number of divorces in a given year for every 1,000 people in the population. From November 2004 to November 2005, there were 3.6 divorces for every 1,000 Americans. There were also 7.5 marriages per 1,000 people in the population, returning us to right around our "one divorce for every two marriages" (Munson and Sutton, 2006).

Crude divorce or marriage rates have certain problems. Obviously, when calculating the crude divorce rate, counting every 1,000 people in the population means including many unmarried people, children, the elderly, the already divorced, and so on. These people cannot become divorced. It is therefore a statistic that is highly susceptible to the age distribution, proportions of married and single people in the population, and to changes in such population characteristics.

Refined Divorce Rate

Considered the most useful measure of divorce, the **refined divorce rate** measures the number of divorces that occur in a given year for every 1,000 marriages (as measured by married women age 15 and older). In 1998, the refined rate was 19 to 20 divorces per 1,000 married women, meaning that 2% of marriages ended in divorce.

Note that the range of available statistics produces different impressions about the reality of divorce in the United States. The ratio measure gives the most alarming impression, the one most closely approximating "one out of two marriages," or 50% of marriages, ending in divorce. When we use the refined rate of 2% of marriages ending in divorce annually, the picture seems much less bleak. The reality represented by each statistic is the same, but the meanings we attach to each statistic, and therefore the understanding it creates, vary significantly.

Predicting Divorce

Another divorce statistic worth mentioning is the **predictive divorce rate.** This calculation (too complicated for our purposes) allows researchers to estimate how many new marriages will likely end in divorce. The prevailing estimate is that somewhere between 40% and 50% of marriages entered into in a year are likely to become divorces, but some put the estimate as high as 60%.

Estimating future trends is a tricky business. Because this estimate is based on experience of prior birth cohorts (people born between specific years), we cannot be confident that current and future cohorts will make the same choices or face the same circumstances as their predecessors.

But even these predictions need to be more carefully assessed. As we show in subsequent sections describing factors associated with divorce, not everyone faces the same risk of divorce. As articulated by Barbara Dafoe Whitehead and David Popenoe (2004), "The background characteristics of people entering a marriage have major implications for their risk of divorce." They go on to report the decreases in vulnerability to divorce during the first 10 years of marriage that are shown in Table 14.2.

Table 14.2 ■ Vulnerability to Divorce in First 10 Years of Marriage

Factor in Risk of Divorce	Percentage of Decrease
Annual income over $50,000 (versus under $25,000)	−30
Having a baby 7 months or more after marriage (versus before marriage)	−24
Marrying after 25 years of age (versus under 18)	−24
Own family of origin intact (versus divorced parents)	−14
Religious affiliation (versus none)	−14
Some college (versus high school dropout)	−13

Ultimately, Dafoe Whitehead and Popenoe (2005, 19) offer the following, more reassuring assessment of the likelihood of experiencing divorce: "So if you are a reasonably well-educated person with a decent income, come from an intact family and are religious, and marry after age twenty-five without having a baby first, your chances of divorce are low indeed."

Divorce Trends in the United States

If we look at long-term divorce trends, the unmistakable conclusion is that the twentieth century saw dramatic increases in marital breakups. If we look, instead, over the past 25 years, a different picture emerges. In more recent decades, the divorce rate dropped (see Table 14.3). Divorce rates in the United States have "plateaued" and then leveled off after reaching their peak in 1979. As we show shortly, this did not occur equally for all groups.

Both marriage and divorce rates have declined. The marriage rate is at its lowest point since the 1930s, and the 2.22 million marriages in 2005 reflect a recent decline from the 2.38 million marriages performed in 1997 (Munson and Sutton 2003). As to divorce, we can see that after three-quarters of a century of increases (minus, of course, the "time-out" of the 1950s), in more recent years the rate has declined. Most recently, there were 2% fewer divorces in 1998 than in 1997 (when there were 1.16 million divorces) and 7% fewer than the 1.22 million divorces occurring in 1992, which represented the all-time high in numbers of divorce. In addition, the 2005 crude divorce rate of 3.6 per 1,000 people is lower than it has been since the 1970s. There are multiple stories to tell about trends in divorce and causes of divorce.

Factors Affecting Divorce

Sometimes it is easy to point to the cause of a particular divorce. Perhaps one spouse was unfaithful or abusive and the marriage was brought to a quick end. In other instances, even the divorcing parties can't identify the exact cause or causes that led to divorce. Researchers have looked at factors affecting wider divorce

Table 14.3 ■ Divorce and Marriage through the Twentieth Century and Beyond

Year	Marriages	Rate per 1,000	Divorces	Rate per 1,000	Rate per 1,000 married women
1900	709,000	9.3	55,751	0.7	3
1920	1,274,476	12.0	170,506	1.6	8
1940	1,595,879	12.1	264,000	2.0	9
1960	1,523,000	8.5	393,000	2.2	9.2
1970	2,158,802	10.6	708,000	3.5	14.9
1980	2,406,708	10.6	1,189,000	5.2	22.6
1985	2,413,000	10.2	1,178,000	5.0	21.7
1990	2,448,000	9.8	1,182,000	4.7	20.9
1995	2,336,000	8.9	1,169,000	4.4	19.8
1998	2,244,000	8.4	1,135,000	4.2	NA*
2001	2,327,000	8.4	NA	4.0	NA

*NA means data not available.

rates, as well as divorce decisions. Some analyses address the complex sets of changes that make divorce rates hard to predict. For example, Heaton (2002) notes that there have been increases in the prevalence of premarital sex, premarital births, cohabitation, and both racial and religious intermarriage. All of these tend to be associated with higher likelihood of marital instability, especially divorce. Yet there have been increases in age at marriage and in educational attainment, which tend to be associated with higher rates of stable marriage. These latter trends are among the factors that have counterbalanced the former trends, leading to declining rates of divorce. In this section, we look at both the larger societal or demographic factors and the individual and couple characteristics that may be related to the likelihood of divorce.

Societal Factors

As seen earlier, even the reduced divorce rates starting in the late 1990s were *six times* the rate at the beginning of the twentieth century. They were twice as high as the rates in 1960. In addition, divorce rates in the United States are higher than rates elsewhere in the industrialized world (see Tables 14.4).

Changed Nature of the Family

The shift from an agricultural society to an industrial one undermined many of the family's traditional functions. Schools, the media, and peers are now important sources of child socialization and childcare. Hospitals and nursing homes manage birth and care

Table 14.4 ■ International Variation in Refined Divorce Rate

Divorces per 1,000 Married Women			
Country	1980	1990	1995
United States	23	21	20
Canada	10	11	11
Denmark	11	13	12
France	6	8	9
Germany	6	8	9
Italy	1	2	2
Japan	5	6	6
Sweden	11	12	14
United Kingdom	12	13	13

SOURCE: U.S. Census Bureau 1998, Table 1,346.

for the sick and aged. Because the family pays cash for goods and services rather than producing or providing them itself, its members are no longer interdependent.

As a result of losing many of its social and economic underpinnings, the family is less of a necessity.

It is now simply one of many choices we have: We may choose singlehood, cohabitation, marriage, or divorce—and if we choose to divorce, we enter the cycle of choices again: singlehood, cohabitation, or marriage and possibly divorce for a second time. A second divorce leads to our entering the cycle for a third time, and so on.

Social Integration

Social integration—the degree of interaction between individuals and the larger community—is a potentially important factor related to the incidence of divorce. The social integration approach regards such factors as urban residence, church membership, and population change as especially important in explaining divorce rates (Breault and Kposowa 1987; Glenn and Shelton 1985; Glenn and Supancic 1984).

Among African Americans, the lowest divorce rate is found among those born and raised in the South; African Americans born and raised in the North and West have the highest divorce rates. Similarly, those who live in urban areas, where the divorce rate is higher than in rural areas are less likely to be subject to the community's social or moral pressures. They are more independent and have greater freedom of personal choice.

Individualistic Cultural Values

American culture has traditionally been individualistic. We highly value individual rights, we cherish images of an individual battling nature, and we believe in individual responsibility. It should not be surprising that many view the individual as having priority over the family when the two conflict. Since the 1950s, perhaps as a reaction to the alienation and stifling conformity of the time, we have increasingly valued self-fulfillment and personal growth (Guttman 1993).

As marriage and the family lost many of their earlier social and economic functions, their meaning shifted. Marriage and family are viewed as paths to *individual* happiness and fulfillment. We marry for love and then expect marriage and our partners to bring us happiness. When individual needs conflict with family demands, however, we no longer automatically submerge our needs to those of the family. We often struggle to balance individual and family needs. But if we are unable to do so, divorce has emerged as an alternative to an unhappy or unfulfilling marriage and as an escape from a mean-spirited or violent marriage.

Demographic Factors

A number of demographic factors appear to have a correlation with divorce, including employment status, income, education level, ethnicity, and religion.

Employment Status

Among Caucasians, a higher divorce rate is more characteristic of low-status occupations, such as factory worker, than of high-status occupations, such as executive (Greenstein 1985; Martin and Bumpass 1989). Unemployment, which contributes to marital stress, is also related to increased divorce rates. Studies conflict as to whether employed wives are more likely than nonemployed wives to divorce; overall, however, the findings seem to suggest that female employment contributes to the likelihood of divorce since the wife is less dependent on her husband's earnings (White 1991). Wives' employment may also lead to conflict about the traditional division of household labor, childcare stress, and other work spillover problems that, in turn, create marital distress.

Employment also creates more opportunities for spouses to meet someone else and to embark on an extramarital sexual relationship. The presence and numbers of attractive alternative partners positively influences the risk of divorce. Scott South, Katherine Trent, and Yang Shen (2001) call this the *macrostructural opportunity perspective,* calling attention to the importance of attention to the opportunities for spouses to form potentially destabilizing opposite-sex relationships that are embedded within macrosocial structures, such as the workplace.

Also related to employment effects are the hours worked. Harriet Presser (2000) estimated that among men married less than 5 years and with young children, working night shifts increased their likelihood of divorce or separation by six times compared to men with similar families who worked days. Women with similar families who work nights face three times the likelihood of separation and divorce compared to those who work days. In the absence of children, the same effects are not found.

Income

The higher the family income, the lower the divorce rate for both Caucasians and African Americans. It is interesting, however, that the higher a woman's individual income, the greater her chances of divorce, perhaps because with greater incomes women are not economically dependent on their husbands or because conflict over inequitable work and family roles increases marital tension.

Each spouse's income alone does not explain divorce, nor does the relative income earned by each spouse. Stacy Rogers (2004) found that the highest risk of divorce occurred in marriages in which wives contributed between 50% and 60% of the family's resources if spouses were at low or moderate levels of happiness. However, "happier spouses have little incentive to divorce, irrespective of spouses' relative economic contributions" (Rogers 2004, 71). Thus, neither higher-earning wives nor lower-earning husbands are automatically prone to divorce.

Educational Level

The decline in divorce that occurred in the 1980s and 1990s happened mostly for college-educated women and men (Martin, 2004). The positive effect of education appears to be greatest in early marriage. During the first 3 years of marriage, the predicted risk of divorce among married women with less than 12 years of education is more than twice that for high school–educated women, and nearly four times the risk faced by women who have been to college (South 2001).

Of course, educational attainment is usually linked with other factors that affect marital success. For example, men and women pursuing higher education tend to delay marriage and children until they're older. Plus, increased education may lead to acquiring values more conducive to marital success (Heaton 2002).

One way in which education can affect divorce is by shaping attitudes toward divorce. One study concluded that college graduates had the most restrictive attitudes toward divorce, believing that "it should be more difficult to obtain a divorce than it is now." Women who haven't completed high school have the least restrictive attitudes (Martin and Parashar 2006).

Ethnicity

About a third of first marriages end in separation or divorce within the first ten years of marriage for white (32%) and Hispanic (34%) women, for non-Hispanic black women the figure reaches nearly half of first marriages (47%) (Phillips and Sweeney, 2005). Bulanbda and Brown (2004) estimate that blacks face a risk of divorce nearly 1.5 times that of whites (Bulanda and Brown, 2004).

In Julie Phillips and Megan Sweeney's (2005) carefully controlled, multivariate analysis of the risk of divorce among a sample of more than 4,500 white, black, and Mexican American women, black women have a 54% greater risk of experiencing a marital separation or divorce than do white women. Foreign-born Mexican women have a 76% reduced risk compared to white women. U.S.-born Mexican American women had risks of divorce that fell between those of Caucasian and African American women. These differences persist even when comparing women with similar experience in premarital cohabitation, with similar family backgrounds, and of similar education, employment, and age at marriage.

Religion

According to sociologists Vaughn Call and Timothy Heaton (1997, 391), "No single dimension of religion adequately describes the effect of *religious* experience on marital stability." Both *religiosity* (strength of religious commitment and participation) and religious affiliation have been linked to risk of divorce. Frequency of attendance at religious services (not necessarily the depth of beliefs) tends to be associated with the divorce rate. The greater the involvement in religious activities, the less the likelihood of divorce. But interestingly, a difference between spouses in frequency of attendance is a risk factor, too. Marriages in which wives attend services weekly and husbands don't attend have a greater risk of divorce than those marriages in which neither spouse attends religious services. The lowest risk is found among couples in which both spouses attend services regularly (Call and Heaton 1997).

Since all major religions discourage divorce, highly religious men and women are less likely to accept divorce because it violates their values. It may also be that a shared religion and participation in organized religious life affirms the couple relationship (Guttman 1993; Wineberg 1994; Call and Heaton 1997). Religiosity even seems to influence the likelihood of divorce when marital problems arise, suggesting that religion plays a role in the decision of whether or not to seek a divorce (Lowenstein 2005).

By religion, the lowest divorce rate is for Jews, followed by Catholics and then Protestants. The highest rates are found among those with no religious affiliation and those couples in religious intermarriages. However, compared with attendance, the effect of religious affiliation on divorce is a modest one, especially among marriages in which spouses are of the same religious affiliation (Call and Heaton 1997). Because the Roman Catholic Church only "allows" divorce through the use of annulments and no longer excommunicates divorced people by refusing them the sacraments, the annulment rate increased greatly over the last decades of the twentieth century (Woodward, Quade, and Kantrowitz 1995).

Life Course Factors

Different aspects of the life course may affect the probability of divorce for some individuals, including age at time of marriage, premarital pregnancy and childbirth, cohabitation, remarriage and intergenerational transmission.

Age at Time of Marriage

The age at which people marry is "the most consistent predictor of marital stability identified in social science research" (Heaton 2002). Young, especially adolescent, marriages are more likely to end in divorce than are marriages that take place when people are in their 20s or older. Close to 50% of those who marry before age 18 and 40% of those who marry before turning 20 divorce. Younger partners are less likely to be emotionally mature, younger marriages may be more likely to involve premarital pregnancy, and marrying "young" may be associated with curtailment of education, which has economic consequences that can undermine marital stability. Only 25% of those who marry when older than 25 end up divorced. The effect of age at marriage is not the same for all ethnic groups, however. Marrying in their teens has a "destabilizing effect" on Caucasian and African American women, but not on Mexican American women (Phillips and Sweeney 2005).

■ *Marrying young, especially in one's teens, significantly increases one's risk of divorce.*

© Grace/zefa/CORBIS

Premarital Pregnancy and Childbirth

Premarital pregnancy or birth significantly increases the likelihood of divorce, the risks being 1.2 to 1.3 times greater than for women without such experiences (Kposowa). Risks are especially high if the pregnant woman is an adolescent, drops out of high school, and faces economic problems following marriage. If a woman gives birth before marriage, the likelihood for divorce in a subsequent marriage increases, especially in the early years. This negative, "destabilizing" effect of a premarital conception on marriage is stronger for African Americans than for Caucasians (Phillips and Sweeney 2005).

Cohabitation

As shown in an earlier chapter, premarital cohabitation is associated with a higher risk of a later divorce. Whether this is an effect of cohabitation—say, by altering people's attitudes toward marriage and divorce—or a reflection of the less traditional attitudes toward marriage and family, including attitudes toward divorce, that cohabitants bring with them into cohabitation is unclear.

Remarriage

You might expect that having been married and divorced (at least) once would make people better at making a subsequent marriage succeed. That may seem as intuitively sensible as would an expectation that cohabitation would create more successful marriages, yet the assumption that cohabitation would lead to success turned out to be quite off the mark. So would the expectation that people learn from and avoid the same mistakes the second (or third, or fourth, or . . .) time around. The divorce rate among those who remarry is *higher* than it is for those who enter first marriages.

It is not entirely clear why there is a higher divorce rate in remarriages. Some researchers suggest that the cause may lie in a "kinds-of-people" explanation. The probability factors associated with the kinds of people who divorced in first marriages—everything from low levels of education to unwillingness to settle for unsatisfactory marriages—are present in subsequent marriages, increasing their likelihood of divorce. Similarly, people bring their same personality problems to any new relationship. Others argue that the unique dynamics of subsequent marriages, especially the presence of stepchildren, increase the chances of divorce. In fact, subsequent marriages that involve stepchildren have twice the likelihood of divorce as first marriages (Schoen 2002).

Intergenerational Transmission

Those whose parents divorce are subject to **intergenerational transmission**—the increased likelihood that divorce will later occur to them (Raschke 1987; Amato 1996). It is now estimated that parental divorce increases the chance of a daughter's marriage ending within the first 5 years by as much as 70%. If both the husband's and the wife's parents have been divorced, the odds of divorce increase by 189%. How can we explain this intergenerational cycle?

Paul Amato (1996) notes that children of divorced parents are more likely to marry younger, cohabit, and experience higher levels of economic hardship. They become more pessimistic about lifelong marriage and develop more liberal attitudes toward divorce. In addition, females whose parents divorce develop less traditional attitudes about women's family roles, value self-sufficiency, and possess stronger attachments to paid employment. Each of these could raise susceptibility to divorce. Interestingly, parental "marital discord" in the absence of divorce has been found to have little consequence for their children's risk of divorce. Furthermore high-discord marriages that ended in divorce only minimally raised their children's risk of divorce. However, where low-discord marriages ended in divorce the children were especially vulnerable to divorce themselves (Amato and DeBoer 2001).

Using survey data from more than 1,300 individuals from the Study of Marriage over the Life Course, Amato examined the relative role of these factors. He found that the major effects of parental divorce that led to later divorce were acquired "problematic behaviors" (such as anger, jealousy, infidelity) and life course variables (such as age at marriage). On the other hand, the intergenerational connection was not well explained by people's attitudes toward divorce.

Amato (1996) draws other interesting conclusions:

- The increased risk of divorce holds in second marriages, as well as first marriages.
- The effects are especially pronounced in "offspring marriages" (marriages by children of divorced parents) of short duration but are not present in marriages of long duration.
- The effects are strongest when parents divorce early in their children's lives (age 12 or younger).

Keep in mind that, as with intergenerational cycles of family violence, this relationship is neither automatic nor inevitable. It is, however, an important factor that can undermine marital success. Perhaps children of divorce need to more consciously guard against behaviors that might undermine their marriages.

One way in which parental divorce may be assumed to affect children's risk of divorce is in shaping their attitudes toward divorce. Children of divorced parents, especially daughters of divorced parents, are more likely to possess pro-divorce attitudes (Kapinus 2004). Research that examined the effect of parents' attitudes on more than 400 children of divorce (Kapinus 2004) concludes:

- There appears to be a "critical period," namely, the late teens, when parents' attitudes toward divorce have special salience to their children.
- Parental divorce affects sons' and daughters' attitudes toward divorce differently. Daughters of divorce are more likely to express "pro-divorce attitudes" than are sons of divorce.
- Diminished relationships with fathers after divorce and continued postdivorce conflict between parents may lead sons toward negative attitudes toward divorce. Yet postdivorce conflict between parents does not have the same effect on daughters.

Family Processes

The actual day-to-day marital processes of communication—handling conflict, showing affection, and other marital interactions—may be the most important factors holding marriages together or dissolving them (Gottman 1994).

Marital Happiness

Although it seems reasonable that there would be a strong link between marital happiness (or, rather, the lack of happiness) and divorce, this is true only during the earliest years of marriage. Low levels of liking and trusting a partner are associated with long-term outcomes such as reduced satisfaction and elevated risk of divorce. The strength of the relationship between low marital happiness and divorce decreases in later stages of marriage, however (White and Booth 1991).

Eventually, alternatives to marriage and barriers to divorce appear to influence divorce decisions more strongly than does marital happiness. With nothing better to leave for, or if there are too many obstacles to overcome in leaving, a couple might stay married even if unhappy. Although the opposite is also true— even if happy one partner might leave for a more attractive alternative—it is probably less common. The presence of alternatives to a spouse has an effect on marital stability that can be observed among both high- and low-risk couples (that is, among those with other predisposing factors and those without).

The importance of the availability of attractive alternatives to a spouse has sometimes been overlooked as a factor accounting for divorce. Scott South, Katherine Trent, and Yang Shen (2001, 753) note that "satisfied and dissatisfied spouses alike remain, consciously or not, in the marriage market." As explained earlier, the workplace is a central component of such a market.

Children

Although 60% of divorces involve children, couples with children divorce less often than couples without children. The birth of the first child reduces the chance of divorce to almost nil in the year following the birth (White 1991). Furthermore, couples with two children divorce less often than couples with one child or no children (Diekmann and Schmidheiny 2004). This does not mean that having children will spare parents from a divorce or that troubled spouses should become parents so that their troubles will disappear. It may well be that troubled spouses hold off having children or, if they have a child, resist having more because of their troubles. Thus, the quality of the marriage may lead to childbearing more than vice versa.

There are some situations in which the presence of children may be related to higher divorce rates. Premaritally conceived (during adolescence) children and physically or mentally limited children are associated with divorce, as are children from prior marriages or relationships. Children in general can contribute to marital dissatisfaction and possibly divorce, according to one researcher (Raschke 1987): "It could be expected that normal children at least contribute to strains in an already troubled marriage, given the consistent findings that children, especially in adolescent years, lower marital satisfaction." At the same time, however, women without children have considerably higher divorce rates than women with children.

Marital Problems

If you ask divorced people to give the reasons for their divorce, they are not likely to say, "I blame the changing nature of the family" or "It was demographics." They are more likely to respond, "She was on my case all the time" or "He just didn't understand me"; if they are charitable, they might say, "We just weren't right for each other." Personal characteristics leading to conflicts are important factors in the dissolution of relationships.

Studies of divorced men and women cite such problems as alcoholism, drug abuse, marital infidelity, sexual incompatibility, and conflicts about gender roles as relationship factors leading to their divorces. They also often cite external events—problems with in-laws or the effect of jobs (Amato and Previti 2003). Paul Amato and Denise Previti (2003) found the most common reasons given by their sample to be infidelity, incompatibility, alcohol or drug use, growing apart, personality problems, lack of communication, and abuse (physical or mental).

Gender differences in reasons for divorce indicate that, in general, women cite emotional or relationship reasons, incompatibility, infidelity, unhappiness, and insufficient love, as well as aspects of their former husband's personality or behaviors (such as abusiveness, neglect of children or home, and substance use). They are less likely to blame themselves. Men more often cite external factors or claim ignorance—they say they do not know what happened (Amato and Previti 2003).

People of high socioeconomic status are more likely to stress communication problems, incompatibility of or changes in values or interests, and their former spouse's self-centeredness. People of low socioeconomic status more often mention such things as financial problems, physical abuse, going out with the boys or girls, employment problems, neglect of home responsibilities, and drinking.

We know from studying enduring marriages that marriages often continue in the face of such problems. Recent research (Amato and Rogers 1997) on the connections between marital problems and divorce reveals that reports of marital problems in 1980 were associated with later divorce between 1980 and 1992. Based on interviews with almost 2,000 people, Paul Amato and Stacy Rogers (1997) found the following:

- Although men's and women's reports differed in the particular problems they emphasized, both predicted divorce equally well.

- Certain problems such as jealousy, moodiness, anger, poor communication, and drinking increased the odds of later divorce; sexual infidelity was an especially strong predictor of divorce.

- People who later divorce report a higher number of problems as early as 9 to 12 years before their divorce. Thus, their assessments of problems are not after-the-fact justifications concocted to account for or justify their divorce.

- Marital problems are **proximal causes** of later divorce. They are features of the relationship that directly raise the probability of divorce. There are also background characteristics, such as age at marriage, prior cohabitation, education, income, church attendance, and parental divorce that operate as more **distal causes.** These are brought by each spouse to the relationship and raise the likelihood that marital problems will later arise.

No-Fault Divorce

Since 1970, beginning with California's Family Law Act, all 50 states have adopted **no-fault divorce**—the legal dissolution of a marriage in which guilt or fault by one or both spouses does not have to be established. It is unclear exactly how or how much no-fault divorce has affected divorce rates. Some contend that liberalization of divorce law led to increases in the divorce rate in both the United States and in other countries (for example, Scotland, England, and Wales) (Lowenstein 2005). It is debatable that this has, by itself, affected the divorce rate. Unambiguously, however, liberalization of divorce law has altered the process of divorce by decreasing the time involved in the legal process and it has altered the grounds for determining postdivorce financial responsibility.

The Stations of the Divorce Process

Divorce is not a single event. You don't wake up one morning and say, "I'm getting a divorce," and then leave. It's a far more complicated process (Kitson and Morgan 1991). It may start with little things that at first you hardly notice—a rude remark, thoughtlessness, an unreasonable act, a "closedness." Whatever the particulars may be, they begin to add up. Other times, however, the sources of unhappiness are more blatant—yelling, threatening, or battering. For whatever reasons, the marriage eventually becomes unsatisfactory; one or both partners become unhappy.

We know less about the process of marital breakdown and divorce than we ought to, especially given its prevalence in the United States. We understand more about falling in love and courtship than we do about falling out of love and divorce (Furstenberg and Cherlin 1991).

Anthropologist Paul Bohannan (1970b) developed one of the more influential descriptive models of the divorce process. (For a discussion of other models, see Guttman 1993.) Bohannan's model consists of six **stations of divorce:** emotional, legal, economic, coparental, community, and psychic. As people divorce, they undergo these stations, or "divorces," although they neither have a particular order nor begin and end simultaneously. The level of intensity of these different divorces varies at different times and for different couples.

- *The emotional divorce.* The emotional divorce, when one spouse (or both) begins to disengage from the marriage, to feel "something isn't quite right," begins well before the legal divorce. But even as divorce papers are filed, the partners may find themselves feeling ambivalent. Because the emotional divorce is not complete, they may try to reconcile. The partners may undermine each other's self-esteem with indifference or destructive criticism. From the outside, the marriage may appear to be functioning adequately, but its heart is missing.

- *The legal divorce.* The legal divorce is the court-ordered termination of a marriage. Although we tend to associate "divorce" with the legal divorce, by the time someone is "officially" legally divorced much has happened. Furthermore, long after the legal decree couples may still be working their way through the other dimensions of divorce. The legal decree permits divorced spouses to remarry and conduct themselves in a way that is legally independent of each other. The legal divorce also sets the terms for the division of property and child custody, issues that may lead to bitterly contested divorce battles. Many of the unresolved issues of the emotional divorce, such as feelings of hurt and betrayal, may be acted out during the legal divorce. No-fault divorce was intended to minimize these issues.

- *The economic divorce.* The economic basis of marriage often becomes most painfully apparent during the economic divorce. Most property acquired during a marriage is considered joint property and is divided between the divorcing spouses. The property settlement is based on the assumption that each spouse contributes to the estate. This contribution may be nonmonetary, as in the case of traditional homemakers whose "moral assistance and domestic services" permitted their husbands to work outside the home. As part of the economic divorce, alimony and child support may be ordered by the court. As the partners go their own ways, husbands and wives often experience different consequences in their standards of living as they set up separate households and no longer pool their resources. Women usually experience a decline in their standards of living, men sometimes see theirs increase.

- *The coparental divorce.* Marriages end, but parenthood does not. Spouses may divorce each other, but they do not divorce their children. (Even those parents who never see their children remain fathers and mothers.) This may be the most complicated aspect of divorce, because it also gives rise to single-parent families and, in most cases, stepfamilies, considered in more detail in Chapter 15. As parents divorce, issues of child custody, visitation, and support must be dealt with. The effect of divorce on children must be understood, negative consequences must be minimized as much as possible, and new ways of relating to the children and former spouses must be developed, keeping the children's best interest foremost in mind.

- *The community divorce.* When people divorce, their social world changes. In-laws become ex-laws; often they lose (or stop) contact. (This is particularly troublesome when in-laws are also grandparents.) Old friends may choose sides or drop out; they may not be as supportive as desired. New friends may replace old ones as divorced men and women begin dating again. They may enter the singles subculture, where activities center on dating. Single parents may feel isolated from such activities because childrearing often leaves them no leisure and diminished income leaves them no money.

- *The psychic divorce.* The psychic divorce is accomplished when a former spouse becomes irrelevant to a sense of self and emotional well-being. For example, people are psychically divorced when they learn that an ex-spouse has gotten a promotion; married someone smarter and better looking; bought a 4 × 4; and received an honorary doctorate—and they don't care. As part of the psychic divorce, each partner develops a sense of independence, completeness, and stability. Navigating through the psy-

chic station may be more difficult and take a good deal longer than it does to experience the other stations of divorce.

The divorce process, as you can see, is complex. It takes place on many different levels. Those who go through divorce experience both pain and liberation but eventually emerge as new women and men living a dramatically different life.

Reflections

From what you know about divorce, either from your own experience as a child or partner or from the experiences of friends or other family members, how well does Bohannan's six-station model describe the experience? Are some stages more difficult than others? Why?

Uncoupling: The Process of Separation

Perhaps the crucial event in a marital breakdown is the act of separation. Although separation generally precedes divorce, not all separations lead to divorce. Furthermore, those that do may first involve attempts at reconciliation, in that about one-third of the divorced women become divorced after attempting at least one marital reconciliation (Wineberg 1999). A statistic now more than a decade old indicates that perhaps 1 in 10 marriages experiences a separation and reconciliation (Wineberg and McCarthy 1993). Those who reconcile may have separated to dramatize their complaints, create emotional distance, or dissipate their anger (Kitson 1985).

People do not suddenly separate or divorce. Instead, they gradually move apart through a set of fairly predictable stages. Sociologist Diane Vaughan (1986) calls this process *uncoupling.* The process appears to be the same for married or unmarried couples and for gay or lesbian relationships. The length of time together does not seem to affect the process.

"Uncoupling begins," Vaughan observes, "as a quiet, unilateral process." Usually one person, the **initiator,** is unhappy or dissatisfied but keeps such feelings to himself or herself. Because the dissatisfied partner is unable to find satisfaction within the relationship, he

■ *In the early phases of the process of separation, estrangement can grow before both parties are fully aware of what has happened.*

© Walter Hodges/Getty Images/Stone

or she begins turning elsewhere. This is not a malicious or intentional turning away; it is done to find self-validation without leaving the relationship. In doing so, however, the dissatisfied partner "creates a small territory independent of the coupled identity" (Vaughan 1986).

Eventually, the initiator decides that he or she can no longer go on. She or he may go through a process of mourning the demise of what is still an intact marriage (Emery 1994, cited in Amato 2000).

After the relationship ends, initiators have better adjustment to divorce and carry less postdivorce attachment to their former spouses (Wang and Amato 2000).

Reflections

From your experience, how well does "uncoupling" describe the process of separating from someone you care about? Are there missing elements or elements that should be emphasized? What about separation distress? In your experience, what was it like? What things were you able to do to alleviate it? What advice would you give others about it?

Uncoupling does not end when the end of a relationship is announced or even when the couple physically separates. Acknowledging that the relationship cannot be saved represents the beginning of the last stage of uncoupling. Diane Vaughan (1986) describes the process:

Partners begin to put the relationship behind them. They acknowledge that the relationship is unsaveable. Through the process of mourning they, too, eventually arrive at an account that explains this unexpected denouement. "Getting over" a relationship does not mean relinquishing that part of our lives that we shared with another but rather coming to some conclusion that allows us to accept and understand its altered significance. Once we develop such an account, we can incorporate it into our lives and go on.

The New Self: Separation Distress and Postdivorce Identity

Examining the experiences of those who divorce may be as good a way as any to see how much our married self becomes part of our deepest self. When people separate or divorce, many feel as if they have "lost an arm or a leg." This analogy, as well as the traditional marriage rite in which a man and a woman are pronounced "one," reveals an important truth of marriage: The constant association of both partners makes each almost a physical part of the other. This dynamic is true even if two people are locked in conflict; they, too, are attached to each other (Masheter 1991).

Separation Distress

Most newly separated people do not know what to expect. There are no divorce ceremonies or rituals to mark this major turning point. Yet people need to understand divorce to alleviate some of its pain and burden. Except for the death of a spouse, divorce is the greatest stress-producing event in life (Holmes and Rahe 1967). The changes that take place during separation are crucial because at this point a person's emotions are at their rawest and most profound. Men and women react differently during this period. Many people experience **separation distress,** situational anxiety caused by separation from an attachment figure. Researchers have considerable knowledge about the negative consequences accompanying marital separation, some of which we discuss here. In looking at this negative effect, however, we need to keep in mind that eventually the negative aspects of separation may be balanced by positive aspects, such as the possibility of finding a more compatible partner, constructing a better (or different) life, developing new dimensions of the self, enhancing self-esteem, and marrying a better parent for the children. These positive consequences may follow, or be intertwined with, separation distress. In the pain of separation, we may forget that a new self is being born.

Almost everyone suffers separation distress when a marriage breaks up. The distress is real but, fortunately, does not last forever (although it may seem so). The distress is situational and is modified by numerous external factors. About the only men and women who do not experience distress are those whose marriages were riddled by high levels of conflict. In these cases, one or both partners may view the separation with relief (Raschke 1987).

During separation distress, almost all attention is centered on the missing partner and is accompanied by apprehensiveness, anxiety, fear, and often panic. "What am I going to do?" "What is he or she doing?" "I need him . . . I need her . . . I hate him . . . I love him . . . I hate her . . . I love her."

Sometimes, however, the immediate effect of separation is not distress but euphoria. This usually results from feeling that the former spouse is not necessary, that the old fights and the spouse's criticism are gone forever, and that life will now be full of possibilities and excitement. That euphoria is soon gone. Almost everyone falls back into separation distress.

Whether a person had warning and time to prepare for a separation affects separation distress. An unexpected separation is probably most painful for the partner who is left. Separations that take place during the first 2 years of marriage, however, are less difficult for the husband and wife to weather. Those couples who separate after 2 years find separation more difficult because it seems to take about 2 years for people to become emotionally and socially integrated into marriage and their marital roles (Weiss 1975). After that point, additional years of marriage seem to make little difference in the spouses' reaction to separation.

As the separation continues, separation distress slowly gives way to loneliness. Eventually, loneliness becomes the most prominent feature of the broken relationship. Old friends can sometimes help provide stability for a person experiencing a marital breakup, but those who give comfort need to be able to tolerate the other person's loneliness.

Establishing a Postdivorce Identity

A person goes through two distinct phases in establishing a new identity following marital separation: *transition* and *recovery* (Weiss 1975). The transition period begins with the separation and is characterized by separation distress and then loneliness. In this period's later stages, most people begin functioning in an orderly way again, although they still may experience bouts of upset and turmoil. The transition period generally ends within the first year. During this time, individuals have already begun making decisions that provide the framework for new selves. They have entered the role of single parent or noncustodial parent, have found a new place to live, have made important career and financial decisions, and have begun to date. Their new lives are taking shape.

The recovery period usually begins in the second year and lasts between 1 and 3 years. By this time, the separated or divorced individual has already created a reasonably stable pattern of life. The marriage is becoming more of a distant memory, and the former spouse does not arouse the intense passions she or he once did. Mood swings are not as extreme, and periods of depression are fewer. Yet the individual still has self-doubts that lie just beneath the surface. A sudden reversal, a bad time with the children, or doubts about a romantic involvement can suddenly destroy a divorced person's confidence. By the end of the recovery period, the distress has passed. It may take some people longer than others to recover because each person experiences the process in his or her own way. But

To some degree, gender influences how individuals respond to divorce. Research indicates that divorced men experience greater emotional distress and report more suicidal thoughts than do women (Riesman and Gerstel 1985; Rosengren, Wedel, and Wilhelmensen 1989; Wallerstein and Kelly 1980). Because women are more likely to initiate divorce, research suggests that they experience fewer postdivorce psychological problems. This may be because they have begun the detachment process earlier than men (Lawson and Thompson 1996). Furthermore, divorced men exhibit higher rates of auto accidents, alcohol abuse, diabetes, heart disease, and mental illness than do divorced women. Higher rates of mortality have been found to exist among divorced men and women, especially if they have remarried or are cohabiting (Hemstrom 1996).

The immediate effect of divorce on women is economic. This is especially true if they become the primary custodial parent. Many women who are granted child support do not receive the full amount, and as many as one in four receive nothing. A combination of lowered earning power, increased expenses, and lack of financial support results in a decreased standard of living for the divorced mother and her children.

The psychological responses experienced among partners are numerous, ranging from anger to depression to ambivalence. Although some men suffer little distress following divorce (Albrecht 1980), generally men seem to experience the greater emotional distress, possibly because of their more frequent social isolation (Reismann 1990). In addition, men report greater attachment to their former spouses and are more likely to desire to rekindle the marriage (Bloom, and Kindle 1985).

Almost 60% of divorces involve children (Kitson and Morgan 1991), and because most children of divorced parents end up in the physical custody of their mothers, fathers must face new emotional territory regarding these issues and their relationships with their children.

Single parenting for the mother involves added responsibility to an already overburdened workload. Noncustodial parenting raises new role expectations concerning the quality of the parent–child relationship, normative behaviors, and discipline.

Social support is positively correlated with lower distress and positive adjustment. Additionally, as with other stressors in a person's life, it is often the individual's perception of the event, not the stress itself that influences how a person adjusts to change. If those experiencing separation and divorce can begin to view and accept their changing circumstances as presenting new challenges and opportunities, there is a greater likelihood that the physiological and psychological symptoms of stress that follow divorce can be reduced.

most are surprised by how long the recovery takes—they forget that they are undergoing a major discontinuity in their lives.

Dating Again

A new partner reduces much of the distress caused by separation. A new relationship prevents the loneliness caused by emotional isolation. It also reinforces a person's sense of self-worth. It will not necessarily eliminate separation distress caused by the disruption of intimate personal relations with the former partner, children, friends, and relatives, but it "often produces a decline in depression, health complaints, and visits to the doctor, and an increase in self-esteem. When someone loves you and values you, you begin thinking that you are worth caring about" (Hetherington and Kelly 2002, 78–79).

Initiating this process may be stressful. A first date after years of marriage and subsequent months of singlehood evokes some of the same emotions felt by inexperienced adolescents (Spanier and Thompson 1987).

For many divorced men and women, the greatest problem is how to meet other unmarried people. They believe that marriage has put them "out of circulation," and many are not sure how to get back in. Because of the marriage squeeze, separated and divorced men in their 20s and 30s are at a particular disadvantage: considerably fewer women are available than men. The squeeze reverses itself at age 40 when there are significantly fewer single men available. The problem of meeting others is most acute for single mothers who are full-time parents in the home because they lack opportunities to meet potential partners. Divorced men, having fewer childcare responsibilities and more income than divorced women, tend to have more active social lives.

Several features of dating following separation and divorce differ from premarital dating. First, dating does not seem to be a leisurely matter. Divorced people are often too pressed for time to waste it on a first date that might not go well. Second, dating may be less spontaneous if the divorced woman or man has primary responsibility for children. The parent must make arrangements about childcare; he or she may wish not to involve the children in dating. Third, finances may be strained; divorced mothers may have income only from low-paying or part-time jobs or TANF benefits yet have many childcare expenses. In some cases, a father's finances may be strained by paying alimony or child support. Finally, separated and divorced men and women often have a changed sexual ethic based on the simple fact that there are few divorced virgins (Spanier and Thompson 1987).

Sexual relationships are often an important component in the lives of separated and divorced men and women. Engaging in sexual relations for the first time following separation may help people accept their newly acquired single status. Because sexual fidelity is an important element in marriage, becoming sexually active with someone other than an ex-spouse is a dramatic symbol that the old marriage vows are no longer valid.

Consequences of Divorce

Most divorces are not contested; between 85% and 90% are settled out of court through negotiations between spouses or their lawyers. But divorce, whether it is amicable or not, is a complex legal process involving highly charged feelings about custody, property, and children (who are sometimes treated by angry partners as property to be fought over).

Economic Consequences of Divorce

Probably the most damaging consequences of the no-fault divorce laws are that they systematically impoverish divorced women and their children. Following divorce, women are primarily responsible for both child-rearing *and* economic support (Maccoby et al. 1993). As a result, women are at a greater risk for poverty than they were during their marriage. Even if a woman is not plunged into poverty, she often experiences a dramatic downward turn in her economic condition (Garrison 1994; Morgan 1991). A single mother's income shows about a 27% decline (Peterson 1996; Smock 1993).

Husbands typically enhance their earning capacity during marriage. In contrast, wives generally decrease their earning capacity because they either quit or limit their participation in the workforce to fulfill family roles. This withdrawal from full participation limits their earning capacity when they reenter the workforce. Divorced homemakers have outdated experience, few skills, and no seniority. Thus, they may not be "equal" to their former husbands at the point of divorce. Rules that treat a woman as if she is equal to her husband simply serve to deprive her of the financial support she needs (Weitzman 1985).

Although it is often claimed that unlike women, men experience enhanced financial well-being following divorce, this outcome depends on the division of wage earning that characterized the failed marriage. For white men who contributed less than 60% of their marital standard of living, divorce precipitates a decline in their living standards. On the other hand, men whose share of the household income was greater than 80% experience significant increases in their living standards after their marriages end (McManus and DiPrete 2001).

Another factor that leads to women's economic slide is lack of child support. When marriage ends, many women face the triple consequences of gender, ethnic, and age discrimination as they seek to support themselves and their children. Because the workplace favors men in terms of opportunity and income, separation and divorce does not affect them as adversely. Whereas the disparities in income between Caucasian and African American women are significant during marriage, following a divorce Caucasian women suffer a relatively greater decline in their standards of living and the income levels of Caucasian and African American women converge (Morgan 1991). Mexican American women suffer relatively less decline in economic status than do Anglo American women because Latinas are already more economically disadvantaged. But because their lives have prepared them for greater economic adversity, Latinas' emotional well-being appears to suffer less than does that of Anglo American women following divorce (Wagner 1993).

Alimony and Child Support

Alimony is the money payment a former spouse makes to the other to meet his or her economic needs. It is *not* intended to be punitive. It is instead designed to address the economic vulnerability that a spouse may find himself or herself in after the end of the marriage. Alimony is paid until the receiving spouse remarries or dies. Death of the paying spouse may not end alimony obligations, however. The deceased's estate may be required to continue to honor the alimony decision even after the paying spouse dies (http://www.answers.com/topic/alimony).

Alimony is different from **child support**—the monetary payments made by the noncustodial spouse to the custodial spouse to assist in childrearing expenses. For many women, their source of income changes upon divorce from primarily joint wages earned during marriage to their own wages, supplemented by child support payments, alimony, help from relatives, and welfare.

The legal criteria around both alimony and child support have undergone some notable changes in the past two decades. The Child Support Enforcement Amendments, passed in 1984, and the Family Support Act of 1988 require states to deduct delinquent support from fathers' paychecks, authorize judges to use their discretion when support agreements cannot be met, and mandate periodic reviews of award levels to keep up with the rate of inflation. In addition, all states implemented automatic wage withholding of child support in 1994. Chien-Chung Huang and colleagues (2005) contend that nearly every year for the past two decades, Congress has passed new laws designed to strengthen child support enforcement. Furthermore, spending by both state and federal governments on child support enforcement increased from less than $1 billion a year in 1978 to $5.2 billion in 2002 (Huang, Mincy, and Garfinkel 2005). Recent research has shown that enforcement has had a beneficial effect on compliance with child support orders (Meyer and Bartfeld 1996).

Data also indicate that most children entitled to child support from their fathers do not receive it (Huang, Mincy, and Garfinkel 2005). One determinant of fathers' compliance with their support obligations is their ability to pay. When child support obligations exceed 35% of a father's income, he is less likely to comply (Meyer and Bartfeld 1996). In general, lower-income fathers are required to pay greater shares of their income in child support. Compliance

by these fathers would "moderately improve" if their child support obligations were in line with those of higher-earning fathers. Yet, reducing the amount fathers have to pay would result in a "net loss" of about 38% to children (Huang, Mincy, and Garfinkel 2005).

People are generally more approving, at least in principle, of child support, than they are of alimony. In the past, alimony represented the continuation of the husband's responsibility to support his wife. Currently, laws determine that alimony be awarded on the basis of need to those women or men who would otherwise be indigent. At the same time, some assert that alimony represents the return of a woman's "investment" in marriage (Oster 1987; Weitzman 1985). Lenore Weitzman (1985) argues that a woman's homemaking and childcare activities must be considered important contributions to her husband's present and future earnings. If divorce rules do not give a wife a share of her husband's enhanced earning capacity, then the "investment" she made in her spouse's future earnings is discounted. According to Weitzman, alimony and child support awards should be made to divorced women in recognition of the wife's primary childcare responsibilities and her contribution to her ex-husband's work or career. Such awards will help raise divorced women and children above the level of poverty to which they have been cast as a result of no-fault divorce's specious equality.

Reflections

Why are alimony and child support often such emotional issues in divorce? On what basis should alimony be awarded? Child support? Why do many noncustodial parents fail to pay child support? What could be done to improve their likelihood of supporting their children?

Employment

The economic effect of divorce on women with children is especially difficult because their employment opportunities are often constrained by the necessity of caring for children (Maccoby et al. 1993). Childcare costs may consume a third or more of a poor single mother's income. Women may work fewer hours because of the need to care for their children.

Separation and divorce dramatically change many mothers' employment patterns (Morgan 1991). If a mother was not employed before separation, she is

Issues and Insights Lesbians, Gay Men, and Divorce

Although there are no reliable studies, it is estimated that about one-fifth of gay men and one-third of lesbians have been married to someone of the opposite sex. Estimates of bisexual men and women who are married run into the millions. Relatively few gay men, lesbians, and bisexuals are consciously aware of their sexual orientation at the time they marry. Those who are aware may not disclose their feelings to their prospective partners. When married lesbians and gay men acknowledge their gayness to themselves, they often feel that they are "living a lie" in their marriage. Although they may deeply love their spouses, most eventually divorce.

How is it that lesbians and gay men marry heterosexuals in the first place? As shown in Chapter 6, the gay or lesbian identity process is difficult and complex. Because of fear and denial, some gay men and lesbians are unable to acknowledge their sexual feelings. They believe or hope they are heterosexual and do their utmost to suppress their same-sex fantasies or behaviors. They often believe that their homosexuality is

just a "phase." Typically they hold negative stereotypes about homosexuality and cannot bring themselves to believe or accept that they might be "one of them." Marriage is one way of convincing themselves that they are heterosexual. In addition to "curing" or denying their gayness, their motivations to marry are no different from heterosexuals (Bozett 1987). Like heterosexuals, gay men and lesbians marry because of pressure from family, friends, and fiancé, genuine love for the fiancé, the wish for companionship, and the desire to have children.

When husbands or wives discover their partner's homosexuality or bisexuality, they may initially experience shock; others experience temporary relief. Mysteries are explained: why the spouse disappears for periods of time, why mysterious phone calls occur, the spouse's lack of sexual interest. But whether shocked or relieved, inevitably the heterosexual spouse feels deceived or stupid. Many feel shame (Hays and Samuels 1987). One woman, who felt ashamed to tell anyone of her distress, recalled, "His coming out of the closet in some ways put the family in the closet" (Hill 1987). At the same time, the gay, lesbian, or bisexual spouse often feels deeply grieved (Voeller 1980): Many people date, marry, and be-

come parents, only to realize too late the error they made. They then find themselves deeply pained, fearful of losing their children through lawsuits, of losing spouses they care for but are ill suited to, of depriving their spouses and themselves of more deeply appropriate and meaningful relationships, and of causing their friends and other relatives deep pain.

When gay men, lesbians, or bisexuals disclose their orientation to their spouses, separation and divorce are the usual outcomes. Many gay men and lesbians are also parents when they separate from their spouses. It is generally important for them to affirm their identities both as gay or lesbian and as a parent (Bozett 1989c). This is especially important because negative stereotypes portray gay men and lesbians as "antifamily." Men and women begin to fuse their identities as gay or lesbian with their parental role.

A study of gay fathers reported that gay men usually do not reveal their orientation to their children unless the parents are separating or the gay father develops a gay love relationship (Bozett 1989c). As with divorced fathers in general, gay fathers usually do not have custody of the children, but lesbians, like other divorced women, are more likely to have custody (Bozett 1989b).

likely to seek a job following the split-up. The reason is simple: if she and her children relied on alimony and child support alone, they would soon find themselves on the street. Most employed single mothers are still on the verge of financial disaster, however. On the average, they earn less than married fathers. This is partly because women tend to earn less than men and partly because they work fewer hours, primarily because of childcare responsibilities (Garfinkel and McLanahan 1986). The general problems of women's lower earning capacity and lack of adequate childcare are

particularly severe for single mothers. Gender discrimination in employment and lack of societal support for childcare condemn millions of single mothers and their children to poverty.

Noneconomic Consequences of Divorce

In comparison to married people, the picture of divorced individuals is fairly bleak. Reviewing the research literature of the 1990s, Paul Amato (2000) notes the following: Compared with married people, divorced individuals experience more psychological distress, poorer self-concepts, lower levels of psychological well-being, lower levels of happiness, more social isolation, less satisfying sex lives, and more negative life events. They also have greater risks of mortality and report more health problems. Compared to married women and men, major depression is three times higher for separated or divorced women and *nine times higher for separated or divorced men*. British data reveal a similar story. Marital separation is accompanied by significant increases in heavy drinking during the period of separation (Power, Rogers, and Hope 2000). Also, Terrance Wade and David Pevalin (2004) found that for those exiting a marriage through separation and divorce there is a much higher prevalence of mental health problems. Of note, they also found that such problems are evident before the marital disruption, indicating that the relationship between mental health and divorce goes both ways.

Linda Waite and Maggie Gallagher, in their book *The Case for Marriage* (2000), take on the question of whether being married makes people happier or whether it is happier people who get married and *stay* married. Citing research that compared the emotional health of a sample of people over time—some who married and stayed married, some who never married or remained divorced, and others who married and divorced—they report the following: "When people married, their mental health improved—consistently and substantially. Meanwhile, over the same period, when people separated and divorced, they suffered substantial deterioration in mental and emotional well-being, including increases in depression and declines in reported happiness. . . . Those who dissolved a marriage also reported less personal mastery, fewer positive relations with others, less purpose in life, and less self-acceptance than their married peers did."

Waite and Gallagher (2000) also note that compared to married people, divorced (and widowed) women and men were three times as likely to commit suicide. Among the divorced, as among the general population, more men than women commit suicide. However, divorced women are "the most likely to commit suicide, followed by widowed, never-married and married, in that order." As parents, divorced individuals have more difficulty raising children. They display more role strain, whether they are custodial or noncustodial parents, and they display less authoritative parenting styles (Amato 2000).

Despite the stark picture that surfaces, for some people divorce is associated with positive consequences. These include higher levels of personal growth, greater autonomy, and—for some women—improvements in self-confidence, career opportunities, social lives, and happiness, as well as a stronger sense of control (Amato 2000). In addition, we would be remiss if we didn't point out evidence suggesting that remaining unhappily married is worse than divorcing. People who find themselves in "long-term low-quality marriages" are less happy than those who divorce and remarry. They also have lower overall life satisfaction, lower overall health, and lower self-esteem than those who divorce and remain single (Hawkins and Booth 2005).

Children and Divorce

Slightly more than half of all divorces involve children. Popular images of divorce depict "broken homes," but it is important to remember that an intact nuclear family, merely because it is intact, does not guarantee children an advantage over children in a single-parent family or a stepfamily. A traditional family wracked with spousal violence, sexual or physical abuse of children, alcoholism, neglect, severe conflict, or psychopathology creates a destructive environment likely to inhibit children's healthy development. Living in a two-parent family with severe marital conflict is often more harmful to children than living in a tranquil single-parent family or stepfamily. Children living in happy two-parent families appear to be the best adjusted, and those from conflict-ridden two-parent families appear to be the worst adjusted. Children from single-parent families are in the middle. The key to children's adjustment following divorce is a lack of conflict between divorced parents (Kline, Johnston, and Tschann 1991).

Telling children that their parents are separating is one of the most difficult and unhappy events in life. Whether or not the parents are relieved about the separation, they often feel extremely guilty about their children. Many children may not even be aware of parental discord (Furstenberg and Cherlin 1991). Even those that are may be upset by the separation, but their distress may not be immediately apparent.

Qualitative research by Heather Westberg, Thorana Nelson, and Kathleen Piercy (2002) indicates that

■ *Notifying children of a decision to divorce is difficult for both the parents and children.*

children's reaction is influenced by how the news is disclosed and is shaped by the perception that life will be relatively better or relatively worse afterward. For those to whom the news was disclosed long before the divorce occurred, by the time it "finally" happened it was experienced as relief.

As psychologist Judith Wallerstein suggested in her book, *Second Chances* (Wallerstein and Blakeslee 1989), divorce is differently experienced within the family. For at least one of the divorcing spouses, divorce is welcomed as an escape from an unpleasant or unfulfilling relationship. Both spouses may come to appreciate the "second chance" they receive with divorce: the opportunity to make a better choice and build themselves a better relationship. Children may not see the breakup of their parents' marriage as an "opportunity." However, under certain circumstances—especially "in households where parents engage in a long-term process of overt, unresolved conflict," children are at risk of developing emotional and developmental problems so long as their parents stay together (Booth and Amato 2001). For such children, divorce may, indeed, come as a relief.

Lisa Strohschein (2005) found that children's antisocial behaviors such as bullying and lying were reduced after divorce of parents who had been experiencing high levels of dysfunction. The stress relief that comes with divorce may, however, become apparent only after enough time passes (Strohschein 2005).

Conversely, when parental conflict is limited and kept from the children, the risk of developmental and emotional problems is low. But for those children from low-conflict parental marriages, divorce may represent "an unexpected, unwelcome, and uncontrollable event." They face the loss of one parent, the emotional distress of the remaining parent, and perhaps a decline in standard of living (Booth and Amato 2001).

The Three Stages of Divorce for Children

Part of the difficulty in determining the effect of divorce on children is a failure to recognize that, just as it is for adults, divorce is a process as opposed to a single event. Divorce comprises a series of events and changes in life circumstances. Many studies focus on only one part of the process and identify that part with divorce itself. Yet at different points in the process, children are confronted with different tasks and adopt different coping strategies. Furthermore, the diversity of children's responses to divorce is the result, in part, of differences in temperament, gender, age, and past experiences.

A study by psychologist Judith Wallerstein (1997) found that children from divorced families suffered both emotionally and developmentally. Young children fared worse than older children. Depending on the point in the process, boys tend to do less well than girls. In the "crisis period" of the 2 years following separation, boys' suffering is especially evident. This may be because they must internalize different gendered styles of reacting to distress. It is also the case, however, that after separation most boys live with their mothers and not their fathers. This, too, can exacerbate their suffering (Furstenberg and Cherlin 1991).

According to Wallerstein, children experience divorce as a three-stage process. Studying 60 California families during a 5-year period, she argued that divorce consisted of the initial, transition, and restabilization stages:

- *Initial stage.* The initial stage, following the decision to separate, was extremely stressful; conflict escalated, and unhappiness was endemic. The children's aggressive responses were magnified by the parents' inability to cope because of the crisis in their own lives.

- *Transition stage.* The transition stage began about a year after the separation, when the extreme emotional responses of the children had diminished or disappeared. The period was characterized by restructuring of the family and by economic and social changes: living with only one parent and visiting the other, moving, making new friends and losing old ones, financial stress, and so on. The transition period lasted between 2 and 3 years for half the families in the study.

- *Restabilization stage.* Families had reached the restabilization stage by the end of 5 years. Economic and social changes had been incorporated into daily living. The postdivorce family, usually a single-parent family or stepfamily, had been formed.

Children's Responses to Divorce

Decisive in children's responses to divorce are their age and developmental stage (Guttman 1993). A child's age affects how the response to one parent leaving home, changes (usually downward) in socioeconomic status, moving from one home to another, transferring schools, making new friends, and so on.

Developmental Tasks of Divorce

Judith Wallerstein suggested that children must undertake six developmental tasks when their parents divorce (Wallerstein 1983). The first two tasks need to be resolved during the first year. The other tasks may be worked on later; often they may need to be reworked because the issues often recur. How children resolve these tasks differs by age and social development. The tasks are as follows:

- *Acknowledging parental separation.* Children often feel overwhelmed by feelings of rejection, sadness, anger, and abandonment. They may try to cope with them by denying that their parents are "really" separating. They need to accept their parents' separating and to face their fears.

- *Disengaging from parental conflicts.* Children need to psychologically distance themselves from their parents' conflicts and problems. They require such distance so that they can continue to function in

■ *Children react differently to divorce depending on their age. Most feel sad, but the eventual outcome for children depends on many factors, including having competent and caring custodial parent, siblings, and friends and their own resiliency. The postdivorce relationship between parents and the custodial parent's economic situation are also important factors.*

© Kindra Clineff/Index Stock

their everyday activities without being overwhelmed by their parents' crisis.

- *Resolving loss.* Children lose not only their familiar parental relationship but also their everyday routines and structures. They need to accept these losses and focus on building new relationships, friends, and routines.

- *Resolving anger and self-blame.* Children, especially young ones, often blame themselves for the divorce. They are angry with their parents for disturbing their world. Many often "wish" their parents would divorce, and when their parents do, they feel responsible and guilty for "causing" it.

- *Accepting the finality of divorce.* Children need to realize that their parents will probably not get back together. Younger children hold "fairy tale" wishes that their parents will reunite and "live happily ever after." The older the child is, the easier it is for him or her to accept the divorce.

- *Achieving realistic expectations for later relationship success.* Children need to understand that their parents' divorce does not condemn them to unsuccessful relationships as adults. They are not damaged by witnessing their parents' marriage; they can have fulfilling relationships themselves.

YOUNGER CHILDREN. Younger children react to the initial news of a parental breakup in many different ways. Feelings range from guilt to anger and from sorrow to relief, often vacillating among all of these. Preadolescent children, who seem to experience a deep sadness and anxiety about the future, are usually the most upset. Some may regress to immature behavior, wetting their beds or becoming excessively possessive. Most children, regardless of their age, are angry because of the separation. Very young children tend to have more temper tantrums. Slightly older children become aggressive in their play, games, and fantasies—for example, pretending to hit one of their parents.

A recent study using longitudinal data collected over a 12-year period examines parent–child relationships before and after divorce. Researchers found that marital discord may exacerbate children's behavior problems, making them more difficult to manage (Amato and Booth 1996). Because discord between parents often preoccupies and distracts them from the tasks of parenting, they appear unavailable and unable to deal with their children's needs. This study reinforced a growing body of evidence showing that many problems assumed to be caused by divorce are present before marital disruption.

School-age children may blame one parent and direct their anger toward him or her, believing the other one innocent. But even in these cases the reactions are varied. If the father moves out of the house, the children may blame the mother for making him go or they may be angry at the father for abandoning them, regardless of reality. Younger schoolchildren who blame the mother often mix anger with placating behavior, fearing she will leave them. Preschool children often blame themselves, feeling that they drove their parents apart by being naughty or messy. They beg their parents to stay, promising to be better. It is heartbreaking to hear a child say, "Mommy, tell Daddy I'll be good. Tell him to come back. I'll be good. He won't be mad at me anymore." A study of 121 white children between the ages of 6 and 12 found that about 33% initially blamed themselves for their parents' divorce. After a year, the figure dropped to 20% (Healy, Stewart, and Copeland 1993). The largest factor in self-blaming was being caught in the middle of parental conflict. Children who blamed themselves displayed more psychological symptoms and behavior problems than those who did not blame themselves.

When parents separate, children want to know with whom they are going to live. If they feel strong bonds with the parent who leaves, they want to know when they can see him or her. If they have brothers or sisters, they want to know if they will remain with their siblings. They especially want to know what will happen to them if the parent they are living with dies. Will they go to their grandparents, their other parent, an aunt or uncle, or a foster home? These are practical questions, and children have a right to answers. They need to know what lies ahead for them amid the turmoil of a family split-up so that they can prepare for the changes.

Some parents report that their children seemed to do better psychologically than they themselves did after a split-up. Children often have more strength and inner resources than parents realize. The outcome of separation for children, Robert Weiss (1975) observes, depends on several factors related to the children's age. Young children need a competent and loving parent to take care of them; they tend to do poorly when a parenting adult becomes enmeshed in constant turmoil, depression, and worry. With older, preadolescent children, the presence of brothers and sisters helps because the children have others to play with and rely on in addition to the single parent. If they have good friends or do well in school, this contributes to their self-esteem. Regardless of the child's age, it is important that the absent parent continue to play a role in the child's life. The children need to know that they have not been abandoned and that the absent parent still cares (Wallerstein and Kelly 1980b). They need continuity and security, even if the old parental relationship has radically changed.

ADOLESCENTS. Many adolescents find parental separation traumatic. Studies indicate that much of what appear to be negative results of divorce (personal changes, parental loss, economic hardships, and psychological adjustments) are often more likely the result of parental conflict that precedes and surrounds the divorce (Amato and Keith 1991; Morrison and Cherlin 1995; Amato and Booth 1996). A study by Youngmin Sun found that such problems as poor psychological well-being, academic difficulties, and behavioral problems are present among adolescents from divorced families *at least a year before* the divorce (Sun 2001).

Reflections

As you look at the adjustments that children must make when their parents divorce, are there others you would add? Which ones do you believe are the most important? Most difficult? If you were a divorcing parent, what strategies would you use to help your children adjust to divorce? How would your strategies differ according to the age of the child or adolescent? What do you think the experience might be of adult children whose parents divorce?

Adolescents may try to protect themselves from the conflict preceding separation by distancing themselves. Although they usually experience immense turmoil within, they may outwardly appear cool and detached. Unlike younger children, they rarely blame themselves for the conflict. Rather, they are likely to be angry with both parents, blaming them for upsetting their lives. Adolescents may be particularly bothered by their parents' beginning to date again. Some are shocked to realize that their parents are sexual beings, especially when they see a separated parent kiss someone or bring someone home for the night. The situation may add greater confusion to the adolescents' emerging sexual life. Some may take the attitude that if their mother or father sleeps with a date, why can't they? Others may condemn their parents for acting "immorally."

Kathleen Boyce Rodgers and Hillary Rose (2002) assert that the negative effects of divorce on adolescents can be tempered. They suggest that strong peer support, a strong attachment to school, and high levels of support and monitoring by parents can lessen the negative consequences adolescents otherwise encounter.

Helping Children Adjust

Helen Raschke's (1987) review of the literature on children's adjustment after divorce found that the following factors were important:

- Before separation, open discussion with the children about the forthcoming separation and divorce and the problems associated with them.
- The child's continued involvement with the noncustodial parent, including frequent visits and unrestricted access.
- Lack of hostility between the divorced parents.
- Good emotional and psychological adjustment to the divorce on the part of the custodial parent.
- Good parenting skills and the maintenance of an orderly and stable living situation for the children.

Continued involvement with the children by both parents is important for the children's adjustment. The greatest danger is that children may be used as pawns by their parents after a divorce. The recently divorced often suffer from a lack of self-esteem and a sense of failure. One means of dealing with the feelings caused by divorce is to blame the other person. To prevent further hurt or to get revenge, divorced parents may try to control each other through their children. A recent study has shown that children are likely to suffer long-term psychological damage—well into adulthood—if the parents do not consider their emotional needs during the divorce process (Wallerstein 1997).

Betwixt and Between: Children Caught in the Middle

One of the presumed consequences of divorce for children is the sense of being caught in the middle, forced to choose sides, and being pulled in different directions by their parents. Some have even suggested that feeling caught between parents may be one of the factors that differentiate children's reactions to divorce, explaining why some do better and some do worse. Such feelings may also lead to adolescent depression and deviant behavior. Evidence indicates that older adolescents are more likely than younger adolescents and children to feel caught. In addition, such feelings may extend well into adulthood, although reduced contact with both parents may lessen the intensity of such feelings.

When caught in the middle, children may opt for one of three strategies: try to maintain positive relationships with both parents, form an alliance with one parent over and against the other, or reject both parents. Trying to remain close to two embattled parents may exact costs that outweigh the benefits of such relationships. Choosing sides comes at the expense of a relationship with one parent and can trigger guilt toward the abandoned parent and resentment toward the. Rejecting both parents means losing closeness to both—a steep price to pay.

Paul Amato and Tamara Afifi (2006) also found that parents put more pressure on daughters than on sons to take sides in their disputes, and feeling caught in the middle is of more negative consequence for mothers and daughters than for mothers and sons.

Multiple Perspectives on the Long-Term Effects of Divorce on Children

There are multiple perspectives on how and why divorce affects children (Amato 1993). Specified outcomes range from negative through neutral to positive (Whitehead 1996; Coontz 1997). There is enough divergent information that we could selectively cite research to make either a more pessimistic or a more optimistic generalization. We review some of these mixed findings here.

A variety of studies reviewed by Barbara Dafoe Whitehead, in her strongly anti-divorce book, *The Divorce Culture* (1997), suggest multiple ways in which children suffer after their parents divorce. First, across racial lines, children of divorce suffer substantial reduction in family income as a direct result of divorce. Second, most children experience a weakening of ties with their fathers, suffering damage when and after fathers leave. She suggests that separation and later divorce induce a "downward spiral" in father–child relationships, wherein distance between them grows, and children eventually lose their fathers' "love, support, and substantial involvement." Third, children suffer a loss of "residential stability," often having to move from the family home because of drops in their economic standing.

Whitehead goes on to detail other measurable ways in which children suffer: reduced school performance, increased likelihood of dropping out, worsened and increased behavioral problems, a greater likelihood of becoming teen parents. Many of these same outcomes were identified as among the "risks and problems associated with stepfamily life" (Whitehead 1996).

In her more optimistic book, *The Way We Really Are: Coming to Terms with America's Changing Families* (1997), Stephanie Coontz tempers some of this distressing news. While acknowledging the "agonizing process" that accompanies divorce and the ways in which children, especially, can be hurt by divorce, Coontz qualifies the more pessimistic interpretations. In a subtle but important comparison, she notes that research shows "*not* that children in divorced families have more problems but that *more* children of divorced parents have problems" (Coontz 1997, emphases in original). In other words, all children of divorce do not suffer the negative consequences identified by researchers and reported by people such as Whitehead. Coontz reminds us that although more children in divorced homes drop out or become pregnant than do children whose parents stay married, "divorce does not account for the majority of such social problems as high school dropout rates and unwed teen motherhood" (Coontz 1997). Finally, Coontz goes even further in an optimistic direction, noting that there are some measures on which large proportions of children of divorced homes score higher than do average children from homes with two parents. She reports that children of single parents (usually single mothers) spend more time talking with their custodial parent, receive more praise for their academic successes, and face fewer pressures toward conventional gender roles. Thus, she argues, in some ways, single-parent households may be beneficial environments within which to be raised (Coontz 1997).

Just How Bad Are the Long-Term Consequences of Divorce?

The message about the long-term consequences varies according to the research examined. Influential longitudinal research conducted by Judith Wallerstein highlights fairly extensive, long-term trauma and distress that stays with and affects children of divorce well into adulthood. Beginning with *Surviving the Breakup: How Children and Parents Cope With Divorce* (Wallerstein and Kelly 1980), through *Second Chances: Men, Women, and Children a Decade After Divorce* (Wallerstein and Blakeslee 1989), and culminating with *The Unexpected Legacy of Divorce: A 25-Year Landmark Study* (Wallerstein, Lewis, and Blakeslee 2000), Wallerstein has followed a sample of (originally) 60 families, with 131 children among them, as they divorced and went through the subsequent adjustment processes at 18 months, 5 years, 10 years, 15 years, and ultimately 25 years. Seventy-five percent of the original families, and 71% of the 131 children were studied for all three books.

Wallerstein found that at the 5-year mark, more than a third of the children were struggling in school, experiencing depression, had difficulty with friendships, and continued to long for a parental reconciliation. At the 10-year follow-up, she indicated that almost half of the children carried lingering problems and they had become worried, sometimes angry, underachieving young adults. Three-fifths of the children of divorce retained a lingering sense of rejection by one or both parents and suffered especially poor relationships with their fathers. Finally, at the quarter-century point, Wallerstein asserted that the effects of divorce on children reached their peak in adulthood, where the ability to form and maintain committed intimate relationships was negatively affected (see Amato 2003).

A more moderate view of the long-term effects of divorce emerges from other studies (Hetherington and Kelly 2002 and Amato 2003). E. Mavis Hetherington undertook the Virginia Longitudinal Study of Divorce and Remarriage, which initially consisted of following a sample of 144 families with a 4-year-old "target child." Half of the sample families were divorced, half were married. Initially they were to be followed and restudied at 2 years to compare how those who divorced fared in comparison to those who did not. Eventually, the sample was expanded, and subsequent research was conducted at 2, 6, 11, and 20 years after divorce. As the "target children" (that is, the initial 4-year-olds) married, had a child, or cohabited for more than 6 months, they were further studied (Hetherington and Kelly 2002). Meanwhile, families were added to the sample at each wave, to reach a final sample of 450, evenly split between nondivorced, divorced, and remarried families.

Throughout the research, a variety of qualitative and quantitative data were collected on personalities of parents and children, adjustment, and relationships within and outside the family (Hetherington 2003).

The impression that Hetherington's research leaves is more encouraging than the one received from Wallerstein's studies. For example, most adults and children adapt to the divorce within 2 to 3 years. Although at the 1-year mark, 70% of the divorced parents were wrestling with animosity, loneliness, persistent attachment, and doubts about the divorce, by 6 years, most were moving toward building new lives. More than 75% of the sample said that the divorce had been a good thing, more than 50% of the women and 70% of the men had remarried, and most had embarked on the postdivorce paths they would continue to take (Hetherington 2003).

In considering the effects of divorce on children, Hetherington reports that 20% of her sample of youths from divorced and remarried families was troubled and displayed a range of problems, including depression and irresponsible, antisocial behavior. They had the highest dropout rate, had the highest divorce rate (as they themselves married), and were the most likely to be struggling economically. But perhaps more important, "80 percent of children from divorced homes eventually are able to adapt to their new life and become reasonably well adjusted" (Hetherington and Kelly 2002, 228). Given that 10% of youths from nondivorced homes also were struggling, the difference for children from divorced as opposed to nondivorced homes was fairly small (10%).

As Hetherington points out, the optimal outcome for adults and their children is to be in a happily married household. Nevertheless, her research indicates that we may overstate the risks and fail to recognize the resilience of men, women, and children of divorce.

Paul Amato (2003) suggests that much of the divorce research supports Wallerstein's claims that divorce is "disruptive and disturbing" in the lives of children, but he fails to find the same strength and pervasiveness of the supposed effects. Using still other longitudinal data gathered as part of the Marital Instability over the Life Course Study, Amato reports that 90% of children with divorced parents achieve the same level of adult well-being as children of "continuously married parents" (Amato 2003). Amato further suggests that children who experience multiple family transitions (parental divorce, remarriages, subsequent divorces, and so on) are the ones who most suffer. He found that children who experienced only a single parental divorce (without any additional parental transitions) were no different in their psychological well-being than children of continuously married parents.

Child Custody

Of all the issues surrounding separation and divorce, custody issues are particularly poignant because they represent continued versus strained or even severed ties between one parent and his or her children. When the court awards custody to one parent, the decision is generally based on one of two standards: the *best interests of the child* or the *least detrimental of the available alternatives*. In practice, however, custody of the children is awarded to the mother in about 90% of the cases. Three reasons can be given for this: (1) women usually prefer custody, and men do not; (2) giving custody to the mother is traditional; and (3) the law reflects a bias that assumes women are naturally better able to care for children.

Sexual orientation has also been a traditional basis for awarding custody (Baggett 1992; Beck and Heinzerling 1993). In the past, a parent's homosexuality has been sufficient grounds for denying custody, but increasingly, courts are determining custody on the basis of parenting ability rather than sexual orientation. Interviews with children whose parents are gay or lesbian testify to the children's acceptance of their parents' orientation without negative consequences (Bozett 1987).

Types of Custody

The major types of custody are sole, joint, and split. In **sole custody,** the child lives with one parent, who has sole responsibility for physically raising the child and making all decisions regarding his or her upbringing. There are two forms of joint custody: legal and physical. In **joint legal custody,** the children live primarily with one parent, but both share jointly in decisions about their children's education, religious training, and general upbringing. In **joint physical custody,** the children live with both parents, dividing time between the two households. Even though joint custody does not necessarily mean that the child's time is evenly divided between parents, it gives children the chance for a more normal and realistic relationship with each parent (Arnetti and Keith 1993). Under **split custody,** the children are divided between the two parents; the mother usually takes the girls and the father, the boys. Split custody often has harmful effects on sibling bonds and should be entered into only cautiously (Kaplan, Hennon, and Ade-Ridder 1993).

Parental satisfaction with court imposed custody arrangements depends on many factors (Arditti 1992; Arditti and Allen 1992). These include how hostile the divorce was, whether the noncustodial parent perceives visitation as lengthy and frequent enough, and how close the noncustodial parent feels to his or her children. In addition, the amount of support payments affects satisfaction. If parents feel they are paying too much or were "cheated" in the property settlement, they are also likely to feel that the custody arrangements are unfair. Unfortunately, custodial satisfaction is not necessarily related to the best interests of the child.

The anger and conflict surrounding custody arrangements helped give rise to a fathers' rights movement and remain key rallying points among "men's rights" advocates (Coltrane and Hickman 1992). The fathers' rights movement depicts its participants as caring fathers who want equal treatment regarding child custody, visitation, and support (Bertoia and Drakich 1993). Given the nature of changing gender roles and the reality of economic hardships, more mothers are relinquishing their children to the fathers.

This trend of fathers seeking and gaining custody of their children comes despite many judges' traditional attitudes about gender and established childcare patterns. Research concerning the effects of a father's custody on the psychological well-being of children reveals no conclusive evidence to preclude or prefer it. The chances of a father gaining custody are improved when the children are older at the time of the divorce, the oldest is male, and the father is the plaintiff in the divorce (Fox and Kelly 1995). Regardless of who is awarded custody, however, it is important when possible for children to maintain close ties with both parents following a divorce (Howell, Brown, and Eichenberger 1992).

Sole Custody

Most children continue to live with their mothers after divorce. This occurs for several reasons. First, because women have traditionally been responsible for childrearing, sole custody by mothers has seemed the closest approximation to the traditional family, especially if the father is given free access. Second, many men have not had the day-to-day responsibilities of childrearing and do not feel (or are not perceived to be) competent in that role. Sole custody does not mean that the noncustodial parent is prohibited from seeing his or her children.

Judith Wallerstein and Joan Kelly (1980b) believe that if one parent is prohibited from sharing important aspects of the children's lives, he or she will withdraw from the children in frustration and grief. Children experience such withdrawal as a rejection and suffer as a result.

It is generally considered in the best interests of the children for them to have easy access to the noncustodial parent. Changes in the noncustodial parent's relationship with his or her children may be related to the difficulties and psychological conflicts arising from visitation and divorce, the noncustodial parent's ability to deal with the limitations of the visiting relationship, and the age and gender of the child (Wallerstein and Kelly 1980a).

Joint Custody

Joint custody, in which both parents continue to share legal rights and responsibilities, has become a preferred form of legal custody. A number of advantages accrue to this type of arrangement. First, it allows both parents to continue their parenting roles. Second, it avoids a sudden termination of a child's relationship with one of his or her parents. Joint-custody fathers tend to be more involved with their children; they spend time with them and share responsibility and decision making (Bowman and Ahrons 1985). Third, dividing the labor lessens many of the burdens of constant childcare experienced by most single parents.

Joint physical custody, however, requires considerable energy from the parents in working out both the logistics of the arrangement and their feelings about each other. Many parents with joint custody find it difficult, but they nevertheless feel that it is satisfactory.

The children do not always like joint custody as much as the parents do. In practice, children rarely split their time evenly between parents (Little 1992).

Any custody arrangement has both benefits and drawbacks, and joint custody is no exception. Although it may be in the best interests of the parents for each of them to continue parenting roles, it may not necessarily be in the best interests of the child. For parents who choose joint custody, it appears to be a satisfactory arrangement. But when joint custody is mandated by the courts over the opposition of one or both parents, it may be problematic. Joint custody may force two parents to interact (*cooperate* is too benign a word) when they would rather never see each other again, and the resulting conflict and ill will may be detrimental to the children. Parental hostility may

make joint custody the worst form of custody (Opie 1993).

Reflections

What form of custody do you believe is the most advantageous to a child? What factors would you consider important in deciding which is the best type of custody for a particular child? If two parents constantly battled over their children, what are some of the consequences you might expect for the children? How do children cope in such circumstances?

Noncustodial Parents

Only recently is research emerging about noncustodial parents. Popular images of noncustodial parents depict them as absent and noncaring, as reflected in the widespread popular use of the term *deadbeats*, which refers specifically to noncustodial parents who fail to maintain their support obligations. A more accurate picture depicts varying degrees of involvement (Bray and Depner 1993; Depner and Bray 1993). Noncustodial parent involvement exists on a continuum in terms of caregiving, decision making, and parent–child interaction. Involvement also changes depending on whether the custodial family is a single-parent family or a stepfamily (Bray and Berger 1993).

Noncustodial fathers often suffer grievously from the disruption or disappearance of their father roles following divorce. They feel depressed, anxious, and guilt ridden; they feel a lack of self-esteem (Arditti 1990). The change in status from full-time father to noncustodial parent leaves fathers bewildered about how they are to act; there are no norms for an involved noncustodial parent. Men often act irresponsibly after a divorce, failing to pay child support and possibly becoming infrequent parts of their children's lives. This lack of norms makes it especially difficult if the relationship between the former spouses is bitter. Without adequate norms, fathers may become "Disneyland Dads," who interact with their children only during weekends, when they provide treats such as movies and pizza, or they may become "Disappearing Dads," absenting themselves from all contact with their children. For many concerned noncustodial fathers, the question is simple but painful: "How can I be a father if I'm not a father anymore?"

Noncustodial fathers often weigh the costs of continued involvement with their children, such as emotional pain and role confusion, against the bene-fits, such as emotional bonding (Braver et al. 1993a, 1993b). Those fathers who maintain their connections are generally older and remarried; they have little or no conflict with their ex-spouses and no significant problems with their children (Wall 1992). For others, however, the costs outweigh the benefits. They are not successful in being noncustodial fathers and abandon the role. A study of noncustodial parents in a support group found that common themes included children rejecting parents and parents rejecting children (Greif and Kristall 1993).

Children often eventually have little contact with the nonresidential parent. This reduced contact seems to weaken the bonds of affection. A study of 18- to 22-year-olds whose parents were divorced found that almost two-thirds had poor relationships with their fathers and one-third had poor relationships with their mothers—about twice the rate of a comparable group from nondivorced families (Zill, Morrison, and Coiro 1993). Divorced fathers are less likely to consider their children sources of support in times of need (Amato 1994; Cooney 1994). Although perhaps better than Frank Furstenberg and Christine Nord's (1985) claim of more than two decades ago that "marital dissolution involves either a complete cessation of contact between the nonresidential parent and child or a relationship that is tantamount to a ritual form of parenthood," noncustodial parents certainly see their relationships suffer considerably.

■ *It is usually important for a child's postdivorce adjustment that he or she have continuing contact with the noncustodial parent. Noncustodial parents are involved with their children in varying degrees.*

Custody Disputes

As many as one-third of all postdivorce legal cases involve children. Vagueness of the "best interests" and "least detrimental alternative" standards by which parents are awarded custody may encourage custody fights by making the outcome of custody hearings uncertain and increasing hostility. Any derogatory evidence or suspicions, ranging from dirty faces to child abuse, may be considered relevant evidence. As a result, child custody disputes are fairly common in the courts. They are often quite nasty.

Divorce Mediation

The courts are supposed to act in the best interests of the child, but they often victimize children by their emphasis on legal criteria rather than on the children's

psychological well-being and emotional development (Schwartz 1994). There is increasing support for the idea that children are better served by those with psychological training than by those with legal backgrounds (Miller 1993). Growing concern about the effect of litigation on children's well-being has led to the development of divorce mediation as an alternative to legal proceedings (Walker 1993).

Divorce mediation is the process in which a mediator attempts to assist divorcing couples in resolving personal, legal, and parenting issues in a

In 1996, as a way of trying to strengthen marriage and reduce divorce rates, Louisiana became the first state in the United States to establish a two-tiered system of marriage.

Marrying couples could choose either a "standard marriage" or a covenant marriage (Hewlett and West 1998; see also Chapter 9 of this book). Following Louisiana's lead, other states have enacted their own covenant marriage legislation. Regardless of the state in question, covenant marriage usually consists of something close to the following, which is drawn from the Louisiana law:

> We do solemnly declare that marriage is a covenant between a man and a woman who agree to live together as husband and wife for so long as they both may live. We have chosen each other carefully and disclosed to one another everything which could adversely affect the decision to enter into this marriage.
>
> We have received premarital counseling on the nature, purposes, and responsibilities of marriage. We have read the Covenant Marriage Act, and we understand that a Covenant Marriage is for life. If we experience marital difficulties, we commit ourselves to take all reasonable efforts to preserve our marriage, including marital counseling.
>
> With full knowledge of what this commitment means, we do hereby declare that our marriage will be bound by Louisiana law on Covenant Marriages and we promise to love, honor, and care for one

another as husband and wife for the rest of our lives.

This is supplemented by an affidavit by the parties that they have discussed with a religious representative or counselor their intent to enter a covenant marriage. Included is their agreement to seek marital counseling in times of marital difficulties, and their agreement to the grounds for terminating the marriage.

We cannot say whether covenant marriage will "work" to reduce the prevalence of divorce. It may have no effect, because the people who elect to enter such a marriage may already perceive marriage as a relationship to keep "till death do us part."

This certainly seems to be the case based on recent research by Laura Sanchez and colleagues (2002). After interviews with three Louisiana focus groups of about a dozen participants each that represented different views on marriage and divorce, the researchers suggest that advocates and opponents of covenant marriage have different perceptions of marriage, marriage reform, divorce, and children's well-being.

The six conservative Christian couples they interviewed, married 11 to 56 years, saw a dangerous decline of traditional two-parent families, a decline in the value placed on motherhood, a general unwillingness to sacrifice for spouse and children, and the emergence of a "culture of divorce." They had converted their marriages to covenant marriages just months before they were interviewed.

The second focus group, a dozen feminist activists (11 females and 1 male, ages 20 to 50), saw traditional marriage as "inherently patriarchal" and detrimental to women's independence and rights. They also suggested that marriage (from courtship through weddings) is a commercial-

ized competition for men, with "victory" (that is, marriage) celebrated with indulgent and conspicuous consumption. They were strongly suspicious of and against covenant marriage.

The third focus group consisted of 10 low-income women (9 black, 1 white), all residents of public housing.

Of the 10, 2 women were married (18 years and 26 years each), a few were divorced, a few cohabited, and some never married. These women were chosen to explore issues related to poverty and welfare and how attitudes about marriage might affect or might be affected by their socioeconomic status.

This group had more practical and less politically ideological views of marriage. They valued marriage and saw numerous disadvantages faced by unmarried women. They perceived no-fault divorce as a source of a reduced commitment to marital responsibility, allowing people easy opportunities to leave rather than fix marriages. They also felt that divorce and single parenthood harmed children. Marriage was portrayed as an ideal worth aspiring toward, but they also acknowledged the problems of "falling out of love, growing apart, and modern strains on women and men in marriage" (Sanchez et al. 2002, 103).

The values expressed by the three focus groups suggest that in the short run, covenant marriage will appeal to those who already endorse its assumptions about marriage. To those who have concerns about inequalities in traditional marriage or worry about women's rights in families, covenant marriage will be unappealing.

To do more than "preach to the choir"—appealing to those who already share the covenant marriage philosophy—will be more difficult for proponents of such reform.

cooperative manner. More than two-thirds of U.S. states offer or require mediation through the courts over such legal issues as custody and visitation. Mediators act as facilitators to help couples arrive at mutually agreed-upon solutions. Although some mediators are attorneys by profession, in the divorce process they neither act as lawyers for nor give advice to either party. Mediators can be either private or court ordered. Mediators generally come from marriage counseling, family therapy, and social work backgrounds, although increasing numbers are coming from other backgrounds and are seeking training in divorce mediation (DeWitt 1994).

Mediation has many goals. A primary goal is to encourage divorcing parents to see shared parenting as a viable alternative and to reduce anxiety about shared parenting (Kruk 1993). Their role is not to save the marriage but to see that couples exit the marriage with less conflict, feeling that their interests were represented. Data on satisfaction indicates that those who use mediators as part of their divorce process have greater levels of satisfaction than those who divorce through adversarial means. They also spend less to end their marriages, because divorce mediation is less financially costly than divorce that relies on litigation alone (http://www.divorceinfo.com/doesmediation-work.htm).

When mediation is court mandated, topics are generally limited to custody and visitation issues. Divorcing parents often find mediation helpful for these issues. In contrast to court settings, mediation provides an informal setting to work out volatile issues. Men and women both report that mediation is more successful at validating their perceptions and feelings than is litigation. Furthermore, women, the poor, and those from ethnic groups are less likely to experience bias in mediation than in a courtroom setting (Rosenberg 1992).

Some courts order parents to participate in seminars covering the children's experience of divorce, as well as problem solving and building coparent relationships (Petersen and Steinman 1994). Parents report that these seminars help them become more aware of their children's reactions and give them more options for resolving child-related disputes.

Divorcing parents also report that mediation helps decrease behavioral problems in their children (Slater, Shaw, and Duquesnel 1992). If parents can work through their differences apart from their children, the children are less likely to react to the anger and fear they might otherwise observe.

What To Do about Divorce

As the previous pages have illustrated, divorcing is a painful process for those involved, and it leaves families and individuals changed forever.

Most people will agree that we would be better off reducing the rate of divorce, but how can that goal be achieved? First, we must decide on the most important cause of the high divorce rates in the United States.

If we believe that divorce rates rose partly because we made it easier and more acceptable to divorce, should we restigmatize divorce? Make exiting a marriage more difficult? If divorce rates rose with the increasing economic independence of women, how can we reduce divorce? Do we need to encourage employed women to stay home? How then do their families survive without their incomes (see Chapter 12)? If part of the explanation for rising divorce rates is in the increasing importance given to self-fulfillment and the decline of both familistic self-sacrifice and religious constraints, how can we reduce divorce? Can we change people's values? Finally, if increases in divorce result from the weakening of all but the emotional function of marriage and the reduction, especially, of the family's economic role, can *anything* be done about divorce?

Part of the dilemma has to do with how we perceive divorce. Is divorce the *problem,* or is it a *solution* to other problems? Do we want to impose restrictions on divorce that require people to remain in unfulfilling, possibly dangerous relationships? The societal reactions to reducing divorce have been largely of two kinds: cultural and legal. From a cultural perspective, some commentators bemoan the popular cultural denigration of marriage (Whitehead 1993, 1997; Popenoe 1993). They suggest that we "dismantle the divorce culture" we have constructed by more consistently championing and effectively demonstrating the benefits of stable, lifelong marriage.

Instead of celebrating "family diversity" and glorifying single-parent households, they believe we should consistently reiterate the idea that marriage is a lifelong commitment involving considerable sacrifice. If that means we must "restigmatize divorce," then that is what we should do (Whitehead 1997).

The other emphasis has been a legal one. Believing that marriage was weakened and divorce increased by no-fault divorce legislation, some have argued that we make divorce *harder to obtain*. Some states have contemplated repealing no-fault divorce legislation or raising marriage ages. Some states have enacted a two-tiered system of marriage in which couples are allowed and encouraged to consider *covenant marriage*—marriage under laws that require couples to undergo premarital counseling, swear to the lifelong commitment of marriage, and promise to divorce only under extraordinary circumstances and only after seeking marriage counseling (see the "Issues & Insights" box on covenant marriages and Chapter 9). Too new to yet evaluate, the covenant marriage system has appealed to both those who wish to reduce divorce and those who wish to establish a more traditional, even religious, understanding of marriage commitments.

The difficulty behind both cultural and legal efforts is that in attempting to make divorce harder or less attractive, they do little to make staying married easier. This, too, could be done. It might entail enacting some work–family policy initiatives to ease the stress and strain facing two-earner households. On the subject of financial resources, because we know that divorce hits hardest at lower- and working-class levels, bolstering the economic stability and security of low-income families might also lead to less divorce.

If we can't reduce or eliminate divorce, we should at least do what we can to protect those who go through divorce, especially children (Coontz 1997; Furstenberg and Cherlin 1991). We should devote resources that will help custodial parents raise their children more effectively. This means, among other things, ensuring their access to quality childcare when they are at work, guaranteeing their receipt of financial obligations (such as child support and alimony) from their former spouses, and helping them avoid the devastating plunge into poverty. In addition, ex-spouses must be instructed in how to display more amicable relationships with each other and should be expected to do so. Because at least some effects of divorce are tied to the level of postdivorce conflict and adjustment, taking steps to reduce conflict and ensure more effective adjustment will benefit children and their parents. Early and aggressive intervention into the postdivorce family (such as teaching anger management or instructing fathers about the vital roles they can still play) constitutes such intervention (Coontz 1997; Furstenberg and Cherlin 1991).

There is no denying that separation or divorce is typically filled with pain for all involved—husband, wife, and children. Furthermore, as we have seen, both the process and its outcomes are often different for husbands and wives and for parents and children. Hopefully, this chapter has increased your understanding of how much divorce there is, the multiple factors that have led to shifts in the divorce rate and that expose individuals to greater or lesser risk of divorce, and the different perspectives on what we can and should do about divorce. Keep in mind that as one family ends, new family forms emerge. These include new relationships and possibilities, new circumstances and responsibilities, and new families with unique relationships: the single parent or the stepfamily. These are the families that we explore in the next chapter.

Summary

- Divorce is an integral part of the contemporary American marriage system, which values individualism and emotional gratification. The divorce rate increased significantly in the 1960s but leveled off in the early 1990s. Between 40% and 50% of all current marriages end in divorce.

- Researchers are increasingly viewing divorce as part of the family life cycle rather than as a form of deviance. Divorce creates the single-parent family, remarriage, and the stepfamily.

- Among the statistics researchers use to measure divorce are the *ratio* of marriages to divorces, the *crude rate* of divorces per 1,000 people in a population, the *refined rate* of divorces per 1,000 marriages, and the *predictive rate* of the future likelihood of divorce within a cohort.

- The likelihood of divorce is lower for those who earn more than $50,000, marry after age 25, come from an "intact" parental marriage, have some religious affiliation, and have attended college.

- The trend in divorce has been downward since the 1980s.

- Compared to other countries, the U.S. divorce rate is among the highest.

- A variety of societal, demographic, and life course factors can affect the likelihood of divorce. The most important factors may be family processes: marital happiness, presence of children (in some cases), and marital problems.

- *No-fault divorce* revolutionized divorce by eliminating fault finding and the adversarial process and by treating husbands and wives as equals. An unintended consequence of no-fault divorce is the growing poverty of divorced women with children.

- Divorce can be viewed as a process involving six *stations* or processes: emotional, legal, economic, coparental, community, and psychic. As people divorce, they undergo these stations simultaneously, but the intensity level of these stages varies at different times.

- Uncoupling is the process by which couples drift apart in predictable stages. It is differently experienced by the initiator and his or her partner. Uncoupling ends when both partners acknowledge that the relationship cannot be saved.

- In establishing a new identity, newly separated people go through transition and recovery. They may experience *separation distress,* often followed by loneliness. The more personal, social, and financial resources a person has at the time of separation, the easier the separation generally will be.

- Women generally experience downward mobility after divorce. The economic effect on men is more mixed and depends on what proportion of the marital income they were responsible for before the divorce.

- *Child support* often goes unpaid, despite a number of legal initiatives to increase compliance by parents who owe support. A major determinant of compliance is what percentage of the parent's income is expected in support.

- Psychological distress, reduced self-esteem, less happiness, more isolation, and less satisfying sex lives are among noneconomic consequences of divorce. For some, the consequences of divorce are more positive than negative and include higher levels of personal growth, more autonomy, and—for women—improvements if their social lives, career opportunities and self-confidence.

- Remaining in an unhappy marriage reduces life satisfaction, mental and physical health, and self-esteem.

- Children are typically told about the divorce by mothers. Children's overall reactions are usually negative. For those to whom the news is told long before the actual divorce, the divorce itself may be experienced as relief.

- Consequences for children depend on the nature of their parents' marriage. In highly dysfunctional, high-conflict households, children may experience parental divorce as relief. However, in low-conflict marriages, even when parents lack commitment and happiness, divorce will likely be experienced as "unexpected, unwelcome, and uncontrollable."

- Children in the divorce process go through three stages: (1) the initial stage, lasting about a year, when turmoil is greatest; (2) the transition stage, lasting up to several years, in which adjustments are being made to new family arrangements, living and economic conditions, friends, and social environment; and (3) the restabilization stage, when the changes have been integrated into the children's lives.

- A significant factor affecting the responses of children to divorce is their age. Young children tend to act out and blame themselves, whereas adolescents tend to remain aloof and angry at both parents for disrupting their lives. Adolescents may be bothered when their parents date again. Many problems assumed to be caused by divorce are present before marital disruption.

- Factors affecting a child's adjustment to divorce include (1) open discussion before divorce, (2) continued involvement with noncustodial parent, (3) lack of hostility between divorced parents, (4) good psychological adjustment to divorce by custodial parent, and (5) stable living situation and good parenting skills. Continued involvement with the children by both parents is important for the children's adjustment.

- Although divorce has been said to put children in the middle of parental conflict, this seems to occur more in intact, high-conflict parental marriages.

- Longitudinal studies following children of divorce over decades have come to different conclusions about how bad the long-term consequences of divorce are and how long they last.

- Custody is generally based on one of two standards: the best interests of the child or the least detrimental of the available alternatives. The major types of custody are *sole, joint,* and *split.* Physical custody is generally awarded to the mother. Joint custody has become more popular because men are becoming increasingly involved in parenting.

- Noncustodial parent involvement exists on a continuum from absent to intimately and regularly involved. Noncustodial parents often feel deeply grieved about the loss of their normal parenting role.

- *Divorce mediation* is a process in which a mediator attempts to assist divorcing couples in resolving personal, legal, and parenting issues in a cooperative manner.

- Recent legislative initiatives such as *covenant marriage* are attempts to reduce the divorce rate by strengthening the marriage commitment.

Key Terms

alimony 506

child support 506

crude divorce rate 492

distal causes 500

divorce mediation 517

initiator 501

intergenerational transmission 497

joint custody 516

joint legal custody 515

joint physical custody 515

no-fault divorce 500

predictive divorce rate 492

proximal causes 500

ratio measure of divorce 492

refined divorce rate 492

separation distress 503

social integration 494

sole custody 515

split custody 515

stations of divorce 500

Resources on the Internet

Companion Website for This Book

http://www.thomsonedu.com/sociology/strong

Gain an even better understanding of this chapter by going to the companion website for additional study resources. Take advantage of the Pre- and Post-Test quizzing tool, which is designed to help you grasp difficult concepts by referring you back to review specific pages in the chapter for questions you answer incorrectly. Use the flash cards to master key terms and check out the many other study aids you'll find there. Visit the Marriage and Family Resource Center on the site. You'll also find special features such as access to Info-Trac© College Edition (a database that allows you access to more than 18 million full-length articles from 5,000 periodicals and journals), as well as GSS Data and Census information to help you with your research projects and papers.

CHAPTER 15

New Beginnings: Single-Parent Families, Remarriages, and Blended Families

Outline

What Do YOU Think?

Are the following statements TRUE or FALSE?
You may be surprised by the answers (see answer key on the following page).

T F   **1** Researchers are increasingly viewing stepfamilies as normal families.

T F   **2** Divorce does not end families.

T F   **3** Single parent families today are as likely to be headed by fathers as by mothers.

T F   **4** Second marriages are significantly happier than first marriages.

T F   **5** More than half of all marriages are remarriages for both spouses.

T F   **6** Children tend to have greater power in single-parent families than in traditional nuclear families.

T F   **7** Becoming a stepfamily is a process.

T F   **8** Stepmothers generally experience less stress in stepfamilies than stepfathers because stepmothers are able to fulfill themselves by nurturing their stepchildren.

T F   **9** Researchers are increasingly finding that remarried families and intact nuclear families are similar to each other in many important ways.

T F   **10** People who remarry and those who marry for the first time tend to have similar expectations.

When Paige was 6 and Daniel 8, their parents separated and divorced. The children continued to live with their mother, Sophia, in a single-parent household while spending weekends and holidays with their father, David. After a year, David began living with Jane, a single mother who had a 5-year-old daughter, Lisa. Three years after the divorce, Sophia married John, who had joint physical custody of his two daughters, Sally and Mary, aged 7 and 9. Some eight years after their parents divorced, Paige and Daniel's family included: two biological parents, two stepparents, three stepsisters, one stepbrother, and two half-brothers. In addition, they had assorted grandparents, stepgrandparents, biological and stepaunts, uncles, and cousins.

Today's families mark a definitive shift from the traditional family system, based on lifetime marriage and the intact nuclear family, to a pluralistic family system, including families created by divorce, remarriage, and births to single women. This new pluralistic family system consists of three major types of families: (1) intact nuclear families, (2) single-parent families (either never married or formerly married), and (3) stepfamilies. **Single-parent families** are families consisting of one parent and one or more children; the parent can be divorced, widowed, or never married. **Stepfamilies** are families in which one or both partners have children from a previous marriage or relationship. Stepfamilies are sometimes referred to as **blended families.**

In fact, a third of Americans are expected to marry, divorce, and remarry, at some point in their lives (Sweeney, 2002). In more than 40% of current marriages, one or both spouses are remarrying (Goldscheider and Sassler, 2006). A third of all children are likely to live in a married or cohabiting stepfamily sometime before they reach adulthood (White and Gilbreth, 2001).

To better understand the world Paige and David live in, a world that you may or may not know well, we need to examine some major patterns in our evolving pluralistic family system. In this chapter we examine single-parent families, binuclear families, remarriage, and stepfamilies. Because of this shift to a pluralistic family system, researchers are beginning to reevaluate these family types, view them as normal rather than deviant family forms (Coleman and Ganong 1991; Pasley and Ihinger-Tallman 1987). If we shift our perspective from structure to function, the important question is no longer whether a particular family form is deviant. (If we measure "deviant" by the statistical prevalence of a family form, the traditional nuclear family may soon become deviant.) The important question becomes whether a specific family—regardless of whether it is a traditional family, a single-parent family, or a stepfamily—succeeds in performing its functions. In a practical sense, as long as a family is fulfilling its functions, it is a kind of normal family. This chapter considers these versions of normal families.

Reflections

What effect does it have on your views of single-parent families and stepfamilies to think of them as "normal" families? As "abnormal" or "deviant" families? If you were reared in a single-parent family or stepfamily, did your friends, relatives, schools, and religious groups treat your family as normal? Why?

Single-Parent Families

In the United States, as throughout the world, single-parent families have increased and continue to grow in number (Burns and Scott 1994). Although no other family type has increased in number as rapidly, single-parent families may not be accurately or adequately understood. All too often, they are still treated negatively in the popular imagination, negated as either "broken homes" or as headed by women, especially teens, who casually bear children "out of wedlock." These images are clearly inadequate, based on ideas and stereotypes that misdirect us from a more accurate understanding. The "broken home" image is based on the ideal of the "happy" traditional family; the assumed irresponsibility of single mothers is based on moralism, occasionally mixed with racism, condemning women for bearing children outside of marriage;

Answer Key for What Do You Think

1 True, see p. 526;   2 True, see p. 532;   3 False, see p. 528;   4 False, see p. 536;   5 False, see p. 534;
6 True, see p. 530;   7 True, see p. 538;   8 False, see p. 541;   9 False, see p. 537;   10 False, see p. 533.

and the "promiscuous teenage mother" stereotype ignores the reality that most births to single mothers are to women older than 20. Finally, although more than 80% of single, custodial parents are female, these images overlook the situations and experiences of single fathers.

Between 1970 and 2002, the percentage of children living in single-parent families more than doubled, increasing from 13% to 28% (Fields 2003).

In previous generations, the life pattern most women experienced was (1) marriage, (2) motherhood, and (3) widowhood. Single-parent families existed in the past, but they were typically the result of widowhood rather than either divorce or births to unmarried women. Significant numbers were headed by men. But a new marriage and family pattern has taken root. Its greatest effect has been on women and their children. Divorce and births to unmarried mothers are the key factors creating today's single-parent family.

The life pattern many married women today experience is (1) marriage, (2) motherhood, (3) divorce, (4) single parenting, (5) remarriage, and (6) widowhood. For those who are not married at the time of their child's birth, the pattern may be (1) dating or cohabitation, (2) motherhood, (3) single parenting with the later possibility but no certainty of (4) marriage, and (5) widowhood. Finally, some who marry, divorce, and remarry, may experience subsequent divorces and or remarriages; they embody the characteristics that comprise serial monogamy.

Characteristics of Single-Parent Families

Single-parent families share a number of characteristics, including the following: creation by widowhood, divorce, or births to unmarried women; usually female headed; significance of ethnicity; poverty; diversity of living arrangements; and transitional character. In addition, some single-parent families are created intentionally through planned pregnancy, artificial insemination, and adoption. Others are headed by lesbians and gay men (Miller 1992). Finally, many single-parent households contain two cohabiting adults and are therefore not *single-adult* households (Fields 2003).

Creation by Divorce or Births to Unmarried Women

Single-parent families today are usually created by marital separation, divorce, or births to unmarried women rather than by widowhood. Throughout the world, including the United States, single-parent families created through births to unmarried women are increasing at a higher rate than are single-parent families created through divorce (Burns and Scott 1994). In 2002, 34% of all births were to unmarried women. The number of children living with an unmarried couple more than tripled between 1980 and 2000. Today, 19 million children under age 18 live in 9.4 million households with either the mother only or the father only (Fields 2003; U.S. Census Bureau 2002, Table 58).

In comparison to single parenting by widows, single parenting by divorced or never-married mothers receives considerably less social support. Widowed mothers often receive social support from their husband's relatives. A divorced mother usually receives little assistance from her own kin and considerably less (or none) from her former partner's relatives. Our culture is still ambivalent about divorce and tends to consider divorce-induced, single-parent families as somewhat deviant (Kissman and Allen 1993). It is even less supportive of families formed by never-married mothers. Conservatives have recently returned to earlier forms of stigmatization by characterizing children of never-married women as "illegitimate" and their mothers as "unwed mothers."

■ *Unmarried adolescent mothers are empowered to build successful families when they have emotional and financial support from their families, educational and employment opportunities, and childcare.*

Headed by Mothers (and Sometimes Fathers)

More than 80% of single-parent families are headed by women (Zhan and Pandey 2004). This has important economic ramifications because of gender discrimination in wages and job opportunities, as discussed in Chapter 12. Still, at least 1.9 million men are custodial single parents, raising one or more children. Like women, men take different paths to single parenthood. They either divorce or separate from their children's mothers or they raise children from relationships in which they were never married.

Significance of Ethnicity

Ethnicity remains an important demographic factor in single-parent families. In 2002, among Caucasian children, 20% lived in single-parent families; among African American children, 53% lived in such families; among Hispanics, 30% lived in single-parent families, and among Asian and Pacific Islander children, 15% lived in such households (Fields 2003). White single mothers were more likely to be divorced than their African American or Latino counterparts, who were more likely to be unmarried at the time of the birth or widowed.

Poverty

Married women usually experience a sharp drop in their income when they separate or divorce (as discussed in Chapter 14). Among unmarried single mothers, poverty and motherhood often go hand in hand. Because they are women, because they are often young, and because they are often from ethnic minorities, single mothers have few financial resources. They are under constant economic stress in trying to make ends meet (McLanahan and Booth 1991). They work for low wages, endure welfare, or both. They are unable to plan because of their constant financial uncertainty. They move more often than two-parent families as economic and living situations change, uprooting themselves and their children. They accept material support from kin but often at the price of receiving unsolicited "free advice," especially from their mothers.

Both mother-only and father-only families are more likely to be poor than are two-parent families; in 2000, 5% of married-couple families lived in poverty compared to 12% of single-father families and 25% of single-mother families. Clearly, however, the association between single parenthood and poverty is greater for mothers than for fathers (Zhan and Pandey 2004).

Compared to *married fathers,* single fathers are substantially less well off. They are younger, less educated, less likely to have jobs, and more likely to receive public assistance and to live in poverty. They are also more likely to be African American. Min Zhan and Shanta Pandey show that the gap between married and single fathers has grown since 1980.

Matter of Fact

Among children in divorced single-parent families, 32.4% live in poverty (U.S. Bureau of Statistics 1996).

Diversity of Living Arrangements

There are many different kinds of single-parent households. Children under age 18 are nearly five times as likely to live with a single mother as with a single father (23% to 5%) (Fields 2003). Single-parent families also show great flexibility in managing childcare and housing with limited resources. In doing so, they rely on a greater variety of household arrangements than is suggested by the umbrella heading "single-parent household." For example, many young African American mothers live with their own mothers in a three-generation setting.

Of perhaps more interest is that many "single-parent households" actually contain the parent and his or, more often her, unmarried partner. In 2002, for example, 11% (1.8 million) of the 16.5 million children living with single mothers also lived with their mothers' unmarried partners. A third (1.1 million) of the 3.3 million children living with an unmarried father also lived with their fathers' unmarried partners (Fields 2003).

Even in the absence of parents' *live-in* partners, parents' romantic partners may play important roles in their children's lives. For example, many children of single mothers and nonresidential biological fathers have a **social father**—a male relative, family associate, or mothers' partner—"who demonstrates parental behaviors and is like a father to the child" (Jayakody and Kalil 2002).

Along these same lines, single parents, especially mothers, often rely on a combination of state or federal assistance and **private safety nets:** support from their social networks on which they can fall back in times of economic need (Hamer and Marchioro 2002; Harknett 2006).

Social support, whether from family or friends, can lead to enhanced well-being and self-esteem among economically disadvantaged single mothers. These, in return, may lead to more effective parenting, even under difficult and highly stressful conditions. Without such support, mothers raising children on their own in economically distressed, potentially dangerous, urban neighborhoods are more likely to experience psychological distress, which then negatively affects their parenting behavior (Kotchik, Dorsey, and Heller 2005).

Transitional Form

Single parenting is usually a transitional state. A single mother has strong motivation to marry or remarry because of cultural expectations, economic stress, role overload, and a need for emotional security and intimacy. The increasing presence of social fathers, including mothers' live-in romantic partners, may be part of the reason low-income families increasingly cohabit rather than marry. The presence of such men can reduce the various pushes toward marriage or remarriage (Jayakody and Kalil 2002).

Intentional Single-Parent Families

For many single women in their 30s and 40s, single parenting has become a more accepted, intentional, and less transitional lifestyle (Seltzer 2000; Gongla and Thompson 1987; Miller 1992). Some older women choose unmarried single parenting because they have not found a suitable partner and are concerned about declining fertility. They may plan their pregnancies or choose donor insemination or adoption. If their pregnancies are unplanned, they decide to bear and rear the child. Others choose single parenting because they do not want their lives and careers encumbered by the compromises necessary in marriage. Still others choose it because they don't want a husband but they do want a child.

Lesbian and Gay Single Parents

There may be 2.5 million to 3.5 million lesbian and gay single parents. Most were married before they were aware of their sexual orientation or married with hopes of "curing" it. They became single parents as a result of divorce. Others were always aware of being lesbian or gay; they chose adoption or donor insemination to have children. Said one gay adoptive father, "I always knew I wanted to be a father." A lesbian who was artificially inseminated said, "I started to get this baby hunger. I just needed to have a child" (Miller 1992).

Children in Single-Parent Families

Children born outside of marriage tend to suffer economic disadvantages that may then lead to other educational, social, and behavioral outcomes. Their disadvantages tend to be worse than those experienced by children of divorced parents or by children in two-parent, married households (Seltzer 2000). They are more likely to engage in high-risk, "health compromising" behaviors such as cigarette smoking, drug and alcohol use, and unprotected sex; are less likely to graduate from high school and college; are more likely to have a child outside of marriage and/or during their teens; are more likely to be "idle" (out of school and out of work), have lower earnings, and suffer lower levels of psychological well-being; and are more vulnerable to divorce and marital instability as adults (King, Harris, and Heard 2004).

The bulk of research on the effects divorced, single-parent households have on children points to some negative outcomes in areas such as behavioral problems, academic performance, social and psychological adjustment, and health. The gaps between children in such households and those whose parents remain continuously married are relatively small but consistent. As Paul Amato (2000) reports, especially when exposed to associated negative life events such as having to move or change schools, the effects of living in a divorced, single-parent home can create particular adjustment difficulties. The consequences appear to be linked to the lack of economic resources but also to the reduced money, attention, guidance, and social connections—what researchers call **social capital**—that fathers provide.

Parental Stability and Loneliness

After a divorce, single parents are usually glad to have the children with them. Everything else seems to have fallen apart, but as long as divorced parents have their children, they retain their parental function. Their children's need for them reassures them of their own importance. A mother's success as a parent becomes even more important to counteract the feelings of low self-esteem that result from divorce.

Feeling depressed, the mother knows she must bounce back for the children. Yet after a short period, she comes to realize that her children do not fill the void left by her missing spouse. The children are a chore, as well as a pleasure, and she may resent being constantly tied down by their needs. Thus, minor incidents with the children—a child's refusal to eat or a temper tantrum—may be blown out of proportion. A major disappointment for many new single parents is the discovery that they are still lonely. It seems almost paradoxical. How can a person be lonely amid all the noise and bustle that accompany children? However, children do not ordinarily function as attachment figures; they can never be potential partners. Any attempt to make them so is harmful to both parent and child. Yet children remain the central figures in the lives of single parents. This situation leads to a second paradox: although children do not completely fulfill a person, they rank higher in most single mothers' priorities than anything else.

Changed Family Structure

A single-parent family is not the same as a two-parent family with one parent temporarily absent. The permanent absence of one parent dramatically changes the way in which the parenting adult relates to the children. Generally, the mother becomes closer and more responsive to her children. Her authority role changes, too. A greater distinction between parents and children exists in two-parent homes. Rules are developed by both mothers and fathers. Parents generally have an implicit understanding to back each other up in childrearing matters and to enforce mutually agreed-on rules. In the single-parent family, no other partner is available to help maintain such agreements; as a result, the children may find themselves in a more egalitarian situation with more power to negotiate rules. They can be more stubborn, cry more often and louder, whine, pout, and throw temper tantrums. Any parent who has tried to convince children to do something they do not want to do knows how soon an adult can be worn down.

Additional "handicaps" faced by single-parent families include the following:

- With only one adult in the household, if that adult is distressed, overwhelmed, or angry, the tone of the whole house is affected (Coontz 1997).

- Facing more intense time pressures, single parents are less able to participate in their children's schooling, and spend less time monitoring their children's homework (Coontz 1997).

- Parental depression, especially among custodial mothers, can affect their abilities to parent effectively and thus exposes their children to more "adjustment problems" (Amato 2000).

- Single mothers with higher levels of life stresses and less time for themselves are more likely to be anxious and to transmit their anxiety to their children. Repeated experiences of transmitted anxiety from mother to child can lead to chronic distress in children (Larson and Gillman 1999).

On the "plus side," children in single-parent homes may also learn more responsibility, spend more time talking with their custodial parent, and face less pressure to conform to more traditional gender roles (Coontz 1997). They may learn to help with kitchen chores, to clean up their messes, or to be more considerate. In the single-parent setting, the children are encouraged to recognize the work their mother does and the importance of cooperation.

Although single parents continue to demonstrate love and creativity in the face of adversity, research on their children reveals some negative long-term consequences. In adolescence and young adulthood, children from single-parent families had fewer years of education and were more likely to drop out of high school. They had lower earnings and were more likely to be poor. They were more likely to initiate sex earlier, become pregnant in their teens, and cohabitate but not marry earlier (Furstenberg and Teitler 1994). Furthermore, they were more likely to divorce. These conclusions are consistent for Caucasians, African Americans, Latinos, and Asian Americans. The reviewers note that socioeconomic status accounts for some, but not all, of the effects. Some effects are attributed to family structure.

Harriette Pipes McAdoo (1988, 1996) traces the cause to poverty, not to single parenthood. She notes that African American families are able to meet their children's needs in a variety of structures. "The major problem arising from female-headed families is poverty," she writes (McAdoo 1988). "The impoverishment of Black families has been more detrimental than the actual structural arrangement."

Successful Single Parenting

Single parenting is difficult, but for many single parents, the problems are manageable. Almost two-thirds of divorced single parents found that single parenting grows easier over time (Richards and Schmiege 1993). As we discuss single parenting, it is important to note that many of the characteristics of successful single parents and their families are shared by all successful families.

Characteristics of Successful Single Parents

In-depth interviews with successful single parents found certain themes running through their lives (Olsen and Haynes 1993):

- *Acceptance of responsibilities and challenges of single parenthood.* Successful single parents saw themselves as primarily responsible for their families; they were determined to do the best they could under varying circumstances.
- *Parenting as first priority.* In balancing family and work roles, their parenting role ranked highest. Romantic relationships were balanced with family needs.
- *Consistent, nonpunitive discipline.* Successful single parents realized that their children's development required discipline. They adopted an authoritative style of discipline that respected their children and helped them develop autonomy.
- *Emphasis on open communication.* They valued and encouraged expression of their children's feelings and ideas. Parents similarly expressed their feelings.
- *Fostering individuality supported by the family.* Children were encouraged to develop their own interests and goals; differences were valued by the family.
- *Recognition of the need for self-nurturance.* Single parents realized that they needed time for themselves. They needed to maintain an independent self that they achieved through other activities, such as dating, music, dancing, reading, classes, and trips.
- *Dedication to rituals and traditions.* Single parents maintained or developed family rituals and traditions, such as bedtime stories; family prayer or meditation; sit-down family dinners at least once a week; picnics on Sundays; visits to Grandma's; or watching television or going for walks together.

Single-Parent Family Strengths

Although most studies emphasize the stress of single parenting, some studies view it as building strength and confidence, especially for women (Amato 2000; Coontz 1997). A study of 60 white single mothers and 11 white single fathers (most of whom were divorced) identified five family strengths associated with successful single parenting (Richards and Schmiege 1993):

- *Parenting skills.* Successful single parents the ability to take on both expressive and instrumental roles and traits. Single mothers may teach their children household repairs or car maintenance; single fathers may become more expressive and involved in their children's daily lives.
- *Personal growth.* Developing a positive attitude toward the changes that have taken place in their lives helps single parents, as does feeling success and pride in overcoming obstacles.
- *Communication.* Through good communication, single parents can develop trust and a sense of honesty with their children, as well as an ability to convey their ideas and feelings clearly to their children and friends.
- *Family management.* Successful single parents develop the ability to coordinate family, school, and work activities and to schedule meals, appointments, family time, and alone time.
- *Financial support.* Developing the ability to become financially self-supporting and independent is important to single parents.

Among the single parents in the study, more than 60% identified parenting skills as one of their family strengths. In addition, 40% identified family management as a strength in their families (Richards and Schmiege 1993). About 25% identified personal growth and communication among their family strengths.

Barbara Risman's (1986) research on custodial single fathers showed their abilities to be attentive, nurturing caregivers to their children. Rather than relying on paid help or female social supports, men became the nurturers in their children's lives. They were involved in their personal, social, and academic lives and saw to it that their emotional and physical needs were met. To Risman, they affirmatively answer the question in her title, "Can Men Mother?"

Binuclear Families

One of the most complex and ambiguous relationships in contemporary America is what some researchers call the **binuclear family**—a postdivorce family system with children (Ahrons and Rodgers 1987; Ganong and Coleman 1994). It is the original nuclear family divided in two. The binuclear family consists of two nuclear families—the maternal nuclear family headed by the mother (the ex-wife) and the paternal one headed by the father (the ex-husband). Both single-parent families and stepfamilies are forms of binuclear families.

Divorce ends a marriage but not a family. It dissolves the husband–wife relationship but not necessarily the father–mother, mother–child, or father–child relationship. The family reorganizes itself into a binuclear family. In this new family, ex-husbands and ex-wives may continue to relate to each other and to their children, although in substantially altered ways. The significance of the maternal and paternal components of the binuclear family varies. In families with joint physical custody, the maternal and paternal families may be equally important to their children. In single-parent families headed by women, the paternal family component may be minimal.

To clarify the different relationships, researchers Constance Ahrons and Roy Rodgers (1987) divide the binuclear family into five subsystems: former spouse, remarried couple, parent–child, sibling (stepsiblings and half-siblings), and mother/stepmother–father/stepfather.

Mother/Stepmother–Father/Stepfather Subsystems

The relationship between new spouses and former spouses often influences the remarried family. The former spouse can be an intruder in the new marriage and a source of conflict between the remarried

■ *Entered into with great enthusiasm, blending families is a complex process. In addition to new spousal roles, families must craft new parent-child relationships and new sibling relationships.*

Margaret Joanne Cotton Photography

couple. Other times, the former spouse is a handy scapegoat for displacing problems. Much of current spouse–former spouse interaction depends on how the ex-spouses feel about each other.

Recoupling: Courtship in Repartnering

Certain norms governing courtship before first marriage are fairly well understood. As courtship progresses, individuals spend more time together; at the same time, their family and friends limit time and energy demands because "they're in love." Courtship norms for second and subsequent marriages, however, are not so clear (Ganong and Coleman 1994; Rodgers and Conrad 1986).

For example, when is it acceptable for formerly married (and presumably sexually experienced) men and women to become sexually involved? What type of commitment validates "premarital" sex among postmarital men and women? How long should courtship last before a commitment to marriage is made? Should the couple cohabit? Without clear norms, courtship following divorce can be plagued by uncertainty about what to expect.

Remarriage courtships tend to be short, unless preceded by cohabitation. If we consider postdivorce cohabitation as an end point, even an intermediate one, in the "courtship process," the process is shorter than would be indicated by marriage dates. Research on how postdivorce cohabitation affects the timing of remarriage shows that postdivorce cohabitation tends to lead to a longer waiting time until remarriage than is experienced by those who don't cohabitate before remarrying (Xu, Hudspeth, and Bartkowski 2006).

As noted earlier, almost one-third of divorced individuals marry within a year of their divorces. This may indicate, however, that they knew their future partners before they were divorced. If neither partner has children, courtship for remarriage may resemble courtship before the first marriage, with one major exception: The memory of the earlier marriage exists as a model for the second marriage. Courtship may trigger old fears, regrets, habits of relating, wounds, or doubts. At the same time, having experienced the day-to-day living of marriage, the partners may have more realistic expectations. Their courtship may be complicated if one or both are noncustodial parents. In that event, visiting children present an additional element.

Cohabitation

Increases in the rates of cohabitation in the United States include many divorced women and men who cohabit before or instead of remarrying. As great an increase as has occurred in premarital cohabitation, *postdivorce* cohabitation is even more common (Xu, Hudspeth, and Bartkowski 2006). Thus, although remarriage rates have declined in recent years, "recoupling" through cohabitation remains common (Coleman, Ganong, and Fine 2000).

Larry Ganong and Marilyn Coleman (1994) describe cohabitation as "the primary way people prepare for remarriage," making it a major difference between first-time marriages and remarriages. This may reflect the desire to test compatibility in a "trial marriage" to prevent later marital regrets (Buunk and van Driel 1989). However, couples who lived together before remarriage did not discuss stepfamily issues any more than did those who did not cohabit (Ganong and Coleman 1994).

- Remarital happiness is about 28% lower for postdivorce cohabiters than for noncohabiters.

- Remarital instability is around 65% greater for cohabiters than for noncohabiters.

- As of now, it is impossible to determine whether postdivorce cohabitation or the types of individuals who cohabit (the selection effect) are responsible for the effect cohabitation has on remarriages. This should be familiar; we posed the same question about the effects of cohabitation on first marriages.

Matter of Fact

Recent research has found that having children in the home has a strong positive effect on economic distress and a strong negative effect on income (Shapiro 1996).

Courtship and Children

Courtship before remarriage differs considerably from that preceding a first marriage if one or both members in the dating relationship are custodial parents. Single parents are not often a part of the singles world because such participation requires leisure and money, which single parents generally lack. Children rapidly consume both of these resources.

Although single parents may wish to find a new partner, their children usually remain the central figures in their lives. This creates a number of new problems. First, the single parent's decision to go out at night may lead to guilt feelings about the children. If a single mother works and her children are in day care, for example, should she go out in the evening or stay at home with them? Second, a single parent must look at a potential partner as a potential parent. A person may be a good companion and listener and be fun to be with, but if he or she does not want to assume parental responsibilities, the relationship will often stagnate or be broken off. A single parent's new companion may be interested in assuming parental responsibilities, but the children may regard him or her as an intruder and try to sabotage the new relationship.

A single parent may also have to decide whether to permit a lover to spend the night when children are in the home. This is often an important symbolic act. For one thing, it brings the children into the parent's new relationship. If the couple has no commitment, the parent may fear the consequences of the children's emotional involvement with the lover; if the couple breaks up, the children may be adversely affected.

Remarriage

The eighteenth-century writer Samuel Johnson described **remarriage**—a marriage in which one or both partners have been previously married—as "the triumph of hope over experience." Americans are a hopeful people. Many newly divorced men and women express great wariness about marrying again, yet they are actively searching for mates. Women often view their divorced time as important for their development as individuals, whereas men, who often complain that they were pressured into marriage before they were ready, become restless as "born-again bachelors" (Furstenberg 1980).

Remarriage Rates

More than 40% of all marriages in the United States are marriages in which at least one partner has been previously married (Coleman, Ganong, and Fine 2000; Goldscheider and Sassler 2006). Of those, 20% remarry other divorced men and women, and approximately 22% marry never-married individuals (U.S. Census Bureau 1996). One out of ten marriages is a third marriage for one or both partners (Goldscheider and Sassler 2006).

Remarriage is common among divorced people, especially men, who have higher remarriage rates than women (Coleman, Ganong, and Fine 2000). Still, 54% of divorced women remarry within 5 years, and 75% remarry within 10 years (Bramlett and Mosher 2002).

In recent years, the remarriage rate has slightly declined. The decline may be partly the result of the desire on the part of divorced men and women to avoid the legal responsibilities accompanying marriage. Instead of remarrying, many are choosing to cohabit.

Paul de Graaf and Matthijs Kalmijn (2003) report that nearly all research indicates that the likelihood of remarriage is negatively affected by the presence of children and by the adult's age. The age effect, however, appears to be stronger for women. Remarriage is more likely among white divorced women and among younger women—women 25 years or younger at the time of divorce. Eighty percent of these younger women remarry within 10 years, compared to 68% of women older than 25 years at the breakup of their marriage.

African American women are less likely than Caucasian or Hispanic women to remarry. Within 5 years after a divorce, approximately 33% of black women, 44% of Hispanic women, and nearly 60% of white women had entered a remarriage (Bramlett and Mosher 2002).

In addition to age and ethnicity, socioeconomic variables such as education may affect remarriage rates, although research that has identified effects is not consistent. Education appears to work differently for women's and men's likelihood of remarriage, raising a man's likelihood of remarriage but reducing a woman's (Coleman, Ganong, and Fine 2000).

Gender

There are a number of reasons that more men than women remarry. First, divorced women tend to be older than never-married women. Given the tendency for men to marry women younger than themselves and that older women are seen as less attractive and therefore less desirable as spouses, women face more competition and possess fewer "resources" to bring to a remarriage. They are also more likely to have custody of children, which can reduce both the ease with which they socialize or date and their appeal as potential spouses.

Presence of Children

Children lower the probability of remarriage for both women and men, but especially for women (Coleman, Ganong, and Fine 2000). The effects are most marked when a woman has three or more children. Most research, however, is 15 to 20 years old, and the increased incidence of single-parent families and stepfamilies may have decreased some of the negative effect of children. Whereas researchers generally speculate that children are a "cost" in remarriage, some point out that some men may regard children as a "benefit" in the form of a ready-made family (Ganong and Coleman 1994). Some research suggests that the stepparent with no biological children experiences the most negative effect (MacDonald and DeMaris 1995).

Initiator Status

Research suggests that initiators will be more likely to remarry than noninitiators (see Chapter 14). In their decisions about seeking a divorce, initiators may factor in the prospect for reentering marriage. They also may be "better prepared emotionally" than noninitiators to remarry. The advantage initiators have over noninitiators may be temporary because noninitiators lag behind initiators in the process of adjusting to and accepting the ending of their marriages (Sweeney 2002). Indeed, Megan Sweeney found that initiators enter new relationships "substantially more quickly than noninitiators," with the effect operating for the first 3 years after separation for men's remarriage patterns.

Need, Attractiveness, or Opportunity?

For women, the highest remarriage rate takes place in the 20s; it declines by a quarter in the 30s and by two-thirds in the 40s. What's going on that accounts for the changing probabilities? First, they may have less drive to remarry. Second, they are more likely to have characteristics that affect their suitability to potential partners. Finally, the pool of eligible and available partners is smaller for remarriage, and grows smaller as women age. More potential partners of their same age will be already married. As a result of these processes, men and women may be willing to "settle for less." They

may choose someone they would not have chosen when they were younger (Ganong and Coleman 1994).

The Remarriage Marketplace

There are three main contexts from which divorced women and men might find another partner: in the workplace, through leisure activities, and through their social network. Women and men who are employed and who are socially integrated are more likely to find a new partner. Employment affects their opportunities to remarry by adding the workplace as a venue in which they are likely to meet potential partners.

> **Matter of Fact**
>
> Research has concluded that remarriage indeed offers enhanced psychological well-being (Shapiro 1996).

> **Reflections**
>
> If you were seeking a marital partner, would you consider a previously married person? Why or why not? Would it make a difference if he or she already had children?

Characteristics of Remarriage

Remarriage is different from first marriage in a number of ways. First, the new partners get to know each other during a time of significant changes in life relationships, confusion, guilt, stress, and mixed feelings about the past (Keshet 1980). They have great hope that they will not repeat past mistakes, but there is also often some fear that the hurts of the previous marriage will recur (McGoldrick and Carter 1989). The past is still part of the present. A Talmudic scholar once commented, "When a divorced man marries a divorced woman, four go to bed."

Remarriages occur later than first marriages. People are at different stages in their life cycles and may have different goals. Divorced people may have different expectations of their new marriages. A woman who already has had children may enter a second marriage with strong career goals. In her first marriage, raising children may have been more important.

In an early study of second marriages in Pennsylvania, Frank Furstenberg (1980) discovered that three-fourths of the couples had a different conception of love than couples in their first marriages. Two-thirds thought they were less likely to stay in an unhappy marriage; they had already survived one divorce and knew they could make it through another. Four out of five believed their ideas of marriage had changed.

Marital Satisfaction and Stability in Remarriage

According to various studies, remarried people are about as satisfied or happy in their second marriages as they were in their first marriages. As in first marriages, marital satisfaction appears to decline with the passage of time (Coleman and Ganong 1991). Yet although marital happiness and satisfaction may be similar in first and second marriages, remarried couples are more likely to divorce. As Marilyn Coleman, Larry Ganong, and Mark Fine (2000) note, "serial remarriages are increasingly common."

How do we account for this paradox? Researchers have suggested several reasons for the higher divorce rate in remarriage. (See Ganong and Coleman 1994 for a discussion of various models explaining the greater fragility of remarriage.)

First, people who remarry after divorce often have a different outlook on marital stability and are more likely to use divorce to resolve an unhappy marriage (Booth and Edwards 1992). Frank Furstenberg and Graham Spanier (1987) note that they were continually struck by the willingness of remarried individuals to dissolve unhappy marriages: "Regardless of how unattractive they thought this eventuality, most indicated that after having endured a first marriage to the breaking point they were unwilling to be miserable again simply for the sake of preserving the union."

Second, despite its prevalence, remarriage remains an "incomplete institution" (Cherlin 1981). Society has not evolved norms, customs, and traditions to guide couples in their second marriages. There are no rules, for example, defining a stepfather's responsibility to a child: Is he a friend, a father, a sort of uncle, or what? Nor are there rules establishing the relationship between an individual's former spouse and his or her present partner: Are they friends, acquaintances, rivals, or strangers? Remarriages don't receive the same family and kin support as do first marriages (Goldenberg and Goldenberg 1994).

Third, remarriages are subject to stresses that are not present in first marriages. The vulnerability of remarriage to divorce is especially real if children from a prior relationship are in the home (Booth and Edwards 1992). Children can make the formation of the husband–wife relationship more difficult because they compete for their parents' love, energy, and attention. In such families, time together alone becomes a precious and all-too-rare commodity. Furthermore, although children have little influence in selecting their parent's new husband or wife, they have immense power in "deselecting" them Marilyn Ihinger-Tallman and Kay Pasley (1987):

> Children can create divisiveness between spouses and siblings by acting in ways that accentuate differences between them. Children have the power to set parent against stepparent, siblings against parents, and stepsiblings against siblings.

The divorce-proneness of remarriages seems to lessen and become more like that of first marriages as people age. People who enter remarriage after turning 40 may face a lower divorce likelihood than that found among first marriages (Coleman, Ganong, and Fine 2000).

Blended Families

Remarriages that include children are different from those that do not. These *blended families* that emerge from remarriage with children are traditionally known as *stepfamilies*. They are also sometimes called *reconstituted, restructured,* or *remarried families* by social scientists—names that emphasize their structural differences from other families. Attempting to focus more on the positive aspect of blending (and striving to steer clear of the negative connotations of "steps" as in "evil stepmother"), some refer to new stepchildren or stepparents as "bonus" children or "bonus" parents. A website for Bonus Families (http://www.bonusfamilies .com/), a nonprofit organization whose goal is to promote "peaceful coexistence between divorced or separated parents and their new families," suggests that at different phases different terms may be more appropriate or acceptable:

> At first you may not feel like a family. The label stepfamily seems just fine because no one really knows

their place and may hate being there, but as you get to know each other, you blend a little. Now you are at the second level, a blended family. The ultimate goal, however, is to become a bonus family. In a bonus family you feel appreciated for who you are even though you are not biologically related to everyone in the family. You play an active role in the new family *and* your family has developed a way to solve conflicts where everyone feels respected and cared for.

Satirist Art Buchwald, however, called them "tangled families." In alluding to the complexity of relationships that result, his term comes close to the truth in some cases. Whatever we decide to call them, there soon may be more stepfamilies in America than any other family form (Pill 1990). If we care about families, we need to understand and support stepfamilies.

A Different Kind of Family

When we enter a stepfamily, many of us expect to recreate a family identical to an intact family. The intact nuclear family becomes the model against which we judge our successes and failures. But researchers believe that blended families are significantly different from intact families (Ganong and Coleman 1994; Papernow 1993; Pill 1990). If we try to make our feelings and relationships in a stepfamily identical to those of an intact family, we are bound to fail. But if we recognize that the stepfamily works differently and provides different satisfactions and challenges, we can appreciate the richness it brings us and have a successful stepfamily.

Structural Differences

Six structural characteristics make the stepfamily different from the traditional first-marriage family (Visher and Visher 1979, 1991). Each one is laden with potential difficulties.

1. *Almost all the members in a stepfamily have lost an important primary relationship.* The children may mourn the loss of their parent or parents, and the spouses may mourn the loss of their former mates. Anger and hostility may be displaced onto the new stepparent.

2. *One biological parent typically lives outside the current family.* In stepfamilies that form after divorce, the absent former spouse may either support or interfere with the new family. Power struggles may occur between the absent parent and the custodial parent, and there may be jealousy between the absent parent and the stepparent.

3. *The relationship between a parent and his or her children predates the relationship between the new partners.* Children have often spent considerable time in a single-parent family structure. They have formed close and different bonds with the parent. A new husband or wife may seem to be an interloper in the children's special relationship with the parent. A new stepparent may find that he or she must compete with the children for the parent's attention. The stepparent may even be excluded from the parent–child system.

4. *Stepparent roles are ill defined.* No one knows quite what he or she is supposed to do as a stepparent. remarried families tend to model themselves after traditional nuclear families, so stepparents often expect that their role will be similar to the parent role. However, some are reluctant to assume an active parenting role, and some attempt to assume such a role too quickly. Children may resist the efforts made by stepparents to become involved in their lives. Most stepparents try role after role until they find one that fits.

5. *Many children in stepfamilies are also members of a noncustodial parent's household.* Each home may have differing rules and expectations. When conflict arises, children may try to play one household against the other. Furthermore, as Emily and John Visher (1979) observe:

   The lack of clear role definition, the conflict of loyalties that such children experience, the emotional reaction to the altered family pattern, and the loss of closeness with their parent who is now married to another person create inner turmoil and confused and unpredictable outward behavior in many children.

6. *Children in stepfamilies have at least one extra pair of grandparents.* Children gain a new set of step-grandparents, but the role these new grandparents are to play is usually not clear. A study by Graham Spanier and Frank Furstenberg (1980) found that step-grandparents were usually quick to accept their "instant" grandchildren.

Numerous researchers have found that children in stepfamilies exhibit about the same number of adjustment problems as children in single-parent families and more problems than children in original, two-parent families (Furstenberg and Cherlin 1991; McLenahan and Sandefor 1994; Nicholson, Fergusson, and Horwood 1999; Coleman, Ganong, and Fine 2000). Others suggest that stepfamily life may be more difficult for children than living in a single-parent household.

In addition, research reveals that relations between stepparents and their stepchildren are often of "low quality," characterized by less frequent activities together than between biological parents and children, less warmth and support from stepparents to stepchildren, and less involvement by stepparents in monitoring and controlling their stepchildren's activities (Stewart 2005).

A new partner is "a second pair of eyes and hands" who can share in the various, often burdensome, tasks of childrearing. Likewise, new partners can be sources of emotional and social support, strengthening the mother's authority in the household, assisting her with difficult decisions, comforting her when parenting is stressful, and potentially inhibiting her from acting in negative or hurtful ways toward her children. Certainly, these effects will be for the better for children (Thomson et al. 2001).

The Developmental Stages of Blended Families

Individuals and families blend into and become a stepfamily through a process—through a series of developmental stages. Each person—the biological parent, the stepparent, and the stepchild (or children)—experiences the process differently. For family members, it involves seven stages, according to a study of stepfamilies by Patricia Papernow (1993). The early stages are fantasy, immersion, and awareness; the middle stages are mobilization and action; and the later stages are contact and resolution.

It takes most stepfamilies about 7 years to complete the developmental process. Some may complete it in 4 years, and others take many, many years. Some only go through a few of the stages and become stuck. Others split up with divorce. But many are successful. Becoming a stepfamily is a slow process that moves in small ways to transform strangers into family members.

Early Stages: Fantasy, Immersion, and Awareness

The early stages in becoming a stepfamily include the courtship and early period of remarriage, when each individual has his or her fantasy of their new family. It is a time when the adults (and sometimes the children) hope for an "instant" nuclear family that will fulfill their dreams of how families should be. They have not yet realized that stepfamilies are different from nuclear families.

FANTASY STAGE. During the fantasy stage, biological parents hope that the new partner will be a better spouse and parent than the previous partner. They want their children to be loved, adored, and cared for by their new partners. They expect their children to love the new parent as much as they do.

New stepparents fantasize that they will be loving parents who are accepted and loved by their new stepchildren. They believe that they can ease the load of the new spouse, who may have been a single parent for years. One stepmother recalled her fantasy: "I would meet the children and they would gradually get to know me and think I was wonderful. . . . I just knew they would love me to pieces. I mean, how could they not?" (Papernow 1993). Of course, they did not.

The children, meanwhile, may have quite different fantasies. They may still feel the loss of their original families. Their fantasies are often that their parents will get back together. Others fear they may "lose" their parent to an interloper, the new stepparent. Some fear that their new family may "fail" again. Still others are concerned about upheavals in their lives, such as moving, going to new schools, and so on.

IMMERSION STAGE. The immersion stage is the "sink-or-swim" stage in a stepfamily. Reality replaces fantasy. "We thought we would just add the kids to this wonderful relationship we'd developed. Instead we spent three years in a sort of Cold War over them," recalled one stepparent (Papernow 1993).

For children, a man's transformation from "Mom's date" to stepfather may be the equivalent of the transformation from Dr. Jekyll to Mr. Hyde. Suddenly an outsider becomes an insider—with authority, as described by one 12-year-old (whose new stepmother also had children): "In the beginning it's fun. Then you realize that your whole life is going to change. Everything changes . . . now there's all these new people and new rules" (Papernow 1993). Children may also feel disloyal to their absent biological parent if they show affection to a stepparent. (Biological parents can make a difference: They can let their children know it's okay to love a stepparent.)

AWARENESS STAGE. The awareness stage in stepfamily development is reached when family members "map" the territory. This stage involves individual and joint family tasks. The individual task is for each member to identify and name the feelings he or she experiences in being in the new stepfamily. A key feeling for stepparents to acknowledge is feeling like an outsider. They need to become aware of feelings of aloneness; they must discover their own needs; and they must set some distance between themselves and their stepchildren. They need to understand why their stepchildren are not warmly welcoming them, as they had expected.

Biological parents need to become aware of unresolved feelings from their earlier marriages and from being single parents. They may feel pulled from the multiple demands of their children and their partners. Biological parents may feel resentment toward their children, their partners, or both (Papernow 1993).

Children in the awareness stage often feel "bumped" from their close relationship with the single parent. They miss cuddling in bed in the morning, the bedtime story, the wholehearted attention. When a new stepparent moves in, their feelings of loss over their parents' divorce are often rekindled. Loyalty issues resurface. If they are not pressured into feeling "wonderful" about their new family, however, they can slowly learn to appreciate the benefits of an added parent and friend who will play with them or take them places.

Middle Stages: Mobilization and Action

In the middle stages of stepfamily development, family members are more clear about their feelings and relationships with one another. They have given up many of their fantasies. They understand more of their own needs. They have mapped the new territory. The family, however, remains biologically oriented. Parent–child relationships are central. In this stage, changes involve the emotional structure of the family.

MOBILIZATION STAGE. In the mobilization stage, family members recognize differences. Conflict becomes more open. Members mobilize around their unmet needs. A stepmother described this change: "I started realizing that I'm different than Jim [the husband] is, and

I'm going to be a different person than he is. I spent years trying to be just like him and be sweet and always gentle with his daughter. But I'm not always that way. I think I made a decision that what I was seeing was right" (Papernow 1993). The challenge in this stage is to resolve differences while building the stepfamily's sense of family.

Stepparents begin to take a stand. They stop trying to be the ideal parent. They no longer are satisfied with being outsiders. Instead, they want their needs met. They begin to make demands on their stepchildren: to pick up their clothes, be polite, do the dishes. Similarly, they make demands on their partners to be consulted; they often take positions regarding their partners' former spouses. Because stepparents make their presence known in this stage, the family begins to change. The family begins to integrate the stepparent into its functioning. In doing so, the stepparent ceases being an outsider and the family increasingly becomes a real stepfamily.

For biological parents, the mobilization stage can be frightening. The stepparents' desire for change leaves biological parents torn. Biological parents feel they must protect their children and yet satisfy the needs of their partners.

Children often attempt to resolve loyalty issues at this stage. They have been tugged and pulled in opposite directions by angry parents too long. Often the adults paid no attention to them. Finally, the children have had enough and can articulate their feelings. After hearing her parents squabble one time too many, one girl reflected: "I thought, this stinks. It's horrible. After the 50 millionth time I said, 'That's your problem. Talk to each other about it,' and they didn't do it again" (Papernow 1993).

ACTION STAGE. In the action stage, the family begins to take major steps in reorganizing itself as a stepfamily. It creates new norms and family rituals. Although members have different feelings and needs, they begin to accept each other. Most important, stepfamily members develop shared, realistic expectations and act on them.

Stepcouples begin to develop their own relationship independent from the children. They also begin working together as a parental team. Stepparents begin to take on disciplinary and decision-making roles; they are supported by the biological parents. Stepparents begin to develop relationships with their stepchildren independent of the biological parents. Stepparent–stepchild bonds are strengthened.

Later Stages: Contact and Resolution

The later stages in stepfamily development involve solidifying the stepfamily. Much of the hard work has been accomplished in the middle stages.

CONTACT STAGE. In the contact stage, stepfamily members make intimate contact with one another. Their relationships become genuine. They communicate with a sense of ease and intimacy. The couple relationship becomes a sanctuary from everyday family life. The stepparent becomes an "intimate outsider" with whom stepchildren can talk about things "too hot" for their biological parents, such as sex, drugs, their feelings about the divorce, and religion.

For the stepparent, a clear role finally emerges—what is now called the **stepparent role.** The role varies from stepparent to stepparent and from stepfamily to stepfamily because, as shown earlier, it is undefined in our society. It is mutually suitable to both the individual and the different family members.

RESOLUTION STAGE. The stepfamily is solid in its resolution stage. It no longer requires the close attention and work of the middle stages. Family members feel that earlier issues have been resolved.

Not all relationships in stepfamily are necessarily the same; they may differ according to the personalities of each individual. Some relationships develop more closely than others. But in any case, there is a sense of acceptance. The stepfamily has made it and has benefited from the effort.

Reflections

If you are a member of a stepfamily, what were your experiences at the different stages? If you are not, ask friends or relatives who are members what their experiences were at the different stages. If you were to become a stepparent, how would you handle each stage?

Problems of Women and Men in Stepfamilies

Most people go into stepfamily relationships expecting to recreate the traditional nuclear family: they are full of love, hope, and energy. Although women and men may enter stepfamilies equally hopeful, they do not experience the same things.

Women in Stepfamilies

Stepmothers tend to experience more problematic family relationships than do stepfathers (Santrock and Sitterle 1987; Kurdek and Fine 1993; Hetherington and Stanley-Hagan 1999). To various degrees, women enter stepfamilies with certain feelings and hopes. Stepmothers generally expect to do the following (Visher and Visher 1979, 1991):

- Make up to the children for the divorce or provide children whose mothers have died with a maternal figure
- Create a happy, close-knit family and a new nuclear family
- Keep everyone happy
- Prove that they are not wicked stepmothers
- Love the stepchild instantly and as much as their biological children
- Receive instant love from their stepchildren

Needless to say, most women are disappointed. Expectations of total love, happiness, and the like would be unrealistic in any kind of family, be it a traditional family or a stepfamily. The warmer a woman is to her stepchildren, the more hostile they may become to her because they feel she is trying to replace their "real" mother. If a stepmother tries to meet everyone's needs—especially her stepchildren's, which are often contradictory, excessive, and distancing—she is likely to exhaust herself emotionally and physically. It takes time for her and her children to become emotionally integrated as a family.

One thing that makes stepmothering more difficult than stepfathering is the role women typically play in childrearing. Women are expected to and expect to become nurturing, primary caregivers, although this role may not be adequately acknowledged or appreciated by their stepchildren. Consequently, there are more opportunities for them to encounter stress and experience conflict with their stepchildren, and thus poorer relationships with their stepchildren may occur.

Stepchildren tend to view relationships with stepmothers as more stressful than relationships with stepfathers. If their biological mothers are still living, they may feel their stepmothers threaten their relationships with their birth mothers (Hetherington and Stanley-Hagan 1999). Stepmothers married to men who have their children full-time often experience greater problems than stepmothers whose children are with them part-time or occasionally (Furstenberg and Nord

1985). Bitter custody fights may leave children emotionally troubled and hostile to stepmothers. In other instances, children (especially adolescents) may have moved into their father's home because their mother could no longer handle them. In either case, the stepmother may be required to parent children who have special needs or problems. Stepmothers may find these relationships especially difficult. Typically, stepmother–stepdaughter relationships are the most problematic (Clingempeel et al. 1984). Relationships become even more difficult when the stepmothers never intended to become full-time stepparents.

Men in Stepfamilies

Different expectations are placed on men in stepfamilies. Because men are generally less involved in childrearing, they usually have few "cruel stepparent" myths to counter. Nevertheless, men entering stepparenting roles may find certain areas particularly difficult at first (Visher and Visher 1991). A critical factor in a man's stepparenting is whether he has children of his own. If he does, they are more likely to live with his ex-wife. In this case, the stepfather may experience guilt and confusion in his stepparenting because he feels he should be parenting his own children. When his children visit, he may try to be "Superdad," spending all his time with them and taking them to special places. His wife and stepchildren may feel excluded and angry.

A stepfather usually joins an already established single-parent family. He may find himself having to squeeze into it. The longer a single-parent family has been functioning, the more difficult it usually is to reorganize it. The children may resent his "interfering" with their relationship with their mother. His ways of handling the children may be different from his wife's, resulting in conflict with her or with her children (Marsiglio 2004; Wallerstein and Kelly 1980b).

Working out rules of family behavior is often the area in which a stepfamily encounters its first real difficulties. Although the mother usually wants help with discipline, she often feels protective if the stepfather's style is different from hers. To allow a stepparent to discipline a child requires trust from the biological parent and a willingness to let go. Disciplining often elicits a child's testing response: "You're not my real father. I don't have to do what you tell me." Homes are more positive when parents include children in decision making and are supportive (Barber and Lyons 1994). Nevertheless, disciplining establishes legitimacy, because only a parent or parent figure is expected to

discipline in our culture. Disciplining may be the first step toward family integration, because it establishes the stepparent's presence and authority in the family.

In comparison to birth parents, stepfathers tend to have more limited and less positive relationships with their stepchildren. They communicate less, display less warmth and affection, and are typically less involved. Some research also indicates that among divorced, noncustodial fathers, remarriage and stepfathering may lead to development of closer relationships with stepchildren than with their biological children.

The new stepfather's expectations are important. Although the motivations to stepparent are often quite different from those of biological parents, research from the 1987–1988 National Survey of Families and Households shows that 55% of stepfathers found it somewhat or definitely true that having stepchildren was just as satisfying as having their own children (Sweet, Bumpass, and Call 1988). Despite this, stepparents tend to view themselves as less effective than natural fathers view themselves (Beer 1992).

However, the process of **paternal claiming,** embracing stepchildren as if they were biological children and becoming involved in the processes of nurturing, providing for, and protecting them, is a two-way process. Stepfathers must build an appropriate identity, but both birth mothers and the stepchildren also help create or hinder the development of a sense of familial "we-ness" (Marsiglio 2004). The complex role that the stepfather brings to his family often creates role ambiguity and confusion that takes time to work out. However, the potential for deep, mutually gratifying, and meaningful relationships between stepfathers and stepchildren is there, as illustrated in the *Real Families* feature, "Claiming Them as Their Own: Stepfather-Stepchild Relationships."

Conflict in Stepfamilies

Achieving family solidarity in the stepfamily is a complex task. When a new parent enters the former single-parent family, the family system is thrown off balance. Where equilibrium once existed, there is now disequilibrium. A period of tension and conflict usually marks the entry of new people into the family system. Questions arise about them: Who are they? What are their rights and their limits? Rules change. The mother may have relied on television as a babysitter, for example, permitting the children unrestricted viewing in the afternoon. The new stepfather, however, may want

Real Families

Claiming Them as Their Own: Stepfather-Stepchild Relationships

Sometimes I feel like I'm on the outside looking in because—sometimes I wish she was mine. I guess because we're just that close . . . in my heart, I feel like I'm her father. . . . I know in reality, I'm not but, I'm going to give her all the benefit that a father should. I'm going to make sure she gets those benefits. Even though her dad is giving them to her, she is given a little extra and I figure that extra go a long way. . . .

Sociologist William Marsiglio conducted interview research with a diverse group of 36 stepfathers, including the 35-year-old stepfather just quoted. Of the men, 25 were married, 7 more cohabited with their female partners, and 4 lived apart from their partners. They ranged in age from 20–54, with an average of 36 years of age. Educationally, 16 of the men were college graduates, 12 more graduated high school and attended some college, and 8 had either just completed or failed to complete high school. Racially, 27 of the men were Caucasian and 9 were African American. In addition 22 men had biological children of their own, and 11 were living with at least one of "their own" biological children. Marsiglio (2004, 34) wanted to uncover men's experiences of "claiming stepchildren" and identified 10 properties of the claiming process.

Among these properties is the *degree of deliberativeness*—how much thought men give to their relationships with their stepchildren and how conscious and deliberate they are in coming to orient themselves to their stepchildren as "their" children.

Although some men experience the paternal claiming process gradually, as events unfold that may include key turning points, some men, like 41-year-old Terry, decide at the outset that the relationship is to be "all or nothing." As he told Marsiglio:

It was like, if I'm coming into this relationship, then I'm coming in a hundred percent. I'm either going to be an all husband and an all father or nothing at all. I can't have like half a relationship. I can't be half a father. Where do you draw the line? . . . If I'm going to love you, I'm going to be your father. I'm going to be there all the way.

Other properties include the degree to which they have and use opportunities to be involved across a range of paternal behaviors; the extent to which they find themselves thinking about, mindful of, or daydreaming about their stepchildren in ways that biological fathers do; and the degree to which they seek and are publicly acknowledged as a father figure by others: schoolteachers, coaches, neighbors, and—in the case of adoptions—the law.

Marsiglio (2004) also identified five conditions that encourage men to perceive their stepchildren as their own: the stepfather's identification with the stepchild, the stepfather's personality, the birth mother's involvement, the stepchildren's perceptions and reactions, and the biological father's presence and involvement.

Here is how the first condition—the degree to which stepfathers identify with their stepchildren, seeing similarities in personalities, interests, or personalities—was expressed by a man Marsiglio calls Thomas:

They're my kids. I look at them like they're my boys, I tell everybody,

they're my boys. And I don't want to take nothing away from (Danny's) dad, but I've raised them for so long now, I mean . . . you have a child in your home for the amount of time that I have, you feed them and long enough, they'll start acting and looking just like you, you know what I'm saying? They just do. They just call me, call me "dad."

For men who have biological children, perceiving that stepchildren are their own may mean coming to feel similarly toward their stepchildren and biological children. Marsiglio's interview with 30-year-old Brandon, revealed how this was experienced:

I really don't (feel differently toward stepchildren). I mean, I thought initially when we first, we all moved in together that maybe—I was a little worried, how am I going to feel towards them? But now . . . I consider them my kids even though I'm not the biological father. I don't really try to step in to take—for them to call me dad or anything like that—but I don't really see them as any different. I mean . . . I'll do my best to protect them and treat them fairly.

These sorts of reactions may not be commonplace, but neither are they aberrations. Some stepfather–stepchild ties become quite powerful, becoming the equivalent of relationships between biological parents and children. Thus, when we read or hear generalizations about distance or deficiency in stepparent–stepchild relationships, we would do well to remember the words and sentiments expressed by Marsiglio's interviewees. We would also be well advised to consider some factors that might enhance or facilitate the paternal claiming process.

to limit the children's afternoon viewing, and this creates tension. To the children, everything seemed fine until this stepfather came along. He has disrupted their old pattern. Chaos and confusion will be the norm until a new pattern is established, but it takes time for people to adjust to new roles, demands, limits, and rules.

Conflict takes place in all families: traditional nuclear families, single-parent families, and stepfamilies. If some family members do not like each other, they will bicker, argue, tease, and fight. Sometimes they have no better reason for disruptive behavior than that they are bored or frustrated and want to take it out on someone. These are fundamentally personal conflicts. Other conflicts are about definite issues: dating, use of the car, manners, television, or friends, for example. These conflicts can be between partners, between parents and children, or among the children themselves. Certain types of stepfamily conflicts, however, are of a frequency, intensity, or nature that distinguishes them from conflicts in traditional nuclear families. Recent research on how conflict affects children in stepfather households found that parental conflict does not account for children's lower level of well-being (Hanson, McLanahan, and Thompson 1996). These conflicts are about favoritism; divided loyalties; discipline; and money, goods, and services.

Favoritism

Favoritism exists in families of first marriages, as well as in stepfamilies. In stepfamilies, however, the favoritism often takes a different form. Whereas a parent may favor a child in a biological family on the basis of age, sex, or personality, in stepfamilies favoritism tends to run along kinship lines. A child is favored by one or the other parent because he or she is the parent's biological child. If a new child is born to the remarried couple, they may favor him or her as a child of their joint love. In American culture, where parents are expected to treat children equally, favoritism based on kinship seems particularly unfair.

Divided Loyalties

"How can you stand that lousy, low-down, sneaky, nasty mother (or father) of yours?" demands a hostile parent. It is one of the most painful questions children can confront, because it forces them to take sides against someone they love. One study (Lutz 1983) found that about half of the adolescents studied confronted situations in which one divorced parent talked

negatively about the other. Almost half of the adolescents felt themselves "caught in the middle." Three-quarters found such talk stressful.

Divided loyalties put children in no-win situations, forcing them not only to choose between parents but also to reject new stepparents. Children feel disloyal to one parent for loving the other parent or stepparent. But as shown in the last chapter, divided loyalties, like favoritism, can exist in traditional nuclear families as well. This is especially true of conflict-ridden families in which warring parents seek their children as allies.

Reflections

Think about conflicts involving favoritism, loyalty, discipline, and the distribution of resources. Do you experience them in your family of orientation? If so, how are they similar to, or different from, stepfamily conflicts? If you are in a stepfamily, do you experience them in your current family? How are these conflicts similar or different in your original family versus your current family? If you are a parent or stepparent, how are these issues played out in your current family?

Discipline

Researchers generally agree that discipline issues are among the most important causes of conflict among remarried families (Ihinger-Tallman and Pasley 1987). Discipline is especially difficult to deal with if the child is not the person's biological child. Disciplining a stepchild often gives rise to conflicting feelings within the stepparent. Stepparents may feel that they are overreacting to the child's behavior, that their feelings are out of control, and that they are being censured by the child's biological parent. Compensating for fears of unfairness, the stepparent may become overly tolerant.

The specific discipline problems vary from family to family, but a common problem is interference by the biological parent with the stepparent (Mills 1984). The biological parent may feel resentful or overreact to the stepparent's disciplining if he or she has been reluctant to give the stepparent authority. As one biological mother who believed she had a good remarriage stated (Ihinger-Tallman and Pasley 1987):

> Sometimes I feel he is too harsh in disciplining, or he doesn't have the patience to explain why he is punishing and to carry through in a calm manner,

which causes me to have to step into the matter (which I probably shouldn't do). . . . I do realize that it was probably hard for my husband to enter marriage and the responsibility of a family instantly . . . but this has remained a problem.

As a result of interference, the biological parent implies that the stepparent is wrong and undermines his or her status in the family. Over time, the stepparent may decrease involvement in the family as a parent figure.

Money, Goods, and Services

Problems of allocating money, goods, and services exist in all families, but they can be especially difficult in stepfamilies. In first marriages, husbands and wives form an economic unit in which one or both may produce income for the family; husband and wife are interdependent. Following divorce, the binuclear family consists of two economic units: the custodial family and the noncustodial family. Both must provide separate housing, which dramatically increases their basic expenses. Despite their separation, the two households may nevertheless continue to be extremely interdependent. The mother in the custodial single-parent family, for example, probably has reduced income. She may be employed but still dependent on child support payments or TANF (see Chapter 12). She may have to rely more extensively on childcare, which may drain her resources dramatically. The father in the noncustodial family may make child support payments or contribute to medical or school expenses, which depletes his income. Both households have to deal with financial instability. Custodial parents can't count on always receiving their child support payments, which makes it difficult to undertake financial planning.

When one or both of the former partners remarry, their financial situation may be altered significantly. Upon remarriage, the mother receives less income from her former partner or lower welfare benefits. Instead, her new partner becomes an important contributor to the family income. At this point, a major problem in stepfamilies arises. What responsibility does the stepfather have in supporting his stepchildren? Should he or the biological father provide financial support? Because there are no norms, each family must work out its own solution.

Stepfamilies typically have resolved the problem of distributing their economic resources by using a one-pot or two-pot pattern (Fishman 1983). In the *one-pot* pattern, families pool their resources and distribute them according to need rather than biological relationship. It doesn't matter whether the child is a biological child or a stepchild. One-pot families typically have relatively limited resources and consistently fail to receive child support from the noncustodial biological parent. By sharing their resources, one-pot families increase the likelihood of family cohesion.

In *two-pot* families, resources are distributed by biological relationship; need is secondary. These families tend to have a higher income, and one or both parents have former spouses who regularly contribute to the support of their biological children. Expenses relating to children are generally handled separately; usually there are no shared checking or savings accounts. Two-pot families maintain strong bonds among members of the first family. For these families, a major problem is achieving cohesion in the step-family while maintaining separate checking accounts.

Just as economic resources need to be redistributed following divorce and remarriage, so do goods and services (not to mention affection). Whereas a two-bedroom home or apartment may have provided plenty of space for a single-parent family with two children, a stepfamily with additional residing or visiting stepsiblings can experience instant overcrowding. Rooms, bicycles, and toys, for example, need to be shared; larger quarters may have to be found. Time becomes a precious commodity for harried parents and stepparents in a stepfamily. When visiting stepchildren arrive, duties are doubled. Stepchildren compete with parents and other children for time and affection.

It may appear that remarried families are confronted with many difficulties, but traditional nuclear families also encounter financial, loyalty, and discipline problems. We need to put these problems in perspective. (After all, half of all current marriages end in divorce, which suggests that first marriages are not problem free.) When all is said and done, the problems that remarried families face may not be any more overwhelming than those faced by traditional nuclear families (Ihinger-Tallman and Pasley 1987).

Family Strengths of Blended Families

Because we have traditionally viewed stepfamilies as deviant, we have often ignored their strengths. Instead, we have seen only their problems. We end this chapter by focusing on the strengths of blended families.

Family Functioning

Although traditional nuclear families may be structurally less complicated than stepfamilies, stepfamilies are nevertheless able to fulfill traditional family functions. A binuclear single-parent, custodial, or non-custodial family may provide more companionship, love, and security than the particular traditional nuclear family it replaces. If the nuclear family was ravaged by conflict or violence, for example, the single-parent family or stepfamily that replaces it may be considerably better, and because children now see happy parents, they have positive role models of marriage partners (Rutter 1994). Second families may not have as much emotional closeness as first families, but they generally experience less trauma and crisis (Ihinger-Tallman and Pasley 1987).

New partners may have greater objectivity regarding old problems or relationships. Opportunity presents itself for flexibility and patience. As family boundaries expand, individuals grow and adapt to new personalities and ways of being. In addition, new partners are sometimes able to intervene between former spouses to resolve long-standing disagreements, such as custody or childcare arrangements.

Effect on Children

As shown, blended families are often associated with problematic outcomes for children. But potentially, blended families can offer children benefits that can compensate for the negative consequences of divorce and of living with a single parent. Remember the notion of "bonus families" introduced earlier? Here are some ways in which stepfamilies offer children some bonuses:

- Children gain multiple role models from which to choose. Instead of having only one mother or father after whom to model themselves, children may have two mothers or fathers: the biological parents and the stepparents.

- Children gain greater flexibility. They may be introduced to new ideas, different values, or alternative politics. For example, biological parents may be unable to encourage certain interests, such as music or model airplanes, whereas a stepparent may play the piano or be a die-hard modeler. In such cases, that stepparent can assist the stepchildren in pursuing their development. In addition, children often have alternative living arrangements that enlarge their perspectives.

- Stepparents may act as a sounding board for their children's concerns. They may be a source of support or information in areas in which the biological parents feel unknowledgeable or uncomfortable.

- Children may gain additional siblings, either as stepsiblings or half-siblings, and consequently gain more experience in interacting, cooperating, and learning to settle disputes among peers.

- Children gain an additional extended kin network, which may become at least as important and loving as their original kin network.

- A child's economic situation is often improved, especially if a single mother remarried.

- Children may gain parents who are happily married. Most research indicates that children are significantly better adjusted in happily remarried families than in conflict-ridden nuclear families.

It is clear that the American family is no longer what it was through most of the last century. The rise of the single-parent family and stepfamily, however, does not imply an end to the nuclear family. Rather, these forms provide different paths that contemporary families take as they strive to fulfill the hopes, needs, and desires of their members, and they are becoming as American as Beaver Cleaver's family and apple pie.

Summary

- Many of today's families depart from the traditional family system, based on lifetime marriage and the intact nuclear family.

- Our pluralistic family system consists of three major types of families: (1) intact nuclear families, (2) single-parent families (either never married or formerly married), and (3) stepfamilies.

- Single-parent families tend to be created by divorce or births to unmarried women, are generally headed by women, are predominantly African American or Latino, are usually poor, involve a variety of household types, and are usually a transitional stage.

- Because of gender discrimination and inequality in wages or job opportunities, many female-headed families face economic hardship.

- Single, custodial fathers take different paths to single parenthood. Most likely reasons fathers obtain custody are because mothers are financially unable to provide adequate care for children, mothers are physically or psychologically unfit, or mothers do not want full-time responsibility for raising children.

- Both mother-only and father-only families are more likely to be poor than are two-parent families.

- Many "single-parent households" actually contain the parent and his or, her unmarried partner. Even in the absence of parents' live-in partners, parents' romantic partners may play important roles in children's lives.

- Many children of single mothers and nonresidential biological fathers have a *social father*—a male relative, family associate, or mothers' partner who behaves like a father to the child.

- Single parents, especially mothers, often come to rely on a combination of state or federal assistance and *private safety nets*: support from their social networks on which they can fall back in times of economic need. These can be the sources of emergency transportation, financial help, childcare, and emotional support, all of which may make a difference between success or failure in finding and keeping a job and in raising a child with less distress.

- Children of single parents are more likely to engage in high risk, "health compromising" behaviors and to suffer a variety of educational, economic and personal costs. These consequences appear to be linked to the lack of economic resources and to reduced money, attention, guidance, and social connections—what researchers call *social capital*—from fathers.

- Relations between the parent and his or her children change after divorce: the single parent generally tends to be emotionally closer but to have less authority. Family strengths associated with successful single parenting include parenting skills, personal growth, communication, family management, and financial support.

- The *binuclear family* is a postdivorce family system with children. It consists of two nuclear families: the mother-headed family and the father-headed family.

- Courtship for second marriage lacks clear norms. Courtship is complicated by the presence of children because remarriage involves the formation of a stepfamily.

- Cohabitation is more common in the "courtship" process leading to remarriages. As with cohabitation in first marriages, cohabitation before remarriages leads to higher rates of marital instability.

- Remarriage rates are lower for those who have children and as adults age. More men than women remarry. Those who initiate the divorce are more likely to remarry within 3 years than noninitiators.

- Explanations of remarriage focus on factors such as need, attractiveness, and opportunity.

- Remarriage differs from first marriage in several ways.

- Remarried couples are more likely to divorce than couples in their first marriages. This may be because of their willingness to use divorce as a means of resolving an unhappy marriage or because remarriage is an "incomplete institution." Stresses accompanying stepfamily formation may also be a contributing factor.

- The stepfamily or blended family differs from the original family because almost all members have lost an important primary relationship, one bi-

ological parent lives outside the current family, the relationship between a parent and his or her children predates the new marital relationship, *stepparent roles* are ill defined.

- Traditionally, researchers viewed stepfamilies from a "deficit" perspective, assuming that stepfamilies are very different from traditional nuclear families. More recently, stepfamilies have been viewed as normal families.

- Research in the United States and a number of other countries reveals some hazards of stepfamilies for children including: academic difficulties; higher risk of physical and mental health problems, earlier onset of sexual activity; greater risk of dropping out of school and of involvement in substance use and criminal activity. Some research indicates that girls adjust less well than boys to stepfamily life.

- Relations between stepparents and their stepchildren have been characterized as "disengaged."

- Becoming a stepfamily is a process—a series of developmental stages. Each person—the biological parent, the stepparent, and the stepchild (or children)—experiences the process differently. For family members, it involves seven stages. The early stages are fantasy, immersion, and awareness; the middle stages are mobilization and action; the later stages are contact and resolution.

- Although both often experience difficulty in being integrated into the family, stepmothers tend to experience greater stress in stepfamilies than do stepfathers. The warmer a woman is to her stepchildren, the more hostile they may become to her because they feel she is trying to replace their "real" mother.

- Men are generally less involved in childrearing, they usually have few "cruel stepparent" myths to counter. A stepfather usually joins an already established single-parent family. The longer a single-parent family has been functioning, the more difficult it usually is to reorganize it.

- Despite the aforementioned difficulties, many men attempt paternal claiming of stepchildren, embracing them as though they were their own children.

- A key issue for stepfamilies is family solidarity—the feeling of oneness with the family. Conflict in stepfamilies is often over favoritism; divided loyalties; discipline; and money, goods, and services. The addition of a new baby into a stepfamily neither solidifies nor divides the family.

- Stepfamily strengths may include improved family functioning and reduced conflict between former spouses. Children may gain multiple role models, more flexibility, concerned stepparents, additional siblings, additional kin, improved economic situation, and happily married parents.

Key Terms

binuclear family 532

blended families 526

paternal claiming 541

private safety nets 529

remarriage 534

single-parent families 526

social capital 529

social father 528

stepfamilies 526

stepparent role 540

Resources on the Internet

Companion Website for This Book

http://www.thomsonedu.com/sociology/strong

Gain an even better understanding of this chapter by going to the companion website for additional study resources. Take advantage of the Pre- and Post-Test quizzing tool, which is designed to help you grasp difficult concepts by referring you back to review specific pages in the chapter for questions you answer incorrectly. Use the flash cards to master key terms and check out the many other study aids you'll find there. Visit the Marriage and Family Resource Center on the site. You'll also find special features such as access to Info-Trac© College Edition (a database that allows you access to more than 18 million full-length articles from 5,000 periodicals and journals), as well as GSS Data and Census information to help you with your research projects and papers.

Appendix A
Sexual Structure and the Sexual Response Cycle

The Female Reproductive System

External Genitalia

The female external genitalia are known collectively as the vulva, which includes the mons veneris, labia, clitoris, urethra, and introitus. The *mons veneris* (literally, "mountain of Venus") is a protuberance formed by the pelvic bone and covered by fatty tissue. The *labia* are the vaginal lips surrounding the entrance to the vagina. The *labia majora* (outer lips) are two large folds of spongy flesh extending from the mons veneris along the midline between the legs. The outer edges of the labia majora are often darkly pigmented and are covered with pubic hair beginning in puberty. Usually the labia majora are close together, giving them a closed appearance. The *labia minora* (inner lips) lie within the fold of the labia majora. The upper portion folds over the clitoris and is called the *clitoral hood.* During sexual excitement, the labia minora become engorged with blood and double or triple in size. The labia minora contain numerous nerve endings that become increasingly sensitive during sexual excitement.

The *clitoris* is the center of erotic arousal in the female. It contains a high concentration of nerve endings and is highly sensitive to erotic stimulation. The clitoris becomes engorged with blood during sexual arousal and may increase greatly in size. Its tip, the clitoral glans, is especially responsive to touch.

Between the folds of the labia minora are the urethral opening and the *introitus.* The introitus is the opening to the vagina; it is often partially covered by a thin perforated membrane called the *hymen,* which may be torn accidentally or intentionally before or during first intercourse. On either side of the introitus is a tiny *Bartholin's gland* that secrets a small amount of moisture during sexual arousal.

Internal Genitalia

The *vagina* is an elastic canal extending from the vulva to the cervix. It envelops the penis during sexual intercourse and is the passage through which a baby is normally delivered. The vagina's first reaction to sexual arousal is "sweating," that is, producing lubrication through the vaginal walls.

A few centimeters from the vaginal entrance, on the vagina's anterior (front) wall, there is, according to some researchers, an erotically sensitive area that they have dubbed the "Grafenberg spot" or "G-spot." The spot is associated with female ejaculation, the expulsion of

Figure A.1 ■ External Female Genitalia

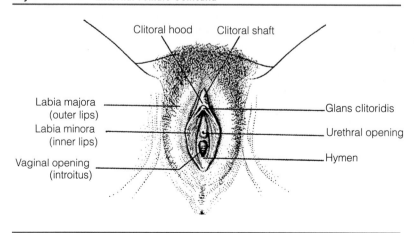

Clitoral hood   Clitoral shaft

Labia majora (outer lips)
Labia minora (inner lips)
Vaginal opening (introitus)

Glans clitoridis
Urethral opening
Hymen

Figure A.2 ■ Cross Section of the Female Reproductive System

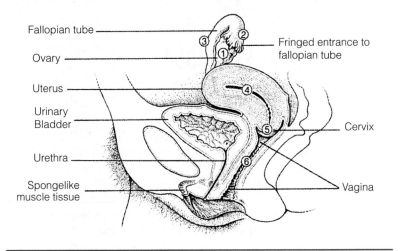

1. A follicle matures in the ovary and releases an ovum. 2. The fimbriae trap the ovum and move it into the fallopian tube. 3. The ovum travels through the fallopian tube to the uterus. 4. If the ovum is fertilized, the resulting blastocyst descends into the uterus. 5. If not fertilized, the ovum is discharged through the cervix into the vagina along with the shed uterine lining during the menstrual flow. 6. The vagina serves as a passageway to the body's exterior.

clear fluid from the urethra, which is experienced by a small percentage of women.

A female has two *ovaries,* reproductive glands (gonads) that produce *ova* (eggs) and the female hormones *estrogen* and *progesterone.* At the time a female is born, she already has all the ova she will ever have— more than forty thousand of them. About four hundred will mature during her lifetime and be released during ovulation; ovulation begins in puberty and ends at menopause.

The Path of the Egg

The two *fallopian tubes* extend from the uterus up to, but not touching the ovaries. When an egg is released from an ovary during the monthly *ovulation,* it drifts into a fallopian tube, propelled by waving *fimbriae* (the fingerlike projections at the end of each tube). If it is fertilized by sperm, fertilization usually takes place within the fallopian tube. The fertilized egg will then move into the uterus.

The *uterus* is a hollow, muscular organ within the pelvic cavity. The pear-shape uterus is normally about 3 inches long, 3 inches wide at the top, and 1 inch at the bottom. The narrow, lower part of the uterus projects into the vagina and is called the *cervix.* If an egg is fertilized, it will attach itself to the inner lining of the uterus, the *endometrium.* Inside the uterus it will develop into

an embryo and then into a fetus. If an egg is not fertilized, the endometrial tissue that developed in anticipation of fertilization will be shed during *menstruation.* Both the unfertilized egg and the inner lining of the uterus will be discharged in the menstrual flow.

The Male Reproductive System

The Penis

Both urine and semen pass through the penis. Ordinarily, the penis hangs limp and is used for the elimination of urine because it is connected to the bladder by the urinary duct (urethra). The penis is usually between 2.5 and 4 inches in length. When a man is sexually aroused, it swells to about 5 to 8 inches in length, is hard, and becomes erect (hence, the term *erection*). When the penis is erect, muscle contractions temporarily close off the urinary duct, allowing the ejaculation of semen.

The penis consists of three main parts: the root, the shaft, and the glans penis. The *root* connects the penis to the pelvis. The *shaft,* which is the spongy body of the penis, hangs free. At the end of the shaft is the *glans penis,* the rounded tip of the penis. The opening at the tip of the glans is called the *urethral meatus.* The glans penis is especially important in sexual arousal because

Figure A.3 ■ External Male Genitalia

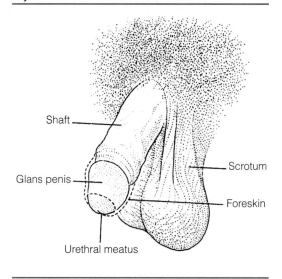

Shaft

Glans penis

Urethral meatus

Scrotum

Foreskin

it contains a high concentration of nerve endings, making it erotically sensitive. The *frenulum,* a small area of skin on the underside of the penis where the glans and shaft meet, is especially sensitive. The glans is covered by a thin sleeve of skin called the *foreskin.* Circumcision, the surgical removal of the foreskin, may damage the frenulum.

When the penis is flaccid, blood circulates freely through its veins and arteries, but as it becomes erect, the circulation of blood changes dramatically. The ar-

teries expand and increase the flow of blood into the penis. The spongelike tissue of the shaft becomes engorged and expands, compressing the veins within the penis so that the additional blood cannot leave it easily. As a result, the penis becomes larger, harder, and more erect.

The Testes

Hanging behind the male's penis is his *scrotum,* a pouch of skin holding his two *testes* (singular *testis;* also called *testicles*). The testes are the male reproductive glands (also called *gonads*), which produce both sperm and the male hormone *testosterone.* The testes produce sperm through a process called *spermatogenesis.* Each testis produces between 100 million and 500 million sperm daily. Once the sperm are produced, they move into the *epididymis,* where they are stored prior to ejaculation.

The Path of the Sperm

The epididymis merges into the tubular *vas deferens* (plural *vasa deferentia*). The vasa deferentia can be felt easily within the scrotal sac. Extending into the pelvic cavity, each vas deferens widens into a flasklike area called the *ampulla* (plural *ampullae*). Within the ampullae, the sperm mix with an activating fluid from the *seminal vesicles.* The ampullae connect to the *prostate gland* through the *ejaculatory ducts.* Secretions from the prostate account for most of the milky, gelatinous

Figure A.4 ■ Cross Section of the Male Reproductive System

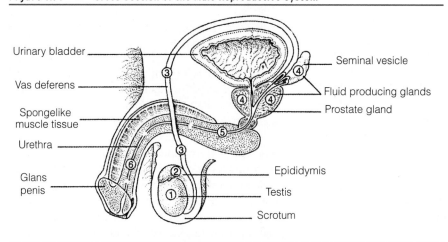

Urinary bladder

Vas deferens

Spongelike muscle tissue

Urethra

Glans penis

Seminal vesicle

Fluid producing glands

Prostate gland

Epididymis

Testis

Scrotum

1. The testis produces sperm. 2. Sperm mature in the epididymis. 3. During ejaculation, sperm travel through the vas deferens. 4. The seminal vesicles and the prostate gland provide fluids. 5. Sperm mix with the fluids, making semen. 6. Semen leaves the penis by way of the urethra.

liquid that makes up the *semen* in which the sperm are suspended. Inside the prostate, the ejaculatory ducts join to the urinary duct from the bladder to form the urethra, which extends to the tip of the penis. The two *Cowper's glands,* located below the prostate, secrete a clear, sticky fluid into the urethra that appears as small droplets on the meatus during sexual excitement.

If the erect penis is stimulated sufficiently through friction, an ejaculation usually occurs. *Ejaculation* is the forceful expulsion of semen. The process involves rhythmic contractions of the vasa deferentia, seminal vesicles, prostate, and penis. Altogether, the expulsion of semen may last from three to fifteen seconds. It is also possible to have an orgasm without the expulsion of semen.

The Sexual Response Cycle

Psychological and Physiological Aspects

When we respond sexually, we begin what is known as the *sexual response cycle.* Helen Singer Kaplan (1979) developed a model to describe the sexual response cycle. According to this model, the cycle consists of three phases: the desire phase, the excitement phase, and the orgasmic phase. The desire phase represents the psychological element of the sexual response cycle; the excitement and orgasmic phases represent its physiological aspects.

Sexual Desire

Desire can exist separately from overtly physical sexual responses. It is the psychological component that motivates sexual behavior. We can feel desire but not be physically aroused. It can suffuse our bodies without producing explicit sexual stirrings. We experience sexual desire as erotic sensations or feelings that motivate us to seek sexual experiences. These sensations generally cease after orgasm.

Physiological Responses: Excitement and Orgasm

A person who is sexually excited experiences a number of bodily responses. Most of us are conscious of some of these responses: a rapidly beating heart, an erection or lubrication, and orgasm. Many other responses may take place below the threshold of awareness, such as curling of the toes, the ascent of the testes, the withdrawal of the clitoris beneath the hood, and a flush across the upper body.

The physiological changes that take place during sexual response cycle depend on two processes: vasocongestion and myotonia. *Vasocongestion* occurs when body tissues become engorged with blood. For example, blood fills the genital regions of both males and females, causing the penis and clitoris to enlarge. *Myotonia* refers to increased muscle tension as orgasm approaches. Upon orgasm, the body undergoes involuntary muscle contractions and then relaxes. (The word *orgasm* is derived from the ancient Sanskrit *urja,* meaning "vigor" or "sap.")

Excitement Phase

In women, the vagina becomes lubricated and the clitoris enlarges during the excitement phase. The vaginal barrel expands, and the cervix and uterus elevate, a process called "tenting." The labia majora flatten and rise; the labia minora begin to protrude. The breasts may increase in size, and the nipples may become erect. Vasocongestion causes the outer third of the vagina to swell, narrowing the vaginal opening. This swelling forms the *orgasmic platform;* during sexual intercourse, it increases the friction against the penis. The entire clitoris retracts but remains sensitive to touch.

In men, the penis becomes erect as a result of vasocongestion, and the testes begin to rise. The testes may enlarge to as much as 150 percent of their unaroused size.

Orgasmic Phase

Orgasm is the release of physical tensions after the buildup of sexual excitement; it is usually accompanied by ejaculation of semen in physically mature males. In women, the orgasmic phase is characterized by simultaneous rhythmic contractions of the uterus, orgasmic platform, and rectal sphincter. In men, muscle contractions occur in the vasa deferentia, seminal vesicles, prostate, and the urethral bulb, resulting in the ejaculation of semen; contractions of the rectal sphincter also occur. Ejaculation usually accompanies male orgasm, but ejaculation and orgasm are separate processes.

Following orgasm, one of the most striking differences between male and female sexual response occurs

Figure A.5 ▪ Stages of Female Sexual Response (internal left; external right)

INTERNAL
Excitement

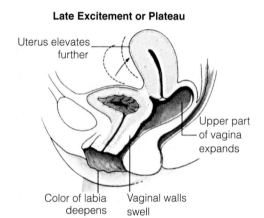

Uterus elevates

Pubic bone
Bladder

Vaginal lubrication appears

Clitoris enlarges

Inner labia swell

Outer labia

EXTERNAL
Unaroused

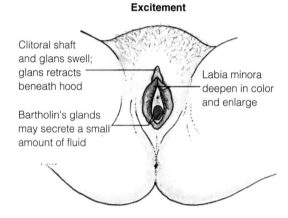

Clitoral hood

Urethra

Labia minora

Labia majora

Anus

Late Excitement or Plateau

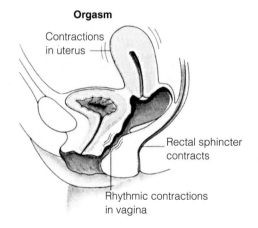

Uterus elevates further

Upper part of vagina expands

Color of labia deepens

Vaginal walls swell

Excitement

Clitoral shaft and glans swell; glans retracts beneath hood

Labia minora deepen in color and enlarge

Bartholin's glands may secrete a small amount of fluid

Orgasm

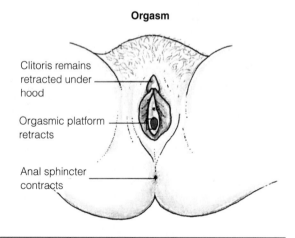

Contractions in uterus

Rectal sphincter contracts

Rhythmic contractions in vagina

Orgasm

Clitoris remains retracted under hood

Orgasmic platform retracts

Anal sphincter contracts

Excitement

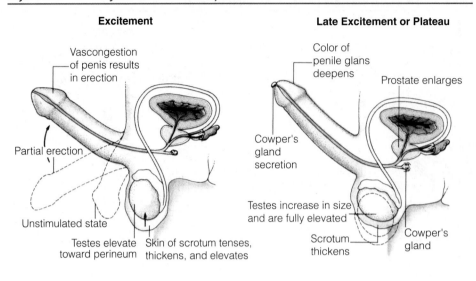

Vascongestion of penis results in erection

Partial erection

Unstimulated state

Testes elevate toward perineum

Skin of scrotum tenses, thickens, and elevates

Late Excitement or Plateau

Color of penile glans deepens

Prostate enlarges

Cowper's gland secretion

Testes increase in size and are fully elevated

Scrotum thickens

Cowper's gland

Orgasm

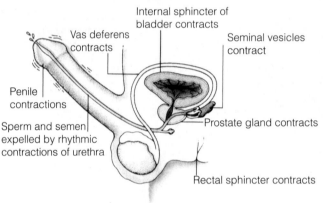

Internal sphincter of bladder contracts

Vas deferens contracts

Seminal vesicles contract

Penile contractions

Sperm and semen expelled by rhythmic contractions of urethra

Prostate gland contracts

Rectal sphincter contracts

as males experience a *refractory period*. The refractory period denotes the time following orgasm during which male arousal levels return to prearousal or excitement levels. During the refractory period, additional orgasms are impossible. Females do not have any comparable period. As a result, they have greater potential for multiple orgasms—that is, for having a series of orgasms. Although most women have the potential for multiple orgasms, only about 13 to 16 percent regularly experience them. For multiple orgasms, women generally require continued stimulation of the clitoris. Most women (or their partners), however, do not seek additional orgasms after the first one because our culture uses the first orgasm (usually the male's) as a marker to end sexual activities.

In sexual intercourse, orgasm has many functions. For men, it serves a reproductive function by causing ejaculation of semen into a women's vagina. For both men and women, it is a source of erotic pleasure, whether it is an autoerotic or relational context; it is intimately connected with our sense of well-being. We may measure both our sexuality and ourselves in terms of orgasm. Did I have one? Did my partner have one? When we measure our sexuality by orgasm, however, we discount activities that do not necessarily lead to orgasm, such as touching, caressing, and kissing. We discount erotic pleasure as an end in itself.

Men tend to be more consistently orgasmic than women, especially in sexual intercourse. If all women are potentially orgasmic, why do smaller proportion of women have orgasms than men? An answer may be found in our dominant cultural model that calls for female orgasm to occur as a result of penile thrusting during heterosexual intercourse in the face-to-face, male-above position. This traditional American model calls for a "no-hands" approach. The women is supposed to be orgasmic without manual or oral stimulation by her partner or herself. If she is orgasmic during masturbation or cunnilingus, such orgasms are usually discounted because they aren't considered "real" sex—that is, heterosexual intercourse.

The problem for women in sexual intercourse is that the clitoris frequently does not receive sufficient stimulation from penile thrusting alone to permit orgasm. In an influential study on female sexuality, Shere Hite (1976) found that only 30 percent of her three thousand respondents experienced orgasm regularly through sexual intercourse "without more direct manual clitoral stimulation being provided at one time of orgasm." Hite concludes that many women need manual stimulation during intercourse to be orgasmic. They also need to be assertive. There is no reason why a women cannot be manually stimulated by herself or her partner to orgasm before or after intercourse. But to do so, a woman has to assert her own sexual needs and move away from the idea that sex is centered around male orgasm.

Appendix B
Pregnancy, Conception, and Fetal Development

Once fertilization of the ovum by a sperm occurs, the birth will take place in approximately 266 days, if the pregnancy is not interrupted. Traditionally, physicians count the first day of the pregnancy as the day on which the woman began her last menstrual period; thus, they calculate the gestation (pregnancy) period to be 280 days, which is also 10 lunar months.

Following fertilization, which normally occurs within the fallopian tube, the fertilized ovum, or *zygote,* undergoes a series of divisions during which the cells replicate themselves. After four or five days, the zygote contains about a hundred cells and is called *blastocyst.* On about the fifth day, the blastocyst arrives in the uterine cavity, where it floats for a day or two before implanting itself in the soft, blood-rich uterine wall (endometrium), which has spent the past three weeks preparing for its arrival. This process of *implantation* takes about a week. The hormone human chorionic gonadotropin (HCG), which is secreted by the blastocyst, maintains the uterine environment in an "embryo-friendly" condition and prevents the shedding of the endometrium that would normally occur during menstruation.

The blastocyst, or pre-embryo, rapidly grows into an *embryo* (which will, in turn, be referred to as a *fetus* around the eighth week of development). During the first two or three weeks of development, the embryonic membranes, including the *amnion*—a membranous sac that will contain the embryo and *amniotic fluid*—and the yolk sac are formed.

During the third week, extensive cell migration occurs and the stage is set for the development of the organs. The first body segments and the brain begin to be formed. The digestive and circulatory systems begin to develop in the fourth week; the heart begins to pump blood. By the end of the first month, the spinal cord and nervous system have also begun to develop.

The fifth week sees the formation of arms and legs. In the sixth week, the eyes and ears form. At seven weeks, the reproductive organs begin to differentiate in the males; female reproductive organs continue to develop. At eight weeks, the fetus is about the size of a thumb, although the head is nearly as large as the body. The brain begins to function to coordinate the development of the internal organs. Facial features begin to form, and bones begin to develop.

Arms, hands, fingers, legs, feet, toes, and eyes are almost fully developed at twelve weeks. At fifteen weeks, the fetus has a strong heartbeat, fair digestion, and active muscles. Most bones are developed by then, and the eyebrows appear. At this stage, the fetus is covered with a fine, downy hair called *lanugo.* (Figure B.1 and Figure B.2 shows the actual size of the developing embryo and fetus through its first sixteen weeks.)

Throughout its development, the fetus is nourished through the *placenta.* The placenta begins to develop from part of thee blastocyst following implantation. This organ grows larger as the fetus does, passing nutrients from the mother's bloodstream to the fetus, to which it is attached by the umbilical cord. The placenta blocks blood corpuscles and large molecules.

By five months, the fetus is 10 to 12 inches long and weighs between one-half and one pound. The internal organs are well developed, although the lungs cannot function well outside the uterus. At six months, the fetus is 11 to 14 inches long and weighs more than a pound. At seven months, it is 13 to 17 inches long, weighing about three pounds. At this point, most healthy fetuses are viable—capable of surviving outside the womb. (Although some fetuses are viable at five or six months, they require specialized care to survive.) The fetus spends the final two months of gestation growing rapidly. At term (nine months), it will be about 20 inches long and weigh about seven pounds.

Figure B.1 ■ Embryonic and Fetal Development

Fetal development, or gestation, takes approximately 266 days from fertilization of the ovum to birth. These photographs chronicle various stages of the process.

(a) After ejaculation, several million sperm move through the cervical mucus toward the fallopian tubes; an ovum has descended into one of the tubes. En route to the ovum, millions of sperm are destroyed in the vagina, uterus, or fallopian tubes. Some go the wrong direction in the vagina and others swim into the wrong tube.

(b) The ovum has divided for the first time following fertilization; the mother's and father's chromosomes have united. In subsequent cell divisions the genes will be identified. After about a week the blastocyst will implant itself into the uterine lining.

(c) The embryo is five weeks old and is two-fifths of an inch long. It floats in the embryonic sac. The major divisions of the brain can be seen as well as an eye, hands, arms, and a long tail.

(d) The embryo is now seven weeks old and is almost an inch long. Its outer and inner organs are developing. It has eyes, a nose, a mouth, lips, and a tongue.

(e) At twelve weeks, the fetus is over three inches long and weighs almost an ounce.

(f) At sixteen weeks, the fetus is more than six inches long and weighs about seven ounces. All organs have been formed. The time that follows is now one of simple growth.

(a)

(b)

(c)

(d)

(e)

(f)

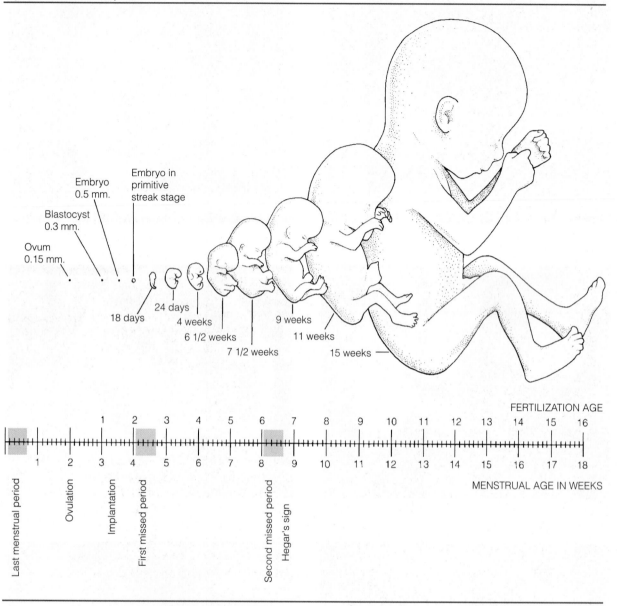

Ovum
0.15 mm.

Blastocyst
0.3 mm.

Embryo
0.5 mm.

Embryo in
primitive
streak stage

18 days

24 days

4 weeks

6 1/2 weeks

7 1/2 weeks

9 weeks

11 weeks

15 weeks

FERTILIZATION AGE

1 2 3 4 5 6 7 8 9 10 11 12 13 14 15 16

1 2 3 4 5 6 7 8 9 10 11 12 13 14 15 16 17 18

MENSTRUAL AGE IN WEEKS

Last menstrual period

Ovulation

Implantation

First missed period

Second missed period

Hegar's sign

Appendix C
The Budget Process

A budget is a plan for spending and saving. It requires you to estimate your available income for a particular period of time and decide how to allocate this income toward your expenses. A working budget can help you implement your money management plan. A well-planned budget does several things for you and your household. It can help you do the following:

- Prevent impulse spending
- Decide what you can or cannot afford
- Know where your money goes
- Increase savings
- Decide how to protect against the financial consequences of unemployment, accidents, sickness, aging, and death

A working budget need not be complicated or rigid. However, preparing one takes planning, and following one takes determination. You must do several things to budget successfully.

First, communicate with other members of your household, including older children. Consider each person's needs and wants so that all family members feel they are a part of the plan. Everyone may work harder to make the budget a success and be less inclined to overspend if they realize the consequences. When families fail to communicate about money matters, it is unlikely that a budget will reflect a workable plan.

Second, be prepared to compromise. This is often difficult. Newlyweds, especially, may have problems. Each may have been living on an individual income and not be accustomed to sharing or may have been in school and dependent on parents. If, for example, one wants to save for things and the other prefers buying on credit, the two will need to discuss the pros and cons of both methods and decide on a middle ground that each can accept. A plan cannot succeed unless there is a financial partnership.

Third, exercise willpower. Try not to indulge in unnecessary spending. Once your budget plan is made,

opportunities to overspend will occur daily. Each household member needs to encourage the others to stick to the plan

Fourth, develop a good record-keeping system. At first, all members of the household may need to keep records of what they spend. This will show how well they are following the plan and will allow intermediate adjustments in the level of spending. Record keeping is especially important during the first year of a spending plan, when you are trying to find a budget that works best for you. Remember: a good budget is flexible, requires little clerical time, and, most important, works for you.

Choosing a Budget Period

A budget may cover any convenient period of time—one month, three months, or a year. Make sure the period you choose is long enough to cover the bulk of household expenses and income. Remember: Not all bills are due monthly, and every household experiences some seasonal expenses. Most personal budgets are for twelve months. You can begin the twelve-month period at any time during the year. If this is your first budget, you may want to set up a trial plan for a shorter time to see how it works.

After setting up your plan, subdivide it into more manageable operating periods. For a yearly budget, divide income and expenses by 12, 24, 26, or 52, depending on your pay schedule or when your bills are due. Most paychecks are received weekly or every two weeks. Although most bills are due once a month, not all are due at the same time in the month. Try using each paycheck to pay your daily expenses and expenses that will be due within the next week or two. This way, you will be able to pay your bills on time. You may also want to allocate something from each paycheck toward large expenses that will be due soon.

Worksheet 1 ■ Estimating Your Income

Source	January	February	March	April	May	June
Net salary:*						
Household member 1						
Household member 2						
Household member 3						
Household member 4						
Social Security payments						
Pension payments						
Annuity payments						
Veterans' benefits						
Assistance payments						
Unemployment compensation						
Allowances						
Alimony						
Child support						
Gifts						
Interest						
Dividends						
Rents from real estate						
Other						
Monthly Totals						

*Net salary is the amount that comes into the household for spending and saving after taxes, Social Security, and other deductions.

Developing a Successful Budget

Step 1: Estimate Your Income

Total the money you expect to receive during the budget period. Use Worksheet 1 as a guide in estimating your household income. Begin with regular income that you and your family receive—wages, salaries, income earned from a farm or other business, Social Security benefits, pension payments, alimony, child support, veterans' benefits, public assistance payments, unemployment compensation, allowances, and any other income. Include variable income, such as interest from bank accounts and investments, dividends

from stock and insurance, rents from property you own, gifts, and money from any other sources.

If your earnings are irregular, it may be difficult to estimate your income. It is better to underestimate than to overestimate income when setting up a budget. Some households have sufficient income, but its receipt does not coincide with the arrival of bills. For these households, planning is very important.

Step 2: Estimate Your Expenses

After you have determined how much your income will be for the planning period, estimate your expenses. You may want to group expenses into one of three cat-

July	August	September	October	November	December	Yearly Totals

egories: fixed, flexible, or set-asides. Fixed expenses are payments that are basically the same amounts each month. Fixed regular expenses include such items as rent or mortgage payments, taxes, and credit installment payments. Fixed irregular expenses are large payments due once or twice a year, such as insurance premiums. Flexible expenses vary from one month to the next, such as amounts spent on food, clothing, utilities, and transportation. Set-asides are variable amounts of money accumulated for special purposes, such as for seasonal expenses, savings and emergency funds, and intermediate and long-term goals.

Use old records, receipts, bills, and canceled checks to estimate future expenses, if you are satisfied with what your dollars have done for you and your family in the past. If you are not satisfied, now is the time for change. Consider which expenses can be cut back and which expenses need to be increased. If you spent a large amount on entertainment, for example, your new budget may reallocate some of this money to a savings account to contribute to some of your future goals.

If you do not have past records of spending, or if this is your first budget, the most accurate way to find out how much you will need to allow for each expense is to keep a record of your household spending. Carry a pocket notebook in which you jot down expenditures during a week or pay period, and total the amounts at the end of each week. You may prefer to keep an account book in a convenient place at home and make entries in it. Kept faithfully for a month or two, the record can help you find out what you spend for categories such as food, housing, utilities, household operation, clothing, transportation, entertainment, and personal items. Use this record to estimate expenses in your plan for future spending. You also need to plan for new situations and changing conditions that increase or decrease expenses. For example, the cost of your utilities may go up.

Total your expenses for a year and divide to determine the amounts that you will have to allocate toward each expense during the budgeting period. Record your estimate for each budgetary expense in the space provided on Worksheet 2. Begin with the regular fixed expenses that you expect to have. Next, enter those fixed expenses that are due once or twice a year. Many households allocate a definite amount each budget period toward these expenses to spread out the cost.

One way to meet major expenses is to set aside money regularly before you start to spend. Keep your set-aside funds separate from other funds so you will not be tempted to spend them impulsively. If possible, put them in an account where they will earn interest. You may also plan at this point to set aside a certain amount toward the long-term and intermediate goals you listed on Worksheet 1. Saving could be almost as enjoyable as spending, once you accept the idea that saving money is not punishment, but instead a systematic way of reaching your goals. You do without some things now in anticipation of buying what will give you greater satisfaction later.

You may want to clear up debts now by doubling up on your installment payments or putting aside an extra amount in your savings fund to be used for this purpose. Also, when you start to budget, consider designating a small amount of money for emergencies. Extras always come up at the most inopportune times. Every household experiences occasional minor crises too small to be covered by insurance but too large to be absorbed into the day-to-day budget. Examples may be a blown-out tire or an appliance that needs replacing. Decide how large a cushion you want for meeting emergencies. As your fund reaches the figure you have allowed for emergencies, you can start saving for something else. Now, record money allocated for occasional major expenses, future goals, savings, emergencies, and any other set-asides in the space provided for them on Worksheet 2.

After you have entered your fixed expenses and your set-asides, you are ready to consider your flexible expenses. Consider including here a personal allowance, or "mad money," for each member of the household. A little spending money that does not have to be accounted for gives everyone a sense of freedom and takes some of the tedium out of budgeting.

Step 3: Balance

Now you are ready for the balancing act. Compare your total expected income with the total of your planned expenses for the budget period. If your planned budget equals your estimated future income, are you satisfied with this outcome? Have you left enough leeway for emergencies and errors? If your expenses add up to more than your income, look again at all parts of the plan. Where can you cut down? Where are you overspending? You may have to decide which things are most important to you and which ones can wait. You may be able to do some trimming on your flexible expenses.

Once you have cut back your flexible expenses, scan your fixed expenses. Maybe you can make some sizable reductions here, too. Rent is a big item in a budget. Some households may want to consider moving to a lower-priced apartment or making different living arrangements. Others may turn in a too-expensive car and seek less expensive transportation. Look back at Worksheet 1. You may need to reallocate some of this income to meet current expenses. Perhaps you may have to consider saving for some of your goals at a later date.

If you have cut back as much as you think you can or are willing to do and your plan still calls for more than you make, consider ways to increase your income. You may want to look for a better-paying job, or a part-time second job may be the answer. If only one spouse is employed, consider becoming a duel-earner family. The children may be able to earn their school lunch and extra spending money by doing odd jobs in your neighborhood, such as cutting grass or baby-sitting. Older children can work part-time on weekends to help out. Another possibility, especially for short-term problems, is to draw on savings. These are decisions each individual household has to make.

If your income exceeds your estimate of expenses—good! You may decide to satisfy more of your immediate wants or to increase the amount your family is setting aside for future goals.

Carrying Out Your Budget

After your plan is completed, put it to work. This is when your determination must really come into play. Can you and your family resist impulse spending?

Become a Good Consumer

A vital part of carrying out a budget is being a good consumer. Learn to get the most for your money, to recognize quality, to avoid waste, and to realize time costs as well as money costs in making consumer decisions.

Keep Accurate Records

Accurate financial records are necessary to keep track of your household's actual money inflow and outgo. A successful system requires cooperation from everyone in the household. Receipts can be kept and entered at the end of each budget period in a "Monthly Expense Record." It is sometimes a good idea to write on the back of each receipt what the purchase was for, who made it, and the date. Decide which family member will be responsible for paying bills or making purchases, and decide who will keep the record system up to date.

The household business record-keeping system does not need to be complex. The simpler it is, the more likely it will be kept current. Store your records in one spot—a set of folders in a file drawer or other fire-resistant box is a good place. You can assemble a folder for each of several categories, including budget, food, clothing, housing, insurance, investments, taxes, health, transportation, and credit. Use these folders for filing insurance policies, receipts, warranties, cancelled checks, bank statements, purchase contracts, and other important papers. Many households also rent a safe deposit box at the bank for storing deeds, stock certificates, and other valuable items.

Evaluating Your Budget

The information on Worksheet 2 can help you determine whether your actual spending follows your plan. If your first plan did not work in all respects, do not be discouraged. A budget is not something you make once and never touch again. Keep revising until the results satisfy you.

Dealing with Unemployment

Step 1: Take Time to Talk

Come right out and let your family know what is going on. Lay-off Plant closing? Depressed economy? Business down? Explain what happened. Break down the big words so that everyone understands, especially the kids.

Fill in everyone at a family meeting or on a one-to-one basis. The important thing is not to leave anyone in the dark. If a family meeting seems out of the question, take time to talk when cleaning up after meals, cutting or raking the lawn, or taking trips to the store. Don't sugar coat the facts or tell "fairy tales." Living with less money will force your family to make hard changes. Yet let your kids know that even though there's less money, they can still count on a loving family—maybe more loving than ever.

Step 2: Take Time to Listen

Let everyone have a say about what these changes mean to him or her. Especially now, kids should be seen and heard.

Listen to words and actions. Is someone suddenly having a lot of crying spells, sleeping in late all the time, acting mean, drinking heavily, withdrawing, abusing drugs, complaining of stomach pains?

Step 3: Find Out Who's Hurting

Let everyone say what he or she is really feeling from time to time.

Just repeat whatever you hear, right when it's said. Then look for a nod to see if you heard it right. Is someone feeling helpless, sad, unloved, confused, worried, frightened, angry, like a burden to the family?

Try not to say "You shouldn't fell that way" because someone may be in real pain. The best you can do is let your loved ones have their say and get it off their chests.

worksheet 2 ■ Expense Estimate and Budget Balancing Sheet, Fixed Expenses (Prepare for Each Month)

Month:	Amount Estimated	Amount Spent	Difference
Rent			
Mortgage			
Installments:			
Credit card 1			
Credit card 2			
Credit card 3			
Automobile loan			
Personal loan			
Student loan			
Insurance:			
Life			
Health			
Property			
Automobile			
Disability			
Set-asides:			
Emergency fund			
Major expenses			
Goals			
Savings and investments			
Allowances			
Education:			
Tuition			
Books			
Transportation:			
Repairs			
Gas and oil			
Parking and tolls			
Bus and taxi			
Recreation			
Gifts			
Other			
Total Fixed Expenses for Month			

Step 4: Let Your Feelings Out, Together or Alone

Give everyone in your family a space and time to let deep feelings out. Don't bottle them up or hide them from yourselves. If you're not comfortable showing others how you feel or fear you may strike someone who's dear to you, consider getting out of the house for a run or a brisk walk; having a good cry, alone; hitting a cushion or pillow; going to your room, shutting the door, and screaming; or all of the above.

Step 5: Solve Problems Together

Every week, look at the changes taking place in your household, and work out ways to deal with them. Working together as a team, your family can do more than survive. It can grow together and come through stronger.

Decide together things like these: what we can't afford now; what things we can do for family fun that will not cost a lot of money; who will do what chores around the house; how we will all get by with less. If your discussions break down, go back to Step 1.

If you have a lot of trouble going through these steps, professional help may be what you need. Call and make an appointment with the family service agency nearest you. Whether or not you have money to pay for the services, the agency will do its best to help your family. Remember: You are not alone.

GLOSSARY

A

abstinence Refraining from sexual intercourse, often on religious or moral grounds.

accommodation According to Jean Piaget, the process by which a child makes adjustments in his or her cognitive framework in order to incorporate new experiences.

acquaintance rape Rape in which the assailant is personally known to the victim, usually in the context of a dating relationship. Also known as *date rape*.

acquired immunodeficiency syndrome (AIDS) An infection caused by the human immunodeficiency virus (HIV), which suppresses and weakens the immune system, leaving it unable to fight opportunistic infections.

adolescence The social and psychological state occurring during puberty.

affiliated kin Unrelated individuals who are treated as if they were related.

agape [AH-ga-pay] According to sociologist John Lee's styles of love, altruistic love.

AIDS See acquired immunodeficiency syndrome.

alimony Court-ordered monetary support to a spouse or former spouse following separation or divorce.

anal eroticism Sexual activities involving the anus.

anal intercourse Penetration of the anus by the penis.

anonymity A state or condition requiring that no one, including the researcher can connect particular responses to the individuals who provided them.

anti-gay prejudice Strong dislike, fear, or hatred of gay men and lesbians because of their homosexuality. See also *homophobia*.

assimilation In Jean Piaget's cognitive developmental theory, the process through which the developing child makes new information compatible with his or her world understanding.

attachment theory of love A theory maintaining that the degree and quality of an infant's attachment to his or her primary caregiver is reflected in his or her love relationships as an adult.

authoritarian child rearing A parenting style characterized by the demand for absolute obedience.

authoritative child rearing A parenting style that recognizes the parent's legitimate power and also stresses the child's feelings, individuality, and need to develop autonomy.

autoeroticism Erotic behavior involving only the self; usually refers to masturbation but also includes erotic dreams and fantasies.

B

basic conflict Pronounced disagreement about fundamental roles, tasks, and functions. Cf. *nonbasic conflict*.

battering A violent act directed against another, such as hitting, slapping, beating, stabbing, shooting, or threatening with weapons.

bias A personal leaning or inclination.

binuclear family A postdivorce family with children, consisting of the original nuclear family divided into two families, one headed by the mother, the other by the father; the two "new" families may be either single-parent or stepfamilies.

bipolar gender role model The traditional view of masculinity and femininity in which male and female gender roles are seen as polar opposites, with males possessing exclusively instrumental traits and females possessing exclusively expressive traits.

birth rate The number of births per year per thousand people in a given community or group. Cf. *fertility rate*.

bisexuality Sexual involvement with both sexes, usually sequentially rather than during the same time period.

blended family A family in which one or both partners have a child or children from an earlier marriage or relationship; a stepfamily. See also *binuclear family*.

boomerang generation Individuals who, as adults, return to their family home and live with their parents.

bundling A colonial Puritan courtship custom in which a couple slept together with a board separating them.

C

caregiver role In family caregiving, the role of the person who provides the most ongoing physical work and decision making relating to the one who is being cared for.

case-study method In clinical research, the in-depth examination of an individual or small group in some form of psychological treatment in order to gather data and formulate hypotheses.

child-free marriage A marriage in which the partners have chosen not to have children.

child sexual abuse Any sexual interaction, including fondling, erotic kissing, oral sex, or genital penetration, that occurs between an adult (or older adolescent) and a prepubertal child.

child support Court-ordered financial support by the non-custodial parent to pay or assist in paying child-rearing expenses incurred by the custodial parent.

clan A group of families related along matrilineal or patrilineal descent lines, regarded as the basic family unit in some cultures.

clinical research The in-depth examination of an individual or small group in clinical treatment in order to gather data and formulate hypotheses. See also *case-study method.*

closed field A setting in which potential partners may meet, characterized by a small number of people who are likely to interact, such as a class, dormitory, or party. Cf. *open field.*

cognition The mental processes, such as thought and reflection, that occur between the moment we receive a stimulus and the moment we respond to it.

cognitive developmental theory A theory of socialization associated with Swiss psychologist Jean Piaget in which the emphasis was placed on the child's developing abilities to understand and interpret their surroundings.

cohabitation The sharing of living quarters by two heterosexual, gay, or lesbian individuals who are involved in an ongoing emotional and sexual relationship. The couple may or may not be married.

coitus The insertion of the penis into the vagina and subsequent stimulation; sexual intercourse.

coming out For gay, lesbian, and bisexual individuals, the process of publicly acknowledging one's sexual orientation.

common couple violence Sociologist Michael Johnson's term for the more routine forms of partner violence that results from disputes and disagreements, and for which there is a high degree of gender symmetry.

companionate love A form of love emphasizing intimacy and commitment.

companionate marriage A marriage characterized by shared decision making and emotional and sexual expressiveness.

complementary needs theory A theory of mate selection suggesting that we select partners whose needs are different from and/or complement our own needs.

conceptualization The specification and definition of concepts used by the researcher.

conduct of fatherhood Men's actual participation in raising their children.

confidentiality An ethical rule according to which the researcher knows the identities of participants and can connect what was said to who said it but promises not to reveal such information publicly.

conflict theory A social theory that views individuals and groups as being basically in competition with each other. Power is seen as the decisive factor in interactions.

conjugal family A family consisting of husband, wife, and children. See also *nuclear family.*

conjugal relationship A relationship formed by marriage.

consanguineous relationship A relationship formed by common blood ties.

continuous coverage system The responsibility facing new parents in which someone must be available to care for their infant around the clock or provide alternate caregiving arrangements. This new temporal reality introduces conflict over which parent will be most directly involved and how much free time each will retain.

coprovider families Families that are dependent on economic activity from both men and women.

couvade The psychological or ritualistic assumption of the symptoms of pregnancy and childbirth by the male.

covenant marriage A new antidivorce reform of legal marriage in which couples acknowledge the lifelong nature of their marital commitments. They are required to undergo premarital counseling, promise to seek marital counseling if they experience serious marital difficulties, and pledge to divorce only under extreme hardships via a fault-based divorce.

crude birthrate A statistic reflecting the number of births per thousand people in the population.

crude divorce rate A statistical measure of divorce calculated on the basis of the number of divorces per 1,000 people in the population.

culture of fatherhood Ralph LaRossa's term for the beliefs we have about the roles, responsibilities, and involvement of fathers in raising their children. LaRossa noted that these beliefs have changed more dramatically than has the conduct of fatherhood.

cunnilingus Oral stimulation for female genitals.

cycle of violence According to Lenore Walker's research, the recurring three-phase battering cycle of (1) tension building, (2) explosion, and (3) reconciliation.

D

date rape Rape in which the assailant is personally known to the victim, usually in the context of a dating relationship. Also known as *acquaintance rape.*

Defense of Marriage Act Federal legislation signed into law by President Clinton denying recognition to same-sex couples, should any state legalize same-sex marriage.

dependent variable A variable that is observed or measured in an experiment and may be affected by another variable. See independent variable.

developmental systems approach An approach to human development that recognizes the importance of the individual's interactions within a complex and changing family system and within the numerous systems of the larger society.

division of labor The interdependence of persons with specialized tasks and abilities. Within the family, labor is traditionally divided along gender lines. See also *complementary marriage model.*

divorce mediation The process in which a mediator (counselor) assists a divorcing couple in resolving personal, legal, and parenting concerns in a cooperative manner.

double standard of aging The devaluation of women in contrast to men in terms of attractiveness as they age.

duration-of-marriage effect The accumulation over time of various factors, such as poor communication, unresolved conflicts, role overload, heavy work schedules, and child-rearing responsibilities, that negatively affect marital satisfaction.

dyspareunia Painful sexual intercourse.

E

economic distress The stressful aspects of the economic life of individuals or families, including unemployment, poverty, and worrying about money.

ego In psychoanalytic theory, the part of the personality that is rational and mediates between the demands of the id and the constraints imposed by society. See also *id* and *superego*.

egocentric fallacy The mistaken belief that one's own personal experience and values are those of others in general.

empty nest The experience of parents when the last grown child has left home. The "empty nest syndrome," in which the mother becomes depressed after the children have gone, is believed to be more of a myth than a reality.

endogamy Marriage within a particular group. Cf. *exogamy*.

engagement A pledge to marry.

environmental influences The wider context and external influences on families that are the focus of family ecological theory.

equity theory A theory emphasizing that social exchanges must be fair or equally beneficial over the long run.

erectile dysfunction Inability or difficulty in achieving erection.

eros 1. From the Greek *eros* [love], the fusion of love and sexuality. 2. According to sociologist John Lee's styles of love, the passionate love of beauty.

ethical guidelines Standards agreed upon by professional researchers. These guidelines protect the privacy and safety of individuals who provide information in a research setting.

ethnic group A large group of people distinct from others because of cultural characteristics, such as language, religion, and customs, transmitted from one generation to another. See also *minority group* and *racial group*.

ethnocentric fallacy (also ethnocentrism) The belief that one's own ethnic group, nation, or culture is inherently superior to others. See also *racism*.

exchange theory See *social exchange theory*.

experimental research A research method involving the isolation of specific factors (variables) under controlled circumstances to determine the effects of each factor.

expressive trait A supportive or emotional personality trait or characteristic.

extended family The family unit of parent(s), child(ren), and other kin, such as grandparents, uncles, aunts, and cousins. See also *conjugal extended family* and *consanguineous extended family*.

extended household A household composed of several different families.

extrafamilial sexual abuse Child sexual abuse that is perpetrated by nonrelated individuals. Cf. *intrafamilial sexual abuse*.

extramarital sex Sexual activities, especially sexual intercourse, occurring outside the marital relationship.

F

fallacy A fundamental error in reasoning that affects our understanding of a subject.

familialism A pattern of social organization in which family loyalty and strong feelings for the family are important.

family A unit of two or more persons, of which one or more may be children who are related by blood, marriage, or affiliation and who cooperate economically and may share a common dwelling place.

family life cycle A developmental approach to studying families, emphasizing the family's changing roles and relationships at various stages, beginning with marriage and ending when both spouses have died.

family of cohabitation The family formed by two people living together whether married or unmarried; may include children or stepchildren.

family of orientation The family in which a person is reared as a child. Cf. *family of procreation*.

family of origin See *family of orientation*.

family of procreation The family formed by a couple and their child or children. See also *family of cohabitation*.

family policy A set of objectives concerning family well-being and specific measures initiated by government to achieve them.

family systems theory A theory viewing family structure as created by the pattern of interactions between its various subsystems, and individual actions as being strongly influenced by the family context.

family work The unpaid work that is undertaken by family members to sustain the family, such as housework, laundry, shopping, yard maintenance, budgeting and bill-paying, and care of children, the sick, and the elderly.

feedback In communication, an ongoing process in which participants and their messages produce a result and are subsequently modified by the result.

fellatio Oral stimulation of the male genitals.

feminism 1. The principle that women should have equal political, social, and economic rights with men. 2. The social movement to obtain for women political, social, and economic equality with men.

feminization of poverty The shift of poverty to females, primarily as a result of high divorce rates and births to unmarried women.

fertility rate In a given year, the number of live births per 1,000 women aged 15–44 years. See also *birth rate*.

fictive kin ties The extension of kinshiplike attributes to non-blood relationships (such as friends or neighbors) to demonstrate their importance and to symbolize the mutual reciprocity found within them.

field of eligibles A group of individuals of the same general background and age who are culturally approved potential marital partners.

foreplay Erotic activity prior to coitus, such as kissing, caressing, sex talk, and oral/genital contact; petting.

friendship An attachment between people; the foundation for a strong love relationship.

G

game stage According to symbolic interactionist theories of George Herbert Mead, the stage in self-development in which one can understand a variety of other perspectives simultaneously.

gender The division into male and female, often in a social sense; sex.

gender identity The psychological sense of whether one is male or female.

gender ideology Arlie Hochschild's term for what individuals believe they ought to do as husbands or wives, and how they believe paid and unpaid work should be divided.

gender-rebellion feminism Versions of feminism that emphasize the interconnectedness between multiple inequalities (race, class, sexual orientation, age, and gender), and see gender inequality as only one aspect of wider social inequality.

gender-reform feminism Versions of feminism that stress how similar women and men are and emphasize the need for equal rights and opportunities for both genders.

gender-resistant feminism Versions of feminism that advocate separatist strategies, wherein women establish women-only social institutions and settings.

gender role The culturally assigned role that a person is expected to perform based on male or female gender.

gender-role attitude A personal belief regarding appropriate male and female personality traits and behaviors.

gender-role behavior An actual activity or behavior in which males or females engage according to their gender role.

gender-role stereotype A rigidly held and oversimplified belief that all males and females possess distinctive psychological and behavioral traits as a result of their gender.

gender theory A theory in which gender is viewed as the basis of hierarchal social relations that justify greater power to males.

H

halo effect The tendency to infer positive characteristics or traits based on a person's physical attractiveness.

hegemonic models of gender Dominant models of masculinity and femininity.

heterogamy Marriage between those with different social or personal characteristics. Cf. *homogamy.*

heterosexuality Sexual orientation toward members of the opposite sex.

HIV See *human immunodeficiency virus.*

homemaker role A family role usually allocated to women, in which they are primarily responsible for home management, child rearing, and the maintenance of kin relationships. Traditionally the role is associated with economic dependency and has primacy over other female roles.

homeostasis A social group's tendency to maintain internal stability or balance and to resist change.

homoeroticism Erotic attraction to members of the same sex.

homogamy Marriage between those with similar social or personal characteristics. Cf. *heterogamy.*

homophobia Irrational or phobic fear of gay men and lesbians.

homosexuality Sexual orientation toward members of the same sex. See also *gay male* and *lesbian.*

honeymoon effect The tendency of newly married couples to overlook problems, including communication problems.

hostile environment An environment created through sexual harassment in which the harassed person's ability to learn or work is negatively influenced by the harasser's actions.

human immunodeficiency virus (HIV) The virus causing AIDS.

hypergamy A marriage in which one's spouse is of a higher social class or rank.

hypogamy A marriage in which one's spouse is from a lower social standing.

hypothesis An unproven theory or proposition tentatively accepted to explain a collection of facts.

I

id In psychoanalytic theory, the part of the personality that seeks to gratify pleasurable needs, especially sexual ones. See also *ego* and *superego.*

identity bargaining The process of role adjustment in a relationship, involving identifying with a role, having the role validated by others, and negotiating with the partner to make changes in the role.

ideology of "intensive mothering" The term used by Sharon Hays to refer to beliefs about what mothers ought to provide their children. The key elements of intensive mothering are full-time attention, self-sacrificing devotion, and expert-guided, labor-intensive involvement with the child, whose needs are more pressing than those of mothers.

incest Sexual intercourse between individuals too closely related to marry, usually interpreted to mean father/daughter, mother/son, or brother/sister. See also *intrafamilial sexual abuse.*

independent variable A variable that may be changed or manipulated in an experiment.

infant mortality rate The number of deaths for every 1,000 live births.

instrumental trait A practical or task-oriented personality trait or characteristic.

interaction In communication, a reciprocal act that takes place between at least two people.

intermittent extended family The family that is formed when a family takes in other relatives in times of need.

intervening variable A variable that is affected by the independent variable and in turn affects the dependent variable.

intrafamilial sexual abuse Child sexual abuse that is perpetrated by related individuals, including steprelatives. See also *incest;* cf. *extrafamilial sexual abuse.*

J

jealousy An aversive response occurring because of a partner's or other significant person's real, imagined, or likely involvement with or interest in another person.

joint custody Custody arrangement in which both parents are responsible for the care of the child. Joint custody takes two forms: *joint legal custody* and *joint physical custody.* See also *sole custody* and *split custody.*

joint legal custody Joint custody in which the child lives primarily with one parent but both parents jointly share in important decisions regarding the child's education, religious training, and general upbringing.

joint physical custody Joint custody in which the child lives with both parents in separate households and spends more or less equal time with each parent.

K

kaddish In Judaism, a form of prayer.

kinship system The social organization of the family conferring rights and obligations based on an individual's status.

L

lesbian A female sexually oriented toward other females.

lesbian separatist A lesbian interested in creating a separate "womyn's" culture distinct from both heterosexual and gay culture. The lesbian separatist movement was strongest in the late 1960s through the early 1980s.

life chances Opportunities to enjoy a healthy and fulfilling life that are affected by one's social class standing.

looking-glass self In the symbolic interactionist theory of Charles Cooley, the looking-glass self refers to the influence of others' perceptions of us on how we come to perceive ourselves.

lower-middle class The socioeconomic class made up of white-collar service workers with incomes between $25,000 and $50,000, who own or rent more modest homes than the upper-middle class and purchase more affordable automobiles

ludus [LOO-dus] According to sociologist John Lee's styles of love, playful love.

M

macrosystem In ecological terms, the broadest level of environmental influences, encompassing the laws, customs, attitudes, and belief systems of the wider society, all of which influence individual development and experience.

mania According to sociologist John Lee's styles of love, obsessive love.

marital rape Forced sexual contact by a husband with his wife; legal definitions of marital rape differ among states.

marriage The legally recognized union between a man and woman in which economic cooperation, legitimate sexual interactions, and the rearing of children may take place.

marriage squeeze The phenomenon in which there are greater numbers of marriageable women than marriageable men, particularly among older women and African-American women. See also *marriage gradient.*

masturbation Manual or mechanical stimulation of the genitals by self or partner; a form of autoeroticism.

matriarchal Pertaining to the mother as the head and ruler of a family. Cf. *patriarchal.*

matriarchy A form of social organization in which the mother or eldest female is recognized as the head of the family, kinship group, or tribe, and descent is traced through her. Cf. *patriarchy.*

matrilineal Descent or kinship traced through the mother. Cf. *patrilineal.*

menopause Cessation of menses for at least one year as a result of aging.

minority group A social category composed of people whose status places them at economic, social, and political disadvantage. Cf. *majority group;* see also *ethnic group.*

minority status Social rank having unequal access to economic and political power.

modeling The process of teaching or learning using imitation.

monogamy 1. The practice of having only one husband or wife at a time. 2. [colloq.] Sexual exclusiveness.

N

neurosis A psychological disorder characterized by anxiety, phobias, and so on.

no-fault divorce The dissolution of marriage because of irreconcilable differences for which neither party is held responsible.

nonbasic conflict Pronounced disagreement about nonfundamental or situational issues. Cf. *basic conflict.*

nonmarital sex Sexual activities, especially sexual intercourse, that take place among older single individuals. Cf. *premarital sex* and *extramarital sex.*

nonverbal communication Communication of emotion by means other than words, such as touch, body movement, and facial expression.

nuclear family The basic family building block, consisting of a mother, father, and at least one child; in popular usage, used interchangeably with *traditional family.* Some anthropologists argue that the basic nuclear family is the mother and child dyad.

O

objective statement A factual statement presenting information based on scientifically measured findings, not on opinions or personal values.

objectivity Suspending the beliefs, biases, or prejudices we have about a subject until we have really understood what is being said.

observational research Research method using unobtrusive, direct observation.

open adoption A form of adoption in which the birth mother has an active part in choosing the adoptive parents; there is a certain amount of information exchanged between the birth mother and the adoptive parents, and there may be some form of continuing contact between the birth mother, the child, and the adoptive family following adoption.

open field A setting in which potential partners may not be likely to meet, characterized by large numbers of people who do not ordinarily interact, such as a beach, shopping mall, or large university campus. Cf. *closed field*.

open marriage A marriage in which the partners agree to allow one another to have openly acknowledged and independent sexual relationships outside the marriage.

operationalization The identification and/or development of research strategies to observe or measure concepts.

opinion An unsubstantiated belief or conclusion based on personal values or biases.

outing The act of publicly disclosing the sexual orientation of gays, lesbians, or bisexuals.

P

parental image theory A theory of mate selection suggesting that we select partners similar to our opposite-sex parents.

Parents' Bill of Rights Sylvia Hewlett and Cornel West's recommended policy initiatives and reforms to improve the conditions under which parents attempt to raise children.

passionate love Intense, impassioned love. Cf. *companionate love*.

patriarchal Pertaining to the father as the head and ruler of a family. Cf. *matriarchal*.

patriarchy A form of social organization in which the father or eldest male is recognized as the head of the family, kinship group, or tribe, and descent is traced through him. Cf. *matriarchy*.

patrilineal Descent or kinship traced through the father. Cf. *matrilineal*.

pedophilia Adult sexual attraction to prepubescent children that is intense and recurring; an adult's use of children for sexual purposes.

peer A person of equal status, as in age, class, position, or rank.

peer marriages Marriages built on principles of equity, equality, and "deep friendship," between spouses.

Husbands and wives divide up housework and childcare more equally than is typical (between 50-50 and 60-40), and exhibit high levels of empathy and communication.

permissive child rearing A parenting style stressing the child's autonomy and freedom of expression, often over the needs of the parents.

phenotype A set of genetically determined anatomical and physical characteristics, such as skin and hair color and facial structure.

play stage In Mead's theory of self-development, the stage in which children "play at" being specific other people, taking on one role or viewpoint at a time.

pleasuring The giving and receiving of sensual pleasure through nongenital touching.

polyandry The practice of having more than one husband at the same time. See also *polygamy;* cf. *polygyny*.

polygamy The practice of having more than one husband or wife at the same time; plural marriage. See also *polyandry, polygyny,* and *consanguineous extended family*.

polygyny The practice of having more than one wife at the same time. See also *polygamy;* cf. *polyandry*.

postpartum period A period of about three months following childbirth during which critical family adjustments are made.

power The ability to exert one's will, influence, or control over another person or group.

pragma Practical love, according to sociologist John Lee's styles of love.

predictive divorce rate A statistical calculation of the expected divorce rate of people who enter marriage in a given year.

premarital sex Sexual activities, especially sexual intercourse, prior to marriage, especially among young, never-married individuals.

principle of least interest A theory of power in which the person less interested in sustaining a relationship has the greater power.

profamily movement A social movement emphasizing conservative family values, such as traditional gender roles, authoritarian child rearing, premarital virginity, and opposition to abortion.

prototype In psychology, concepts organized into a mental model.

proximity Nearness to another in terms of both physical space and time.

psychoanalytic theory The Freudian model of personality development, in which maturity is seen as the ability to gain control over one's unconscious impulses.

psychosexual development The growth of the psychological aspects (such as attitudes and emotions) of sexuality that accompany physical growth.

psychosocial theory A theory of human psychological development that emphasizes the role of family and society in such development.

PWA Person with AIDS.

Q

qualitative research Small groups or individuals are studied in an in-depth fashion.

quantitative research Samples taken from a large number of subjects.

R

racial group A large group of people defined as distinct because of their phenotype (genetically transmitted anatomical and physical characteristics, especially facial structure and skin color). Cf. *ethnic group.*

racism The practice of discrimination and subordination based on the belief that race determines character and abilities. See also *ethnocentric fallacy* and *ethnocentrism.*

rape Sexual act against a person's will or consent as defined by law, usually including sexual penetration by the penis or other object; it may not, however, necessarily include penile penetration of the vagina. Also known as *sexual assault.* See also *acquaintance rape* and *marital rape.*

ratio measure of divorce A statistical calculation reflecting the ratio of the number of divorces in a given year to the number of marriages in that same year.

reactive jealousy Jealousy that occurs when a partner's past, present, or anticipated involvement with another is revealed. Cf. *suspicious jealousy.*

refined divorce rate A statistic reflecting the number of divorces in a given year for every thousand married couples.

rejection sensitivity The tendency to anticipate and overreact to rejection.

relative love and need theory A theory of power in which the person gaining the most from a relationship is the most dependent.

remarriage A marriage in which one or both partners have been previously married.

residential propinquity A pattern in which the chances of two people marrying are greater the closer they live to each other.

role The pattern of behavior expected of a person in a group or culture as a result of his or her social position, such as husband or wife in a family.

role conflict See *interrole conflict.*

role interference See *interrole conflict.*

role overload The experience of having more prescribed activities in one or more roles than can be comfortably or adequately performed. See also *role strain.*

role reversed couples Couples in which women are the sole or dominant wage earners, while men stay home and assume domestic and child-care responsibilities traditionally performed by women.

role strain Difficulties, tensions, or contradictions experienced in performing a role, often because of multiple role demands. See also *interrole conflict* and *role overload.*

S

sandwich generation Individuals and families who care for both their own children and their aging parents at the same time.

scientific method A method of investigation in which a hypothesis is formed on the basis of impartially gathered data and is then tested empirically.

secondary data analysis Use of research gathered by public sources of information.

second shift Arlie Hochschild's term for the domestic responsibilities awaiting employed women after their paid work hours are completed.

self-care Children under age 14 caring for themselves at home without supervision by an adult or older adolescent.

self-disclosure The revelation of deeply personal information about oneself to another.

separation distress A psychological state following separation that may be characterized by depression, anxiety, intense loneliness, or feelings of loss.

sex 1. Biologically, the division into male and female. 2. Sexual activities.

sexual coercion Nonconsensual sexual behavior such as rape, sexual assault, and sexual harassment

sexual dysfunction Recurring problems in sexual functioning that cause distress to the individual or partner; may have a physiological or psychological basis.

sexual enhancement Any means of improving a sexual relationship, including developing communication skills, fostering a positive attitude, giving a partner accurate and adequate information, and increasing self-awareness.

sexual harassment Deliberate or repeated unsolicited verbal comments, gestures, or physical contact that is sexual in nature and unwelcomed by the recipient. Two types of sexual harassment involve (1) the abuse of power and (2) the creation of a hostile environment. See also *hostile environment.*

sexual intercourse Coitus; heterosexual penile/vaginal penetration and stimulation.

sexual orientation Sexual identity as heterosexual, gay, lesbian, or bisexual.

sexual script A culturally approved set of expectations as to how one should behave sexually as male or female and as heterosexual, gay, or lesbian.

shift couples Two-earner households in which spouses work different, often nonoverlapping shifts, so that one partner is home while the other is at work.

SIDS See *sudden infant death syndrome.*

single-parent family A family with children, created by divorce or unmarried motherhood, in which only one parent is present. A family consisting of one parent and one or more children.

social classes Groupings of people who share a common economic position by virtue of their wealth, income, power, and prestige, and thus have similar social and familial experiences.

social construct An idea or concept created by society. As applied to gender, it refers to the ways in which we define gender and then act on our beliefs about it.

social exchange theory A theory that emphasizes the process of mutual giving and receiving of rewards, such as love or sexual intimacy, in social relationships, calculated by the equation Reward − Cost = Outcome.

social integration The degree of interaction between individuals and the larger community.

social learning theory A theory of human development that emphasizes the role of cognition (thought processes) in learning.

social mobility A term used to refer to movement up or down the socioeconomic ladder, which can occur within a person's lifetime or between generations.

social role A socially established pattern of behavior that exists independently of any particular person, such as the husband or wife role or the stepparent role.

socialization The shaping of individual behavior to conform to social or cultural norms.

socioeconomic status A term used to refer to the combined effects of income, occupational prestige, wealth, education, and income on a person's lifestyle and opportunities.

sole custody Child custody arrangement in which only one parent has both legal and physical custody of the child. See also *joint custody* and *split custody*.

spirit marriage In Canton, China, a marriage of two deceased persons, arranged by their families to provide family continuity.

split custody Custody arrangement when there are two or more children in which custody is divided between the parents, the mother generally receiving the girls and the father receiving the boys.

spontaneous abortion The natural but fatal expulsion of the embryo or fetus from the uterus; miscarriage.

stepfamily A family in which one or both partners have a child or children from an earlier marriage or relationship. Also known as a *blended family;* see also *binuclear family.*

stepparent role The role a stepparent forges for herself or himself within the stepfamily as there is no such role clearly defined by society.

stereotype A rigidly held, simplistic, and overgeneralized view of individuals, groups, or ideas that fails to allow for individual differences and is based on personal opinion and bias rather than critical judgment.

stimulus-value-role theory A three-stage theory of romantic development proposed by Bernard Murstein: (1) stimulus brings people together; (2) value refers to the compatibility of basic values; (3) role has to do with each person's expectations of how the other should fulfill his or her roles.

storge [STOR-gay] According to sociologist John Lee's styles of love, companionate love.

structural functionalism A sociological theory that examines how society is organized and maintained by examining the functions performed by its different structures. In marriage and family studies, structural functionalism examines the functions the family performs for society, the functions the individual performs for the family, and the functions the family performs for its members.

subsystem A system that is part of a larger system, such as family, and religious and economic systems being subsystems of society and the parent/child system being a subsystem of the family.

sudden infant death syndrome (SIDS) The death of an apparently healthy infant during its sleep from unknown causes.

superego In psychoanalytic theory, the part of the personality that has internalized society's demands and acts as a sort of conscience to control the id. See also *ego* and *id*.

survey research Research method using questionnaires or interviews to gather information from small, representative groups and to infer conclusions that are valid for larger populations.

suspicious jealousy Jealousy that occurs when there is either no reason for suspicion or only ambiguous evidence that a partner is involved with another. Cf. *reactive jealousy.*

symbolic interaction A theory that focuses on the subjective meanings of acts and how these meanings are communicated through interactions and roles to give shared meaning.

T

TANF See *Temporary Assistance to Needy Families.*

taking the role of the other In symbolic interactionist theories of socialization, the ability to see things from someone else's vantage point.

Temporary Assistance to Needy Families (TANF) The current government program designed to financially assist families with children during times of poverty.

theory A set of general principles or concepts used to explain a phenomenon and to make predictions that may be tested and verified experimentally.

total fertility rate A complicated statistic that estimates the number of births a hypothetical group of 1,000 women would have if they experience across their childbearing years the age-specific rates for a given year.

traditional family In popular usage, an intact, married two-parent family with at least one child, which adheres to conservative family values; an idealized family. Popularly used interchangeably with *nuclear family.*

traumatic sexualization The process of developing inappropriate or dysfunctional sexual attitudes, behaviors, and feelings by a sexually abused child.

trial marriage Cohabitation with the purpose of determining compatibility prior to marriage.

Sternberg's triangular theory of love A theory developed by Robert Sternberg emphasizing the dynamic quality of love as expressed by the interrelationship of three elements: intimacy, passion, and decision/commitment.

trust Belief in the reliability and integrity of another.

two-person career An arrangement in which it takes the efforts of two spouses to ensure the career success of one. One spouse, typically the husband, can devote himself or herself fully to career pursuits because of the help and assistance received from his or her spouse. This help and assistance includes taking care of all family and domestic needs, but also often includes unpaid supportive roles (such as entertaining business colleagues).

U

unrequited love Love that is not returned.

upper-middle class A socioeconomic class consisting of college-educated, highly paid professionals (for example, lawyers, doctors, engineers) who have annual incomes that may reach into the hundreds of thousands of dollars.

V

value judgment An evaluation based on ethics or morality rather than on objective observation.

values The social principles, goals, or standards held as acceptable by an individual, family, or group.

variable In experimental research, a factor, such as a situation or behavior, that may be manipulated. See also *independent variable* and *dependent variable.*

Viagra An oral medication for erectile dysfunction that has restored many men's abilities to engage in sexual activity.

violence An act carried out with the intention of causing physical pain or injury to another.

virginity The state of not having engaged in sexual intercourse.

W

wheel theory of love A theory developed by Ira Reiss holding that love consists of four interdependent processes: rapport, self-revelation, mutual dependency, and intimacy fulfillment.

work spillover The effect that employment has on time, energy, activities, and psychological functioning of workers and their families.

working class A socioeconomic class comprised of skilled laborers with high school or vocational educations. The working class lives somewhat precariously, with little savings and few liquid assets should illness or job loss occur.

BIBLIOGRAPHY

Abbey, Antonia. "Acquaintance Rape and Alcohol Consumption on College Campuses: How Are They Linked?" *Journal of American College Health* 39, 4 (January 1991): 165–169.

Abbey, A., and R. J. Harnish. "Perception of Sexual Intent: The Role of Gender, Alcohol Consumption, and Rape Supportive Attitudes." *Sex Roles* 32, 5–6 (March 1995): 297–313.

Abelsohn, David. "A 'Good Enough' Separation: Some Characteristic Operations and Tasks." *Family Process* 31, 1 (1992): 61–83.

Absi-Semaan, Nada, Gail Crombie, and Corinne Freeman. "Masculinity and Femininity in Middle Childhood: Developmental and Factor Analyses." *Sex Roles: A Journal of Research* 28 (1993): 187–207.

Acitelli, Linda K. "Gender Differences in Relationship Awareness and Marital Satisfaction among Young Married Couples." *Personality and Social Psychology Bulletin* 18, 1 (1992): 102–110.

Acker, Joan. "From Sex Roles to Gendered Institutions." *Contemporary Sociology* 21, 5 (1992): 565–570.

Ackerman, Diane. *The Natural History of Love.* New York: Random House, 1994.

Adams, Bert. "The Family Problems and Solutions." *Journal of Marriage and the Family* 47, 3 (August 1985): 525–529.

Adams, David. "Identifying the Assaultive Husband in Court: You Be the Judge." *Response to the Victimization of Women and Children* 13, 1 (1990): 13–16.

Adelmann, Pamela K. "Psychological Well-Being and Home-maker vs. Retiree Identity among Older Women." *Sex Roles* 29, 3–4 (1993): 195–212.

Adler, Nancy, Susan Hendrick, and Clyde Hendrick. "Male Sexual Preference and Attitudes toward Love and Sexuality." *Journal of Sex Education and Therapy* 12, 2 (September 1989): 27–30.

"A Fatal, Unknowing Dose: With GHB, Line Between a High and Death is Narrow." ABC-News.Go.com., March 2, 2000.

Ahlander, N., and K. Bahr, "Beyond Drudgery, Power, and Equity: Toward an Expanded Discourse on the Moral Dimensions of Housework in Families." *Journal of Marriage and the Family* 57,1 (February 1995): 54–68.

Ahn, Helen Noh, and Neil Gilbert. "Cultural Diversity and Sexual Abuse Prevention." *Social Service Review* 66, 3 (September 1992): 410–428.

Ahrons, Constance, and Roy Rodgers. *Divorced Families: A Multidisciplinary View.* New York: Norton, 1987.

Ainsworth, Mary D., M. D. Blehar, E. Waters, and S. Wall. *Patterns of Attachment: A Psychological Study of the Strange Situation.* Hillsdale, NJ: Lawrence Erlbaum, 1978.

Alapack, Richard. "The Adolescent First Kiss." *Humanistic Psychologist* 19, 1 (March 1991): 48–67.

Albrecht, S. L. "Reactions and Adjustments to Divorce: Differences in the Experiences of Males and Females." *Family Relations* 29 (1980): 59–70.

Aldous, Joan. *Family Careers: Developmental Change in Families.* New York: Wiley, 1978.

———. "American Families in the 1980s: Individualism Run Amok?" *Journal of Family Issues* 8, 4 (December 1987): 422–425.

———. "Perspectives on Family Change." *Journal of Marriage and the Family* 52, 3 (August 1990): 571–583.

Aldous, Joan, and Wilfried Dumon. "Family Policy in the 1980s: Controversy and Consensus." In *Contemporary Families: Looking Forward, Looking Back,* edited by A. Booth. Minneapolis: National Council on *Family Relations,* 1991.

Aldous, Joan. *Family Careers: Rethinking the Developmental Perspective.* Thousand Oaks, Calif.: Sage Publications, 1996.

Aldous, Joan. "New Views of Grandparents in Intergenerational Context." *Journal of Family Issues* 16, 1 (January 1995): 104–122.

Aldous, Joan, and Gail M. Mulligan "Fathers' Child Care and Children's Behavior Problems: A Longitudinal Study." *Journal of Family Issues* 23, 5 (July 2002): 624–647.

"Alimony," answers.com, 2006.

Allen, Katherine, Rosemary Blieszner, and Karen Roberto. "Families in the Middle and Later Years: A Review and Critique of Research in the 1990's." *Journal of Marriage and the Family* 62, 4 (November 2000): 911–926.

Allen, Katherine R., David H. Demo, and Mark A. Fine, *Handbook of Family Diversity.* New York: Oxford University Press, 2000.

Allen, Katherine R., and Kristin M. Baber. "Starting a Revolution in Family Life Education: A Feminist Vision." *Family Relations* 41, 4 (October 1992): 378–384.

Allport, Gordon. *The Nature of Prejudice.* Garden City, NY: Doubleday, 1958.

Amato, Paul R. "Who Cares for Children in Public Places? Naturalistic Observation of Male and Female Caretakers." *Journal of Marriage and the Family* 51 (November 1989): 981–990.

———. "Parental Absence During Childhood and Depression in Later Life." *The Sociological Quarterly* 32 (1991): 543–556.

———. "Children's Adjustment to Divorce: Theories, Hypotheses, and Empirical Support." *Journal of Marriage and the Family* 55 1 (February 1993): 23–32.

———. "Explaining the Intergenerational Transmission of Divorce," *Journal of Marriage and the Family* 58 (August 1996): 628–640.

———. "Reconciling Divergent Perspectives: Judith Wallerstein, Quantitative Family Research, and Children of Divorce." *Family Relations* 52, 4 (October 2003) 332–339.

Amato, Paul R. "Tension Between Institutional and Individual Views of Marriage." *Journal of Marriage and Family* 66, 4 (November 2004): 943–958.

Amato, Paul R. "The Future of Marriage." In *Vision 2004: What is the Future of Marriage,* edited by Paul R. Amato and Nancy Gonzalez, Minneapolis, MN: National Council on Family Relations, 2004: 99–102.

Amato, Paul, and Stacy Rogers. "A Longitudinal Study of Marital Problems and Subsequent Divorce." *Journal of Marriage and the Family* 59 (August 1997): 612–624.

Amato, Paul R. "Father-Child Relations, Mother-Child Relations, and Offspring Psychological Well-Being in Early Adulthood." *Journal of Marriage and the Family* 56, 4 (November 1994): 1031–1042.

———. "The Consequences of Divorce for Adults and Children," *Journal of Marriage and the Family* 62, 4 (November 2000): 1269–1288.

Amato, Paul R., and Alan Booth. "A Prospective Study of Divorce and Parent-Child Relationships." *Journal of Marriage and the Family* 58, 2 (May 1996): 356–365.

Amato, Paul R., and Bruce Keith. "Parental Divorce and the Well-Being of Children: A Meta-Analysis." *Psychological Bulletin* 110 (1991): 26–46.

Amato, Paul R., and Danelle De Boer. "The Transmission of Marital Instability Across Generations: Relationship Skills or Commitment to Marriage?" *Journal of Marriage and Family* 63 (November 2001): 1038–1051.

Amato, Paul R., and Denise Previti. "People's Reasons for Divorcing: Gender, Social Class, the Life Course, and Adjustment." *Journal of Family Issues* 24, 5 (July 2003): 602–626.

Amato, Paul R., and Tamara D. Afifi. "Feeling Caught between Parents: Adult Children's Relations with Parents and Subjective Well-Being." *Journal of Marriage and Family* 68, 1 (February 2006): 222–235.

Amato, Paul R., David R. Johnson, Alan Booth, and Stacy J. Rogers. "Continuity and Change in Marital Quality between 1980 and 2000." *Journal of Marriage and Family* 65, 1 (February 2003): 1–22.

Ambert, Anne-Marie, Patricia Adler, Peter Adler, and Daniel Detzner. "Understanding and Evaluating Qualitative Research." *Journal of Marriage & Family* 57, 4 (November 1995): 879–908.

Ambry, Margaret K. "Childless Chances." *American Demographics* 14, 4 (April 1992): 55.

American Academy of Child and Adolescent Psychiatry. "Making Day Care a Good Experience." No. 20 (October 1992).

American Federation of State, County and Municipal Employees. "Fact Sheets: The Family and Medical Leave Act (FMLA)," afscme.org, 2006.

American Psychological Association. *Interim Report of the APA Working Group on Investigation of Memories of Childhood Abuse.* Washington, DC: American Psychological Association, 1994.

American Psychological Association. "Understanding Child Sexual Abuse: Education, Prevention, and Recovery," http://apa.org/releases/sexabuse, 2001.

American Society for Reproductive Medicine. "2003 Assisted Reproductive Technology (ART) Report: Success Rates." In *Assisted Reproductive Technologies: A Guide for Patients, 2003.* Society for Assisted Reproductive Technology, Birmingham, AL, 2003. http://www.sart.org/ARTPatients.html.

American Sociological Association. "24/7 Economy's Work Schedules Are Family Unfriendly and Suggest Needed Policy Changes." ASA News (May 2004): http://www.asanet.org.

———. "Achieving 'Adulthood' Is More Elusive for Today's Youth: Transition to 'Adulthood' Occurring at a Later Age." ASA News (August 2, 2004): http://www.asanet.org.

———. "Data Support Americans' Sense of an Accelerating 'Time Warp'; Balance Between Work and Family Remains Elusive for Many Workers." ASA News (November 2004): http://www.asanet.org.

———. "The Sexual Revolution and Teen Dating Trends is Explored in ASA's Magazine, Contexts," www2.asanet.org/media/cntrisman.html, March 2002.

Anderson, Stephen. "Parental Stress and Coping during the Leaving Home Transition." *Family Relations* 37 (April 1988): 160–165.

Aponte, Robert, with Bruce Beal and Michelle Jiles. "Ethnic Variation in the Family: The Elusive Trend Toward Convergence." In *Handbook of Marriage and the Family,* 2nd ed., edited by M. Sussman, S. Steinmetz, and G. Peterson. New York: Plenum, 1999.

Arditti, Joyce A . "Noncustodial Fathers: An Overview of Policy and Resources." *Family Relations* 39, 4 (October 1990): 460–465.

———. "Factors Related to Custody, Visitation, and Child Support for Divorced Fathers: An Exploratory Analysis." *Journal of Divorce and Remarriage* 17, 3–4 (1992): 23–42.

Arditti, Joyce A., and Katherine R. Allen. "Understanding Distressed Fathers' Perceptions of Legal and Relational Inequities Postdivorce." *Family and Conciliation Courts Review* 31, 4 (1993): 461–476.

Aries, Phillipe. *Centuries of Childhood.* New York: Vintage, 1962.

Arms, Suzanne. *Adoption: A Handful of Hope.* Berkeley, CA: Celestial Arts, 1990.

Armstrong, Penny, and Sheryl Feldman. *A Wise Birth.* New York: Morrow, 1990.

Aron, Arthur, and Elaine Aron. "Love and Sexuality." In *Sexuality in Close Relationships,* edited by K. McKinney and S. Sprecher. Hillsdale, NJ: Lawrence Erlbaum, 1991.

Aron, Arthur, et al. "Experiences of Falling in Love." *Journal of Social and Personal Relationships* 6 (1989): 243–257.

Arrighi, Barbara A., and David J. Maume Jr. "Workplace Subordination and Men's Avoidance of Housework." *Journal of Family Issues* 21, 4 (May 2000): 464–487.

"Artificial Insemination," ivf-infertility.com, 2005.

"At a Glance." *OURS: The Magazine of Adoptive Families* 23, 6 (November 1990): 63.

Atkinson, Alice M. "Rural and Urban Families' Use of Child Care." *Family Relations* 43, 1 (January 1994): 16–22.

Atkinson, Jean. "Gender Roles in Marriage and the Family: A Critique and Some Proposals." *Journal of Family Issues* 8 (1987): 5–41.

Atkinson, Maxine P., and Stephen P. Blackwelder. "Fathering in the 20th Century." *Journal of Marriage and the Family* 55, 4 (November 1993): 975–986.

Atwood, J. D., and J. Gagnon. "Masturbatory Behavior in College Youth." *Journal of Sex Education and Therapy* 13 (1987): 35–42.

Avellar, Sarah, and Pamela Smock. "The Economic Consequences of the Dissolution of Cohabiting Unions." *Journal of Marriage and Family* 67, 2 (May 2005): 315–327.

Aviv, Rachel. "Exes Play Tug-of-War for Pets." *Seattle Times,* August 29, 2004.

Axinn, William G., and Arland Thornton. "Mothers, Children, and Cohabitation: The Intergenerational Effects of Attitudes and Behavior." *American Sociological Review* 58, 2 (1993): 233–246.

Babbie, Earl. *The Practice of Social Research,* 6th ed. Belmont, CA: Wadsworth: 1986.

Babbie, Earl. *Basics of Social Research—with SPSS.* Belmont, CA: Wadsworth. 2002.

"Baby Blues Family Tree," babyblues.com, May 1999.

Bachu, Amara. *Current Population Reports (Series P23, No. 197),* Washington, DC: U.S. Census Bureau, 1999a.

Bachu, Amara. "Is Childlessness among American Women on the Rise?" U.S. Census Bureau, Population Division, Fertility and Family Statistics Branch. Washington, DC: U.S. Census Bureau, 1999b.

Baggett, Courtney R. "Sexual Orientation: Should It Affect Child Custody Rulings?" *Law and Psychology Review* 16 (1992): 189–200.

Baldwin, J. D., S. Whitely, and J. I. Baldwin. "The Effect of Ethnic Group on Sexual Activities Related to Contraception and STDs." *Journal of Sex Research* 29, 2 (May 1992): 189–206.

Balsam, Kimberly F., and Dawn M. Szymanski. "Relationship Quality and Domestic Violence in Women's Same-Sex Relationships: The Role of Minority Stress." *Psychology of Women Quarterly* 29, 3 (September 2005): 258–269.

Bankston, Carl, and Jacques Henry. "Endogamy Among Louisiana Cajuns: A Social Class Explanation." *Social Forces* 77 (4), 1999: 1317–1338.

Barbach, Lonnie. *For Each Other: Sharing Sexual Intimacy.* Garden City, NY: Doubleday, 1982.

Barber, B. L., and J. M. Lyons. "Family Processes and Adolescent Adjustment in Intact and Remarried Families." *Journal of Youth and Adolescence* 23, 4 (August 1994): 421–436.

Barnes, Jessica and Claudette Bennett. "The Asian Population: 2000." *Census 2000 Brief* (February 2002).

Barnes, Terrance, and Claudette Bennett. "We the People: Asians in the United States." U.S. Census Bureau: Census 2000 Special Reports. Washington, D.C.: U.S. Census Bureau CENSR-17 (December 2004).

Bartholomew, Kim. "Avoidance of Intimacy: An Attachment Perspective." *Journal of Social and Personal Relationships* 7, 2 (1990): 147–178.

Basile, Kathleen C., Ileana Arias, Sujata Desai, and Martie Thompson. "The Differential Association of Intimate Partner Physical, Sexual, Psychological, and Stalking Violence and Posttraumatic Stress Symptoms in a Nationally Representative Sample of Women." *Journal of Traumatic Stress* 17, 5 (October 2004): 413–421.

Basow, S. A. Gender: *Stereotyping and Roles.* Pacific Grove, CA: Brooks/Cole, 1992.

Basow, Susan A. *Gender, Stereotypes and Roles,* 4th ed. Pacific Grove, CA. Brooks/Cole: 1993.

Baumeister, Roy F., Sara R. Wotman, and Arlene M. Stillwell. "Unrequited Love: On Heartbreak, Anger, Guilt, Script-lessness, and Humiliation." *Journal of Personality and Social Psychology* 64, 3 (March 1993): 377–394.

Baumrind, Diana. "Current Patterns of Parental Authority." *Developmental Psychology Monographs* 4, 1 (1971): 1–102.

———. "Rejoinder to Lewis's Reinterpretation of Parental Firm Control Effects: Are Authoritative Families Really Harmonious?" *Psychological Bulletin* 94, 1 (July 1983): 132–142.

Baumrind, Diana. "The Influence of Parenting Style on Adolescent Competence and Substance Use." *Journal of Early Adolescence* 11, 1 (February, 1991): 56–95.

Baxter, Janeen. "To Marry or Not to Marry: Marital Status and the Household Division of Labor." *Journal of Family Issues* 26, 3 (April 2005): 300–321.

Baxter, Jean. "Gender Equality and Participation in Housework: A Cross-National Perspective," *Journal of Comparative Family Studies* 28 (Autumn 1997): 220–248.

BBC. "Map: Parenthood Policies in Europe," news.bbc.co.uk, March 2006.

Bearman, Peter, and Hannah Bruckner. "Promising the Future: Virginity Pledges and First Intercourse." American Journal of Sociology 106, 4 (January 2001): 859–913.

Becerra, Rosina. "The Mexican American Family." In *Ethnic Families in America: Patterns and Variations,* 3rd ed., edited by C. Mindel et al. New York: Elsevier North Holland, 1988.

Bechhofer, L., and L. Parrot. "What Is Acquaintance Rape?" In *Acquaintance Rape: The Hidden Crime,* edited by A. Parrott and L. Bechhofer. New York: Wiley, 1991.

Beck, Joyce W., and Barbara M. Heinzerling. "Gay Clients Involved in Child Custody Cases: Legal and Counseling Issues." *Psychotherapy in Private Practice* 12, 1 (1993): 29–41.

Beck, Melinda. "Miscarriages." *Newsweek* (August 15, 1988): 46–49.

Beck, Rubye, and Scott Beck. "The Incidence of Extended Households among Middle-Aged Black and White Women." *Journal of Family Issues* 10, 2 (June 1989): 147–168.

Becker, Howard. *Outsiders.* New York: Free Press, 1963.

Beer, William. *American Stepfamilies.* New Brunswick, NJ: Transaction, 1992.

Beier E. G., and D. P. Sternberg. "Subtle Cues between Newly Weds." *Journal of Communication* 27 (1997): 92–97.

Beitchman, Joseph H., et al. "A Review of the Long-Term Effects of Child Sexual Abuse." *Child Abuse and Neglect* 16, 1 (January 1992): 101.

Belknap, Joanne. "The Sexual Victimization of Unmarried Women by Nonrelative Acquaintances." In *Violence in Dating Relationships: Emerging Social Issues,* edited by M. Pirog-Good, and J. Stets. NY: Praeger Publishers, pp. 205–218, 1989.

Bell, Robert. *Worlds of Friendship.* Beverly Hills, CA: Sage, 1981.

Bellah, Robert, et al. *Habits of the Heart.* Berkeley, CA: University of California Press, 1985.

Bem, Sandra. "Androgyny versus the Tight Little Lives of Fluffy Women and Chesty Men." *Psychology Today* 9, 4 (September 1975a): 58–59 ff.

———. "Sex Role Adaptability: One Consequence of Psychological Androgyny." *Journal of Personality and Social Psychology* 31, 4 (1975b): 634–643.

———. *The Lenses of Gender: Transforming the Debate on Sexual Inequality.* New Haven, CT: Yale University Press, 1993.

Benenson, Joyce, and Athena Christakos. "The Greater Fragility of Females' Versus Males' Closest Same-Sex Friendships." *Child Development* 74, 4 (July 2003): 1123–1129.

Benokraitis, Nijole V., ed. *Feuds About Families: Conservative, Centrist, Liberal, and Feminist Perspectives."* Upper Saddle River, NJ: Prentice Hall, 2000.

Berardo, Donna Hodgkins, Constance L. Shehan, and Gerald R. Leslie. "A Residue of Tradition: Jobs, Careers, and Spouses' Time in Housework." *Journal of Marriage and Family* 49, 2 (May 1987): 381.

Bergen, David J., and John E. Williams. "Sex Stereotypes in the United States Revisited: 1972–1988." *Sex Roles* 24, 7/8 (1991): 413–423.

Berger, C. R. "Planning and Scheming: Strategies for Initiating Relationships." In *Accounting for Relationships: Explanations, Representation and Knowledge,* edited by R. Burnett, P. McChee, and D. Clarke. New York: Methuen, 1987.

Benokraitis, N. V., and J. R. Feagin. *Modern Sexism: Blatant, Subtle, and Covert Discrimination.* Englewood Cliffs, NJ: Prentice-Hall, 1995.

Berger, Peter, and Hansfried Kellner. "Marriage and the Construction of Reality." In *The Family: Its Structure and Functions,* edited by Rose Coser. New York: St. Martin's Press, 1974.

Bernard, J. "The Good-Provider Role: Its Rise and Fall." *American Psychologist* 36 (1981): 1–12.

Bernard, Jessie. *The Future of Marriage,* 2nd ed. New York: Columbia University Press, 1982.

Berrick, J. D., and R. P. Barth. "Child Sexual Abuse Prevention—Research Review and Recommendations." *Social Work Research and Abstracts* 28 (1992): 6–15.

Berscheid, Ellen. "Emotion." In *Close Relationships,* edited by H. H. Kelley et al. New York: Freeman, 1983.

Bertoia, Carl, and Janice Drakich. "The Father's Rights Movement." *Journal of Family Issues* 14, 4 (December 1993): 592–615

Besharov, Douglas and Timothy Sullivan. "Welfare Reform and Marriage," *The Public Interest* 125 (1996): 81–94.

Betchen, Stephen. "Male Masturbation as a Vehicle for the Pursuer/Distancer Relationship in Marriage." *Journal of Sex and Marital Therapy* 17, 4 (December 1991): 269–278.

Bevan, Emma, and Daryl Higgins. "Is Domestic Violence Learned? The Contribution of Five Forms of Child Maltreatment to Men's Violence and Adjustment." *Journal of Family Violence* 17, 3 (September 2002): 223–245.

Billy, John, Nancy Landale, William Grady, and Denise Zimmerle. "Effects of Sexual Activity on Adolescent Social and Psychological Development." *Social Psychology Quarterly* 51, 3 (September 1988): 190–212.

Billy, John, Koray Tanfer, William R. Grady, and Daniel H. Klepinger. "The Sexual Behavior of Men in the United States." *Family Planning Perspectives* 25, 2 (March 1993): 52–60.

Bird, Chloe. "Gender Differences in the Social and Economic Burdens of Parenting and Psychological Distress." *Journal of Marriage and the Family* 59(4) 1997: 809–823.

Bird, Chloe E., and Catherine E. Ross. "Houseworkers and Paid Workers: Qualities of the Work and Effects on Personal Control." *Journal of Marriage and the Family* 55, 4 (November 1993): 913–925.

Blackwell, Debra. "Marital Homogamy in the United States: The Influence of Individual and Paternal Education." *Social Science Research* 27 (2) 1998: 159–164.

Blair, Sampson Lee. "The Sex-Typing of Children's Household Labor: Parental Influence on Daughters' and Sons' Housework." *Youth and Society* 24, 2 (1992): 178–203.

———. "Employment, Family, and Perceptions of Marital Quality among Husbands and Wives." *Journal of Family Issues* 14, 2 (1993): 189–212.

Blair, Sampson, and Daniel Lichter. "Measuring the Division of Household Labor: Gender Segregation and Housework among American Couples." *Journal of Family Issues* 12, 1 (March 1991): 91–113.

Blair-Loy, Mary and Amy Wharton. "Employees Use of Work-Family Policies and the Workplace Social Context." *Social Forces* 80, 3 (March 2002): 813-845.

Blankenhorn, David. *Fatherless America: Confronting Our Most Urgent Social Problem.* New York: Basic Books, 1995.

Block, Jeanne. "Differential Premises Arising from Differential Socialization of the Sexes: Some Conjectures." *Child Development* 54 (December 1983): 1335–1354.

Blood, Robert, and Donald Wolfe. *Husbands and Wives.* Glencoe, IL: Free Press, 1960.

Bloom, L., and R. Kindle. "Demographic Factors in the Continuing Relationship between Former Spouses." *Family Relations* 34 (1985): 375–381.

Blow, Adrian J., and Kelley Hartnett. "Infidelity in Committed Relationships II: A Substantive Review." *Journal of Marital and Family Therapy* 31, 2 (April 2005): 183–216.

Blumenfeld, Warren, and Diane Raymond. *Looking at Gay and Lesbian Life.* Boston: Beacon Press, 1989.

Blumstein, Philip. "Identity Bargaining and Self-Conception." *Social Forces* 53, 3 (1975): 476–485.

Blumstein, Philip, and Pepper Schwartz. *American Couples.* New York: McGraw-Hill, 1983.

Bohannan, Paul, ed. *Divorce and After.* New York: Doubleday, 1970.

Boland, Joseph, and Diane Follingstad. "The Relationship between Communication and Marital Satisfaction: A Review." *Journal of Sex and Marital Therapy* 13, 4 (December 1987): 286–313.

Bonney, Lewis A. "Planning for Postdivorce Relationships: Factors to Consider in Drafting a Transition Plan." *Family and Conciliation Courts Review* 31, 3 (1993): 367–372.

Bonus Families. Home page, bonusfamilies.com, 2006.

Book, Cassandra L., et al. *Human Communication: Principles, Contexts, and Skills.* New York: St. Martin's Press, 1980.

Booth, A., and J. N. Edwards. "Starting Over: Why Remarriages Are More Unstable. *Journal of Family Issues* 13, 2 (June 1992): 179–194.

Booth, Alan, and Paul Amato. "Parental Predivorce Relations and Offspring Postdivorce Well-Being." *Journal of Marriage and Family* 63 (February 2001): 197–212.

Booth, Alan, et al. "Divorce and Marital Instability over the Life Course." *Journal of Family Issues* 7 (1986): 421–442.

Borhek, Mary. "Helping Gay and Lesbian Adolescents and Their Families: A Mother's Perspective." *Journal of Adolescent Health Care* 9, 2 (March 1988): 123–128.

Borisoff, Deborah, and Lisa Merrill. *The Power to Communicate: Gender Differences as Barriers.* Prospect Heights, IL: Waveland, 1985.

Borland, Dolores. "An Alternative Model of the Wheel Theory." The Family Coordinator (July 1975): 289–292.

Boss, Pauline. "Ambiguous Loss Research, Theory, and Practice: Reflections After 9/11." *Journal of Marriage & Family* 66, 3 (August 2004): 551–566.

Bouchard, Genevieve, "Adult Couples Facing a Planned vs. an Unplanned Pregnancy: Two Realities." *Journal of Family Issues* 26, 5 (July 2005): 619–637.

Boushey, Heather, and Joseph Wright. "Working Moms and Child Care." Washington, DC: Center for Economic Policy and Research. www.cepr.net/publications/child_care_2004.htm, May 5, 2004.

Bowker, Lee. *Beating Wife Beating.* Lexington, MA: Lexington Books, 1983.

Bowlby, John. *Attachment and Loss.* New York: Basic Books, 1969.

Bowlby, John. *Attachment and Loss.* 3 vols. New York: Basic Books, 1980.

Bowlby, J. *Attachment and Loss. Vol 2: Separation: Anxiety and Anger.* New York: Basic Books, 1973, reissued in 1999.

Bowman, Madonna, and Constance Ahrons. "Impact of Legal Custody Status on Father's Parenting Post-Divorce." *Journal of Marriage and the Family* 47, 2 (May 1985): 481–485.

Bozett, Frederick W. "Gay Fathers: A Review of the Literature." *Journal of Homosexuality* 18 (1989): 137–62.

Bozett, Frederick W., ed. *Gay and Lesbian Parents.* New York: Praeger, 1987.

Bradbury, Thomas N., and Benjamin Karney. "Understanding and Altering the Longitudinal Course of Marriage." *Journal of Marriage and Family* 66, 4 (November 2004): 862–879.

Bradsher, Keith. "3 Guilty of Manslaughter in Slipping Drug to Girl," *New York Times,* March 15, 2000.

Bramlett, M. D., and Mosher, W. D. "Cohabitation, Marriage, Divorce, and Remarriage in the United States." National Center for Health Statistics. Vital Health Stat 23(22). 2002.

Brantley, Angel, David Knox, and Marty Zussman. "When and Why: Gender Differences in Saying 'I Love You' Among College Students." *College Student Journal* 36 (December 2002).

Braver, Sanford, et al. "A Social Exchange Model of Nonresidential Parent Involvement." In *Nonresidential Parenting: New Vistas in Family Living,* edited by C. E. Depner, and J. H. Bray. Newbury Park, CA: Sage Publications, 1993a.

———. "A Longitudinal Study of Noncustodial Parents: Parents without Children." *Journal of Family Psychology* 7, 1 (June 1993b): 9–23.

Bray, James. "Family Assessment: Current Issues in Evaluating Families." *Family Relations* 44, 4 (October 1995): 469–477.

Bray, James H., and Sandra H. Berger. "Noncustodial Father and Paternal Grandparent Relationship in Stepfamilies." *Family Relations* 39, 4 (October 1990): 414–419.

Bray, James H., and Charlene Depner. "Nonresidential Parents: Who Are They?" In *Nonresidential Parenting: New Vistas in Family Living,* edited by C. E. Depner and J. H. Bray. Newbury Park, CA: Sage, 1993.

Brayfield, April A. "Employment Resources and Housework in Canada." *Journal of Marriage and the Family* 54, 1 (February 1992): 19–30.

Breault, K. D., and Augustine Kposowa. "Explaining Divorce in the United States: A Study of 3,111 Counties, 1980." *Journal of Marriage and the Family* 49, 3 (August 1987): 549–558.

Brennan, Kelly, and Phillip R. Shaver. "Dimensions of Adult Attachment, Affect Regulation, and Romantic Relationship Functioning." *Personality & Social Psychology Bulletin* 21, 3 (1995): 267–283.

Brennan, Robert, Rosalind Barnett, and Karen Gareis. "When She Earns More Than He Does: A Longitudinal Study of Dual-Earner Couples." *Journal of Marriage & Family* 63, 1 (February 2001): 168–183.

Bretschneider, Judy, and Norma McCoy. "Sexual Interest and Behavior in Healthy 80– to 102–Year-Olds." *Archives of Sexual Behavior* 17, 2 (April 1988): 109–128.

Brewster, Karin and Irene Padavic. "Change in Gender-Ideology, 1977–1996: The Contributions of Intracohort Change and

Population Turnover. *Journal of Marriage and the Family* 62, 2 (May 2000): 477–488.

Brewster, Mary P. "Power and Control Dynamics in Prestalking and Stalking Situations." *Journal of Family Violence* 18, 4 (August 2003): 207–217.

Brezosky, Lynn. "Parents Fighting State over Girl's Care Want New Judge." Associated Press: September 28, 2005.

Bridges, Judith S. "Pink or Blue: Gender Congratulations Cards." *Psychology of Women* 17, 2 (1993): 193–205.

Bringle, Robert G., and Glenda J. Bagby. "Self-Esteem and Perceived Quality of Romantic and Family Relationships in Young Adults." *Journal of Research in Personality* 26, 4 (1992): 340–356.

Bringle, Robert, and Bram Buunk. "Jealousy and Social Behavior: A Review of Person, Relationship, and Situational Determinants." In *Review of Personality and Social Psychology, Vol. 6, Self, Situation, and Social Behavior,* edited by P. Shaver. Newbury Park, CA: Sage, 1985.

———. "Extradyadic Relationships and Sexual Jealousy." In *Sexuality in Close Relationships,* edited by K. McKinney and S. Sprecher. Hillsdale, NJ: Lawrence Erlbaum, 1991.

Brittingham, Angela, and G. Patricia de la Cruz. "Ancestry 2000: A Census 2000 Brief. C2KBR-35." Washington, DC: United States Census Bureau: 2004.

Brittingham, Angela, and de la Cruz, G. Patricia. "We the People of Arab Ancestry in the United States." Census 2000 Special Reports #21. US Census Bureau, Washington DC, 2005.

Britton, D. M. "Homophobia and Homosociality: An Analysis of Boundary Maintenance." *Sociological Quarterly* 31, 3 (September 1990): 423–439.

Broman, Clifford. "Satisfaction among Blacks: The Significance of Marriage and Parenthood." *Journal of Marriage and the Family* 50, 1 (February 1988): 45–51.

Bronfenbrenner, Urie. *The Ecology of Human Development.* Cambridge, MA: Harvard University Press, 1979.

Brooks, Jane B. *Parenting in the 90s.* Mountain View, CA: Mayfield, 1994.

Brown, Lyn Mikel, and Carol Gilligan. *Meeting at the Crossroads: Women's Psychology and Girl's Development.* Cambridge, MA: Harvard University Press, 1992.

Brown, Susan, and Alan Booth. "Cohabitation Versus Marriage: A Comparison of Relationship Quality," *Journal of Marriage and the Family* 58 (August) 1996: 668–678.

———. "Stress at Home, Peace at Work: A Test of the Time Bind Hypothesis." *Social Science Quarterly* 83, 4 (December 2002a): 905–919.

———. "Bending the Time Bind: Rejoinder to Hochschild and Goodman." *Social Science Quarterly* 83, 4, (December 2002b): 941–946.

Browne, Angela, and David Finkelhor. "Initial and Long-Term Effects: A Review of the Research." In *Sourcebook on Child Sexual Abuse,* edited by D. Finkelhor. Beverly Hills, CA: Sage, 1986.

Brubaker, Timothy H. "Families in Later Life: A Burgeoning Research Area." *Journal of Marriage & Family* 52, 4 (November 1990): 959–981.

Bryant, Z. Lois, and Marilyn Coleman. "The Black Family as Portrayed in Introductory Marriage and Family Textbooks." *Family Relations* 37, 3 (July 1988): 255–259.

Buffum J. "Prescription Drugs and Sexual Function." *Psychiatr. Med.* 10 (1992): 181–198.

Budig, Michelle, and Paula England. "The Wage Penalty for Motherhood." *American Sociological Review* 66 (April 2001): 204–225.

Bulanda, Jennifer R., and Susan L. Brown. "Race–Ethnic Differences in Marital Quality and Divorce." Working Paper Series 06-08. Center for Family and Demographic Research. Bowling Green, OH: Bowling Green State University, 2004.

Bumpass, Larry, and Hsien-Hen Lu. "Trends in Cohabitation and Implications for Children's Family Contexts in the United States." *Population Studies* 54, 1 (March 2000): 29–41.

Bumpass, Larry, and James Sweet. "Changing Patterns of Remarriage." *Journal of Marriage & Family* 52, 3 (August 1990): 747–756.

Bumpass, Larry L., Teresa C. Martin, and James A. Sweet. "The Impact of Family Background and Early Marital Factors on Marital Disruption." *Journal of Family Issues* 12, 1 (1991): 22–42.

Bumpass, Larry, James Sweet, and Teresa Castro Martin. "Changing Patterns of Remarriage." *Journal of Marriage and the Family* 52, 3 (August 1990): 747–756.

Bunting, Madeline. "Behind the Baby Gap Lies a Culture of Contempt for Parenthood." The Guardian (March 7, 2006).

Burgess, Ernest. "The Family as a Unity of Interacting Personalities." *The Family* 7, 1 (March 1926): 3–9.

Burkett, Elinor. The Baby Boon: *How Family Friendly America Cheats the Childless.* New York: Free Press, 2000.

Burleson, Brant, and Wayne Denton. "The Relationship Between Communication Skill and Marital Satisfaction: Some Moderating Effects." *Journal of Marriage and the Family* 59, 4 (November 1997): 884–902.

Burns, Ailsa, and Cath Scott. *Mother-Headed Families and Why They Have Increased.* Hillsdale, NJ: Erlbaum, 1994.

Bush, Catherine R., Joseph P. Bush, and Joyce Jennings. "Effects of Jealousy Threats on Relationship Perceptions and Emotions." *Journal of Social and Personal Relationships* 5, 3 (August 1988): 285–303.

Buss, D. M. "Marital Assortment for Personality Dispositions: Assessment with Three Different Data Sources." *Behavior Genetics* 14 (1984): 111–123.

Butler, Amy C. "Gender Differences in the Prevalence of Same-Sex Sexual Partnering: 1988–2002." Social Forces 84, 1 (September 2005): 421–449.

Buunk, Bram, and P. Dykstra. "Gender Differences in Rival Characteristics That Evoke Jealousy in Response to Emotional versus Sexual Infidelity." *Personal Relationships* 11 (2004): 395–408.

Buunk, Bram, and Ralph Hupka. "Cross-Cultural Differences in the Elicitation of Sexual Jealousy." *Journal of Sex Research* 23, 1 (February 1987): 12–22.

Buunk, Bram, and Barry van Driel. *Variant Lifestyles and Relationships.* Newbury Park, CA: Sage, 1989.

Byers, E. S., and L. Heinlein. "Predicting Initiations and Refusals of Sexual Activities in Married and Cohabiting Heterosexual Couples." *Journal of Sex Research* 26 (1989): 210–231.

Byrne, Donn, and Karen Murnen. "Maintaining Love Relationships." In *The Psychology of Love,* edited by R. Sternberg and M. Barnes. New Haven, CT: Yale University Press, 1988.

Byrne, Michael, Alan Carr, and Marie Clark. "Power in Relationships of Women with Depression." *Journal of Family Therapy* 26, 4 (November 2004): 407–429.

Caetano, Raul, John Schafer, Catherine Clark, Carol Cunradi, and Kelly Raspberry. "Intimate Partner Violence, Acculturation, and Alcohol Consumption Among Hispanic Couples in the United States." *Journal of Interpersonal Violence* 15, 1 (January 2000): 30–45.

Cahoon, D. E. M., Edmonds, R. M. Spaulding, and J. C. Dickens. "A Comparison of the Opinions of Black and White Males and Females Concerning the Occurrence of Rape." *Journal of Social Behavior and Personality* 10, 1 (March 1995): 91–100.

Call, Vaughn R. A., and Tim B. Heaton. "Religious Influence on Marital Stability." *Journal for the Scientific Study of Religion* 36, 3 (September 1997): 382–392.

Callan, Victor. "The Personal and Marital Adjustment of Mothers and of Voluntarily and Involuntarily Childless Wives." *Journal of Marriage and the Family* 47, 4 (November 1985): 1045–1050.

Camarota, Steven. "Immigrants from the Middle East: A Profile of the Foreign-born Population from Pakistan to Morocco." Center for Immigration Studies (August 2002).

Cancian, F. M. "Gender Politics: Love and Power in the Private and Public Spheres." In *Family in Transition,* edited by A. S. Skolnick and J. H. Skolnick. Glenview, IL: Scott, Foresman, 1989.

Cancian, Francesca. "Gender Politics: Love and Power in the Private and Public Spheres." In *Gender and the Life Course,* edited by A. Rossi. Hawthorne, NY: Aldine, 1985: 253–262.

———. *Love In America: Gender and Self Development.* New York: Oxford University Press, 1987.

Cann, Arnie. "Rated Importance of Personal Qualities Across Four Relationships." *The Journal of Social Psychology* 144, 3 (2004): 322–334.

Canter, Lee, and Marlene Canter. *Assertive Discipline for Parents.* Santa Monica, CA: Canter and Associates, 1985.

Cargan, Leonard, and Matthew Melko. *Singles: Myths and Realities.* Beverly Hills, CA: Sage, 1982.

Carl, Douglas. "Acquired Immune Deficiency Syndrome: A Preliminary Examination of the Effects on Gay Couples and Coupling." *Journal of Marital and Family Therapy* 12, 3 (July 1986): 241–247.

Caron, Sandra L., and Eilean G. Moskey. "Changes over Time in Teenage Sexual Relationships: Comparing the High School Class of 1950, 1975, and 2000." *Adolescence* 37, 147 (Fall 2002): 515–526.

Carpenter, Laura. "Gender and the Meaning and Experience of Virginity Loss in the Contemporary United States." *Gender and Society* 16, 3 (June 2002): 345–365.

Carpenter, Laura. *Virginity Lost: An Intimate Portrait of First Sexual Experiences.* New York: New York University Press, 2005.

Carr, Deborah. "The Desire to Date and Remarry Among Older Widows and Widowers." *Journal of Marriage and Family* 66, 4 (November 2004): 1051–1068.

Carter, Betty, and Monica McGoldrick, eds. *The Changing Family Life Cycle,* 2nd ed. Boston: Allyn and Bacon, 1989.

Carter, D. Bruce. "Sex Role Research and the Future New Directions for Research." In *Current Conceptions of Sex Roles and Sex Typing,* edited by D. Bruce Carter. New York: Praeger, 1987.

Casper, Lynne, and Suzanne Bianchi. *Continuity and Change in the American Family.* Thousand Oaks, CA: Sage Publications, 2002.

Casper, Lynne M., and Kristin E. Smith. "Dispelling the Myths: Self-care, Class, and Race." *Journal of Family Issues* 23, 6 (2002): 716–727.

Cate, Rodney M., and Sally A. Lloyd. *Courtship.* Newbury Park, CA: Sage, 1992.

CDC. *See* Centers for Disease Control and Prevention.

Centers for Disease Control and Prevention. "Statewide Prevalence of Illicit Drug Use by Pregnant Women—Rhode Island." *Morbidity and Mortality Weekly Report* 39, 14 (April 3, 1990): 225–227.

———. "HIV Survey in Childbearing Women." *National AIDS Hotline Training Bulletin* 103 (June 15, 1994a): 2.

———. "Surveillance Report: U.S. AIDS Cases Reported through June 1994." *HIV/AIDS Surveillance Report,* 1994b.

———. "Surveillance Report: U.S. AIDS Cases Reported through December 1996." *HIV/AIDS Surveillance Report,* 1996.

———. "Births, Marriages, Divorces and Deaths: Provisional Data for 2001." *National Vital Statistics Reports.* 50, 14 (September 2002).

———. "Intimate Partner Violence Prevention, Facts-NCIPC," National Center for Injury Prevention and Control, www.cdc.gov, 2006.

———. "Life Expectancy at Birth, at 65 years of Age, and at 75 Years of Age, According to Race and Sex: United States, Selected Years 1900–2001," National Center for Health Statistics, www.cdc.gov, 2003.

———. "Racial/Ethnic Disparities in Infant Mortality-United States, 1995–2002," Morbidity and Mortality Weekly Report, www.cdc.gov, 2005.

Chan, Connie S. "Asian American Adolescents: Issues in the Expression of Sexuality." In *Sexualities: Identities, Behaviors, and Society,* edited by Michael Kimmel and Rebecca Plante. New York: Oxford University Press, 2004: 106–112.

Cherlin, Andrew. *Marriage, Divorce, Remarriage.* Cambridge, MA: Harvard University Press, 1981.

Cherlin, Andrew, and Frank Furstenberg, Jr. *The New American Grandparent.* New York: Basic Books, 1986.

Cherlin, Andrew J. "The Deinstitutionalization of American Marriage." *Journal of Marriage and Family* 66, 4 (November 2004): 848–861.

Cherlin, Andrew J., Linda Burton, Tera Hurt, and Diane Purvin, "The Influence of Physical and Sexual Abuse on Marriage and Cohabitation." *American Sociological Review* 69, 6 (December 2004): 768–789.

Chesser, Barbara Jo. "Analysis of Wedding Rituals: An Attempt to Make Weddings More Meaningful." *Family Relations* 29, 2 (April 1980).

Child Welfare Information Gateway. "Costs of Adopting: A Factsheet for Families," childwelfare.gov, 2004.

Children Now. "Fall Colors: 2003-2004 Prime Time Diversity Report. Oakland CA: Children Now. (2004).

Chinitz, J., and R. Brown. "Religious Homogamy, Marital Conflict, and Stability in Same-Faith and Interfaith Jewish Marriages." *Journal for the Scientific Study of Religion* 40, 4 (December 2001): 723–733.

Chira, Susan. "Years after Adoption, Adults Find Past, and New Hurdles." *New York Times* (August 30, 1993): B1, B6.

Chodorow, Nancy. *The Reproduction of Mothering: Psychoanalysis and the Sociology of Gender.* Berkeley: University of California Press, 1978.

Christensen, Andrew, and Christopher L. Heavey. "Gender and Social Structure in the Demand/Withdraw Pattern of Marital Conflict." *Journal of Personality & Social Psychology* 59, 1 (July 1990): 73–81.

Christopher, F. Scott, and Susan Sprecher. "Sexuality in Marriage, Dating, and Other Relationships: A Decade Review." *Journal of Marriage and the Family* 62, 4. (November 2000): 999–1017.

Christopher, F. S., and M. M. Frandsen. "Strategies of Influence in Sex and Dating." *Journal of Social and Personal Relationships* 7 (1990): 89–105.

Ciabattari, T. "Cohabitation and Housework: The Effects of Marital Intentions." *Journal of Marriage and Family* 66 (2004): 118–125.

Ciancanelli, Penelope, and Bettina Berch. "Gender and the GNP." In *Analyzing Gender,* edited by B. Hess and M. Marx Ferree. Newbury Park, CA: Sage, 1987.

Ciccarelli, Janice C., and Linda J. Beckman. "Navigating Rough Waters: An Overview of Psychological Aspects of Surrogacy." *Journal of Social Issues* 61, 1 (March 2005): 21–43.

Ciccarelli, John K., and Janice C. Ciccarelli. "The Legal Aspects of Parental Rights in Assisted Reproductive Technology." *Journal of Social Issues* 61, 1 (March 2005): 127–137.

Clare, Pamela. "Romancing the Store: Why Not Give Women What They Want?" www.publishersweekly.com/article/CA6285280.html, November 21, 2005.

Clatterbaugh, Kenneth. *Contemporary Perspectives on Masculinity: Men, Women, and Politics in Modern Society,* 2nd ed. Boulder, CO: Westview, 1997.

Clawson, Dan, and Naomi Gerstel. "Caring for Our Young: Child Care in Europe and the United States." *Contexts* 1, 4 (Fall/Winter 2002): 28–35.

Clemmens, Donna A. "Adolescent Mothers' Depression After the Birth of Their Babies: Weathering the Storm." *Adolescence* 37, 147, (Fall 2002): 551–565.

Clingempeel, W. Glenn, et al. "Stepparent-Stepchild Relationships in Stepmother and Stepfather Families: A Multimethod Study." *Family Relations* 33 (1984): 465–473.

Cochran, Susan D., Vickie M. Mays, and Laurie Leung. "Sexual Practices of Heterosexual Asian-American Young Adults: Implications for Risk of HIV Infection." *Archives of Sexual Behavior* 20, 4 (August 1991): 381–394.

Coggle, Frances, and Grace Tasker. "Children and Housework." *Family Relations* 31 (July 1982): 395–399.

Cohan, Catherine, and Stacey Kleinbaum. "Toward a Greater Understanding of the Cohabitation Effect: Premarital Cohabitation and Marital Communication." *Journal of Marriage and the Family* 64, 1 (February 2002): 163–179.

Cohen, T .F. "What Do Fathers Provide? Reconsidering the Economic and Nurturant Dimensions of Men as Parents." In *Men, Work and Family,* edited by J. C. Hood. Newbury Park, CA: Sage , 1993.

Cohen, Theodore. *Men's Family Roles: Becoming and Being Husbands and Fathers.* Doctoral Dissertation, Boston University. (University Microfilms No. 86–09272) 1986.

———. "Remaking Men: Men's Experiences Becoming and Being Husbands and Fathers and Their Implications for Reconceptualizing Men's Lives," *Journal of Family Issues* 8 (1987): 57–77.

Cohen, Theodore, ed. *Men and Masculinity: A Text-Reader.* Belmont, CA: Wadsworth, 2001.

Cohen, Theodore, and John C. Durst. "Leaving Work and Staying Home: The Impact on Men of Terminating the Male Economic Provider Role." In *Men and Masculinity: A Text-Reader,* edited by T. Cohen. Belmont, CA: Wadsworth, 2001.

Cohler, Bertram, and Scott Geyer. "Psychological Autonomy and Interdependence within the Family." In *Normal Family Processes,* edited by F. Walsh. New York: Guilford Press, 1982.

Cole, Robert. "Mental Illness and the Family." In *Vision 2010: Families and Health Care,* edited by B. A. Elliott. Minneapolis: National Council on *Family Relations*, 1993: 18–19.

Coleman, Marilyn, and Lawrence Ganong. "The Cultural Stereotyping of Stepfamilies." In *Remarriage and Stepparenting: Current Research and Theory,* edited by K. Pasley and M. Ihinger-Tallman. New York: Guilford Press, 1987.

———. "Remarriage and Stepfamily Research in the 1980s: Increased Interest in an Old Form." In *Contemporary Families: Looking Forward, Looking Back,* edited by A. Booth. Minneapolis: National Council on Family Relations, 1991.

Coleman, Marilyn, Lawrence Ganong, and Mark Fine. "Reinvestigating Remarriage: Another Decade of Progress." *Journal of Marriage and the Family* 62, 4 (November 2000): 1288–1307.

Coleman, Thomas. "A Week for Singles but Greeting Cards Can't Be Found." Column One: Eye on Unmarried America, unmarriedamerica.org, September 2005.

Collier, J., M. Z. Rosaldo, and S. Yanagisako. "Is There a Family? New Anthropological Views." In *Rethinking the Family: Some Feminist Questions,* edited by B. Thorne and M. Yalom. New York: Longman, 1982.

Coltrane, Scott. *Family Man: Fatherhood, Housework, and Gender Equity.* New York: Oxford University Press, 1996.

———. "Research on Household Labor: Modeling and Measuring the Social Embeddedness of Routine Family Work." *Journal of Marriage and the Family* 62, 4 (November 2000): 1208–1233.

Coltrane, Scott, and Neal Hickman. "The Rhetoric of Rights and Needs: Moral Discourse in the Reform of Child Custody and Child Support Laws." *Social Problems* 39, 4 (1992): 400–420.

"A Comparative Survey of Minority Health." The Commonwealth Fund. New York, 1996.

Condry, J., and S. Condry. "The Development of Sex Differences: A Study of the Eye of the Beholder." *Child Development* 47, 4 (1976): 812–819.

Connell, Robert. *Gender and Power: Society, the Person, and Sexual Politics.* Stanford, CA: Stanford University Press, 1987.

———. *Masculinities.* Berkeley: University of California Press, 1995.

Coombs, Mary. "Transgenderism and Sexual Orientation: More Than a Marriage of Convenience?" *National Journal of Sexual Orientation Law* 3, 1 (1997), ibiblio.org/gaylaw/.

Cooney, Teresa M. "Young Adults' Relations with Parents: The Influence of Recent Parental Divorce." *Journal of Marriage and the Family* 56, 1 (February 1994): 45–56.

Coontz, Stephanie. *The Way We Really Are: Coming to Terms with America's Changing Families.* New York: Basic Books, 1997.

———. "Divorcing Reality: Other Researchers Question Wallerstein's Conclusions," *Children's Advocate,* Action Alliance for Children, January-February, 1998.

———. "The World Historical Transformation of Marriage." *Journal of Marriage and Family* 66, 4 (November 2004): 974–979.

———. Marriage, *A History: From Obedience To Intimacy, Or How Love Conquered Marriage.* New York: Viking, 2005.

Copenhaver, Stacey, and Elizabeth Grauerholz. "Sexual Victimization Among Sorority Women: Exploring the Link Between Sexual Violence and Institutional Practices." *Sex Roles* 24, (1/2) 1991:31–41.

Cortese, Anthony. "Subcultural Differences in Human Sexuality: Race, Ethnicity, and Social Class." In *Human Sexuality: The Societal and Interpersonal Context,* edited by K. McKinney and S. Sprecher. Norwood, NJ: Ablex, 1989.

Council on Family Law. The *Future of Family Law: Law and the Marriage Crisis in North America.* New York: Institute for American Values, 2005.

Cowan, Carolyn Pope, and Philip Cowan. *When Partners Become Parents: The Big Life Change for Couples.* New York: Basic Books, 1992.

Cowan, Carolyn Pope, and Philip Cowan. *When Partners Become Parents: The Big Life Change for Couples.* Mahwah, NJ: Lawrence Erlbaum, 2000.

Cowan, Philip, and Carolyn Cowan. "Becoming a Family: Research and Intervention." In *Methods of Family Research: Biographies of Research Projects,* edited by I. Sigel and G. Brody. Hillsdale, NJ: Lawrence Erlbaum, 1990.

Cramer, David, and Arthur Roach. "Coming Out to Mom and Dad: A Study of Gay Males and Their Relationships with Their Parents." *Journal of Homosexuality* 14, 1–2 (1987): 77–88.

Cramer, Robert E., William T. Abraham, Lesley M. Johnson, and Barbara Manning-Ryan. "Gender Differences in Subjective Distress to Emotional and Sexual Infidelity: Evolutionary or Logical Inference Explanation." *Current Psychology* 20, 4 (Winter, 2001-2002): 327–336.

Creti, L., and E. Libman. "Cognition and Sexual Expression in the Aging." *Journal of Sex and Marital Therapy* 15, 2 (June 1989): 83–101.

Crissey, Sara R. "Race/Ethnic Differences in the Marital Expectations of Adolescents: The Role of Romantic Relationships." Journal of Marriage and Family 67, 3 (August 2005): 697–709.

Crittenden, Ann. *The Price of Motherhood: Why the Most Important Job in the World Is Still the Least Valued.* New York: Metropolitan Books. 2001.

Crohan, Susan. "Marital Quality and Conflict Across the Transition to Parenthood in African American and White Couples." *Journal of Marriage & Family* 58, 4 (November 1996): 933–944

Crohan, Susan E. "Marital Quality and Conflict Across the Transition to Parenthood in African American and White Couples." *Journal of Marriage and Family* 58, 4 (November 1996): 933–944.

Crosby, Faye. *Juggling: The Unexpected Advantages of Balancing Career and Home for Women and Their Families.* New York: Free Press, 1991.

Crouter, Ann, Matthew Bumpus, Melissa Head, and Susan McHale. "Implications of Overwork and Overload for the Quality of Men's Family Relationships." *Journal of Marriage and Family* 63 (May 2001): 404–416.

Crouter, Ann C., and Beth Manke. "The Changing American Workplace: Implications for Individuals and Families." *Family Relations* 43, 2 (April 1994): 117–124.

Crouter, Ann C., and Susan M. McHale. "The Long Arm of the Job: Influences of Parental Work on Child Rearing." In *Parenting: An Ecological Perspective,* edited by T. Luster and L. Okagaki. Hillsdale, NJ: Lawrence Erlbaum, 1993.

Cuber, John F., and Peggy Harroff. *The Significant Americans, a Study of Sexual Behavior among the Affluent.* NY: Appleton-Century, 1965.

Culp, Rex E., Alicia S. Cook, and Pat C. Housley. "A Comparison of Observed and Reported Adult-Infant Interactions: Effects of Perceived Sex." *Sex Roles* 9 (April 1983): 475–479.

Cummings, E. Mark, Marcie C. Goeke-Morey, and Lauren M. Papp. "Children's Responses to Everyday Marital Conflict Tactics in the Home." *Child Development* 74, 6 (November 2003): 1918–1929.

Cunningham, Mick. "Parental Influences of the Gendered Division of Housework." *American Sociological Review* 66 (April 2001): 184–203.

Cunradi, Carol, Raul Caetano, and John Schafer. "Socio-economic Predictors of Intimate Partner Violence among White, Black, and Hispanic Couples in the United States." *Journal of Family Violence* 17, 4 (December 2002): 377–389.

Cupach, William, and J. Comstock. "Satisfaction with Sexual Communication in Marriage." *Journal of Social and Personal Relationships* 7 (1990): 179–186.

Cupach, William, and Sandra Metts, "Sexuality and Communication in Close Relationships." In *Sexuality in Close Relationships,* edited by K. McKinney and S. Sprecher. Hillsdale, NJ: Lawrence Erlbaum, 1991.

Curry, Tim, Robert Jiobu, and Kent Schwirian. *Sociology for the 21st Century.* Prentice Hall, 2002.

Cyr, Mireille, John Wright, Pierre McDuff, and Alain Perron. "Intrafamilial Sexual Abuse: Brother-Sister Incest Does Not Differ from Father-Daughter and Stepfather-Stepdaughter Incest." *Child Abuse & Neglect* 26, 9 (September 2002): 957–974.

Dainton, M. "The Myths and Misconceptions of the Stepmother Identity: Descriptions and Prescriptions for Identity Management. *Family Relations* 42, 1 (January 1993): 93–98.

Daley, Suzanne. "Girl's Self-Esteem Is Lost on Way to Adolescence, New Study Finds." *New York Times* (January 9, 1991): B1.

Daly, Kerry. "Reshaping Fatherhood: Finding the Models." *Journal of Family Issues* 14, 4 (December 1993): 510–530.

Daly, Kerry. "Spending Time with the Kids: Meanings of Family Time for Fathers." *Family Relations* 45, 4 (October 1996): 466–476.

Daly, Kerry J. "Deconstructing Family Time: from Ideology to Lived Experience." *Journal of Marriage & Family* 63, 2 (May 2001): 283–295.

Darling, Nancy. "Parenting Style and Its Correlates." www.athealth.com/Practitioner/ceduc/parentingstyles.html, March 1999.

Darling-Fisher, Cynthia, and Linda Tiedje. "The Impact of Maternal Employment Characteristics on Fathers' Participation in Child Care." *Family Relations* 39, 1 (January 1990): 20–26.

Datzman, Jeanine, and Carol Brooks Gardner. "'In My Mind, We Are All Humans': Notes on the Public Management of Black-White Interracial Romantic Relationships." *Marriage & Family Review* 30, 1/2, (2000): 5–25.

D'Augelli, A. R., and M. L. Rose. "Homophobia in a University Community: Attitudes and Experience of White Heterosexual Freshmen." *Journal of College Student Development* 31, 6 (1990): 484–491.

Davidson, Kenneth, and Carol Darling. "The Stereotype of Single Women Revisited." *Health Care for Women International* 9, 4 (October 1988): 317–336.

Davis, S. "Men as Success Objects and Women as Sex Objects: A Study of Personal Advertisements." *Sex Roles* 23 (July 1990): 43–50.

Davis, Shannon, and Theodore Greenstein. "Cross-national Variations in the Division of Household Labor." *Journal of Marriage & Family* 66, 5 (December 2004): 1260–1271.

Deal, James E., Karen S. Wampler, and Charles F. Halverson. "The Importance of Similarity in the Marital Relationship." *Family Process* 31, 4 (1992): 369–382.

Declercq, E., C. Sakala, M. P. Corry, S. Applebaum, and P. Risher. "Listening to Mothers: Report of the First National U.S. Survey of Women's Childbearing Experiences." New York: Maternity Center Association, October 2002.

DeGenova, Mary Kay. *Families in Cultural Context: Strengths and Challenges in Diversity.* Mountain View, CA: Mayfield, 1997.

Degler, Carl. *At Odds.* New York: Oxford University Press, 1980.

de Graaf, Paul M., and Matthijs Kalmijn. "Alternative Routes in the Remarriage Market: Competing-Risk Analyses of Union Formation after Divorce." *Social Forces* 81, 4 (June 2003): 1459–1498.

de Graaf, Paul M., and Matthijs Kalmijn. "Divorce Motives in a Period of Rising Divorce." *Journal of Family Issues* 27, 4 (April 2006): 483–505.

DeGroot, J. M., et al. "Correlates of Sexual Abuse in Women with Anorexia Nervosa and Bulimia Nervosa." *Canadian Journal of Psychiatry* 37, 7 (September 1992): 516–518.

De Judicibus, Margaret A., and Marita P. McCabe. "Psychological Factors and the Sexuality of Pregnant and Postpartum Women." *Journal of Sex Research* 39, 2 (May 2002): 94–83.

De la Cruz, G. Patricia, and Angela Brittingham. *The Arab Population: 2000. Census 2000 Brief C2KBR-23.* Washington, D.C.: U.S. Bureau of the Census, December 2003.

DeLamater, John D., and Morgan Sill. "Sexual Desire in Later Life." *Journal of Sex Research* 42, 2 (May 2005): 138–149.

Del Carmen, Rebecca. "Assessment of Asian-Americans for Family Therapy." In *Mental Health of Ethnic Minorities,* edited by F. Serafica et al. New York: Praeger, 1990.

DeMaris, Alfred, and K. Vaninadha Rao. "Premarital Cohabitation and Subsequent Marital Stability in the United States: A Reassessment." *Journal of Marriage and the Family* 54, 1 (February 1992): 178–190.

DeMaris, Alfred. "The Influence of Intimate Violence on Transitions out of Cohabitation." *Journal of Marriage and Family* 63 (February 2001): 235–246.

Demo, David, and Alan Acock. "The Impact of Divorce on Children." In *Contemporary Families: Looking Forward, Looking Back,* edited by A. Booth. Minneapolis: National Council on Family Relations, 1991.

Demo, David H., and Alan C. Acock. "Family Diversity and the Division of Domestic Labor: How Much Have Things Really Changed?" *Family Relations* 42, 3 (July 1993): 323–331.

Demo, David, and Martha Cox. "Families with Young Children: A Review of Research in the 1990's." *Journal of Marriage and the Family*, 62, 4 (November 2000): 876–895.

Demos, John. *A Little Commonwealth*. New York: Oxford University Press. 1970.

Demos, Vasilikie. "Black Family Studies in the *Journal of Marriage and the Family* and the Issue of Distortion: A Trend Analysis." *Journal of Marriage and the Family* 52, 3 (August 1990): 603–612.

Denov, Myriam S. "To a Safer Place? Victims of Sexual Abuse by Females and Their Disclosures to Professionals." *Child Abuse & Neglect* 27, 1 (January 2003):47–62.

Dentzer, Susan. "Do the Elderly Want to Work?" *U.S. News and World Report* (May 14, 1990): 48–50.

DeParle, Jason. "Suffering in the Cities Persists as U.S. Fights Other Battles." *New York Times* (January 27, 1991): 15.

Depner, Charlene, and James Bray, eds. *Nonresidential Parenting: New Vistas in Family Living*. Newbury Park, CA: Sage, 1993.

Derlega, Valerian J., Sandra Metts, Sandra Petronio, and S. Margulis. *Self-Disclosure*. Newbury Park, CA: Sage, 1993.

de Vries, B., C. Jacoby, and C. G. Davis. "Ethnic Differences in Later Life Friendship." *Canadian Journal on Aging* 15 (1996): 226–244.

DeWitt, P. M. "Breaking Up Is Hard To Do." *American Demographics*, reprint package (1994): 14–16.

Diamond, Lisa. "Emerging Perspectives on Distinctions Between Romantic Love and Sexual Desire." *Current Directions in Psychological Science* 13, 3 (2004): 116–119.

Diamond, Lisa M. "What Does Sexual Orientation Orient? A Biobehavioral Model Distinguishing Romantic Love and Sexual Desire." *Psychological Review* 110, 1 (January 2003): 173–193.

Dibiase, Rosemarie, and Jaime Gunnoe. "Gender and Culture Differences in Touching Behavior." *Journal of Social Psychology* 144, 1 (February 2004): 49–62.

DiBlasio, Frederick A., and Brent B. Benda. "Gender Differences in Theories of Adolescent Sexual Activity." *Sex Roles* 27, 5/6 (1992): 221–236.

Diekmann, Andreas, and Kurt Schmidheiny. "Do Parents of Girls Have a Higher Risk of Divorce? An Eighteen-Country Study." *Journal of Marriage & Family* 66, 3 (August 2004): 651–660.

Dietz, Tracy. "An Examination of Violence and Gender Role Portrayals in Video Games: Implications for Gender Socialization and Aggressive Behavior," *Sex Roles: A Journal of Research* 38 (March 1998): 5–6.

DiIorio, C., M. Kelley, and M. Hockenberry-Eaton. "Communication about Sexual Issues: Mothers, Fathers, and Friends." *Journal of Adolescent Health* 23, (1999):181–189.

Dillon, Nancy, Austin Fenner, and Alison Gendar. "Bound, Beaten, Starved, Killed: Stepdad & Mom Face Murder Rap in Death of Girl, 7." *New York Daily News*, January 11, 2006.

Dilworth-Anderson, Peggye, and Harriette Pipes McAdoo. "The Study of Ethnic Minority Families: Implications for Practitioners and Policymakers." *Family Relations* 37, 3 (July 1988): 265–267.

Dion, Karen, et al. "What Is Beautiful Is Good." *Journal of Personality and Social Psychology* 24 (1972): 285–290.

Dodson, Fitzhugh. "How to Discipline Effectively." In *Experts Advise Parents*, edited by E. Shiff. New York: Dell, 1987.

Dodson, Lisa, and Jillian Dickert. "Girls' Family Labor in Low-Income Households: A Decade of Qualitative Research." *Journal of Marriage & the Family* 66, 2 (May 2004): 218–332.

"Does Divorce Mediation Work?" divorceinfo.com, 2006.

Doherty, William J. Edward Kouneski, and Martha Erickson. "Responsible Fathering: An Overview and Conceptual Framework." *Journal of Marriage and the Family* 60, 2 (1998): 277–292.

Dolgin, Kim. "Men's Friendships: Mismeasured, Demeaned, and Misunderstood?" In *Men and Masculinity: A Text Reader*, edited by T. Cohen. Belmont, CA: Wadsworth, 2001.

Donnelly, Kathleen. "Breaking the Barriers." *San Jose Mercury News* (September 27, 1993): 1C, 8C.

Downey, Geraldine, C. Bonica, and C. Rincon. "Rejection Sensitivity and Conflict in Adolescent Romantic Relationships." In *Adolescent Romantic Relationships*, edited by W. Furman, B. Brown, and C. Feiring. New York: Cambridge University Press, 1999.

Doyle, James. *The Male Experience*, (3rd ed.). Dubuque, Iowa: W. C. Brown, 1994.

Dreikurs, Rudolph, and V. Soltz. Children: *The Challenge*. New York: Hawthorne Books, 1964.

Driver, Janice, and John Gottman. "Daily Marital Interactions and Positive Affect During Marital Conflict Among Newlywed Couples." *Family Process* 43, 3 (September 2004): 301–314.

Drugger, Karen. "Social Location and Gender-Role Attitudes: A Comparison of Black and White Women." *Gender and Society* 2, 4 (December 1988): 425–448.

Duck, Steve, ed. *Dynamics of Relationships*. Thousand Oaks, CA: Sage, 1994.

Duncan, Greg, and Willard Rodgers. "Longitudinal Aspects of Childhood Poverty." *Journal of Marriage and the Family* 50, 4 (November 1988): 1007–1022.

Duncombe, Jean, and Dennis Marsden. "Love and Intimacy: The Gender Division of Emotion and 'Emotion Work': A Neglected Aspect of Sociological Discussion of Heterosexual Relationships." *Sociology* 27, 2 (May 1993): 221–242.

Dworetsky, John P. *Introduction to Child Development*. 4th ed. St. Paul, MN: West, 1990.

Edin, Kathryn, Maria Kefalas, and Joanna Reed. "A Peek Inside the Black Box: What Marriage Means for Poor Unmarried Parents." *Journal of Marriage and Family* 66, 4 (November 2004): 1007–1014.

Edleson, Jeffrey, et al. "Men Who Batter Women." *Journal of Family Issues* 6, 2 (June 1985): 229–247.

Egeland, Byron. "A History of Abuse Is a Major Risk Factor for Abusing the Next Generation." In *Current Controversies in*

Family Violence, edited by R. Gelles and D. Loseke. Newbury Park, CA: Sage, 1993.

Eggebeen, David, and Chris Knoester. "Does Fatherhood Matter for Men?" *Journal of Marriage and Family* 63 (May 2001): 381–393.

Ehrenreich, Barbara. *The Hearts of Men.* Garden City, NY: Anchor/Doubleday, 1984.

———. "Housework Is Obsolescent." *Time* (October 25, 1993): 92.

Ekerdt, David J., and Stanley DeViney. "Evidence for a Preretirement Process among Older Male Workers." *Journal of Gerontology* 48, 2 (March 1993): S35–S43.

Elliott, Diana M., and John Briere. "Sexual Abuse Trauma Among Professional Women: Validating the Trauma Symptom Checklist (TSC-40)." *Child Abuse and Neglect* 16, 3 (May 1992): 391 ff.

Ellwood, David. *Poor Support: Poverty in the American Family.* New York: Basic Books, 1988.

Emery, Robert E., Sheila G. Matthews, and Katherine M. Kitzmann. "Child Custody Mediation and Litigation: Parents' Satisfaction and Functioning One Year after Settlement." *Journal of Consulting and Clinical Psychology* 62, 1 (1994): 124–129.

Eriksen, Shelley, and Naomi Gerstel. "A Labor of Love or Labor Itself: Care Work Among Adult Brothers and Sisters." *Journal of Family Issues* 23, 7 (October 2002): 836–856.

Erikson, Erik. *Childhood and Society.* New York: Norton, 1963.

Eshelman, J. Ross. *The Family* (8th ed.). Needham Heights, MA: Allyn and Bacon, 1997.

Evans-Pritchard, E. E. *Kinship and Marriage among the Nuer.* Oxford: Oxford University Press, 1951.

Evenson, Ranae J., and Robin W. Simon. "Clarifying the Relationship Between Parenthood and Depression." *Journal of Health & Social Behavior* 46, 4 (December 2005): 341–358.

Faderman, Lillian. *Odd Girls and Twilight Lovers.* New York: Columbia University Press, 1991.

Fagot, Beverly, and Mary Leinbach. "Socialization of Sex Roles within the Family." In *Current Conceptions of Sex Roles and Sex Typing,* edited by D. Bruce Carter. New York: Praeger, 1987.

Faludi, Susan. *Backlash: The Undeclared War Against American Women.* New York: Crown, 1991.

"Family Friendly Hotels," family-friendly-hotels.com, 2006.

Farrell, Warren. *The Myth of Male Power.* New York: Berkley Books, 2001

Farquhar, Dion. "Reproductive Technologies Are Here to Stay." *Sojourner* 20, 5 (January 1995): 6–7.

Faulkner, Rhonda A., Maureen Davey, and Adam Davey. "Gender-Related Predictors of Change in Marital Satisfaction and Marital Conflict." *American Journal of Family Therapy* 33, 1 (January–February 2005): 61–83.

Fay, Robert, Charles Turner, Albert Klassen, and John Gagnon. "Prevalence and Patterns of Same-Gender Sexual Contact among Men." *Science* 243, 4889 (January 20, 1989): 338–348.

Feeney, Judith A., and Patricia Noller. "Attachment Style as a Predictor of Adult Romantic Relationships." *Journal of Personality and Social Psychology* 58, 2 (February 1990): 281–291.

Fehr, Beverly. "Prototype Analysis of the Concepts of Love and Commitment." *Journal of Personality and Social Psychology* 55, 4 (1988): 557–579.

Ferree, Myra Marx. "Beyond Separate Spheres: Feminism and Family Research." In *Contemporary Families: Looking Forward, Looking Back,* edited by A. Booth. Minneapolis: National Council on Family Relations, 1991.

Field, Craig, and Raul Caetano. "Intimate Partner Violence in the U.S. General Population: Progress and Future Directions." *Journal of Interpersonal Violence* 20, 4 (April 2005): 463–469.

Fields, Jason. *America's Families and Living Arrangements: 2003.* Current Population Reports, P20-553. U.S. Census Bureau, Washington, DC,2003.

Field, Tiffany. "Quality Infant Day Care and Grade School Behavior and Performance." *Child Development* 62 (1991): 863–870.

———. "America's Families and Living Arrangements, 2003." *Current Populations Reports* P20-533. Washington, D.C.: United States Census Bureau, 2004.

Fields, Jason, and Lynne Casper. "America's Families and Living Arrangements: March 2000." *Current Population Reports* (Series P20, No. 537). Washington, DC: U.S. Census Bureau, 2001.

Filene, Peter. *Him/Her/Self: Sex Roles in Modern America* (2nd ed.). Baltimore: Johns Hopkins University Press, 1986.

Fincham, Frank, and Steven Beach. "Forgiveness in Marriage: Implications for Psychological Aggression and Constructive Communication." *Personal Relationships* 9, 3 (September 2002): 239–251.

Fincham, Frank D., and Steven R. H. Beach. "Conflict in Marriage: Implications for Working with Couples." *Annual Review of Psychology* 50, 1 (1999): 47–77.

Fincham, F. D., and T. N. Bradbury. "The Impact of Attributions in Marriage: A Longitudinal Analysis." *Journal of Personality and Social Psychology* 53 (1987): 510–517.

Fineman, Martha Albertson, and Roxanne Mykitiuk. *The Public Nature of Private Violence: The Discovery of Domestic Abuse.* New York: Routledge, 1994.

Finkel, Judith A., and Finy J. Hansen. "Correlates of Retrospective Marital Satisfaction in Long-Lived Marriages: A Social Constructivist Perspective." *Family Therapy* 19, 1 (1992): 1–16.

Finkelhor, D., and K. Yllo. *License to Rape: Sexual Abuse of Wives.* New York: Holt, Rinehart, and Winston, 1985.

Finkelhor, David. *Sexually Victimized Children.* New York: Free Press, 1979.

———. "Common Features of Family Abuse." In *The Dark Side of Families,* edited by D. Finkelhor et al. Beverly Hills, CA: Sage, 1983.

———. *Child Sexual Abuse: New Theory and Research.* New York: Free Press, 1984.

———. "Prevention: A Review of Programs and Research." In *Sourcebook on Child Sexual Abuse,* edited by D. Finkelhor. Beverly Hills, CA: Sage Publications, 1986a.

———. "Prevention Approaches to Child Sexual Abuse." In *Violence in the Home: Interdisciplinary Perspectives,* edited by M. Lystad. New York: Brunner/Mazel, 1986b.

———. "Sexual Abuse of Children." In *Vision 2010: Families and Violence, Abuse and Neglect,* edited by R. J. Gelles. Minneapolis: National Council on *Family Relations,* 1995.

Finkelhor, David, and Larry Baron. "High Risk Children." In *Sourcebook on Child Sexual Abuse,* edited by D. Finkelhor. Beverly Hills, CA: Sage, 1986.

Finkelhor, David, and Angela Browne. "Initial and Long-Term Effects: A Conceptual Framework." In *Sourcebook on Child Sexual Abuse,* edited by D. Finkelhor. Beverly Hills, CA: Sage, 1986.

Finkelhor, David, and Richard Ormrod. "Factors in the Underreporting of Crimes against Juveniles." *Child Maltreatment* 6, 3 (2001): 219–229.

Finkelhor, David, G. Hotaling, I. A. Lewis, and C. Smith. *Missing, Abducted, Runaway, and Throwaway Children in America.* Washington, DC: U.S. Department of Justice, 1990.

Fischer, Lucy R. "Mothers and Mothers-in-Law." *Journal of Marriage and the Family* 45, 1 (February 1983): 187–192.

Fishman, Barbara. "The Economic Behavior of Stepfamilies." *Family Relations* 32 (July 1983): 356–366.

Fitting, Melinda, Peter Rabins, M. Jane Lucas, and James Eastham. "Caregivers for Demented Patients: A Comparison of Husbands and Wives." *Gerontologist* 26 (1986): 248–252.

Flaks, David K., Ilda Ficher, Frank Masterpasqua, and G. Joseph. "Lesbians Choosing Motherhood: A Comparative Study of Lesbians and Heterosexual Parents and Their Children." *Developmental Psychology* 31, 1 (January 1995): 105–114.

Flynn, Clifton P. "Sex Roles and Women's Response to Courtship Violence." *Journal of Family Violence* 5, 1 (March 1990): 83–94.

Foa, Uriel G., Barbara Anderson, J. Converse, and W. A. Urbansky, "Gender-Related Sexual Attitudes: Some Cross-Cultural Similarities and Differences." *Sex Roles* 16, 19–20 (May 1987): 511–519.

Foley, L., et al. "Date Rape: Effects of Race of Assailant and Victim and Gender of Subjects." *Journal of Black Psychology* 21, 1 (February 1995): 6–18.

Folk, Karen F., and Andrea H. Beller. "Part-Time Work and Child Care Choices for Mothers of Preschool Children." *Journal of Marriage and the Family* 55, 1 (February 1993): 147–157.

Folk, Karen Fox, and Yunae Yi. "Piecing Together Child Care with Multiple Arrangements: Crazy Quilt or Preferred Pattern for Employed Parents of Preschool Children." *Journal of Marriage and the Family* 56, 3 (August 1994): 669–680.

Follingstad, D. R, E. S. Hause, L. L. Rutledge, and D. S. Polek. "Effects of Battered Women's Early Responses on Later Abuse Patterns." *Violence and Victims* 7 (1992): 109–128.

Follingstad, Diane R., L. L. Rutledge, B. J. Berg, and E. S. Haure. "The Role of Emotional Abuse in Physically Abusive Relationships." *Journal of Family Violence* 5, 2 (June 1990): 107–120.

Forste, Renata, and Tim Heaton. "Initiation of Sexual Activity among Female Adolescents." *Youth and Society* 19, 3 (March 1988): 250–268.

Fox, Greer, and Velma McBride Murry. "Gender and Families: Feminist Perspectives and Family Research." *Journal of Marriage and the Family* 62, 4 (November 2000): 1160-1172.

Fox, Greer L., and Robert F. Kelly. "Determinants of Child Custody Arrangements at Divorce." *Journal of Marriage and the Family* 57, 3 (August 1995): 693–708.

Fraser, Julie, Eleanor Maticka-Tyndale, and Lisa Smylie. "Sexuality of Canadian Women at Midlife." *Canadian Journal of Human Sexuality* 13, 3–4 (2004): 171–187.

French, J. P., and Bertram Raven. "The Bases of Social Power." In *Studies in Social Power,* edited by L. Cartwright. Ann Arbor: University of Michigan Press, 1959.

Friedan, Betty. *The Feminine Mystique.* New York: Dell, 1963.

Friedman, Rochelle, and Bonnie Gradstein. *Surviving Pregnancy Loss.* Boston: Little, Brown, 1982.

Frieze, Irene H. "Female Violence Against Intimate Partners: An Introduction." *Psychology of Women Quarterly* 29 (2005): 229–237.

Fuchs, Dale. "Spanish Socialists' Proposals Opposed by Church." *New York Times,* May 30, 2004.

Furstenberg, Frank F., Jr. and A. J. Cherlin. *Divided Families: What Happens to Children When Parents Part.* Cambridge, MA: Harvard University Press, 1991.

Furstenberg, Frank F., Jr. and J. O. Teitler. "Reconsidering the Effects of Marital Disruption: What Happens to Children of Divorce in Early Adulthood?" *Journal of Family Issues* 15, 2 (June 1994): 173–190.

Furstenberg, Frank F., Jr. "Good Dads-Bad Dads: Two Faces of Fatherhood" In *The Changing Family,* edited by A. Cherlin. New York: Urban Institute Press, 1988.

———. "Reflections on Remarriage." *Journal of Family Issues* 1, 4 (1980): 443–453.

Furstenberg, Frank F., Jr., and Christine Nord. "Parenting Apart: Patterns in Childrearing after Marital Disruption." *Journal of Marriage and the Family* 47, 4 (November 1985): 893–904.

Furstenberg, Frank F., Jr., and Graham Spanier, eds. *Recycling the Family—Remarriage after Divorce.* Rev ed. Newbury Park, CA: Sage, 1987.

Furstenberg, Frank F., Jr., Sheela Kennedy, Vonnie C. Mcloyd, Rubén G. Rumbaut, and Richard A. Settersten Jr. "Growing Up Is Harder To Do." *Contexts* 3, 3 (Summer 2004): 33–41.

Fuwa, Makiko. "Macro-level Gender Inequality and the Division of Household Labor in 22 Countries." *American Sociological Review* 69, 6 (December 2004): 751–767.

Gager, Constance T., Teresa M. Cooney, and Kathleen Thiede. "The Effects of Family Characteristics and Time Use on Teenagers' Household Labor." *Journal of Marriage & Family* 61, 4 (November 1999): 982–994.

Gagnon, John, and William Simon. "The Sexual Scripting of Oral Genital Contacts." *Archives of Sexual Behavior* 16, 1 (February 1987): 1–25.

Ganong, L. H., and M. Coleman. "Effects of Remarriage on Children: A Review of the Empirical Literature." *Family Relations* 33 (1984): 389–406.

Ganong, L., and M. Coleman. "Stepparent: a Pejorative Term?" *Psychological Reports* 52 (1983): 919–922.

Ganong, Lawrence, and Marilyn Coleman. "Gender Differences in Expectations of Self and Future Partner." *Journal of Family Issues* 13, 1 (March 1992): 55–64.

———. *Remarried Family Relation*ships. Newbury Park, CA: Sage Publications, 1994.

Ganong, Lawrence, Marilyn Coleman, and Gregory Kennedy. "The Effects of Using Alternate Labels in Denoting Stepparent or Stepfamily Status." *Journal of Social Behavior and Personality* 5 (1990): 453–463.

Gans, Herbert. "Symbolic Ethnicity: The Future of Ethnic Groups and Cultures in America." In *On the Making of Americans,* edited by H. Gans. Philadelphia: University of Pennsylvania, 1979.

Gao, Ge. "Stability of Romantic Relationships in China and the United States." In *Cross-Cultural Interpersonal Communication,* edited by S. Ting-Toomey and F. Korzenny. Newbury Park, CA: Sage, 1991.

Garbarino, James. *Children and Families in the Social Environment.* Hawthorne, NY: Aldine De Gruyter, 1982.

Garfinkel, Irwin, and Sara McLanahan. *Single Mothers and Their Children: A New American Dilemma.* Washington, DC: Urban Institute Press, 1986.

Garner, Abigail. *Families Like Mine: Children of Gay Parents Tell It Like It Is.* New York: HarperCollins, 2005.

Garnets, L., et al. "Violence and Victimization of Lesbians and Gay Men: Mental Health Consequences." *Journal of Interpersonal Violence* 5 (1990): 366–383.

Gecas, Viktor, and Monica Seff. "Social Class, Occupational Conditions, and Self-Esteem." *Sociological Perspectives* 32 (1989): 353–364.

———. "Families and Adolescents." In *Contemporary Families: Looking Forward, Looking Back,* edited by A. Booth. Minneapolis: National Council on *Family Relations,* 1991.

Gelles, Richard J., and Murray Straus. *Intimate Violence: The Definitive Study of the Causes and Consequences of Abuse in the American Family.* New York: Simon and Schuster, 1988.

Gelles, Richard J. "Child Abuse and Violence in Single-Parent Families: Parent Absence and Economic Deprivation." *American Journal of Orthopsychiatry* 59, 4 (October 1989): 492–501.

———. "Through a Sociological Lens: Social Structure and Family Violence." In *Current Controversies in Family Violence,* edited by R. Gelles and D. Loseke. Newbury Park, CA: Sage, 1993.

Gelles, Richard J., and Jon R. Conte. "Domestic Violence and Sexual Abuse of Children: A Review of Research in the Eighties." In *Contemporary Families: Looking Forward, Looking Back,* edited by A. Booth. Minneapolis: National Council on Family Relations, 1991.

Gelles, Richard J., and Claire Pedrick Cornell. *Intimate Violence in Families.* Newbury Park, CA: Sage, 1985.

Gelles, Richard J., and Claire Pedrick Cornell. *Intimate Violence in Families,* 2nd ed. Newbury Park, CA: Sage, 1990.

Genevie, Lou, and Eva Margolies. *The Motherhood Report: How Women Feel about Being Mothers.* New York: Macmillan, 1987.

Gentile, D., and J. R. Gentile. "Violent Video Games as Exemplary Teachers." Paper presented at the Biennial Meeting of the Society for Research in Child Development, Atlanta, GA, April 9, 2005.

Gentile, D., P. Lynch, J. Linder, and D. Walsch. "The Effects of Violent Video Game Habits on Adolescent Hostility, Aggressive Behaviors, and School Performance." *Journal of Adolescence* 27, 1 (February 2004): 5–22.

Gentile, D., and C. Anderson. "Video Games." In *Encyclopedia of Human Development,* Vol. 3, edited by N. J. Salkind. Thousand Oaks, CA: Sage, 2006: 1303–1307.

Gentile, D., and D. Walsch. "A Normative Study of Family Media Habits." *Journal of Applied Developmental Psychology* 23 (2002): 157–178.

George, Kenneth, and Andrew Behrendt. "Therapy for Male Couples Experiencing Relationship Problems and Sexual Problems." *Journal of Homosexuality* 14, 1–2 (1987): 77–88.

Gerris, Jan, Maja Dekovic, and Jan Janssens. "The Relationship between Social Class and Childrearing Behaviors: Parents' Perspective Taking and Value Orientations." *Journal of Marriage and the Family* 59, 4 (November 1997): 834–847.

Gerson, Kathleen. *Hard Choices: How Women Decide About Work, Career, and Motherhood.* Berkeley: University of California Press, 1985.

———. *No Man's Land: Men's Changing Commitments to Family and Work.* New York: Basic Books, 1993.

Gibson-Davis, Christina, Kathryn Edin, and Sara McLanahan. "High Hopes But Even Higher Expectations: The Retreat From Marriage Among Low-Income Couples." *Journal of Marriage and Family* 67, 3 (December 2005): 1301–1312.

Gilbert, Lucia, et al. "Perceptions of Parental Role Responsibilities: Differences between Mothers and Fathers." *Family Relations* 31 (April 1982): 261–269.

Gies, Frances. and Joseph Gies. *Marriage and the Family in the Middle Ages.* New York: Harper & Row, Publishers, 1987.

Gillen, K., and S. J. Muncher. "Sex Differences in the Perceived Casual Structure of Date Rape: A Preliminary Report." *Aggressive Behavior* 21, 2 (1995): 101–112.

Gillespie, Rosemary. "Childfree and Feminine: Understanding the Gender Identity of Voluntarily Childless Women." *Gender & Society* 17, 1 (February 2003): 122–136.

Gilligan, Carol. *In a Different Voice: Psychological Theory and Women's Development.* Cambridge, MA: Harvard University Press, 1982.

Gilmore, David. *Manhood in the Making.* New Haven, CT: Yale University Press, 1990.

Glazer-Malbin, Nona, ed. *Old Family/New Family.* New York: Van Nostrand, 1975.

Gleick, Elizabeth. "Tower of Psychobabble: Pronouncing on the Differences Between the Sexes Has Made John Gray Master of a Self-Help Universe. But Is He More of a Healer or a Huckster?" *Time,* June 16, 1997.

Glenn, Evelyn N., and Stacey G. H. Yap. "Chinese American Families." In *Minority Families in the United States: A Multicultural Perspective,* edited by R. L. Taylor. Englewood Cliffs, NJ: Prentice Hall, 1994.

Glenn, Norval. "Duration of Marriage, Family Composition, and Marital Happiness." *National Journal of Sociology* 3 (1989): 3–24.

———. "The Recent Trend in Marital Success in the United States," *Journal of Marriage and the Family* 53 (2) May 1991: 261–270.

"Who's Who in the Family Wars: A Characterization of the Major Ideological Factions," In *Feuds About Families: Conservative, Centrist, Liberal, and Feminist Perspectives,* edited by N. Benokraitis. Upper Saddle River, NJ: Prentice Hall, 2000: 2–13.

Glenn, Norval, and Beth Ann Shelton. "Regional Differences in Divorce in the United States." *Journal of Marriage and the Family* 47, 3 (August 1985): 641–652.

Glenn, Norval, and Michael Supancic. "The Social and Demographic Correlates of Divorce and Separation in the United States: An Update and Reconsideration." *Journal of Marriage and the Family* 46, 3 (August 1984): 563–575.

Glick, Paul. "The Family Life Cycle and Social Change." *Family Relations* 38, 2 (April 1989): 123–129.

———. "Fifty Years of Family Demography." *Journal of Marriage and the Family* 50, 4 (November 1988): 861–873.

Gnezda, Therese. *The Effects of Unemployment on Family Functioning.* Prepared Statement to the Select Committee on Children, Youth and Families, House of Representatives, at Hearings on the New Unemployed, Detroit, March 4, 1984. Washington, DC: Government Printing Office, 1984.

Goelman, Hillel, et al. "Family Environment and Family Day Care." *Family Relations* 39, 1 (January 1990): 14–19.

Goetting, Ann. "The Six Stages of Remarriage: Developmental Tasks of Remarriage after Divorce." *Family Relations* 31 (April 1982): 213–222.

———. "Patterns of Support among In-Laws in the United States: A Review of Research." *Journal of Family Issues* 11, 1 (1990): 67–90.

Goetz, Kathryn W. and Cynthia J. Schmiege. "From Marginalized to Mainstreamed: The HEART Project Empowers the Homeless." *Family Relations* 45, 4 (October 1996).

Goldenberg, H., and I. Goldenberg. *Counseling Today's Families,* 2nd ed. Pacific Grove, CA: Brooks/Cole, 1994.

Goldscheider, Frances and Sharon Sassler. "Creating Stepfamilies: Integrating Children into the Study of Union Formation." *Journal of Marriage and Family* 68, 2 (May 2006): 273–291.

Goldstein, David, and Alan Rosenbaum. "An Evaluation of the Self-Esteem of Maritally Violent Men." *Family Relations* 34, 3 (July 1985): 425–428.

Goleman, Daniel. "Spacing of Siblings Strongly Linked to Success in Life." *New York Times* (May 28, 1985): 17–18.

———. "Gay Parents Called No Disadvantage." *New York Times* (March 11, 1992).

Gondolf, Edward. "Male Batterers." In *Family Violence: Prevention and Treatment,* edited by R. L. Hampton et al. Newbury Park, CA: Sage, 1993.

Gondolf, Edward W. "Evaluating Progress for Men Who Batter." *Journal of Family Violence* 2 (1987): 95–108.

———. "The Effect of Batterer Counseling on Shelter Outcome." *Journal of Interpersonal Violence* 3, 3 (September 1988): 275–289.

Gongla, Patricia, and Edward Thompson, Jr., "Single Parent Families." In *Handbook of Marriage and the Family,* edited by M. Sussman and S. Steinmetz. New York: Plenum Press, 1987.

Goode, William. "Force and Violence in the Family." *Journal of Marriage and the Family* 33 (November 1971): 624–636.

Goode, William, ed. *The Family.* 2nd ed. Englewood Cliffs, NJ: Prentice Hall, 1982.

Goode, William. "Why Men Resist." *Dissent* 27, 2 (1980): 181–193.

Goodman, Lisa, and Deborah Epstein. "Refocusing on Women: A New Direction for Policy and Research on Intimate Partner Violence." *Journal of Interpersonal Violence* 20, 4 (April 2005): 479–487.

Gordon, Thomas. *P.E.T. in Action.* New York: Bantam Books, 1978.

Gornick, Janet C., and Marcia K. Meyers. "Helping America's Working Parents: What We Can Learn from Europe and Canada." Work and Family Program research paper. Washington, DC: New America Foundation, November 2004. www.newamerica.net/publications/policy/helping_americas_working_parents.

Gottman, John, James Coan, Sybil Carrere, and Catherine Swanson. "Predicting Marital Happiness and Stability from Newlywed Interactions" *Journal of Marriage and the Family* 60, 1 (February 1998): 5-22.

Gottman, John M. *What Predicts Divorce? The Relationship between Marital Processes and Marital Outcomes.* Hillsdale, NJ: Lawrence Erlbaum, 1994.

Gottman, John M. *Why Marriages Succeed or Fail and How You Can Make Yours Work.* New York: Simon and Schuster, 1995.

Gottschalk, Lorene. "Same-sex Sexuality and Childhood Gender Non-conformity: A Spurious Connection." *Journal of Gender Studies* 12, 1 (March 2003):35–51.

Gough, Kathleen. "Is the Family Universal: The Nayer Case." In *A Modern Introduction to the Family,* edited by N. Bell and E. Vogel. New York: Free Press, 1968.

Graham-Kevan, Nicola, and John Archer. "Investigating Three Explanations of Women's Relationship Aggression." *Psychology of Women Quarterly* 29 (2005): 270–277.

Grall, Timothy. "Custodial Mothers and Fathers and Their Child Support: 2001." *Current Population Reports.* Washington, D.C.: U.S. Census Bureau. (October 2003).

Grall, Timothy. "Support Providers: 2002." Current Population Reports (Series P70, No. 99). Washington, DC: U.S. Census Bureau, 2005.

Gray, John. *Men Are From Mars, Women Are From Venus.* New York: Harper Collins. 1992.

Gray, John. *Men Are from Mars, Women Are from Venus.* New York: HarperCollins. 1993.

Greeff, Abraham P., and Tanya deBruyne. "Conflict Management Style and Marital Satisfaction." *Journal of Sex and Marital Therapy* 26, 4 (October 2000): 321–334.

Greenberger, Ellen. "Explaining Role Strain: Intrapersona." *Journal of Marriage and the Family* 52, 1 (February 1994): 115–118.

Greenberger, Ellen, and Robin O'Neil. "Parents' Concern about Their Child's Development: Implications for Father's and Mother's Well-Being and Attitudes toward Work." *Journal of Marriage and the Family* 52, 3 (August 1990): 621–635.

Greenstein, Theodore. "Marital Disruption and the Employment of Married Women." *Journal of Marriage and the Family* 52 (1990): 657–676.

Greenstein, Theodore N. "Husbands' Participation in Domestic Labor: Interactive Effects of Wives' and Husbands' Gender Ideologies." *Journal of Marriage and the Family* 58, 3 (August 1996): 585–595.

Greeson, Larry E. "Recognition and Ratings of Television Music Videos: Age, Gender, and Sociocultural Effects." *Journal of Applied Social Psychology* 21, 23 (1991): 1908–1920.

Greif, Geoffrey. "Children and Housework in the Single Father Family." *Family Relations* 34, 3 (July 1985): 353–357.

Greif, Geoffrey L., and Joan Kristall. "Common Themes in a Group for Noncustodial Parents." *Families in Society* 74, 4 (1993): 240–245.

Griffin-Shelley, Eric. "The Internet and Sexuality: A Literature Review-1983–2002." *Sexual & Relationship Therapy* 18, 3 (August 2003) 355–371.

Griswold, Robert. *Fatherhood in America: A History.* New York: Basic Books, 1993.

Grosswald, Blanche. "The Effects of Shift Work on Family Satisfaction." *Families in Society* 85, 3 (July–September 2004): 413–423.

Grotevant, Harold, and Julie Kohler. "Adoptive Families." In *Parenting and Child Development in "Nontraditional" Families,* edited by M. E. Lamb. Mahwah, NJ: Lawrence Erlbaum, 1999.

Grotevant, H. D., R. G. McCoy, C. Elde, and D. L. Frave. "Adoptive Family System Dynamics: Variations by Level of Openness in the Adoption." *Family Process* 33 (1994): 125–146.

Groth, Nicholas. *Men Who Rape: The Psychology of the Offender.* New York: Plenum Press, 1980.

Grzywacz, Joseph, and Nadine Marks. "Family, Work, Work-Family Spillover, and Problem Drinking During Midlife." *Journal of Marriage and the Family* 62, 2 (May 2000): 336–348.

Guerrero Pavich, Emma. "A Chicana Perspective on Mexican Culture and Sexuality." In *Human Sexuality, Ethnoculture, and Social Work,* edited by L. Lister. New York: Haworth, 1986.

Guldner, Gregory. "Long-Distance Romantic Relationships: Prevalence and Separation-Related Symptoms in College Students." *Journal of College Student Development* 37, 3 (May-June 1996): 289–296.

Guldner, Gregory T., and Clifford H. Swensen. "Time Spent Together and Relationship Quality: Long-Distance Relationships as a Test Case." *Journal of Social and Personal Relationships* 12, 2 (1995): 313–320.

Gulledge, Andrew, Michelle Gulledge, and Robert Stahmann. "Romantic Physical Affection Types and Relationship Satisfaction." *The American Journal of Family Therapy* 31 (2003): 233–242.

Gullo, Karen. "Parents Juggling Multiple Care Arrangements for Kids, Study Shows." Associated Press: March 8, 2000.

Guttentag, M., and P. Secord. *Too Many Women.* Newbury Park, CA: Sage, 1983.

Guttman, Herbert. *The Black Family: From Slavery to Freedom.* New York: Pantheon, 1976.

Guttman, Joseph. *Divorce in Psychosocial Perspective: Theory and Research.* Hillsdale, NJ: Lawrence Erlbaum, 1993.

Haaga, D. A. "Homophobia?" *Journal of Behavior and Personality* 6 (1991): 171–174.

Haferd, Laura. "Paddling Returns to Child Rearing." *San Jose Mercury News* (December 20, 1986): D12.

Hafstrom, Jeanne, and Vicki Schram. "Chronic Illness in Couples: Selected Characteristics, Including Wife's Satisfaction with and Perception of Marital Relationships." *Family Relations* 33 (1984): 195–203.

Hall, D. R., and J. Z. Zhao. "Cohabitation and Divorce in Canada: Testing the Selectivity Hypothesis." *Journal of Marriage and the Family* 57 (May 1995): 421–427.

Hamer, Jennifer, and Kathleen Marchioro. "Becoming Custodial Dads: Exploring Parenting Among Low-Income and Working-Class African American Fathers." *Journal of Marriage & Family* 64, 1 (February 2002): 116–129.

Hamilton, B. E., J. A. Martin, and P. D. Sutton. "Births: Preliminary Data for 2002." National Vital Statistics Report 51 (11). Hyattsville, MD: National Center for Health Statistics (2003).

Hamilton, B., J. Martin, S. Ventura, P. Sutton, and F. Menacker. "Births: Preliminary Data for 2004." *National Vital Statistics Reports* 54, 8. Hyattsville, MD: National Center for Health Statistics, 2004.

Hansen, Christine H., and Ranald D. Hansen. "The Influence of Sex and Violence on the Appeal of Rock Music Videos." *Communication Research* 17, 2 (1990): 212–234.

Hansen, Gary. "Dating Jealousy among College Students." *Sex Roles* 12, 7–8 (April 1985): 713–721.

———. "Extradyadic Relations during Courtship." *Journal of Sex Research* 23, 3 (August 1987): 383–390.

Hanson, B., and C. Knopes. "Prime Time Tuning Out Varied Cultures." *USA Today* (July 6, 1993).

Hanson, Thomas L., Sara S. McLanahan, and Elizabeth Thomson. "Double Jeopardy: Parental Conflict and Stepfamily Outcomes for Children." *Journal of Marriage and the Family,* 58, 1 (February 1996): 141–154.

Hare-Mustin, Rachel T., and Jeanne Marecek. "On Making a Difference." In *Making a Difference: Psychology and the Construction of Gender,* edited by R. T. Hare-Mustin and J. Marcek. New Haven, CT: Yale University Press, 1990.

Harknett, Kristen. "The Relationship between Private Safety Nets and Economic Outcomes among Single Mothers." *Journal of Marriage and Family* 68, 1 (February 2006): 172–191.

Harknett, Kristen, and Sara McLanahan. "Racial and Ethnic Differences in Marriage After the Birth of a Child." *American Sociological Review* 69 (December 2004): 790–811.

Harriman, Lynda. "Personal and Marital Changes Accompanying Parenthood." *Family Relations* 32, 3 (July 1983): 387–394.

Harrington, Michael. *The Other America: Poverty in the United States.* New York: Macmillan, 1962.

Harris, David, and Hiromi Ono. "How Many Interracial Marriages Would There Be If All Groups Were of Equal Size in All Places? A New Look at National Estimates of Interracial Marriage." *Social Science Research* 34, 1 (2005): 236–251.

Harry, Joseph. "Decision Making and Age Differences among Gay Male Couples." In *Gay Relationships,* edited by J. DeCecco. New York: Haworth, 1988.

Hart, J., E. Cohen, A. Gingold, and R. Homburg. "Sexual Behavior in Pregnancy: A Study of 219 Women." *Journal of Sex Education and Therapy* 17, 2 (June 1991): 88–90.

Hassebrauck, Manfred, and Beverly Fehr. "Dimensions of Relationship Quality." *Personal Relationships* 9 (2002): 253–270.

Hatfield, Elaine, and R. Rapson. *Love, Sex, and Intimacy: The Psychology, Biology, and History.* New York: Harper-Collins, 1993.

Hatfield, Elaine, and Susan Sprecher. *Mirror, Mirror: The Importance of Looks in Everyday Life.* New York: State University of New York, 1986.

Hatfield, Elaine, and G. William Walster. *A New Look at Love.* Reading, MA: Addison-Wesley, 1981.

Hawkins, Alan J., and Tomi-Ann Roberts. "Designing a Primary Intervention to Help Dual-Earner Couples Share Housework and Childcare." *Family Relations* 41, 2 (April 1992): 169–177.

Hawkins, Alan J., Tomi-Ann Roberts, Shawn Christiansen, and C. M. Marshall. "An Evaluation of a Program to Help Dual-Earner Couples Share the Second Shift." *Family Relations* 43, 2 (April 1994): 213–220.

Hawkins, Alan, and David Dollahite, eds. *Generative Fathering: Beyond Deficit Perspectives.* Vol. 3, *Current Issues in the Family.* Thousand Oaks, CA: Sage, 1997.

Hawkins, D., and Alan Booth. "'Unhappily Ever After': Effects of Long-Term Low-Quality Marriages on Well Being." *Social Forces* 84, 1 (September 2005): 451–471.

Haynes, Faustina. "Gender and Family Ideals: An Exploratory Study of Black Middle-Class Americans." *Journal of Family Issues* 21, 7, (October 2000): 811–837.

Hays, Sharon. *Cultural Contradictions of Motherhood.* New Haven, CT: Yale University Press, 1996.

Hays, Dorothea, and Aurele Samuels. "Heterosexual Women's Perceptions of their Marriages to Bisexual or Homosexual Men." *Journal of Homosexuality* 18, 2 (1989): 81–100.

Hazan, Cindy, and Philip Shaver. "Romantic Love Conceptualized as an Attachment Process." *Journal of Personality and Social Psychology* 52 (March 1987): 511–524.

"Health Encyclopedia—Diseases and Conditions: Dyspareunia," healthscout.com, 2001.

Healy, Joseph M., Abigail J. Stewart, and Anne P. Copeland. "The Role of Self-Blame in Children's Adjustment to Parental Separation." *Personality and Social Psychology Bulletin* 19, 3 (1993): 279–289.

Heaton, T. "Factors Contributing to Increasing Marital Stability in the United States." *Journal of Family Issues* 23, 3 (2002): 392–409.

Hecht, Michael L., Peter J. Marston, and Linda Kathryn Larkey. "Love Ways and Relationship Quality in Heterosexual Relationships." *Journal of Social & Personal Relationships* 11, 1 (February 1994): 25–43.

Hefner, R., et al. "Development of Sex-Role Transcendence." *Human Development* 18 (1975): 143–158.

Heilbrun, Carolyn. *Toward a Recognition of Androgyny.* New York: Norton, 1982.

Helms-Erikson, Heather. "Marital Quality Ten Years After the Transition to Parenthood: Implications of the Timing of Parenthood and the Division of Housework." *Journal of Marriage and Family* 63 (November 2001): 1099–1110.

Hemstrom, Orjan. "Is Marriage Dissolution Linked to Differences in Morbidity Risks for Men and Women?" *Journal of Marriage and the Family* 58, 2 (May 1996): 366–378.

Henderson, Stephen. "Weddings: Vows—Rakhi Dhanoa and Ranjeet Purewal." *New York Times* (August 18, 2002).

Hendrick, Clyde, and Susan Hendrick. "A Theory and Method of Love." *Journal of Personality and Social Psychology* 50 (February 1986): 392–402.

———. "Attachment Theory and Close Adult Relationships." *Psychological Inquiry* 5, 1 (1994): 38–41.

Hendrick, Susan. "Self-Disclosure and Marital Satisfaction." *Journal of Personality and Social Psychology* 40 (1981): 1150–1159.

Henneck, Rachel. "Family Policy in the U.S., Japan, Germany, Italy and France: Parental Leave, Child Benefits/Family Allowances, Child Care, Marriage/Cohabitation, and Divorce." www.contemporaryfamilies.org, May 2003.

Henslin, James. *Essentials of Sociology: A Down to Earth Approach,* 3rd ed. Needham Heights, MA: Allyn and Bacon, 2000.

Herbert, Bob. "Scapegoat Time." *New York Times* (December 16, 1994): A19.

Herek, Gregory M. "Heterosexuals' Attitudes Toward Bisexual Men and Women in the United States." *Journal of Sex Research* 39, 4 (November 2002): 264–275.

Herring, Cedric, and Karen R. Wilson-Sadberry. "Preference or Necessity? Changing Work Roles of Black and White Women, 1973–1990." *Journal of Marriage and the Family* 55, 2 (May 1993): 314–325.

Herring, Susan, Inna Kouper, Lois Scheidt, and Elijah Wright. "Women and Children Last: The Discursive Construction of Weblogs." In *Into the Blogosphere: Rhetoric, Community, and Culture of Weblogs,* edited by Laura Gurak, Smiljana Antonijevic, Laurie Johnson, Clancy Ratliff, and Jessica Reyman. blog.lib.umn.edu/blogosphere/women_and_children.html, retrieved August 13, 2004.

Hetherington, E. Mavis. "Intimate Pathways: Changing Patterns in Close Personal Relationships Across Time." *Family Relations* 52, 4 (October 2003): 318–331.

Hetherington, E. Mavis and Margaret Stanley-Hagan. "Stepfamilies." In *Parenting and Child Development in 'Nontraditional' Families* edited by M. Lamb. Mahwah, New Jersey: Lawrence Erlbaum, 1999: 137–160.

Hetherington, E. Mavis and John Kelly. *For Better or Worse: Divorce Reconsidered.* New York: W.W. Norton, 2002.

Heuveline, Patrick, and Jeffrey Timberlake. "The Role of Cohabitation in Family Formation: The United States in Comparative Perspective." *Journal of Marriage & Family* 66, 5 (December 2004): 1214–1230.

Hewlett, Sylvia. *A Lesser Life: The Myth of Women's Liberation in America.* New York: Morrow, 1986.

———. *When the Bough Breaks: The Cost of Neglecting Our Children.* New York: Basic Books, 1991.

Hewlett, Sylvia, and Cornel West. *The War Against Parents: What We Can Do for America's Beleaguered Moms and Dads.* New York: Houghton Mifflin, 1998.

Heyl, Barbara. "Homosexuality: A Social Phenomenon." In *Human Sexuality: The Societal and Interpersonal Context,* edited by K. McKinney and S. Sprecher. Norwood, NJ: Ablex, 1989.

Hibbard, R.A., and G. L. Hartman. "Behavioral Problems in Alleged Sexual Abuse Victims." *Child Abuse and Neglect* 16 (1982): 755–762.

Higgins, C., L. Duxbury, and C. Lee. "Impact of Life-cycle Stage and Gender on the. Ability to Balance Work and Family Responsibilities." *Family Relations* 43 (1994): 144–150.

Higginbottom, Susan F., Julian Barling, and E. Kevin Kelloway. "Linking Retirement Experiences and Marital Satisfaction: A Mediational Model." *Psychology and Aging* 8, 4 (1993): 508–516.

Hill, Ivan. *The Bisexual Spouse.* McLean, VA: Barlina Books, 1987.

Hirschl, Thomas A., Joyce Altobelli, and Mark R. Rank. "Does Marriage Increase the Odds of Affluence? Exploring the Life Course Probabilities." *Journal of Marriage and Family* 65, 4 (November 2003): 927–938.

Hochschild, Arlie. *The Second Shift: Working Parents and the Revolution at Home.* New York: Viking Press, 1989.

Hochschild, Arlie Russell. *The Time Bind: When Work Becomes Home and Home Becomes Work.* New York: Holt, 1997.

Hoffman, Kristi L., K. Jill Kiecolt, and John N. Edwards. "Physical Violence between Siblings: A Theoretical and Empirical Analysis." *Journal of Family Issues* 26, 8 (November 2005): 1103–1130.

Hoffman, Lois Wladis. "Maternal Employment: 1979." *American Psychologist* 34 (1979): 859–865.

Holmes, T., and R. Rahe. "The Social Readjustment Rating Scale." *Journal of Psychosomatic Medicine* 11 (1967): 213–218.

Holtzen, D. W., and A. A. Agresti. "Parental Responses to Gay and Lesbian Children." *Journal of Social and Clinical Psychology* 9, 3 (September 1990): 390–399.

Hook, Misty, Lawrence Gerstein, Lacy Detterich, and Betty Gridley. "How Close Are We? Measuring Intimacy and Examining Gender Differences." *Journal of Counseling and Development* 81 (Fall 2003): 462–472.

Hopper, Joseph. "The Symbolic Origins of Conflict in Divorce." *Journal of Marriage and Family* 63 (May 2001): 430–445.

Hort, Barbara, et al. "Are People's Notions of Maleness More Stereotypically Framed Than Their Notions of Femaleness?" *Sex Roles* 23, 3 (February 1990): 197–212.

Hort, Barbara E., M. D. Leinbach, and B. I. Fagot. "Is There Coherence among the Cognitive Components of Gender Acquisition?" *Sex Roles* 24, 3–4 (1991): 195–207.

Horwitz, Alan, and Helene White, "The Relationship of Cohabitation and Mental Health: A Study of a Young Adult Cohort," *Journal of Marriage and the Family* 60 (May 1998): 505–514.

Hotaling, Gerald T., and David B. Sugarman. "A Risk Marker Analysis of Assaulted Wives." *Journal of Family Violence* 5, 1 (March 1990): 1–14.

Hotaling, Gerald T., et al., eds. *Coping with Family Violence: Research and Perspectives.* Newbury Park, CA: Sage, 1988.

———. *Family Abuse and Its Consequences.* Newbury Park, CA: Sage, 1988.

Houran, James, et al. "Do Online Matchmaking Tests Work? An Assessment of Preliminary Evidence for a Publicized 'Predictive Model of Marital Success.'" *North American Journal of Psychology* 6 (2004): 507–526.

Houseknecht, Sharon K. "Voluntary Childlessness." In *Handbook of Marriage and the Family,* edited by M. B. Sussman and S. K. Steinmetz. New York: Plenum Press, 1987.

Houseknecht, Sharon K., Suzanne Vaughan, and Ann Statham. "The Impact of Singlehood on the Career Patterns of Professional Women." *Journal of Marriage and the Family* 49, 2 (May 1987): 353–366.

Houts, Leslie A. "But Was It Wanted? Young Women's First Voluntary Sexual Intercourse." *Journal of Family Issues* 26, 8 (November 2005): 1082–1102.

Howard, Judith. "A Structural Approach to Interracial Patterns in Adolescent Judgments about Sexual Intimacy." *Sociological Perspectives* 31, 1 (January 1988): 88–121.

Howes, Carollee. "Can the Age of Entry into Child Care and the Quality of Child Care Predict Adjustment in Kindergarten?" *Developmental Psychology* 26 (1990): 292–303.

Huang, Chien-Chung, Ronald B. Mincy, and Irwin Garfinkel. "Child Support Obligations and Low-Income Fathers." *Journal of Marriage and Family* 67, 5 (December 2005): 1213–1225.

Hughes, Jean O'Gorman, and Bernice R. Sandler. "Friends Raping Friends—Could It Happen to You?" Project on the Status and Education of Women, Association of American Colleges, 1987.

"Humor Page," nokidding.net, 2006.

Huston, Ted. "The Social Ecology of Marriage and Other Intimate Unions." *Journal of Marriage and the Family* 62, 2 (May 2000): 298–321.

Huston, Ted, S. M. McHale, and A. C. Crouter. "When the Honeymoon's Over: Changes in the Marriage Relationship over the First Year." In *The Emerging Field of Personal Relationships,* edited by S. Duck and R. Gilmour. Hillsdale, NJ: Lawrence Erlbaum, 1986.

Huston, Ted, et al. "From Courtship to Marriage: Mate Selection as an Interpersonal Process." In *Personal Relationships 2: Developing Personal Relationships,* edited by S. Duck and R. Gilmour. London: Academic Press, 1981.

Huston, Ted, and Heidi Melz. "The Case for (Promoting) Marriage: The Devil Is in the Details." *Journal of Marriage and Family* 66, 4 (November 2004): 943–958.

Huston, Ted L., and Gilbert Geis. "In What Ways Do Gender-Related Attributes and Beliefs Affect Marriage?" *Journal of Social Issues* 49, 3 (1993): 87–106.

Hutchins, Loraine, and Lani Kaahumanu. "Overview." In *Bi Any Other Name: Bisexual People Speak Out,* edited by L. Hutchins and L. Kaahumanu. Boston: Alyson Publications, 1991.

Hynie, Michaela, John E. Lydon, Sylvana Cote, and Seth Wiener. "Relational Sexual Scripts and Women's Condom Use: The Importance of Internalized Norms." *Journal of Sex Research* 35, 4 (November 1998): 370–380.

Ihinger-Tallman, Marilyn, and Kay Pasley. "Divorce and Remarriage in the American Family: A Historical Review." In *Remarriage and Stepparenting: Current Research and Theory,* edited by K. Pasley and M. Ihinger-Tallmann. New York: Guilford Press, 1987a.

———. *Remarriage.* Newbury Park, CA: Sage, 1987b.

Impett, Emily, and Letitia Anne Peplau. "Why Some Women Consent to Unwanted Sex With a Dating Partner: Insights from Attachment Theory." *Psychology of Women Quarterly* 26 (2002): 360–370.

Isensee, Rik. *Love Between Men: Enhancing Intimacy and Keeping Your Relationship Alive.* New York: Prentice Hall, 1990.

Ishii-Kuntz, Masako. "Japanese American Families." In *Families in Cultural Context,* edited by M. K. DeGenova. Mountain View, CA: Mayfield, 1997.

Ishi-Kuntz, Masako. "Racial and Ethnic Diversity in Marriage: Asian American Perspective." In *Vision 2004: What is the Future of Marriage,* edited by Paul R. Amato and Nancy Gonzalez, Minneapolis, MN: National Council on Family Relations, 2004: 36–40.

Jaccard, James, Patricia J. Dittus, and Vivian V. Gordan. "Parent-Adolescent Congruency in Reports of Adolescent Sexual Behavior and in Communications about Sexual Behavior." *Child Development* 69, 1 (February 1998): 247–261.

Jackson, Susan M., Fiona Cram, and Fred W. Seymour. "Violence and Sexual Coercion in High School Students' Dating Relationships." *Journal of Family Violence* 15, 1 (March 2000): 23–36.

Jacobs, Jerry, and Kathleen Gerson. The *Time Divide: Work, Family, and Gender Inequality.* Cambridge, MA: Harvard University Press, 2004.

Jankowiak, W., and E. Fischer. "A Cross-Cultural Perspective on Romantic Love," *Ethnology,* 31, 1992: 149–155.

Janus, Samuel and Cynthia Janus. *The Janus Report.* New York: Wiley, 1993.

Jayakody, Rukmalie, and Ariel Kalil. "Social Fathering in Low-Income, African American Families with Preschool Children." *Journal of Marriage & Family* 64, 2 (May 2002): 504–516.

Jeffrey, T. B., and L. K. Jeffrey. "Psychologic Aspects of Sexual Abuse in Adolescence." *Current Opinion in Obstetrics and Gynecology* 3, 6 (December 1991): 825–831.

Jensen, Larry C., and Janet Jensen. "Family Values, Religiosity, and Gender." *Psychological Reports* 73, 2 (October 1993): 429–430.

Jensen-Scott, Rhonda L. "Counseling to Promote Retirement Adjustment." *Career Development Quarterly* 1, 3 (1993): 257–267.

Joesch, Jutta. "The Effects of the Price of Child Care on AFDC Mothers' Paid Work Behavior." *Family Relations* 40, 2 (April 1991): 161–166.

Johnson, Catherine B., Margaret S. Stockdale, and Frank E. Saal. "Persistence of Men's Misperceptions of Friendly Cues across a Variety of Interpersonal Encounters." *Psychology of Women Quarterly* 15, 3 (September 1991): 463–475.

Johnson, David, Lynn White, John Edwards, and Alan Booth. "Dimensions of Marital Quality: Toward Methodological and Conceptual Refinement." *Journal of Family Issues* 7 (1986): 31–49.

Johnson, Julia Overturf, and Barbara Downs. "Maternity Leave and Employment Patterns: 1961–2000." *Current Population Reports* (Series P70, No. 103). Washington, DC: U.S. Census Bureau, 2005.

Johnson, Leanor Boulin. "Perspectives on Black Family Empirical Research: 1965–1978. In *Black Families,* edited by H. Pipes McAdoo. Newbury Park, CA: Sage, 1988.

Johnson, M. "An Exploration of Men's Experience and Role at Childbirth" *Journal of Men's Studies* 10, 2 (Winter 2002).

Johnson, Michael and Janel Leone. "The Differential Effects of Intimate Terrorism and Situational Couple Violence: Findings from the National Violence Against Women Survey." *Journal of Family Issues* 26, 3 (April 2005): 322–349.

Johnson, Michael. "Patriarchal Terrorism and Common Couple Violence: Two Forms of Violence Against Women." *Journal of Marriage and the Family* 57 (2) (May 1995): 283–294.

Johnson, Michael, and Kathleen Ferraro. "Research on Domestic Violence in the 1990's: Making Distinctions," *Journal of Marriage and the Family*, 62, 4 (November 2000): 948–963.

Johnston, Deirdre, and Debra Swanson. "Invisible Mothers: A Content Analysis of Motherhood Ideologies and Myths in Magazines." *Sex Roles* 49 (July 2003).

Jones, Carl. *Mind Over Labor.* New York: Penguin, 1988.

Jones, T. S., and M. S. Remland. "Sources of Variability in Perceptions of and Responses to Sexual Harassment." *Sex Roles* 27, 3–4 (August 1992): 121–142.

Jorgensen, Stephen, and Russell Adams. "Predicting Mexican-American Family Planning Intentions: An Application and Test of a Social Psychological Model." *Journal of Marriage and the Family* 50, 1 (February 1988): 107–119.

Julian, T. W., P. McKenry, and L. McKelvey. "Cultural Variations in Parenting: Perceptions of Caucasian, African American, Hispanic, and Asian American Parents." *Family Relations* (43), 1994: 30–37.

Julien, D., H. J. Markman, and K. Lindahl. "A Comparison of a Global and Microanalytic Coding System: Implications for Future Trends in Studying Interactions." *Behavioral Assessment* 11 (1989): 81–100.

Juni, Samuel, and Donald W. Grimm. "Marital Satisfaction and Sex-Roles in a New York Metropolitan Sample." *Psychological Reports* 73, 1 (1993): 307–314.

Jurich, Anthony, and Cheryl Polson. "Nonverbal Assessment of Anxiety as a Function of Intimacy of Sexual Attitude Questions." *Psychological Reports* 57 (3, Pt. 2), (December 1985): 1247–1243.

Kachadourian, Lorig K., Frank Fincham, and Joanne Davila. "The Tendency to Forgive in Dating and Married Couples: The Role of Attachment and Relationship Satisfaction." *Personal Relationships* 11, 3 (September 2004): 373–393.

Kagan, Jerome, and N. Snidman. "Temperamental Factors in Human Development." *American Psychologist* 46, 8 (1991): 856–862.

Kalis, Pamela, and Kimberly Neuendorf. "Aggressive Cue-Prominence and Gender Participation in MTV." *Journalism Quarterly* 66, 1 (March 1989): 148–154, 229.

Kalmijn, Matthijs. "Intermarriage and Homogamy: Causes, Patterns, Trends." *Annual Review of Sociology* 24 (1998): 395–421.

Kalmijn, Matthijs. "Marriage Rituals as Reinforcers of Role Transitions: An Analysis of Weddings in The Netherlands." *Journal of Marriage & Family* 66, 3 (August 2004): 582–594.

Kalmijn, Matthijs, and Henk Flap. "Assortative Meeting and Mating: Unintended Consequences of Organized Settings for Partner Choices." *Social Forces* 79, 4 (June 2001): 1289–1312.

Kantrowitz, Barbara. "Gay Families Come Out." *Newsweek* (November 4, 1996): 51–57.

Kapinus, Carolyn. "The Effects of Parents' Attitudes Toward Divorce on Offspring's Attitudes: Gender and Parental Divorce as Mediating Factors." *Journal of Family Issues* 25 1 (January 2004): 112–135.

Kaplan, Helen Singer. *Disorders of Desire.* New York: Simon and Schuster, 1979.

Kaplan, L., C. Hennon, and L. Ade-Ridder. "Splitting Custody between Parents: Impact on the Sibling System." *Families in Society: the Journal of Contemporary Human Services CEU Article* 30 (1993):131–144.

Kassop, Mark. "Salvador Minuchin: A Sociological Analysis of His Family Therapy Theory." *Clinical Sociological Review* 5 (1987): 158–167.

Katchadourian, Herant. *Midlife in Perspective.* San Francisco: Freeman, 1987.

Katz, Jennifer, Stephanie Washington Kuffel, and Amy Coblentz. "Are There Gender Differences in Sustaining Dating Violence? An Examination of Frequency, Severity, and Relationship Satisfaction." *Journal of Family Violence* 17, 3 (September 2002): 247–271.

Katz, Lynn Fainsilber, and Erica M. Woodin. "Hostility, Hostile Detachment, and Conflict Engagement in Marriages: Effects on Child and Family Functioning." *Child Development* 73, 2 (March 2002): 636–690.

Katz, Michael B. *The Undeserving Poor: From War on Poverty to War on Welfare.* New York: Pantheon, 1990.

Katzev, A. R., R. L. Warner, and A. C. Acock. "Girls or Boys? Relationship of Child Gender to Marital Instability." *Journal of Marriage and the Family* 56 (February 1994): 89–100.

Kaufman, Joan, and Edward Zigler. "The Intergenerational Transmission of Abuse Is Overstated." In *Current Controversies in Family Violence,* edited by R. Gelles and D. Loseke. Newbury Park, CA: Sage, 1993.

Kawamoto, Walter T., and Tamara C. Cheshire. "American Indian Families." In *Families in Cultural Context,* edited by M. K. DeGenova. Mountain View, CA: Mayfield, 1997.

Keene-Reid, Jennifer, and John Reynolds. "Gender Differences in the Job Consequences of Work-to-Family Spillover." *Journal of Family Issues* 26, 3 (2005): 275–299.

Keillor, Garrison. "It's Good Old Monogamy That's Really Sexy." *Time* (October 17, 1994): 71.

Keith, Pat, and Robbyn Wacker. "Grandparent Visitation Rights: An Inappropriate Intrusion or Appropriate Protection?" *International Journal of Aging and Human Development* 54, 3 (2002): 191–204.

Keller, David, and Hugh Rosen. "Treating the Gay Couple within the Context of Their Families of Origin." *Family Therapy Collections* 25 (1988): 105–119.

Kellett, J. M. "Sexuality of the Elderly." *Sexual and Marital Therapy* 6, 2 (1991): 147–155.

Kelley, Douglas L., and Judee K. Burgoon. "Understanding Marital Satisfaction and Couple Type as Functions of Relational Expectations." *Human Communication Research* 18, 1 (1991): 40–69.

Kelley, Harold. "Love and Commitment." In *Close Relationships*, edited by H. Kelley et al. San Francisco: Freeman, 1983.

Kelley, Harold, et al., eds. *Close Relationships*. San Francisco: Freeman, 1983.

Kempe, C. H., F. Silverman, B. Steele, W. Droegmueller, and H. Silver. "The Battered Child Syndrome." *Journal of the American Medical Association* 181 (1962):17–24.

Kennedy, Gregory E. "College Students' Expectations of Grandparent and Grandchild Role Behaviors." *Gerontologist* 30, 1 (1990): 43–48.

Kennedy, Gregory E., and C. E. Kennedy. "Grandparents: A Special Resource for Children in Stepfamilies." *Journal of Divorce and Remarriage* 19, 3–4 (1993): 45–68.

Keshet, Jamie. "From Separation to Stepfamily." *Journal of Family Issues* 1, 4 (December 1980): 517–532.

Kiecolt, K. Jill. "Satisfaction with Work and Family Life: No Evidence of a Cultural Reversal." *Journal of Marriage and Family* 65 (February 2003): 23–35.

Kikumura, Akemi, and Harry Kitano. "The Japanese American Family." In *Ethnic Families in America: Patterns and Variations*, 3rd ed., edited by C. Mindel et al. New York: Elsevier North Holland, 1988.

Kilborn, Peter T. "Day Care: Key to Welfare Reform." *San Francisco Chronicle* (June 1, 1997): A-4.

Kilmartin, Christopher. *The Masculine Self*. New York: Macmillan, 1994.

Kim, Janna L., and L. Monique Ward. "Pleasure Reading: Associations between Young Women's Sexual Attitudes and Their Reading of Contemporary Women's Magazines." *Psychology of Women Quarterly* 28, 1 (March 2004): 48.

Kim, Jungsik, and Elaine Hatfield. "Love Types and Subjective Well-Being: A Cross-Cultural Study." *Social Behavior and Personality* 32, 2 (2004): 173–182.

———. *The Gendered Society*. New York: Oxford University Press, 2000.

———. *Manhood in America: A Cultural History*. New York: Free Press, 1994.

Kimmel, Michael. "Masculinity as Homophobia: Fear, Shame, and Silence in the Construction of Gender Identity." In *Theorizing Masculinities*, edited by H. Brod and M. Kaufman. Thousand Oaks, CA: Sage, 1995.

———. *Manhood in America: A Cultural History*. Berkeley: University of California Press, 1996.

Kimmel, Michael, and Michael Messner, eds. *Men's Lives*, 4th ed. Needham Heights, MA: Allyn and Bacon, 1998.

Kimmel, Michael, and Rebecca Plante. "The Gender of Desire: The Sexual Fantasies of College Women and Men." In *Sexualities: Identities, Behaviors, and Society*, edited by Michael Kimmel and Rebecca Plante. New York: Oxford University Press, 2004.

King, Laura A. "Emotional Expression, Ambivalence over Expression, and Marital Satisfaction." *Journal of Social and Personal Relationships* 10, 4 (1993): 601–607.

King, Valarie, Kathleen Mullan Harris, and Holly E. Heard. "Racial and Ethnic Diversity in Nonresident Father Involvement." *Journal of Marriage and the Family* 66 (2004): 1–21.

King, Valarie, and Mindy E. Scott. "A Comparison of Cohabiting Relationships among Older and Younger Adults." *Journal of Marriage and Family* 67, 2 (May 2005): 271–285.

Kinsey, Alfred, Wardell Pomeroy, and Clyde Martin. *Sexual Behavior in the Human Male*. Philadelphia: Saunders, 1948.

Kinsey, Alfred, Wardell Pomeroy, Clyde Martin, and P. Gebhard. *Sexual Behavior in the Human Female*. Philadelphia: Saunders, 1953.

Kissman, Kris, and JoAnn Allen. *Single Parent Families*. Newbury Park, CA: Sage, 1993.

Kitano, K., and H. Kitano. The Japanese-American Family" In *Ethnic Families in America: Patterns and Variations*, 4th ed., edited by C. Mindel, R. Habenstein, and R. Wright, Jr. Upper Saddle River, NJ: Prentice Hall, 1998: 311–330.

Kito, Mie. "Self-Disclosure in Romantic Relationships and Friendships among American and Japanese College Students." *Journal of Social Psychology* 145, 2 (April 2005): 127–140.

Kitson, Gay and Leslie Morgan. "The Multiple Consequences of Divorce: A Decade Review. In *Contemporary Families: Looking Forward, Looking Back*, edited by A. Booth. Minneapolis: National Council on *Family Relations*, 1991b.

Kitson, Gay. "Marital Discord and Marital Separation: A County Survey." *Journal of Marriage and the Family* 47 (August 1985): 693–700.

Kitson, Gay C., Richard D. Clark, Norman B. Rushforth, Paul M. Brinich, Howard S. Sudak, and Stephen J. Zyranski. "Research on Difficult Family Topics: Helping New and Experienced Researchers Cope with Research on Loss," *Family Relations* 45, 2 (1996): 183–188.

Kitson, Gay, and Leslie Morgan. "Consequences of Divorce." In *Contemporary Families: Looking Forward, Looking Back* edited by A. Booth. Minneapolis: National Council on *Family Relations*, 1991a.

Kitzinger, Sheila. *The Complete Book of Pregnancy and Childbirth*. New York: Knopf, 1989.

Klein, David M., and James M. White. *Family Theories: An Introduction*. Thousand Oaks, CA: Sage, 1996.

Klein, Fritz. "The Need to View Sexual Orientation as a Multivariate Dynamic Process: A Theoretical Perspective." In *Homosexuality/Heterosexuality: Concepts of Sexual Orientation*, edited by David McWhirter, Stephanie Sandovers, and June Machover Reinisch. New York: Oxford University Press, 1990.

Kline, Marsha, Janet Johnson, and Jeanne Tschann. "The Long Shadow of Marital Conflict: A Model of Children's Postdivorce Adjustment." *Journal of Marriage and the Family* 53 (2) May 1991: 297–309.

Klinetob, Nadya, and David Smith. "Demand-Withdraw Communication in Marital Interaction: Tests of Interspousal Contingency and Gender Role Hypothesis." *Journal of Marriage and the Family*, 58, 4 (November 1996): 945–957.

Klonoff-Cohen, Hillary S., S. L. Edelstein, and E. S. Lefkowitz, "The Effect of Passive Smoking and Tobacco Exposure through Breast Milk on Sudden Infant Death Syndrome." *Journal of the American Medical Association* 273 (1995): 795–798.

Kluwer, Esther, Jose Heesink, and Evert Van de Vliert. "Marital Conflict About the Division of Household Labor and Paid Work." *Journal of Marriage and the Family* 58, 4 (November 1996): 958–969.

———. "The Marital Dynamics of Conflict Over the Division of Labor." *Journal of Marriage and the Family* 59, 3 (August 1997): 635–654.

———. "The Division of Labor across the Transition to Parenthood: A Justice Perspective." Journal of Marriage and Family 64, 4 (November 2002): 930–943.

Knafo, D., and Y. Jaffe. "Sexual Fantasizing in Males and Females." *Journal of Research in Personality* 18 (1984): 451–462.

Knapp, J., and R. Whitehurst. "Sexually Open Marriage and Relationships: Issues and Prospects." In *Marriage and Alternatives: Exploring Intimate Relationships,* edited by R. Libby and R. Whitehurst. Glenview, IL: Scott, Foresman, 1977.

Knoester, Chris, and Alan Booth. "Barriers to Divorce." *Journal of Family Issues* 21, 1 (January 2000): 78–99.

Knox, David, Marty Zussman, and Vivian Daniels. "College Students Attitudes Toward Interreligious Marriage." *College Student Journal* 36, 1 (March 2002): 84–87.

Knox, David, Marty E. Zusman, Vivian Daniels, and Angel Brantley. "Absence Makes the Heart Grow Fonder? Long Distance Dating Relationships Among College Students." *College Student Journal* 36, 3 (2002): 364–366.

Knox, David, and Caroline Schacht. "Sexual Behaviors of University Students Enrolled in a Human Sexuality Course." *College Student Journal* 26, 1 (March 1992): 38–40.

———. *Choices in Relationships: An Introduction to Marriage and the Family,* 6th Edition. Belmont, CA: Wadsworth Pub Co, 2000.

Knox, David, and K. Wilson. "Dating Behaviors of University Students." *Family Relations* 30 (1981): 83–86.

Knox, David, and Tim Britton. "Age Discrepant Relationships Reported by University Faculty and Their Students." *College Student Journal* 31, 3 (September 1997): 280.

Knudson-Martin, Carmen, and Anne Rankin Mahoney. "Language and Processes in the Construction of Equality in New Marriages." *Family Relations* 47, 1 (January 1998): 81–91.

Kohlberg, Lawrence. "The Cognitive-Development Approach to Socialization." In *Handbook of Socialization Theory and Research,* edited by A. Goslin. Chicago: Rand McNally, 1969.

Komarovsky, Mirra. *Blue Collar Marriage.* New York: Vintage, 1962.

———. *Blue-Collar Marriage.* 2nd ed. New Haven, CT: Yale University Press, 1987.

Konner, Melvin. *Childhood.* Boston: Little, Brown, 1991.

Kortenhaus, Carole M., and Jack Demarest. "Gender Role Stereotyping in Children's Literature: An Update." *Sex Roles: A Journal of Research* 28, 3–4 (1993): 219–323.

Kortenhaus, C. M., and J. Demarest. "Gender Role Stereotyping in Children's Literature: An Update." *Sex Roles* 28 (1993): 219–232.

Koss, M., L. Goodman, L. Fitzgerald, N. Russo, G. Keita, and A. Browne. *No Safe Haven: Male Violence Against Women at Home, at Work, and in the Community.* Washington, D.C.: American Psychological Association, 1994.

Koss, Mary. "Hidden Rape: Sexual Aggression and Victimization in a National Sample of Students in Higher Education." In *Rape and Sexual Assault,* vol. 2, edited by A. W. Burgess. New York: Garland, 1988.

Koss, Mary P., and Sarah L. Cook. "Facing the Facts: Date and Acquaintance Rape Are Significant Problems for Women." In *Current Controversies in Family Violence,* edited by R. Gelles and D. Loseke. Newbury Park, CA: Sage, 1993.

Kposowa, Augustine J. "The Impact of Race on Divorce in the United States." *Journal of Comparative Family Studies* 29, 3 (Autumn 1998): 529–548.

Kranichfeld, Marion. "Rethinking Family Power." *Journal of Family Issues* 8, 1 (March 1987): 42–56.

Krause, Neal. "Race Differences in Life Satisfaction among Aged Men and Women." *Journals of Gerontology* 48, 5 (1993): S235–S244.

Kreider, Rose M. "Adopted Children and Stepchildren: 2000." *Census 2000 Special Reports.* Washington, DC: U.S. Census Bureau, 2003.

Kreider, Rose M. "Number, Timing, and Duration of Marriages and Divorces: 2001." *Current Population Reports,* P70–97. Washington, DC, U.S. Census Bureau, 2005.

Kruk, Edward. "Promoting Co-operative Parenting after Separation: A Therapeutic/Interventionist Model of Family Mediation." *Journal of Family Therapy* 15, 3 (1993): 235–261.

Kubey, Robert. "Media Implications for the Quality of Family Life" in *Media, Children and the Family: Social Scientific, Psychodynamic and Clinical Perspectives,* edited by D. Zillman, J. Bryant, and A. C. Huston. Hillsdale, NJ: Lawrence Erlbaum, 1994.

Kübler-Ross, Elisabeth. *On Death and Dying.* New York: Macmillan, 1969.

———. *Working It Through.* New York: Macmillan, 1982.

———. *AIDS: The Ultimate Challenge.* New York: Macmillan, 1987.

Kudson-Martin, C. and A. Mahoney. "Language Processes in the Construction of Equality in Marriages. *Family Relations,* 47 (1998): 81–91.

Kurdek, L. A. "Conflict Resolution Styles in Gay, Lesbian, Heterosexual Nonparent, and Heterosexual Parent Couples." *Journal of Marriage and Family* 56, 3 (August 1994): 705–722.

Kurdek, L. A., and M. A. Fine. "The Relation Between Family Structure and Young Adolescents' Appraisals of Family Climate and Parenting Behavior." *Journal of Family Issues* 14 (June 1993): 279–290.

Kurdek, Lawrence. "Correlates of Negative Attitudes Toward Homosexuals in Heterosexual College Students." *Sex Roles* 18 (1988): 727–738.

Kurdek, Lawrence, and P.J. Smith. "Partner Homogamy in Married, Cohabiting, Heterosexual, Gay and Lesbian Couples." *Journal of Sex Research* 23 (1987): 212–232.

Kurz, Demie. "Caring for Teenage Children." *Journal of Family Issues* 23, 6 (September 2002): 748–767.

Lakoff, Robin. *Language and Women's Place.* New York, 1975.

Lamanna, Mary Ann, and Agnes Riedmann. *Marriages and Families; Making Choices in a Diverse Society.* Belmont, CA: Wadsworth, 1997.

Lamb, K. A., G. Lee, and A. DeMaris. "Union Formation and Depression: Selection and Relationship Effects." *Journal of Marriage and the Family* 65, 4 (2003): 953–962.

Lamb, Michael. "Book Review." *Journal of Marriage and the Family* 55, 4 (November 1993): 1047–1049.

Lamb, Michael (ed.) *The Role Of The Father In Child Development.* New York : Wiley, 1997.

Lamb, Michael (ed.) *The Father's Role: Applied Perspectives.* New York : J. Wiley, 1986.

Lamb, Michael E. "Nonparental Childcare." *Parenting and Child Development in "Nontraditional" Families,* edited by M. Lamb. Mahwah, NJ: Lawrence Erlbaum, 1999.

Lamb, Michael E., ed. *Parenting and Child Development in "Nontraditional" Families.* Mahwah, NJ: Lawrence Erlbaum Associates, 1999.

Laner, Mary R. "Violence or Its Precipitators: Which Is More Likely to Be Identified as a Dating Problem." *Deviant Behavior* 11, 4 (October 1990): 319–329.

Langman, L. "Social Stratification." In *Handbook of Marriage and the Family,* edited by M.G. Sussman and S.K. Steinmetz. New York: Plenum Press, 1987, pp. 211–246.

Laner, Mary R., and N. Ventrone. "Dating Scripts Revisited." *Journal of Family Issues* 21, 4 (May 2000): 488–500.

Lannutti, Pamela J., and Kenzie A. Cameron. "Beyond the Breakup: Heterosexual and Homosexual Post-Dissolutional Relationships." *Communication Quarterly* 50, 2 (2002), 153–170.

Lantz, Herman. "Family and Kin as Revealed in the Narratives of Ex-Slaves." *Social Science Quarterly* 60, 4 (March 1980): 667–674.

Lareau, Annette. "Invisible Inequality: Social Class and Childrearing in Black Families and White families." *American Sociological Review* 67, 5 (October 2002): 747–776.

Lareau, Annette. *Unequal Childhoods.* Berkeley, CA: University of California Press, 2003.

LaRossa, Ralph. "Fatherhood and Social Change." *Family Relations* 37 (1988): 451–458.

LaRossa, Ralph, and Maureen Mulligan LaRossa. *The Transition to Parenthood: How Infants Change Families.* Beverly Hills, CA: Sage, 1981.

LaRossa, Ralph, Charles Jaret, Malati Gadgil, and Robert Wynn. "Gender Disparities in Mother's Day and Father's Day Comic Strips: A 55 Year History." *Sex Roles* 44, 11/12, (June 2001): 693–719.

LaRossa, Ralph, Charles Jarret, Malati Gadgil, and G. Robert Wynn, "The Changing Culture of Fatherhood in Comic-Strip Families: A Six-Decade Analysis." *Journal of Marriage and Family* 62, 2 (May 2000): 375–387.

Larsen, Andrea S., and David H. Olson. "Predicting Marital Satisfaction Using PREPARE: A Replication Study." *Journal of Marital and Family Therapy* 15, 3 (1989): 311–322.

Larson, Jeffry H., Stephan M. Wilson, and Rochelle Beley. "The Impact of Job Insecurity on Marital and Family Relationships." *Family Relations* 43, 2 (April 1994): 138–143.

Larson, Reed, and Mark Ham. "Stress and "Storm and Stress." In "Early Adolescence: The Relationship of Negative Events with Dysphoric Affect" *Developmental Psychology* 29, 1 (January 1993): 130–140.

Larson, Reed, and Maryse Richards. "Family Emotions: Do Young Adolescents and Their Parents Experience the Same States?" *Journal of Research on Adolescence* 4, 4 (1994): 567–583.

Larson, Reed, and Sally Gillman. "Transmission of Emotions in the Daily Interactions of Single-Mother Families." *Journal of Marriage & Family* 61, 1 (February 1999): 21–373.

LaSala, Michael. "Extradyadic Sex and Gay Male Couples: Comparing Monogamous and Nonmonogamous Relationships." *Families in Society: The Journal of Contemporary Social Services* 85, 3 (2004): 405–412.

Lasch, Christopher. *Haven in a Heartless World.* New York: Basic Books, 1977.

Lauer, Jeanette, and Robert Lauer. *'Til Death Do Us Part: How Couples Stay Together.* New York: Haworth Press, 1986.

Laumann, Edward, John Gagnon, Robert Michael, and Stuart Michaels. *The Social Organization of Sexuality: Sexual Practices in the United States.* Chicago: University of Chicago Press, 1994.

Lavee, Y., and D. H. Olson. "Seven Types of Marriage: Empirical Typology Based on ENRICH." *Journal of Marital and Family Therapy* 19 (1993): 325–340.

Lawson, Erma Jean, and Aaron Thompson. "Black Men's Perceptions of Divorce-Related Stressors and Strategies for Coping with Divorce." Journal of *Family Relations* 17, 2 (March 1996): 249–273.

Layne, Linda. "Breaking the Silence: An Agenda for a Feminist Discourse of Pregnancy Loss." *Feminist Studies* 23, 2 (1997): 289–315.

Lederer, William, and Don Jackson. *Mirages of Marriage.* New York: Norton, 1968.

Lee, John A. *The Color of Love.* Toronto: New Press, 1973.

———. "Love Styles." In *The Psychology of Love,* edited by R. Sternberg and M. Barnes. New Haven, CT: Yale University Press, 1988.

Lee, Thomas, Jay Mancini, and Joseph Maxwell. "Contact Patterns and Motivations for Sibling Relations in Adulthood." *Journal of Marriage and the Family* 52, 2 (May 1990): 431–440.

Lehr, Sally, Alice Demi, Colleen DiIorio, Jeffrey Facteau, and Sally T. Lehr. "Predictors of Father–Son Communication about Sexuality." *Journal of Sex Research* 42, 2 (May 2005): 119–129.

Lehrer, E., and C. Chiswick. "The Religious Composition of Unions." *Demography* 30 (1993): 385–404.

Leifer, Myra. *Psychological Effects of Motherhood: A Study of First Pregnancy.* New York: Praeger, 1990.

Leitenberg, H., M. J. Detzer, and D. Srebnik. "Gender Differences in Masturbation and the Relation of Masturbation Experience in Preadolescence and or Early Adolescence to Sexual Behavior and Sexual Adjustment in Young Adulthood." *Archives of Sexual Behavior* 22, 2 (April 1993): 87–98.

Leland, Nancy, and Richard Barth. "Characteristics of Adolescents Who Have Attempted to Avoid HIV and Who Have Communicated with Parents About Sex." *Journal of Adolescent Research* 8, 1 (1993): 58–76.

Leslie, Leigh, and Bethany Letiecq. "Marital Quality of African American and White Partners in Interracial Couples." *Personal Relationships* 11 (2004): 559–574.

Lesnick-Oberstein, M., and L. Cohen. "Cognitive Style, Sensation Seeking, and Assortative Mating." *Journal of Personality and Social Psychology* 46 (1984): 112–117.

Levenson, Robert W., Laura L. Carstensen, and John M. Gottman. "Long-Term Marriage: Age, Gender, and Satisfaction." *Psychology and Aging* 8, 2 (1993): 301–313.

———. "The Influence of Age and Gender on Affect, Physiology, and Their Interrelations: A Study of Long-Term Marriages." *Journal of Personality & Social Psychology* 67, 1 (July 1994): 56–68.

Levin, Irene. "The Model Monopoly of the Nuclear Family." Paper presented at the National Conference on *Family Relations*, Baltimore, November 1993.

Levine, James. *Working Fathers: Strategies for Balancing Work and Family.* Reading, MA: Addison Wesley Longman, 1997.

Levine, Martin. "The Life and Death of Gay Clones." In *Gay Culture in America: Essays from the Field,* edited by G. Herdt. Boston: Beacon, 1992.

Levinger, G. "A Social Psychological Perspective on Marital Dissolution." *Journal of Social Issues* 32, 1 (1976), 21–47.

Levy-Shiff, R. "Individual and Contextual Correlates of Marital Change Across the Transition to Parenthood." *Developmental Psychology,* 30, 4 (1994): 591–601.

Levy-Shiff, Rachel, Ilana Goldshmidt, and Dov Har-Even. "Transition to Parenthood in Adoptive Families." *Developmental Psychology* 27, 1 (1991): 131–140.

Lewin, Tamar. "Financially Set, Grandparents Help Keep Families Afloat, Too." *New York Times* (July 14, 2005): A1.

Lewin, Tamar. "Up From the Holler: Living in Two Worlds, At Home in Neither." *New York Times,* May 19, 2005.

Lewis, Oscar. "The Culture of Poverty." *Scientific American,* 215, 4 (1966): 19–25.

Libman, E. "Sociocultural and Cognitive Factors in Aging and Sexual Expression: Conceptual and Research Issues." *Canadian Psychology* 30, 3 (July 1989): 560–567.

Lichter, Daniel, and Martha Crowley. "Poverty in America: Beyond Welfare Reform." *Population Bulletin* 57, 2 (2002): 3–36.

Lieberman, B. "Extrapremarital Intercourse: Attitudes toward a Neglected Sexual Behavior." *Journal of Sex Research* 24 (1988): 291–299.

Lieberson, Stanley, and Mary Waters. *From Many Strands: Ethnic and Racial Groups in Contemporary America.* New York: Russell Sage Foundation, 1988.

Liebow, Elliot. *Tally's Corner; A Study Of Negro Streetcorner Men.* Boston, Little, Brown, 1967.

Lin, Chin-Yau Cindy, and Victoria Fu. "A Comparison of Child-rearing Practices among Chinese, Immigrant Chinese, and Caucasian-American Parents." *Child Development* 61, 2 (April 1990): 429–434.

Lin, I-Fen, Nora Cate Schaeffer, Judith A. Seltzer, and Kay L. Tuschen "Divorced Parents' Qualitative and Quantitative Reports of Children's Living Arrangements." *Journal of Marriage & Family* 66, 2 (May 2004): 385–397.

Lindsey, Linda. *Gender Roles: A Sociological Perspective,* 3rd ed. Upper Saddle River, NJ: Prentice Hall, 1997.

Lino, Mark. "Expenditures on a Child by Husband-Wife Families." *Family Economics Review* 3, 3 (1990): 2–12.

———. *Expenditures on Children by Families: 2002.* Washington, DC: U.S. Department of Agriculture, Center for Nutrition Policy and Promotion, 2003.

———. "Expenditures on Children by Families: 2003." Family Economics and Nutrition Review 16, 1 (Winter 2004): 31–38.

Lips, Hilary. *Sex and Gender.* 3rd ed. Mountain View, CA: Mayfield, 1997.

Little, Margaret A. "The Impact of the Custody Plan on the Family: A Five-Year Follow-up: Executive Summary." *Family and Conciliation Courts Review* 30, 2 (1992): 243–251.

Livernois, Joe. "County Gears Up for Huge Welfare Shift." *Monterey County Herald* (June 1, 1997): A-1, A-8.

Lloyd, Sally A. "Physical and Sexual Violence During Dating and Courtship." In *Vision 2010: Families and Violence, Abuse and Neglect,* edited by R. J. Gelles. Minneapolis: National Council on *Family Relations,* 1995.

Lloyd, Sally A., and Beth C. Emery. "The Dynamics of Courtship Violence." Paper presented at the annual meeting of the National Council on Family Relations, Seattle, WA, November 1990.

London, Richard, James Wakefield, and Richard Lewak. "Similarity of Personality Variables as Predictors of Marital Satisfaction." *Personality and Individual Differences* 11, 1 (1990): 39–43.

Long, Vonda O., and Estella A. Martinez. "Masculinity, Femininity, and Hispanic Professional Women's Self-Esteem and Self-Acceptance." *Journal of Counseling & Development* 73, 3 (November/December 1994): 183–186.

Longman, Phillip. "The Cost of Children." U.S. News and World Report (March 30, 1998): 51–58.

LoPresto, C., M. Sherman, and N. Sherman. "The Effects of a Masturbation Seminar on High School Males' Attitudes, False Beliefs, Guilt, and Behavior." *Journal of Sex Research* 21 (1985): 142–156.

Lorber, Judith. *Paradoxes of Gender.* New Haven, CT: Yale University Press, 1994.

———. *Gender Inequality: Feminist Theories and Politics.* Los Angeles: Roxbury, 1998.

"Louisiana Civil Code Art. 98. Mutual Duties Of Married Persons," marriagedebate.com, 1999.

Loulan, JoAnn. *Lesbian Sex.* San Francisco: Spinsters Books, 1984.

Low, David. "Nintendo Reveal Sales Figures: Including World-wide DS Sales up to September." palgn.com.au/article .php?id=3605&sid=4cea9e5e814470cb7ea6fd462d04a13e, December 19, 2005.

Lowenstein, Ludwig. "Causes and Associated Features of Divorce as Seen by Recent Research." *Journal of Divorce & Remarriage* 42, 3/4 (2005): 153–171.

Lutz, Patricia. "The Stepfamily: An Adolescent Perspective." *Family Relations* 32, 3 (July 1983): 367–375.

Maccoby, Eleanor E., Christy M. Buchanan, Robert H. Mnookin, and Sanford M. Dornbusch. "Postdivorce Roles of Mothers and Fathers in the Lives of Their Children." *Journal of Family Psychology* 7, 1 (1993): 24–38.

MacDonald, W. L., and A. DeMaris. "Remarriage, Stepchildren, and Marital Conflict. Challenges to the Incomplete Institutionalization Hypothesis. *Journal of Marriage and the Family* 57, 2 (May 1995): 387–398.

Mace, David, and Vera Mace. "Enriching Marriage." In *Family Strengths,* edited by N. Stinnet et al. Lincoln: University of Nebraska Press, 1979.

———. "Enriching Marriages: The Foundation Stone of Family Strength." In *Family Strengths: Positive Models for Family Life,* edited by N. Stinnet et al. Lincoln: University of Nebraska Press, 1980.

Madden, Mary, and Amanda Lenhart. "On-line Dating." Pew Internet and American Life Project, www.pewinternet.org, March 5, 2006.

Mahdi, Akbar. "Role Reversal and its Consequences in the Iranian Immigrant Family." In *Men and Masculinity: A Text-Reader,* edited by Theodore Cohen, Belmont, CA: Wadsworth Publishing, 2001.

Makepeace, James. "Dating, Living Together, and Courtship Violence." In *Violence in Dating Relationships: Emerging Social Issues,* edited by M. Pirog-Good and J. Stets. New York: Praeger, 1989.

Maloney, Lawrence. "Behind Rise in Mixed Marriages." *U.S. News and World Report* (February 10, 1986): 68–69.

Mancini, Jay, and Rosemary Bliszner. "Research on Aging Parents and Adult Children." *Journal of Marriage and the Family* 51, 2 (May 1989): 275–290.

———. "Aging Parents and Adult Children: Research Themes in Intergenerational Relations." In *Contemporary Families: Looking Forward, Looking Back,* edited by A. Booth. Minneapolis: National Council on *Family Relations,* 1991.

———. "Social Provisions in Adulthood: Concept and Measurement in Close Relationships." *Journal of Gerontology* 47, 1 (1992): 14–20.

Manning, Wendy. "Children and the Stability of Cohabiting Couples." *Journal of Marriage & Family* 66, 3 (August 2004): 674–689.

Manning, Wendy D., and Kathleen Lamb. "Adolescent Well-Being in Cohabiting, Married, and Single-Parent Families." *Journal of Marriage & Family* 65, 4 (November 2003): 876–893.

Manning, Wendy D., and Pamela Smock. "First Comes Cohabitation and Then Comes Marriage?" *Journal of Family Issues* 23, 8 (November 2002): 1065–1087.

Marecek, Jeanne, et al. "Gender Roles in the Relationships of Lesbians and Gay Men." In *Gay Relationships,* edited by J. DeCecco. New York: Haworth Press, 1988.

Margolin, Gayla, Linda Gorin Sibner, and Lisa Gleberman. "Wife Battering." In *Handbook of Family Violence,* edited by V. B. Van Hasselt et al. New York: Plenum Press, 1988.

Margolin, Leslie. "Sexual Abuse by Grandparents." *Child Abuse and Neglect* 16, 5 (September 1992): 735.

Marin, Rick. "At-Home Fathers Step Out to Find They Are Not Alone." *New York Times* (January 2, 2000): 1, 16.

Markman, Howard. "Prediction of Marital Distress: A Five-Year Followup." *Journal of Consulting and Clinical Psychology* 49 (1981): 760–761.

———. "The Longitudinal Study of Couples' Interactions: Implications for Understanding and Predicting the Development of Marital Distress." In *Marital Interaction,* edited by K. Hahlweg and N. S. Jacobsen. New York: Guilford Press, 1984.

Markman, Howard, et al. "The Prediction and Prevention of Marital Distress: A Longitudinal Investigation." In *Understanding Major Mental Disorders: The Contribution of Family Interaction Research,* edited by K. Hahlweg and M. Goldstein. New York: Family Process Press, 1987.

Marks, Nadine. "Flying Solo at Midlife: Gender, Marital Status, and Psychological Well-Being." *Journal of Marriage & Family* 58, 4 (November 1996): 917–932.

Marks, Nadine, James Lambert, and Heejeong Choi. "Transitions to Caregiving, Gender and Psychological Well-Being: A Prospective U.S. National Study," *Journal of Marriage and the Family* 64, 3 (August 2002): 657–667.

Marks, Stephen. *Three Corners: Integrating Marriage & The Self.* Lexington, MA: Lexington Books, 1986.

Marks, Stephen. "What Is a Pattern of Commitment?" *Journal of Marriage and the Family* 52, 1 (February 1994): 112–115.

Marmor, Judd. "Homosexuality and the Issue of Mental Illness." In *Homosexual Behavior,* edited by J. Marmor. New York: Basic Books, 1980a.

———. "The Multiple Roots of Homosexual Behavior." In *Homosexual Behavior,* edited by J. Marmor. New York: Basic Books, 1980b.

———, ed. *Homosexual Behavior.* New York: Basic Books, 1980c.

"Marriage Rights and Benefits," nolo.com, 2006.

Marsiglio, William. "Paternal Engagement Activities with Minor Children." *Journal of Marriage & Family* 53, 4 (November 1991): 973–986.

———. *Procreative Man.* New York: New York University, 1998.

———. *Stepdads : Stories Of Love, Hope, And Repair.* Lanham, MD: Rowman & Littlefield, 2004.

Martin, Douglas. "Many Dads Struggle to Fit New Roles." *New York Times* (June 20, 1993): 11.

Martin, Karin A. "Giving Birth Like a Girl," *Gender & Society* 17, 1 (February 2003): 54–72.

Martin, S. P. "Growing Evidence for a Divorce Divide? Education and Marital Dissolution Rates in the United States Since the 1970s." Russell Sage Foundation Working Papers: Series on Social Dimensions of Inequality. New York: Russell Sage Foundation, 2004.

Martin, Steven, and Sangeeta Parashar. "Women's Changing Attitudes Toward Divorce, 1974–2002: Evidence for an Educational Crossover." *Journal of Marriage and Family* 68, 1 (February 2006): 29–40.

Martin, T. C., and L. L. Bumpass. "Recent Trends in Marital Disruption." *Demography* 26 (February 1989): 37–51.

Marwell, G., et al. "Legitimizing Factors in the Initiation of Heterosexual Relationships." Paper presented at the First International Conference on Personal Relationships, Madison, WI, July 1982.

Masheter, Carol. "Postdivorce Relationships Between Ex-Spouses: The Roles of Attachment and Interpersonal Conflict." *Journal of Marriage and the Family* 53 (1) April 1991: 103–110.

Masters, John, et al. "The Role of the Family in Coping with Childhood Chronic Illness." In *Coping with Chronic Disease,* edited by T. Burish and L. Bradley. New York: Academic Press, 1983.

Masters, William, and Virginia Johnson. *Human Sexual Inadequacy.* Boston: Little, Brown, 1970.

Mathis, Richard D., and Zoe Tanner. "Cohesion, Adaptability, and Satisfaction of Family Systems in Later Life." *Family Therapy* 18, 1 (1991): 47–60.

Maticka-Tyndale, Eleanor. "Sexual Scripts and AIDS Prevention: Variations in Adherence to Safer-Sex Guidelines." *Journal of Sex Research* 28, 1 (February 1991): 145–166.

Matiella, Ana Consuelo. *Positively Different: Creating a Bias-Free Environment for Children.* Santa Cruz, CA: Network Publications, 1991.

Mattessich, Paul, and Reuben Hill. "Life Cycle and Family Development." In *Handbook of Marriage and the Family,* edited by M. Sussman and S. Steinmetz. New York: Plenum Press, 1987.

Matthews, Sarah, and Tana Rosner. "Shared Filial Responsibility: The Family as the Primary Caregiver." *Journal of Marriage and the Family* 50, 1 (February 1988): 185–195.

Matthews, T., and B. Hamilton. "Trend Analysis of the Sex Ratio at Birth in the United States." *National Vital Statistics Reports* 53, 20–37. Hyattsville, MD: National Center for Health Statistics, 2006.

Mays, Vickie M., S. D. Cochran, G. Bellinger, and R. G. Smith. "The Language of Black Gay Men's Sexual Behavior: Implications for AIDS Risk Reduction." *Journal of Sex Research* 29, 3 (August 1992): 425–434.

Max, Sarah. "Living With the In-Laws: Multigenerational Households, Though Still Uncommon, Seem to Be Growing in Popularity." Money/CNN.com (April 22, 2004).

Mazor, Miriam. "Barren Couples." *Psychology Today* (May 1979): 101–108, 112.

Mazor, Miriam, and Harriet Simons, eds. *Infertility: Medical, Emotional and Social Considerations.* New York: Human Sciences Press, 1984.

McAdoo, Harriette Pipes. "Changes in the Formation and Structure of Black Families: The Impact on Black Women." Working paper no. 182, Center for Research on Women, Wellesley College, Wellesley, MA, 1988a.

———, ed. *Black Families,* 2nd ed. Beverly Hills, CA: Sage, 1988b.

———. *Black Families,* 3rd ed. Thousand Oaks, CA: Sage, 1996.

McCaghy Charles, and Timothy Capron. *Deviant Behavior: Crime, Conflict and Interest Groups.* Boston, MA: Allyn & Bacon, 1997.

McCaghy, Charles, Timothy Capron, and J. D. Jamieson. *Deviant Behavior: Crime, Conflict, and Interest Groups,* 5th ed. Needham Heights, MA: Allyn and Bacon, 2000.

McFarlane, Judith, Jacquelyn Campbell, and Kathy Watson. "Intimate Partner Stalking and Femicide: Urgent Implications for Women's Safety." *Behavioral Sciences and the Law* 20 (2002): 51–68.

McGill, Michael. *The McGill Report on Male Intimacy.* New York: Henry Holt, 1985.

McGoldrick, Monica, and B. Carter, eds. *The Changing Family Life Cycle,* 2nd ed. Boston: Allyn and Bacon, 1989.

McHale, Susan, and Ted Huston. "The Effect of the Transition to Parenthood on the Marriage and Family Relationship: A Longitudinal Study." *Journal of Family Issues* 6, 4 (December 1985): 409–433.

McHugh, Maureen, Nicole Livingston, and Amy Ford. "A Postmodern Approach to Women's Use of Violence: Developing Multiple and Complex Conceptualizations." *Psychology of Women Quarterly* 29, 4 (December 2005): 323–336.

McKim, Margaret K. "Transition to What? New Parents' Problems in the First Year." *Family Relations* 36, 1 (January 1987): 22–25.

McKinnon, Jesse. "The Black Population in the United States: March 2002." *U.S. Census Bureau Current Population Reports.* Series P20-541. Washington, D.C. (2003).

McKinnon, Jesse, and C. Bennett. "We the People: Blacks in the United States: Census 2000 Special Reports." *U.S. Bureau of the Census,* 2005.

McLanahan, Sara, and Karen Booth. "Mother-Only Families: Problems, Prospects, and Politics." In *Contemporary Families: Looking Forward, Looking Back,* edited by A. Booth. Minneapolis: National Council on *Family Relations,* 1991.

McLanahan, Sara, and Karen Booth. "Mother-Only Families: Problems, Prospects, and Politics." *Journal of Marriage and the Family* (51) 1989: 557–580.

McLeer, S. V., et al. "Sexually Abused Children at High Risk for Post-Traumatic Stress Disorder." *Journal of the American*

Academy of Child and Adolescent Psychiatry 31, 5 (September 1992): 875–879.

McLoyd, Vonnie, Ana Marie Cauce, David Takeuchi, and Leon Wilson. "Marital Processes and Parental Socialization in Families of Color: A Decade Review of Research." *Journal of Marriage and the Family* 62, 4 (November 2000): 1070–1093.

McLoyd, Vonnie, and Julia Smith. "Physical Discipline and Behavior Problems in African American, European American, and Hispanic Children: Emotional Support as a Moderator" *Journal of Marriage and the Family* 64, 1 (February 2002): 40–53.

McMahon, Kathryn. "The Cosmopolitan Ideology and the Management of Desire." *Journal of Sex Research* 27, 3 (August 1990): 381–396.

McManus, Patricia, and Thomas DiPrete. "Losers and Winners: The Financial Consequences of Separation and Divorce for Men." *American Sociological Review* 66 (April 2001): 246–268.

McRoy, R. G., H. D. Grotevant, and S. Ayers-Lopez. *Changing Patterns in Adoption.* Austin, TX: Hogg Foundation for Mental Health, 1994.

Mead, Margaret. *Male and Female.* New York: Morrow, 1975.

Mederer, H., and R. J. Gelles. "Compassion or Control: Intervention in Cases of Wife Abuse" *Journal of Interpersonal Violence* 4, 1 (March 1989): 25–43.

Medora, Nilufer P., Jeffry H. Larson, Nuran Hortaçsu, and Parul Dave. "Perceived Attitudes towards Romanticism: A Cross-Cultural Study of American, Asian-Indian, and Turkish Young Adults." *Journal of Comparative Family Studies* 33, 2 (Spring 2002): 155–178.

Melito, Richard. "Adaptation in Family Systems: A Developmental Perspective." *Family Processes* 24 (1985): 89–100.

Menaghan, Elizabeth, and Toby Parcel. "Parental Employment and Family Life Research in the 1980s." In *Contemporary Families: Looking Forward, Looking Back,* edited by A. Booth. Minneapolis: National Council on *Family Relations,* 1991.

Merriam-Webster Online Dictionary. "Cybersex," m-w.com, 2006.

Mertensmeyer, C., and Coleman, M. " Correlates of Inter-Role Conflict in Young Rural and Urban Parents." *Family Relations* 36 (1976), 425-429. Also in R. Marotz-Baden, C. Hennon, and T. Brubaker (Eds.) *Families in Rural America: Stress, Adaptation, and Revitalization,* Minneapolis: National Council on Family Relations.

Messerschmidt, James. *Masculinities and Crime.* Nanham, MD: Rowman and Littlefield, 1993.

Metts, Sandra, and William Cupach. "The Role of Communication in Human Sexuality." In *Human Sexuality: The Social and Interpersonal Context,* edited by K. McKinney and S. Sprecher. Norwood, NJ: Ablex, 1989.

Metz, Michael E., and Norman Epstein. "Assessing the Role of Relationship Conflict in Sexual Dysfunction." *Journal of Sex and Marital Therapy* 28, 2 (March 2002): 139–164.

Meyer, Daniel R., and Judi Bartfeld. "Compliance with Child Support Orders in Divorce Cases." *Journal of Marriage and the Family* 58, 1 (February 1996): 201–212.

Meyer, Daniel R., Elizabeth Phillips, and Nancy L. Maritato. "The Effects of Replacing Income Tax Deductions with Children's Allowance." *Journal of Family Issues* 12, 4 (December 1991): 467–491.

Michael, Robert, John Gagnon, Edward Laumann, and Gina Kolata. *Sex in America: The Definitive Survey.* Boston: Little, Brown, 1994.

Migliaccio, Todd. "Abused Husbands: A Narrative Analysis." *Journal of Family Issues* 23, 1 (January 2002): 26–52.

Milhausen, Robin, and Edward S Herold. "Does the Sexual Double Standard Still Exist? Perceptions of University Women." *The Journal of Sex Research* 36, 4 (November 1999): 361–368.

Miller, Glenn. "The Psychological Best Interests of the Child." *Journal of Divorce and Remarriage* 19, 1–2 (1993): 21–36.

Miller, Richard B., Jeremy B. Yorgason, Jonathan G. Sandberg, and Mark B. White. "Problems That Couples Bring To Therapy: A View Across the Family Life Cycle." *American Journal of Family Therapy* 31, 5 (October–December 2003): 395.

Miller, Ron. "Black and White Television." *San Jose Mercury News* (June 15, 1992): D1, D5.

Mills, C. Wright. *The Sociological Imagination.* New York, Oxford University Press, 1959.

Mills, David. "A Model for Stepparent Development." *Family Relations* 33 (1984): 365–372.

Mindel, Charles H., Robert W. Habenstein, and Roosevelt Wright Jr., eds. *Ethnic Families in America: Patterns and Variations,* 3rd. ed. New York: Elsevier North Holland, 1976, 1981, 1988.

Minkler, Meredith, and Kathleen M. Roe. *Grandmothers as Caregivers: Raising Children of the Crack Cocaine Epidemic.* Family Caregiver Applications Series 2. Newbury Park, CA: Sage, 1993.

Mintz, Steven. *Huck's Raft: A History of American Childhood.* Cambridge, MA: Belknap Press of Harvard University; New Ed edition: 2004.

Mintz, Steven, and Susan Kellogg. *Domestic Revolutions: A Social History of American Family Life.* New York: Free Press, 1988.

Minuchin, Salvador. *Family Therapy Techniques.* Cambridge, MA: Harvard University Press, 1981.

Miracle, Tina, A. Miracle, and R. Baumeister, *Human Sexuality: Meeting Your Basic Needs.* Upper Saddle River, NJ: Prentice Hall, 2003.

Mirowsky, John. "Parenthood and Health: The Pivotal and Optimal Age at First Birth." *Social Forces* 81, 1 (2002): 315.

"Missouri Revised Statutes: Chapter 451—Marriage, Marriage Contracts, and Rights of Married Women (Section 451.030)," www.moga.state.mo.us/statutes/C400-499/4510000030.htm, August 2005.

Moffatt, Michael. *Coming of Age in New Jersey: College and American Culture.* New Brunswick, NJ: Rutgers University Press, 1989.

Money, John. *Love and Lovesickness.* Baltimore: Johns Hopkins University Press, 1980.

Montagu, Ashley. *Touching,* 3rd ed. New York: Columbia University Press, 1986.

Montgomery, Marilyn J., and Gwendoly T. Sorell. "Differences in Love Attitudes Across Family Life Stages." *Family Relations* 46, 1 (January 1997): 55–61.

Moore, M. M. "Nonverbal Courtship Patterns in Women: Context and Consequences." *Ethology and Sociobiology* 6, 2 (1985): 237–247.

Moore, Teresa. "Study Reveals Deep Scars of Divorce." *San Francisco Chronicle* (June 3, 1997): Al.

Morgan, S. Philip, Diane Lye, and Gretchen Condran. "Sons, Daughters, and the Risk of Marital Disruption." *American Journal of Sociology* 94 (July 1988): 110–129.

Morrison, Donna Ruane, and Amy Ritualo. "Routes to Children's Economic Recovery After Divorce: Are Cohabitation and Remarriage Equivalent?" *American Sociological Review* 65 (August 2000): 560–580.

Morrison, Donna R., and Andres J. Cherlin. "The Divorce Process and Young Children's Well-Being: A Prospective Analysis." *Journal of Marriage and the Family* 57, 3 (August 1995): 800–812.

Morse, Katherine A., and Steven L. Neuberg. "How Do Holidays Influence Relationship Processes and Outcomes? Examining the Instigating and Catalytic Effects of Valentine's Day." *Personal Relationships* 11, 4 (December 2004): 509–527.

Mosher, D. L., and S. S. Tomkins. "Scripting the Macho Man: Hypermasculine Socialization and Enculturation." *Journal of Sex Research* 25 (February 1988): 60–84.

Moynihan, Daniel Patrick. *The Negro Family: The Case for National Action.* Washington, DC: U.S. Government Printing Office, 1965.

Muehlenhard, Charlene. "Misinterpreted Dating Behaviors and the Risk of Date Rape." *Journal of Social and Clinical Psychology* 9, 1 (1988): 20–37.

Muehlenhard, Charlene L., and S. W. Cook. "Men's Self-Reports of Unwanted Sexual Activity." *Journal of Sex Research* 24 (1988): 58–72.

Muehlenhard, Charlene L., and L. C. Hollabaugh. "Do Women Sometimes Say No When They Mean Yes? The Prevalence and Correlates of Women's Token Resistance to Sex." *Journal of Personality and Social Psychology* 54 (May 1988), 872–879.

Muehlenhard, Charlene L., and M. Linton. "Date Rape and Sexual Aggression in Dating Situations." *Journal of Consulting Psychology* 34 (April 1987): 186–196.

Muehlenhard, Charlene L., and M. L. McCoy. "Double Standard/Double Bind." *Psychology of Women Quarterly* 15 (1991): 447–461.

Muehlenhard, Charlene L., I. G. Ponch, J. L. Phelps, and L. M. Giusti. "Definitions of Rape: Scientific and Political Implications." *Journal of Social Issues* 48, 1 (Spring 1992): 23–44.

Muehlenhard, Charlene L., and J. Schrag. "Nonviolent Sexual Coercion." In *Acquaintance Rape: The Hidden Crime,* edited by A. Parrot and L. Bechhofer. New York: Wiley, 1991.

Mullen, Paul E. "The Crime of Passion and the Changing Cultural Construction of Jealousy." *Criminal Behavior and Mental Health* 3, 1 (1993): 1–11.

Munson, M. L., and P. D. Sutton. "Births, Marriages, Divorces, and Deaths: Provisional Data for November 2005." *National Vital Statistics Reports* 54, 18. Hyattsville, MD: National Center for Health Statistics, 2006.

"Muppet Gender Gap." *Media Report to Women* 21, 1 (1993): 8.

Muram, David, K., Miller, and A. Cutler. "Sexual Assault of the Elderly Victim." *Journal of Interpersonal Violence* 7, 1 (March 1992): 70–76.

Murdock, George. *Social Structure.* New York: Free Press, 1967.

Murnen, S. K., A. Perot, and D. Byrne. "Coping with Unwanted Sexual Activity: Normative Responses, Situational Determinants, and Individual Differences." *Journal of Sex Research,* 26 (1989): 85–106.

Murphy-Berman, Virginia, Helen Levesque, and John Berman. "U.N. Convention on the Rights of the Child." *American Psychologist* 51, 12 (December 1996): 1257–1262.

Murstein, Bernard. *Who Will Marry Whom: Theories and Research in Marital Choice.* New York: Springer Publishing, 1976.

———. *Paths to Marriage: Family Studies Text Series,* vol. 5. Beverly Hills, CA: Sage, 1986.

———. "A Clarification and Extension of the SVR Theory of Dyadic Pairing." *Journal of Marriage and the Family* 49 (1987): 929–933.

Myers, Jane and Matthew Shurts. "Measuring Positive Emotionality: A Review of Instruments Assessing Love." *Measurement and Evaluation in Counseling and Development* 34 (January 2002): 238–254.

Nadelson, Carol, and Maria Sauzier. "Intervention Programs for Individual Victims and Their Families." In *Violence in the Home: Interdisciplinary Perspectives,* edited by M. Lystad. New York: Brunner/Mazel, 1986.

Nanda, Serena. *Neither Man Nor Woman: The Hijras of India.* Belmont. CA: Wadsworth, 1990.

National Center for Health Statistics. "Advance Report of Final Natality Statistics, 1991." *Monthly Vital Statistics Report* 42, 3 (Supplement, September 9, 1993).

———. *Healthy People 2000 Review.* Hyattsville, Maryland: Public Health Service, 1994.

National Center for Health Statistics, 2000.

———. "Births, Marriages, Divorces and Deaths: Provisional Data for 2001" *National Vital Statistics Report* 50, 14 (September 2002).

National Climatic Data Center. "Climate of 2005: Summary of Hurricane Katrina," www.ncdc.noaa.gov, December 29, 2005.

National Committee on Pay Equity. "The Wage Gap by Gender and Race," infoplease.com, 2006.

Nelsen, Jane. *Positive Discipline.* New York: Ballantine, 1987.

Neuman, W. Lawrence. *Social Research Methods: Qualitative and Quantitative Approaches,* 4th ed. Needham Heights, MA: Allyn and Bacon, 2000.

Newcombe, Nora. *Child Development: Change Over Time,* 8th ed. New York: HarperCollins, 1996.

Newcomer, Susan, and Richard Udry. "Oral Sex in an Adolescent Population." *Archives of Sexual Behavior* 14, 1 (February 1985): 41–46.

Newman, David. *Sociology of Families.* Thousand Oaks, CA: Pine Forge, 1999.

Newman, Katherine. *Falling from Grace.* New York: Free Press, 1988.

Ney, Philip G. "Transgenerational Abuse." In *Intimate Violence: Interdisciplinary Perspectives,* edited by E. C. Viano. Washington, DC: Hemisphere, 1992.

Nichols, William C., and Mary A. Pace-Nichols. "Developmental Perspectives and Family Therapy: The Marital Life Cycle." *Contemporary Family Therapy: An International Journal* 15, 4 (1993): 299–315.

Nicholson, Jan M., David Fergusson, and L. John Horwood. "Effects on Later Adjustment of Living in a Stepfamily During Childhood and Adolescence." *Journal of Child Psychology* 40, 3, (1999): 405–416.

Nielsen, Joyce McCarl, Russell K. Endo, and Barbara L. Ellington. "Social Isolation and Wife Abuse: A Research Report." In *Intimate Violence,* edited by E. Viano. Washington, DC: Hemisphere Publishing, 1992.

"Nielsen Reports Americans Watch TV at Record Levels," www.nielsenmedia.com, 2005.

Niolon, Richard. "Sibling Sexual Abuse," psychpage.com, 2000.

Noller, Patricia. *Nonverbal Communication and Marital Interaction.* Oxford, England: Pergamon, 1984.

Noller, Patricia, and Mary Anne Fitzpatrick. "Marital Communication." In *Contemporary Families: Looking Forward, Looking Back,* edited by A. Booth. Minneapolis: National Council on Family Relations, 1991.

Nomaguchi, Kei, Melissa Milkie and Suzanne Bianchi. "Time Strains and Psychological Well-Being: Do Dual Earner Mothers and Fathers Differ?" *Journal of Family Issues* (September 2005): 756–791.

Noonan, Mary. "The Impact of Domestic Work on Men's and Women's Wages." *Journal of Marriage and Family* 63 (November 2001) 1134–1145.

Norton, Arthur J. "Family Life Cycle: 1980." *Journal of Marriage and the Family* 45 (1983): 267–275.

Notarius, Clifford, and Jennifer Johnson. "Emotional Expression in Husbands and Wives." *Journal of Marriage and the Family* 44, 2 (May 1982): 483–489.

Notman, Malkah T., and Eva P. Lester. "Pregnancy: Theoretical Considerations." *Psychoanalytic Inquiry* 8, 2 (1988): 139–159.

Nurmi, Jari Erik. "Age Differences in Adult Life Goals, Concerns, and Their Temporal Extension: A Life Course Approach to Future-Oriented Motivation." *International Journal of Behavioral Development* 15, 4 (1992): 487–508.

Nwoye, Augustine. "A Framework for Intervention in Marital Conflicts over Family Finances: A View from Africa." *American Journal of Family Therapy* 28, 1 (January–March 2000): 75–87.

Nye, F. Ivan. "Fifty Years of Family Research." *Journal of Marriage and the Family* 50, 2 (May 1988): 305–316.

Oakley, Ann ed. *Sex, Gender, and Society.* Rev. ed. New York: Harper and Row, 1985.

Ogunwole, Stella. *The American Indian and Alaskan Native Population: 2000. Census 2000 Brief C2KBR/01-15.* Washington, D.C.: U.S. Bureau of the Census (February 2002).

Oggins, Jean, Joseph Veroff, and Douglas Leber. "Perceptions of Marital Interaction among Black and White Newlyweds." *Journal of Personality and Social Psychology* 65, 3 (1993): 494–511.

Olds, S. W. *The Eternal Garden: Seasons of Our Sexuality.* New York: Times Books, 1985.

O'Leary, K. Daniel. "Through a Psychological Lens: Personality Traits, Personality Disorders, and Levels of Violence." In *Current Controversies in Family Violence,* edited by R. Gelles and D. Loseke. Newbury Park, CA: Sage, 1993.

Olenick, I. "Odds of Spousal Infidelity Are Influenced by Social and Demographic Factors." *Family Planning Perspectives* 32, 3 (May/June 2000): 148.

Oliver, Mary Beth, and Janet Shibley Hyde. "Gender Differences in Sexuality: A Meta-analysis." *Psychological Bulletin* 114, 1 (1993): 29–51.

Olson, David H., and John DeFrain. *Marriage and the Family; Diversity and Strengths,* 2nd ed. Mountain View, CA: Mayfield, 1997.

O'Neil, Robin, and Ellen Greenberger. "Patterns of Commitment to Work and Parenting: Implications for Role Strain." *Journal of Marriage and the Family* 52, 1 (February 1994): 101–115.

Ono, Hiromi. "Marital History Homogamy between the Divorced and the Never Married among non-Hispanic Whites." *Social Science Research* 34, 2 (2005): 333.

Opie, Anne. "Ideologies of Joint Custody." *Family and Conciliation Courts Review* 31, 3 (1993): 313–326.

Oropesa, R. S., and Nancy S. Landale. "The Future of Marriage and Hispanics." *Journal of Marriage and Family* 66, 4 (November 2004): 901–920.

Oster, Sharon. "A Note on the Determinants of Alimony." *Journal of Marriage and the Family* 49, 1 (February 1987): 81–86.

O'Sullivan, Lucia, and E. Sandra Byers. "College Students' Incorporation of Initiator and Restrictor Roles in Sexual Dating Interactions." *Journal of Sex Research* 29, 3 (August 1992): 435–446.

Over, Ray, and Gabriel Phillips. "Differences between Men and Women in Age Preferences for a Same-Sex Partner." *Behavioral and Brain Sciences* 20 (1997): 137–143.

Overturf Johnson, Julia. "Who's Minding the Kids? Child Care Arrangements: Winter 2002." Current Population Reports. P70-101. U.S. Census Bureau, Washington, D.C. (2005).

Overturf Johnson, Julia, and Barbara Downs. "Maternity Leave and Employment Patterns: 1961-2000." *Current Population Report P70-103.* U.S. Census Bureau, Washington, D.C. (2005).

Padilla, E. R., and K. E. O'Grady. "Sexuality among Mexican Americans: A Case of Sexual Stereotyping." *Journal of Personality and Social Psychology* 52 (1987): 5–10.

Palkovitz, Robert. "Fathers' Birth Attendence, Early Contact, and Extended Contact with Their Newborns: A Critical Review." *Child Development* 56, 2 (April 1985): 392–407.

———. "Reconstructing 'Involvement': Expanding Conceptualizations of Men's Caring in Contemporary Families." In *Generative Fathering: Beyond Deficit Perspectives,* Vol. 3, *Current Issues in the Family,* edited by A. Hawkins and D. Dollahite. Thousand Oaks, CA: Sage, 1997: 200–216.

Paludi, M. A. "Sociopsychological and Structural Factors Related to Women's Vocational Development." *Annals of New York Academy of Sciences,* 602 (1990): 157–168.

Papanek, Hannah. "Men, Women, and Work: Reflections on the Two-Person Career." *American Journal of Sociology* 78, 4 (January 1973): 90–110.

Papernow, Patricia L. *Becoming a Stepfamily.* San Francisco: Jossey-Bass, 1993.

Parents, Families, and Friends of Lesbians and Gays. "What Is Marriage, Anyway?" pflag.org, 2006.

Park, Kristin. "Stigma Management among the Voluntarily Childless." *Sociological Perspectives* 45, 1 (Spring 2002): 21–45.

Pasley, Kay. "Family Boundary Ambiguity: Perceptions of Adult Stepfamily Family Members." In *Remarriage and Stepparenting: Current Research and Theory,* edited by K. Pasley and M. Ihinger-Tallman. New York: Guilford Press, 1987.

Pasley, Kay, and Marilyn Ihinger-Tallman, eds. *Remarriage and Stepparenting: Current Research and Theory.* New York: Guilford Press, 1987.

Patterson, Charlotte. "Lesbian and Gay Parents and Their Children: Summary of Research Findings." In *Lesbian and Gay Parenting American Psychological Association* (2005).

Paul, L., and J. Galloway. "Sexual Jealousy: Gender Differences in Response to Partner and Rival." *Aggressive Behavior* 20, 33 (1994): 203–211.

Paulson, Sharon, and Cheryl Somers. "Students' Perceptions of Parent-Adolescent Closeness and Communication about Sexuality: Relations with Sexual Knowledge, Attitudes, and Behaviors." *Journal of Adolescence* 23, 5 (October, 2000): 629–644.

Peoples, James and Garrick Bailey. *Humanity: An Introduction to Cultural Anthropology,* 7th ed. Belmont, CA: Wadsworth Publishing, 2006.

Peplau, Letitia. "Research on Homosexual Couples." In *Gay Relationships,* edited by J. DeCecco. New York: Haworth Press, 1988.

Peplau, Letitia, and Susan Cochran. "Value Orientations in the Intimate Relationships of Gay Men." *Gay Relationships,* edited by J. DeCecco. New York: Haworth Press, 1988.

Peplau, Letitia Anne, and Steven Gordon. "The Intimate Relationships of Lesbians and Gay Men." In *Gender Roles and Sexual Behavior,* edited by E. Allgeier and N. McCormick. Palo Alto, CA: Mayfield, 1982.

Peplau, Letitia Anne, Charles T. Hill, and Zick Rubin. "Sex Role Attitudes in Dating and Marriage: A 15-Year Follow-Up of the Boston Couples Study." *Journal of Social Issues* 49, 3 (1993): 31–53.

Peplau, Letitia Anne, Rosemary Veniegas, and Susan Miller Campbell. "Gay and Lesbian Relationships." (pp. 200–215) In *Sexualities: Identities, Behaviors, and Society,* edited by Michael Kimmel, Rebecca Plante. New York: Oxford University, 2004.

Perkins, Kathleen. "Psychosocial Implications of Women and Retirement." *Social Work* 37, 6 (1992): 526–532.

Perlman, Daniel, and Steve Duck, eds. *Intimate Relationships: Development, Dynamics, and Deterioration.* Beverly Hills, CA: Sage, 1987.

Perry-Jenkins, Maureen, and Karen Folk. "Class, Couples, and Conflict: Effects of the Division of Labor on Assessments of Marriage in Dual-Earner Marriages." *Journal of Marriage and the Family* 56, 1 (February 1994): 165–180.

Perry-Jenkins, Maureen, Rena Repetti, and Ann Crouter. "Work and Family in the 1990's." *Journal of Marriage and the Family,* 62, 4, (November 2000): 981–998.

Peters, Stefanie, et al. "Prevalance." In *Sourcebook on Child Sexual Abuse,* edited by D. Finkelhor. Beverly Hills, CA: Sage, 1986.

Petersen, Larry. "Interfaith Marriage and Religious Commitment among Catholics." *Journal of Marriage and the Family* 48, 4 (November 1986): 725–735.

Petersen, Virginia, and Susan B. Steinman. "Helping Children Succeed after Divorce: A Court-Mandated Educational Program for Divorcing Parents." *Family and Conciliation Courts Review* 32, 1 (1994): 27–39.

Peterson, Gary W., and Boyd C. Rollins. "Parent-Child Socialization." In *Handbook of Marriage and the Family,* edited by M. B. Sussman and S. K. Steinmetz. New York: Plenum Press, 1987.

Peterson, Marie Ferguson. "Racial Socialization of Young Black Children." In *Black Children: Social, Educational, and Parental Environments,* edited by H. Pipes McAdoo and J. McAdoo. Beverly Hills, CA: Sage, 1985.

Peterson, Richard R. "A Reevaluation of the Economic Consequences of Divorce." *American Sociological Review* 61, 3 (June 1996).

Peyrot, Mark, et al. "Marital Adjustment to Adult Diabetes: Interpersonal Congruence and Spouse Satisfaction." *Journal of Marriage and the Family* 50, 2 (May 1988): 363–376.

Phillips, Julie A., and Megan M. Sweeney. "Premarital Cohabitation and Marital Disruption among White, Black, and Mexican American Women." *Journal of Marriage and Family* 67, 2 (May 2005): 296–314.

———. "Can Differential Exposure to Risk Factors Explain Recent Racial and Ethnic Variation in Marital Disruption?" *Social Science Research* 35, 2 (June 2006): 409–434.

Pill, Cynthia. "Stepfamilies: Redefining the Family." *Family Relations* 39, 2 (April 1990): 186–193.

Pillemer, Karl, and J. Jill Suitor. "Elder Abuse." In *Handbook of Family Violence,* edited by V. B. Van Hasselt et al. New York: Plenum Press, 1988.

Pina, Darlene, and Vern Bengston. "The Division of Household Labor and Wives' Happiness: Ideology, Employment, and

Perceptions of Support." *Journal of Marriage and Family* 55 (November 1993): 901–912.

Piotrkowski, Chaya, Robert Rapoport, and Rhona Rapoport. "Families and Work" In *Handbook of Marriage and the Family,* edited by M. Sussman and S. Steinmetz. New York: Plenum, 1987: 251–283.

Pistole, M. Carole. "Attachment in Adult Romantic Relationships: Style of Conflict Resolution and Relationship Satisfaction." *Journal of Social Personal Relationships* 6, 4 (1989): 505–512.

Pleck, Joseph. *Working Wives/Working Husbands.* Beverly Hills, CA: Sage, 1985.

Pogrebin, Letty Cottin. *Family Politics.* New York: McGraw-Hill, 1983.

Polatnik, M. Rivka. "Too Old for Child Care? Too Young for Self Care? Negotiating After-School Arrangements for Middle School." *Journal of Family Issues* 23, 6 (September 2002): 728–747.

Pollack, William. *Real Boys: Rescuing Our Sons from the Myths of Boyhood.* New York: Random House, 1998.

Popenoe, David. "American Family Decline, 1960–1990: A Review and Appraisal." *Journal of Marriage and the Family* 55, 1993: 527–555.

———. *Life Without Father: Compelling New Evidence That Fatherhood and Marriage Are Indispensable for the Good of Children and Society.* New York: Free Press, 1996.

Popovich, Paula M., et al. "Perceptions of Sexual Harassment as a Function of Sex of Rater and Incident Form and Consequence." *Sex Roles* 27 (December 1992): 609–625.

Population Resource Center. Executive Summary: A Demographic Profile of Hispanics in the U.S. Washington, D.C. (2001).

Power, Chris, and Bryan Rodgers. "Heavy Alcohol Consumption and Marital Status: Disentangling the Relationship in a National Study of Young Adults." *Addiction* 94, 10 (October 1999): 1477–1487.

Presser, Harriet. "Nonstandard Work Schedules and Marital Instability." *Journal of Marriage and the Family,* 62, 1 (February 2000): 93–110.

Presser, Harriet. *Working in a 24/7 Economy: Challenges for American Families.* New York: Russell Sage Foundation, 2003.

Price, John. "North American Indian Families." In E*thnic Families in America: Patterns and Variations,* 2nd ed., edited by C. Mindel and R. Habenstein. New York: Elsevier-North Holland, 1981.

Priest, Ronnie. "Child Sexual Abuse Histories among African-American College Students: A Preliminary Study." *American Journal of Orthopsychiatry* 62, 3 (July 1992): 475.

Proctor, B., and J. Dalaker. *Poverty in the United States: 2002. Current Population Reports, Series P60-222.* U.S. Bureau of the Census: Washington, D.C. (2003).

Quiles, Jose. "Romantic Behaviors of University Students: A Cross-Cultural and Gender Analysis in Puerto Rico and the United States." *College Student Journal* 37, 3 (September 2003): 354.

Ramirez Barranti, Chrystal. "The Grandparent/Grandchild Relationship: Family Resource in an Era of Voluntary Bonds." *Family Relations* 34, 3 (July 1985): 343–353.

Rank, Mark R. and Li-Chen Cheng. "Welfare Use Across Generations: How Important Are the Ties That Bind?" *Journal of Marriage and the Family* 57, 3 (August 1995): 673–684.

Rao, Kavitha, et al. "Child Sexual Abuse of Asians Compared with Other Populations." *Journal of the American Academy of Child and Adolescent Psychiatry* 31, 5 (September 1992): 880 ff.

Rapp, Carol, and Sally Lloyd. "The Role of 'Home as Haven' Ideology in Child Care Use." *Family Relations* 38, 4 (October 1989): 427–430.

Raschke, Helen. "Divorce." In *Handbook of Marriage and the Family,* edited by M. Sussman and S. Steinmetz. New York: Plenum Press, 1987.

Real, Terrence. *I Don't Want to Talk About It: Overcoming the Secret Legacy of Male Depression.* New York: Fireside, 1997.

Reed, David, and Martin Weinberg. "Premarital Coitus: Developing and Establishing Sexual Scripts." *Social Psychology Quarterly* 47, 2 (June 1984): 129–138.

Reedy, M., et al. "Age and Sex Differences in Satisfying Love Relationships across the Adult Life Span." *Human Development* 24 (1981): 52–86.

Reeves, Terrance, and Claudette Bennett. "The Asian and Pacific Island Population in the United States: March 2002." *Current Populations Reports.* Washington, DC: Government Printing Office. www.census.gov, 2003.

Reeves, Terrance J., and Claudette E. Bennett. *We the People: Asians in the United States, Census 2000 Special Reports, CENSR-17.* U.S. Census Bureau, Washington, DC, 2004.

Regan, Pamela. *The Mating Game: A Primer on Love, Sex, and Marriage.* Thousand Oaks, CA: Sage, 2003.

Regan, Pamela C., E. R. Koca, and T. Whitlock. "Ain't Love Grand! A Prototype Analysis of Romantic Love." *Journal of Social and Personal Relationships* 15 (1999): 411–420.

Reilly, Mary E., et al. "Tolerance for Sexual Harassment Related to Self-Reported Sexual Victimization." *Gender and Society* 6, 1 (March 1992): 122–138.

Reisman, C. *Divorce Talk: Women and Men Make Sense of Personal Relationships.* New Brunswick, NJ: Rutgers University Press, 1990.

Reiss, Ira. *Family Systems in America.* 3rd ed. New York: Holt, Rinehart, and Winston, 1980.

Renzetti, Claire. "Violence in Gay and Lesbian Relationships." In *Vision 2010: Families and Violence, Abuse and Neglect,* edited by R. J. Gelles. Minneapolis: National Council on *Family Relations,* 1995.

Renzetti, Claire, and Daniel Curran. *Living Sociology.* Needham Heights, MA: Allyn and Bacon, 1998.

———. *Women, Men and Society,* 4th ed. Boston: Allyn and Bacon, 1999.

———. *Women, Men, and Society,* 5th ed. Needham Heights, MA: Allyn and Bacon, 2003.

Rexroat, Cynthia. "Race and Marital Status Differences in the Labor Force Behavior of Female Family Heads: The Effect of

Household Structure." *Journal of Marriage and the Family* 52, 3 (August 1990): 591–601.

Rexroat, Cynthia, and Constance Shehan. "The Family Life Cycle and Spouses' Time in Housework." *Journal of Marriage and the Family* 49, 4 (November 1987): 737–750.

Rice, F. Phillip, and Kim Gale Dolgin. *The Adolescent: Development, Relationships and Culture,* 10th ed. Boston: Allyn and Bacon, 2002.

Rice, Susan. "Sexuality and Intimacy for Aging Women: A Changing Perspective." *Journal of Women and Aging* 1, 1–3 (1989): 245–264.

Rich, M., E. Woods, E. Goodman, J. Emans, and R. DuRant. "Aggressors or Victims: Gender and Race in Music Video Violence." *Pediatrics* 101, 4 (April 1998): 669–674.

Richards, Leslie N., and Cynthia J. Schmiege. "Problems and Strengths of Single-Parent Families: Implications for Practice and Policy." *Family Relations* 42, 3 (July 1993): 277–285.

Ridley, Jane, and Michael Crowe. "The Behavioural-Systems Approach to the Treatment of Couples." *Sexual and Marital Therapy* 7, 2 (1992): 125–140.

Riggs, David S. "Relationship Problems and Dating Aggression: A Potential Treatment Target." *Journal of Interpersonal Violence* 8, 1 (1993): 18–35.

Riggs, Janet Morgan. "Impressions of Mothers and Fathers on the Periphery of Childcare." *Psychology of Women Quarterly* 29, 1 (2005): 58–62.

Riseden, Andrea D., and Barbara E. Hort. "A Preliminary Investigation of the Sexual Component of the Male Stereotype." Unpublished paper, 1992.

Risman, Barbara. "Can Men Mother? Life as a Single Father." *Family Relations* (35) 1986: 95–102.

Risman, Barbara. "Can Men Mother? Life as a Single Father." In *Gender in Intimate Relationships,* edited by B. Risman and P. Schwartz. Belmont, CA: Wadsworth, 1989.

Risman, Barbara. "Intimate Relationships From a Microstructural Perspective:Men Who Mother." Gender & Society 1, 1 (March 1987): 6–32.

———. *Gender Vertigo: American Families in Transition.* New Haven, CT: Yale University Press, 1998.

Risman, Barbara, and Danette Johnson-Sumerford. "Doing It Fairly: A Study of Post Gender Marriages." *Journal of Marriage and the Family* 60 (February 1998): 23–40.

Roberts, Michael C., Kristi Lekander, and Debra Fanurik. "Evaluation of Commercially Available Materials to Prevent Child Sexual Abuse and Abduction." *American Psychologist* 45, 6 (June 1990): 782–783.

Roberts, Nicole, and Robert Levenson. "The Remains of the Workday: Impact of Job Stress and Exhaustion on Marital Interaction in Police Couples." *Journal of Marriage and Family* 63, 4 (November 2001): 1052–1067.

Robinson, I., K. Ziss, B. Ganza, and S. Katz. "20 Years of the Sexual Revolution, 1965–1985—An Update." *Journal of Marriage and the Family* 53, 1 (February 1991): 216–220.

Rodgers, Kathleen Boyce, and Rose, Hilary A. "Risk and Resiliency Factors among Adolescents Who Experience Marital Transitions." *Journal of Marriage and Family* 64, 4 (November 2002): 1024–1037.

Rodgers, Roy. Family *Interaction and Transaction: The Developmental Approach.* Englewood Cliffs, NJ: Prentice Hall, 1973.

Rodgers, Roy, and L. Conrad. "Courtship for Remarriage: Influences on Family Reorganization after Divorce." *Journal of Marriage and the Family* 48 (1986): 767–775.

Roehling, Patricia V., Lorna Hernandez Jarvis, and Heather Swope. "Variations in Negative Work-Family Spillover Among White, Black, and Hispanic American Men and Women: Does Ethnicity Matter?" *Journal of Family Issues* 26, 6 (September 2005): 840–865.

Roenrich, L., and B. N. Kinder. "Alcohol Expectancies and Male Sexuality: Review and Implications for Sex Therapy." *Journal of Sex and Marital Therapy* 17 (1991): 45–54.

Rogers, Stacy. "Dollars, Dependency, and Divorce: Four Perspectives on the Role of Wives' Income." *Journal of Marriage and the Family* 66, 1 (February 2004): 59–74.

Rogers, Susan M., and Charles F. Turner. "Male-Male Sexual Contact in the U.S.A.: Findings from Five Sample Surveys, 1970–1990." *Journal of Sex Research* 28, 4 (November 1991): 491–519.

Romance Writers of America. Home page, rwanational.org, 2006.

Rose, Amanda. "Co-Rumination in the Friendships of Girls and Boys." *Child Development* 73, 6 (December 2002): 1830–1843.

Rosen, M. P., et al. "Cigarette Smoking: An Independent Risk Factor for Atherosclerosis in the Hypograstric-Cavernous Arterial Bed of Men with Arteriogenic Impotence." *Journal of Urology* 145, 4 (April 1991): 759–776.

Rosenberg, Joshua D. "In Defense of Mediation." *Family and Conciliation Courts Review* 30, 4 (1992): 422–467.

Rosengren, A., H. Wedel, and L. Wilhelmsen. "Marital Status and Mortality in Middle-Aged Swedish Men." *American Journal of Epidemiology* 129, 1 (January 1989): 54–64.

Ross, Catherine. "Reconceptualizing Marital Status as a Continuum of Social Attachment," *Journal of Marriage and the Family* 57 (1) February 1995: 129–140.

Ross, Catherine, John Mirowsky, and Karen Goldsteen. "The Impact of the Family on Health." In *Contemporary Families: Looking Forward, Looking Back,* edited by A. Booth. Minneapolis: National Council on *Family Relations,* 1991.

Rossi, Alice. "Transition to Parenthood." *Journal of Marriage and the Family* 30 (1) February 1968: 26–39.

Rubin, A. M., and J. R. Adams. "Outcomes of Sexually Open Marriages." *Journal of Sex Research* 22 (1986): 311–319.

Rubin, Lillian. *Worlds of Pain: Life in the Working Class Family.* New York: Basic Books, 1976.

———. *Intimate Strangers: Men and Women Together.* New York: Perennial, 1983.

———. *Just Friends: The Role of Friendship in Our Lives.* New York: Harper and Row, 1985.

———. *Erotic Wars.* New York: Farrar, Straus and Giroux, 1990.

———. *Families on the Faultline: America's Working Class Speaks about the Family, the Economy, Race, and Ethnicity.* New York: HarperCollins, 1994.

Rubin, Zick. *Liking and Loving: An Invitation to Social Psychology.* New York: Holt, Rinehart, and Winston, 1973.

Rudd, J., and S. Herzberger. "Brother-sister Incest—Father-daughter Incest: A Comparison of Characteristics and Consequences." *Child Abuse and Neglect* 23, 9 (September 1999): 915–928.

Runyan, William. "In Defense of the Case Study Method." *American Journal of Orthopsychiatry* 52, 3 (July 1982): 440–446.

Russell, Diana. *Sexual Exploitation: Rape, Child Sexual Abuse, and Workplace Harassment.* Beverly Hills, CA: Sage, 1984.

Russell, Diana E. H. *The Secret Trauma: Incest in the Lives of Girls and Women.* New York: Basic Books, 1986.

———. *Rape in Marriage.* Rev. ed. Bloomington, IN: Indiana University Press, 1990.

Russell, Graeme. "Problems in Role-Reversed Families." In *Reassessing Fatherhood,* edited by C. Lewis and M. O'Brien. London: Sage, 1987: 161–179.

Rust, Paula. "Two Many and Not Enough: The Meanings of Bisexual Identities." In *Sexualities: Identities, Behaviors, and Society,* edited by Michael Kimmel and Rebecca Plante. New York: Oxford University Press, 2004.

Rutter, V. "Lessons from Stepfamilies." *Psychology Today* (May 1994): 30–33, 60.

Rydell, R., A. McConnell, and R. Bringle. "Jealousy and Commitment: Perceived Threat and the Effect of Relationship Alternatives." *Personal Relationships* 11 (2004): 451–468.

Sadker, Myra, and David Sadker. *Failing at Fairness: How American Schools Cheat Girls.* New York: C. Scribner's Sons, 1994.

Safilios-Rothschild, Constantina. "The Study of the Family Power Structure." *Journal of Marriage and the Family* 32, 4 (November 1970): 539–543.

———. "Family Sociology or Wives' Sociology? A Cross-Cultural Examination of Decisionmaking." *Journal of Marriage and the Family* 38 (1976): 355–362.

Sagrestano, Lynda, Christopher Heavey, and Andrew Christensen. "Perceived Power and Physical Violence in Marital Conflict." *Journal of Social Issues* 55, 1 (Spring 1999): 65–79.

Sahlstein, Erin. "Relating at a Distance: Negotiating Being Together and Being Apart in Long-distance Relationships." *Journal of Social & Personal Relationships* 21, 5 (October 2004): 689–710.

Saitoti, Tepelit Ole. *The Worlds of a Masai Warrior: An Autobiography.* Berkeley: University of California Press, 1986.

Salholz, E. "The Future of Gay America." *Newsweek* (March 12, 1990): 20–25.

Sampson, Ronald. *The Problem of Power.* New York: Pantheon, 1966.

Sanchez, Laura. "Feminism and Families." *Journal of Marriage and the Family* 59, 4 (1997): 1031–1032.

Sanchez, Laura, and Constance T. Gager. "Hard Living, Perceived Entitlement to a Great Marriage, and Marital Dissolution." *Journal of Marriage and Family* 62, 3 (August 2000): 708–722.

Sanchez, Laura, Steven Nock, James D. Wright, and Constance Gager. "Setting the Clock Forward or Back?" *Journal of Family Issues,* 23, 1 (January 2002): 91–120.

Sander, William. "Catholicism and Intermarriage in the United States." *Journal of Marriage and the Family* 55, 4 (November 1993): 1037–1041.

Sanford, Keith. "Problem–Solving Conversations in Marriage: Does It Matter What Topics Couples Discuss?" *Personal Relationships* 10, 1 (March 2003): 97–112.

Santrock, John, and Karen Sitterle. "Parent-Child Relationships in Stepmother Families." In *Remarriage and Stepparenting: Current Research and Theory,* edited by K. Pasley and M. Ihinger-Tallman. New York: Guilford Press, 1987.

Sassler, Sharon. "The Process of Entering into Cohabiting Unions." *Journal of Marriage and Family* 66, 2 (May 2004): 491–505.

Satir, Virginia. *The New Peoplemaking.* Rev. ed. Mountain View, CA: Science and Behavior Books, 1988.

———. *Peoplemaking.* Palo Alto, CA: Science and Behavior Books, 1972.

Sayer, Liana C., Suzanne M. Bianchi, and John P. Robinson. "Are Parents Investing Less in Children? Trends in Mothers' and Fathers' Time with Children." *American Journal of Sociology* 110, 1 (2004): 1–43.

Scanzoni, John. *Sexual Bargaining.* 2nd ed. Englewood Cliffs, NJ: Prentice Hall, 1980.

———. "Reconsidering Family Policy: Status Quo or Force for Change?" *Journal of Family Issues* 3, 3 (September 1982): 277–300.

Schaap, Cas, Bram Buunk, and Ada Kerkstra. "Marital Conflict Resolutions." In *Perspectives on Marital Interaction,* edited by P. Noller and M. A. Fitzpatrick. Philadelphia: Multilingual Matters, 1988.

Schoen, Robert. "Union Disruption in the United States." *International Journal of Sociology* 32, 4 (Winter 2002/2003): 36–50.

Schoen, Robert, and Yen-Hsin Alice Cheng. "Partner Choice and the Differential Retreat From Marriage." *Journal of Marriage and Family* 68, 1 (February 2006): 1–10.

Schoeni, R., and K. Ross. "Family Support during the Transition to Adulthood." Transitions to Adulthood: Research Brief No. 1, October 2003. MacArthur Foundation Research Network on Transitions to Adulthood and Public Policy. Philadelphia: University of Pennsylvania, 2003.

Schooler, Carmi. "Psychological Effects of Complex Environments during the Life Span: A Review and Theory." In *Cognitive Functioning and Social Structure over the Life Course,* edited by C. Schooler and K. Warner Schaie. Norwood, NJ: Ablex, 1987.

Schwartz, Pepper. *Peer Marriage: How Love Between Equals Really Works.* New York: Free Press, 1994.

Schwebel, Andrew I. Ryan L. Dunn, Barry F. Moss, and Maureena A. Renner. "Factors Associated with Relationship Stability in Geographically Separated Couples." *Journal of College Student Development* 33, 3 (1992): 222–230.

Scott, Janny. "Low Birth Weight's High Cost." *Los Angeles Times,* (December 24, 1990): 1.

Scott, Joan. "Gender: A Useful Category of Historical Analysis." *American Historical Review* 91 (1986): 1053–1075.

Seccombe, Karen. "Families in Poverty in the 1990's: Trends, Causes, Consequences, and Lessons Learned." *Journal of Marriage and the Family,* 62, 4 (November 2000): 1094–1113.

Seidman, Steven. "Revisiting Sex Role Stereotyping in MTV Videos." *International Journal of Instructional Media* 26, 1 (Winter 1999).

Seltzer, Judith. "Families Formed Outside of Marriage." *Journal of Marriage and the Family,* 62, 4 (November 2000): 1247–1268.

Sennett, Richard, and Jonathan Cobb. *The Hidden Injuries of Class.* New York: Vintage, 1972.

Serovich, Julianne M., Sharon J. Price, and Steven F. Chapman. "Former In-Laws as a Source of Support." *Journal of Divorce and Remarriage* 17, 1–2 (1991): 17–25.

Shapiro, Adam D. "Explaining Psychological Distress in a Sample of Remarried and Divorced Persons. *Journal of Family Issues* 17, 2 (March 1996): 186–203.

Shapiro, Jerrold Lee. *The Measure of a Man: Becoming the Father You Wish Your Father Had Been.* New York: Delacorte, 1993.

Shapiro, Joanna. "Family Reactions and Coping Strategies in Response to the Physically Ill or Handicapped Child: A Review." *Social Science and Medicine* 17 (1983): 913–931.

Sharpsteen, Don J. "Romantic Jealousy as an Emotion Concept: A Prototype Analysis." *Journal of Social and Personal Relationships* 10, 1 (1993): 69–82.

Shaver, Phillip, Cindy Hazan, and D. Bradshaw. "Love as Attachment: The Integration of Three Behavioral Systems." In *The Psychology of Love,* edited by R. Sternberg and M. Barnes. New Haven, CT: Yale University Press, 1988.

Shelton, Beth A., and Daphne John. "Does Marital Status Make a Difference? Housework among Married and Cohabiting Men and Women." *Journal of Family Issues* 14, 3 (September 1993): 401–420.

Sheppard, V. J., E. S. Nelson, and V. Andreoli-Mathie. "Dating Relationships and Infidelity: Attitudes and Behaviors." *Journal of Sex and Marital Therapy* 21, 3 (Fall 1995): 202–212.

Sheridan, Lorraine, Raphael Gillett, and Graham M. Davies. "Stalking: Seeking the Victim's Perspective." Paper presented at the Seventh Annual Conference of the British Psychological Society Division of Criminological and Legal Psychology, University of Cambridge, September 29–October 1, 1997.

Sheridan, Lorraine, and Graham M. Davies. "Stalking: The Elusive Crime." *Legal and Criminological Psychology* 6, 2 (September 2001): 133–147.

Shibley-Hyde, J., and S. Jaffee. "Becoming a Heterosexual Adult: The Experiences of Young Women." *Journal of Social Issues* 56, 2 (2000): 283–296.

Shon, Steven, and Davis Ja. "Asian Families." In *Ethnicity and Family Therapy,* edited by M. McGoldrick et al. New York: Guilford Press, 1982.

Shook, Nancy, Deborah Gerrity, Joan Jurich, and Allen Segrist. "Courtship Violence Among College Students: A Comparison of Verbally and Physically Abusive Couples." *Journal of Family Violence* 15, 1 (2000): 1–23.

Shostak, Arthur B. "Tomorrow's Family Reforms: Marriage Course, Marriage Test, Incorporated Families, and Sex Selection Mandate." *Journal of Marital and Family Therapy* 7, 4 (October 1981): 521–528.

———. "Singlehood." In *Handbook of Marriage and the Family,* edited by M. Sussman and S. Steinmetz. New York: Plenum Press, 1987.

Shurtleff, Mark, and Terry Goddard. The Primer: Helping Victims of Domestic Violence and Child Abuse in Polygamous Communities. Utah Attorney General's Office and Arizona Attorney General's Office, http://www.attorneygeneral.utah.gov/polygamy/The Primer.pdf. (2005).

Shweder, Richard A. "Hey, You Still Just Don't Understand." *New York Times* (August 14, 1994).

Sigelman, C. K., et al. "Courtesy Stigma: The Social Implications of Associating with a Gay Person." *Journal of Social Psychology* 131 (1991): 45–56.

Signorielli, N. *Reflections of Girls in the Media: A Content Analysis across Six Media.* Menlo Park, CA: Kaiser Family Foundation.

Silver, Donald, and B. Kay Campbell. "Failure of Psychological Gestation." *Psychoanalytic Inquiry* 8, 2 (1988): 222–223.

Silverstein, Judith L. "The Problem with In-Laws." *Journal of Family Therapy* 14, 4 (1992): 399–412.

Silverstein, Meril, Xuan Chen, and Kenneth Heller. "Too Much of a Good Thing? Intergenerational Social Support and the Psychological Well-Being of Older Parents." *Journal of Marriage and the Family* 58, 4 (November 1996): 970–982.

Simmons, Tavia, and Grace O'Neil. Households *and Families: 2000. A Census 2000 Brief. C2KBR/01-8.* Washington. D.C.: U.S. Census Bureau (September 2001).

Simmons, Tavia, and Jane Lawler Dye, "Grandparents Living with Grandchildren, 2000," *Census 2000 Brief* (October 2003).

Simons, Ronald, and Les Whitbeck. "Sexual Abuse as a Precursor to Prostitution and Victimization Among Adolescent and Adult Homeless Women." *Journal of Family Issues* 12 (3) September 1991: 361–379.

Skitka, L. J., and C. Maslach. "Gender Roles and the Categorization of Gender-Relevant Information." *Sex Roles* 22 (1990): 3–4.

Slater, Alan, Jan A. Shaw, and Joseph Duquesnel. "Client Satisfaction Survey: A Consumer Evaluation of Mediation and Investigative Services: Executive Summary." *Family and Conciliation Courts Review* 30, 2 (1992): 252–259.

"Sleeping on Back Saves 1,500 Babies." *San Mateo County Times* (June 25, 1996): A-4.

Sluzki, Carlos. "The Latin Lover Revisited." In *Ethnicity and Family Therapy,* edited by M. McGoldrick et al. New York: Guilford Press, 1982.

Small, Stephen, and Dave Riley. "Toward a Multidimensional Assessment of Work Spillover into Family Life." *Journal of Marriage and the Family* 52, 1 (February 1990): 51–61.

Smith, Anthony, Doreen Rosenthal, and Heidi Reichler. "High Schoolers' Masturbatory Practices: Their Relationship to Sexual Intercourse and Personal Characteristics." In *Sexualities: Identities, Behaviors, and Society,* edited by Michael Kimmel and Rebecca Plante. New York: Oxford University Press, 2004.

Smith, Carolyn, Marvin Krohn, Rebekah Chu, and Oscar Best. "African American Fathers: Myths and Realities about Their Involvement with Their Firstborn Children." *Journal of Family Issues* 26, 7 (October 2005): 975–1001.

Smith, Herbert L., and S. Philip Morgan. "Children's Closeness to Father as Reported by Mothers, Sons and Daughters." *Journal of Family Issues* 15, 1 (March 1994): 3–29.

Smith, Ken, and Cathleen Zick. "The Incidence of Poverty among the Recently Widowed: Mediating Factors in the Life Course." *Journal of Marriage and the Family* 48 (1986): 619–630.

Smith, T. W. "Changing Racial Labels: From Negro to Black to African American." *Public Opinion Quarterly* 56, 4 (1992): 496–514.

Smits, Jeroen, Wout Ultee, and Jan Lammers. "Effects of Occupational Status Differences Between Spouses on the Wife's Labor Force Participation and Occupational Achievement: Findings from 12 European Countries." *Journal of Marriage and the Family* 58, 1 (February 1996): 101–115.

Smock, P. J. "The Economic Costs of Marital Disruption for Young Women over the Past Two Decades." *Demography 30,* 3 (August 1993): 353–371.

Smock, Pamela J. "Cohabitation in the United States: An Appraisal of Research Themes, Findings, and Implications." *Annual Review of Sociology* 26, Summer 2000.

Smock, Pamela J. "The Wax and Wane of Marriage: Prospects for Marriage in the 21st Century." *Journal of Marriage and Family* 66, 4 (November 2004): 966–973.

Snarey, John, et al. "The Role of Parenting in Men's Psychosocial Development." *Developmental Psychology* 23, 4 (July 1987): 593–603.

Soboleski, Juliana M., and Valerie King. "The Importance of the Coparental Relationship for Nonresident Fathers' Ties to Children." *Journal of Marriage and Family* 67, 3 (December 2005): 1196–1212.

Solot, Dorian, and Marshall Miller. "Common Law Marriage Fact Sheet." www.unmarried.org/common.html, August 2005.

Sommers-Flanagan, Rita, John Sommers-Flanagan, and Britta Davis. "What's Happening on Music and Television? A Gender Role Content Analysis." *Sex Roles* 28, 11–12 (June 1993): 745–753.

South, Scott. "Time Dependent Effects of Wives' Employment on Marital Dissolution." *American Sociological Review* 66 (April 2001): 226–245.

South, Scott, Katherine Trent, and Yang Shen. "Changing Partners: Toward a Macrostructural-Opportunity Theory of Marital Dissolution." *Journal of Marriage and Family* 63, 3 (August 2001): 743–754.

Spaht, Katherine. "Model Marriage Obligations Statute," marriagedebate.com.

Spanier, Graham B., and R. L. Margolis. "Marital Separation and Extramarital Sexual Behavior." *Journal of Sex Research* 19 (1983): 23–48.

Spanier, Graham B., and Linda Thompson. *Parting: The Aftermath of Separation and Divorce.* Rev. ed. Newbury Park, CA: Sage, 1987.

Spector, I. P., and M. P. Carey. "Incidence and Prevalence of the Sexual Dysfunctions—A Critical Review of the Empirical Literature." *Archives of Sexual Behavior* 19, 4 (August 1990): 389–408.

Spence, Janet, and L. L. Sawin. "Images of Masculinity and Femininity." In *Sex, Gender, and Social Psychology,* edited by V. O'Leary et al. Hillsdale, NJ: Lawrence Erlbaum, 1985.

Spitalnick, Josh, and Lily McNair. "Couples Therapy with Gay and Lesbian Clients: An Analysis of Important Clinical Issues." *Journal of Sex & Marital Therapy* 31, 1 (January/February 2005): 43–56.

Spitz, Mary-Ann Leitz. "Stalking: Terrorism at Our Doors—How Social Workers Can Help Victims Fight Back." *Social Work* 48, 4 (October 2003): 504–512.

Spitzberg, Brian H., and William R. Cupach. "The Inappropriateness of Relational Intrusion." *Inappropriate Relationships: The Unconventional, the Disapproved, and the Forbidden.* Mahway, NJ: Lawrence Erlbaum, 2002.

Spitze, Glenna and John Logan. "Employment and Filial Relations: Is There a Conflict?" *Sociological Forum* 6 (December 1991): 681–698.

Sprecher, Susan. "Influences on Choice of a Partner and on Sexual Decision Making in the Relationship." In *Human Sexuality: The Social and Interpersonal Context,* edited by K. McKinney and S. Sprecher. Norwood, NJ: Ablex, 1989.

Sprecher, Susan, and Kathleen McKinney. *Sexuality.* Newbury Park, CA: Sage, 1993.

Sprecher, Susan, and Pamela Regan. "Liking Some Things (in Some People) More Than Others: Partner Preferences in Romantic Relationships and Friendships." *Journal of Social and Personal Relationships* 19, 4 (2002): 436–481.

Sprecher, Susan, and Susan Hendrick. "Self-Disclosure in Intimate Relationships: Associations with Individual and Relationship Characteristics over Time." *Journal of Social and Clinical Psychology* 23, 6 (December 2004): 836–856.

Springer, D., and D. Brubaker. *Family Caregivers and Dependent Elderly: Minimizing Stress and Maximizing Independence.* Beverly Hills, CA: Sage, 1984.

St. Petersburg Times Special Report. "9/11 For the Record." (September 8, 2002). http://www.sptimes.com/2002/09/08/911/911__For_the_record.shtml.

Stacey, Judith. "Good Riddance to the Family: A Response to David Popenoe," *Journal of Marriage and the Family* 55 (3) (August 1993): 545–547.

Stacey, Judith, and Timothy Biblarz "(How) Does the Sexual Orientation of Parents Matter?" *American Sociological Review* 66, 2 (2001): 159–183.

Stack, Carol B. *All Our Kin: Strategies for Survival in a Black Community.* New York: Harper and Row, 1974.

Stafford, Laura, and James R. Reske."Idealization and Communication in Long-Distance Premarital Relationships." *Family Relations: Journal of Applied Family and Child Studies* 39, 3 (1990): 274–279.

Stallones, Lorann, Martin B. Marx, Thomas F. Garrity, and Timothy P. Johnson. "Pet Ownership and Attachment in Relation to the Health of U.S. Adults, 21 to 64 Years of Age." *Anthrozoos* 4, 2 (1990): 100–112.

Staples, Robert. "The Sexual Revolution and the Black Middle Class." In *The Black Family,* edited by R. Staples. 4th ed. Belmont, CA: Wadsworth, 1991.

Starrels, Marjorie E. "Gender Differences in Parent-Child Relations." *Journal of Family Issues* 15, 1 (March 1994): 148–165.

Starrels, Marjorie E., Sally Bould, and Leon J. Nicholas. "The Feminization of Poverty in the United States: Gender, Race, Ethnicity, and Family Factors." *Journal of Family Issues* 15, 4 (December 1994): 590–607.

Steelman, Lala Carr, and Brian Powell. "The Social and Academic Consequences of Birth Order: Real, Artificial, or Both?" *Journal of Marriage and the Family* 47 (1985): 117–124.

Stein, Peter. "Men and Their Friendships." In *Men In Families,* edited by R. Lewis and R. Salt. Beverly Hills, CA: Sage, 1986: 261–270.

Steinberg, Laurence, and Susan Silverberg. "Marital Satisfaction in Middle Stages of Family Life Cycle." *Journal of Marriage and the Family* 49, 4 (November 1987): 751–760.

Steinmetz, Suzanne. "Family Violence." In *Handbook of Marriage and the Family,* edited by M. Sussman and S. Steinmetz. New York: Plenum Press, 1987.

Steinmetz, Suzanne, Sylvia Clavan, and K. Stein. *Marriage and Family Realities.* New York: Harper and Row, 1990.

Sternberg, Robert, and Michael Barnes, eds. *The Psychology of Love.* New Haven, CT: Yale University Press, 1988.

Stevens, Daphne, Gary Kiger, and Pamela J. Riley. "Working Hard and Hardly Working: Domestic Labor and Marital Satisfaction among Dual-Earner Couples." *Journal of Marriage and Family* 63, 2 (May 2001): 514–526.

Stevens, Gillian, and Robert Schoen. "Linguistic Intermarriage in the United States." *Journal of Marriage and the Family* 50, 1 (February 1988): 267–280.

Stevenson, M. "Tolerance for Homosexuality and Interest in Sexuality Education." *Journal of Sex Education and Therapy* 16 (1990): 194–197.

Stewart, Susan. "Boundary Ambiguity in Stepfamilies." *Journal of Family Issues* 26, 7 (October 2005): 1002–1029.

Stewart, Susan. "How the Birth of a Child Affects Involvement with Stepchildren." *Journal of Marriage & Family* 67, 1 (May 2005): 461–473.

Stockard, Janice. *Daughters of the Canton Delta: Marriage Patterns and Economic Strategies in South China, 1860–1930.* Stanford, CA: Stanford University Press, 1989.

Stockdale, M. S. "The Role of Sexual Misperceptions of Women's Friendliness in an Emerging Theory of Sexual Harassment." *Journal of Vocational Behavior* 42, 1 (February 1993): 84–101.

Storaasli, Ragnar D., and Howard J. Markman. "Relationship Problems in the Early Stages of Marriage: A Longitudinal Investigation." *Journal of Family Psychology* 4, 1 (September 1990): 80–98.

Straus, Murray A. "Physical Assaults by Wives: A Major Social Problem." In *Current Controversies in Family Violence,* edited by R. Gelles and D. Loseke. Newbury Park, CA: Sage, 1993.

Straus, Murray, and Carolyn Field. "Psychological Aggression by American Parents: National Data on Prevalence, Chronicity, and Severity." *Journal of Marriage and the Family* 65, 4 (November 2003): 795–808.

Straus, Murray, Richard Gelles, and Suzanne Steinmetz. *Behind Closed Doors.* Garden City, NY: Anchor Books, 1980.

Straus, Murray, and Carrie Yodanis. "Corporal Punishment in Adolescence and Physical Assaults on Spouses Later in Life: What Accounts for the Link?" *Journal of Marriage and the Family* 58, 4 (November 1996): 825–841.

Strohschein, Lisa. "Parental Divorce and Child Mental Health Trajectories." *Journal of Marriage and Family* 67, 5 (December 2005): 1286–1300.

Strom, Robert, Pat Collinsworth, Shirley Strom, and D. Griswold. "Strengths and Needs of Black Grandparents." *International Journal of Aging and Human Development* 36, 4 (1992–1993): 255–268.

Strong, Bryan, and Christine DeVault. *Human Sexuality: Diversity in Contemporary America,* 2nd ed. Mountain View, CA: Mayfield, 1997.

Suitor, J. Jill. "Marital Quality and Satisfaction with Division of Household Labor." *Journal of Marriage and the Family* 53, 1 (February 1991): 221–230.

Sulloway, Frank J. *Born to Rebel; Birth Order, Family Dynamics, and Creative Lives.* New York: David McKay Company, 1996.

Sun, Yong Min. "Family Environment and Adolescents' Well-Being Before and After Parents' Marital Disruption: A Longitudinal Analysis." *Journal of Marriage and Family* 63, 3 (August 2001): 697–713.

Sun, Yongmin. "The Well-Being of Adolescents in Households With No Biological Parents." *Journal of Marriage & Family* 65, 4 (November 2003): 894–909.

Sung, B. L. *Mountains of Gold.* New York: Macmillan, 1967.

Surra, Catherine. "Research and Theory on Mate Selection and Premarital Relationships in the 1980s." In *Contemporary*

Families: Looking Forward, Looking Back, edited by A. Booth. Minneapolis: National Council on *Family Relations,* 1991.

Surra, Catherine, P. Arizzi, and L. L. Asmussen. "The Association between Reasons for Commitment and the Development and Outcome of Marital Relationships." *Journal of Social and Personal Relationships* 5 (1988): 47–63.

"Surrogacy," ivf-infertility.com, 2005.

Swain, Scott. "Covert Intimacy: Closeness in Men's Friendships." In *Gender in Intimate Relationships: A Microstructural Approach,* edited by B. Risman and P. Schwartz. Belmont, CA: Wadsworth, 1989.

Sweeney, Megan M. "Remarriage and the Nature of Divorce: Does It Matter Which Spouse Chose to Leave?" *Journal of Family Issues* 23, 3 (April 2002): 410–440.

Sweeney, Megan, and Julie Phillips. "Understanding Racial Differences in Marital Disruption: Recent Trends and Explanations." *Journal of Marriage & Family* 66, 3 (August 2004): 639–650.

Sweet, J. A., L. L. Bumpass, and V. R. A. Call. *The Design and Content of the National Survey of Families and Households.* Working Paper NSFH-1. Madison: University of Wisconsin, Center for Demography and Ecology, 1988.

Swift, Carolyn. "Preventing Family Violence: Family-Focused Programs." In *Violence in the Home: Interdisciplinary Perspectives,* edited by M. Lystad. New York: Brunner/Mazel, 1986.

Symons, Donald. *The Evolution of Human Sexuality.* New York: Oxford University Press, 1979.

Szinovacz, Maximiliane. "Family Power." In *Handbook of Marriage and the Family,* edited by M. Sussman and S. Steinmetz. New York: Plenum Press, 1987.

Szinovacz, Maximiliane, and Adam Davey. "Retirement and Marital Decision Making: Effects on Retirement Satisfaction." *Journal of Marriage & Family* 67, 2 (May 2005): 387–398.

Takagi, Diana Y. "Japanese American Families." In *Minority Families in the United States: A Multicultural Perspective,* edited by R. L. Taylor. Englewood Cliffs, NJ: Prentice Hall, 1994.

Tanfer, K., and L. dc Cubbins, "Coital Frequency among Single Women: Normative Constraints and Situational Opportunities." *Journal of Sex Research* 29, 2 (1992): 221–250.

Tannen, Deborah. "Sex, Lies and Conversation: Why Is It So Hard for Men and Women to Talk to Each Other?" *Washington Post* (June 24, 1990).

Tannen, Deborah. *You Just Don't Understand: Women and Men in Conversation.* New York: Morrow, 1990.

Tanner, Litsa, et al. "Images of Couples and Families in Animated Disney Movies." *American Journal of Family Therapy* 31 (2003): 355–374.

Tashiro, Ty, and Patricia Frazier. "'I'll Never Be in a Relationship Like That Again': Personal Growth Following Romantic Relationship Breakups." *Personal Relationships* 10, 1 (March 2003): 113–128.

Taylor, Robert J. "Black American Families." In *Minority Families in the United States: A Multicultural Perspective,* edited by R. L. Taylor. Englewood Cliffs, NJ: Prentice Hall, 1994a.

———. "Minority Families in America." In *Minority Families in the United States: A Multicultural Perspective,* edited by R. L. Taylor. Englewood Cliffs, NJ: Prentice Hall, 1994b.

Taylor, Robert J., Linda M. Chatters, Belinda Tucker, and Edith Lewis. "Developments in Research on Black Families." In *Contemporary Families: Looking Forward, Looking Back,* edited by A. Booth. Minneapolis: National Council on Family Relations, 1991.

Teachman, Jay D., and Karen A. Polonko. "Cohabitation and Marital Stability in the United States." *Social Forces* 69, 1 (September 1990): 207–220.

Teachman, Jay D., R. Vaughn, A. Call, and Karen P. Carver. "Marital Status and Duration of Joblessness Among White Men." *Journal of Marriage and the Family* 56, 2 (May 1994): 415–428.

"Teen Dad Terrell Pough Killed," People.com, 2005.

Ten Kate, N. "Choosing and Using Contraception." *American Demographics* 20, 6 (June 1998).

Tessina, Tina. *Gay Relationships: For Men and Women. How to Find Them, How to Improve Them, How to Make Them Last.* Los Angeles: Jeremy P. Tarcher, 1989.

Testa, Ronald J., Bill N. Kinder, and G. Ironson. "Heterosexual Bias in the Perception of Loving Relationships of Gay Males and Lesbians." *Journal of Sex Research* 23, 2 (May 1987): 163–172.

Thayer, Leo. *On Communication.* Norwood, NJ: Ablex, 1986.

Thoits, Peggy A. "Identity Structures and Psychological Well-Being: Gender and Marital Status Comparisons." *Social Psychology Quarterly* 55, 3 (1992): 236–256.

Thompson, Anthony. "Emotional and Sexual Components of Extramarital Relations." *Journal of Marriage and the Family* 46, 1 (February 1984): 35–42.

Thompson Elizabeth, Jane Mosley, Thomas Hanson, and Sara McLanahan. "Remarriage, Cohabitation, and Changes in Mothering Behavior." *Journal of Marriage and Family* 63 (May 2001): 370–380.

Thompson, Linda. "Family Work: Women's Sense of Fairness," *Journal of Family Issues* 12 (2) June 1991: 181–196.

———. "Conceptualizing Gender in Marriage: The Case of Marital Care." *Journal of Marriage and the Family* 55, 3 (August 1993): 557–569.

Thompson, Linda, and Alexis J. Walker. "Gender in Families: Women and Men in Marriage, Work, and Parenthood." *Journal of Marriage & Family* 51, 4 (November 1989): 845–871.

Thompson, Linda, and Alexis J. Walker. "The Place of Feminism in Family Studies." *Journal of Marriage & Family* 57, 4 (November 1995): 847–865.

Thornton, Arland. "Changing Attitudes toward Family Issues in the United States." *Journal of Marriage and the Family* 51, 4 (November 1989): 873–893.

Tienda, Marta, and Ronald Angel. "Headship and Household Composition among Blacks, Hispanics, and Other Whites." *Social Forces* 61 (1982): 508–531.

Tienda, Marta, and Jennifer Glass. "Household Structure and Labor Force Participation of Black, Hispanic, and White Mothers." *Demography* 22 (1985): 281–394.

Ting-Toomey, Stella. "An Analysis of Verbal Communication Patterns in High and Low Marital Adjustment Groups." *Human Communications Research* 9, 4 (June 1983): 306–319.

Toews, Michelle, Beth Catlett, and Patrick McKenry. "Women's Use of Aggression Toward Their Former Spouses During Marital Separation." *Journal of Divorce and Remarriage* 42, (2005): 1–14.

Tolman, Richard M. "Treatment Program for Men Who Batter." In *Vision 2010: Families and Violence, Abuse and Neglect,* edited by R. J. Gelles. Minneapolis: National Council on Family Relations, 1995.

Topham, Glade L., Jeffry H. Larson, and Thomas B. Holman. "Family-of-Origin Predictors of Hostile Conflict in Early Marriage." *Contemporary Family Therapy* 27, 1 (March 2005): 101–121.

Toufexis, Anastasia. "Older—But Coming on Strong." *Time* (February 22, 1988): 76–79.

Tracy, Kathleen. *The Secret Story of Polygamy.* Naperville, IL: Sourcebooks, 2002.

Tran, Than Van. "The Vietnamese American Family." In *Ethnic Families in America: Patterns and Variations,* 3rd ed., edited by C. H. Mindel et al. New York: Elsevier, 1988.

Treas, Judith, and Vern L. Bengtson. "The Family in Later Years." In *Handbook of Marriage and the Family,* edited by M. Sussman and S. Steinmetz. New York: Plenum Press, 1987.

Treas, Judith, and Deirdre Giesen, "Sexual Infidelity Among Married and Cohabiting Americans." *Journal of Marriage and the Family* 62, 1 (2000): 48–60.

Treichler, Paula. "Feminism, Medicine, and the Meaning of Childbirth." In *Body/Politics: Women and the Discourses of Science,* edited by Mary Jacobus, Evelyn Fox Keller, and Sally Shuttleworth. New York: Routledge, 1990: 113–138.

Troiden, Richard. *Gay and Lesbian Identity: A Sociological Analysis.* New York: General Hall, 1988.

Troll, L. E. "Family Embedded vs. Family-Deprived Oldest-Old: A Study of Contrasts." *International Journal of Aging and Human Development* 38 (1994), 51–63.

Troll, Lillian. "The Contingencies of Grandparenting." In *Grandparenthood,* edited by V. Bengtson and J. Robertson. Beverly Hills, CA: Sage 1985.

Tucker, Judith Stadtman. "The Mother and the Magazine." *The Mothers Movement Online,* mothersmovement.org, 2003.

Tucker, Raymond K., M. G. Marvin, and B. Vivian. "What Constitutes a Romantic Act." *Psychological Reports* 89, 2 (October 1991): 651–654.

Turner, R. Jay, and William R. Avison. "Assessing Risk Factors for Problem Parenting: The Significance of Social Support." *Journal of Marriage and the Family* 47, 4 (November 1985): 881–892.

Turner, Robert L., and M. E. Fakouri. "Androgyny and Differences in Fantasy Patterns." *Psychological Reports* 73, 3 (1993): 1164–1166.

Twohey, Denise, and Micheal Ewing. "The Male Voice of Emotional Intimacy." *Journal of Mental Health Counseling* 17, 1 (January 1995): 54–63.

Tyre, Peg. "The Trouble With Boys." *Newsweek* (January 30, 2006): 44–51.

Tyre, Peg. "The Trouble with Boys." *Newsweek* (January 10, 2006).

Udry, J. Richard. *The Social Context of Marriage.* Philadelphia: Lippincott, 1974.

United Nations. Demographic Yearbook, 2002. New York: United Nations, 2005.

U.S. Bureau of Justice Statistics, *Violence by Intimates: Analysis of Data on Crimes by Current or Former Spouses, Boyfriends, and Girlfriends.* Washington, DC: Department of Justice, 1998.

U. S. Bureau of Labor Statistics. A *Profile of the Working Poor: 2003.* Report 983. Washington, D.C. (March 2005).

U.S. Census Bureau, *Current Population Report,* 1998. www.census.gov.

U. S. Census Bureau. Table 13.2 Total Money Income in 2003 of Families by Type, and by Hispanic Origin Type of Householder: 2004. Washington, D.C. : U.S. Census Bureau, Population Division, Ethnic & Hispanic Statistics Branch

U.S. Census Bureau, Historical Abstracts of the United States, Colonial Times to 1970, Series A. Washington, DC: U.S. Government Printing Office, 160–171, 1989.

———. "Marital Status of the Population 15 Years and Over, by Sex and Race: 1950 to Present," MS-1. Washington, DC: U.S. Government Printing Office, 2001a.

———. "Annual Social and Economic Supplement," Current Population Reports, Series P20-553. Washington, DC: U.S. Government Printing Office, 2003a.

———. "Census of Population and Housing Characteristics of American Indians and Alaska Natives by Tribe and Language: 2000, PHC-5. Washington, DC: U.S. Government Printing Office, 2003b.

———. "Annual Social and Economic Supplement." Current Population Reports. Washington, DC: U.S. Government Printing Office, 2004a.

———. Income, Poverty, and Health Insurance Coverage in the United States: 2004. Report P60, n. 229. Washington, DC: U.S. Government Printing Office, Table B-2, pp. 52–57, 2004b.

———. "2005 American Community Survey," www.census.gov, 2005a.

———. "Facts for Features: Unmarried and Single Americans Week," www.census.gov, 2005b.

———. "Hispanic Population in the United States: 2004 March CPS," www.census.gov, 2005c.

———. Statistical Abstract of the United States. Washington, DC: U.S. Government Printing Office, 2006.

U.S. Census Bureau News. "Families and Living Arrangements: Americans Marrying Older, Living Alone More, See

Households Shrinking, Census Bureau Report (2006) http://www.census.gov/Press-Release/www/releases/archives/families_households/006840.html

———. "Race and Hispanic or Latino." U.S. Census Bureau, Census 2000 Summary File 1, Matrices P3, P4, PCT4, PCT5, PCT8, and PCT11, Washington, DC: U.S. Government Printing Office 2000a.

———. *Infant Mortality.* International Data Base, 1996a.

———. *Statistical Abstract of the United States.* Washington, DC: U.S. Government Printing Office, 1996b.

———. *Statistical Abstract of the United States.* Washington, DC: U.S. Government Printing Office, 2000b.

———. *Statistical Abstract of the United States.* Washington, DC: U.S. Government Printing Office, 2001b.

———. *Statistical Abstract of the United States.* Washington, DC: U.S. Government Printing Office, 2002.

———. *Statistical Abstract of the United States.* Washington, DC: U.S. Government Printing Office, 2003c.

U.S. Department of Justice. "Sexual Offenses and Offenders." Washington, DC: Bureau of Justice Statistics, 1997.

U.S. Department of Labor. "The Family and Medical Leave Act of 1993," Public Law 103-3, www.dol.gov/esa/regs/statutes/whd/fmla.htm, 1993.

Vaillant, Caroline O., and George E. Vaillant. "Is the U-Curve of Marital Satisfaction an Illusion? A 40–Year Study of Marriage." *Journal of Marriage and the Family* 55, 1 (February 1993): 230–240.

Valentine, Deborah. "The Experience of Pregnancy: A Developmental Process." *Family Relations* 31, 2 (April 1982): 243–248.

Van Buskirk. "Soap Opera Sex: Tuning In, Tuning Out." *US* (August 1992): 64–67.

Vande Berg, Leah R., and Diane Streckfuss. "Prime-Time Television's Portrayal of Women and the World of Work: A Demographic Profile." *Journal of Broadcasting and Electronic Media* (March 1992): 195–207.

Van Horn, K. Roger, Angela Arnone, Kelly Nesbitt, Laura Desilets, Tanya Sears, Michelle Giffin, and Rebecca Brudi "Physical Distance and Interpersonal Characteristics in College Students' Romantic Relationships." *Personal Relationships* 4,1 (1997): 25–34.

Vaquera, Elizabeth, and Grace Kao. "Private and Public Displays of Affection among Interracial and Intra-Racial Adolescent Couples." *Social Science Quarterly* 86, 2 (June 2005): 484–508.

Vaselle-Augenstein, Renata, and Annette Ehrlich. "Male Batterers: Evidence for Psychopathology." In *Intimate Violence: Interdisciplinary Perspectives,* edited by E. C. Viano. Washington, DC: Hemisphere, 1992.

Vasquez-Nuttall, E., et al. "Sex Roles and Perceptions of Femininity and Masculinity of Hispanic Women: A Review of the Literature." *Psychology of Women Quarterly* 11 (1987): 409–426.

Vasta, Ross. "Physical Child Abuse: A Dual-Component Analysis." *Developmental Review* 2, 2 (June 1992): 125–149.

Vaughan, Diane. *Uncoupling: Turning Points in Intimate Relationships.* New York: University Press, 1986.

Vaughan, Diane. *Uncoupling : Turning Points In Intimate Relationships.* New York: Vintage Reprint Edition, 1990.

Veevers, Jean. *Childless by Choice.* Toronto: Butterworth, 1980.

Vega, William. "Hispanic Families." In *Contemporary Families: Looking Forward, Looking Back,* edited by A. Booth. Minneapolis: National Council on Family Relations, 1991.

Vemer, Elizabeth, et al. "Marital Satisfaction in Remarriage: A Meta-analysis." *Journal of Marriage and the Family* 53, 3 (August 1989): 713–726.

Ventura, Jacqueline N. "The Stresses of Parenthood Reexamined." *Family Relations* 36, 1 (January 1987): 26–29.

Vermont Secretary of State. "The Vermont Guide to Civil Unions," sec.state.vt.us/otherprg/civilunions/civilunions.html, August 2006.

Visher, Emily B., and John S. Visher. *Stepfamilies: A Guide to Working with Stepparents and Stepchildren.* New York: Brunner/Mazel, 1979.

———. *How to Win as a Stepfamily.* New York: Brunner/Mazel, 1991.

Voeller, Bruce. "Society and the Gay Movement." In *Homosexual Behavior,* edited by J. Marmor. New York: Basic Books, 1980.

Vogel, D. A., M. A. Lake, and S. Evans. "Children's and Adults' Sex-Stereotyped Perceptions of Infants." *Sex Roles* 24 (1991): 605–616.

Vogel, David L., Stephen R. Wester, and Martin Heesacker. "Dating Relationships and the Demand–Withdraw Pattern of Communication." *Sex Roles* 41, 3–4 (1999): 297–306.

Voydanoff, Patricia. *Work and Family Life.* Newbury Park, CA: Sage, 1987.

———."Economic Distress and Family Relations: A Review of the Eighties." In *Contemporary Families: Looking Forward, Looking Back,* edited by A. Booth. Minneapolis: National Council on *Family Relations,* 1991.

Voydanoff, Patricia, and Brenda Donnelly. "Work and Family Roles and Psychological Distress." *Journal of Marriage and the Family* 51, 4 (November 1989): 933–941.

Vredevelt, P. *Empty Arms: Emotional Support for Those Who Have Suffered Miscarriage or Stillbirth.* Sisters, OR: Questar, 1994.

Waehler, C.A. *Bachelors: The Psychology of Men Who Haven't Married.* Westport, CT: Praeger, 1996.

Wade, Terrance J., and David J. Pevalin. "Marital Transitions and Mental Health." *Journal of Health and Social Behavior* 45, 2 (June 2004): 155–170.

Wagner, Cynthia G. "Homosexual Relationships." *Futurist* 40, 3 (May/June 2006): 6.

Wagner, Roland M. "Psychosocial Adjustments during the First Year of Single Parenthood: A Comparison of Mexican-American and Anglo Women." *Journal of Divorce and Remarriage* 19, 1–2 (1993): 121–142.

Waite, Linda. "Does Marriage Matter?" *Demography* 32, 4 (November 1995): 483–507.

Waite, Linda, and Maggie Gallagher. *The Case for Marriage: Why Married People are Happier, Healthier, and Better off*

Financially. New York: Doubleday, 2000; Broadway Books, 2001.

Waldman, Steven, and Lucy Shackelford. "Welfare Booby Traps." *Newsweek* (December 12, 1994): 34–35.

Walker, E. A., et al. "Medical and Psychiatric Symptoms in Women with Childhood Sexual Abuse." *Psychosomatic Medicine* 54, 6 (November 1992): 658–664.

Walker, Karen. "Men, Women, and Friendship: What They Say, What They Do," *Gender and Society* 8, 2 (June 1994).

Walker, Lenore. *The Battered Woman Syndrome.* New York: Harper Colophon, 1979.

———. *The Battered Woman.* New York: Springer Publishing, 1984.

———. "The Battered Woman Syndrome Is a Psychological Consequence of Abuse." In *Current Controversies in Family Violence,* edited by R. Gelles and D. Loseke. Newbury Park, CA: Sage Publications, 1993.

Wall, Jack C. "Maintaining the Connection: Parenting as a Noncustodial Father." *Child and Adolescent Social Work Journal* 9, 5 (1992): 441–456.

Waller, Willard, and Reuben Hill. *The Family: A Dynamic Interpretation.* New York: Dryden Press, 1951.

Wallerstein, Judith, and Sandra Blakeslee. *Second Chances: Men, Women, and Children a Decade after Divorce.* New York: Ticknor & Fields, 1989.

———. *Second Chances: Men, Women, and Children a Decade after Divorce* (paperback, revised). New York: Houghton Mifflin, 1995.

Wallerstein, Judith, and Joan Kelly. "Effects of Divorce on the Visiting Father-Child Relationship." *American Journal of Psychiatry* 137, 12 (December 1980a): 1534–1539.

———. *Surviving the Breakup: How Children and Parents Cope with Divorce.* New York: Basic Books, 1980b.

Wallerstein, Judith, Julia Lewis, and Sandra Blakeslee. *The Unexpected Legacy of Divorce: A Twenty-Five Year Landmark Study.* New York: Hyperion Press, 2000.

Wallerstein, Judith (with Julia Lewis, Sandra Blakeslee). *The Unexpected Legacy Of Divorce: A 25 Year Landmark Study.* New York : Hyperion, 2000.

Walzer, Susan, and Thomas P. Oles. "Accounting for Divorce: Gender and Uncoupling Narratives." *Qualitative Sociology* 26, 3 (Fall 2003): 331–349.

Walzer, Susan. *Thinking About the Baby: Gender and Transitions into Parenthood.* Philadelphia: Temple University, 1998.

Wang, H., and P. Amato. "Predictors of Divorce Adjustment: Stressors, Resources, and Definitions." *Journal of Marriage and the Family* 62 (August 2000): 655–668.

Warren, Jennifer A. and Phyllis J. Johnson. "The Impact of Workplace Support on Work-Family Role Strain." *Family Relations* 44, 2 (April 1995): 163–169.

Watkins, William G., and Arnon Bentovim. "The Sexual Abuse of Male Children and Adolescents: A Review of Current Research." *Journal of Child Psychology and Psychiatry and Allied Disciplines* 33, 1 (January 1992): 197–248.

Weeks, Jeffrey. *Sexuality and Its Discontents.* London: Routledge, 1985.

Weinberg, M. S., and C. J. Williams. *Male Homosexuals: Their Problems and Adaptations.* New York: Penguin, 1974.

Weinberg, Martin S., C. J. Williams, and Douglas W. Pryor. *Dual Attraction: Understanding Bisexuality.* New York: Oxford University Press, 1994.

Weiss, David. "Open Marriage and Multilateral Relationships: The Emergence of Nonexclusive Models of the Marital Relationship." In *Contemporary Families and Alternative Lifestyles,* edited by E. Macklin and R. Rubin. Beverly Hills, CA: Sage, 1983.

Weiss, Robert. *Marital Separation.* New York: Basic Books, 1975.

Weitzman, Lenore. *The Marriage Contract : Spouses, Lovers, and the Law.* New York: Macmillan, 1981.

———. *The Divorce Revolution: The Unexpected Social and Economic Consequences for Women and Children in America.* New York: Free Press, 1985.

Weitzman, Susan. *Not to People Like Us: Hidden Abuse in Upscale Marriages.* New York: Basic Books, 2000.

Weizman, R., and J. Hart. "Sexual Behavior in Healthy Married Elderly Men." *Archives of Sexual Behavior* 16, 1 (February 1987): 39–44.

Werner, Carol M., Barbara B. Brown, Irwin Altman, and Brenda Staples. "Close Relationships in Their Physical and Social Contexts: A Transactional Perspective." *Journal of Social and Personal Relationships* 9, 3 (1992): 411–431.

West, Candace, and Don Zimmerman. "Doing Gender." *Gender and Society* 1 (1987): 125–151.

Westberg, Heather, Thorana S. Nelson, and Kathleen W. Piercy. "Disclosure of Divorce Plans to Children: What the Children Have To Say." *Contemporary Family Therapy* 24, 4 (December 2002): 525–542.

Whealin, Julia. "Men and Sexual Trauma: A National Center for PTSD Fact Sheet," ncptsd.va.gov, 2006.

Whisman, Mark, Amy Dixon, and Benjamin Johnson. "Therapists' Perspectives of Couple Problems and Treatment Issues in Couple Therapy." *Journal of Family Psychology* 11, 3 (September 1997): 361–366.

Whitbourne, Susan, and Joyce Ebmeyer. *Identity and Intimacy in Marriage: A Study of Couples.* New York: Springer-Verlag, 1990.

White, G., and P. Mullen. *Jealousy: Theory Research and Clinical Strategies.* New York: Guilford, 1989.

White, Jacquelyn W. "Feminist Contributions to Social Psychology." *Contemporary Social Psychology* 17, 3 (September 1993): 74–78.

White, James, and D. Klein. *Family Theories,* 2nd ed. Thousand Oaks, CA:Sage, 2002.

White, Joseph, and Thomas Parham. *The Psychology of Blacks: An African-American Perspective,* 2nd ed. Englewood Cliffs, NJ: Prentice Hall, 1990.

White, Lynn. "Determinants of Divorce." In *Contemporary Families: Looking Forward, Looking Back,* edited by A. Booth. Minneapolis: National Council on Family Relations, 1991.

White, Lynn, and Alan Booth. "The Transition to Parenthood and Marital Quality." *Journal of Family Issues* 6 (1985): 435–449.

———. "Divorce over the Life Course: The Role of Marital Happiness." *Journal of Family Issues* 12, 1 (March 1991): 5–22.

White, Lynn, and Bruce Keith. "The Effect of Shift Work on the Quality and Stability of Marital Relations." *Journal of Marriage and the Family* 52, 2 (May 1990): 453–462.

Whitehead, Barbara Dafoe. "Dan Quayle Was Right," *Atlantic Monthly.* April 1993: 47–84.

Whitehead, Barbara Dafoe. *The Divorce Culture: Rethinking Poor Commitments to Marriage and the Family.* New York: Knopf Publishing Group, 1996.

———. *The Divorce Culture.* New York: Knopf, 1997.

Whitehead, Barbara Dafoe, and David Popenoe. "Social Indicators of Marital Health and Well Being." The State of Our Unions, 2005: The Social Health of Marriage in America. The National Marriage Project. New Brunswick, NJ: Rutgers University, 2005.

Whitehead, Barbara Dafoe, and David Popenoe. "Social Indicators of Marital Health and Well Being." The State of Our Unions, 2004: The Social Health of Marriage in America. The National Marriage Project. New Brunswick, NJ: Rutgers University, 2004.

Widmer, Eric D., Judith Treas, and Robert Newcomb. "Attitudes toward Nonmarital Sex in 24 Countries." *Journal of Sex Research* 35 (November 1998): 349–358.

Wilcox, Allen, et al. "Incidents of Early Loss of Pregnancy." *New England Journal of Medicine* 319, 4 (July 28, 1988): 189–194.

Wilkie, Colleen F., and Elinor W. Ames. "The Relationship of Infant Crying to Parental Stress in the Transition to Parenthood." *Journal of Marriage and the Family* 48, 3 (August 1986): 545–550.

Wilkinson, Doris Y. "American Families of African Descent." In *Families in Cultural Context,* edited by M. K. DeGenova. Mountain View, CA: Mayfield, 1997.

Wilkinson, Doris, et al., eds. "Transforming Social Knowledge: The Interlocking of Race, Class, and Gender." *Gender and Society* [Special issue] (September 1992).

Willetts, Marion. "An Exploratory Investigation of Heterosexual Licensed Domestic Partners." *Journal of Marriage and the Family* 65, 4 (November 2003): 939–952.

Williams, John, and Arthur Jacoby. "The Effects of Premarital Heterosexual and Homosexual Experience on Dating and Marriage Desirability." *Journal of Marriage and the Family* 51 (May 1989): 489–497.

Willing, Richard. "Research Downplays Risk of Cousin Marriages." *USA Today* (April 4, 2002).

Wilson, Pamela. "Black Culture and Sexuality." *Journal of Social Work and Human Sexuality* 4, 3 (March 1986): 29–46.

Wilson, S. M., and N. P. Medora. "Gender Comparisons of College Students' Attitudes toward Sexual Behavior." *Adolescence* 25, 99 (September 1990): 615–627.

Wilson, William Julius. *The Truly Disadvantaged: The Inner City, the Underclass, and Public Policy.* Chicago: University of Chicago Press, 1987.

Wineberg, H. "Marital Reconciliation in the United States: Which Couples Are Successful?" *Journal of Marriage and the Family* 56, 1 (February 1994): 80–88.

Wineberg, H. "The Timing of Remarriage among Women Who Have a Failed Marital Reconciliation in the First Marriage." *Journal of Divorce and Remarriage* 30, 3/4 (June 1999): 57–69.

Wineberg, H., and J. McCarthy. "Separation and Reconciliation in American Marriages." *Journal of Divorce & Remarriage* 20, 1-2 (1993): 21–42.

Winslow, Sarah. "Work–Family Conflict, Gender, and Parenthood, 1977–1997." *Journal of Family Issues* 26, 6 (2005): 727–755.

Winton, Chester. *Frameworks for Studying Families.* Guilford, CT: Dushkin, 1995.

———. *Children as Caregivers: Parental and Parentified Children.* Boston: Pearson Allyn and Bacon, 2003.

Wise, T. N., S. Epstein, and R. Ross. "Sexual Issues in the Medically Ill and Aging." *Psychiatric Medicine* 10 (1992): 169–180.

Wolf, Michelle, and Alfred Kielwasser. "Introduction: The Body Electric: Human Sexuality and the Mass Media." *Journal of Homosexuality* 21, 1/2 (1991): 7–18.

Wolf, Rosalie S. "Abuse and Neglect of the Elderly." In *Vision 2010: Families and Violence, Abuse and Neglect,* edited by R. J. Gelles. Minneapolis: National Council on Family Relations, 1995.

Wolfinger, Nicholas H. "Parental Divorce and Offspring Marriage: Early or Late?" *Social Forces* 82, 1 (September 2003): 337–353.

Wong, Morrison G. "The Chinese American Family." In *Ethnic Families in America: Patterns and Variations,* 3rd ed., edited by C. H. Mindel et al. New York: Elsevier, 1988.

Woodward, K. L., V. Quade, and B. Kantrowitz. "Q: When Is a Marriage Not a Marriage?" *Newsweek* (March 13, 1995): 58–59.

Woodward, Kenneth. "New Rules for Making Love and Babies." *Newsweek* (March 23, 1987): 42–43.

Wright, Julia. "Getting Engaged: A Case Study and a Model of the Engagement Period as a Process of Conflict-Resolution." *Counselling Psychology Quarterly* 3, 4 (1990): 399–408.

Wrigley, Julia, and Dreby, Joanna. "Fatalities and the Organization of Childcare." *American Sociological Review* 70, 5 (October 2005): 729–757.

Wu, Zheng, and Randy Hart. "Union Disruption in Canada." *International Journal of Sociology* 32, 4 (Winter 2002/2003): 51–75.

Wu, Zheng, and Lindy MacNeill. "Education, Work, and Childbearing after Age 30." *Journal of Comparative Family Studies* 33, 2 (Spring 2002): 191–213.

Wu, Zheng, Margaret J. Penning, Michael S. Pollard, and Randy Hart. "In Sickness and in Health." *Journal of Family Issues* 24, 6 (September 2003): 811–838.

Wyatt, G., D. Gutherie, and C. Notgrass "The Differential Effects of Women's Child Sexual Abuse and Subsequent Sexual Revictimization." *Journal of Consulting and Clinical Psychology* 60, 2 (1992): 67–73.

Wyatt, Gail, and Sandra Lyons-Rowe. "African American Women's Sexual Satisfaction as a Dimension of Their Sex Roles," *Sex Roles* 22, 7–8 (April 1990): 509–524.

Wyatt, Gail, Stephanie Peters, and Donald Guthrie. "Kinsey Revisited I: Comparisons of the Sexual Socialization and Sexual Socialization and Sexual Behavior of Black Women over 33 Years." *Archives of Sexual Behavior* 17, 3 (June 1988a): 201–239.

———. "Kinsey Revisited II: Comparison of the Sexual Socialization and Sexual Socialization and Sexual Behavior of Black Women over 33 Years." *Archives of Sexual Behavior* 17, 4 (August 1988b): 289–332.

Wyatt, Gail Elizabeth, et al. "Differential Effects of Women's Child Sexual Abuse and Subsequent Sexual Revictimization." *Journal of Consulting and Clinical Psychology* 60, 2 (April 1992): 167.

Wyche, Karen F. "Psychology and African-American Women: Findings from Applied Research." *Applied and Preventive Psychology* 2, 3 (June 1993): 115–121.

Xu, Xiaohe, Clark D. Hudspeth, and John P. Bartkowski. "The Timing of First Marriage: Are There Religious Variations?" *Journal of Family Issues* 26, 5 (July 2005): 584–618.

Xu, Xiaohe, Clark D. Hudspeth, and John P. Bartkowski. "The Role of Cohabitation in Remarriage." *Journal of Marriage and Family* 68, 2 (May 2006): 261–274.

Yellowbird, Michael, and C. Matthew Snipp. "American Indian Families." In *Minority Families in the United States: A Multicultural Perspective,* edited by R. L. Taylor. Englewood Cliffs, NJ: Prentice Hall, 1994.

Yeung, W. Jean, John Sandberg, Pamela Davis-Kean, and Sandra Hofferth. "Children's Time With Fathers in Intact Families." *Journal of Marriage and Family* 63 (February 2001): 136–154.

Ÿllo, Kersti. "Through a Feminist Lens: Gender, Power, and Violence." In *Current Controversies in Family Violence,* edited by R. Gelles and D. Loseke. Newbury Park, CA: Sage , 1993.

———. "Marital Rape." In *Vision 2010: Families and Violence, Abuse and Neglect,* edited by R. J. Gelles. Minneapolis: National Council on Family Relations, 1995.

Young, L. "Sexual Abuse and the Problem of Embodiment." *Child Abuse and Neglect* 16, 1 (1992): 89–100.

Zaidi, A. U., and M. Shuraydi. "Perceptions of Arranged Marriages by Young Pakistani Muslim Women Living in a Western Society." *Journal of Comparative Family Studies* 33, 4 (2002): 495–514.

Zhan, Min, and Shanta Pandey. "Economic Well-Being of Single Mothers: Work First or Postsecondary Education?" *Journal of Sociology and Social Welfare* 31, 3 (September 2004): 87–112.

Zilbergeld, Bernie. *The New Male Sexuality.* New York: Bantam Books, 1992.

Zill, Nicholas, Donna R. Morrison, and Mary J. Coiro. "Long-Term Effects of Parental Divorce on Parent-Child Relationships, Adjustment, and Achievement in Young Adulthood." *Journal of Family Psychology* 7, 1 (1993): 91–103.

Zimmerman, Shirley L. "The Welfare State and Family Breakup: The Mythical Connection." *Family Relations* 40, 2 (April 1991): 139–147.

Zink, Therese, and Frank Putnam. "Intimate Partner Violence Research in the Health Care Setting: What are Appropriate and Feasible Methodological Standards?" *Journal of Interpersonal Violence* 20, 4 (April 2005): 365–372.

Zinn, Maxine Baca. "Family, Feminism, and Race." *Gender and Society* 4 (1990): 68–82.

———. "Adaptation and Continuity in Mexican-Origin Families." In *Minority Families in the United States: A Multicultural Perspective,* edited by R. L. Taylor. Englewood Cliffs, NJ: Prentice Hall, 1994.

"Zits: About the Comic," kingfeatures.com/features/comics/zits/about.htm, 2006.

Zucker, Alyssa. "Disavowing Social Identities: What it Means When Women Say 'I'm Not a Feminist But....'" *Psychology of Women Quarterly* 28 (December 2004): 423–435.

Zvonkovic, Anisa N., Kathleen M. Greaves, Cynthia J. Schmiege, and Leslie D. Hall. "The Marital Construction of Gender through Work and Family Decisions: A Qualitative Analysis." *Journal of Marriage and the Family* 58, 1 (February 1996): 91–100.

Photo credits

Name Index

Martin, S. P., 495
Martin, T. C., 287, 495
Martinez, E. A., 426
Marvin, M. G., 215
Marwell, G., 292
Marx, M. B., 14
Masheter, C., 502
Maslach, C., 120
Masterpasqua, F., 410
Masters, W., 214
Mathis, R. D., 351
Maticka-Tyndale, E., 211
Maume, D., 427
Maume, D. J., Jr., 434, 435, 448
Max, S., 20
Maxwell, J., 21
Mays, V. M., 215
McAdoo, H. P., 73, 96, 530
McCabe, M. P., 369, 379
McCaghy, C., 460
McCarthy, 501
McClennen, 466
McConnell, A., 178
McCormick, J., 473
McCoy, M. L., 469
McCoy, N., 214
McDuff, P., 479
McFarlane, J., 176, 177
McGill, M., 160
McGoldrick, M., 343, 349, 535
McHale, S. M., 245, 337, 397, 425
McHugh, M., 457, 462, 463, 464
McKelvey, L., 98
McKenny, 458
McKenry, P., 98
McKim, M. K., 383
McKinney, K., 292
McKinnon, J., 97, 98
McLanahan, S., 92, 326, 327, 491, 507, 528, 538, 543
McLeer, S. V., 480
McLoyd, V., 97, 98, 100, 102, 104, 105, 134, 136, 403, 412
McMahon, K., 196
McManus, P., 505
McNair, L., 205
McRoy, R. G., 377
McVey, 481
Mead, G. H., 401
Mead, M., 118
Meckel, 72
Mederer, H., 471
Medora, N. P., 154, 155, 156, 215
Melito, R., 51
Melko, M., 305
Melz, H., 323, 325, 326, 327, 335, 336, 337, 341, 342
Menacker, F., 362
Menaghan, E., 131, 437
Merrill, L., 241
Mertensmeyer, C., 445
Messerschmidt, J., 117
Messner, 143
Messner, M., 52
Metts, S., 250, 292
Metz, M. E., 268
Meyer, D. R., 450, 475, 506

Meyers, M. K., 451, 452
Michael, R., 164, 202, 208, 213, 214, 215, 216, 223, 224, 285, 297
Michaels, S., 164, 202, 208, 214, 215, 216, 223, 224, 297
Migliaccio, T., 464, 472
Milhausen, R., 194
Milkie, M., 423
Miller, 195
Miller, G., 517
Miller, M., 310
Miller, P. A., 213
Miller, R., 527, 529
Miller, R. B., 266
Mills, C. W., 110
Mills, D., 543
Milner, 57
Min, P. G., 261
Mincy, R. B., 506
Mindel, C. H., 180, 287
Minkler, M., 352, 414
Mintz, S., 21, 24, 70, 72, 73, 74, 75, 76, 77, 78, 79, 80, 81, 84, 154
Minuchin, S., 50, 57
Miracle, A., 194, 195, 201, 203, 205, 213, 215, 218, 220, 226, 227
Miracle, T., 194, 195, 201, 203, 205, 213, 215, 218, 220, 226, 227
Mirowsky, J., 14, 366, 422, 423, 430
Mnookin, R. H., 505, 506
Moffatt, M., 193, 196, 215
Money, J., 164
Montagu, A., 239
Montgomery, M., 182
Montgomery, M. J., 166
Moore, M. M., 239, 292
Moore-Hirschl, 221
Morgan, 505, 506
Morgan, L., 500, 504
Morgan, S. P., 392
Morris, 451
Morrison, D. R., 512, 517
Morse, K., 300
Mosher, D. L., 207
Mosher, W. D., 325, 534
Moskey, E., 198
Moss, B. F., 64
Mostashari, 109
Moultrup, 221, 222
Moynihan, D. P., 96
Muehlenhard, C. L., 465, 468, 469
Mullen, P., 177, 178
Mulligan, G. M., 394
Muncher, S. J., 469
Munson, M. L., 80, 492, 493
Murdock, G., 12, 18
Murdock, R., 7
Murnen, K., 184
Murphy, 57
Murphy-Berman, V., 96
Murray, 271
Murray, C., 449
Murry, V., 122
Murry, V. M., 15, 259
Murstein, B., 154, 291, 292
Mustaine, 177
Myers, J., 166, 167, 169

Mykitiuk, 463

N
Nadelson, C., 482
Nanda, S., 117
Nelsen, J., 402
Nelson, T., 508
Nesbitt, K., 64
Neuberg, S., 300
Neuendorf, K., 133
Neuman, W. L., 32, 38
Newcomb, R., 221
Newcombe, N., 400
Newcomer, S., 195
Newman, D., 461
Newman, K., 94, 447
Ney, P. G., 459, 477
Nicholas, L. J., 90
Nichols, W. C., 335
Nicholson, J. M., 538
Niolon, R., 479
Nock, S., 518
Noller, P., 171, 237, 238, 245, 246, 251
Nomaguchi, K., 423
Noonan, M., 425, 434
Nord, C., 517, 540
Norton, A. J., 240
Notarius, C., 250
Notgrass, 481
Notman, M. T., 368
Nurmi, J. E., 335
Nye, F. I., 57

O
Oakley, A., 125, 428
Oggins, J., 343
Ogunwole, S., 105
Okazaki, 192
Olds, S. W., 220
O'Leary, K. D., 458
Olenik, I., 221
Oleshansky, 130
Oliver, 293
Oliver, M. B., 214
Olsen, 531
Olson, D., 356
Olson, D. H., 335, 337, 338
O'Neil, G., 20
O'Neil, R., 426
Ono, H., 285, 288, 289, 290
Opie, A., 516
Ormrod, R., 398
Oropesa, R. S., 323, 324, 325, 326
Ortega, 287
Oster, S., 506
O'Sullivan, 195
O'Sullivan, L., 218
Over, R., 288
Overturf Johnson, J., 431

P
Pace-Nichols, M. A., 335
Padavic, I., 134
Palkovitz, R., 376, 392

Paludi, M. A., 443
Pandey, S., 528
Papanek, H., 90
Papernow, P. L., 537, 538, 539
Papp, 270
Parashar, S., 495
Parcel, T., 131, 437
Parham, T., 250
Park, K., 364, 365
Parrot, L., 468
Pasch, 271
Pasley, K., 7, 526, 536, 543, 545
Patterson, C., 164, 409, 410, 411
Paul, 180
Paulson, 195
Pence, 445
Peoples, J., 155
Peplau, A., 313
Peplau, L., 199, 205
Peplau, L. A., 130, 164
Perlman, D., 291
Perron, A., 479
Perry-Jenkins, M., 423, 426, 433, 440
Person, 47
Peters, S., 193
Petersen, L., 287
Petersen, V., 519
Peterson, 352
Peterson, G. W., 400
Peterson, M. F., 409
Peterson, R. R., 505
Petronio, S., 250
Pevalin, D., 508
Phelps, J. L., 469
Phillips, E., 450
Phillips, G., 288
Phillips, J., 496
Phillips, J. A., 308, 310, 316, 497
Piaget, J., 124, 399, 400
Piercy, K., 508
Pill, C., 537
Pillemer, K., 477
Pina, D., 433
Pingree, 34
Piotkowski, C., 437
Pistol, 171
Pistole, M. C., 263
Plante, R., 213
Pleck, 394
Pleck, J., 437
Pogrebin, L. C., 406
Polatnik, M. R., 398, 399
Pollack, W., 129
Polonko, K. A., 313
Polson, C., 215
Pomeroy, W., 202
Ponch, I. G., 469
Popenoe, D., 24, 27, 80, 135, 332, 391, 492, 493, 519
Potter, E. H., III, 478
Pouton, 480
Power, C., 508
Presser, H., 438, 439, 495
Previti, D., 499
Price, J., 105
Price, J. H., 213
Price, S. J., 345
Priest, R., 480

Proctor, B., 90
Pryor, D. W., 210
Purvin, D., 461, 478, 480, 481
Putnam, F., 471

Q

Quade, V., 496
Quindlen, A., 388

R

Rabins, P., 415
Rahe, R., 503
Ramsey, P. G., 6
Rank, M. R., 89, 333
Rannik, E., 174
Rao, 339
Rao, K., 480
Rapkin, 468
Rapoport, R., 437
Rapp, C., 445
Rapson, R., 156
Raschke, H., 98, 491, 497, 499, 503, 512
Raspberry, K., 465
Raven, B., 256
Ravo, 340
Raymond, D., 202
Real, T., 135
Reamy, 379
Rebecca, 130
Reed, J., 326, 327, 332, 333, 340
Reeves, T. J., 102, 103, 104
Regan, P., 152, 153, 160, 161, 164, 165, 166, 167, 170, 171, 175, 176, 177, 247, 271, 282, 292, 293, 294, 297, 299
Reisman, C. K., 504
Reiss, I., 171, 172, 173, 174, 182
Renner, M. A., 64
Rentfrow, J., 295
Renzetti, C., 85, 88, 117, 119, 125, 126, 127, 129, 130, 140, 141, 142, 143, 219, 461, 466
Repetti, R., 423, 426
Reske, J., 64
Rexroat, C., 435, 444
Reynolds, J., 425
Ribeau, 250
Rice, F. P., 39
Rich, M. E., 133
Richards, E. S., 39
Richards, L. N., 531
Ridley, J., 179
Riedmann, A., 350
Riggs, 180
Riggs, J. M., 393
Riley, D., 424
Riley, P., 436, 438
Rincon, C., 170
Riseden, A. D., 191
Risman, B., 122, 130, 131, 132, 133, 138, 141, 161, 198, 440, 531
Roach, A., 204
Roberto, K., 351, 352, 412
Roberts, 194, 195
Roberts, M. C., 483
Roberts, N., 424
Roberts, T.-A., 433, 434

Robinson, I., 218
Rodgers, B., 508
Rodgers, K. B., 512
Rodgers, R., 46, 47, 532, 533
Rodgers, W., 90
Rodin, 481
Roe, K. M., 352, 414
Roehling, P. V., 424, 425
Roenrich, L., 223
Roger, K., 64
Rogers, 480
Rogers, S., 329, 495, 499
Rogers, S. M., 202
Rollins, B. C., 400
Romero-Garcia, 136
Rosaldo, M. Z., 19
Rose, A., 157
Rose, H., 512
Rose, M. L., 206
Rosen, H., 164
Rosen, M. P., 223
Rosenbaum, A., 463
Rosenberg, J. D., 519
Rosengren, A., 504
Ross, 223
Ross, C., 14, 91, 315, 422, 423, 430
Ross, C. E., 428
Ross, K., 412
Rossi, A., 381
Rowe, 354
Ruane, 281, 373
Rubin, L., 17, 88, 91, 92, 94, 110, 430, 435, 439, 440, 447
Rubin, Z., 130, 134, 160, 165, 168, 193, 241, 415, 427
Rudd, J., 479
Rumbaut, R. G., 412
Runyan, W., 57
Rushforth, N. B., 36, 57
Russell, D. E. H., 466, 479, 480
Russell, G., 135, 441
Russell, J. N., 152
Russo, N., 443
Rust, P., 207, 208
Rutledge, L. L., 180
Rutter, V., 545
Rydell, R., 178

S

Saal, F. E., 469
Sadker, D., 128, 129
Sadker, M., 128, 129
Safilios-Rothschild, C., 257
Sagrestano, L., 246, 247, 255, 256, 258
Sahlstein, E., 64
Samuels, A., 507
Sanchez, 52
Sanchez, L., 100, 518
Sandberg, J., 394, 396
Sandberg, J. G., 266
Sandefur, G., 538
Sander, W., 287
Sandler, B. R., 469
Sanford, K., 248
Santrock, J., 47, 540
Sassler, S., 309, 526, 534
Satir, V., 50, 57, 248, 275

SUBJECT INDEX

Boldface numbers indicate definitions of terms.

B

Bargaining, in conflict resolution, 272
Basic conflicts, **259**–260
Battered child syndrome, **472**
Battered women and the law, 471–472
Battering, **462**–463. *See also* Child abuse and neglect; Violence
Beauty and the Beast, 158
Beginning marriages, 335–341
   cohabitation and, 339
   engagement and, 338–339
   establishing boundaries and, 343–346
   establishing marital roles and, 342–343
   predicting marital success and, 335–338
   weddings and, 339–340
Behavior modification, in childrearing, 403
Belgium, same-sex marriage in, 9
Berdaches, 117
Betrayal, child sexual abuse and, 482
Bias, **36**
Bifurcation of fatherhood, **393**
Bifurcation of working time, **423**
Binuclear families, **17, 532**
   mother/stepmother-father/stepfather subsystems, 532–533
   recouping, 533–534
Biological determinism, 131
Biological influences in child development, 404
Bipolar gender role, **118**
Birth. *See* Childbirth
Birth control, 226, 366
Birthrate
   contemporary marriages and families, 79–81
   demographics, 83–84
Bisexuality, **200**, 207–210
Black Americans. *See* African American(s); African-American families
Blamers, 248
Blended families, **526**, 536–546. *See also* Stepfamilies
"Blogging" about sex, 197
Boomerang generation, **350**–351
Boundaries of new families, establishing, 343–346
Brazelton, T. Berry, 401
Breadwinning mothers, 440–442
Breaking up, 297–302
Brother-sister sexual abuse, 479
Buber, Martin, 184
Bulgaria, housework survey in, 61
Bundling, **71**

C

California
   domestic partnerships/civil unions, 12
   extended families, 20
   same-sex couple benefits, 10
Canada, same-sex marriage in, 9
Caregivers, 416
Caring, intimate love and, 184
Case-study method, **57**
Catalyst effect, 300
Centrists, ideological position of, **25**
Channeling, socialization through, 126

Child abuse and neglect, 472–474
   homicides, 477
   interventions, 475–476
   at risk families, 474–475
   sexual. *See* Child sexual abuse
Childbirth, 370–372
   divorce, factors affecting, 497
   feminist approach to, 371
   medicalization of, **370**–371
   men and, 376
   single-parent families, creation of, 527
   view of by mothers, after, 371–372, 374–375
Childcare, 394–399
   active, 394–396
   day care, 396–398
   lack of adequate care, 443–445
   mental, 396
   nonparental, 396–399
   school-age childcare, 398–399
   self-care, **398**–399
   supplemental, 396–398
Child custody, 515
   disputes, 517
   noncustodial parents, 516–517
   types of, 515–516
Child development, 404–406
   basic needs and, 404–406
   biological influences, 404
   individual temperament and, 404
   psychosexual, in family context, 405–406
   self-esteem and, 405
   socialization and. *See* Socialization
Child-free marriages, **364**
Child less, 364
Childrearing
   authoritarian, **403**
   authoritative, **403**
   contemporary strategies, 402–403
   dual-earner marriages/families, 436
   expert advice on, 401–403
   indulgent, **403**
   permissive, **403**
   socialization. *See* Socialization
   uninvolved parenting, **403**
Children, 362
   adult children at home, 350–351
   age at which to have, 366
   in blended families, 545–546
   caring for parents, 414–415
   in cohabitation *versus* marriage, 315
   divorce and, 508–514
   family processes affecting divorce, 499
   fertility patterns in the United States, 362–364
   forgoing, 364–365
   marital satisfaction and, 344, 347–348
   middle-aged marriages with, 348–350
   parentified, 414–415
   remarriage and, 535
   self-care of, **398**–399
   in single-parent families, 529–531
   socialization of. *See* Socialization
   in stepfamilies, 545–546. *See also* Stepfamilies
   unmarried parenthood, 363–364
   violence against parents and, 476–477
Child sexual abuse, **478**–479

brother-sister sexual abuse, 479
   effects of, 480–482
   father-daughter sexual abuse, 479
   preventing, 483–484
   at risk children, 480
   trauma, 482
   treatment programs, 482–483
   uncle-niece sexual abuse, 480
Child support, **506**
China
   love in, 156
   Na of China, 8
   spirit marriages, **20**
Chlamydia, 228
Choice-shaping factors, 287
Christianity, marriage in, 9
Christian Promise Keepers, 134
Church of Jesus Christ of Latter-day Saints, polygamy and, 13
Civil unions, 9
Clans, **7**
Clarifying, marital satisfaction and, 262
Clarity, in childrearing, 402
Clinical psychologists, 273
Clinical research, 57
Clockspring variation on Reiss's Wheel Theory, 174
Closed fields, in romantic relationships, 292
Close relationship model of legal marriage, **332**
Coercive power, 256
Coercive sex, 468–469
Coexistence, in conflict resolution, 272
Cognitive development theory, **124**, 399–400
Cohabitation, **82**, 306–317
   binuclear families and, 533
   common-law marriage, 310–311
   divorce, factors affecting, 497
   domestic partners, **311**, 312
   extrarelational sex in, 297
   family life cycle stage, 339
   gay and lesbian, 311–313
   marital communication, later, 245
   marital success and, 316–317
   marriage compared with, 313–316
   rise of, 307–308
   types of, 308–310
Coitus, **216**
College, gender development in, 130
Colonial era, marriage and families during, 70–73
Comadres, 102
Comics strips and family life changes, 86–87
Coming out, **204**–205
Commitment, **155**
   changes in, with the passage of time, 181
   in cohabitation *versus* marriage, 313
   intimate love, 182
   and love, 171–172
Common couple violence, **457**
Common-law marriage, **310**–311
Communication
   in childrearing, 402
   cohabitation and later marital communication, 245
   demand-withdrawal communication, **246**
   effectiveness, 251
   ethnicity and, 250

feedback, **252**–254
gender differences in, 241, 242–244, 246–247
ineffective communication, 249
marital communication patterns and satisfaction, 245
marriage and, 241–247
miscommunication, 248–249
mutual affirmation, **254**–255
nonverbal, **237**–241
premarital communication patterns and marital satisfaction, 244–245
self-awareness obstacles, 249–250
self-disclosure and, 250–252
sexual communication, 247
topic-related difficulty, 248
trust and, **252**
verbal, **237**
Community divorce, 501
Community marriages, 341
Compadres, 7, 102
Companionate grandparenting, 414
Companionate love, 156, 168–169, 170
marital sex and, 220
Companionate marriages, **76**–77, 154
Complaining, marital satisfaction and, 262
Complementary needs theory, **290**
"Computers" as communication style, 248
Concepts, **37**
Conceptualization, **37**
in a disaster, 40–41
Concerted cultivation, 92
Conduct of fatherhood, **392**
Confidentiality, **54**–55
Conflict, 259–274
agreement as a gift, 271
anger, dealing with, 260
bargaining, 272
basic conflicts, **259**–260
beneficial nature of, 270–271
child well-being, 270
coexistence, 272
counseling, 273
familial well-being, 270
forgiveness, **273**–274
frequency studies, 266–267
gender differences in handling, 260–262
hostile conflict, **263**
housework and, 269
intimacy and, 259–260
marital satisfaction and resolution of, 262–265
mental health consequences of, 270
money conflicts, 268–269
nonbasic conflicts, **260**
physical health consequences of, 270
professional assistance, 273
resolution of, 262–265, 271–274
sex, fighting about, 265–268
Conflict-habituated marriages, 355
Conflict Resolution Styles Inventory (CRSI), 265
Conflict theory, **49**–50, 54
Confrontation, marital satisfaction and, 262
Conjugal families, **155**
Conjugal model of legal marriage, **332**
Conjugal relationships, **20**
Connecticut

colonial era families, 71–72
domestic partnerships/civil unions, 9, 10, 12
Consanguineous relationships, **20**
Consequences, in childrearing, 402
Conservatives, ideological position of, **25**
Consistency, in childrearing, 402
Consummate love, 168–169
Contact stage of blended families, 540
Contempt, **238**
Continuous coverage system, **382**
Contraception, 226, 366
Conventional sequential work/family role staging, 442
Cooperativeness, **263**
Coparental divorce, 501
Coparental marriages, 341
Coprovider families, **429**
Co-rumination, **157**
Couvade, 371
Covenant marriage, 518, 520
Critical thinking, 33
CRSI (Conflict Resolution Styles Inventory), 265
Crude birthrate, **362**
Crude divorce rate, **492**
Culture of fatherhood, **392**
Culture(s). *See also specific groups;* Ethnicity
contemporary families and, 84
love and, 155, 156
reasons women stay in violent relationships, 470–471
singlehood and, 305
Cunnilingus, **215**
Custody, child. *See* Child custody
Cyrano style of unrequited love, 175
Czech Republic, housework survey in, 61

D

Date rape, **468**–469
Dating, 217–219, 291, 293. *See also* Romantic relationships
binuclear families and, 533–534
breaking up, 297–302
extrarelational sex, 297
problems in, 294–297
scripts, 294–297, **296**
after separation or divorce, 504–505
violence and, 466–468
Day care, 396–398
Deadbeat parents, 516–517
Deductive research, **37**
Deep friendship, **161**
Defense of Marriage Act, 10, 12
Defensiveness, marital satisfaction and, 262
Deinstitutionalization of marriage, **328**
Demand-withdrawal communication, **246**
Demographic factors affecting divorce, 495–496
Demographics
contemporary families and, 83–84
in retreat from marriage, 324–326
Denmark, domestic partnerships/civil unions in, 9
Dependent variable, **37**, 38
Destructive parentification, 415

Developmental approach to marriage, 334–335
Developmental stages of blended families, 538–540
Developmental systems approach to socialization, **400**–401
Developmental tasks of divorce, 510–512
Devitalized marriages, 355
Discipline in blended families, 543–544
Discrimination
against women in the workplace, 442–443
antigay, 205–207
Displays of love, 155
Distal causes of divorce, **500**
Distractors, 248–249
Diverse families, issues of, 407–411
Division of labor, familial. *See* Family work
Divorce, 490–491
alimony and, **506**
child custody and. *See* Child custody
children and, 508–514
child support and, **506**
community, 501
coparental, 501
covenant marriage as response to, 518, 520
demographics, 83, 495–496
distal causes, **500**
economic, 500–501
economic consequences of, 505–507
emotional, 500
family processes affecting, 498–500
gender and stress, 504
identity, postdivorce, 503–504
intergenerational transmission, 497–498
legal, 500
life course factors affecting, 496–498
measuring, 492–493
no-fault, **500**
noneconomic consequences of, 507–508
proximal causes, **500**
psychic, 501
rates in contemporary marriages, 79–81
ratio measure of, **492**
reducing, 519–520
single-parent families, creation of, 527
societal factors affecting, 494–495
stations of process of, 500–501
trends in U.S., 493
Divorce mediation, **517**–519
Divorce rate
contemporary patterns, 82
crude, **492**
predictive, **492**–493
refined, **492**
Domestic marriages, 341
Domestic partners, **311**, 312
Domestic partnerships, 9
Domestic violence. *See* Child abuse and neglect; Child sexual abuse; Violence
Don Quixote style of unrequited love, 176
Drugs, date rape and, 468–469
Dual-earner marriages/families, 141, 432–436
childrearing, 436
coping in, 440
coprovider families, **429**
emotion work, **436**
housework and, 434–436

Flexible work times and environments, 445–446

Flirting, 239

FMLA (Family and Medical Leave Act), 451–452

Foreplay, **191**

Forgiveness, **273**–274

France
domestic partnerships/civil unions, 9
same-sex marriage, 9

Freud, Sigmund, 399, 400

Friedan, Betty, 84, 140

Friendship
deep friendship, **161**
gender and, 157–158
intimacy of, 152–153

Fulfillment of intimacy needs, **173**

Fundamentalist Church of Jesus Christ of Latter-day Saints, 13

G

Gambler's fallacy, 36

Game stage of symbolic interaction theory, **401**

Gay bashing, 205

Gays, **201**
anal eroticism, **216**
breakups, 301
cohabitation, 311–313
coming out, **204**–205
discrimination, 205–207
divorce, 507
domestic partners, **311**, 312
heterosexual relationships compared, 205
identifying oneself as, 202–205
love and, 164
marriage gender restrictions, 331
"meeting," 293
parenthood and, 409–411
prejudice, **205**–207
same-sex marriage, 9–12
sexual coercion in relationships, 468
sexuality in relationships, 219
sexually open relationships, 222–223
singlehood and, 306
single-parent families, 529
stages in acquiring identity, 204
violence in relationships, 466

Gender, 22, **52**, **116**–118
communication differences in, 241, 242–244, 246–247
conflict, differences in handling, 260–262
dating script differences, 294–297
degendering marriage and family, 140–141
divorce stress and, 504
"doing gender," 122
friendship and, 157–158
Great Depression and World Wars, 78
and intimacy, 155–161
jealousy differences, 178–179
love and, 159–160, 164
marriage restrictions, 331
physical attractiveness in mate selection, 283
remarriage and, 534–535
role changes and opportunity for women, 84

sexuality and, 164
and sexual orientation, **121**
social construction of, 121–122
as social structure, 122
violence and, in the exchange-social control model, 459–460
wage gaps, 119, 120

Gender attribution, 120

Gender display or presentation, 121

Gender identity, **117**

Gender ideology, **435**

Gender polarization, 119

Gender-rebellion feminism, **142**

Gender-reform feminism, **141**

Gender-resistant feminism, **141**–142

Gender role(s), **116**–144
change, resistance to, 139
cognitive development theory of, **124**
contemporary roles, constraints of, 138–139
family experiences, 134–138
gender theory of, **121**–122
learning, 124–133
masculinity and femininity, 118–121
men's roles in families and work, 134–136
movements and the family, 139–144
social learning theory of, 122–**123**
women's roles in families and work, 136–138

Gender-role attitudes, **116**

Gender-role behaviors, **116**

Gender-role stereotypes, **116**

Gender stratification, 120

Gender theory, **121**–122

Genital herpes, 228

Genital warts, 228

Germany
domestic partnerships/civil unions, 9
same-sex marriage, 9

Ghetto poor, 89

Giselle style of unrequited love, 175–176

Gonorrhea, 228

Grandparenting, 412–414

Gray, John, 119

Great Britain
domestic partnerships/civil unions, 9
housework survey, 61

Great Depression, 77–78

Grounded theory, 38

H

Haeckel, Ernst, 39

Halo effect, **281**

Hatfield and Sprecher's Passionate Love Scale, 167–168

Hawaii
domestic partnerships/civil unions, 12
extended families, 20
same-sex marriage, 9–10

Health
AIDS/HIV, 227–229
caregivers, 416
in cohabitation *versus* marriage, 315
conflict, consequences of, 270
mental. *See* Mental health

Hegemonic models of gender, **117**

Hendrick and Hendrick's Love Attitude Scale, 167

Hepatitis, 229

Heterogamy, **285**

Heterosexual, **200**–201

Hispanics. *See* Latino families; Latinos/Latinas

History
colonial era, marriage and families during, 70–73
nineteenth-century marriages and families. *See* Nineteenth-century marriages and families
twentieth-century marriages and families. *See* Twentieth-century marriages and families

HIV (human immunodeficiency virus), **227**–229

Homemaker role, **428**

Homeostasis, **51**

Homoeroticism, **202**

Homogamy, **154**, **285**–290

Homophobia, **205**

Homosexual, **200**–201. *See also* Gays; Lesbians

Honeymoon effect, **245**
marital success and, 337

Hostile conflict, **263**

Hostile environment, **443**

Households, **6**

Housework. *See also* Family work
and conflict, 269
survey, 60–62

Hua of Papua, New Guinea, 117

Human immunodeficiency virus (HIV), **227**–229

Hungary, housework survey in, 61

Hurricane Katrina, 40–41

Hypergamy, **288**

Hypogamy, **287**

Hypotheses, 37

I

Iceland, domestic partnerships/civil unions in, 9

Id, **399**

Idealization, 64

Identity bargaining, **343**

Identity, postdivorce, 503–504

Ideology of intensive mothering, **136**, **391**

Ie, 7

Immersion stage of blended families, 538–539

Immigration, 75–76

Incest, **479**

Income. *See also* Economic factors; Poverty; Socioeconomic status
birthrates and, 83–84
divorce and, 495

Independent variables, **37**, 38

India
Hijra of, 117
kin rights and obligations, 20
love in, 156
Nayar of India, 20

Individualistic cultural values affecting divorce, 494–495

Male(s). *See also* Fathers and fatherhood; Men; Parent(s)
   gay. *See* Gays
   traditional gender role, 134–136
Male gender roles
   contemporary, 138–139
   traditional, 134–136
Mandatory arrest policies, **471**
Mania, **167**
Manipulation, socialization through, 125–126
Marianismo, 136
Marital commitments, **347**
Marital decline perspective, **329**
Marital history homogamy, 288–**289**
Marital paradigm, **263**
Marital power
   feminist views on, 258–259
   household labor, division of, 437
   principle of least interest, **257**
   relative love and need theory, 257
   resource theory of power, **257**–258
   sources of, 256–257
Marital rape, **465**–466
Marital resilience perspective, **329**
Marital roles, establishing, 341–343
Marital satisfaction, 344–348
   children and, 344, 347–348
   commitments and, **347**
   conflict resolution, 262–265
   decline in, 344
   duration-of-marriage effect and, 344
   family processes affecting divorce, 498–499
   household labor, division of, 437–438
   individual changes, 348
   marital communication patterns, 245
   measuring, 349
   premarital communication patterns and, 244–245
   remarriage, 536
   social and psychological satisfaction, 344–345
   social context and stress, 346–347
Marital separation
   contemporary patterns, 82
   dating again, 504–505
   distress, 502–**503**
   identity, postdivorce, 503–504
   process of, 501–502
Marital status, in child-parent relationships, 407–408
Marital success
   background factors influencing, 336
   cohabitation and, 316–317
   personality factors, 337
   predicting, 335–338
   relationship factors influencing, 337–338
Marital tasks, 342–343
Marketplace of relationships, 280–281
Marriage, **8**–9, 322–323. *See also* Marital *entries*
   age restrictions, 330
   beginning. *See* Beginning marriages
   benefits of, 333
   between blood relatives, 330
   cohabitation compared with, 313–316
   communication and, 241–247

companionate marriages, **76**–77, 154
contemporary patterns, 82
deinstitutionalization of marriage, **328**
developmental approach to, 334–335
divorce and. *See* Divorce
dual-earner. *See* Dual-earner marriages/families
early marriage. *See* Early marriage
economic cooperation, 15–16
enduring marriages, 354–356
experience *versus* expertise, 4–5
forms of, 12–13
gender development in, 130
gender of spouses, 331
individualized marriage, 328–329
intimacy in, 13–15
later-life marriages. *See* Later-life marriages
learning about, 32–33
legal marriage, 329–332
love and, 165
middle-aged marriages. *See* Middle-aged marriages
modified polygamy, **13**
monogamy, **12**–13
number of spouses, 331
ongoing social controversies, 5–6
peer marriages, 160–**161**
polyandry, **12**
polygamy, **12**–13
polygyny, **12**
rates in contemporary marriages, 79–81
reasons to marry, 332–334
rejection of, 327
religion and, 327–328
reproduction and, 16
rights and benefits of, 11
same-sex marriages, 9–12
serial monogamy, **13**
slave marriages, 9
social class and, 165
social roles and status, 17–18
socialization, **16**–17
spirit marriage, **20**
stations of, 340–341
thinking critically about, 33–37
time of, 328
Marriage and family counselors, 273
Marriage debate, **323**–329
   deinstitutionalization of marriage, **328**
   individualized marriage, 328–329
   marital decline perspective, **329**
   marital resilience perspective, **329**
   religion and, 327–328
   retreat from marriage, **324**–327
Marriage squeeze, **283**
Martin, Don, 58
Masculinity. *See* Gender *entries*
Massachusetts
   colonial era families, 71–72
   same-sex marriage, 9–12
Mass media as socialization agent, 129–130
Masturbation, **214**
Mate selection, 280–291
   field of eligibles, **284**–289
   marketplace of relationships, **280**–281
   physical attractiveness and, 281–284
   theories and stages of, 290–291

Mating gradient, **283**–284
Matriarchal societies, **117,** 137
Matrilineal, **70**
Mead, Margaret, 118
Media
   families in, 34–36
   sexuality, influence on, 196–198
Mediation, of divorce, **517**–519
Medicalization of childbirth, **370**–371
Memories of child sexual abuse, 481–482
Men. *See also* Fathers and fatherhood; Parent(s)
   in blended families, 541
   childbirth and, 376
   dating script differences, 294–297
   dual-earner marriages/families and housework, 434–436
   family work, 427–428
   fatherhood, 391–394
   gender movements, 143–144
   gender roles in families and work, 134–136
   pregnancy and, 369–370
   violence, victims and perpetrators of, 463–464
   virginity and, 200
   wage gap, 442–**443**
Men Are From Mars, Women Are From Venus (Gray), 119, 243–244
Menopause, **211**
Mental health
   conflict, consequences of, 270
   parenthood, effects on, 406–407
   relationship quality and, 315
Mental labor, **396**
Mesosystems, 39
Mexican Americans. *See* Latino families; Latinos/Latinas
Microsystems, 39
Middle age and sexuality, 210–212
Middle-aged marriages, 348–351
   with adolescents, 349–350
   as launching centers, 350–351
   reevaluation and, 351
   with young children, 348–349
Middle class, 88, 93. *See also* Social class
Middle Eastern families, 107–110
Million Man March, 134
Minority groups, **96.** *See also specific groups*
Minority status, **110**
Miscarriages, 372
Miscommunication, 248–249
Mobilization stage of blended family development, 539
Modeling, **123**
Modified extended families, **20**
Modified polygamy, **13**
Money conflicts, 268–269
Monogamy, **12**–13, 220
Mormons, polygamy and, 13
Mother(s). *See also* Parent(s); Women
   active childcare by, 394–396
   breadwinning, 440–442
   lesbians. *See* Lesbians.
Motherhood, 389–391, 432. *See also* Parenthood
   gender role and, 136–138
   intensive mothering ideology, **136, 391**

Multiple masculinities and femininities, **117**
Music videos and gender, 133
Mutual affirmation, **254**–255
Mutual dependency, **173**
Mutual violent control, **457**

N

Na of China, 8
Nambikwara of Africa, 15
National Singles Week, 302
Native American(s)
  affection, displays of, 164
  fertility rates, 362–363
  intermarriage, 285–287
  unmarried parenthood, 363–364
Native-American families, 105–107
  clans, **7**
  colonial era marriage and families, 70–71
  grandparenting and, 413
Nayar of India, 20
Neglect, child. *See* Child abuse and neglect
Neopatriarchal gender ideals, 137
Netherlands
  housework survey, 61
  same-sex marriage, 9
Neuroses, **399**
New Deal, 77
New Jersey
  domestic partnerships/civil unions, 12
  same-sex couple benefits, 11
New York, domestic partnerships/civil unions in, 12
New Zealand, domestic partnerships/civil unions in, 9
9/11, 40–41
1950s, families of, 79
Nineteenth-century marriages and families, 73
  African-American families, 74–75
  childhood and adolescence, 74
  immigration, 75–76
  industrialization, 73–74
  power of love, 74
  women, changing roles of, 74
Nisei, 105
No-drop prosecution policies, **471**
No-fault divorce, **500**
Nonbasic conflicts, **260**
Nonlove, 168–169
Nonmarital sex, **217**–219
Nonparental childcare, 396–399
Nonparental households, **408**
Nonverbal communication, **237**–241
  emotions, expressing, 238
  eye contact, **239**
  functions of, 238
  importance of, 238
  interpersonal attitudes, conveying, 238
  proximity, **238**–239
  touch, 239–241
Norway
  domestic partnerships/civil unions, 9
  same-sex marriage, 9
Nuclear family, **7**, 19
Nurturant fathers, 392

O

Objectivity, **33**–36
Observational research, 57–63
Obsessive relational intrusion, 176–177
Ohio, same-sex marriage in, 12
Older adulthood. *See* Elderly people; Later-life marriages
One-pot stepfamilies, 545
Open adoption, **377**
Open fields, in romantic relationships, **292**
Operationalization, **37**
Opinions, **36**
Opportunity-determining factors, 287
Oral-genital sex, 215–216
Outing, **204**–205

P

Paradigms, 38
Paraphrasing, marital satisfaction and, 262
Parent(s). *See also* Fathers and fatherhood; Mother(s)
  aging parents, care for, 415–416
  becoming one, 378–383
  care for by children, 414–416
  needs of, 406–407
  roles and responsibilities, taking on, 381–382
  as socialization agents, 124–128
  societal indifference to needs of, 407
Parental image theory, 291
Parental influence on sexuality, 194–195
Parenthood. *See also* Childbirth; Fathers and fatherhood; Motherhood
  age at which to have children, 366
  child custody. *See* Child custody
  deferring, 365–366
  ethnicity and, 408–409
  fatherhood, 391–394, 395
  forging, 364–365
  gays and lesbians, 409–411
  gender development in, 130–131
  motherhood, 389–391, 432
  noncustodial parents, 516–517
  preparation for, with humor, 389
  stresses of, 382–383
  unmarried parenthood, 363–364
  violence toward parents, 476–477
Parentified children, 414–415
Parenting, 391
  of adult children, 412
Partners
  choosing. *See* Mate selection
  counseling, 273
Passionate love, **156**
  changes in, with the passage of time, 181
  instability of, 180–182
Passive-congenial marriages, 355
Pastoral counselors, 273
Patriarchal societies, **117**
Patriarchy, **71**
Patrilineal, **70**
PBS survey on families, 24
Pedophilia, **478**–479
Peer(s), **129**
  influence on sexuality, 195–196
  as socialization agent, **129**

Peer marriages, 160–**161**, 439–**440**
Pelvic inflammatory disease (PID), 229
Peoplemaking (Satir), 248
Perceptual accuracy, 251
Performance anxiety, **223**
Permissive childrearing, **403**
Personality
  marital success and, 337
  parentified children, 414–415
Personal validation, features of intimacy, 151
Pets
  custody, 15
  intimacy and, 14–15
Phenotype, **95**
Physical attractiveness in mate selection, 281–284
Physical health consequences of conflict, 270
Physical punishment, in childrearing, 402–403
Piaget, Jean, 124, 399–400
"Pickup lines," 292
PID (pelvic inflammatory disease), 229
Placaters, 248
Play stage of symbolic interaction theory, **401**
Pleasuring, **214**
Poland, housework survey in, 61
Polyandry, **12**
Polygamy, **12**–13
Polygyny, **12**
Popular culture as socialization agent, 129–130
Portugal, domestic partnerships/civil unions in, 9
Positive assortative mating, 285
Postdivorce identity, 503–504
Postgender marriages, 160–**161**, 439–**440**
Postgender relationships, **138**
Postpartum period, **379**–380
Posttraditionalists, 141
Poverty
  lower class, 88–90
  single-parent families and, 528
  unemployment and, 448
  welfare reform and, 448–451
Poverty line, 88
Power, **255**. *See also* Marital power
  child sexual abuse and, 482
  intimacy and, 255–256
  violence and, in the exchange-social control model, 460
Power bases, 255
Power outcomes, 255
Power processes, 255
Pragma, **167**
Predictive accuracy, 251
Predictive divorce rate, **492**–493
Pregnancy, 366–367. *See also* Childbirth
  divorce, factors affecting, 497
  emotional and psychosocial changes during, 367–369
  men and, 369–370
  sexuality during, 369
Pregnancy loss, 372
  coping with, 373–374
  infant mortality, 373
  spontaneous abortion, **372**–373

V

Vaginitis, 229
Valentine's Day, 300
Validating, marital satisfaction and, 262
Value judgments, **36**
Value stage of romantic relationships, 291
Value theory, **290**
Vanatinai of South Pacific, 8
Variables, **37,** 63
Verbal appellation, socialization through, 126
Verbal communication, **237**
Vermont, domestic partnerships/civil unions in, 9, 10, 12
Viagra, **226**
Video games and gender, 132–133
Violence, 456–**457**
 child abuse and neglect. *See* Child abuse and neglect
 class and, 464–465
 coercive sex, 468–469
 costs of, 471
 date rape, **468**–469
 dating violence and abuse, 466–468
 gay and lesbian relationships, 466
 interventions for, 471–472
 intimate partner abuse, **457**
 intimate partner violence, **457**
 marital rape, **465**–466
 men and women and, 462–464
 models of, 458–460
 mythology of, 461
 parents as victims, 476–477
 prevalence of, 460–461
 race and, 465
 rape, **465**–466
 reasons women stay in violent relationships, 470–471
 reducing, 478
 sibling violence, 476
 types of intimate, 457–458
Violent resistance, **457**
Virginity, **199**–200
Vital marriages, 355–356

W

Wage gap, 442–**443**
Washington, domestic partnerships/civil unions in, 12
Weblog, 197
Weddings, 339–340
Welfare reform, 448–451
West Germany, housework survey in, 61
Wheel theory of love, **172**–175
White(s)
 African American families compared, 96–99

Asian American families compared, 102–105
Asian American sexuality compared, 192
child sexual abuse, effects of, 480
cohabitation and, 308–310
divorce, economic consequences of, 505
divorce rates, 496
European ethnic families, 110–111
familial division of labor, 427
fertility rates, 362–363
financial support for children by nonresident fathers, 395
grandparenting, 414
intermarriage, 285–287
intermittent extended families, 352
jealousy and, 180
Latino families compared, 101
love styles, 167
lower class families, 88
marital status, 97, 325
men's roles in families and work, 134
miscarriages, 372
oral-genital sex and, 215
parenthood stresses, 383
poor women and children, 90
premarital sexual involvement, 218
ratio of unmarried men to unmarried women, 284
remarriage, 534
sexual scripts and, 193
single-parent families, 528
socioeconomic status, 99
unmarried parenthood, 363–364
wage gaps, 120
women in the workplace, 429–430
women's roles in families and work, 136
Widowhood, 352–354
Wishful never-married singles, 305
Women. *See also* Motherhood; Mother(s); Parent(s)
 battered women and the law, 471–472
 in blended families, 540–541
 dating script differences, 294–297
 discrimination against women in the workplace, 442–443
 double standard of aging, **283**
 dual-earner marriages/families and housework, 434–436
 economic discrimination against, 442–443
 family work, 428
 feminization of poverty, **90**
 gender roles in families and work, 136–138
 motherhood, 389–391
 nineteenth-century marriages and families, 74
 role changes and opportunity for, 84
 sexual harassment, **443**

violence, victims and perpetrators of, 463–464
virginity and, 200
wage gap, 442–**443**
working women. *See* Working women
World War II and, 77–78
Work
 cohabitation and, 315–316
 divorce, consequences of, 506–507
 employment status, factors affecting divorce, 495
 family. *See* Family work
 family linkages and, 423
 family-to work spillover, 424–425
 marriage contracts, 431
 men's roles, 134–136
 role conflict, **425**–426
 role overload, 425–426
 role reversal, **440**–442
 role strain, **425**–426
 unemployment and families, 446–448
 women's roles, 136–138
Working class, **88.** *See also* Social class
Working poor, 88–89
Working women, 15–16, 429–432. *See also* Dual-earner marriages/families
 employment patterns, 431–432
 marriage contracts, 431
 motherhood, price for, 432
 reasons for entering workforce, 430–431
 role reversal, **440**–442
Workplace
 childcare, lack of, 443–445
 discrimination against women in, 442–443
 family policy and, **451**–452
 gender development in, 131
 inflexible work environments and the time bind and, 445–446
 sexual harassment, **443**
Work spillover, **424**
World Wars, 77–78

Y

You Just Don't Understand: Women and Men in Conversation (Tannen), 242–244
Young adulthood, psychosexual development in, 194–200
 adolescent sexual behavior, 198–199
 media influence, 196–198
 parental influence, 194–195
 patterns for women and men, 200
 peer influence, 195–196
 sexual developmental tasks, 198
 virginity, **199**–200

TO THE OWNER OF THIS BOOK:

I hope that you have found *The Marriage and Family Experience,* Tenth Edition, useful. So that this book can be improved in a future edition, would you take the time to complete this sheet and return it? Thank you.

School and address:_____

Department:_____

Instructor's name:_____

1. What I like most about this book is:_____

_____

_____

2. What I like least about this book is:

_____

_____

3. My general reaction to this book is:

_____

_____

4. The name of the course in which I used this book is:

_____

5. Were all of the chapters of the book assigned for you to read?_____

   If not, which ones weren't?_____

6. In the space below, or on a separate sheet of paper, please write specific suggestions for improving this book and anything else you'd care to share about your experience in using this book.

_____

_____

_____

_____

THOMSON ™

WADSWORTH

BUSINESS REPLY MAIL
FIRST-CLASS MAIL   PERMIT NO. 34   BELMONT CA

POSTAGE WILL BE PAID BY ADDRESSEE

NO POSTAGE
NECESSARY
IF MAILED
IN THE
UNITED STATE

Attn: Chris Caldeira

Wadsworth/Thomson Learning
10 Davis Dr
Belmont CA 94002-9801

OPTIONAL:

Your name:_____ Date: _____

May we quote you, either in promotion for *The Marriage and Family Experience,* Tenth Edition, or in future publishing ventures?

Yes: _____ No: _____

Sincerely yours,

Chris Caldeira